ESSENTIALS OF BIOCHEMISTRY

Dr. Nisha Khalsa presently posted as Professor of Biochemistry, Jodhpur Dental College, Jodhpur has retired as Professor and Head of Biochemistry Department, Dr. S.N. Medical College, Jodhpur. Dr. Nisha Khalsa has 40 years experience of teaching medical biochemistry to MBBS undergraduate and postgraduate students at various Government Medical Colleges of Rajasthan as well as at Nepal. She has published many research papers in national and international medical Journals and has been guide for M.D. (Biochemistry), M.Sc. (Medicine) and Ph.D. (Medicine) Biochemistry candidates of Rajasthan University since 1981. She has been examiner of Medical undergraduate and postgraduate students at various Medical Colleges of India and Nepal.

ESSENTIALS OF BIOCHEMISTRY

Nisha Khalsa Ph.D.

Aavishkar Publishers, Distributors
Jaipur 302 003 (Raj.) India

Published by
Akshay Jain for
Aavishkar Publishers, Distributors
807, Vyas Building, Chaura Rasta
Jaipur 302 003 (Raj.) India
Phone : 2578159
e-mail : aavishkarbooks@hotmail.com

2025

ISBN 978-81-7910-255-8

Price : ₹ 3000.00

Printed at
Sheetal Printers
Jaipur (Raj.)

This Book is dedicated to
my daugther Meenal

FOREWORD

It gives me immense pleasure to give brief introduction of the book "Essentials of Biochemistry" written by Dr. Nisha Khalsa, Professor of Biochemistry (Retired) Dr. S.N. Medical College, Jodhpur, presently posted as Professor of Biochemistry in Jodhpur Dental College, Jodhpur. Dr. Khalsa has vast experience spanning over 40 years of teaching Medical Biochemistry to undergraduate and post graduate MBBS and Dental Students. I am sure this book will be welcomed warmly both by students and teachers of various medical and dental institutions.

A major thrust has been given by the author to make the book more student friendly, in easy to understand the difficult and complex subject matter, while maintaining a balance between basic Fundamental Biochemistry with emerging advances in Molecular Biology, Genetic Engineering and other recent applied aspects of Clinical Biochemistry. Apart from this, the book also covers many relevant applied areas, that are global medical problems such as diabetes, hypercholesterolemia, cancer, obesity and AIDs etc.

I feel the author has taken enormous pains in writing this book, that will benefit and meet the requirements of students of various health sciences.

I am pleased to introduce this book to MBBS and BDS students in particular and to the fraternity of health sciences in general.

DR. P.K. GUPTA
Dean
University of Medical Sciences
Rajasthan

ACKNOWLEDGEMENTS

This book is the outcome of tremendous response of medical students over the years that made me realize the importance of teaching "Fundamental aspects of Medical Biochemistry" in a simple, easy to understand and lucid style. It is my proud privilege to show my gratitude to one and all of my students and colleagues for their support and love throughout my teaching career. Here in particular I would like to thank Manisha and Purvi for their contribution at every stage of planning this book, for helping me in preparation of figures and diagrams and assisting me in proof reading. I would like to thank and acknowledge the patience and support of my family, whilst I had to spend long hours in the preparation of drafts and typing of the manuscript at the computer centre and in my study room.

Finally I would like to thank the publishing team at Aavishkar Publishers, Distributors, Jaipur for their encouragement and persistence. It was a privileged and rewarding experience to work with Akshay Jain and Vipin Jain. My special thanks are to Mr. Jitendra for being patient and helping me in the preparation of figures and diagrammes on computer.

PREFACE

This book "Essentials of Biochemistry" is written by me with a view to provide fundamentals of Medical Biochemistry in concise and easily assimilable form for the students of health sciences in general and MBBS and BDS students in particular.

During my 40 years experience of teaching Medical and now dental students, I have observed students facing the problem of lengthy and difficult presentation of the subject in Western and Indian texts available. In this book an attempt has been made to provide concise yet comprehensive coverage of basic principles of Medical Biochemistry, as well as recent advances in molecular genetics in a balanced and easy to understand format.

In view of the rapid expansion and tremendous advances in the field of Medical Biochemistry and Molecular Genetics, it has become necessary to update the subject to enable the students of health sciences to understand and utilize this vast knowledge to solve health problems at molecular level. The main aim and objective of writing this book has been to present difficult and complex subject matter in lucid, simple and interesting style to make it more student friendly and at the same time to provide recent advances in Clinical Biochemistry and Genetic Engineering that are relevant to medicine and necessary for students of various health sciences, to understand causes and rational treatment of various diseases, that are major concern of physicians and other health workers.

Text of each chapter in the book has been designed to provide easy flow of information to facilitate students to understand and master the subject. Informative statements have been made at the beginning of most of the paragraphs to highlight the key points of concerned subject matter.

Chapters on burning global medical problems such as diabetes, cancer and AIDS are included in this book with a view to provide knowledge of applied biochemistry at molecular level to the students of health sciences.

The book presents an exhaustive and thorough exposition of fundamentals and essentials of Medical Biochemistry specially to meet the curriculum for MBBS and BDS students as recommended by Medical Council of India (MCI) and Dental Council of India (DCI).

—Nisha Khalsa

CONTENTS

SECTION II : BIOENERGETICS & METABOLISM

SECTION III : PROTEIN METABOLISM

SECTION IV : MOLECULAR BIOLOGY & GENETICS

SECTION V : NUTRITIONAL ASPECTS

APPENDICES

BIOCHEMISTRY & MEDICINE

Biochemistry is the chemical basis of life. It is a science that deals with the study of chemistry of life processes and metabolic reactions occurring in a living organism at molecular level. The discipline of biochemistry is an insight into the intricate complexities of the life processes and tries to unravel the mysteries of life in molecular terms. Biochemistry encompasses large areas including chemical nature of living cell, molecular biology, molecular genetics and it describes and explains all life processes in the language of chemical reactions.

The study of Biochemistry and Molecular biology has demonstrated that all living organisms are very similar at molecular level and the major chemical processes occurring in them are almost similar. This shows that a unity exists amongst all living organisms in the vast and diversified living kingdom.

A fundamental knowledge of biochemistry is essential for all the students of health sciences to understand the basic functions of body during health and the alterations that occur in disease state. The normal biochemical processes are main basis of health and development of an individual.

All diseases are manifested in abnormalities in molecular constituents and chemical reactions of the body. This demarcation between the normal biochemical reactions during health and the abnormal changes occurring during disease process is the basis of Clinical Biochemistry.

Modern medical practice is heavily dependent on biochemical analytical procedures for the diagnosis of diseases, as well as for prognosis and effective treatment of the diseases. The laboratory biochemical investigations of body fluids such as blood, urine, CSF etc. reflects the alterations in the chemical composition as clinical manifestations of the disease process and are immensely useful to physicians in diagnosis and prognosis of diseases.

The tremendous advancement in the biochemical techniques and laboratory investigation procedures in recent years had great impact on modern clinical practice. Also the advanced biochemical techniques are the major tools in the research projects of all medical sciences. This reciprocal relationship between biochemistry and medicine has stimulated mutual advancements in both the fields.

The major burning problem faced by the clinicians is that as they find treatment to overcome the existing life threatening diseases, new ones are emerging and assuming frightening heights like AIDS, hepatitis B, new kinds of cancers etc. This has stimulated several arrays of studies and research projects world over on various aspects of biochemical problems including protein synthesis, enzymology and molecular genetics. These studies would help health workers to understand the progress and functional control of the microorganisms responsible for causing the dreaded diseases that are life threatening.

One major prerequisite for good health is **optimum nutrition**. Now a days more emphasis is being exerted on preventive medicine to forstall diseases. Biochemical studies and research has resulted in an enormous impact on our understanding about good nutrition and preventive medicine.

Clinical biochemistry explains several aspect of disease processes and this vast biochemical knowledge is of immense help to physicians in solving various clinical problems to diagnose the diseases, as well as to plan and execute logical treatment. Biochemical investigation procedures are very important in solving clinical problems and play great role in revealing several aspects of disease process. Some important of them are listed below :-

1. Biochemical research reveals the fundamental root cause and mechanisms underlying various diseases.
2. This knowledge help physicians to plan rational and logical treatment of the disease concerned.
3. Biochemical laboratory investigations assists in the diagnosis and differential diagnosis of specific diseases. For example estimation of total and conjugated serum bilirubin levels and specific enzyme estimations (Chapter 34) facilitate differential diagnosis of various types of Jaundice.
4. Screening tests for early diagnosis of the dreaded killer diseases are of great help in management of the patient to prevent severe irreversible clinical complications which can be life threatening.
5. Prenatal (before birth) or just after birth, screening tests facilitate early diagnosis of certain serious genetic diseases. This would help physician to manage and monitor the affected infant, so that severe irreversible clinical complications can be prevented.
6. Biochemical investigations are of paramount importance in proper monitoring of the patient and to assess the response to a treatment or therapy, particularly in killer diseases such as cancer, AIDS, heart diseases, hepatitis B etc.

Biochemistry is the most rapidly developing subject in medical sciences. The molecular genetics has advanced tremendously in last 3-4 decades. The great advancement in recombinant DNA technology and genetic engineering in recent years has revolutionized the modern medical practice.

Synthesis of gene in laboratory with the help of genetic engineering was a landmark in the history of biochemistry and has resulted in immense progress in the field of molecular genetics and its use in medical practice. With the advancement in recombinant DNA technology, it has become possible to synthesize commercially, certain proteins (e.g. insulin) and vaccines in pure form without any contamination for their use in the medical practice.

By genetic engineering it has now become possible to transfer genes from one organism to another, for induction of better qualities in the offsprings. Genes can be clonned in the laboratories to obtain large population of identical copies of parent DNA (gene) to be stored in **Gene Libraries** for their future medicinal uses.

Gene therapy has now become possible to manage several life threatening genetic diseases and inborn errors of metabolic disorders. In this therapy an abnormal (mutant) gene responsible for a specific genetic disease is replaced in the affected subject by introducing a normal gene by special technique so that child born may not be affected by the genetic disease (details in Chapter 23)

Efforts to understand the basic cause of familial hypercholesterolemia, which results in severe atherosclerosis and heart diseases at an early stage, have lead to dramatic progress in our knowledge about **cell membrane receptors** and mechanisms of uptake and metabolism of different lipoprotein fractions.

The advancement in research studies of oncogene activation and cancer development have resulted in discovery of several factors involved in the control of normal cellular growth at molecular level. This knowledge regarding mechanisms triggering a normal cell to become a cancerous cell has resulted in the use of chemotherapeutic drugs in control and treatment of cancerous tumors.

1

PROTEINS AND AMINO ACIDS

Proteins are nitrogenous biopolymers formed by the condensation of many molecules of α–amino acids linked together by peptide bonds. Proteins are primary compounds that are sole basis of life. Life is possible only because of the unique physiochemical properties of thousands of different proteins present in a living cell. Proteins are not only the essential structural component of living protoplasm, but are of paramount importance as they perform wide range of physiological functions that are absolutely necessary for the life.

CLASSIFICATION OF PROTEINS

Chemically proteins are classified into 3 major groups as described here under:

A. **Simple Proteins**—The simple proteins are naturally occurring proteins that are polymers of α -amino acids and on hydrolysis produce a mixture of around 20 α amino acids only.

B. **Conjugated Proteins**—These are compound proteins and contain some non protein groups or substances in addition to the protein part.

C. **Derived Proteins**—These are derived as a result of denaturation or partial hydrolysis of natural simple or compound proteins.

A brief account of classification of proteins along with important examples of each class is given here under:

Proteins

- Simple Proteins
  - Albumin
  - Globulin
  - Glutelin
  - Scleroproteins examples are
    - Collagen
    - Elastin
    - Keratin
  - Protamines
  - Histones
  - Gliadin
- Conjugated Proteins
  - Nucleoproteins (RNA, DNA + histones)
  - Chromoproteins (Heme + globin) e.g. - Hemoglobin, Myoglobin
  - Phosphoprotein e.g. - Casein
  - Lipoproteins (Lipid + Protein) e.g. -VLDL, LDL, HDL
  - Glycoproteins Carbohydrate + Protein e.g. -Immunoglobulins
- Derived Proteins
  - Proteoses
  - Peptones
  - Gelatin
  - Polypeptides

BIOLOGICAL FUNCTIONS OF PROTEINS

1. Structural Proteins—Proteins form essential constituent of structural frame work of various cellular components. Different types of tissues are formed in body due to the different types of proteins present in them, for example collagen of connective tissue.

2. Functional Proteins—Various physiological and biochemical functions of the body tissues are carried by different proteins present in them. Some important examples are enumerated in the Table 1.1.

i) All enzymes are proteins in nature, that catalyze various biochemical and metabolic reactions of the body.

ii) Many hormones are chemically proteins or polypeptide in nature and regulate various physiological and metabolic processes of the body.

3. Transport Proteins—Lipids and metals are transported as lipoproteins and metalloproteins via blood to tissues. The example are VLDL, LDL, HDL, transferrin, ceruloplasmin etc.

4. Oxygen Transport Protein—Transport of oxygen and biological oxidation is achieved by hemoglobin, myoglobin and cytochrome system.

5. Carrier Proteins—Certain carrier protein transport substances by active transport mechanism (e.g.active absorption, facilitated absorption) across semipermeable cell membrane, as well as they facilitate selective reabsorption of important nutrients and electrolytes by renal tubules from glomerular filtrate.

6. Storage form of Proteins—Ferritin is a storage protein that store iron in tissues.

7. Immuno Proteins—γ Globulins act as antibodies against foreign invaders and pathogens to provide immunity to body against infections.

8. Receptor Proteins—The various hormones of the body bind with specific receptors to bring about their diversified physiological action in the target tissues, for example insulin receptor, LDL receptors etc.

The structural and functional relationship of proteins is summarized in Table-1.1

Table 1.1 : Various types of proteins with specific functions

S.No.	Biological Role	Proteins/Examples
1.	Structural proteins	Collagen, Keratin
2.	Enzymes	Pepsin, Amylase, Lipase
3.	Hormones	Insulin, Growth hormone, Glucagon
4.	Transport proteins	Lipoprotein, Transferrin
5.	Oxygen transport proteins	Hemoglobin, Myoglobin
6.	Electron transport chain	Cytochromes
7.	Metalloproteins	Ceruloplasmin,
8.	Receptor proteins	Hormone insulin receptor, LDL receptor
9.	Storage Proteins	Ferritin
10.	Immune function	Gammaglobulins, Immunoglobulins
11.	Contractile proteins	Myosin & actin are involved in muscle contraction

PROTEIN STRUCTURE

Protein are macromolecules with high molecular weight and are formed by the polymerization of many molecules of α -amino acids linked by peptide bonds. The α -amino acids are the building blocks of protein molecule and have an amino group on α C–atom of a carboxylic acid as shown under:

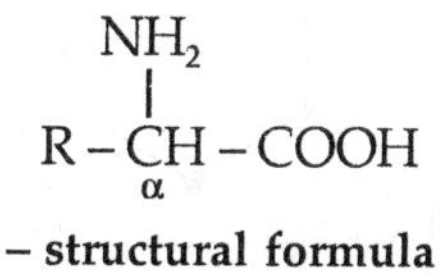

α amino acid – structural formula (R-side chain)

Peptide Bond—When carboxylic group of one amino acid condenses with the α–amino group of another amino acid, a molecule of water is lost and a peptide bond (CO – NH) is formed between two amino acids to from a dipeptide, as shown in the Fig 1.1

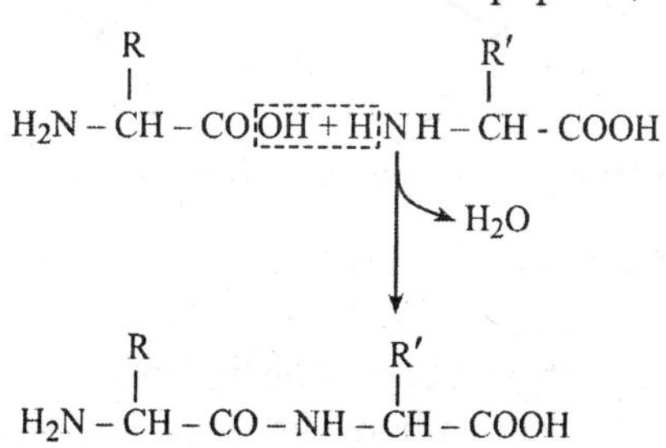

Dipeptide

Fig. 1.1 : Peptide bond formation between two α amino acids to form a dipeptide

POLYPEPTIDE FORMATION

The free – COOH group of the dipeptide formed as described above then condenses with α amino group of third amino acid to form another peptide bond and a tripeptide is produced, which in turn can be linked to many more amino acids sequentially by peptide bonds. In this manner many amino acid molecules are linked together by peptide bonds to produce a long polypeptide chain and a protein molecule is formed (Fig. 1.2).

$$\begin{array}{l} \quad\quad\ R_1 \quad\quad\quad\quad\quad\ \ R_2 \quad\quad\quad\quad\quad\ \ R_3 \quad\quad\quad\quad\quad R_4 \quad\quad\quad\quad\quad\ R_n \\ \quad\quad\ \ | \quad\quad\quad\quad\quad\quad\ | \quad\quad\quad\quad\quad\quad\ | \quad\quad\quad\quad\quad\quad | \quad\quad\quad\quad\quad\quad\ | \\ H_2N—CH—CO—NH—CH—CO—NH—CH—CO—NH—CH\text{-}CO \ldots NH—CH—COOH \end{array}$$

Fig. 1.2 : α – Amino acids linked by peptide bonds in a polypeptide chain

Various types of proteins with specific physiological functions are produced in a living cell by polymerization of the 20 naturally occurring α amino acids, just by varying the **number** and **specific sequence** in which they are linked together in long polypeptide chain.

STRUCTURAL ORGANIZATION OF PROTEINS

There are four levels of structural organization of protein molecule structure as described here under :

1. Primary Structure —The primary structure of protein molecule comprises the number and specific sequence of α–amino acids in a long polypeptide chain. This specific sequence of linear arrangement of amino acids in a long polypeptide chain is determined by the genetic code on genes of DNA molecule during protein biosynthesis.

2. Secondary Structure—The long polypeptide chain then gets folded in a specific but regular manner, that is guided by the primary structure and free groups and radicals present on the polypeptide chain. The two most common forms in which the polypeptide chain is folded are α - helix and β – pleated sheet as shown in the Fig 3.

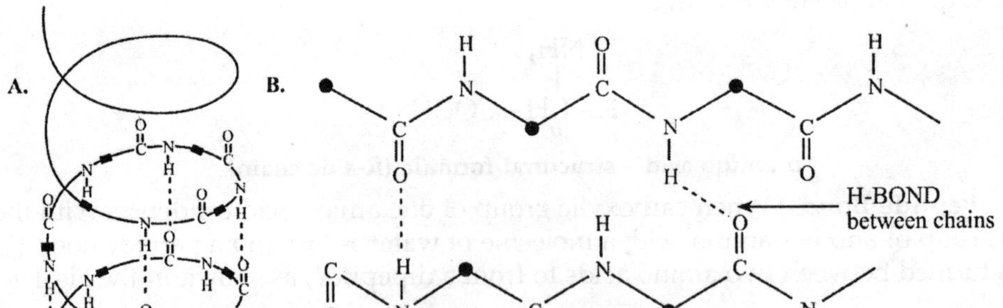

Fig. 1.3 : Folding of polypeptide chain as α– helix (A) and β– pleated form (B)

Bonds maintaining secondary protein structure

a) **Hydrogen Bonds**—The specific folded structures as mentioned above, are maintained by hydrogen bonds. The hydrogen bonds are formed by sharing a hydrogen atom between carbonyl group of one peptide bond and NH group of another peptide bond of the adjacent fold of polypeptide chain. The hydrogen bonds are weak, loose, non covalent bonds which can be easily broken.

b) **Disulphide Linkage**—The disulphide bonds (-S-S-) formed between two cysteine residues of adjacent fold or loops of polypeptide chain are responsible for maintaining the secondary structure of protein molecule. The disulphide linkages are strong, covalent chemical bonds and are not broken easily.

3. Tertiary Structure —The tertiary level of protein structure is formed by packaging compactly, the α helix or β pleated sheet of polypeptides into a globular or fibrous shape or other regular compact molecular shapes as depicted in the Fig. 1.4.

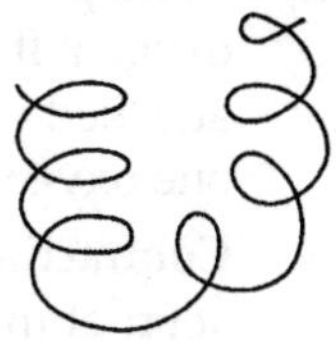

Fig. 1.4 : Tertiary molecular structure of proteins.

Certain physical forces are responsible for maintaining the tertiary level, three dimensional structure of a protein molecule. The secondary and tertiary structure are specified by the sequence of amino acids in the polypeptide chain, which after synthesis gets folded spontaneously in a regular three dimensional biologically active conformation. The forces responsible for maintaining the tertiary level of protein molecular structure are mainly physical hydrophobic interactions and multiple weak Van der waal forces between tightly packed aliphatic (hydrophobic) side chains in the interior of protein molecule.

4. Quaternary Structure—Some proteins are biologically and physiologically functional only, when many monomeric subunits (one polypeptide chain properly folded in three dimensional secondary and tertiary levels) combine in a fourth level (quaternary) in a particular special arrangement. For example hemoglobin is a tetrameric protein having four monomeric subunits i.e. two α and two β polypeptide chains, each attached to a heme unit and are arranged together in a tetrahedron structure as tetramer functional protein.

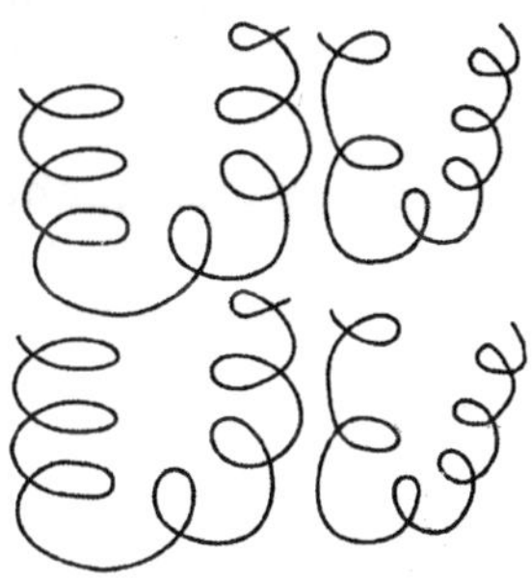

Fig. 1.5 : Quaternary structure of protein with four monomeric units (tetramer)

The quaternary level of protein molecular structure is maintained by certain covalent and non covalent interactions. Many interchain electrostatic bonds stabilize the quaternary structural integrity of the protein molecule. Disintegration of quaternary molecular structure into monomeric subunits results in the loss of biological activity of the protein (Fig. 1.5).

Structure - Functional Relationship—The characteristic biological functions of proteins are related to their tertiary and quaternary level of structural integrity. This refers to the three dimensional conformation of the regularly folded polypeptide chains, which assemble in specific manner to form **special domains**. The specific domains is a section of protein molecule which facilitates the protein to perform a physiological task. For instance a specific substrate binds to enzyme protein at the active site to catalyze a biochemical reaction. A specific ligand or inhibitor interacts and binds on specific domain to modulate its function.

To illustrate the structure-functional relationship following examples of proteins can be taken into consideration.

1. **Enzymes have Two Specific Domains**—These are 1) catalytic domains for specific substrate binding which have certain groups or radicals required for catalysis and 2) allosteric sites on which certain modulators bind to increase or inhibit enzyme catalysis.
2. **Transport Protein**—Hemoglobin has specific domain to bind and transport oxygen. It is a tetrameric protein molecule with four monomeric subunits, each attached with a heme unit containing ferrous iron in a special pocket that binds one oxygen molecule.
3. **Contractile Proteins**—The typical example of contractile proteins is myosin and actin of muscle fibers, that participate in contraction of muscles.
4. **Structural Proteins**—Collagen fiber is the most abundant fibrous protein component of connectives tissue of skin, bones, cartilages, tendon, blood vessel walls etc. It has a strong superhelical, cable like tertiary structure and is rich in hydroxy proline and hydroxy lysyl residues. These form cross linkages so that a fine net work (mesh) is formed, which is then filled up with ground cementing substances (mucopolysaccharide) to give strength to these tissues to make them strong.
5. **Receptor and Carrier Proteins**—These proteins have a specific domain for recognition and binding of only one particular ligand (compound) for its transport across biomembranes or to bring about its action inside the target tissues (e.g. hormone receptors)
6. **Immunoglobulins**—The γ-globulins have specific variable domains to bind specifically to an antigen (foreign protein, virus or bacteria) to destroy them and thus provide immunity to body against infections.

PHYSICAL PROPERTIES OF PROTEINS

Proteins being macromolecules, form colloidal solution in water and exhibit characteristic colloidal properties such as scattering the beam of light and exert osmotic pressure etc. The protein colloidal particles remain in solution and do not aggregate because of the presence of a shell of hydration on protein molecular surface and having similar ionic charges which repel each other.

Amphoteric Nature

The protein molecules have free NH_2 as well as free COOH group of basic and dicarboxylic amino acids respectively, that are not involved in peptide linkages. These free NH_2 and COOH groups ionize depending upon the pH of solution into NH^+_3 (positive charge) or COO^- (negative charge) ions and thus behave as both acid or bases and the protein molecule will move in an electric field either towords cathode or anode.

Isoelectric pH

At isoeletric point (pH) the free NH_2 group of a protein particle receives a H^+ ions from COOH group (intramolecularly) and thus the net charge on protein molecule becomes zero and it is called **zwitter ion**. At this pH the protein particle (zwitter ions) will not move in the electric field (as net charge on particle is zero) and it is known as isoelectric point.

Denaturation

Certain physical agents such as heat, X-rays, high pressure, vigorous shaking and chemicals like phenol, salicylate etc. disintegrate the quaternary, tertiary and secondary level of protein molecular structure by breaking loose physical forces and weak noncovalent hydrogen bonds. Due to this the natural (native) protein polypeptide chain gets uncoiled and unfolded in an irregular shape called denatured protein with altered properties and result in loss of biological activity. The process is known as denaturation of proteins and is depicted in the Fig. 1.6.

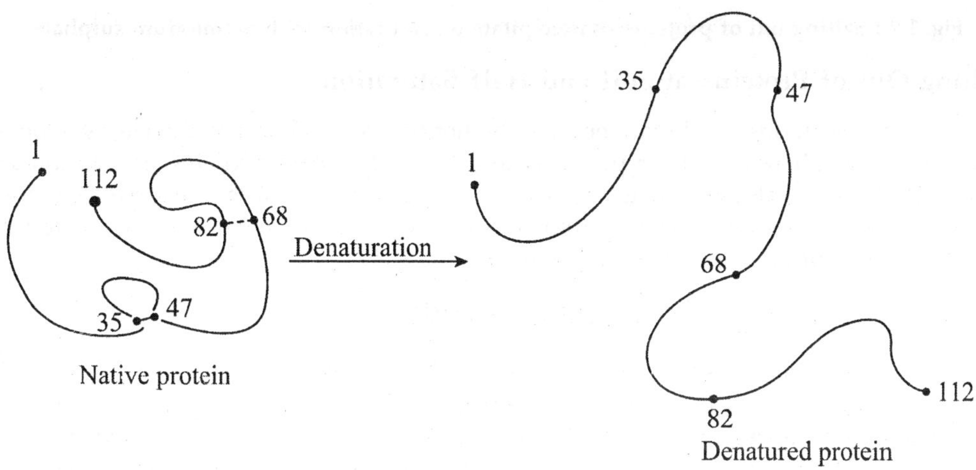

Fig. 1.6 : Denaturation of proteins molecule to unfolded polypetide chain

The denatured protein may or may not be renatured back, depending on the degree of damage to the protein molecular structure. Protein solutions kept for long period at room temperature may denaturate and loose their biological properties. But if protein solution is freezed at low temperature, its molecular structural integrity and biological properties can be preserved.

Heat Coagulation

Some protein solutions such as albumin and globulin, when heated are **denatured irreversibly** and exhibit altered properties, for example reduced solubility and form thick white flocculation called coagulum.

Heat Coagulation Test

This property of coagulation of protein solution by heat forms the basis of heat coagulation test commonly utilized as detection test for the presence of albumin in abnormal urine sample. (details in Chapter 36.)

Salting Out of Protein

The colloidal protein solution is stabilized by two factors. These are : 1) shell of hydration and 2) similar ionic charges on protein particles. Any factor which neutralizes the charges or removes the shell of hydration will cause precipitation of proteins. For instance when a saturated solution of salt or solid ammonium sulphate or sodium sulphate is added to a protein solution and mixed by vigorous shaking, they get precipitated. The process is known as "salting out of protein" as shown in the Fig. 1.7.

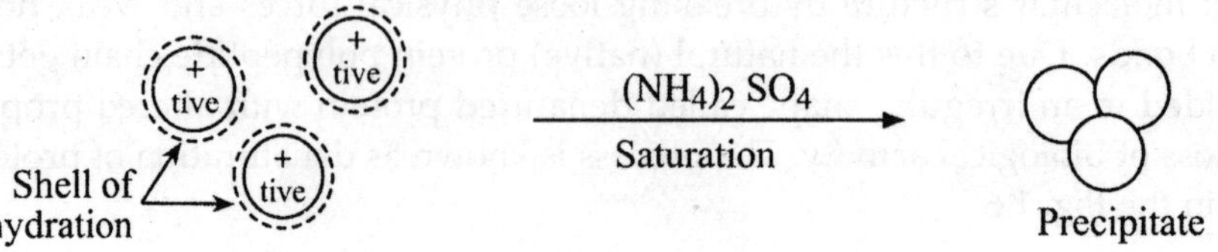

Fig. 1.7 : Salting out of proteins as precipitate on saturation with ammonium sulphate

Salting Out of Proteins at Full and Half Saturation

It has been observed that generally higher the molecular weight (size) of protein, the amount of salt required for precipitation is less. So some proteins in solution are precipitated at half saturation (e.g. – globulin) and some need full saturation of salt to get precipitated such as albumin. A mixture of proteins thus can also be separated by precipitation at different salt concentrations.

AMINO ACIDS

The α–amino acids are building blocks of protein molecule. There are 20 naturally occurring α amino acids, from which all types of proteins are synthesized in a living cell. α–Amino acids have an amino group attached to the α carbon of a carboxylic acid as shown below. As α C-atom of amino acids is asymmetric, D and L isomers of amino acids are possible. However only L-isomer of α–amino acids are utilized in our body for various metabolic purposes and for proteins synthesis.

$$\begin{array}{c} NH_2 \\ | \\ R-CH-COOH \end{array}$$

L-α-amino acid

Classification of α-Amino acids

The naturally occurring 20 α–amino acids differ in the structure of side chain or 'R' group and accordingly are classified into six major groups. In addition to protein synthesis many amino acids serve very important functions in our body. The classification of amino acids along with there chemical structure is summarized in Table 1.2.

Table 1.2 : Classification of Amino acids with chemical nature

S.No.	Classes of Amino Acids	Chemical Structure
1.	**Aliphatic amino acids**	
	(a) Glycine	$\begin{array}{c} H-CH-COO^- \\ \vert \\ NH_3^+ \end{array}$

contd. ...

... contd.

S.No.	Classes of Amino Acids	Chemical Structure
	(b) Alanine	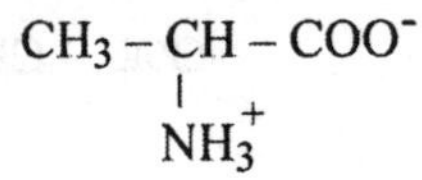
	Branched chain amino acids	
	(c) Valine	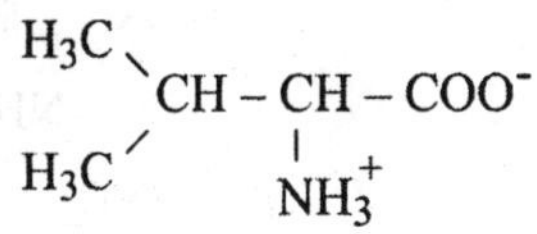
	(d) Leucine	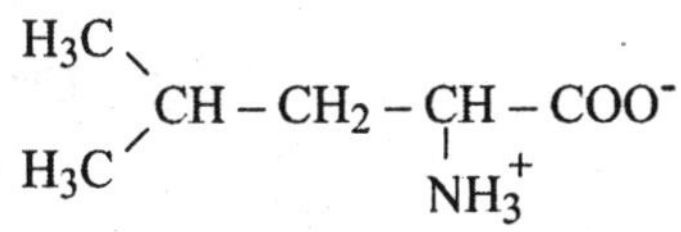
	(e) Isoleucine	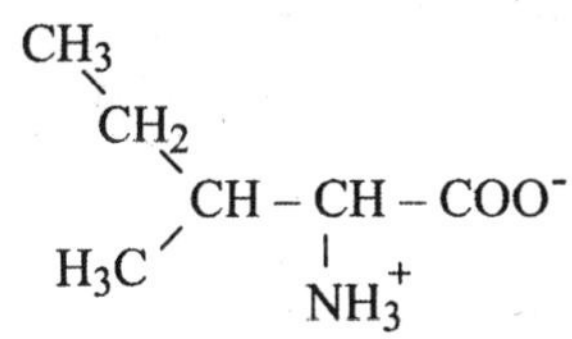
2.	**Hydroxy amino acids**	
	(a) Serine	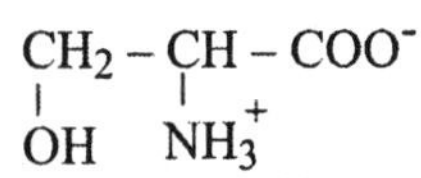
	(b) Threonine	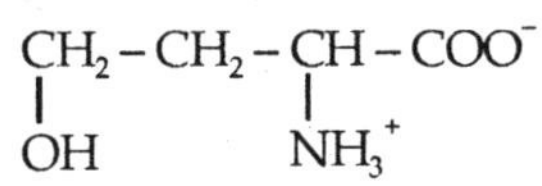
3.	**Sulphur containing amino acids**	
	(a) Methionine	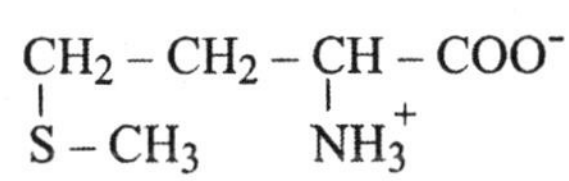
	(b) Cysteine	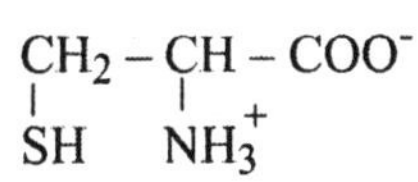
	(c) Cystine	
4.	**Dicarboxylic amino acids**	
	(a) Aspartate	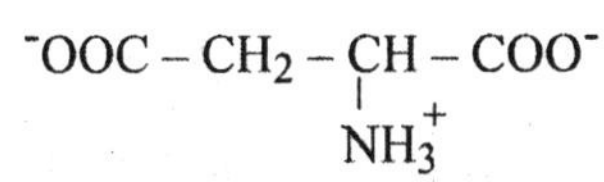

contd. ...

... contd.

S.No.	Classes of Amino Acids	Chemical Structure
	(b) Glutamate	$^{-}OOC-CH_2-CH_2-CH(NH_3^+)-COO^-$
5.	**Dibasic amino acids**	
	(a) Arginine	$H-N(C(=NH_2^+)NH_2)-CH_2-CH_2-CH_2-CH(NH_3^+)-COO^-$
	(b) Lysine	$CH_2(NH_3^+)-CH_2-CH_2-CH_2-CH(NH_3^+)-COO^-$
6.	**Aromatic amino acids**	
	(a) Phenylalanine	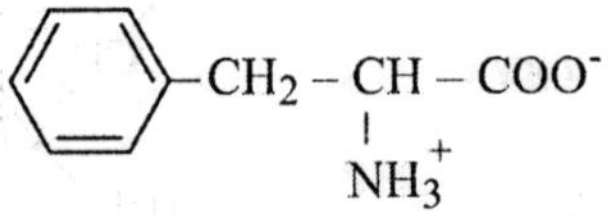
	(b) Tyrosine	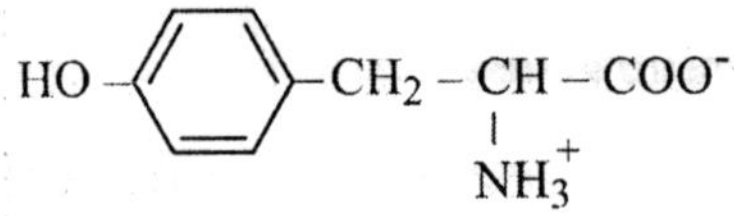
	(c) Tryptophan	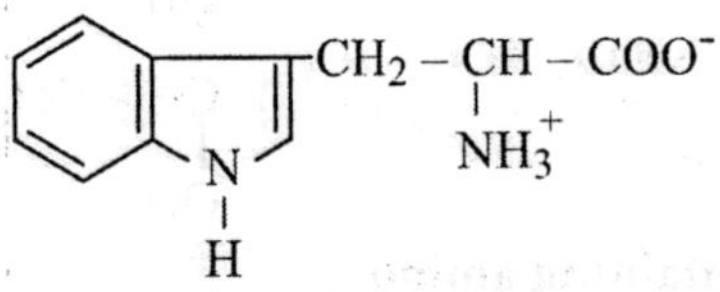
	Imino acids	
	(d) Histidine	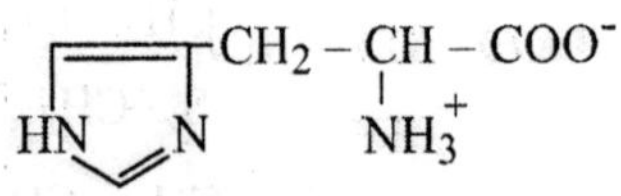
	(e) Proline	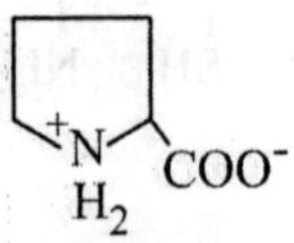

General Characteristics of Amino Acids

1. Isomerism—All naturally occurring amino acids, except glycine have at least one asymmetric C – atom (α C – atom) and show stereoisomerisms. Example are D and L isomers as shown ahead:

L-alanine D-alanine

2. Amphoteric nature—The α–amino acids have both acidic i.e. carboxylic (COOH) and basic amino (NH_2) groups and thus they show both acidic and basic characteristics depending on the pH of medium. The carboxylic and amino groups of amino acids ionize as either positively charged ions or negatively charged ions as shown below:

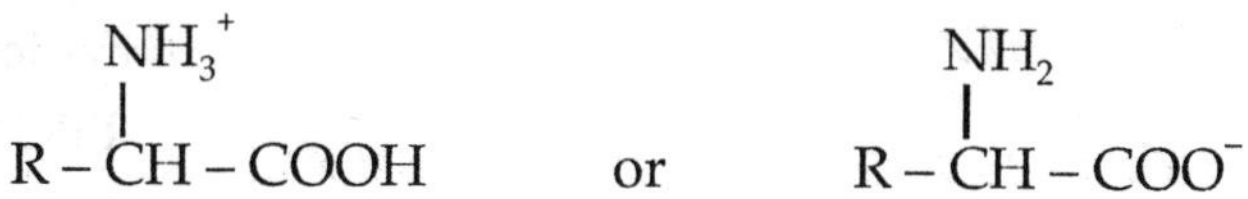

Acidic pH Basic pH

3. Isoelectric pH—At a particular pH the proton (H^+) from COOH group of an amino acid is taken up by NH_2 group intramolecularly with a result, both positive and negative charges are equal on the particle and net charge is zero. This ion will not move in an electric field and is called as zwitter ion. This pH is known as isoelectric point and the events occuring are as illustrated in the Fig.

Isoelectric pH Zwitter ion

Formation of Peptide Bond

The most important characteristic of amino acids is that they condense together by formation of a peptide bond between carboxylic group of one amino acid and amino group of another amino acids, with the removal of elements of water as shown below:

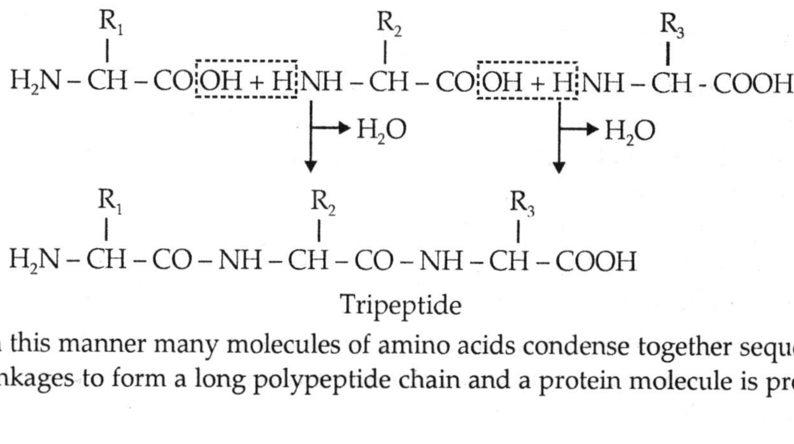

Tripeptide

In this manner many molecules of amino acids condense together sequentially by peptide linkages to form a long polypeptide chain and a protein molecule is produced.

Some important biologically active polypeptides along with their functions are summarized in the Table 1.3.

Table 1.3 : Biologically important polypeptides with their functions

S.No.	Name	Peptide	Occurence	Function
1.	Glutathione	Tripeptide	gly – cys – glu | SH Erythrocytes	In RBC's it provides sulfhydryl buffer and detoxify H_2O_2, superoxide anion and other toxic oxy free radical and maintain RBC fragility.
2.	Oxytocin	Octapeptide	Posterior pituitary hormone,	Causes uterine contraction during labour and stimulate secretion of milk
3.	Vassopressin/ ADH	Octapeptide	Posterior pituitary hormone,	Causes reabsorption of water from glomerular filtrate, regulates blood volume
4.	ACTH, β MSH	Polypeptides	Anterior pituitary hormones	Stimulate secretion of adrenal corticoid hormones
5.	Gastrin, secretin, pancreozymin	Polypeptides	Gastrointestinal hormones	Stimulate GIT secretions

Essential Amino Acids

Out of the twenty naturally occurring amino acids, ten α–amino acids are known as essential amino acids, as they cannot be synthesized in our body and have to be taken in the diet in adequate amount to maintain normal health and growth of an individual. These essential amino acid in our diet are supplied by nutritionally good quality proteins such as albumin, globulin(egg), casein (milk), glutelin (wheat) etc.

Out of these ten essential amino acids, eight are indispensable or essential amino acids while two are considered as semi essential and are as enumerated in the Table 1.4.

Arginine and histidine are considered as semi essential because, though they are synthesized in our body and ase sufficient for health of an adult but quantity is inadequate during growth and development, so must be supplied in the diets of children.

Table 1.4 : Essential Amino Acids

Dietary Indispensable Amino Acids	Semi Essential Amino Acids
Methionine	Arginine
Threonine	Histidine
Tryptophan	
Valine	
Leucine	
Isoleucine	
Phenylalanine	
Lysine	

ESSENTIAL AMINO ACIDS —BIOLOGICAL VALUE OF PROTEINS

As dietary proteins are the main source of essential amino acids for the maintenance of health, growth and development of an individual, both the quality and quantity of proteins in our diet is very important. The nutritional quality (biological value) of protein is determined by the amount and proportions of essential amino acid contents. Accordingly the proteins containing proper quantities of each essential amino acid are considered as good quality proteins (high biological value) and others as fair quality proteins, which lack one or other essential amino acids. The nutritional value of some dietary proteins is summarized in Table 1.5.

Table 1.5 : Nutritional values of proteins

S.No.	Quality of Proteins	Source	Essential Amino Acids Content
1.	**High Quality**		
	Albumin	Egg & Poultry	Rich in all essential amino acids
	Globulin	Egg & Poultry	
	Casein	Milk & Milk Products	
2.	**Good Quality**		
	Meat Proteins	Liver, meat	Appropriate amount of essential amino acids
3.	**Fair Quality**		
	Pulses	Moong, arhar, channa	Lack one or other essential amino acids
	Cereals	Wheat, rice	
	Beans	Peas, soyabean	

DIETARY REQUIREMENT OF PROTEINS

Recommended daily allowance of protein in diet depends on both quality as well as quantity of dietary proteins. Normally 45 to 56 g/day proteins of good quality are recommended to maintain good health of an adult.

Deficiency of Essential Amino Acids

Deficiency of good quality protein or any one essential amino acid in our diet would adversely affect our health, as protein synthesis in our body would be impaired. Protein deficiency is generally observed during growth of infants and during childhood, as requirement of rich amount of essential amino acids (good quality proteins) is necessary for the growth and development of child.

Deficiency State—Clinical Manifestation

Initially lack of good quality protein or deficiency of any single essential amino acid in our diet would result in **negative nitrogen balance,** which is characterized by less N_2 (proteins / amino acids) retention in our body. This results in more catabolism of amino acids (proteins) and more nitrogenous waste product excretion (urea) in urine. The protein synthesis in our body is impaired which results in following clinical manifestations:

- i) **Hypoproteinemia (↓ plasma proteins)** which causes osmotic pressure changes, water retention and edema on face, legs etc.
- ii) **Anemia**—Decreased hemoglobin level in the blood.
- iii) **Protein malnutrition**—When deficiency of good quality proteins in diet is severe enough, it can result in clinical syndrome known as **kwashiorkar**. The clinical symptoms includes stunted growth, fatty liver, jaundice, anemia, edema, skin disorders, dermatitis etc. (details in the Chapter 27.)

❑❑❑

2

Enzymes

Enzymes are biocatalysts that accelerate the rate of biochemical reactions going on in a living organism, which otherwise would not occur in the mild conditions of temperature (37° C) pressure and pH (neutral 7.4) prevailing in a living cell. Life is possible only due to the presence of enzymes, which catalyze and regulate various metabolic processes necessary for life. Almost all functions of body, such as digestion of food, catabolism of nutrients for energy generation to carry out vital activities of living organism and biosynthesis of specialized compounds like proteins, RNA, DNA etc, are catalyzed and controlled by specific enzymes.

DEFINITION

Enzymes are **biocatalyst,** synthesized by living cell, which **accelerate** the rate of biochemical reactions. Chemically they are protein in nature, highly sensitive and very specific in their action.

CHARACTERISTICS OF ENZYMES

1. Efficiency—Unlike ordinary inorganic catalyst, enzymes are very efficient. Hydrolysis of protein in laboratory by boiling with dilute HCl may take many hours, but inside body, in the presence of proteolytic enzymes (eg. pepsin, trypsin), hydrolysis of dietary protein occurs very fast, within a very short time at body temperature (37°C) to completely digest proteins in gastrointestinal tract.

2. Enzyme Sensitivity—Enzymes being protein in nature, their activity is affected very much and are highly sensitive to any change in temperature, pH and other conditions of reaction medium.

3. Enzyme Specificity—Enzymes are highly specific in their action unlike ordinary inorganic catalyst. Enzyme specificities are of various kinds :

(a) **Absolute Specificity:** Some enzymes act on one substrate only e.g. urease act only on urea to decompose it into NH_3 and CO_2.

(b) **Reaction Specificity:** Some enzymes are specific to catalyze only one type of reaction. For example transaminases catalyze transfer of amino group from one amino acid to an α-keto acid.

(c) **Stereo-Specificity:** Enzymes usually act only on **one isomer** to catalyze a reaction. For instance, L-amino acid oxidases, oxidize only L-isomers of amino acids and glucokinase acts on only D-isomer of glucose.

(d) **Optical Specificity:** Enzymes are specific to the extent of acting only on either dextro or levo rotatory isomers.

4. Enzyme activity can be regulated—Unlike ordinary catalyst, enzyme catalysis can be minutely and very efficiently be controlled by the presence of certain activators or inhibitors. This is very important for life, as by regulating activity of key enzyme of a metabolic cycle, the metabolic rate can be increased or decreased by the presence of certain activator or inhibitors, depending on the cellular need at a particular moment.

NOMENCLATURE AND CLASSIFICATION

Enzymes are named by adding suffix "ase" after the name of substrate on which they act, or after the type of reaction they catalyze. For example, urease is called so, as it acts on urea and the enzymes which catalyze transfer of amino group are named transaminases. Previously enzyme were named randomly, as they were discovered.

As enzymes are highly specific in their action, thousands of different enzymes are present in a living organism and it will be very difficult to remember them. So to make it easy, convenient and rational, a **precise classification** of enzymes was developed by **International union** of **biochemists (IUB system)**. According to the IUB system, all enzymes are placed in **6 major classes**, based on type of reactions they catalyze and were given a 4 digit code number for complete information about an enzyme. The 4 digits in code denotes: Class, Subclass, Subsub class and specific enzyme respectively. Classification and selected examples for six major classes of enzymes are given in Table 2.1.

Table 2.1 : Classification of enzymes along with specificities

S.No.	Enzyme Class with examples	Type of Reaction Catalyzed	Example	4 Digit Code	Reaction Catalyzed
1	**Oxidoreductases** Oxidases, dehydrogenase, L- Amino acid oxidases.	Transfer of H_2 / electrons, Addition of oxygen	Alcohol dehydrogenase	1.1.1.1.	Ethanol → Acetaldehyde + + NAD^+ NADH+ H^+
2	**Transferases** Transaminase Transmethylase	Transfer of NH_2, phosphate, acyl, methyl groups etc.	**Hexokinase** ATP- Glucose Phopho - transferase	2.7.1.1	Glucose ⟶ Glu.6 PO_4 + + ATP ADP

contd. ...

... contd.

S.No.	Enzyme Class with examples	Type of Reaction Catalyzed	Example	4 Digit Code	Reaction Catalyzed
3	**Hydrolases** Pepsin, Trypsin, Choline estrase.	Hydrolysis of peptide, glucosidic, ester bonds by adding H_2O	**Lipase** (triacyl glycerol hydrolase)	3.1.1.3	TAG + H_2O ⟶ DAG + FA
4	**Lyases** Fumrase, Histidase	Breaking of bonds without adding H_2O	Aldolase	4.1.2.7.	Fruct 1,6 bisphosphate ↓ DHAP+glyceraldehyde 3-PO_4
5	**Isomerases** Retinal Isomerase	Interchange of one isomer into other	Triosephosphate isomerase	5.3.1.1	Di OH acetone PO_4 ↓ Glyceraldehyde 3-PO_4
6	**Ligases** Acyl CoA carboxylase	Formation of bonds , C-N, C-C by utilizing ATP	Glutamine synthetase	6.3.1.2	Glutamate→Glutamine+ +NH_3 ATP ADP + Pi

CHEMICAL NATURE OF ENZYMES

All enzymes are chemically protein in nature. However, recently a few RNA molecules have been shown to have highly substrate specific catalytic activities and are called as **ribozymes**. Each enzyme protein molecule has its own primary, secondary and tertiary structure and specific molecular configuration, which is essential for its catalytic activity. Many enzymes have **oligomeric functional units** such as dimer, tetramer etc, with quaternary level of structural integrity. For example lactate dehydrogenase and hexokinase are tetramers with four subunits in an active enzyme molecule.

Multienzyme complex—Several enzymes of a particular metabolic cycle exist as multienzyme complex, containing many polypeptide units, each with a specific catalytic activity. They catalyze conversion of substrate in a sequential series of reactions to produce different intermediate products, and then to form final metabolic product. Examples of multienzymes are fatty acid synthase, pyruvate dehydrogenase etc.(Chapters 5 & 11)

Cofactors and Coenzymes

Many enzymes require certain cofactors and coenzymes for their catalytical activity. The complete biologically functional enzyme is called **holoenzyme**, which consists of an **apoenzyme** (the protein part) and a **nonprotein prosthetic** group called the **coenzymes.** The coenzymes are bound to the protein part called apoenzyme, to form an active holoenzyme.

Holoenzyme = Apoenzyme + Coenzyme
(active enzyme) Protein part Nonprotein prosthetic group.

Cofactors are nonprotein prosthetic groups that bind with protein part (apoenzyme) to form an active holoenzyme. If the prosthetic group or coenzyme is removed from the apoenzyme, it looses its catalytical activity. The cofactors may be organic nonprotein prosthetic groups, the coenzyme or inorganic metallic ions, such as Mg^{++}, Ca^{++}, Fe^{++} etc.

Coenzymes are nonprotein organic prosthetic group, of low molecular weight, dialyzable compounds associated with enzyme catalytic functions. Coenzyme takes part in the reaction but reverts back to the original compound after completion of enzyme catalysis and can be regarded as **cosubstrate** or second substrate. Coenzymes are required necessarily by the enzymes belonging to group transfer or H_2/ electron transfer reactions as well as by isomerases and ligases. (Table 2.2).

Vitamin B Complex members are activated as Coenzymes—Most of the Vitamin B complex members are activated in body as important **coenzymes** necessary for **enzymatic reactions** of important metabolic processes. Some important coenzymes along with their biochemical role and reactions catalyzed with specific enzymes are summarized in the Table 2.2.

Table 2.2 : Important coenzymes with specific biochemical role

S.No.	B complex Vitamin	Coenzyme	Enzymes	Reaction Catalyzed/ Metabolic Pathway
1.	Thiamine	Thiamine pyro-phosphate (TPP)	Pyruvate and α keto glu. dehydrogenase. Transketolase	Glycolysis/Citric acid cycle, HMP shunt
2.	Riboflavin	FMN FAD	Dehydrogenase, L&D amino acid oxidases	Biological Oxidation
3.	Niacin	NAD NADP	Dehydrogenases Glu. 6. PO_4 dehydrogenase	Biological Oxidation HMP shunt
4.	Pyridoxal	B_6PO_4 (PLP)	Transaminases, Amino acid decarboxylase	Amino acid metabolism, heme synthesis.
5.	Pantothenate	Coenzyme A	AcylCo A	Fatty acid metabolism
6.	Biotin	Biotin	Carboxylases	CO_2 fixation reaction to increase C-chain
7.	Folic acid	THF (FH_4)	One C - moiety carrier	Purine, pyrimidine & DNA, RNA synthesis
8.	Cyanocobal-amine	Methylcobal-amine, Adenosyl-cobalamine	Methyl malonyl CoA mutase Isomerases	Methylation Isomerization

MECHANISM OF ENZYME CATALYSIS

Enzyme catalysis in biological system is basically similar to inorganic catalysis. For any chemical reaction to proceed, two conditions are necessary:

1. Molecules to react should **collide** with each other.
2. Reactants should possess sufficient **kinetic** energy to overcome the **energy barrier** of the reaction. Most of the reactants in test tube would not react **spontaneously,** as they do not possess suffient energy to overcome the energy barrier of the reaction. Each chemical reaction has its own energy barrier, which has to be crossed by reactants, for a reaction to proceed.

In **ordinary chemical reactions,** by increasing **temperature, pressure** and by providing **low or high pH** to the reaction medium, we can provide sufficient energy to the reactants to cross energy barrier of reaction and reaction rate increases.

But in **biological system** most of the metabolic reactions would not occur at the **mild conditions of temperature** (37° C), **pressure** and **pH** (neutral at 7.4), prevailing in the living cell, if enzymes were not present.

The catalysts and enzymes help in fulfilling the above mentioned conditions and rate of reaction is accelerated.

Enzymes lower the energy barrier of reaction

Enzymes have catalytic site on their molecular surface, which is **specific** for the reactants (substrate), on which they **bind** to form **enzyme-substrate complex,** which is an **alternative transition state.** This **enzyme-substrate transition** state is at **higher energy level,** so the reactants cross the energy barrier of reaction and reaction proceeds at a **fast rate**. The events are as illustrated in the Fig. given below:

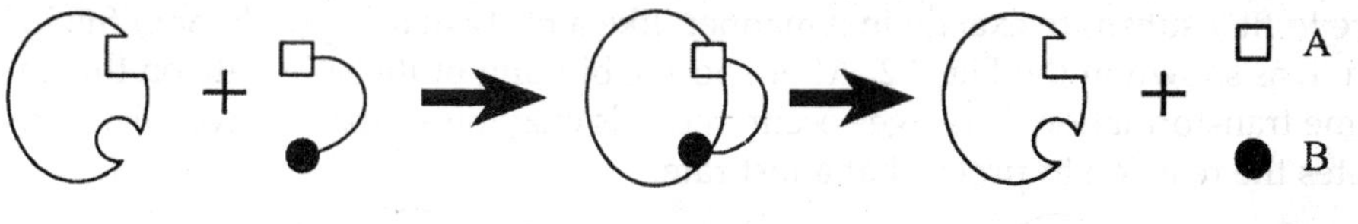

Enzyme | Substrate AB | Enzyme substrate Comple | Enzyme | Products A&B

Groups and radicals at catalytic site—At catalytic sites of enzymes certain groups and radicals, such as sulph hydral (SH), phosphate, OH etc. are present which activate the reactants (substrate), so that enzyme-substrate transition state is raised to an **higher energy level**, which overcomes the energy barrier of the reaction and reaction rate is accelerated.

Gibbs free energy of activation—The enzyme-substrate complex (transition state) has highest free energy in the reaction pathway. The difference in free energy (ΔG), between reaction without enzyme and the transition state (ES complex) is termed as **Gibbs free energy of activation** (ΔG°). The enzyme reduces (lower) energy barrier of the reaction, by lifting substrates bound on catalytic site to a higher energy level, so that they react at a fast rate, as shown in the Fig. 2.1.

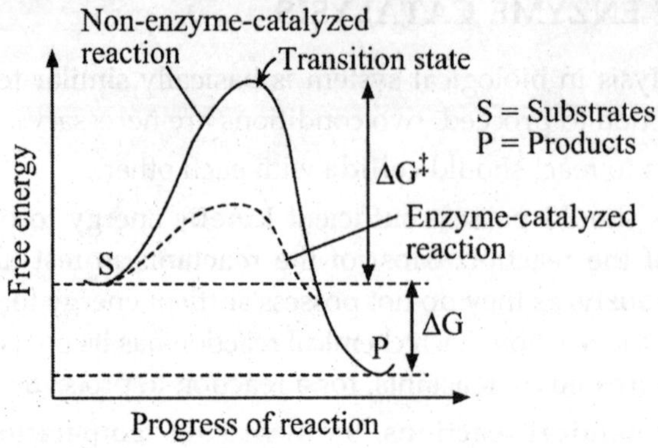

Fig. 2.1 : Effect of enzyme on energy of activation of substrates

Catalytic site - A three dimensional entity—A catalytic site or active site on enzyme molecular surface is present as a three dimensional structure, as **cleft** or **crevices,** with a shape and size complementary to exactly and **specifically fit** the substrate molecules. On catalytic site, with predominantly **non polar** environment, a specific substrate gets attached and is **bound** by **multiple weak forces,** such as H_2 bonds, Van der Waal forces etc. and in some cases by **reversible covalent bonds** with certain groups and radicals of amino acid residues, such as SH, OH, PO_4 groups etc. This binding of substrate on catalytic site with specific groups and radicals, **transforms** the enzyme-substrate complex to a **transition** state, which is at higher energy level (free energy), so that barrier of reaction is crossed and reaction proceeds at a fast rate to form product.

Two models have been proposed to explain the mechanism of **enzyme catalysis** and their **high specificity**—

1. Emil Fischer's lock and key model : According to this lock and key model, the catalytic sites on enzyme surface are present as **prefixed, rigid** clefts or crevices with a shape and size to fit a substrate exactly in a manner, like a particular key(only one) fits in a lock to open it, as shown in the Fig. 2.2. After specific binding of the substrate on this catalytic site, some transformational changes occur and this enzyme - substrate complex formation facilitiates the reaction to proceed at a fast rate.

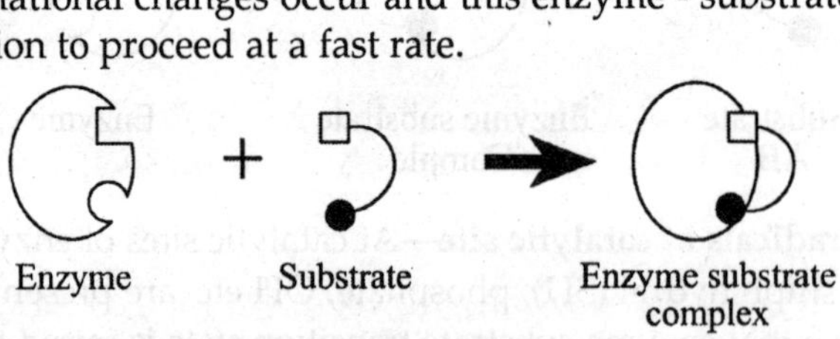

Fig. 2.2 : Lock and key model of Emil Fischer to explain mechanism of enzyme catalysis

However, as enzyme (protein) molecular tertiary and quaternary level of structural integrity is in **fluid state,** it does not allow **prefixed, rigid catalytic** sites, so another theory, an **Induced Fit Hypothesis** was proposed by scientist Koshland.This model is widely accepted, as it does not have prefixed catalytic sites and it explains the mechanism of enzyme catalysis well.

2. Induce Fit Model of Koshland : According to this hypothesis, presence of a substrate in the reaction medium **induces** certain **conformational changes** on the **enzyme**

molecular surface. This brings some distantly placed amino acids with specific groups and radicals to come near to each other to **form** a catalytic site of shape and size to fit only one specific substrate. The substrate then binds on the catalytic site to form an **enzyme-substrate complex, which** catalyze the reaction as shown in the Fig. 2.3.

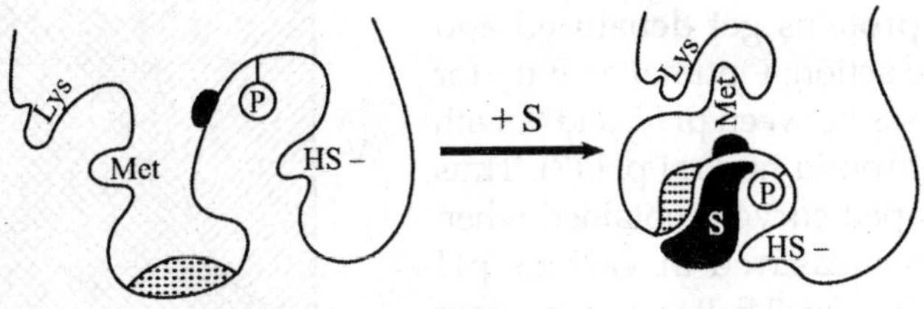

Fig. 2.3 : Induce fit model of Koshland to explain mechanism of enzyme catalysis

Factors affecting enzyme catalysis

Multiple factors affect the **rate** of enzyme catalyzed reactions, mostly because enzymes are **protein** in nature and thus are highly sensitive to any change in environment of reaction medium. Some important factors which influence enzyme catalysis are describe here.

1. Temperature : The **rate** of **enzyme catalyzed** reaction **initially increases** as temperature rises from 0°C, due to increase in kinetic energy of reacting molecules, upto 40 to 50°C rise of temperature, then the **rate declines.** This is because above 50°C, denaturation of enzyme proteins takes place and they loose their catalytic activity. When a graph is plotted depicting rate of enzyme catalysis against temperature, a bell shaped curve is obtained as shown in the Fig. 2.4.

Temperature coefficient (Q_{10})—The temperature coefficient is defined as an increase in rate of enzyme catalyzed reaction in biological system, when temperature of medium is raised by 10°C. For most of the enzymes of body $Q_{10} = 2$. That is the rate of reaction doubles with 10°C rise of temperature between 0°C to 40°C after which rate decline due to the denaturation of enzymes.

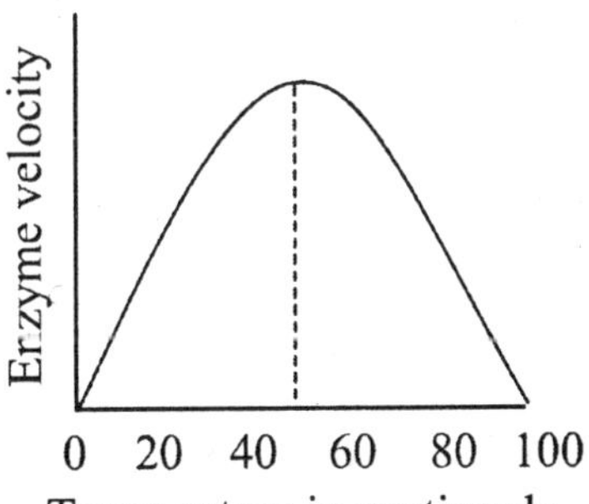

Fig. 2.4 : Effect of temperature on enzyme velocity

Optimum temperature—It is the temperature at which enzyme catalyzed rate is maximum. For majority of enzymes, optimum temperature is around 40°C, as is evident from the curve in Fig. 2.4. As our body temperature is around 37°C, it ensures maximum activity of enzymes in the body tissues. However a few enzymes are active even at 100°C, for instance enzymes of hot spring bacteria and venom enzyme phosphokinase etc.

2. Effect of pH—Alteration in the pH of reaction medium affects enzyme activity considerably. It has been observed that for most of the enzymes, very low pH (below 4) or high pH (>9), enzyme proteins get denatured and loose their catalytic action. Optimal activity for most of the enzymes is between pH 5 and 9, with maximum activity around neutral pH (7). Thus an inverted bell shaped curve is obtained when enzyme activity is measured at various pH values as shown in the Fig. 2.5. It is evident from the curve that enzyme activity is optimum between pH 5 and 9, as they possess **proper structural conformation** as well as correct **ionic charge** to bind maximally with the substrate (SH) as illustrated in the Fig. 2.5.

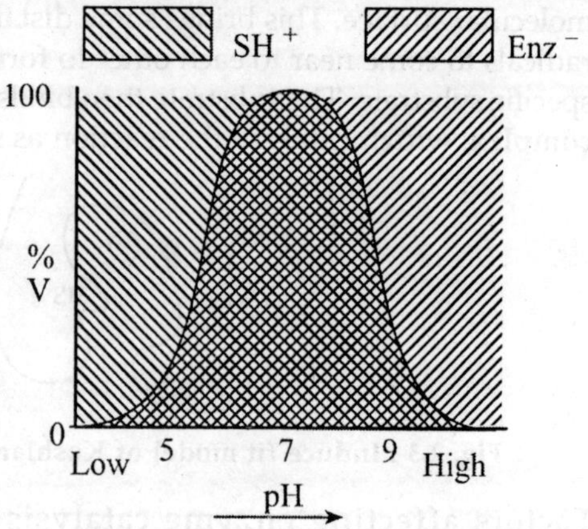

Fig. 2.5 : Effect of pH on velocity of enzyme catalyzed reaction

Optimum pH is that pH at which enzyme activity is maximum. Optimum pH for most of the enzymes is around 7 and as our body tissues and blood pH is maintained around pH 7.4, it is ensured that the enzymes are optimally active.

However there are **few exceptions,** where some enzymes are active at lower or high pH e.g. pepsin is active at pH 1.5 (acidic pH) and alkaline phosphatase is active at alkaline pH - 9.

3. Enzyme concentration—Initial rate of enzyme catalyzed reaction is proportional to enzyme concentration till an equilibrium of reaction is reached, as many biochemical reactions (enzyme catalyzed) are reversible.

4. Substrate concentration—As enzymes are biocatalyst and normally present in less amount as compared to substrate concentration, so enzyme concentration becomes rate limiting factor in most of the enzyme catalyzed biological reactions. When velocity of enzyme catalysis is plotted against substrate concentration, a characteristic hyperbolic (saturation) curve is obtained, as shown in the Fig. 2.6.

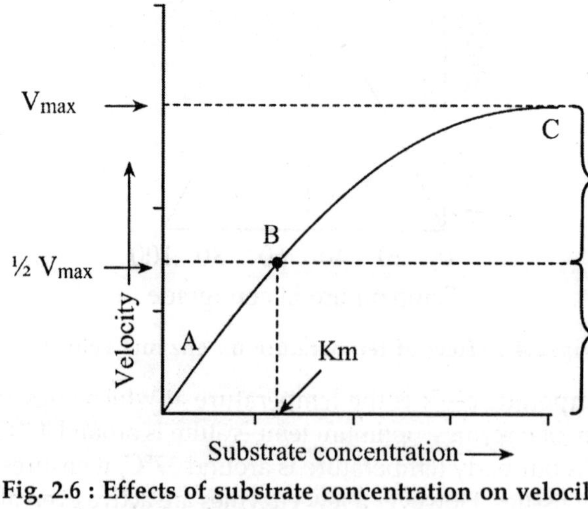

Fig. 2.6 : Effects of substrate concentration on velocily of enzyme catalyzed reaction and Km value

Three phases in the curve are distinct at point A, B and C as evident from the Fig. 2.6.

1. At low substrate concentration-[S]: Initial velocity of enzyme catalyzed reaction is proportional to the substrate concentration, as indicated at point A.
2. At point C, the velocity is maximum with no further increase by increasing substrate concentration [S]. This is because active sites of all enzyme molecules are saturated with substrate and enzyme concentration becomes rate limiting factor at this point.
3. **Point B** in the Fig. 2.6 depicts a more theoritical phase, where the velocity of enzyme catalyzed reaction is just half of maximum velocity (Vmax/2). At this point (B), it is indicated that the half of enzyme molecules are bound with substrate. This substrate concentration is known as **Km value** or **Michaelis Menten constant.**

Michaelis Menten constant—It is defined as the substrate concentration where velocity of enzyme catalyzed reaction is just half of maximum velocity.

Michaelis Menten equation—Michaelis constant (Km) is expressed well by Michaelis Menten equation as given under:

$$\text{Initial velocity } v = \frac{V\max[S]}{Km+[S]}$$

where Km is Michaelis constant, v is Initial velocity, V max is Maxium velocity and [S] is Substrate concentration

Michaelis Menten expression describes behavior of most of the enzymes of our body well. The Michaelis Menten expression explains the 3 phases of the saturation curve i.e., A,B and C as depicted in the Fig. 2.6. At **point B** where [S] is equal to **Km** value, the Michaelis Menten expression can be converted as described below:

$$v = \frac{V\max[S]}{[S]+[S]} = \frac{V\max[\cancel{S}]}{2[\cancel{S}]} = \frac{V\max}{2}$$

This confirms that when [S] is equal to **Km** value, the velocity of enzyme catalyzed reaction is just half of maximum or in other words km is that concentration of [S] where velocity v is half of Vmax.

Physiological Significance of Km

In biological system, mostly the substrate concentration is maintained near Km value (physiological concentration of substrate). This is advantageous because when [S] is near Km value, velocity is highly responsive to even a small change in [S] at point B. The rate of most of the enzyme catalyzed reactions, thus can be increased or decreased **several folds,** (steeply) just by small increase or decrease in [S] respectively, as required by immediate metabolic need of the cell.

Biomedical Significance of Km

Michaelis Menten constant is very significant in biological system as: 1) Km value is constant and characteristic of a particular enzyme and thus is beneficial in the **identification** of enzyme, 2) It is important for finding the **affinity** of enzyme with substrate, as low Km value shows high affinity of enzyme to the substrate and 3) Km value explains the mechanism of various types of enzyme inhibitors well.

Lineweaver Burk plot—Many enzymes do not follow the ideal saturation curve and often require impractically high substrate concentration to get a hyperbolic curve. To overcome this problem, a linear or straight line graphic representation can be obtained by taking reciprocals of Michaelis Menten equation as described here under.

Reciprocal of Michaelis Menten expression is:

$$\frac{1}{v} = \frac{Km + [S]}{Vmax[S]}$$

$$\text{or} \quad \frac{1}{v} = \frac{Km}{Vmax} \times \frac{1}{[S]} + \frac{\cancel{[S]}}{Vmax \times \cancel{[S]}}$$

$$\text{or} \quad \frac{1}{v} = \frac{Km}{Vmax} \times \frac{1}{[S]} + \frac{1}{Vmax}$$

This equation is quite similar to a straight line equation **i.e. y= ax+b** and can be represented as linear plot known as Lineweaver Burk's plot as shown in the Fig. 2.7.

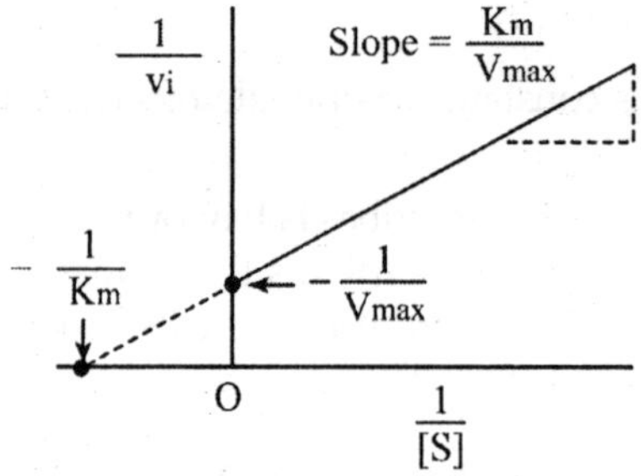

Fig. 2.7 : Lineweaver-Burk or double reciprocal plot

ENZYME INHIBITORS

Many substances inhibit or decrease the rate of enzyme catalyzed reactions. They can be broadly classified into 3 categories:

1. Reversible competitive inhibitors
2. Reversible non competitive inhibitors
3. Irreversible inhibitors (Enzyme poisons)

Reversible Competitive Inhibitors—The chemical structure of **competitive inhibitor** closely resembles to that of the substrate and thus is called as **substrate analogue.** Because of similar chemical structure, the inhibitor (I) can also bind at the catalytic site on enzyme molecule, but binding is **reversible.** So when a substrate analogue (inhibitor) is present, it **competes** with the substrate to bind at catalytic site to form a **EnzI** complex rather than EnzS complex. EnzI complex cannot carry reaction forward to form product, so it revert back to free enzyme after some time. The action of competitive inhibitors may be depicted as given under :

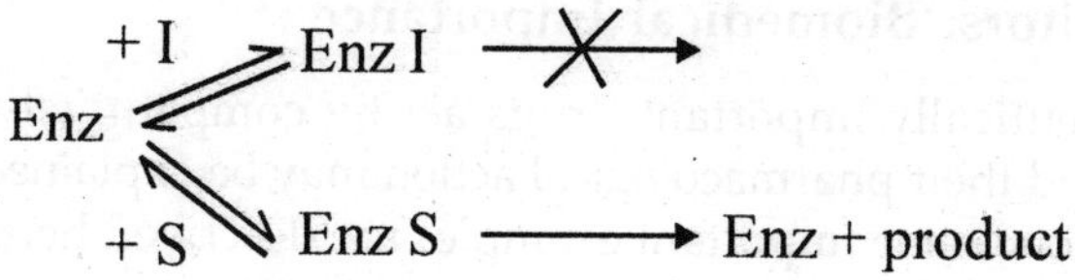

Competitive Inhibitors decrease rate of reaction

The rate of product formation (enzyme catalysis) depends very much on the amount of EnzS formation. In presence of inhibitor, some enzyme molecules will be occupied with formation of EnzI complex, so less enzymes will be available for the formation of EnzS complex and thus the rate of reaction is slowed down and Km value is increased as shown above in the Fig.

Classic example of competitive inhibition : Malonate is a structural analogue of succcinate (an intermediate of CAC) and thus competes for binding on catalytic site of enzyme **succinate dehydrogenase** and thus decrease the rate of the reaction as shown here under.

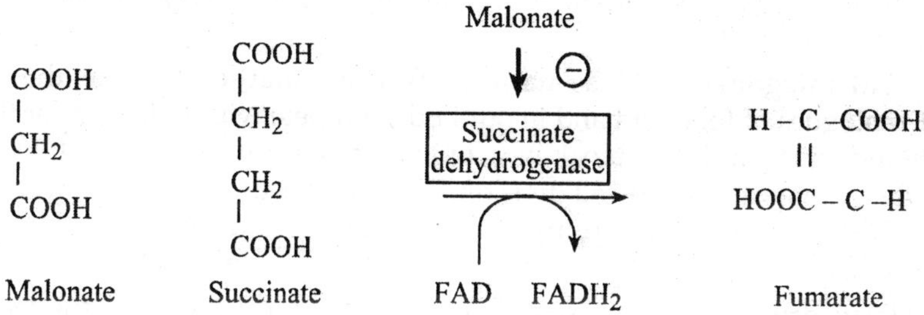

At high substrate concentration but constant inhibitor concentration, the chances of EnzS complex formation increases. Accordingly at infinitely high [S], Vmax of enzyme catalyzed reaction can be attained, as formation of EnzI would be negligible. This can be illustrated by Lineweaver Burk plot as shown in the Fig. 2.8.

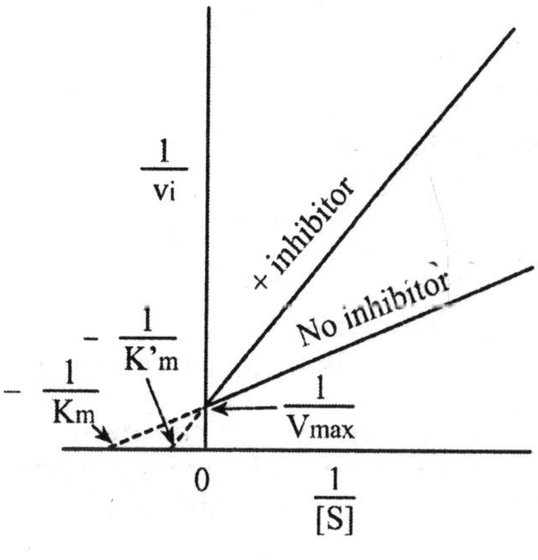

Fig. 2.8 : Lineweaver Burk plot of classic competitive inhibition

Classic competitive inhibition double reciprocal plot illustrates following points :

1. Competitive inhibitor increases apparent km value.
2. Vmax remains same at infinite high [S].

Competitive Inhibitors: Biomedical Importance

Many therapeutically important drugs act by competitively inhibiting the key regulatory enzymes and their pharmacological action may be explained by the mechanism of competitive inhibition. Some important examples are described here.

Sulphonamide and other sulpha drugs are potent bacteriocidal agents that kill bacteria and check infection by blocking folic acid synthesis in the microorganisms, which is required necessarily for bacterial multiplication. Bacteria synthesize folic acid from para aminobenzoic acid (PABA) in the presence of the enzyme **folic acid synthase.** Sulphonamide is a structure analogue of PABA and thus competes for binding on catalytic site of enzyme folic acid synthase and thus block folic acid synthesis in bacteria, so they are killed and infection is checked.

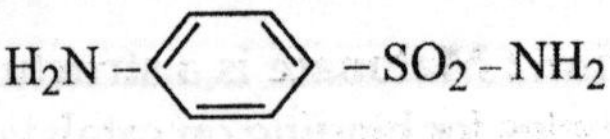

Sulphanilamide

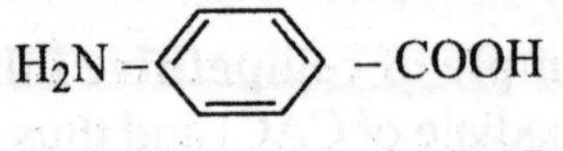

p-amino benzoic acid (PABA)

The sulpha drugs are nontoxic to human cells as we do not synthesize folic acid and take it directly in our diet.

Folic acid antagonists such as **aminopterine** and **methotrexate** on the other hand are structural analogues of folic acid and in our body compete with folic acid for the enzyme **dihydrofolate reductase** and thus block its activation to tetrahydrofolate (THF) coenzyme, necessary for DNA replication and cell division. These drugs thus reduce cancerous growth (Chapter 25). The folic acid antagonists, methotrexate and aminopterin thus act as **anticancerous drugs** and are widely used in the treatment of leukemia and other cancers.

Lovastatin drug is widely used to lower cholesterol levels in the treatment of hypercholesterolemia and thus are protective and reduce the risk of heart diseases. Lovastatin exerts its therapeutic action by competitively inhibiting the key regulatory enzyme of cholesterol biosynthesis, the HMG Co A reductase and thus lowers blood cholesterol level. (Chapter 12).

Dicoumarol is a structural analogue of vitamin K and competes for its activity in the process of activation of blood clotting factors. Therefore it is used therapeutically as **anticoagulant drug** to prevent intravascular thrombosis (Chapter 25.).

Physostigmine is a structural analogue of acetylcholine and competitively inhibits **acetyl cholinesterase** enzyme and thus block breakdown of acetylcholine causing prolonged neural impulse. The drug is used in the treatment of myasthenia gravis, in which neural impulses are weak and patient looses control on voluntary movements of muscles.

Some important drugs which act by competitively inhibiting enzymes to cure and alleviate disease state are summarized in the Table 2.3 along with their mechanism of action.

Table 2.3 : Therapeutic uses of competitive inhibitors

S.No.	Competitive Inhibitor	Enzyme	Substrate	Therapeutic use/ Mechanism
1.	Allopurinol	Xanthine oxidase	Hypoxanthine	Reduce excess production of uric acid and thus beneficial in control of Gout

contd. ...

... contd.

S.No.	Competitive Inhibitor	Enzyme	Substrate	Therapeutic use/ Mechanism
2.	Antibiotics sulphonamide & sulpha drugs	Folic acid synthetase	Para amino benzoic acid (PABA)	-Kill bacteria/**cure infections** by blocking folic acid synthesis in bacteria
3.	Dicumarol	-Clotting Factors activation	Vitamin K	Anticoagulant drugs. Prevent intravascular thrombosis
4.	Aminopterin, Methotrexate	Dihydrofolate reductase	Dihydrofolic acid	**Anticancerous drugs** used in treatment of leukemia and other cancers

Reversible Noncompetitive inhibitors

The noncompetitive inhibitors are chemically **different** from substrate and have affinity to bind at **some other site** (domain) than the catalytic site on enzyme surface. Therefore formation of both **EnzI** complex and **EnzS** as well as **EnzIS** complex formation is possible. Since EnzIS may catalyze formation of product, but at a slower rate than EnzS complex, therefore it decreases the rate of reaction. Thus Vmax is lowered in presence of non competitive inhibitor but Km value remains unaffected as illustrated in the Fig. 2.9.

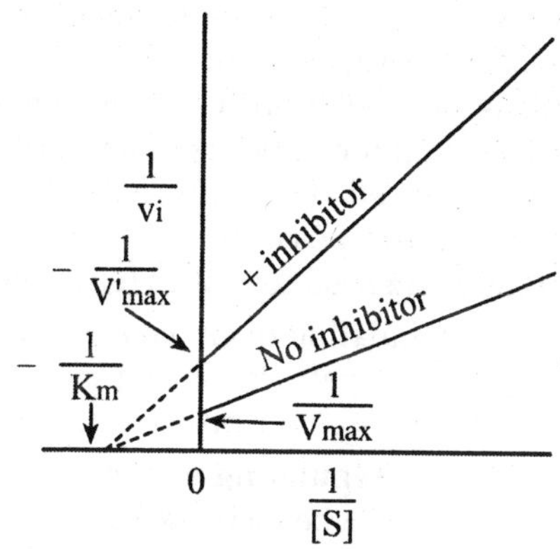

Fig. 2.9 : Double reciprocal plot for reversible non competitive inhibition

Since S has equal affinity to bind both free Enz and with EnzI complex, it is assumed that there is no significant alteration of active site conformation when I bind on the other site of enzyme. The events occuring due to the presence of non compentitive inhibitor can be depicted as—

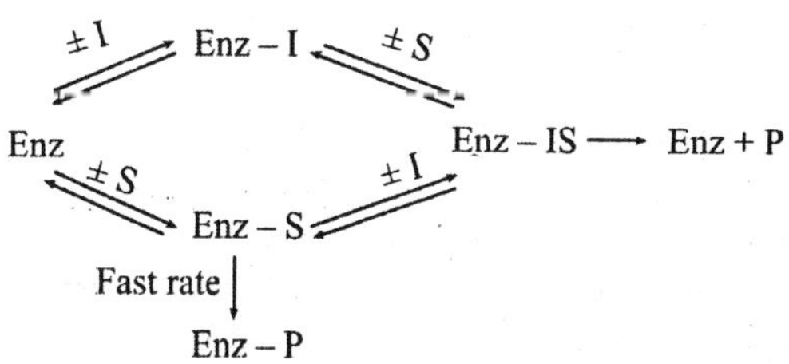

Irreversible inhibitors : Enzyme poisons

Many toxic substances bind **irreversibly** by covalent bonds with SH or OH groups of amino acid residues at the catalytic site of enzymes, thereby inhibit **catalysis** and are toxic to body. Some important enzyme poisons are listed in Table 2.4. along with enzymes which are inactivated by them.

Table 2.4 : Enzyme poisons along with their toxic effects

S. No.	Enzyme poisons	Enzyme Inhibition by binding/precipitation	Toxic effect
1.	Heavy metals Ag, Hg	Precipitate/ inactivate most of enzymes of the body	Heavy metal poisoning may result in death
2.	Iodoacetate	SH group of gly 3-PO_4 dehydrogenase	Block glycolysis/can be fatal
3.	Nerve gas (DFP) di-isopropyl fluorophosphate	OH group of serine at active site of enzymes e.g. acetylcholine esterase	Block nerve impulse conduction, resulting in **paralysis and death**

Regulation of Enzyme Catalysis

One of the most important characteristic feature of a living organism, in order to **survive** is to adapt itself to the ever changing internal as well as external environmental conditions. To achieve this, nature has provided an **unique ability** to an organism to **regulate** various metabolic processes by **controlling** activity of a **rate limiting enzyme.** Thus the enzyme action can be regulated and controlled as and when required by the immediate need of the cell. This feature differentiate the enzymes from ordinary inorganic catalysts and **make life possible** in the ever changing environmental conditions, sometimes even during dangerous calmaties (environmental), such as famine, floods etc.

Regulation of enzyme action—Active regulation of catalytic capacity of key enzyme of a metabolic pathway is of great significance in living organism. This is accomplished by two general mechanisms as described under:

1. **Regulating enzyme synthesis at gene level:** Enzyme synthesis at gene level can be increased or decreased as required by the immediate metabolic need of a living cell, by certain **inducers** or **suppressors** respectively. This process takes **some time** and thus is considered as a **long term effect** on regulation of a metabolic cycle. (Chapter 22)

   Several **hormones** and **II messengers** influence and regulate various metabolic processes by affecting gene expression and rate of enzyme synthesis. The catalytic capacity can be influenced by changing the concentration of enzyme (by regulating synthesis) or by altering the intrinsic catalytic efficiency of enzyme directly.

2. **Rapid regulation of enzyme catalytic efficiency:** Alterations in the intrinsic catalytic efficiency of key enzymes provides rapid means for regulation of metabolic cycles according to the immediate need of a living cell. Enzyme activity can be regulated quickly as required, by various mechanisms such as: (1) by allosteric effectors, (2)by covalent modification and (3) by feedback inhibitions

Allosteric Enzymes

Most of the key enzymes (rate limiting steps) of metabolic processes in the body are **allosteric enzymes,** whose catalytic efficiency (catalysis) can be increased or decreased by the presence of certain **allosteric modifiers.**

Characteristics of Allosteric Enzymes

The allosteric enzymes, in addition to the catalytic site, usually possess some **other sites** called **allosteric sites,** on which an allosteric activator (+ve effector) or an inhibitor (-ve effector) binds to regulate enzyme activity and thus metabolic process as a whole, can be controlled as required.

Conformational alteration—When an allosteric effector is attached on **allosteric site** of a key enzyme, some conformational changes occurs on enzyme surface which **modify catalytic site** of enzyme in **+ve** (positive) or **-ve way** to activate or inhibit its activity respectively.

Multimeric forms of Allosteric Enzymes

Allosteric enzymes are usually oligomeric having 2 or more monomeric units in their quaternary structure. Mostly the allosteric enzymes are **dimers,** which exists in two conformational **states,** the **tout T form** (inactive) and **relaxed R form** (active enzyme) with **proper catalytic** site as shown in the Fig. 2.10.

Allosteric T form Dimer (inactive) cannot bind substrate $\underset{\text{Inhibitor}}{\overset{\text{Activator}}{\rightleftharpoons}}$ R form (active enzyme) can bind **substrate** for catalysis

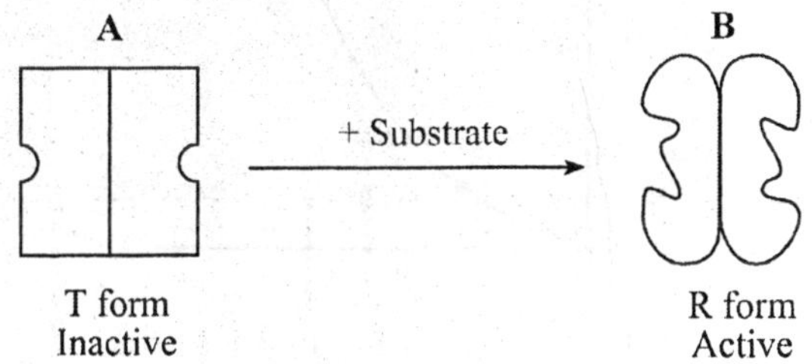

Fig. 2.10 : Diagramatic representation of allosteric enzyme with T and R form and effect of substrate

Co-operativity in allosteric activation—Allosteric (dimers) enzymes are activated rapidly by the presence of substrate, which **induce** rapid conformational change of T form to R form, that binds substrate at properly formed catalytic site. The allosteric effectors by inducing change in conformation of one subunit of dimeric enzyme to R form, automatically changes other subunit to R form by cooperativity. This results in rapid change in T form of the enzymes into active R form to steeply increase rate of catalysis and the effect is known as cooperative binding of substrate as shown in the Fig. 2.12.

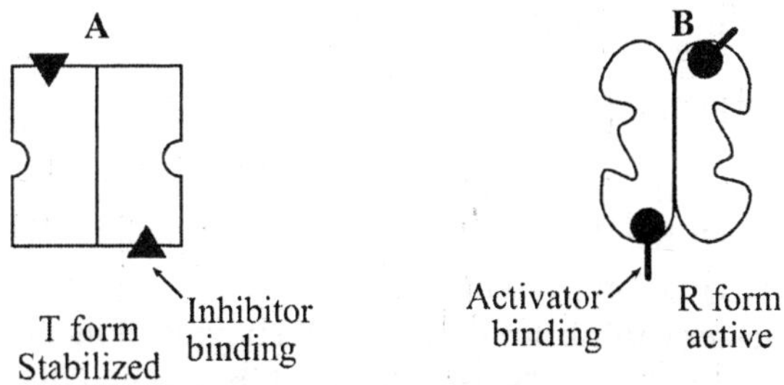

Fig. 2.11 : Effect of modulators on allosteric enzymes. T form is stabilized by binding with inhibitor (A) and binding of an activator changes T form to R form (B)

Influence of Allosteric Effectors—In general the allosteric inhibitor, when gets attached to T form of enzyme at allosteric site, it stabilizes the T form and would not favour the substrate binding, so the enzyme catalysis is inhibited. Whereas the allosteric activator and substrates favours the conversion of T form to R form and enzyme catalysis is stimulated as shown in the Fig. 2.11.

Sigmoid curve : A **sigmoid** or **S shaped** curve is obtained when rate of allosteric enzyme activity is plotted against substrate concentration. Initially, at low [S], (point A) most of the enzymes are in T form and rate of reaction is very low. (less catalysis). When [S] is increased, due to induction of conformational changes and cooperativity, rapid change of T form to R form occurs, with drastic increase in the rate of enzyme catalysis and a steep rise in the velocity of reaction is observed as is evident from the curve (point B) in the Fig. 2.12.

At high [S] concentration (point C) all enzymes are in R form and are saturated with substrate at catalytic sites, so there is no further increase in velocity by increasing [S]. Thus a sigmoid (S) shaped curve is obtained as depicted with points A,B and C in the Fig.2.12.

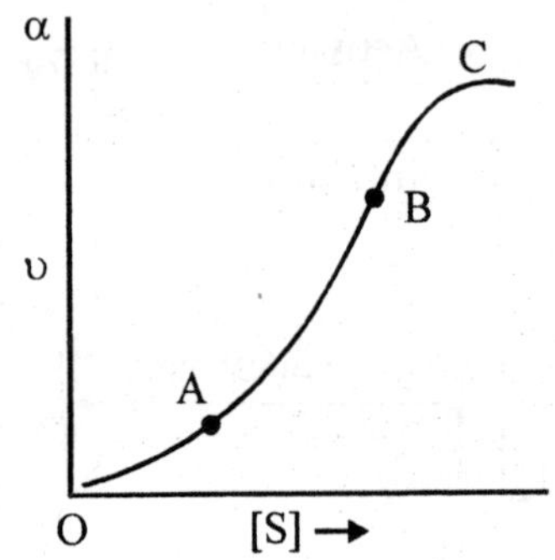

Fig. 2.12 : Sigmoid saturation curve

Allosteric Enzymes : Metabolic Significance

Allosteric enzymes are very important in the body for the regulation of various metabolic pathways. These regulatory enzymes in a particular pathway are termed as **key enzymes or rate limiting enzymes** as the whole metabolic pathway is constrained here as a bottle neck and this is metabolically highly significant. Some important examples are: 1) δ -amino levulenic acid synthase (ALA Synthase) in heme biosynthesis 2) HMG CoA reductase of cholesterol biosynthesis,3) Phosphofructokinase I (glycolysis) 4) aspartate transcarbamoylase (pyrimidine nucleotide synthesis) etc. Details along with allosteric modifiers are given in the Chapter 5.

FEED BACK INHIBITION

In biosynthetic pathways when the concentration of the product is more than required, the product itself **feedback inhibits** an initial, rate limiting key enzyme of its own **biosynthetic pathway.** This is represented in the Fig. 2.13 where the product D (in excess) itself is acting as an allosteric **inhibitor** of the initial enzyme of its own biosynthetic pathway to reduce the rate of its synthesis. The feed back inhibition usually accurs at initial step so as to prevent wastage of energy.

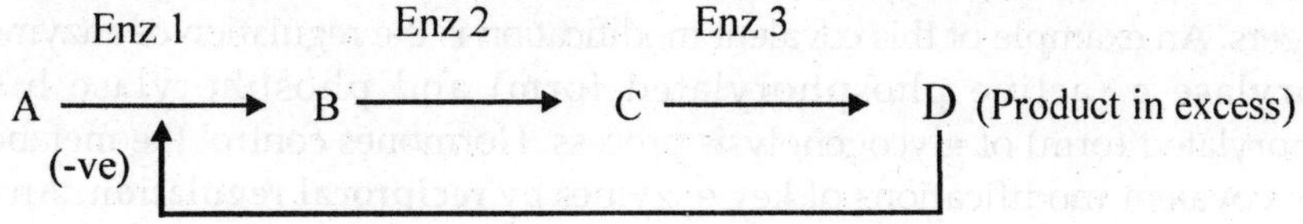

Fig. 2.13 : Feed back inhibition of initial key enzyme by the product

Multiple feedback loops—In a branched chain biosynthetic pathways where many products are synthesized, multiple feed back inhibitions are observed by the **different products** as required by the cell. A classic example is the regulation of biosynthesis of pyrimidine nucleotides, where the final products CTP, UDP and TMP feedback inhibit the key enzyme of their own branch (synthetic) as well as they can feedback inhibit (partially) the first initial enzyme of the common synthetic pathway as illustrated in the Fig. 2.14 (details in the Chapter 18.)

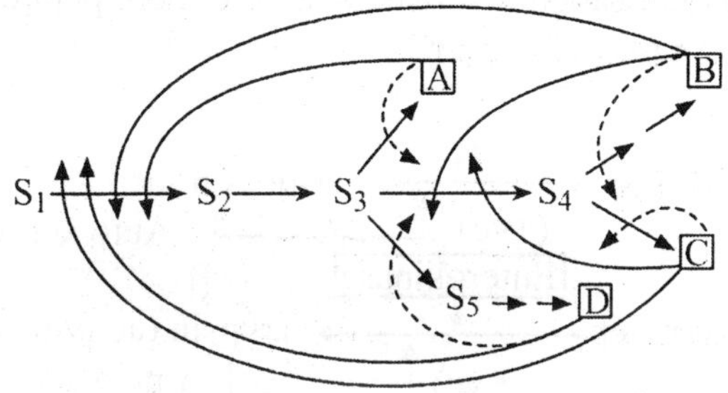

Fig. 2.14 : Multiple loop feedback inhibition by the products

In branched biosynthetic pathways (as mentioned above), the negative feed back inhibitory effects of two or more end products on the first initial regulatory enzyme of common pathway is **strictly additive** and is called as **cumulative feed back inhibition**. Classic example is pyrimidine nucleotides biosynthesis as described in the chapter 8.

COVALENT MODIFICATION

Certain key enzymes of major metabolic pathways are activated or inhibited by covalent modification depending on the need of body. These enzyme exist in active and inactive forms and are interconvertible reversibly. The interconversion is brought about by reversible covalent modification, namely phosphorylation of OH group of serine residue at the catalytic site, or by dephosphorylation (details in chapter 5. Fig. 2.15).

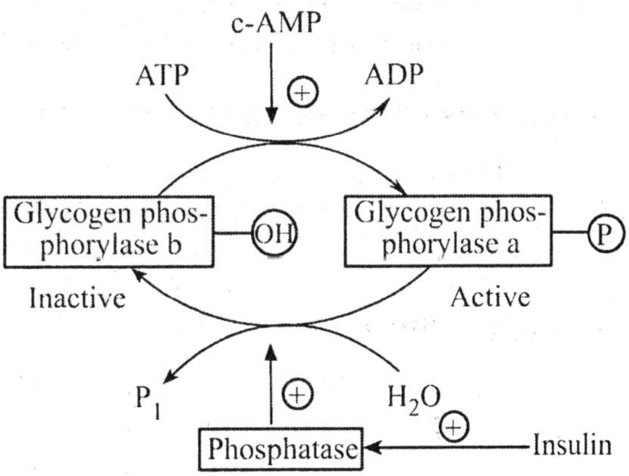

Fig. 2.15 : Reversible activation/inactivation by phosphorylation/dephosphorylation of rate limiting enzymes of glycogen metabolism.

This regulation is exerted by hormones through the mediation of their

II messengers. An example of this covalent modification is the regulation of enzyme glycogen phosphorylase a (active phosphorylated form) and phosphorylase b (inactive, dephosphorylated form) of glycogenolysis process. Hormones control the metabolic cycles, mainly by covalent modifications of key enzymes by **reciprocal regulation**. An important example is glycogen synthesis and breakdown as illustrated in the Fig. 2.15 (details in chapter 5).

Activation of Zymogens

Some exocrine glands such as gastric and pancreatic glands secrete proteolytic enzymes in inactive zymogen form, for instance pepsinogen, trypsinogen etc. These proenzymes are secreted in gastrointestinal secretions and are activated later on, when food comes in stomach or in intestine, to active enzymes **pepsin** and trypsin respectively, by **irreversible covalent** modification by breaking one or more peptide bonds of as shown in the Fig. given below: (Details in Chapter 14)

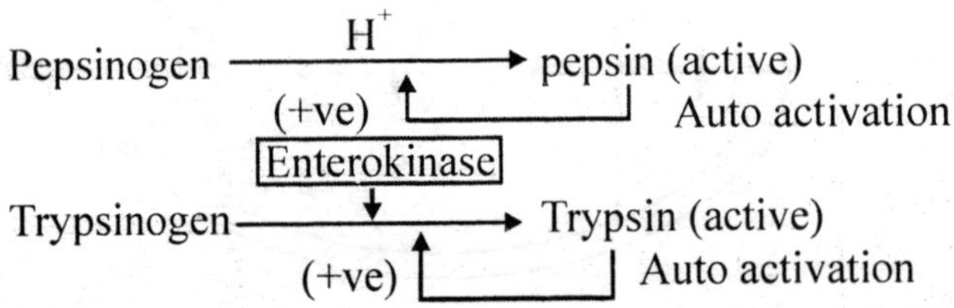

Isoenzymes

Physically distinct forms of same enzyme exist in our body and are called **Isoenzymes**. They catalyze **same reaction** but are different in their electrophoretic mobility, chemical structure and immunological properties. They are synthesized in different tissues and differ in their subunit composition in the multimeric forms (e.g. tetramer, dimer etc) Important example are lactate dehydrogenase, creatine phosphokinase (CPK) etc.

Isoenzymes : Biomedical Importance

Human sera contain several isozymes and their respective levels are raised in different pathological conditions. This observation attracted medical interest in estimation of isoenzymes for differential diagnosis of several diseases.

Separation and Identification of Isozymes

Human serum contain 5 lactate dehydrogenase (LDH) isoenzymes and they can be separated and identified by electrophoresis. Serum sample is subjected to agar gel electrophoresis (Chapter 32.) at pH 8.6, for about 2 hours. The LDH isozyme have different electrophoretic mobility and are thus separated in 5 different bands, which may be visualized by staining with a specific assay mixture. This assay mixture contains the substrates of LDH isozymes to catalyze reaction and a colourless dye is reduced to a coloured product as illustrated in the Fig. 2.16.

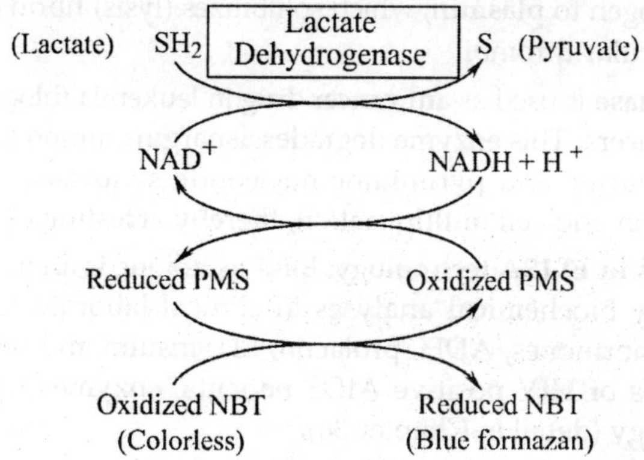

Fig. 2.16 : LDH Isoenzyme assay with complete reactions at 5 regions on electrophoretogram

Subsequently other important isoenzymes were also separated and identified with specific techniques to facilitate differential diagnosis of diseases. Important examples are lactate dehydrogenase, creatine phosphokinase (CPK), alkaline phosphatase etc.

Diagnostic importance—Depending on the tissue/organ involved in the disease process, the respective levels of different isoenzymes increase in serum, as a result of necrosis of infected tissue. Thus their determination in serum is beneficial in differential diagnosis of diseases. Some important isoenzymes with clinical significance are listed in Table 2.5.

Table 2.5 : Some important Isoenzymes with clinical significance

S.No.	Isoenzymes	Types	Diagnostic significance	Plasma level raised
1.	Lactate dehydrogenase (LDH)	5 types LDH 1	Heart disease (MI)	LDH_1 & LDH_2
		LDH 2		
		LDH 3		
		LDH 4		
		LDH 5	Liver disease	LDH_5
2.	Creatine phosphokinase (CPK)	3 types	Heart diseases	
		CPK-MB,	Myocardial Infection	CPK-MB
		CPK BB &		
		CPKMM	Muscular dystrophy	CPK Total

Enzymes as Therapeutic Agents

Recently importance of enzymes as drugs has been demonstrated and nowadays therapeutic uses of enzymes in the treatment of certain diseases has gained importance in clinical medicine. Some important examples include.

1. **Streptokinase :** This enzyme is obtained from streptococcus and is used as drug for clearing (lysis) of intravascular blood clots. Sreptokinase activates

plasminogen to **plasmin,** which solubilizes (lysis) fibrin clot and thus dissolves intravascular thrombi.

2. **Asparginase** is used as **anticancer drug** in leukemia (blood cancer) and in certain other cancers. This enzyme degrades aspargine amino acid, which is necessary for the purine and pyrimidine nucleotide synthesis, thus block rapid DNA replication and cell multiplication, thereby arresting cancerous growth.
3. **Enzymes in ELISA technology:** Elisa technique is an important diagnostic tool for many biochemical analyses in clinical laboratory,(Chapter 37) such as thyroid hormones, ADH, prolactin, LH, insulin and other hormone assay. In diagnosis of HIV positive AIDS patients, enzymes have great role in Elisa technology (detail in Chapter 38).

DIAGNOSTIC IMPORTANCE OF SERUM ENZYMES

Estimation of certain enzyme levels in serum is of great **clinical importance** for **diagnosis** as well as in **prognosis** of many diseases. These enzymes are normally present in very small amounts in blood. They do not have any functions in blood and are released in plasma as a result of normal **process** of tissue wear and tear, which synthesize them. But if that tissue (organ) which produce a particular enzyme is involved in some **disease** process with extensive **necrosis** of the tissue (due to infection), it results in elevation of that enzyme in serum. Their estimation in plasma, thus have great diagnostic value. A brief account of some important enzyme estimations useful in diagnosis of specific diseases is given in Table 2.6 along with their normal serum levels.

Table 2.6 : Some important serum enzymes investigations used in clinical diagnosis

S.No.	Enzyme	Normal serum level	Increased serum level in disease
1.	Amino transferase		
	a. AST (SGOT)	2-45 IU/L	Myocardial infarction
	b. ALT (SGPT)	3-40 IU/L	Viral hepatitis
2.	Digestive enzymes		
	a. Amylase	80-180 Somogy's units/dl	Acute pancreatic
	b. Lipase	0.2-1.5 IU/L	Acute apendicitis
3.	Creatine phospho-Kinase NAC (Total)	10-50 IU/L	Muscle dystrophy
	CPK MB	2-20 IU/L	Myocardial Infarction
4.	Acid phosphatase	0.5-3 KA unit/dl	Prostate carcinoma
	Alkaline phosphatase	3-13 KA unit/dl	Bone diseases, rickets obstructive jaundice etc.
5.	LDH_1 & LDH_2		Myocardial infarction
	LDH_5		Liver diseases

Enzymes Estimation

As enzymes are present in very low concentration in serum, they are estimated indirectly as their catalytic capacity and are expressed in terms of enzyme activity, in form of international units, IU/Liter, or in other units named after the method used for enzymes estimations. The international unit is a standard unit and is a defined as the amount of the enzyme activity that would catalyze conversion of one micromole of the substrate per minute/ liter of the sample.

❑❑❑

3

CARBOHYDRATES OF PHYSIOLOGICAL IMPORTANCE

Carbohydrates are essential constituents of plants and animal organisms. They form major portion of the structural frame work of tissues and serve as storage form of nutrients. Carbohydrates are the most economical and major source of energy in the body to carryon life processes. They are synthesized in plants from CO_2 and water in the presence of green pigment chlorophyll and sunlight by the process known as photosynthesis.

Chemical nature

The carbohydrates are also known as saccharides and are defined as aldehydic or ketonic derivatives of polyhydroxy alcohols and the compounds which produce these derivatives on hydrolysis.

Classification

The carbohydrates are classified as monosaccharides, disaccharides and polysaccharides, depending upon the number of monomeric saccharide units present in the molecule. The classification of carbohydrates with important examples is summarized in the Fig. 3.1.

Monosaccharides—The monosaccharides are simple, monomeric units which cannot be hydrolyzed further. Monosaccharides are **subdivided,** according to the number of carbon atoms present, as trioses (3C-atoms), tetroses (4C) pentoses (5C) and hexoses (6C). They are further subdivided as **aldoses** that contain aldehyde group or **ketoses** with a keto group. The general formula of monosaccarides is $Cn(H_2O)n$.

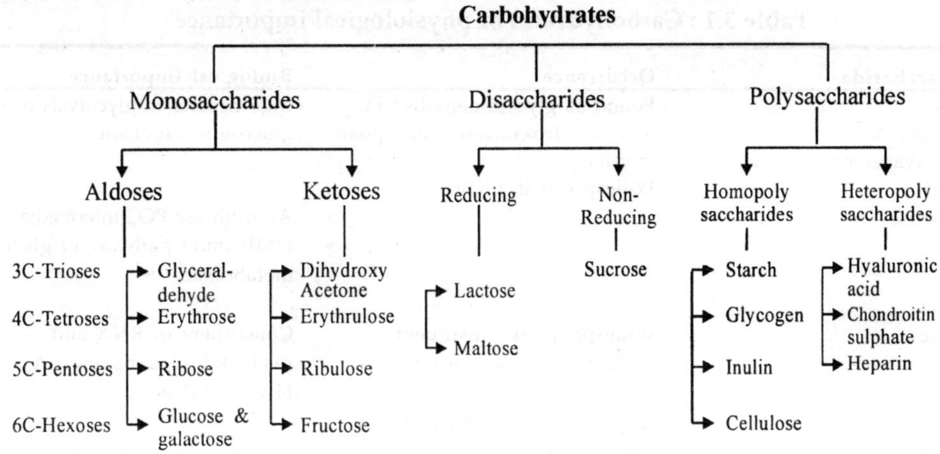

Fig. 3.1 : Classification of carbohydrates with examples of major classes

Disaccharides—The disaccharides are formed by joining together two monosaccharide units by a glycosidic linkage. The general formula is $Cn(H_2O)n\text{-}1$. Disaccharides of physiological importance are maltose, lactose and sucrose, which on hydrolysis produce two molecules of same or different monosaccharides as shown in the reactions given below:

$$\text{Maltose} \xrightarrow[H_2O \ \text{dil HCl/ Boiling}]{\text{Hydrolysis}} \text{2 glucose molecules}$$

$$\text{Lactose} \xrightarrow[H_2O \ \text{dil HCl/ Boiling}]{\text{Hydrolysis}} \text{1 glucose + 1 galactose}$$

$$\text{Sucrose} \xrightarrow[H_2O \ \text{dil HCl/ Boiling}]{\text{Hydrolysis}} \text{1 glucose + 1 fructose}$$

Polysaccharides—The polysaccharides (Greek word, poly means many) are polymers of many monosaccharide units joined together by glycosidic bonds. Polysaccharides are macromolecules with high molecular weight. They are usually nonsugar (not sweet) and form colloidal solution in water. Polysaccharides are further subdivided into two groups :

Homopolysaccharides—The homopolysaccharides are polymers of many molecules of same monosaccharide units. For instance -starch, glycogen, cellulose etc., which yield only glucose molecules on hydrolysis, while inulin produces many fructose molecules on hydrolysis.

A brief account of some physiologically important carbohydrates of each class along with their sources and functions is given in the Table 3.1.

Heteropolysaccharides—These are polymers formed by the condensation of many molecules of two or more different types of monosaccharide units or their derivatives, linked together by glycosidic bonds. Examples are hyaluronic acid, chondroitin sulphates, heparin and blood group substances (details later in the section).

Table 3.1 : Carbohydrates of physiological importance

Monosaccharide	Occurrence	Biological Importance
Trioses Glyceraldehyde Dihydroxyacetone	Found as glyceraldehyde-3 PO_4 and dihydroxyacetone phosphate in cells	Intermediate in glycolysis of glucose metabolism.
Tetroses	Widespread in plants	
D-Erythrose		As erythrose PO_4, intermediate in HMP shunt pathway of glucose metabolism.
Pentoses		
D-Ribose	Widespread as constituent of nucleic acids and nucleotides	Constituent of RNA and nucleotides, coenzymes, ATP, NAD, FAD etc.
Deoxyribose	Constituent of all types of DNA	
D-Ribulose		Intermediate of HMP shunt
D-Xylulose	Constituent of glycoproteins and gums	Intermediate of HMP shunt
Hexoses		
D-Glucose	Main sugar of body, blood and other body fluids, constituent of starch, glycogen, cellulose and disaccharides.	Major source of readily available energy for life processes
D-Galactose	Constituent of milk sugar lactose, Mucopolysaccharides, glycolipids, cerebrosides and gangliosides of brain	As glycolipids of white and grey matter of brain.Chondroitin sulphate as ground cementing substance of connective tissues.
D-Fructose	Widespread in fruits, honey, constituent of sucrose.	As phosphate, intermediate in glycolysis
Disaccharides		
Sucrose	Constituent of cane sugar, beet sugar & fruits.	Common table sugar
Lactose	Milk sugar	Exclusive carbohydrate source in **Milk**, provide glucose for energy, & galactose for synthesis of brain glycolipids in infants
Maltose	Product of starch digestion in GIT by enzyme amylase	
Homopolysaccharides :		
Starch – Made up of many glucose units linked by $\alpha 1-4$ and $\alpha 1-6$ glycosidic bond	Cereals, vegetables, tuber roots etc.	Carbohydrate reserve of plants, important dietary source of fuel, as glucose and energy
Glycogen – animal starch made up of glucose units	Liver, muscles. Animal polysaccharide	Storage reserve polysaccharide of body.
Cellulose – Made up of many glucose units linked by $\beta 1-4$ glycosidic bond	Polysaccharide of plants cell wall and thus essentially present in vegetable and other plant foodstuffs	Not digested by human, constitute dietary fiber. Important in prevention of constipation and colic diseases
Inulin – polymer of fructose	Plants, dahlia bulb, garlic, onion etc.	Not digested by humans, used for renal function test to measure glomerular filtration rate (GFR)

MONOSACCHARIDES

Chemical Nature : Structural Aspects

The monosaccharides are aldehyde or ketonic derivatives of polyhydroxy alcohols.

Isomerism—Monosaccharides contain one or more asymmetric carbon atom, i.e. a carbon atom having four different atoms or group attached to it and thus they exhibit stereoisomerism. Isomers are the compounds which are identical in composition and structural formula but differ in spatial configuration of groups or radicals and are called stereoisomers.

Number of possible stereoisomers depends on the number of asymmetric carbon atoms present in a compound and is given by Van hoff's formula of 2^n, where n is the number of asymmetric carbon atoms. For instance: glucose(hexose) has 4 asymmetric C-atoms, so $4^2 = 16$ isomers of glucose are possible.

Optical activity—Monosaccharides having asymmetric carbon atoms can rotate the beam of plane polarized light and are said to be optically active. An isomer which rotates the plane polarised light to right is known as **dextro rotatory** and is designated as d or by +sign, where as the isomer which rotates the plane polarized light to left is known as **levo rotatory** and is designated as l or (-) isomer. When equal amount of d and l isomer are present in a solution, the resulting mixture has no net optical activity and is called racemic mixture or dl mixture.

Stereoisomerism—Monosaccharides show geometric or D and L isomerism. The orientation of H and OH groups around the carbon atom adjacent to the terminal primary alcohol group determines whether the sugar belongs to D or L series. When OH group is on right side, the compound is D isomer and when OH group on this carbon is on left side, the compound is L isomer, as shown in the Fig. 3.2.

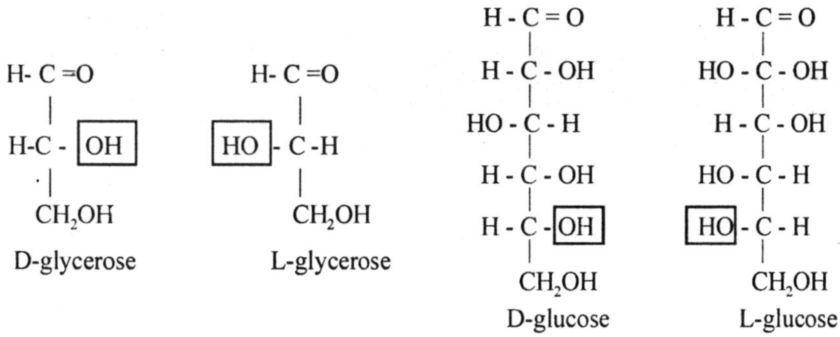

Fig. 3.2 : D and L isomerism in monosaccharides

D and L isomers are **mirror image** of each other. Most of the monosaccharides occuring in mammals are of **D-configuration** and the enzyme machinery of animals are specific to metabolize D series of monosaccharides only. So **humans utilize only D-isomers** of carbohydrates in the body for metabolic purposes.

Ring structure—Although the monosaccharides are commonly represented as straight chain projection given by Fischer (Fig.3.3), they exist mainly as ring structures. These ring structures are produced by intra molecular reaction of the aldehyde or keto group of the sugar with its own-OH group attached to C-atom 4 or 5, to form a **hemiacetal linkage.** By the formation of this intramolecular linkage, two types of rings are formed known as pyranose and furanose rings.

Pyranose and Furanose Ring Structures

Six membered ring form of hexose sugars is called as **pyranose** and has 5 carbon atoms and one oxygen atom in the ring. These are formed by the reaction of aldehyde group of the sugar (on C-atom 1) with OH group attached to fifth carbon atom of molecule as illustrated in the Fig. 3.3.

Similarly sugar with 5 membered ring of four carbon and one oxygen atom is known as **furanose** and is formed by the reaction between aldehyde or keto group with OH group attached to fourth carbon of the sugar. (Fig. 3.3)

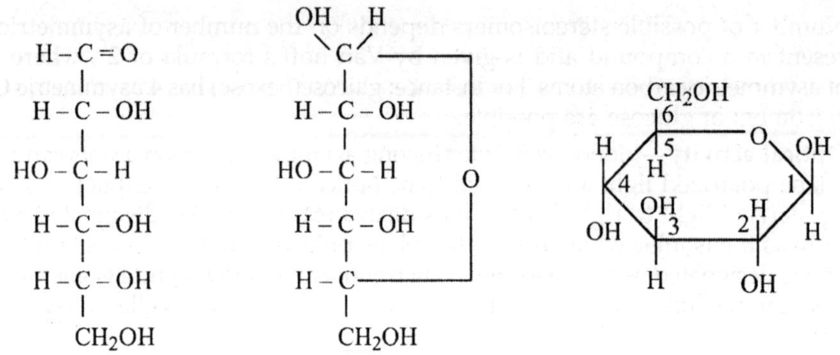

Trace amounts of intermediary — D–Glucose (Open chain); β–D–Glucose — Fischer's formula; β–D–Glucopyranose — Haworth formula

Fig. 3.3 : Ring structures of D-glucopyranose as given by Fischer and Haworth

Ketoses such as fructose exist in form of a five membered ring called **furanose.** The hemiacetal linkage is formed between the keto group present on the second carbon and OH group at fifth carbon atom of the sugar as shown in the Fig. 3.4.

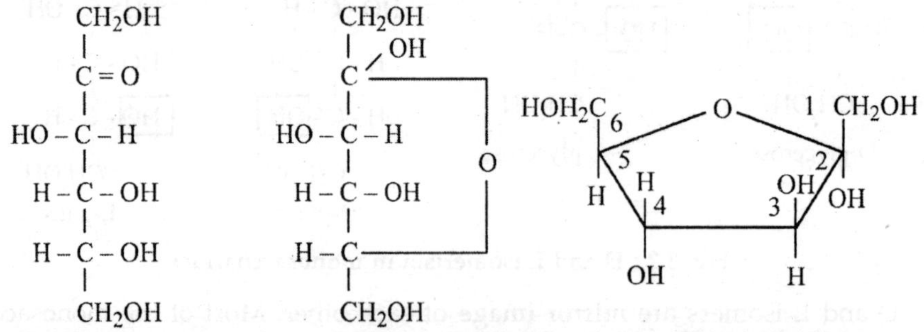

D–Fructose — Open chain projection formula; α–D–Fructose — Fischer formula; α–D–Fructo furanose — Haworth formula

Fig. 3.4 : Ring structures of D fructofuranose as given by Fischer (A) and Haworth (B)

Anomers

Formation of the ring structure of sugar results in creation of an additional asymmetric carbon called **anomeric carbon** i.e. C_1 in aldoses and C_2 in ketoses. The presence of anomeric asymmetric carbon atom results in the formation of two more isomers known as **α** and **β** isomers. These two isomers differ from each other in having different configuration of alcoholic group around anomeric carbon atom. When –OH group is on right side of the anomeric C-atom (C-1 in the Fig. 3.5) it is α isomer and when the –OH group is on left side, it is called β isomer as depicted in the Fig. 3.5.

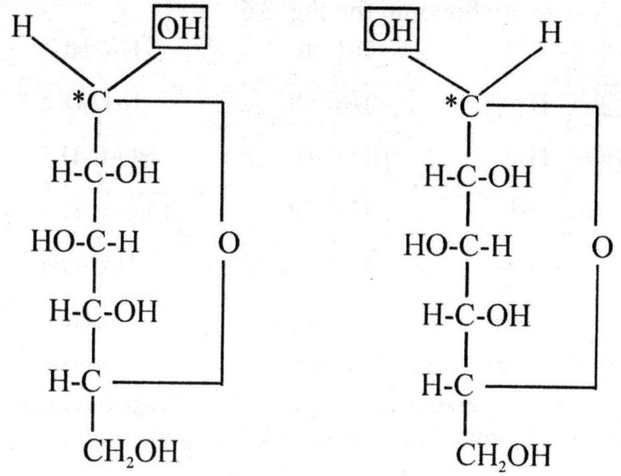

α D-glucose β D-glucose

Fig. 3.5 : α and β Anomers of D-glucose and anomeric C*-atom with OH group on opposite sides (shown in the block)

Mutarotation—The α and β anomers of glucose have different degree of optical rotation, though it is dextro (+) rotatory in both the cases. The solid (powder) glucose exists mainly as α anomer and when is added to water (fresh solution) it shows an optical rotation of +112.2°. The α anomer in the solution, then slowly is converted to β isomer and after sometime an equilibrium is attained between the two α and β anomers with a resultant optical rotation of +52.5°. This change in degree of optical rotation is known as **mutarotation** and the whole event can be summarized as :

$$\underset{+112.2^\circ}{\alpha\text{ D-glucose}} \rightleftharpoons \underset{+52.5^\circ}{\overset{\text{Equilibrium}}{\alpha\ \&\ \beta\text{ isomers}}} \rightleftharpoons \underset{+18.7^\circ}{\beta\text{ D-glucose}}$$

Mutarotation is defined as the change in degree of optical rotation from +112.2° (freshly prepared α-D glucose solution) to +52.5°C due to the interconversion of α anomer to β anomer of glucose till an equilibrium is reached. The equilibrium mixture contains 63% β anomers and 36% of α anomers with traces (1%) of open chain glucose molecules. β anomers predominant at equilibrium because they have more stable configuration.

Mechanism—The mechanism of interconversion of α anomer to β anomer probably involves opening of the α-D ring of glucose by breaking of hemiacetal linkage to a straight chain structure, which then recyclize in different direction(plane) to generate β anomer.

Epimers

The isomers formed due to alteration in the configuration of –H and –OH groups around **a single carbon atom** in a sugar molecule are called **epimers.** For example mannose and galactose are two epimers of glucose. Galactose has a different configuration of groups around fourth carbon from glucose and mannose differs in configuration of groups around second carbon from glucose as shown in the Fig. 3.6.

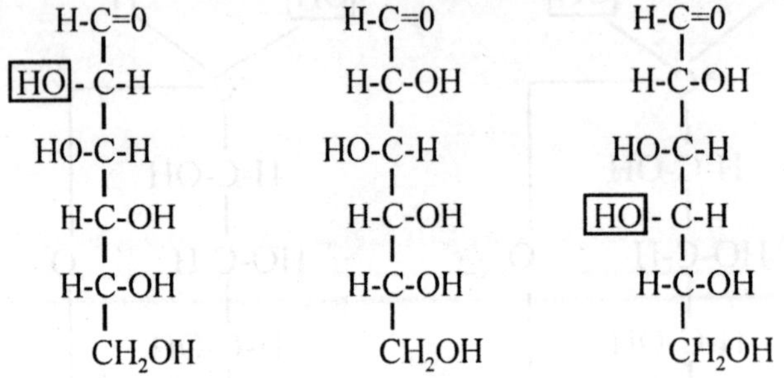

D-Mannose **D-glucose** **D-galactose**

Fig. 3.6 : Epimers of glucose, i.e. mannose and galactose have different configuration of OH group at C-atom 2 and 4 respectively (shown in block)

Aldose-ketose Isomerism

The aldoses and ketoses are the functional group isomers, i.e. sugars having same molecular formula but different chemical structures due to the different functional groups. For example glucose and fructose are isomers with same general formulas of Cn (H_2O)n but they differ as glucose has an aldehyde group, where as fructose has a keto group (Fig. 3.4).

Chemical Properties

1. **Oxidation** : Depending on oxidizing agent used, the terminal aldehyde (or keto group) at anomeric C_1-atom or terminal primary alcoholic group at 6^{th}C, or both, of the sugar can be oxidized to produce sugar acids, such as gluconic, glucuronic and saccharic acids respectively.

   Glucuronic acid—Oxidation of terminal alcoholic group at 6th C-atom of glucose, can be achieved only after protecting the anomeric aldehyde group by UDP nucleotide in the body and then oxidation by specific dehydrogenase enzyme (uronic acid path way, Fig. 7).

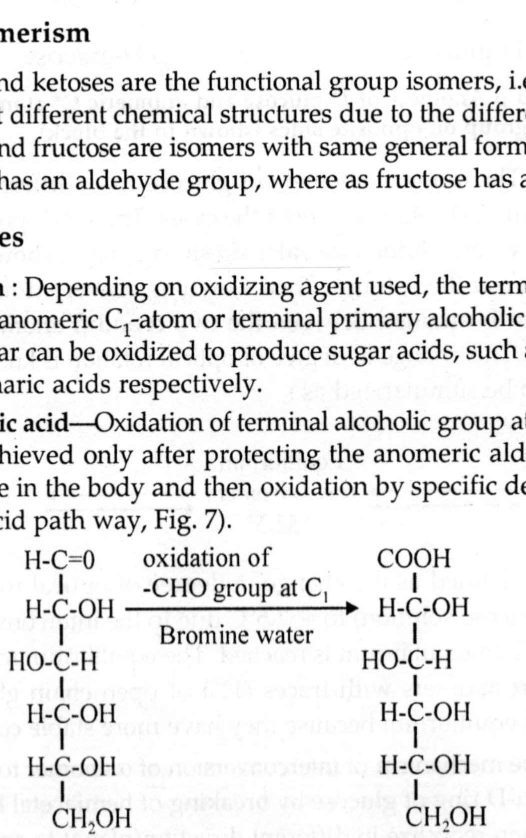

Glucose Gluconic acid

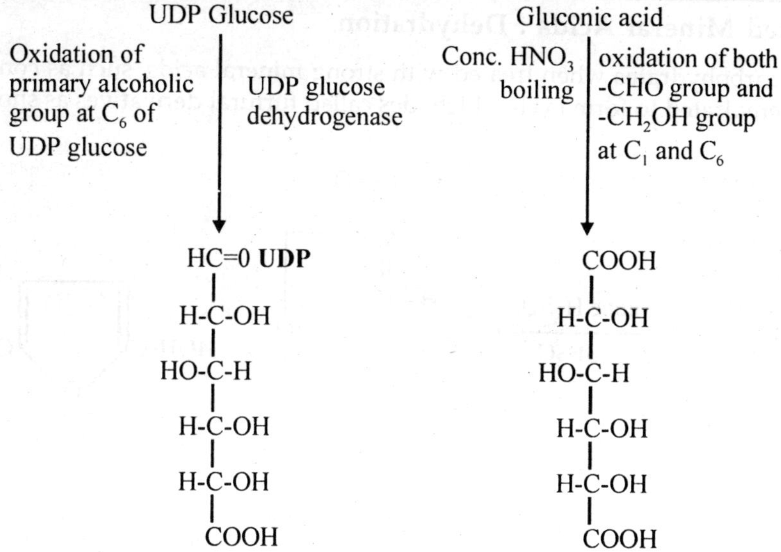

UDP-Glucuronic acid   Glucosaccharic acid

Biomedical Importance : Glucuronic Acid

By an alternative metabolic pathway(uronic acid pathway) glucose is oxidized to glucuronic acid in our body for some specialized functions, which are as listed below :

i) **Synthesis of mucopolysaccharides** : Glucuronic acid is a major constituent of various mucopolysaccharides produced in the body.

ii) **Conjugation reaction in liver :** Insoluble bile pigment bilirubin is conjugated in liver as bilirubin mono and diglucuronide (soluble) for its excretion via bile into GIT.(Chapter 34.)

iii) Hormones(steroid) are conjugated with glucuronic acid for their excretion.

iv) Many drugs are conjugated and detoxified by conjugation with glucuronic acid to be excreted out.

2. Reduction of sugars to form sugar alcohols—The aldehyde and keto group of monosaccharides can be reduced to alcoholic group, to produce corresponding sugar alcohols by reducing agents such as Na-amalgam as shown in the reaction:

$$\text{D-glucose} \xrightarrow{2H} \text{D-sorbitol} \xleftarrow{2H} \text{D-fructose}$$

3. Action of acids :

Dilute mineral acids—Disaccharides and polysaccharides are hydrolyzed into their constituent monosaccharides by boiling with dilute mineral acids as given below.

$$\text{Starch} \xrightarrow[\text{boiling}]{\text{dil HCl}} \text{(glucose)n}$$

$$\text{Sucrose} \xrightarrow[\text{boiling}]{\text{dil HCl}} \text{glucose + fructose}$$

Concentrated Mineral Acids : Dehydration

The carbohydrates when treated with strong mineral acids, such as concentrated H_2SO_4, are dehydrated to form cyclic aldehydes called furfural derivatives as shown in the Fig. 3.7.

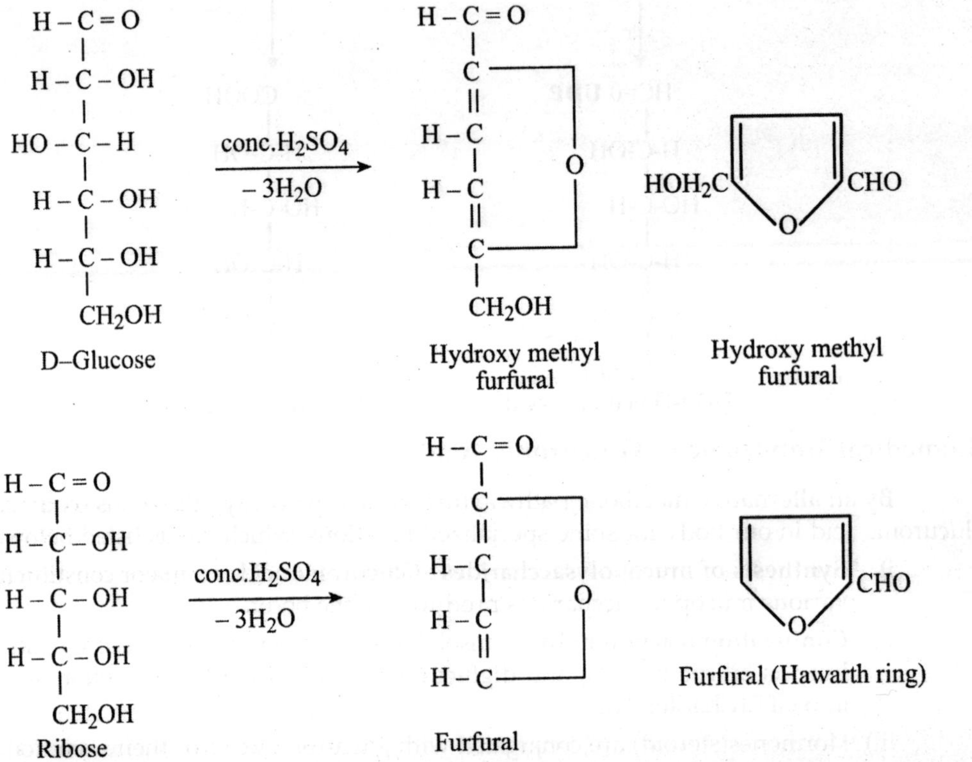

Fig. 3.7 : Action of conc. H_2SO_4 on carbohydrates to produce furfural derivatives

Practical application—This dehydration reaction forms the basis of certain tests used for the detection of carbohydrates. The furfural products formed by the dehydration with strong mineral acids can condense with certain organic phenols to form compounds having characteristic colour and are used for the identification of the carbohydrates.

Molisch's test—With α-naphthol, furfural derivatives gives violet colour. This test is given by all carbohydrates and used as group detection test.

Selivanoff's test—With resorcinol the furfural derivative condense to produce a cherry red colour. This test is given rapidly by fructose and thus is utilized for the detection of fructose. Procedure of the test : To 3 ml of selivanoff's reagent (0.5% resorcinol in dilute HCl), 0.5 ml of sugar solution are added and boiled for 30 seconds. A red colour indicates the presence of fructose.

4. Reducing properties of sugars

Monosaccharides and disaccharides with potential aldehyde and keto group reduce certain metallic ions, for example the cupric ions in hot alkaline medium to give a red precipitate of cuprous oxide.

$$\text{Monosaccharide/reducing dissacharides} + Cu^{+2} \xrightarrow[\text{medium}]{\text{hot alkaline}} \underset{\substack{\text{Cuprous oxide}\\ \text{(Red ppt)}}}{Cu_2O} + \text{Aldonic acid}$$

Practical application—The reducing properties of sugars are utilized for both qualitative as well as for the quantitative determinations of sugars. Reagents containing Cu^{+2} ions are most commonly used. Examples are Fehlings and Benedict's Test,

Benedicts reaction—The Benedict test is commonly used in clinical biochemistry to detect the presence of sugars in the urine sample. The cupric ions of the Benedict's reagent (5 ml) are reduced by boiling with sugar solution (0.5 ml) for two minutes to produce a green, yellow, orange or red precipitate depending on the degree of reduction and concentration of sugar present in the urine sample (Chapter 36). Therefore this test can be used as semiquantitative test.

Barfoed's reaction—In Barfoed's test, cupric ion reduction is carried out in hot acidic (acetic acid) medium and the test is given positive by monosaccharides rapidly, because they are strong reducing compounds as compared to the reducing disaccharides. Therefore this test differentiate monosaccharides from reducing disaccharides.

5. Osazone formation

Monosaccharides and reducing disaccharides with potential aldehyde or ketonic group form characteristic osazones with phenylhydrazine when kept at boiling temperature. The sugar solution with osazone mixture (phenyl hydrazine + sodium acetate) when placed in boiling water bath, forms osazones of characteristic crystalline shape. The osazones of the sugars have characteristic melting points, crystalline structures, and precipitation time and thus are useful in the identification of sugars in biological fluids.

Mechanism of reaction—The reaction involves only first two carbon i.e. the carbonyl carbon and the next adjacent carbon atom. The glucose and fructose differ in their configuration at I[st] two carbon atoms only and thus form the same type of osazone crystals, long needle shape called glucosazones. The reaction involved in osazone formation is described in the Fig. 3.8.

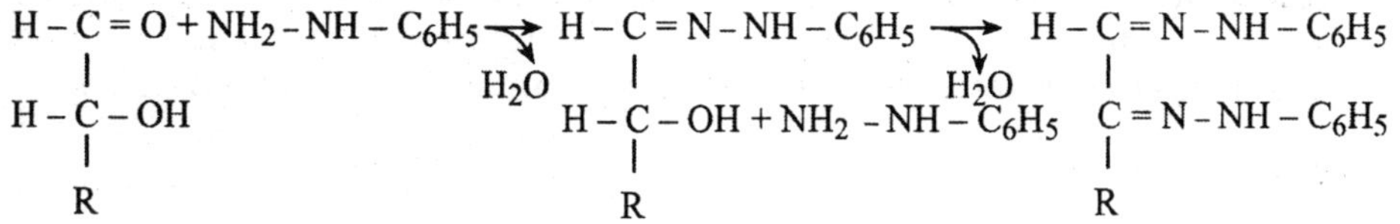

Glucose Phenyl hydrazine Glucose-hydrazone Glucose-osazone

Fig. 3.8 : Osazone formation by reaction of first two C-atoms of reducing sugar with phenyl hydrazine

Characteristics of Osazones

1. Glucosazone crystals are long yellow needles shaped, which are insoluble and precipitate within 5-10 minutes, as fan shaped aggregates or sheaves, commonly described as "bundle of hay" as shown in Fig. 3.9.
2. Lactosazone crystals are fine small needle shaped, that aggregate as powder puff or cotton ball in shape.
3. Maltosazone crystals are broad plate like arranged in sunflower shape.

The lactosazones and maltosazones are formed after boiling for 30-45 minutes and are soluble, so will crystallize after cooling at room temperature. The characteristic shape of the different osazones is depicted in the Fig. 3.9.

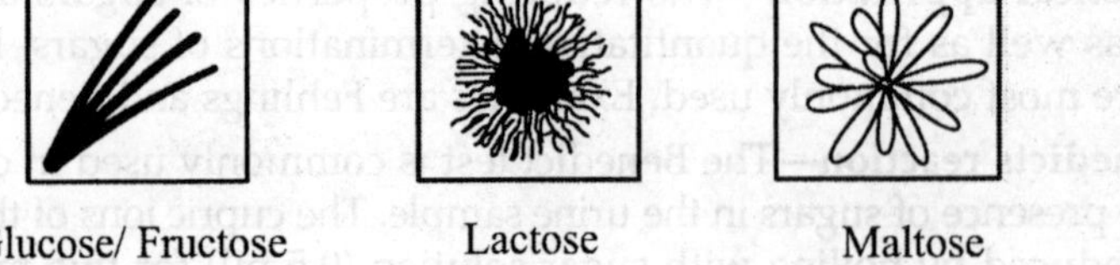

Fig. 3.9 Characteristic osazone crystals of some important sugars.

6. Sugar Derivatives of Biomedical Importance

Deoxy sugars—Deoxy sugars are those in which the oxygen of a –OH group has been removed leaving behind a hydrogen atom. Deoxy sugars of biological importance are :a)2-deoxy ribose which is constituent of nucleic acid DNA (Chapter 17) and b) 6-deoxy-L-galactose, found in glycoproteins and blood group substances.

Amino sugars (hexosamines)—Amino group may be substituted for OH group of sugars to produce amino sugars in our body. Usually the amino group is added the at the second C-atom of hexoses. Example are as given below :

Glycosamine and Galactosamine—The alcoholic –OH group at 2C-of glucose and galactose is replaced by $-NH_2$ group to form glucosamine and galactosamine respectively.

Biomedical importance—Glucosamine and galactosamine are main constituents of various glycosaminoglycans of our body tissues, for example: 1) Glucosamine is constituent of hyaluronic acid, heparin, blood group substance and chitin etc., 2) Galactosamine is a constituent of chondroitin sulphates of the connective tissues and 3) Certain antibiotics such as erythromycin, carbomycin etc. contains amino sugars.

Amino sugar acids—Neuraminic acid is unstable and occurs in nature in form of acetylated derivatives known as sialic acid or N-acetyl neuraminic acid (NANA)

Biomedical importance—Neuraminic acid and sialic acids are constituent of many mucopolysaccharides and glycolipids, such as gangliosides of brain. A number of nitrogenous oligosaccharides which contain neuraminic acid are found in human milk.

7. Glycoside Formation

A very important property of carbohydrates is to form **glycosidic** linkage or bond, which is formed when anomeric OH group (at C atom 1 or 2) of a monosaccharide condenses with a hydroxyl group of **another carbohydrate** or of a non-carbohydrate (aglycon) compound such as methyl alcohol, glycerol, phenol, sterol etc. The glycoside bond formation is very important property of sugars. The monosaccharide join together to form disaccharides, oligosaccharides and polysaccharides by glycosidic linkages.

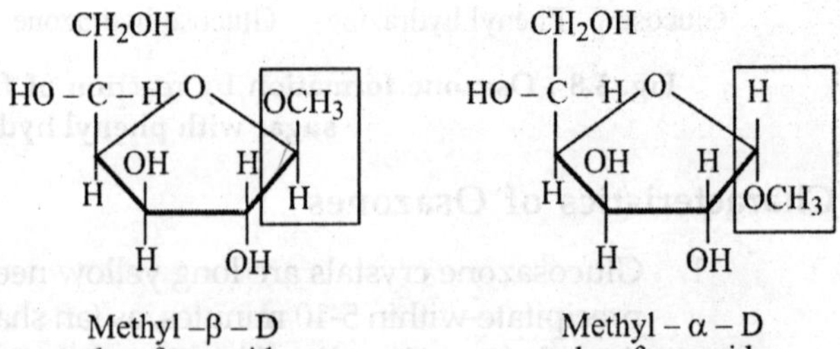

Physiologically Important Glycosides : Biomedical Importance

1. **Cardiac glycosides** are of great biomedical importance because of their potent stimulatory action on the heart and are used as drugs in cases of cardiac insufficiency. Cardiac glycosides have **sterol** as **aglycon** component. Important example is **digitonin.**
2. **Ouabain** is a glycoside that impairs active transport of Na^+ in cardiac muscle as it inhibits Na^+/K^+ ATPase and acts as Na^+/K^+ pump inhibitor.
3. **Phlorhizin**—It block transport of sugar across the mucosal cells of small intestine and renal tubular epithelium. Chemically phlorhizin is a glycoside and on administration causes renal glucosuria.
4. **Streptomycin**—Potent antibiotic drug streptomycin is also a glycoside, chemically.

DISACCHARIDES

Disaccharides are formed by two similar or dissimilar monosaccharide units joined together by a glycosidic linkage. The disaccharide can be a) **Reducing** – with free aldehyde or a keto group, for example maltose and lactose and b) **Non-reducing** –which does not contain free aldehyde or keto groups e.g. sucrose.

Lactose—It is a reducing disaccharide and forms characteristic lactosazones. Lactose is made up of glucose and galactose linked by β1-4 glycosidic linkage. It is also known as **milk sugar** as it is the main sugar present in milk and is secreted by lactating mammary gland. Lactose is necessary for infants as it provides galactose, which is essential for the development of brain. Galactose is an important constituent of brain glycolipids, such as cerebroside and gangliosides. Lactose is produced by formation of β 1-4 glycosidic bond between anomeric OH group of galactose and OH group at 4th carbon of glucose as shown in the Fig. 3.10.

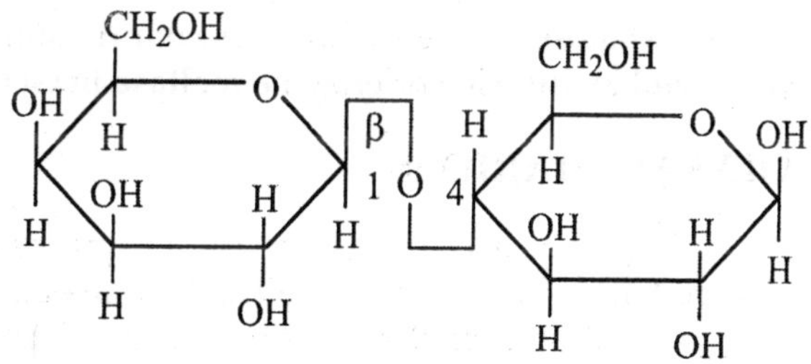

Fig. 3.10 : Lactose structure

Lactose is hydrolyzed by boiling with dilute HCl or by enzyme **lactase,** present on brush border of intestine to produce galactose and glucose.

Sucrose—It is the main sweetening agent used in our food and is known as cane sugar or table sugar. Sucrose is made up of glucose and fructose linked by α 1-2 glycosidic linkage. The aldehyde group of glucose at C_1 is linked by

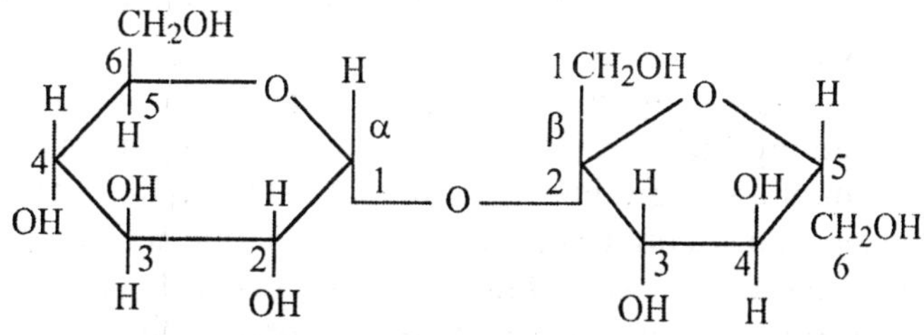

Fig. 3.11 : Structure of Sucrose

glycosidic bond with the keto group of fructose at C_2 in sucrose molecule, so no free reducing groups are left behind. Thus sucrose is a non-reducing disaccharide and will not form osazone.

Inversion—Sucrose is a dextrorotatory with an optical rotaion of +66.5°. Hydrolysis of sucrose yield equal amounts of glucose and fructose. Fructose being strong levorotatory (-92°) than dextrorotary glucose (+52.5°), the resultant mixture has a net levorotation after hydrolysis of sucrose. Therefore there is a change in the sign of optical rotation from dextrorotation to levorotation on hydrolysis of sucrose and is known as inversion of sugar. Sucrose is hydrolyzed by boiling with dil. HCl or by enzyme **sucrase** also known as **invertase** to produce glucose and fructose (Fig. 3.11).

Maltose—It is made up of two glucose units joined by glycosidic bond between–OH group at the first carbon atom of a glucose to the –OH of the fourth carbon atom of another glucose molecule, forming an **α1 – 4 glycosidic linkage** as shown the Fig. 3.12. Maltose is formed in GIT as hydrolysis product of starch and dextrins by the action of enzyme amylase. Maltose is a reducing disaccharide and forms characteristic maltosazones. It is hydrolyzed by dil. HCl or enzyme **maltase** in GIT to produce glucose.

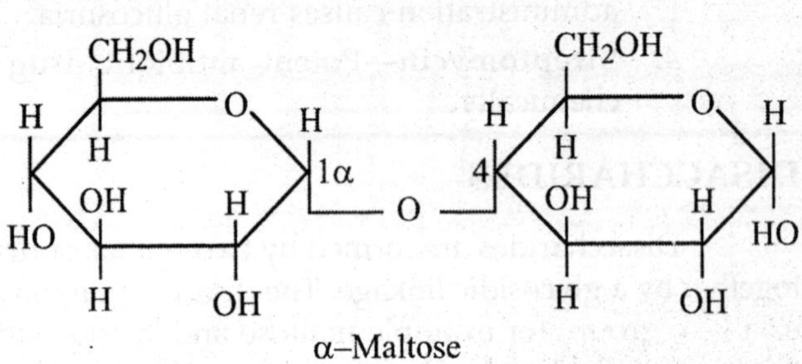

Fig. 3.12 : Structure of α-Maltose

POLYSACCHARIDES

The polysaccharides are the polymers formed by the condensation of many monosaccharide units. or their derivatives joined together by the glycosidic bonds. They are linear as well as branched chain polymers. The polysaccharides primarily have two important functions. 1) They form structural framework of the living cells and 2) the polysaccarides serves as the main storage form of nutrients and energy in the living organisms. The polysaccharide are of two types i.e. homopolysaccharides and hetero polysaccharides.

Homopolysaccharides—These on hydrolysis produce many molecules of single type of monosaccharide and thus are polymers of identical sugar units, for example starch, glycogen, dextrin, cellulose, inulin etc.

Starch—It is a storage carbohydrate reserve of plants and is the major dietary source of carbohydrates for humans. High content of starch is found in cereals, vegetables, tuber roots etc.

Chemically starch molecule contains two polysaccharides components, the **amylose** and **amylopectin**. Amylose consists of D-glucose units linked by α 1-4 glycosidic linkages in form of a straight linear chain, that is folded in a helical structure (Fig. 3.13). Whereas amylopectin is a branched chain polysaccharide and contains in addition to α 1-4 glycosidic bonds, the 1-6 glycosidic linkage to form branching points. Branching occur after every 24-30 glucose units in a linear chain, as shown in the Fig. 3.13.

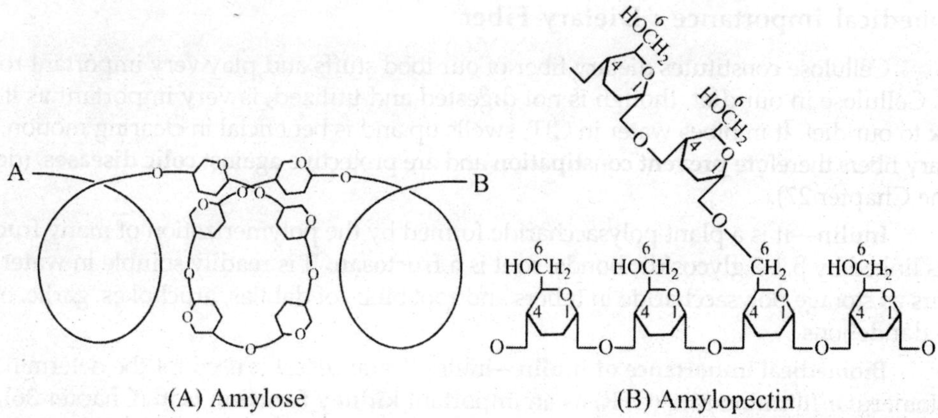

Fig. 3.13 : Structure of amylose (A) and amlylopectin components of starch

Amylose is low molecular weight polysaccharide and forms 10-20% of starch granules. It is made up of about 300 glucose units, joined by α 1-4 glycoside bonds as unbranched, linear polymer. It is soluble in water and gives blue color with iodine solution.

Amylopectin is high molecular weight polysaccharide and forms 80-90% of starch granules. It is formed by about 1000 glucose units jointed together as highly branched polymer. It is insoluble in cold water and gives violet color with iodine solution.

Starch is a non-reducing polysaccharide, as free reducing groups are negligible in the molecule. Starch is hydrolyzed by boiling with dil. HCl to produce glucose or by enzyme **amylase** in GIT to produce many molecules of α maltose.

Glycogen—The glycogen is the storage polysaccharide of animal body and is often called as animal starch. It is stored mainly in liver (6%) and in muscles (up to 1%). The function of liver glycogen is mainly to provide glucose for the maintenance of blood glucose level, particularly in between meals. After 12-18 hours of fasting, the liver becomes almost depleted of glycogen stores.

The function of muscle glycogen is to provide readily available source of glucose 6-phosphate for energy production, required for muscle contraction.

Glycogen is an extensively branched polysaccharide and has structure similar to amylopectin component of the starch. Glycogen is composed of glucose units linked by α 1-L4 glycosidic bonds in the straight chains portions and α 1-6 glycosidic linkage at the branch points. Branching occurs after every 12-18 glucose units, making it a highly branched and compact tree like structure as shown in the Fig. 5.3. Glycogen gives a wine red colour with iodine solution.

Cellulose—Cellulose is a plant polysaccharide and is main constituent of plant structural framework. Thus cellulose form the major component of our vegetable food stuffs, but it is not digested or utilized by humans. Cellulose is a homopolysaccharide formed by the polymerization of D-glucose units linked by **β1-4 glycosidic linkage** (unlike starch which have α 1-4 and α 1-6 glycosidic bonds). Cellulose cannot be digested in our digestive system as we do not have a hydrolyzing enzyme that can break β1-4 glycoside bonds (amylase enzyme specifically breaks only α1-4 glycosidic linkage).

Biomedical importance : Dietary Fiber

Cellulose constitutes dietary fiber of our food stuffs and play very important role in GIT. Cellulose in our diet, though is not digested and utilized, is very important as it add **bulk** to our diet. It imbibes water in GIT, swells up and is beneficial in clearing motion. The dietary fibers therefore **prevent constipation** and are protective against **colic diseases.** (details in the Chapter 27).

Inulin—It is a plant polysaccharide formed by the polymerization of many fructose units linked by β 1-2 glycosidic bonds, so it is a **fructosan.** It is readily soluble in water and occurs as storage polysaccharide in tubers and root bulbs of dahlias, artichokes, garlic, onion and dandelions.

Biomedical importance of inulin—Inulin clearance test is used for the determination of glomerular filtration rate (GFR) as an important **kidney function test.** (Chapter 36).

Dextrins—These are partial digestive intermediates formed in the course of hydrolysis of starch in GIT. **Limit dextrins** are first produced during action of amylase on starch, which hydrolyze glucose units from straight chain portions till the branching points are reached. Dextrins are finally hydrolyzed into maltose and then to glucose in the intestine (Chapter 4).

Heteropolysaccharides : Glycosaminoglycans (GAG)

Mucopolysaccharides (Glycosaminoglycans) are heteropolymers of repetitive units of many sugar units and their derivatives, such as **aminosugars** and **uronic acids.** Acetylated amino groups, sulphates etc are generally present in GAG. The mucopolysaccharides are also found to occur attached to proteins and are called **proteoglycans.** Mucopolysaccharides are essential components of body tissues and form major ground cementing substance of connective tissues of bone, skin, cartilage, blood vessels etc. The mucopolysaccharides held together the collagen and elastin fibers in these tissues, to give them strength. Some important examples of mucopolysaccharides are described here.

Hyaluronic acid—It is an important constituent of **synovial fluid** of bone joints, **vitreous humour** of eye, skin and umlical cord etc. Hyaluronic acid is viscous in nature as it can hold rich amount of water and thus is important as **lubricant** and **shock absorber** in bone joints and in eye ball. It is a polymer made up of many repeating units of D-glucuronic acid and N-acetylglucosamine joined by β 1-3 glycoside bond as shown in the (Fig. 3.13.).

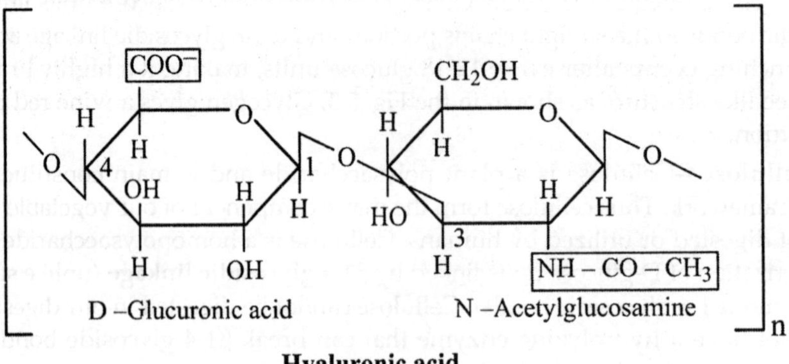

Fig. 3.13 : Hyaluronic acid structure

Chondroitin sulphates—These are polymers of many repeating units of glucuronic or iduronic acid and N-acetylgalactosamine, esterified with sulphuric acid (Fig 3.14). Chondriotin sulphates are the principal mucopolysaccharides present as ground cementing substance of connective tissues, cartilages, tendon, skin and bones etc. and provide strength to these tissues.

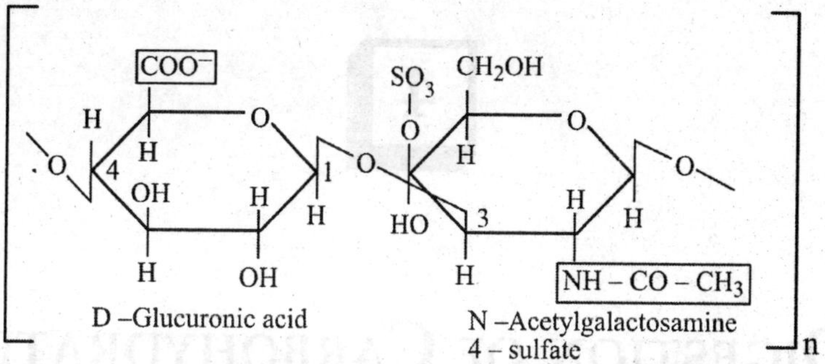

Fig. 3.14 : Structure of chondroitin 4-sulphate

Heparin—It is polymer made up of many repeating units of glucosamine sulfate and sulfated glucuronic acid or iduronic acid. It is an natural anticoagulant present in blood and prevents intravascular blood clotting. Heparin is an intracellular component of mast cells that line arteries especially in liver, lungs and skin.

Biomedical Importance—It is a potent anticoagulant widely used during dialysis of blood in kidney disease with renal failure, as well as in collecting blood for clinical use. Heparin is also used therapeutically in thromboembolic conditions to prevent intravascular thrombus formation in various pathological conditions, such as cardiovascular diseases.

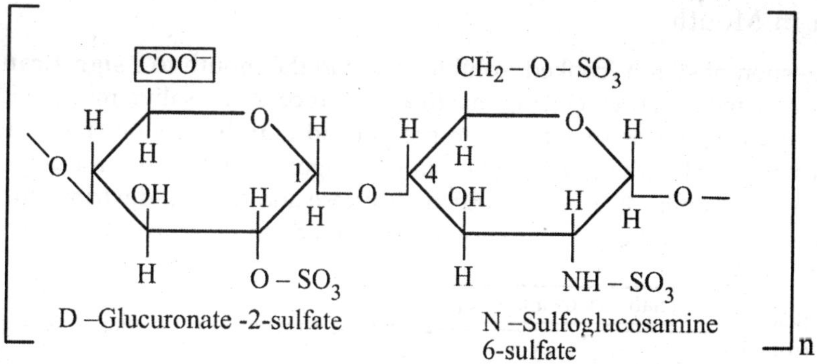

Fig. 3.15 : Structure of heparin

Blood group substances—The blood group antigens are present on the plasma membranes of erythrocytes and differentiate the type of blood as A, B, AB or O and Rh blood groups. These blood group antigens are chemically mucopolysaccharides associated with certain proteins. The blood group antigens are complex heteropolysaccharides formed by the condensation of many units of a variety of different monosaccharides and their derivatives such as N-acetyl galactosamine, galactose, fucose, sialic acid etc.

❑❑❑

4

DIGESTION OF CARBOHYDRATES

DIETARY CARBOHYDRATES

The principal dietary carbohydrates that are **digested** and **utilized** by human beings includes: (a) The polysaccharides – **starch** and glycogen, (b) Disaccharides – **Surcose** and **lactose** (milk sugar) and (c) small amounts of free monosaccharides namely **glucose,** fructose and pentoses obtained from fruits.

Digestion in Mouth

Digestion of starch (cooked, soluble) starts in the mouth to a **significant extent** by the salivary enzyme amylase. During **mastication process,** the saliva mixes with food and salivary **α–amylase,** that is secreted from the **parotid glands,** acts on starch. The enzyme hydrolyzes starch by breaking α 1-4 glycosidic linkages randomly to produce a mixture of small carbohydrate units, such as maltose, maltotriose and α–limit dextrins with branching point having 1-6 glycosidic bond as shown in the reaction given below :

$$\underset{\text{(soluble)}}{\text{Starch}} \xrightarrow[\text{pH 6.8}]{\boxed{\text{Salivary } \alpha\text{–amylase}}} \text{maltose} + \text{maltotriose} + \alpha\text{–limit dextrins}$$

Salivary α–amylase is most active at **optimum pH 6.8** (pH of saliva) and about **30 to 40% dietary starch** is digested by this enzyme. This is because, the activity (hydrolysis) of **α–amylase** continues for an appreciable period, till the food mixed with saliva from mouth reaches via esophagus to stomach.

Digestion in Stomach

Activity of salivary amylase is arrested, when food along with saliva reaches to stomach and is mixed with **gastric juice.** The acidic content of gastric secretion (HCl) lowers the pH of medium, which falls to around **pH 4** at which salivary amylase is inactivated.

No further digestion of **carbohydrates** occur in stomach, as carbohydrate digestive enzymes are not present in the gastric secretion.

Digestion in Small Intestine

The partially digested food from stomach, as acidic chyme comes in **duodenum** and gets mixed with pancreatic secretion and bile. The alkaline contents of these secretions neutralize the acid of gastric chyme and pH of medium becomes alkaline. This provides optimum pH for the activity of pancreatic enzymes.

Pancreatic α -Amylase—It is a potent amylolytic enzyme and acts on and digest all types of starch, including raw or insoluble starch completely. This enzyme specifically hydrolyzes α 1-4 glycosidic linkages of starch at alkaline pH to produce maltose, maltotriose and a mixture of short, branched (1-6 bond) oligosaccharides (α –limit dextrins) and straight chain oligosaccharides as shown in the Fig. 4.1.

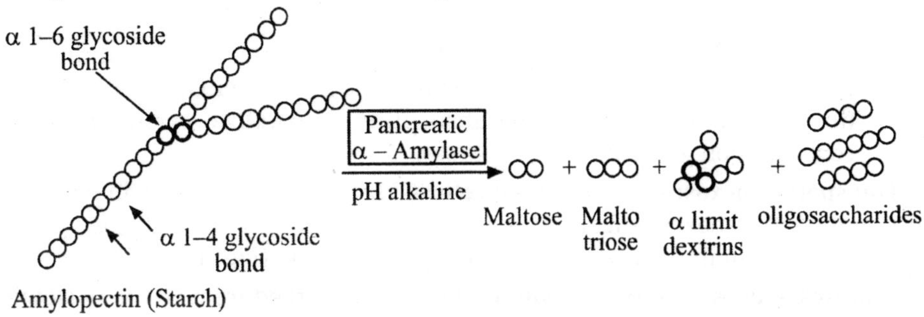

Fig. 4.1 : Diagrammatic representation of hydrolysis of starch by pancreatic α–amylase

Digestion at brush border of intestine

The final digestion of di and oligosaccharides, so produced as described above (end products of starch digestion), as well as dietary disaccharides is achieved by enzyme **oligosaccharidases** and **disaccharidases.** These enzymes are present on the brush borders of **mucosal lining** of **upper jejunum** and hydrolyze the disaccharides into monosaccharides as shown in the Fig. 4.2.

At brush border

Maltose —[Maltase]→ 2 Glucose molecules

Sucrose —[Sucrase]→ 1 Glucose + 1 Fructose

Lactose —[Lactase]→ 1 Glucose + 1 Galactose

Branched α-limit dextrins —[α-limit dextrinase]→ Glucose

Oligosaccharides —[Oligosaccharidase]→ Glucose

Fig. 4.2 : Hydrolysis of disaccharides and oligosaccharides at the brushborder of jejunum by respective enzymes

Absorption of Monosaccharides

The monosaccharides are the end product of carbohydrate digestion in GIT, that are absorbed at the brush border of muscosal lining of jejunum into the portal blood. Two mechanisms are mainly responsible for the absorption of monosaccharides from the intestinal lumen as described below:

A) Active Transport—The glucose and galactose are absorbed from intestine at a very fast rate, **against concentration gradient** by an **active transport system.** This transport system requires a sodium dependent specific carrier protein **(SGLT-I)** which has two specific binding sites – one for **glucose or galactose (epimers)** and another for **sodium** to transport them (both sugar and Na^+) inside mucosal cells. This **active transport system** is linked with Na^+/K^+ pump, that requires ATP (energy), as Na^+ is an extracellular ion and has to be expelled out from mucosal cells, in exchange with K^+ (intracellular ion) against concentration gradient (Fig. 4.3).

B) Facilitated Diffusion—Fructose is absorbed from intestine at **medium rate** by facilitated diffusion process, that requires a specific carrier **GLUT-5.** This process do not utilize energy as the absorption of fructose occurs along with concentration gradient.

Transport of hexoses from mucosal cells: GLUT-2 is a carrier protein required for the transportation of all hexose sugars. This carrier facilitates transport of glucose, galactose as well as fructose from the mucosal cells into the portal blood (Fig. 4.3).

Pentoses—Ribose and other pentose sugars are absorbed from GIT very slowly by **passive diffusion** along with concentration gradient, in which no carrier or energy (ATP) is used.

The mechanisms of absorption of monosaccharides from intestinal lumen into mucosal cells and then into portal blood is depicted in the Fig. 4.3.

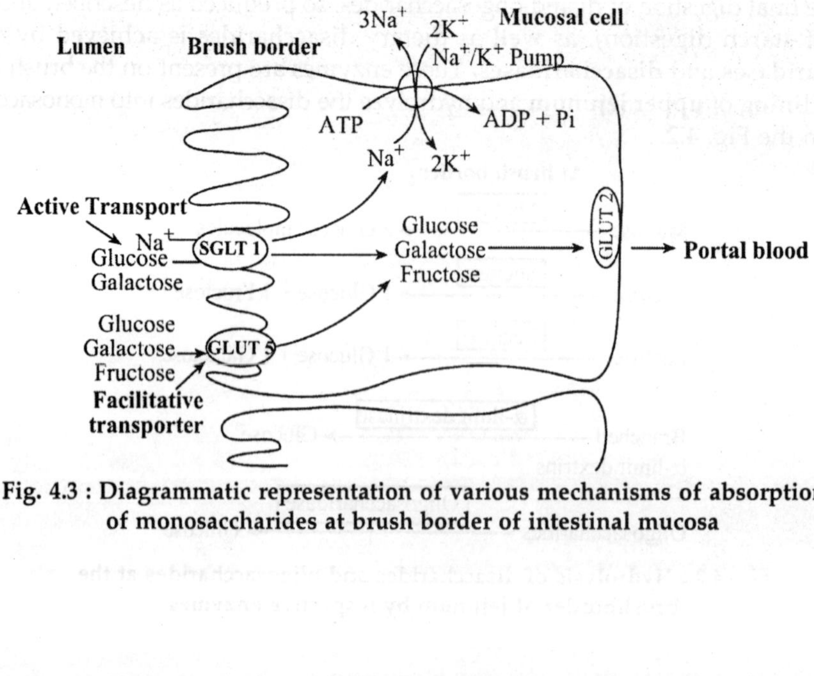

Fig. 4.3 : Diagrammatic representation of various mechanisms of absorption of monosaccharides at brush border of intestinal mucosa

❑❑❑

5

METABOLISM OF CARBOHYDRATES

Carbohydrates are the most economical and major source of energy for the living organisms. Green plants containing active compound chlorophyll synthesize carbohydrates (mainly starch) in the presence of sunlight from carbon dioxide and water. The process is known as **photosynthesis** and the starch so produced is stored in seeds, grains, vegetables, fruits etc. that form our main foodstuffs.

Dietary Carbohydrates and their Fate—The principal carbohydrate in our diet is **starch** and to some extent disaccharides, such as sucrose (cane sugar) and lactose (milk sugar) are also consumed. The dietary carbohydrates are hydrolyzed by the digestive enzymes in gastrointestinal tract to monosaccharides. **Glucose** is the major monosaccharide produced by digestion of carbohydrates, which is absorbed from GIT, along with other monosaccharides, such as galactose and fructose, that are converted into glucose in the liver or in GIT itself (Chapter 4).

Glucose-The Major Sugar of Body—Glucose is the main sugar present in blood, body fluids and tissues and is major source of energy in our body. Glucose plays central role in all carbohydrate metabolism and is well connected with lipid and protein metabolism as is evident from the Fig. 5.1. The major sources and functions of glucose and its role in intermediary metabolism is depicted in the Fig. 5.1.

MAJOR PATHWAYS OF CARBOHYDRATE METABOLISM

A. Glucose oxidation pathways are as given under:

(1) **Glycolysis** and **citric acid cycle** are the major oxidative pathways of glucose to generate energy.

(2) **Hexose Monophosphate** Shunt (HMP Shunt) is a minor pathway, occurring in certain tissues for the generation of reducing power and pentose sugars.

(3) **Uronic acid pathway** is also a minor oxidative pathway of glucose to synthesize glucuronic acid, vitamin C (in lower animals) and for the production of pentose sugars.

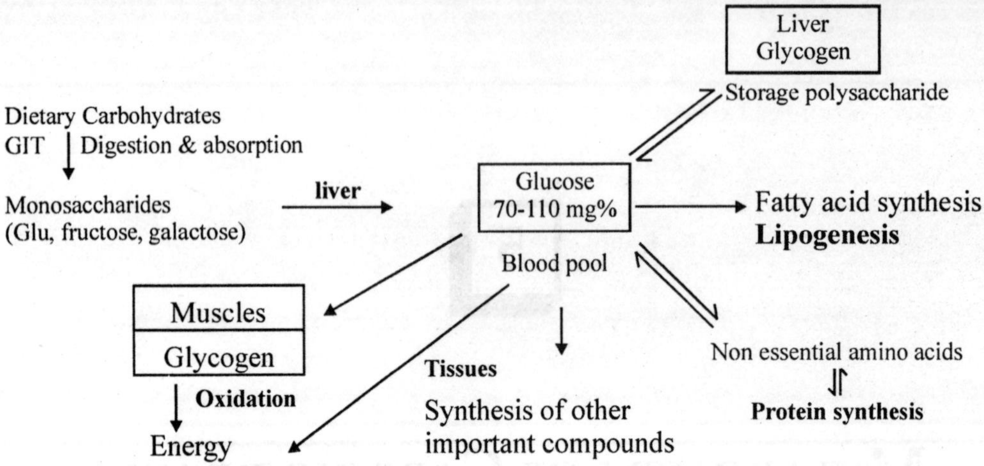

Fig. 5.1 : Sources and functions of glucose and overview of its role in intermediary metabolism

B. Storage of Carbohydrates :

(1) **Glycogenesis** is the synthesis of glycogen from glucose, which is stored in liver and muscles.

(2) Conversion of glucose to other physiologically important compounds.

C. Glucose Production pathways:

(1) **Glycogenolysis** is the breakdown of glycogen into glucose to maintain blood sugar level (BSL) or to provide glucose 6-PO_4 for oxidation purposes.

(2) **Gluconeogenesis :** Denovo synthesis of glucose from non-carbohydrate substances is known as gluconeogenesis and it is a major process, that maintain BSL during fasting and hypoglycaemic conditions.

GLYCOGEN METABOLISM

Glycogen is the storage polysaccharide of animals, similar to the starch of plants. Glycogen is stored mainly in liver, upto 6 – 8 % and in muscles to the extent of 1 – 2 % only. But as muscle mass is more in comparison to liver, the quantity of muscle glycogen is more (250 g), about 3 – 4 times higher than liver glycogen, which is around 75 g.

Function of Glycogen Stores

The primary function of liver glycogen is to provide free glucose to maintain blood sugar level, in between meals and in fasting conditions. While muscle glycogen serves as readily available source of **glucose 6-phosphate** units for energy production purposes during **muscle contraction.**

Glycogen Metabolism include mainly two processes as described here under:

1. **Glycogenesis**—The synthesis of glycogen by the condensation of many glucose units is known as glycogenesis and it requires energy as ATP and UTP for the activation of glucose to **UDP - glucose.**
2. **Glycogenolysis**—Break down of glycogen into glucose 6-PO_4 or free glucose is known as glycogenolysis.

A brief account of glycogen metabolism and its functions is depicted in the Fig. 5.2.

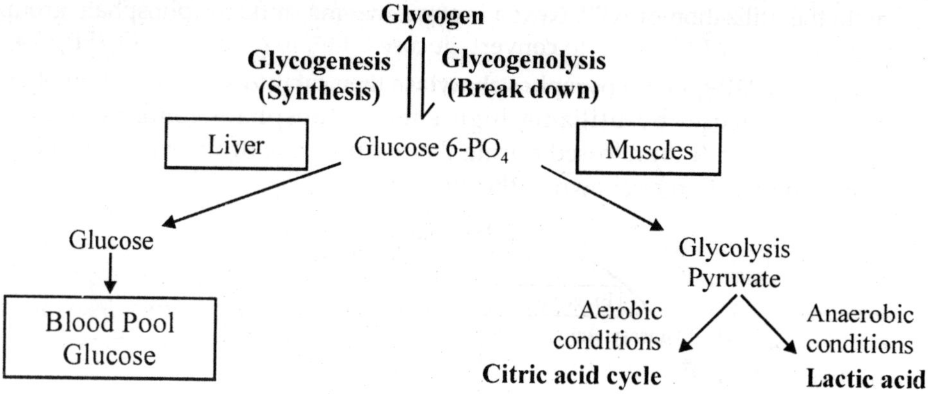

Fig. 5.2 : Overview of glycogen metabolism

Glycogen – Structural Aspects

Glycogen is a homopolysaccharide formed by the condensation of many glucose units (around 1 lac), joined together by glycosidic linkages to produce an extensively branched, tree like structure (Fig. 5.3). In the straight chain portions of branches of glycogen molecule, the glucose units are bonded by α–1, 4 glycosidic linkages, while at branching points, the linkage is α–1, 6 glycosidic bond, as shown in the Fig. 5.3. Glycogen is stored as granules in cytosol of tissues as a reserve storage polysaccharide.

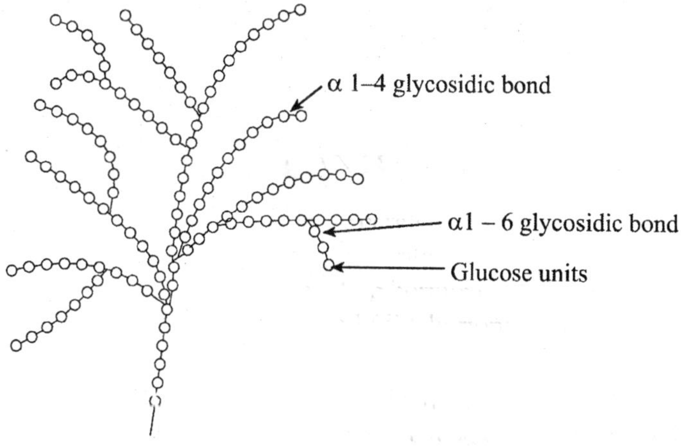

Fig. 5.3 : Tree like glycogen molecule with many glucose units joined by α–1-4 and α–1-6 glycosidic linkages

GLYCOGENSIS

Synthesis of glycogen takes place in cytosol of tissues, primarily in the liver and muscles, where potent enzyme system for glycogenesis are present.

Glycogenesis comprises of following discrete steps and reactions :

1. Activation of Glucose to UDP-Glucose—Glucose is first activated as UDP-glucose, that is prerequisite for glycogen synthesis. For this, glucose is converted into glucose

6-phosphate catalyzed by enzyme **hexokinase** in extrahepatic tissues and by **glucokinase** in liver with the utilization of ATP. Next a **mutase** enzyme shifts the phosphate group from 6C-atom of glucose to 1 C- atom, to convert glucose 6-PO_4 to glucose 1-PO_4 (Fig. 5.4).

Enzyme **UDP-glucose pyrophosphorylase** then catalyzes the activation of glucose-1-PO_4 to UDP-glucose, by utilizing high energy phosphate compound UTP. The pyrophosphate group (PPi) released during the reaction is hydrolyzed to provide energy for the activation reaction as shown in the Fig. 5.4.

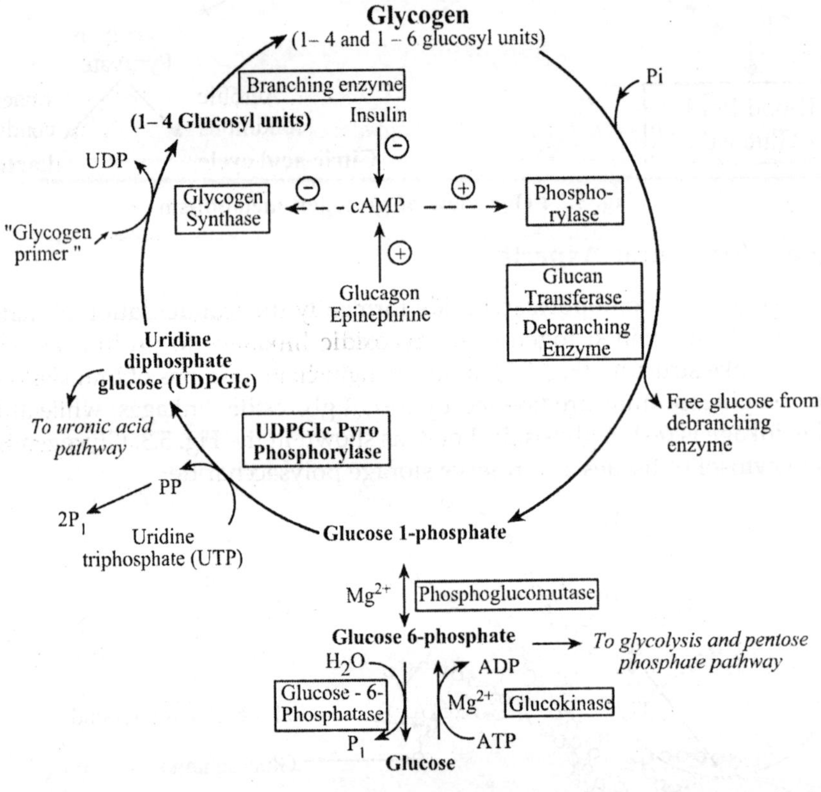

Fig. 5.4 : Reactions of glycogenesis and glycogenolysis pathways and role of c-AMP in reciprocal regulation of the two pathways

UDP-glucose produced as described above is now ready to add glucose units that condense together by glycosidic bonds to synthesize glycogen molecule as described in following steps.

1. Glycogen Synthase—Glycogen synthesis requires the enzyme **glycogen synthase,** that is specific for the formation of α 1-4 glycoside linkage between glucose units to form the straight chain portions of glycogen molecule.

For the synthesis of glycogen, a **glycogen primer** is needed. It can be a small fragment of left over glycogen molecule with a branching point or a preexisting specific protein called **glycogenin** can act as primer. UDP-glucose produced as described above then add glucose units on the short straight branch of the primer by forming α 1-4 glycoside linkage, catalyzed by enzyme **glycogen synthase** and the branches elongates. (Fig. 5.5).

2. Branching Enzyme—For the formation of branching points to introduce new branches in growing tree like glycogen molecule, a branching enzyme is needed. The branching enzyme **glycosyl α transferase** (amylo α 1-4 to 1-6 trans glycosidase), transfers a small fragment (3-4 glucose residues) of a grown up branch (with 11-12 glucose units) to the near by another branch and join the fragment by α 1-6 glycoside linkage to produce a branching point as illustrated in the Fig. 5.5.

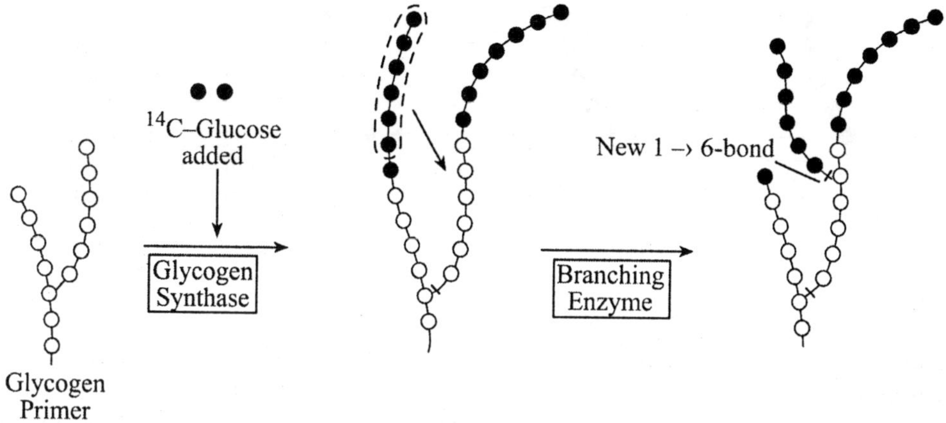

Fig. 5.5 : Formation of branches during glycogenesis by branching enzyme

The newly formed branches are then elongated and new branches are produced by alternative actions of enzymes glycogen synthase and α 1-4 to 1-6 transglycosidase respectively to synthesize a large, extensively branched tree like glycogen molecule.

GLYCOGENOLYSIS

The break down of glycogen molecule into glucose units is known as glycogenolysis. The key enzymes and main reactions of glycogenolysis are basically **different** from glycogen synthesis, as the two processes are **irreversible.** The enzymes and reactions involved in the glycogen breakdown are as described here under:

a) Glycogen phosphorylase a (active)—Phosphorylase-a is the key enzyme of the glycogenolysis process. It is a specific and **unique** enzyme, as it **phosphorylatically hydrolyzes** α 1-4 glycosidic linkage from the straight chain portions of a glycogen branch, by directly utilizing energy of hydrolysis to add inorganic phosphate group to produce **glucose 1-PO_4**. Thus successive glucose units of a glycogen branch, from the non reducing end are phosphorylatically cleaved off by the action of phosphorylase enzyme, till branching point is reached.

b) Debranching enzyme—For the hydrolysis of the branching points, a debranching enzyme is needed. The debranching enzyme is a bifunctional enzyme and it first removes a small fragment of left over branch (3-4 glucose residues) to expose the branching point and transfers it to near by branch by its α 1-4 to 1-4 glucan transferase activity. The second catalytic unit of this enzyme, called amylo α 1-6 glycosidase, then hydrolyzes 1-6 glycosidic bond of the exposed branching point to remove free glucose from the left over branch, as shown in the Fig. 5.6.

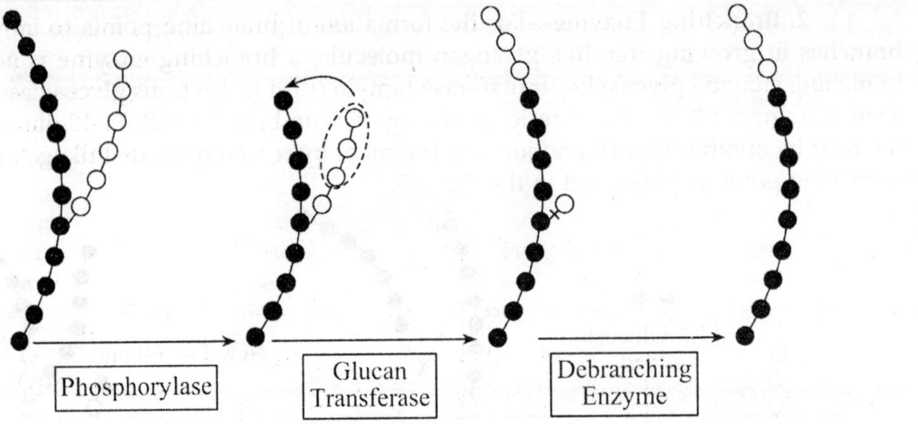

Fig. 5.6 : Action of debranching enzymes during glycogenolysis to remove branching points

During glycogenolysis, by the alternative action of **phosphorylase a** and debranching enzyme, glycogen molecule is completely hydrolyzed to produce glucose 1-PO_4 and some free glucose units. Glucose 1-PO_4 is then converted by a **mutase** enzyme to **glucose 6-PO_4,** which is utilized in muscles for glycolysis to generate energy.

c) Hepatic Glucose 6-phosphatase—However in liver a potent enzyme **glucose 6-phosphatase** (absent in muscles) further hydrolyzes phosphate group from glucose 6-PO_4 to release **free glucose,** which can readily be added to blood to maintain blood sugar level (BSL 70-110 mg %) in fasting conditions (Fig. 5.4).

Reciprocal Regulation of Glycogenesis and Glycogenolysis

At any moment of the time, only **one process** i.e. either glycogen synthesis or glycogen break down proceeds depending on BSL status or the immediate need of the tissues. This is important, otherwise the processes would be futile and energy will be wasted. So the two pathways of glycogen

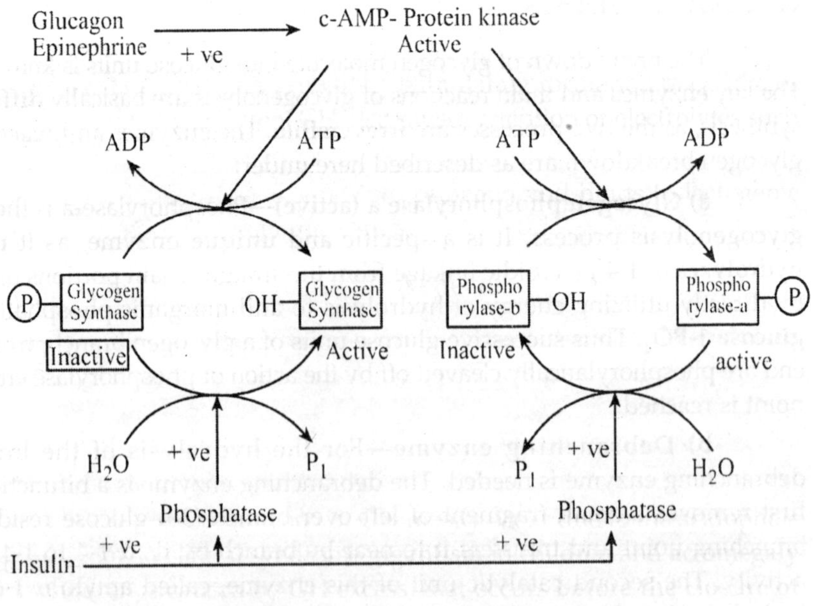

Fig. 5.7 : Reciprocal regulation of key enzymes of glycogen metabolism by c-AMP-protein kinase system

metabolism are regulated **reciprocally** i.e.when one pathway is going on actively in a tissue, the other pathway is stopped or inhibited. This is achieved by a fine coordination and regulation of the activity of the **rate limiting enzymes** of the two pathways, namely **glycogen synthase** and **phosphorylase.**

The reciprocal regulation of the two key enzymes of glycogen metabolism is accomplished by **c-AMP,** which has an activating (+ve) effect on enzyme **phosphorylase** (glycogenolysis) and a **simultaneous inhibiting** (-ve) effect on **glycogen synthase** by covalent modification with simultaneous phosphorylation as shown in the Fig. 5.7.

Cyclic AMP—II Messenger of Hormones

Cyclic AMP is the II messenger through which many hormones exert their potent physiological actions inside the target cell. The c-AMP was discovered by Scientist Earl Sutherland, while working on regulation of glycogenolysis process in vitro.

Cyclic AMP is synthesized from ATP, catalyzed by a cell membrane bound enzyme adenyl cyclase. The hormones such as glucagon and epinephrine on binding with specific hepatic cell membrane receptors, activate adenyl cyclase, to increase c-AMP synthesis as shown in the Fig. 5.8 (details in the chapter 27.). The enzyme phospho- diesterase degrades and inactivates c-AMP. Hormone insulin increases the activity of phosphodiesterase enzyme and thus lowers the c-AMP concentration in the tissues.

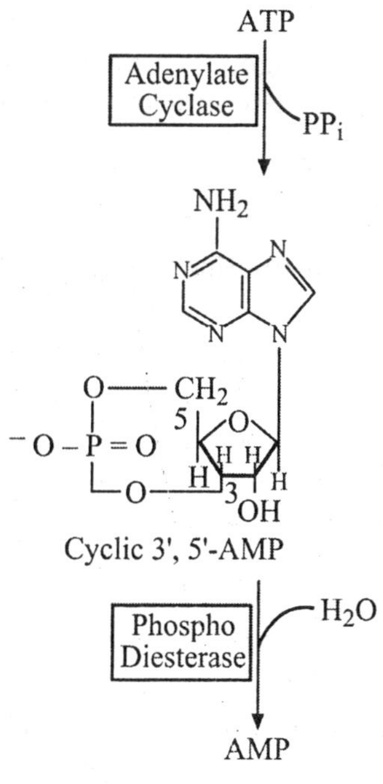

Fig. 5.8 : Synthesis and hydrolysis of c–AMP by respective enzymes

Mechanism of Action of c-AMP

The key enzyme of glycogen metabolism, glycogen phosphorylase and glycogen synthase exists in two forms, an active 'a' form and inactive 'b' form. The c-AMP produced as result of the action of hormones, activates a protein kinase (inactive tetramer as R_2C_2) enzyme by combining with its 'R' subunit, so that active catalytic 'C' subunit becomes free to have its catalytic activity.

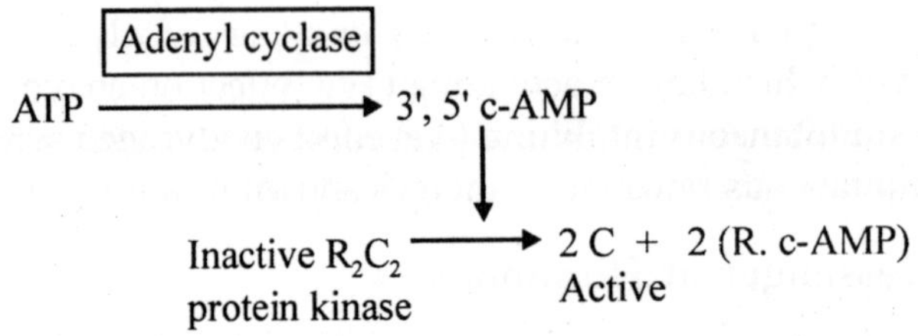

The protein kinase so activated stimulates activation of a cascade system to amplify the hormonal effect, that brings about rapid glycogenolysis as shown in Fig. 5.8.

At the same time c-AMP - protein kinase system inactivates glycogen synthase by phosphorylation and thus inhibits glycogenesis (Fig. 5.8).

Mechanism of Reciprocal Regulation—Salient Features

1. Phosphorylase enzyme is **active** only when OH group of a serine residue, at the catalytic site of the enzyme is phosphorylated by c-AMP-protein kinase system utilizing ATP
2. On the other hand, glycogen synthase is inactivated (inhibited) by phosphorylation of the serine residue of the enzyme at the catalytic site by c-AMP-protein kinase system.

Factors Controlling Reciprocal Regulation

Three factors are involved in the fine control and coordination of reciprocal regulation of glycogenolysis and glycogenesis, as described here under:

1. **Hormonal regulation**—Insulin and glucagon (diabetogenic hormones) have opposite effects on the key enzymes of glycogen metabolism. Glucagon and other diabetogenic hormones increase c-AMP (II messenges) concentration and thus glycogenolysis is stimulated. While insulin inactivates c-AMP and thus glycogenolysis is inhibited (Details in Chapter 7).
2. **Covalent Modification**—c-AMP-protein kinase mediated activation or inactivation of key enzymes by covalent modification i.e. phosphorylation/ dephosphorylation reciprocally regulate the key enzymes of glycogen metabolism as shown in the Fig. 5.8.
3. **Influence of Ca^{++}-Calmodulin System**—Ca^{++}-Calmodulin system stimulates the activation of myophosphorylase several hundred folds during vigorous muscle contraction, to accelerate glycogenolysis at a very fast rate. This is important to provide large amount of energy as ATP needed during excersize.

A brief account of reciprocal regulation of two key enzymes and glycogen metabolic pathways by various factors is depicted in the Fig. 5.8 & Fig. 5.9.

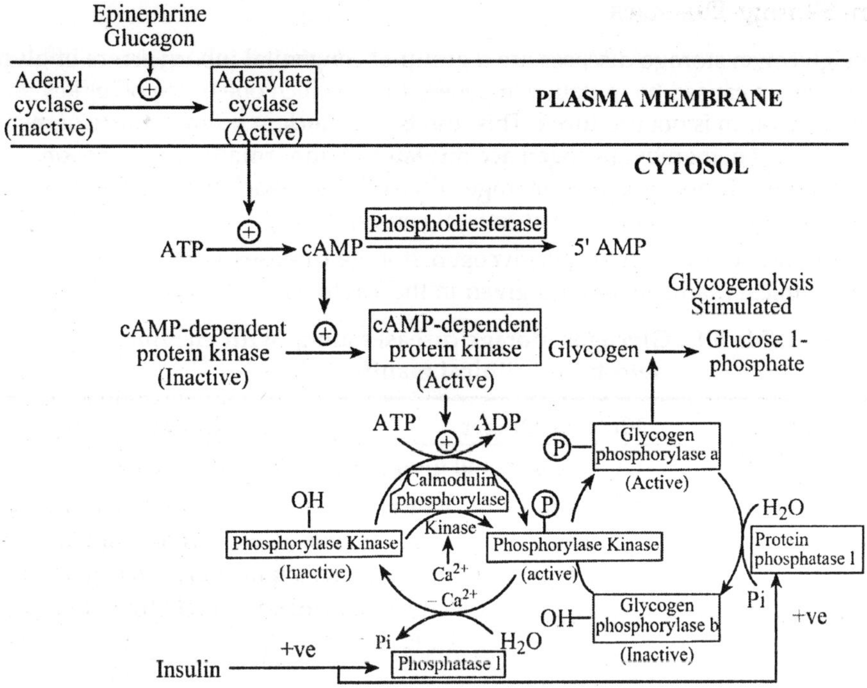

Fig. 5.8 : Regulation of myophosphorylase activity by activation cascade of various factors

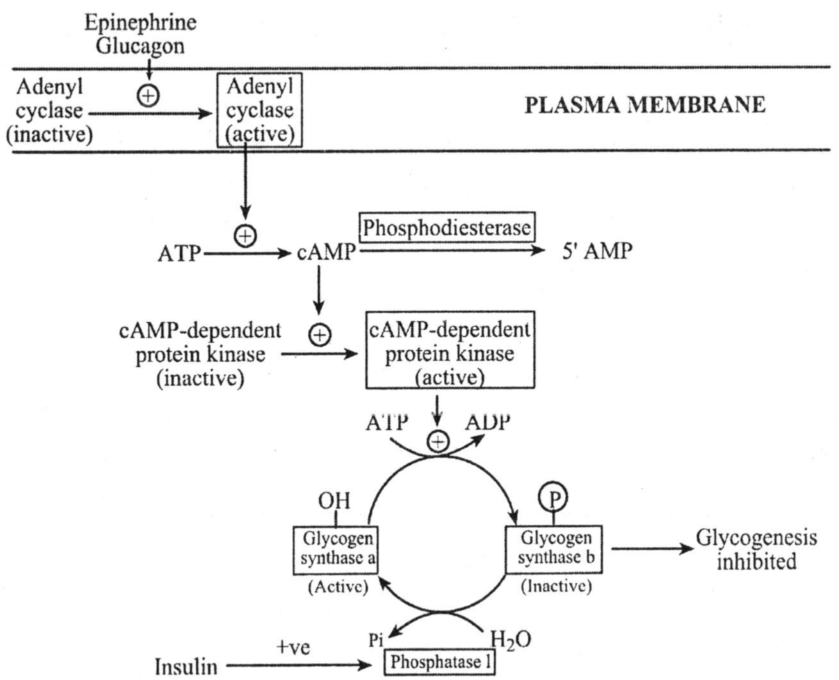

Fig. 5.9 : Hormonal regulation of glycogen synthesis (glycogenesis) by activation cascade of various factors

Glycogen Storage Diseases

The glycogen storage diseases are a group of congenital inborn errors of glycogen metabolism, that are caused by mutation in genes, due to which one or the other key enzyme of glycogen metabolism is not produced. This results in a **block** in normal course of metabolic pathways of glycogen, thereby **glycogen accumulates in abnormally excessive amounts** in liver or in muscles. These glycogen storage diseases are associated with severe clinical manifestations, some of which can lead to fatal consequences.

A brief account of important glycogen storage diseases with the enzyme defect (genetic) and clinical manifestations is given in the Table 5.1.

Table 5.1 : Glycogen storage diseases along with enzyme defects and clinical manifestations

Type	Name	Enzyme Defect	Clinical Manifestations
Type I	Von Gierke's disease	Hepatic & Renal glucose-6- phosphatase deficiency	Hypoglycemia Liver enlargement, cirrhosis & jaundice Renal failure
Type II	Pompe's disease	Cardiac muscle debranching enzyme deficiency	Enlargement of heart, Heart failure
Type III	Forbe's limit dextrinosis	Skeletal muscle debranching enzyme deficiency	Accumulation of limit dextrin in muscles, Energy not released. Difficulty in performing exercise
Type IV	Anderson's disease	Liver & cardiac branching enzyme deficiency	Abnormal glycogen with few branches accumulates. Cirrhosis, jaundice Death due to liver failure
Type V	McArdle's disease	Skeletal muscle phosphorylase deficiency	Muscular exercise may lead to fatal consequences
Type VI	Her's disease	Liver phosphorylase deficiency	Jaundice, liver failure

GLYCOLYSIS

Glycolysis is the major oxidative pathway of glucose, occurring universally in all living organisms for energy production purposes. Main reactions of glycolysis in all living cells are similar, though the end product may be different in different organisms. This pathway is often referred to as **Embden Meyerhoff's** pathway, after the scientists who elucidated and proposed the complete sequence of reactions of glycolysis.

Glycolysis : Salient Features

1. It is an **universal oxidative pathway** of glucose in all living organism for energy production purpose. However in microorganisms, such as yeast, it proceeds in absence of O_2 and end product is ethyl alcohol.

2. Glycolysis is unique, as it can proceed both in the presence of O_2 (**aerobic**) as well as in absence of O_2 (**anaerobic**). **Anaerobic glycolysis** is significant in **vigorously exercising muscles**, where **hypoxia** usually develops, but it continues to provide energy (ATP), though in less quantity and the end product is **lactic acid.**

3. **Anaerobic glycolysis** is also important in tissues lacking **mitochondria,** for instance erythrocytes, cornea and lens of eye etc.

4. Enzymes of glycolysis are present in cytosol of all cells and the process is exclusively cytosolic.

Reactions of Glycolysis

1. **Activation of Glucose**—Before glucose can enter glycolysis for oxidation, it is activated as glucose 6-PO_4 by ATP in the presence of enzyme **hexokinase.** Glucose 6-PO_4 so produced is then isomerized to fructose 6-PO_4 and then one more phosphate group is added by another molecule of ATP, in the presence of enzyme **phosphofructokinase-1** (PFK-1) to produce **fructose 1-6 bisphosphate** as shown in the Fig. 5.10. The enzyme PFK-1 is the key regulatory enzyme, at which the rate of glycolysis is regulated.

2. **Cleavage of Fructose 1-6 Bisphosphate**—Fructose 1-6 bisphosphate (6C-atom compound) is then cleaved in the middle of the molecule into two 3C-atom compounds (trioses), called **glyceraldehyde 3-PO_4** and **dihydroxyacetone phosphate,** catalyzed by enzyme **aldolase.** The triose phosphates so produced are interconvertible and are isomerized in the presence of enzyme **triosephosphate isomerase** to finally produce two molecules of glyceraldehyde 3-phosphate.

3. **Oxidation of triose phosphates**—Two molecules of glyceraldehyde 3-PO_4 are then oxidized by a specific enzyme **glyceraldehyde 3-PO_4 dehydrogenase** by removing 2H ions, that are taken up by coenzyme NAD^+ to generate 2 $NADH+H^+$. At this step the energy of oxidation is very high and is trapped directly as high energy phosphate bond by incorporating inorganic phosphate at **substrate level** to form a high energy compound glycerate l, 3-bisphosphate as illustrate in the Fig. 5.10.

4. **Substrate level phosphorylation and ATP generation**—

a) The high energy phosphate bond of the compound 1,3-bisphosphoglycerate produced as described above is utilized to generate ATP at substrate level. in the presence of enzyme **glycerate bisphosphate kinase.**

b) The term substrate level phosphorylation is referred to as: the generation of ATP at an oxidation step by trapping energy of oxidation directly, without involving oxidative phosphorylation of mitochondrial electron transport chain.

Another high energy phosphate bond and ATP is generated at subtrate level in further steps of glycolysis as described here under:

5. **Oxidation of glycerate 3-PO_4 to phosphoenol pyruvate (PEP)**—The 3 phosphoglycerate left after ATP generation in above step is converted into 2-

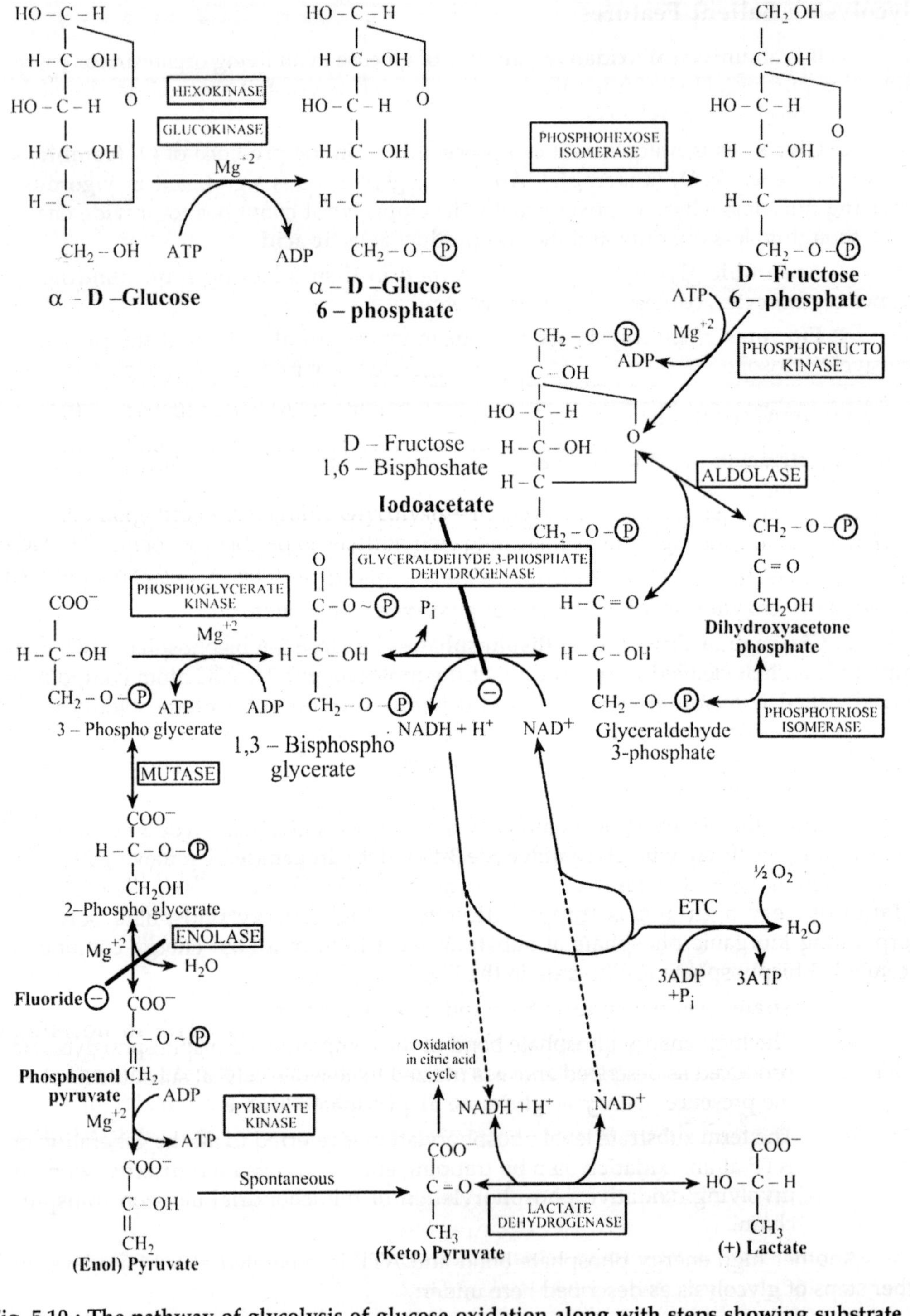

Fig. 5.10 : The pathway of glycolysis of glucose oxidation along with steps showing substrate level ATP generation and fate of reducing equivalents removed⊖shows action of inhibitiors

phosphoglycerate by a **mutase** enzyme and then is dehydrated by the enzyme **enolase** to form phosphoenol pyruvate.

The energy of dehydration is redistributed intramolecularly to raise the phosphate bond at C–atom 2 to produce a very high energy phosphate compound known as **phosphoenol pyruvate (PEP)** as shown in the Fig. 5.10.

6. Conversion of PEP to pyruvate—This high energy compound phosphoenol pyruvate is then utilized in the next step of glycolysis to generate another ATP molecule at substrate level, catalyzed by enzyme **PEP kinase** (Fig. 5.10). The compound enolpyruvate left is unstable and thus spontaneously is converted into pyruvate, the end product of glycolysis process as shown in the Fig. 5.10.

Fate of Pyruvate

Two molecules of pyruvate (3C-compound) so produced at the end of glycolysis of glucose metabolism as described above, has two fates. These are:

a) Oxidatively decarboxylation of pyruvate to acetyl CoA, which enters citric acid cycle (if oxygen is available) for complete oxidation.

b) If hypoxia exist (as in vigorously exerecizing muscles) pyruvate is reduced to lactate by NADH + H^+ to regenerate oxidized NAD^+ coenzymes to continue glycolysis cycle, even in anaerobic conditions as shown in the Fig. 5.10 & 5.11.

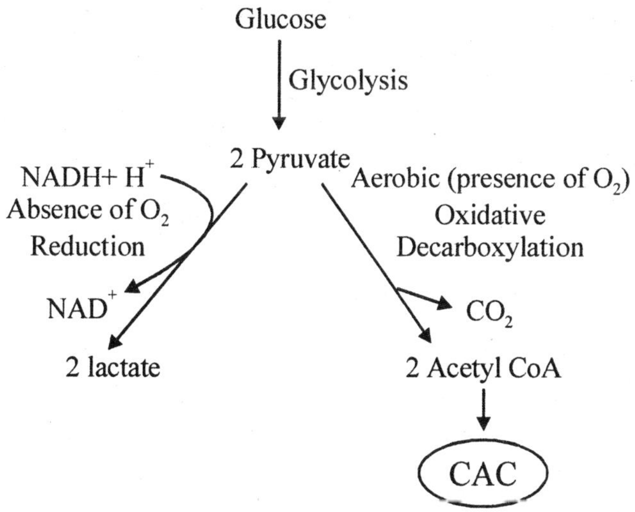

Fig. 5.11 : Fate of pyruvate

Lactate dehydrogenase (LDH)

The enzyme catalyzing the reduction of pyruvate to lactate is called lactate dehydrogenase. In the condition of O_2 deprivation (anaerobic conditions), the availability of **oxidized NAD^+ becomes rate limiting** at the step of oxidation of glyceraldehyde 3-phosphate in glycolysis process. So to regenerate oxidized NAD^+ to continue the glycolysis process, even in anaerobic conditions, pyrurate is reduced to lactic acid (Fig. 5.10).

Energetics of Aerobic Glycolysis

1. Aerobic glycolysis—ATP generation at the steps

(i)	Glyceraldehyde 3-phosphate dehydrogenase →	2 NADH + H^+ → (Biological oxidation on respiratory chain)	6 ATP
(ii)	1 ~ 3-Bisphosphoglycerate → (substrate level ATP generation)		2 ATP
(iii)	Phosphoenolpyruvate → (substrate level ATP generation)		2 ATP
		Total =	10 ATP

Utilization of ATP at the steps :

(i)	Hexokinase (1 ATP utilized for activation)		(-) 1 ATP
(ii)	Phosphofructokinase(1ATP used for activation)		(-) 1 ATP
		Net =	8 ATP

2. Energetics of Anaerobic Glycolysis—In the condition of O_2 deprivation, reduced NADH+ H^+ cannot be oxidized on electron transport chain and to regenerate oxidized NAD^+ to continue glycolysis in vigorously exercizing muscles, NADH + H^+ is utilized to reduce pyruvate to lactate. Thus only 2 ATP are generated in anaerobic glycolysis of glucose oxidation as described below:

ATP generation at steps

I.	1 ~ 3 Bisphosphoglycerate -	substrate level → ATP generation		2 ATP
II	Phosphoenolpyruvate	substrate level → ATP generation		2 ATP
			Total =	4 ATP

ATP utilization at steps:

I.	Hexokinase (activation of glucose)		(-) 1 ATP
II.	PFK (activation of fructose 6-PO_4)		(-) 1 ATP
		Net =	2 ATP

Regulation of Glycolysis

The three steps of glycolysis, where free energy change is very high, are irreversible and thus are rate limiting reactions. These steps are catalyzed by key regulatory enzymes-hexokinase, phosphofructokinase (PFK-1) and pyruvate kinase and their activity can be regulated as described below:

1. **Hexokinase**—This enzyme is inhibited allosterically by the accumulation of glucose 6-PO_4 in the tissues and thus rate of glycolysis decreases.
2. **Phosphofructokinase-1 (PFK-1)**—It is the most important regulatory enzyme of glycolysis, the activity of which is allosterically regulated by activators, such as AMP, fructose 2, 6-bisphosphate, and inhibitors ATP and citrate. (details in the Chapter 7)

3. **Pyruvate Kinase**—The activity of this enzyme is regulated by covalent modification by phosphorylation/dephosphorylation. The enzyme pyruvate kinase is inhibited by ATP, c-AMP and is activated by the hormone insulin.

Details of reciprocal regulation of glycolysis and gluconeogenesis are described in Fig. 5.12 along with the regulation of gluconeogenesis (details in the Chapter 7).

2,3-Bisphosphoglycerate (BPG) Cycle

The 2, 3-BPG cycle is also known as Rapaport Leubering cycle and it is established in erythrocytes. The RBCs are unique, as the only metabolic cycle established in them, is anaerobic glycolysis, in which 2 ATP molecules per mole of glucose oxidized are generated at substrate level. This is because mitochondria are absent in RBCs, so biological oxidative phosphorylation of NADH + H^+ does not occur and thus it is utilized to reduce pyruvate to lactic acid. This is necessary for the regeneration of oxidized NAD^+ to continue the glycolysis process.

2,3-BPG cycle is a **bypass** of main glycolysis process in RBC. 2,3-BPG cycle is advantageous for the economy of RBCs, when ATP is not needed much and one step of substrate level ATP generation is bypassed as described here under:

In 2,3-BPG cycle, the high energy compound 1,3 bisphosphoglycerate is isomerized to 2,3-BPG (a low energy phosphate compound), by shifting the phosphate group from position 1st to 2nd C-atom, catalyzed by a mutase enzyme. The 2,3 BPG later on can be converted to 3-phosphoglycerate by hydrolyzing one phosphate group, to continue the glycolysis process as summarized in the Fig. 5.12.

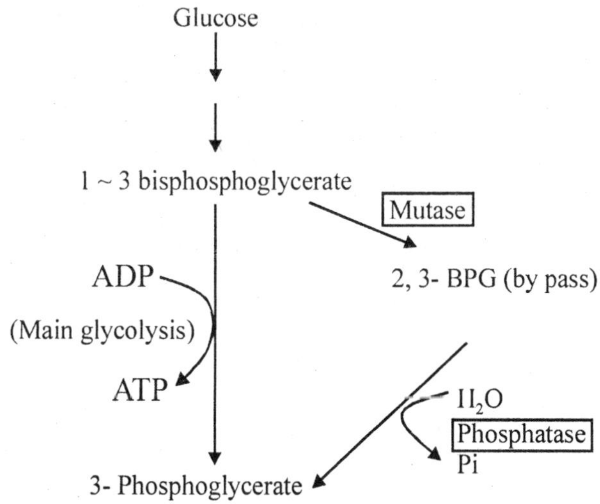

Fig. 5.12 : Reactions of 2,3 BPG cycle as a bypass of main glycolysis process in RBCs

Importance of 2, 3-BPG

2,3-BPG so synthesized in RBCs plays very important role in facilitating unloading of oxygen from oxyhemoglobin at tissue level, even when partial pressure of oxygen is very low. This is because 2, 3-BPG has more affinity to bind with deoxyhemoglobin and thus favours unloading of oxygen from oxy Hb.

Hence it is observed that under hypoxic conditions, the 2,3 BPG production increases in RBCs. This increased level of 2,3 BPG favours release of oxygen from oxyHb, even when pO_2 is low. The compensatory increase in 2,3 BPG at high altitude regions favours O_2 dissociation (details in Chapter 33).

Cori's Cycle

During anaerobic glycolysis, as occurring in vigorously excersizing muscles or in RBCs, lactic acid is the end product. Accumulation of lactic acid in muscles and tissues would cause lowering of pH (lactoacidosis), due to which enzymes of glycolysis are inhibited and muscles get fatigued after some time.

To restore normal pH of muscles, the lactic acid is released into blood and reaches to liver, where by the process gluconeogenesis, it is utilized to synthesize glucose (denovo) and glycogen. These events are depicted in the Fig. 5.13 and are termed as Cori's cycle.

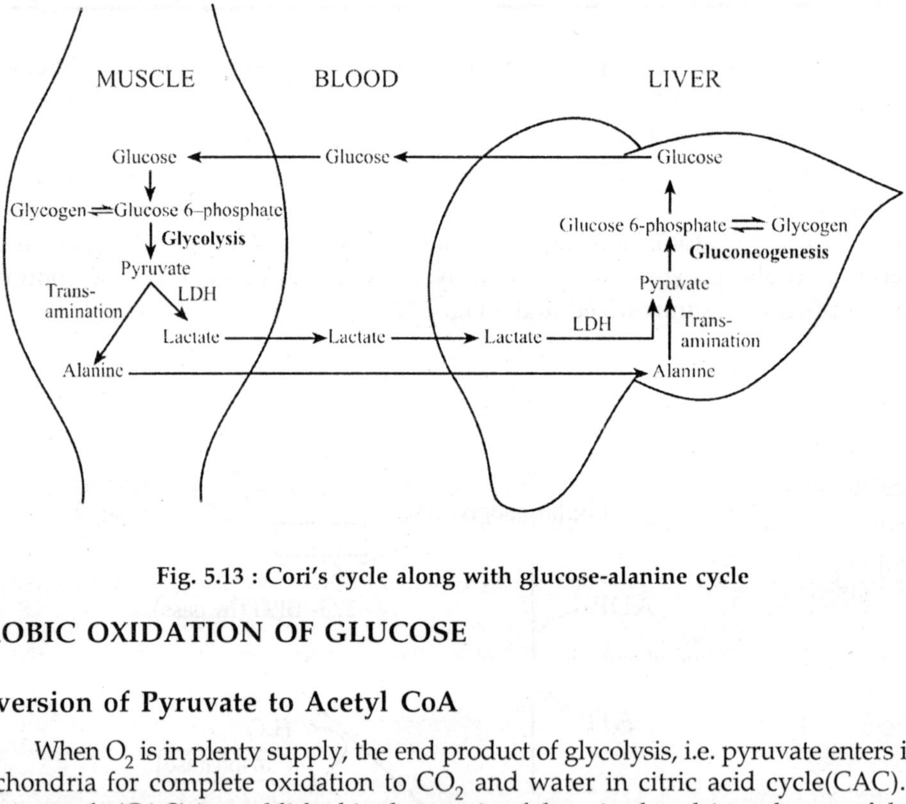

Fig. 5.13 : Cori's cycle along with glucose-alanine cycle

AEROBIC OXIDATION OF GLUCOSE

Conversion of Pyruvate to Acetyl CoA

When O_2 is in plenty supply, the end product of glycolysis, i.e. pyruvate enters inside mitochondria for complete oxidation to CO_2 and water in citric acid cycle(CAC). This oxidative cycle (CAC) is established in the matrix of the mitochondria and most of the free energy of glucose is made available inside mitochondra as described here.

Pyruvate is oxidatively decarboxylated to **acetyl CoA,** catalyzed by a multienzyme complex known as **pyruvate dehydrogenase system (PDH)** present inside mitochondria. The multienzyme **pyruvate dehydrogenase** requires 5 coenzymes, namely TPP, NAD^+, FAD, lipoate and coenzyme A. The overall reaction catalyzed by PDH is as given below:

$$\text{Pyruvate} + NAD^+ + \text{CoA} \xrightarrow{\boxed{PDH}} \text{acetyl CoA} + CO_2 + NADH + H^+$$

Oxidative decarboxylation of pyruvate—Role of thiamine pyrophosphate (TPP)

1. Pyruvate (3 C-compound) is first decarboxylated and the 2 C-atom compound left is taken up by TPP as hydroxyethyl TPP complex. (Fig. 5.14).

2. This complex then transfers the 2 C-atom moiety to oxidized lipoic acid to form acetyl lipoate, catalyzed by dihydrolipoate transacetylase catalyzing unit of PDH multienzyme complex.

3. The acetyl group of dihydrolipoate acetate is then transferred to coenzyme A as shown in the Fig. 5.14.

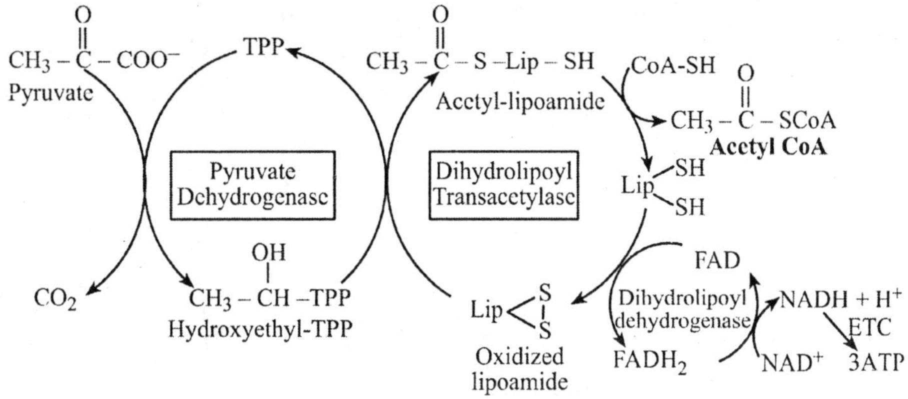

Fig. 5.14 : Reaction sequence of pyruvate dehydrogenase complex to convert pyruvate to acetyl CoA

4. This cycle is completed when reduced lipoate is oxidized back by lipoate dehydrogenase subunit of PDH and reducing equivalents are taken up first by FAD, catalyzed by lipoate dehydrogenase, which are finally transferred to NAD^+ to produce $NADH + H^+$ as shown in the Fig. 5.14.

Salient Features of PDG Catalyzed Reaction

1. This PDH catalyzed reaction links glycolysis process of glucose metabolism to citric acid cycle (aerobic oxidation) for complete oxidation of glucose.

2. One way reaction—PDH catalyzed reaction is irreversible, because of its high energy barrier and so only pyruvate (and hence glucose) can be converted to acetyl CoA, to be utilized for fatty acid synthesis. However fatty acids cannot be converted to glucose, as the reaction (PDH catalyzed) is **irreversible**.

3. Inhibitors of PDH—Arsenic and mercuric ions bind irreversibly with the SH group of lipoic acid and thus they are potent poisons and cause fatal consequences when consumed.

4. Insulin activates enzyme PDH, while glucagon and c-AMP inhibit it by covalent modification by dephosphorylation and phosphorylation respectively.

CITRIC ACID CYCLE

The citric acid cycle, also known as Kreb's cycle or tricarboxylic acid cycle (TCA) is a **common metabolic pathway** for the final and complete oxidation of all nutrients in the body. It is a common furnance in which ultimately all nutrients burn completely to CO_2 and free energy is released, as illustrated in Fig. 5.15.

Location—The citric acid cycle (CAC) is established in the mitochondrial matrix, in close proximity to electron transport chain (ETC) on inner mitochondrial membrane. This facilitates the quick transfer of reducing equivalents removed at various steps of CAC for biological oxidation to generate energy.

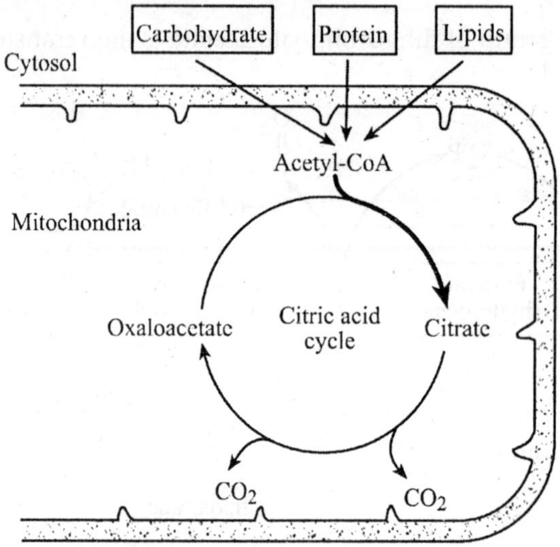

Fig. 5.15 : Common metabolic role and location of citric acid cycle inside mitochondria

Citric Acid Cycle—Salient Features

1. It is a cyclic process occurring in matrix of mitochondria, in which acetyl CoA is completely oxidized to CO_2 and reducing equivalents removed at various steps of the cycle are transported to electron transport chain, to be oxidized by molecular O_2 to liberate energy.

2. The CAC has dual role i.e. it has both oxidative (catabolic) as well as anabolic functions. The glycogenic amino acids after transamination, form one or the other component of CAC, which are finally utilized for glucose synthesis (**gluconeogenesis**) to maintain blood sugar level during fasting conditions.

3. The CAC also plays important role in **fatty acid synthesis** (lipogenesis) to provide and transport acetyl CoA out side mitochondria at the site of fatty acid synthesis (Chapter 11.). Also it is involved in the synthesis of nonessential amino acids required for the protein biosynthesis.

4. Oxaloacetate is the starter of CAC, which is regenerated back after the completion of the cycle.

Reactions Citric Acid Cycle—The citric acid cycle comprizes the following discrete steps and reactions.

1. **Formation of Citrate and Isocitrate**—The citric acid cycle is initiated by oxaloacetate (a 4 C-compound), which condenses with acetyl CoA (2 C-compound) to produce a 6 C-tricarboxylic acid called citric acid. This reaction is catalyzed by the enzyme **citrate synthase** and energy for the condensation is provided by the high energy sulphur bond of acetyl CoA.

Citrate so produced is then isomerized to isocitrate in two steps, catalyzed by enzyme **aconitase.** This enzyme first dehydrates citrate to an unsaturated cis-aconitic acid and then rehydrate it to produce isocitrate. (Fig. 5.16).

2. **Oxidative Decarboxylation of Isocitrate**—Next the isocitrate is dehydrogenated (NADH + H^+ removed) as well as decarboxylated, catalyzed by the enzyme **isocitrate dehydrogenase** to produce α ketoglutarate (5 C-compound) in two steps as shown in Fig. 5.16.

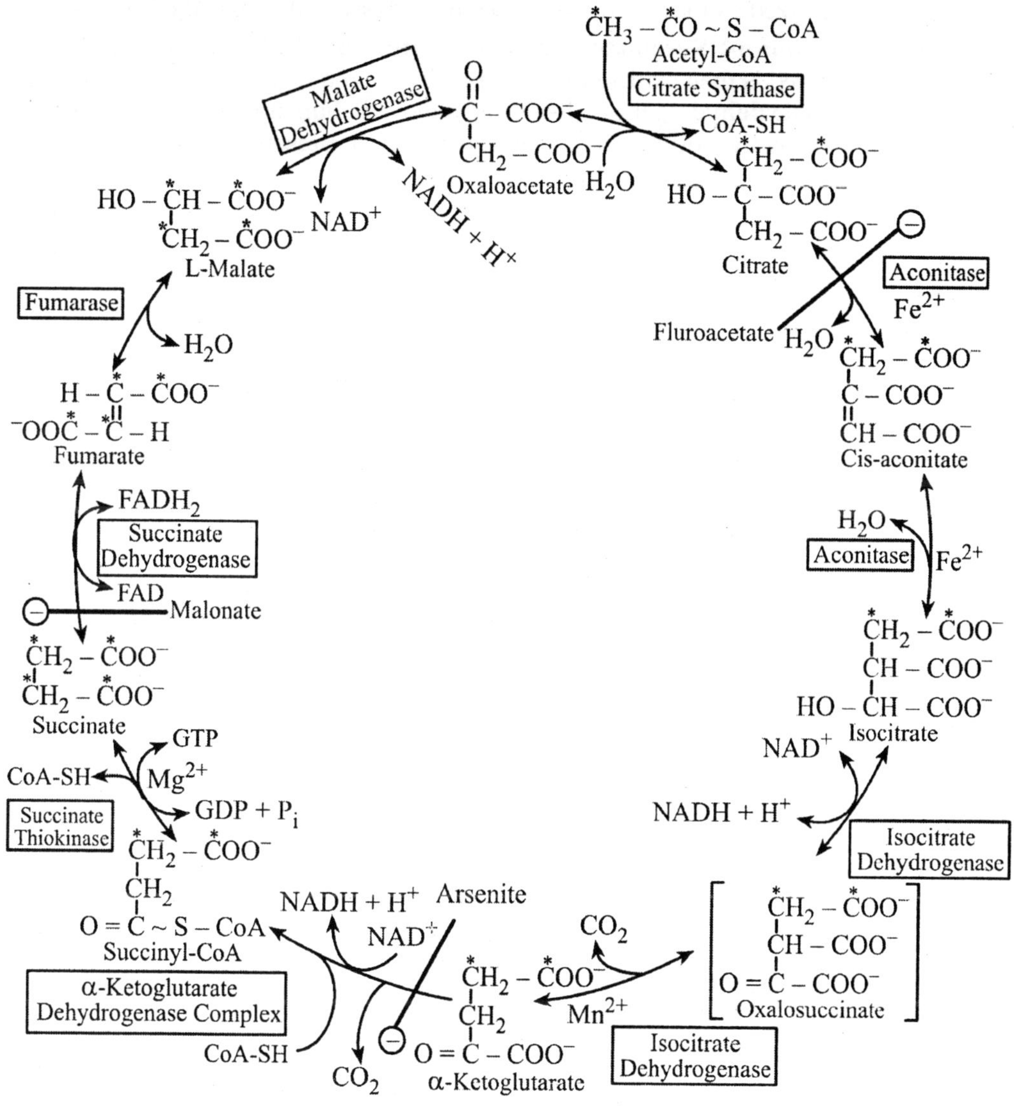

Fig. 5.16 : Reactions of citric acid cycle, along with key enzymes and inhibitors

3. **Oxidative Decarboxylation of α-Ketoglutarate**—In next step of CAC, α-ketoglutarate undergoes a second oxidative decarboxylation reaction, catalyzed by a multienzyme system known as **α-ketoglutarate dehydrogenase** to

produce a 4C-atom compound succinyl CoA (CO_2 and NADH+ H^+ are removed). This reaction is complicated and requires a multienzyme complex, similar to **pyruvate dehydrogenase** with five cofactors, namely TPP, NAD, FAD, lipoate and coenzyme A (details in Fig. 5.14).

4. **Substrate Level ATP Generation**—Succinyl CoA produced in above step has a high energy sulphur bond, which is trapped at substrate level to generate high energy compound GTP and succinate (4C-compound) is liberated. The reaction is catalyzed by the enzyme **succinate thiokinase** as shown in the Fig. 5.16.
5. **Oxidation of Succinate and Regeneration of Oxaloacetate**—Succinate so produced is then oxidized to fumarate by the removal of H^+ ions, catalyzed by **succinate dehydrogenase** ($FADH_2$ is removed). Fumarate is then converted into malate by the addition of elements of water, in presence of enzyme **fumarase.** Malate so produce is finally dehydrogenated (NADH+ H^+ removed) as catalyzed by **malate dehydrogenase** to release back oxaloacetate to complete the cycle as illustrated in the Fig. 5.16.

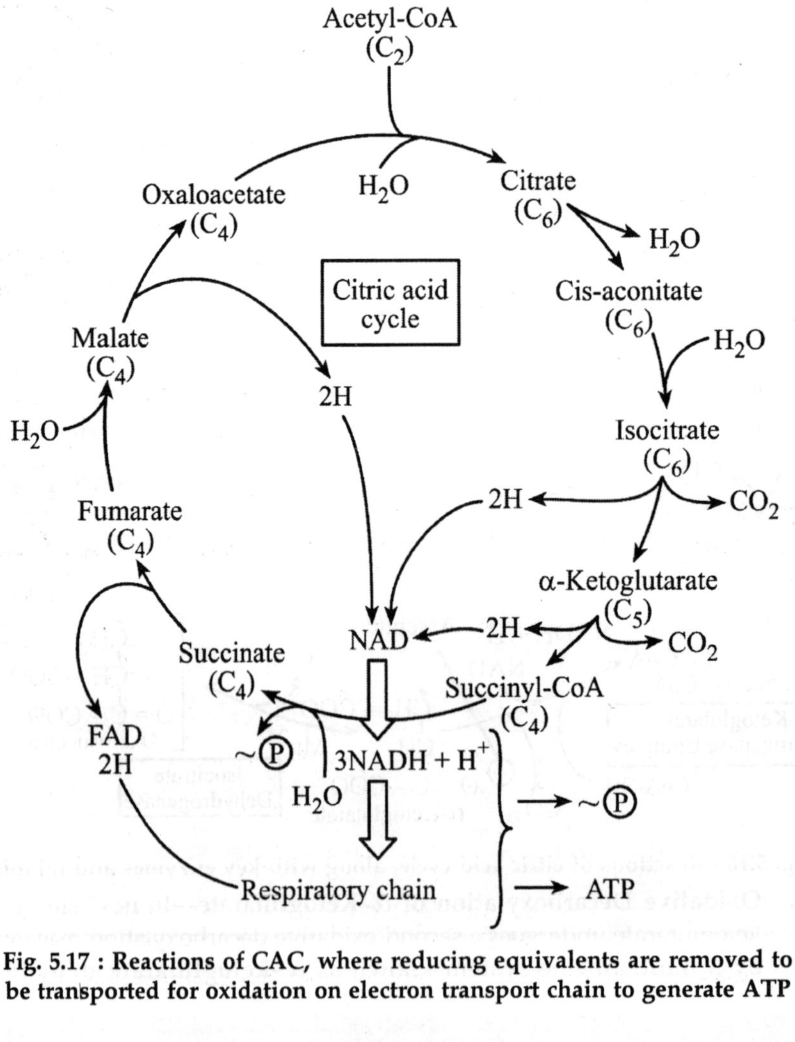

Fig. 5.17 : Reactions of CAC, where reducing equivalents are removed to be transported for oxidation on electron transport chain to generate ATP

Energetics of Citric acid Cycle

	ATP generation at the steps :			
1)	Isocitrate dehydrogenase			
2)	α-Ketoglutarate dehydrogenase	→ 3 NADH + H^+	$\xrightarrow{ETC}$	9 ATP
3)	Malate dehydrogenase			
4)	Succinate dehydrogenase	→ 1 $FADH_2$	$\xrightarrow{ETC}$	2 ATP
5)	Succinate thiokinase (substrate level ATP generation)		→	1 ATP
		Total	=	12 ATP

Thus complete oxidation of acetyl CoA in citric acid cycle to CO_2 and reducing quivalents removed, on oxidation to water on ETC, generates total of **12 ATP molecules.**

Energetics of Complete Oxidation of Glucose—

	ATP generation at steps :		
1.	Aerobic glycolysis of glucose to produce two pyruvate	⟶	8 ATP
2.	2 pyruvate to 2Acetyl CoA → 2 NADH+ H	⟶	6 ATP
3.	2 Acetyl CoA oxidation in 2CAC = 12 x 2	⟶	24 ATP
	So complete oxidation of glucose generates	Total =	38 ATP/mol

Amphibolic Role of Citric Acid Cycle

Citric acid cycle is the major common oxidative pathway, in which all nutrients are oxidized completely in most of the extrahepatic tissues of the body. However in the liver, CAC has both **catabolic** as well as **anabolic functions,** as describe here under:

Anabolic Functions of Citric Acid Cycle

1. Gluconeogenesis—When blood sugar level (BSL) is low, as observed in fasting conditions, glycogenic amino acids are utilized mainly to synthesize glucose denovo to maintain BSL. After transamination the amino acid are converted into one or the other intermediates of CAC, which enter the gluconeogenesis process (Chapter 7.) and are converted to glucose to correct hypoglycaemic state, as shown in the Fig. 5.18.

2. Non essential Amino Acid Synthesis—By transamination reactions catalyzed by specific transaminases, many components of CAC are converted into amino acids, that are required for **protein biosynthesis.**

3. Fatty acid synthesis—Active fatty acid synthesis in cytosolic microsomes requires acetyl CoA as the major precursor. But most of the acetyl CoA is produced inside mitochondria from pyruvate, the end product of glucose oxidation (glycolysis). Only way by which acetyl CoA can come out of mitochondria (mitochondrial membrane is impermeable to acetyl CoA) is through citrate component of CAC. The citric acid comes out of the mitochondria and in the presence of a potent enzyme known as **ATP–citrate lyase** in cytosol, it is cleaved to release acetyl CoA required for the fatty acid synthesis. (Fig. 5.18).

4. Heme Synthesis—Succinyl CoA the main intermediate of CAC is utilized for the synthesis of hemoglobin, myoglobin and cytochromes in the tissues. Succinyl CoA and glycine are the precursors for the synthesis of porphyrins and heme. (details in Chapter 34.)

The anabolic functions of CAC are illustrated in the Fig. 5.18.

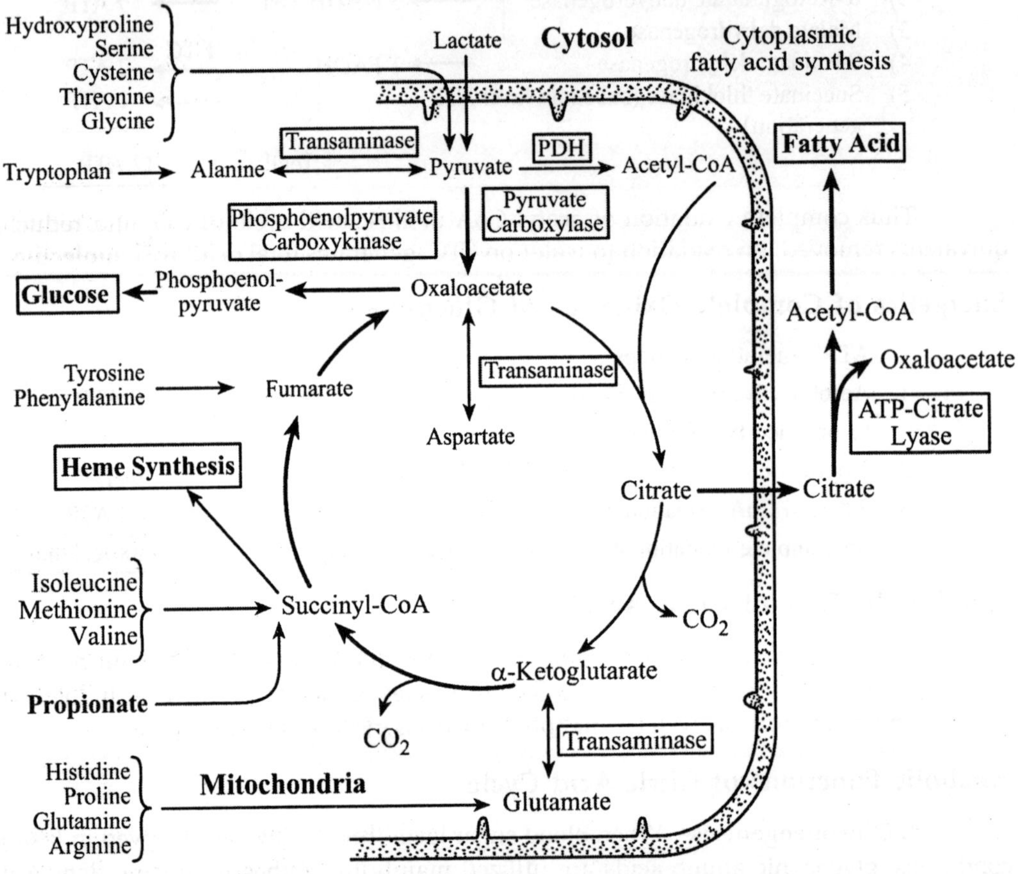

Fig. 5.18 : Anabolic functions of citric acid cycle

□□□

6

Biological Oxidation and Bioenergetics

In order to carry on life processes, a living organism essentially requires a continuous supply of energy. An organism derives this energy from catabolism of nutrients, such as carbohydrates, lipids and proteins in the tissues. And finally by the biological oxidation of reducing equivalents (H^+) removed during the metabolic processes by molecular O_2, the inherent energy of nutrients is made available to carry on life processes.

BIOENERGETICS

For better understanding of biological oxidation, a basic idea of bioenergetics and role of high energy compounds, such as adenosine triphosphate (ATP) is necessary. Bioenergetics or biological thermodynamics deals with the study of all **energy transfer reactions** going on in a living organism.

Usually biological oxidation reactions are accompanied by the liberation of **heat energy,** because during oxidation the reactant moves from high energy level to low energy state **(exergonic reactions).**

Isothermic animals like human beings cannot utilize heat energy directly to carry on vital functions, so the heat energy of oxidation is trapped as **high energy chemical bonds,** for example phosphate or thioester bonds. A convenient method in biological system is to **couple** an exergonic reaction (oxidation), that is accompanied by liberation of heat energy with an endergonic reaction to **trap** the heat energy of oxidation to produce a high energy intermediate (~E). This would then allow ~E to act as **transducer** of utilizable form of energy, termed as **free energy** in various energy requiring processes (Fig. 6.1).

Free energy—In living organisms, ATP (adenosine triphosphate) is the principal high energy compound, that is utilized for performing work in body. Thus ATP is an **utilizable form** of **energy** and the energy of hydrolysis of ATP, used in carrying out life processes is termed as **free energy.**

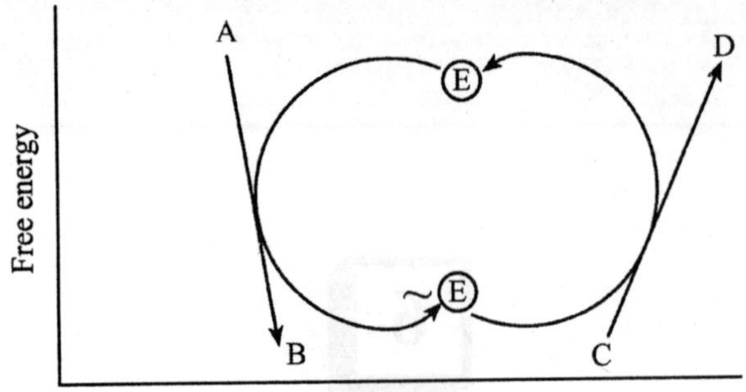

Fig. 6.1 : Formation and utilization of high energy intermediate

Limitations in Capture of Heat Energy

Enthalpy—The total heat energy of an nutrient, that would be evolved when it is completely oxidized to CO_2 and water is known as enthalpy. During biological oxidation of the nutrient in body tissues, this inherent heat energy is evolved slowly at various steps of oxidation which is captured as high energy chemical bonds (ATP).

Free energy—This trapping of heat energy of oxidation as high energy phosphate bonds, that is the utilizable form of energy is termed as free energy and is represented by the symbol 'G'. Trapping of heat energy of oxidation as high energy phosphate bond (free energy) is limited to only 42-48% of the total enthalpy. This is because of the limitations of biological system to capture and convert heat energy into high energy chemical bonds. Thus only some amount of heat energy during biological oxidation is trapped as ATP, while the rest of energy of oxidation is dissipated as heat, and is termed as **entropy** ('S'). In isothermic animals, this entropy (heat energy of oxidation which cannot be trapped as chemical bond) is not a total waste, as it is utilized to maintain the body temperature at 37° C for optimum functioning of enzymes and body organs.

Entropy—It can be defined according to II law of thermodyanamics as the degree of randomness or disorder of a reaction system, that is maximum as the reaction reaches the state of equilibrium. The relationship between changes of enthalpy (ΔH), free energy (ΔG) and entropy (ΔS) can be expressed by the equation:

$$\Delta G = \Delta H - T\Delta S$$ Where T is absolute temperature.

Exergonic reactions are represented by a negative sign of free energy (-'G') and are accompanied by loss of energy and proceed **spontaneously.** While endergonic reactions with gain of energy are represented by +ve 'G', indicating that energy must be supplied to the reactants to complete the reaction.

High Energy Phosphate Compounds

In biological systems, **utilizable form of energy** (free energy) is in form of high energy chemical compounds, which on hydrolysis yield great amount of energy, that can be utilized to carry on life activities. The examples are high energy phosphate compounds that play central role in energy capture and transfer reactions in a living organism. This allows the high energy phosphates to act as **transducer** of energy (energy rich intermediates), for instance **ATP** that yields at least 7 kcal/mol energy on hydrolysis, which is utilized to perform work.

Low energy phosphate bonds—Certain organic phosphate compounds yield less energy on hydrolysis (<7.0 kcal/mol) and are called low energy phosphate compounds. For example glucose 6–PO_4, fructose - 1, 6 bisphosphate etc, where the low phosphate bond is utilized to activate the metabolite, so that it can enter a metabolic cycle.

High energy bonds are usually acid anhydride bonds that on hydrolysis liberate free energy (>7.3 kcal/mol) for carrying out life processes. Scientist Lipmann was first to coin the term high energy phosphate bonds and suggested the use of symbol ~ to represent a high energy bond.

Adenosine triphosphate (ATP)

ATP plays a central role in all energy transfer reactions of the body and is appropriately termed as **energy currency** of living cell. ATP is generated (produced) mainly by two pathways:

1. By **coupling** biological oxidation on **mitochondrial electron transport chain (ETC) with phosphorylation** that capture the energy of oxidation as high energy ATP. This is **major source** of ATP, the utilizable form of energy.
2. By **substrate level phosphorylation**—At certain steps of metabolic cycles, the liberation of energy is very high to enable **trapping of heat energy** directly (without involving ETC on mitochondrial membrane), by incorporation of inorganic phosphate to generate a high energy phosphate bond. For example phosphoenol pyruvate and 1,3 bisphosphoglycerate with a high energy phosphate bond are produced during glycolysis process (Chapter 5.)

ATP utilization—ATP serves as immediate source of utilizable form of energy (free energy) to carry out vital functions of the body, to allow it to act as energy currency of cell, which is illustrated in the Fig. 6.2.

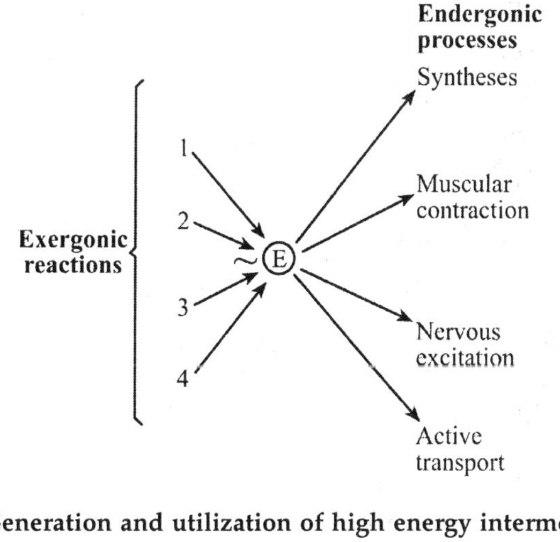

Fig. 6.2 : Generation and utilization of high energy intermediate ~(E)

ATP –Chemical structure—ATP is an adenosine nucleotide made up of adenine, ribose sugar and three phosphate bonds esterified with OH group of ribose unit as shown in the Fig. 6.3. ATP consists of two high energy phosphate bonds and one low energy phosphate bond in its triphosphate unit as adenosine triphosphate (Fig. 6.3).

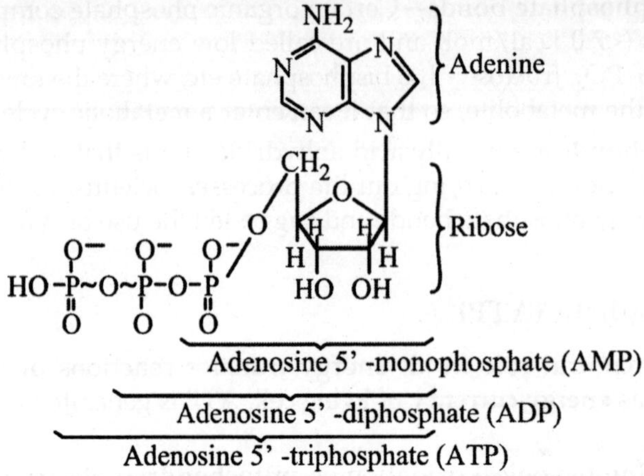

Fig. 6.3 : Chemical structure of ATP

Energy of hydrolysis of ATP—Hydrolysis of the terminal phosphate bond of ATP yield 7.3 Kcal/mol that is the primary source of utilizable form of energy in all energy requiring processes of the body. The reaction involved in hydrolysis of ATP is given below—

$$ATP + H_2O \longrightarrow ADP + Pi + 7.3 \text{ kcal (energy)}$$

Other high energy phosphates along with their energy of hydrolysis are summarized in Table 6.1.

Table 6.1 : High energy phosphate compounds along with their energy of hydrolysis

S.No.	High energy compounds	Energy of hydrolysis kcal/mol
1.	Phosphoenol pyruvate	14.8 kcal/mol
2.	1,3 Bisphosphoglycerate	11.8 kcal/mol
3.	Creatine phosphate (storage form of high phosphate bonds.)	10.3 kcal/mol
4.	Adenosine triphosphate	7.3 kcal/mol
5.	Pyrophosphate (PPi)	6.6 kcal/mol

BIOLOGICAL OXIDATION

Oxidation is defined as loss of electron and reduction as gain of electron. For example ferrous iron (Fe^{2+}) is oxidized to ferric (Fe^{3+}) state by loss of an electron, while ferric iron is reduced to ferrous iron by a gain of electron as illustrated below—

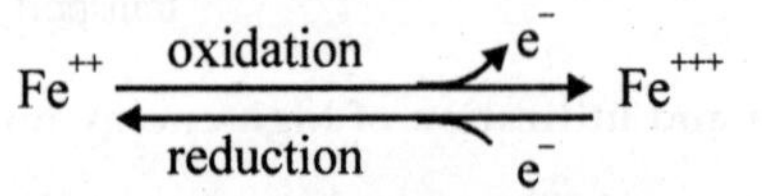

In biological system oxidation can also take place in the body by removal of H^+ from a compound, or by addition of oxygen, in the presence of specific dehydrogenases and oxidase enzymes respectively.

Salient Features of Biological Oxidation

1. Biological oxidation in a living organism is **slow, gradual** and occurs in **many steps,** so as to prevent **explosive** evolution of heat energy, which otherwise can kill the living cell.

2. In biological system, oxidation is **mostly initiated** by **dehydrogenation** i.e. removal of reducing equivalents by coenzymes NAD, FAD etc. catalyzed by specific enzymes called **dehydrogenases,** as shown below—

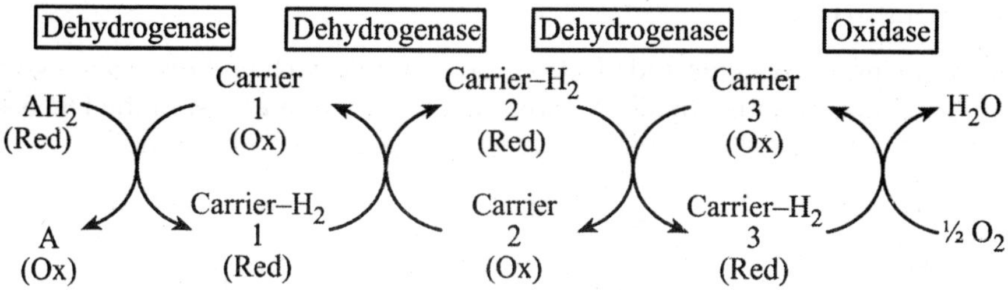

3. Oxidation of reducing equivalents on ETC—In biological system, the free energy of oxidation of substrate is liberated by oxidation, of **C-H bond** and finally by the oxidation of reducing equivalents ($2H^+$) on electron transport chain (ETC) by molecular O_2 to H_2O and not by oxidation of C-C bond by O_2 to produce CO_2.

4. Transfer of reducing equivalents to ETC—The reducing equivalents removed as $NADH+H^+$ and $FADH_2$ from the substrate (nutrient) are then transported to electron transport chain (ETC), on inner mitochondrial membrane for biological oxidation finally by molecular O_2 to H_2O with the liberation of energy at various steps which is captured as high energy phosphate bonds.

Electron Transport Chain : Respiratory Chain

The various components of respiratory chain on inner mitochondrial membrane are arranged according to their increasing order of **redox potential,** from -ve redox potential to +ve redox potential and electrons and reducing equivalents are sequentially transported through them finally to molecular O_2 to be oxidized to water.

Redox potential—It is defined as the tendency or ease at which a substrate can donate or accept electrons. Redox potential is expressed as Eo volts and is directly related to the change in free energy (ΔG) during oxidation/reduction reaction. Redox poential of important components of the respiratory chain are summarize in the Table 6.2.

Table 2 : Redox potential of components of ETC

Redox pair	Eo volts
$2H^+/2H$	- 0.42
$NAD^+/NADH+H^+$	- 0.32
$FMN/FMNH_2$	- 0.30
$FAD/FADH_2$	- 0.22
Coenzyme Q/QH_2	+ 0.10

contd. ...

... contd.

Redox pair	Eo volts
Cytochrome b	+ 0.07
Cytochrome c, c_1	+ 0.23
Cytochrome a, a_3	+ 0.29
½ O_2/H_2O	+ 0.82

Respiratory chain (ETC)

The reducing equivalents and electrons on respiratory chain flow from – ve redox potential to +ve redox potential, finally to molecular O_2 as summarized in the Fig. 6.4. The ETC has a span of 1.14 V and free energy change (ΔG) equal to 53 kcal/mol.

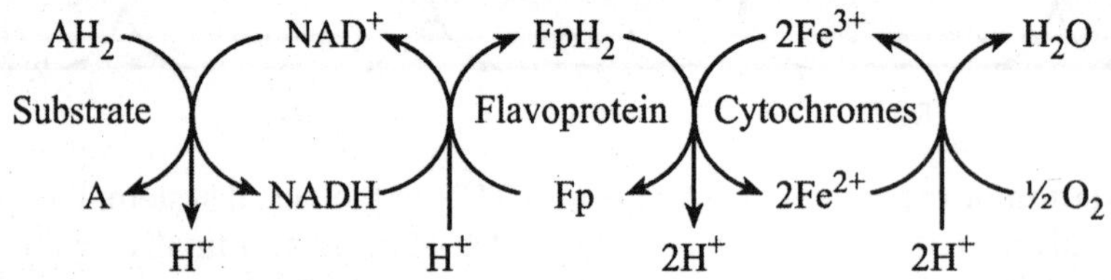

Fig. 6.4 : Major components of respiratory chain established on inner mitochondrial membrane.

Mitochondria - the power house of cell

Mitochondria are appropriately designated as power house of the cell, as they are the centers of metabolic oxidative pathways established in the matrix to generate energy e.g. citric acid cycle, β-oxidation of fatty acids etc.The reducing equivalents removed during metabolic pathways as reduced coenzymes NADH+H^+ and $FADH_2$, are rapidly transferred to ETC on inner mitochondrial membrane for biological oxidation to liberate the energy. The heat energy liberated is then trapped as high energy phosphate bonds (ATP) by coupling oxidation with phosphorylation. Thus mitochondria are major site of ATP generation in the cell.

Components of Electron Transport Chain

There are **five** distinct components or carriers on ETC, that transfer reducing equivalents and electrons from a substrate (nutrient), in various steps through redox carriers (oxido/red. reactions), finally to be oxidized by O_2 to water. Components of ETC are describe here under—

1. Nicotinamide nucleotide (NAD) dependent anaerobic dehydrogenases—NAD is actively involved in most of the dehydrogenation reactions to initiate oxidation of a substrate (nutrient) by specific dehydrogenase enzymes and removes the H^+ ions as NADH + H^+. These enzymes are unable to transfer the H^+ removed from substrate to molecular oxygen directly and are thus termed as anaerobic dehydrogenases.

2. Flavoproteins—(FAD) dependent aerobic dehydrogenases—The flavoprotein nucleotide (FAD) dependent dehydrogenases catalyze removal of H^+ from a substrate, that can be transferred directly to oxygen or to the other substances to produce H_2O_2. These enzymes are thus termed as **aerobic dehydrogenases.** e.g L-amino acid oxidases, xanthine oxidases etc. On ETC the reducing equivalents from NADH+H$^+$ are transferred to a flavoprotein enzyme with FMN as coenzyme as catalyzed by NADH dehydrogenase system. The FMN on ETC requires specific **iron-sulphur protein (FeS)** as cofactor.

3. Iron-sulphur-protein (FeS)—The FeS protein (Non-heme iron, NHI) consists of a ferrous/ferric, oxidation/reduction system to facilitate quick transfer of electrons from FMN to coenzyme Q and from cytochrome b to cytochrome c, the next components of respiratory chain.

4. Coenzymes Q—Next component on ETC is coenzyme Q that is also known as **ubiquinone** because of its universal occurence in all living organisms. It has a quinone/ quinol oxidation/reduction system and has a chemical structure similar to vitamin K with a variable isoprenoid side chain (Fig.6.5). Coenzyme Q is soluble in lipid bilayer of mitochondial membrane and thus can travel across membrane freely to transport the reducing equivalents and electrons to the next component of electron transport chain.

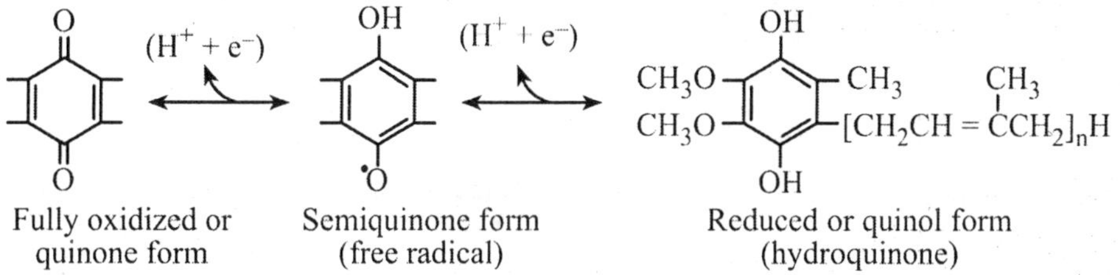

Fig. 6.5 : Chemical structure of coenzyme Q along with quinol/quinone oxidation/reduction system

5. Cytochromes—The cytochromes are essential components of ETC, that link NAD$^+$, FMN coenzyme oxidation/reduction step through coenzyme Q to molecular O_2, the final acceptor of H$^+$ ions. Cytochromes contain a heme like porphyrin cofactor attached to the protein component with **ferrous/ferric oxidation/reduction system,** unlike hemoglobin which always has iron in ferrous state. ETC consists of at least **3 cytochrome** systems, with some additional components as describe below:

(a) **Cytochrome b**—It connects coenzyme Q to the cytochrome system and transfers electron from coenzyme Q to cytochrome c and has **FeS** as essential component to facilitate quick transfer of the electrons.

(b) **Cytochrome c and c_1**—These are central members of ETC with heme porphyrin as cofactor and a Fe^{2+}/Fe^{3+} oxidation/reduction system. Cytochrome c is loosely bound to inner mitochondrial membrane.

(c) **Cytochrome aa_3 system**—Cytochrome aa_3 are terminal component of ETC and are also known as **cytochrome oxidase** as they finally transport electrons and the reducing equivalents to molecular O_2 to be oxidized to water. Cytochrome aa_3, in addition to two heme groups also contain two Cu^{++} ions as essential components, that are necessary to convert molecular O_2 to nascent O^- by transferring electron as shown in the reaction—

A total of 4 electrons are needed for complete reduction of one molecule of O_2 to H_2O.

Organization of Respiratory Chain

The complete respiratory chain (ETC) is established on the inner mitochondrial membrane starting from NADH through a chain of electron carriers, organized sequentially in order of increasing redox potential as shown in the Fig 6.6. The respiratory chain can be disrupted into **4 distinct enzyme complexes** I, II, III and IV, that transport electrons and protons finally to molecular O_2 to be oxidized to water (Fig. 6.6).

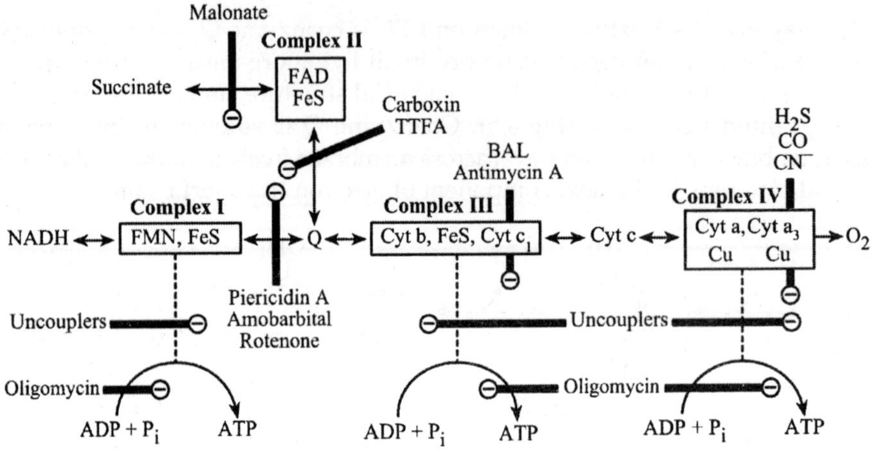

Complex I – NADH coenzyme Q reductase

Complex II – succinate coenzyme Q reductase

Complex III – cytochrome reductase

Complex IV – cytochrome oxidase

Fig. 6.6 : Components of ETC with complexes I-IV along with the cofactors and inhibitors of respiratory chain, at various sites

Sites of ATP Generation—Oxidative Phosphorylation

During transport of electrons and reducing equivalents through various components of respiratory chain, **energy of biological oxidation is released** as heat, slowly and gradually at various steps on ETC. This heat energy of oxidation, released at various steps of ETC is then trapped as high energy phosphate bond of ATP by **coupling (uniting) oxidation with phosphorylation.** The energy of oxidation being trapped as ATP is limited to certain steps only on ETC. These steps which can be coupled with phosphorylation to generate ATP should fulfill following conditions—

1. The steps on ETC, where the free energy change is more than 37 K.joule and redox potential difference of approximately 0.2 volts can be coupled with phosphorylation. This is because, at these steps heat energy liberated is sufficient enough to enable its capture as high energy phosphate bond.
2. The three steps on ETC which fulfill the above mentioned conditions corresponds to complex I, III and IV, and these are the sites of ATP generation on ETC. The three steps (complex I, III an IV), represents **coupling sites** of biological oxidation with phosphorylation for ATP production as shown in Fig. 6.6.

P:O ratio—Experimentally it has been observed that when a substrate is oxidized **by NAD⁺ dependent** dehydrogenase system and reducing equivalents (electrons) are transferred through complete ETC, 3 molecules of phosphate is incorporated per atom of O_2

consumed (1/2 O_2) to generate 3 ATP molecules and thus **P:O ratio is 3:1.** But when a substrate is oxidized by FAD dependent dehydrogenase enzymes, P:O ratio is only 2:1 i.e. only 2 ATP are generated, because in this case **complex** I step is bypassed on the ETC (Fig. 6.6) and complex IV step does not fulfill the conditions necessary to couple oxidation with phosphorylation.

Mechanism of Oxidative Phosphorylation

Several hypotheses were proposed to explain the mechanism of oxidative phosphorylation, such as **chemical coupling** with hypothetical energy rich intermediates (X~I), **conformational** coupling and **chemiosmotic theory.** As the first two hypothesis i.e. chemical and conformational coupling lack experimental evidence, they were rejected.

Chemiosmotic Hypothesis

Chemiosmotic theory proposed by Mitchell (1961) is now widely accepted, as it has enough experimental evidence and proof to explain exact mechanism of oxidative phosphorylation and ATP generation.

According to chemiosmotic theory, at three sites on ETC corresponding to complex I, III and IV, the energy of oxidation liberated is sufficient to **eject two protons outside** the inner mitochondrial membrone (IMM). As the IMM is impermeable to protons, a **proton gradient** develops across the membrane by this ejection of 2 protons outside IMM.

This **Electrochemical potential gradient** so developed drives a **vectorial** (unidirectional i.e. from outside to inside) membrane bound **ATP synthase system** to synthesize ATP and simultaneously **protons are pumped back** inside IMM through F_1-F_o unit to ETC for further oxidation.

F_1-F_O Units on Inner Mitochondial Membrane (IMM)

The IMM is a specialized structure rich in proteins. This membrane is extensively folded to form **cristae,** which greatly increases the surface area of IMM, (Fig. 6.7A).

At the inside of IMM, specialized repeatative **knob** like structures are present projecting towards matrix, termed as F_1 units, which are the **phosphorylating subunits** consisting of active **ATP synthase** system. These knob like F_1 subunits are attached through a stalk (F_O-subunit), which spans the whole IMM, and possesses a proton channel through which protons are pumped back from outside IMM, (Fig. 6.7B) to inside F_1 subunit. This is the only route through which the protons are taken back to transfer them to next component on ETC.

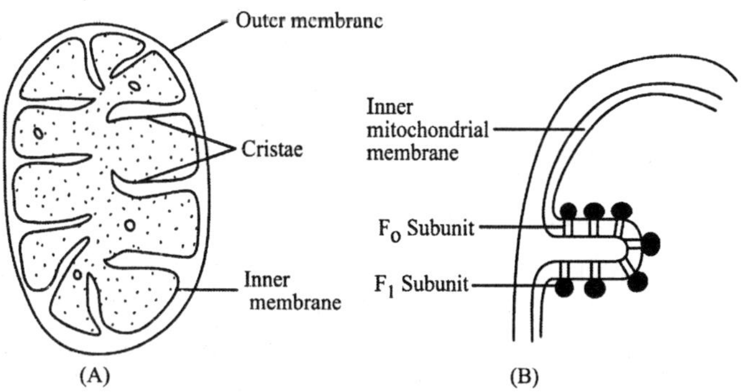

Fig. 6.7 : (A) Structure of mitochondria (B) Enlarged cristae with F_1-F_O subunit

In the F_1 subunit the proton (H^+) activates an ATP synthase enzyme to phosphorylate ADP to generate a high energy phosphate bond to synthesize **ATP** (Fig. 6.8.). At the same time i.e. simultaneously to ATP production, in F_1-F_0 subunit, protons are taken back to next step on ETC to continue the transfer of reducing equivalents to next complex.

Thus biological oxidation on respiratory is **tightly coupled** with phosphorylation and B/O cannot proceed without simultaneous phosphorylation to generate ATP in normal conditions.

Respiratory Chain (ETC)—a Proton Pump

Chemiosmotic theory postulates that the **energy** of biologic oxidation at complex I, III and IV on ETC translocates (eject out) protons H^+ to the outside of inner mitochondrial membrane and ETC acts as **proton pump** creating an electro chemical potential gradient across IMM. This potential difference then drives a vectorial ATP synthase system (F_1- F_O units) to generate ATP.

The flow of protons (proton pump) from Fo channel to F_1 initiates a sequence of conformational changes in F_1 proteins resulting in stimulation of enzyme **ATPsynthase** and ATP is synthesized by the reaction given below—

$$ADP + P_i + H^+ \xrightarrow{\text{ATP synthase}} ATP + H_2O$$

The conformational change of ATPsynthase translocates the protons from F_1 subunit to be released in the mitochondrial matrix and are taken by the next complex on respiratory chain, for the further oxidation.

Three Oxidation/reduction Loops on Respiratory Chain

The chemiosmotic theory further explains that the respiratory chain (ETC) on IMM is folded into three loops corresponding to complex I, III and IV, in such a manner that $2H^+$ are ejected outside the IMM at site I, III and IV complex of ETC. These three loops form the coupling sites for phosphorylation (Fig. 6.8). Each oxidation/reduction loop consist of a H^+ carrier and an electron carrier as illustrated in the Fig. 6.8.

ATP Synthase and F_1 - F_0 Functional Units

The protons that are translocated outside IMM at three sites corresponding to the three loops of ETC with complex. I, III and IV, can reenter back on ETC only through the space of stalk of F_0 subunit to the knob like F_1 functional subunit that possesses an active ATP synthase system. Thus simultaneously to proton reentry through F_1- F_0 system, one molecule of ATP is synthesized as described above (Fig. 6.8) as well as $2H^+$ released by F_1 subunit are taken back to continue biological oxidation on the next complex of ETC.

Evidence Favouring Chemiosmotic Theory

Many experimental evidence favour the chemiosmotic theory to explain mechanism of oxidative phosphorylation. These are listed below:

1. Inner mitochondrial membrane (IMM) is impermeable to many ions including protons.

2. Intact mitochondria (and membranes) are necessary for ATP generation, as in broken membrane proton gradient cannot be generated and ATP will not be produced.

3. By increasing H^+ concentration $[H^+]$ outside IMM by adding acid increases ATP generation, as observed in experimental studies.

4. Uncouplers like 2, 4 DNP increase permeability of IMM to protons (H^+), there by block development of proton gradient across IMM and inhibit ATP production.

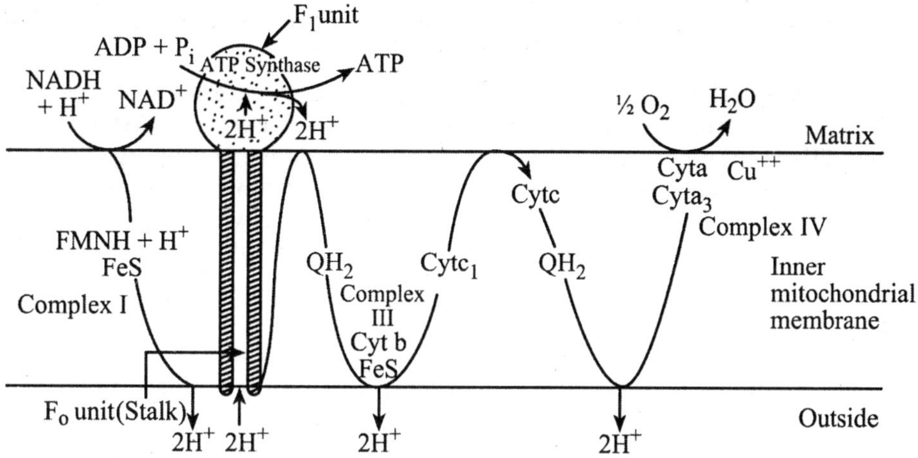

Fig. 6.8 : Diagrammatic representation of ETC with 3 oxi/red loops and F_1- F_O system (at one loop) with ATP synthase coupling as postulated by chemiosmotic theory of Mitchell

Inhibitors of Respiratory Chain (ETC)

A variety of site specific chemical compounds **inhibit** ETC at complex I, III and IV of biological oxidation and ATP generation is impaired. Some example of these inhibitors at their specific site on ETC are shown in the Fig. 6.6.

The respiratory chain inhibitors bind to specific enzyme/coenzymes of complex I, III and IV and block oxidation/reduction steps and transport of electrons and red equivalents is inhibited. Examples are:

1. **At complex I** – Fish poison rotenone, barbiturates and antibiotic piericidine inhibit the enzyme complex NAD^+/FMN reductase.
2. **At complex III** – Antibiotic antimycin A and BAL (Dimercaperol) inhibit enzyme cytochrome reductase.
3. **At complex IV** – Poisons like cyanides, carbon monoxide, H_2S gas have great affinity to get attached to iron of porphyrins of cytochrome a_3 and hemoglobin and prevent transport of O_2 and final oxidation of $2H^+$ by O_2 and thus are **instant** poisons causing death.

Inhibitors of oxidative phosphorylation—Certain non site specific inhibitors inhibit oxidative phosphorylation by different mechanisms. Some important inhibitors of oxidative phosphorylation are listed below—

(a) The antibiotic **oligomycin** and **atractyloside** completely block oxidative phosphorylation and ATP generation by blocking transport of adenine nucleotides across the IMM.

(b) **Uncouplers**—Certain compounds uncouple and separate biological oxidation on ETC from phosphorylation, for instance 2,4 Dinitrophenol (DNP), dinitro cresol, pentachlorophenol and trifluoro carbonylcyano phenylhydrazone (FCCP). The FCCP is 100 times more potent as uncoupler than 2, 4 DNP.

Mechanism of action of uncouplers—The presence of uncouplers make mitochondial membrane permeable to H^+ ions, therefore the electrochemical potential difference will not developed across IMM, but electron transport and biological oxidation will proceed on ETC **without ATP generation** and energy of oxidation would be dissipated as heat.

Physiological uncouplers—Thyroxine hormone in excess uncouples B/O from phosphorylation and thus increases **basal metabolic rate (BMR)**, as seen in **hyperthyroidism.** In certain conditions, uncoupling of oxidative phosphorylation is utilized for biological advantage. For example hibernating animals, newborn infants and animals living in extreme cold climate possess brown adipose tissue, where biological oxidation can proceed without phosphorylation to help them to maintain their body temperature constant.

Brown Adipose Tissue—Physiological Significance

Brown adipose tissue is very significant in hibernating animals, newborn infants and in animals living in excessive cold regions of world (e.g. polar bears). The maintenance of their body temperature is achieved in these conditions by a **specialized brown adipose tissue** present in upper back and neck of these animals. The mitochondria of brown adipose tissue specialize in carrying out B/O without ATP generation (uncoupled from phosphorylation). This results in liberation of total energy of B/O as heat, which help them to maintain their body temperature for adaptation to survive in physiological condition of hibernation and in adverse environmental conditions of extreme cold.

Though brown adipose tissue is not prominent in humans, it is present in some lucky individuals, who may eat as much as they like and not become obese (diet induced thermogenesis).

Mechanism of Thermogenesis

The brown adipose tissue has rich blood supply and high contents of mitochondria and cytochromes but very low activity of **ATP synthase.** A specific protein **thermogenin** is present in this tissue that allows protons to permeate through IMM and thus uncouples B/O from phosphorylation so **more heat** is liberated as illustrated in the Fig. 6.9.

Respiratory Control

Availability of ADP is the most prominent factor, which control biological oxidation and respiration. This is because B/O and phosphorylation is tightly coupled and secondly because in body tissues ATP : ADP ratio is constant. **At resting stage** most cells are in state four (4) of **Chance** and **William's** conditions controlling respiration i.e. availability of ADP only. During resting state ATP stores are saturated (less free ADP) and thus availability of ADP becomes **rate limiting** and **B/O and respiration slows down at rest. During exercise** and performing work, ATP is utilized and more ADP is available for oxidative phosphorylation, therefore O_2 consumption increases and **B/O and respiration rate elevated** (hyperventilation).

Effect of uncouplers—As uncouplers delink B/O from phosphorylation, ADP availability no longer is controlling factor, so B/O and respiration rate becomes uncontrolled in presence of uncouplers.

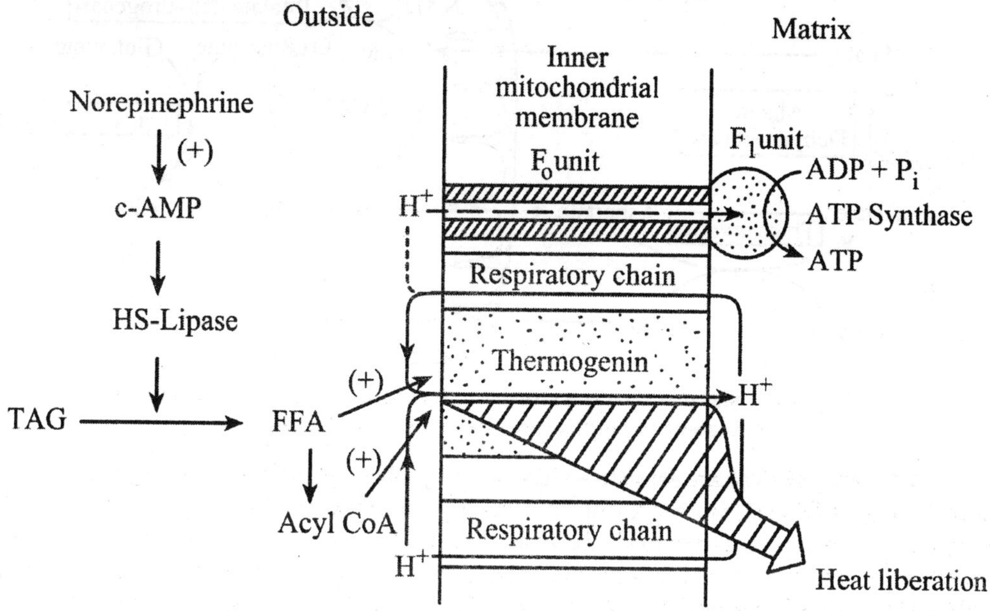

Fig. 6.9 : Mechanism of uncoupling of B/O from phosphorylation by thermogenin in brown adipose tissue

Transport of Reducing Equivalents—Shuttle Pathways

Major oxidative pathways like citric acid cycle and β-oxidation of fatty acids are established inside mitochondrial matrix to facilitate quick transport of reducing equivalents ($NADH+H^+$ and $FADH_2$) to ETC on IMM for B/O and ATP generation. However **Glycolysis** and other oxidative pathways proceed in **cytoplasm** and reducing equivalent removed in these pathways are impermeable through mitochondrial membranes. So special mechanisms are needed for their transport to IMM. Two shuttles are operative for the transport of NADH and $FADH_2$, produced in cytosol to inner mitochondrial membrane for B/O. These are **malate - aspartate shuttle** and **glycerol phosphate** shuttle.

Malate-aspartate shuttle—The malate shuttle is of universal occurance and is well established in most of the body tissues to transport reducing equivalent from cytosol to mitochondria for B/O on IMM. The reducing equivalent produced in cytosol ($NADH+H^+$) are taken up by oxaloacetate and is reduced to malate, that can cross mitochondrial membrane to transport them to ETC on IMM for B/O and ATP generation.

Three ATP molecules are produced by $NADH+H^+$ transported from cytosol through this shuttle. However oxaloacetate is impermeable through mitochondrial membrane so to continue the shuttle (transport of cytosolic $NADH+H^+$), it is first transaminated to aspartate, which is permeable and come out of mitochondria and reproduce oxaloacetate outside by transamination reaction (Fig. 6.10).

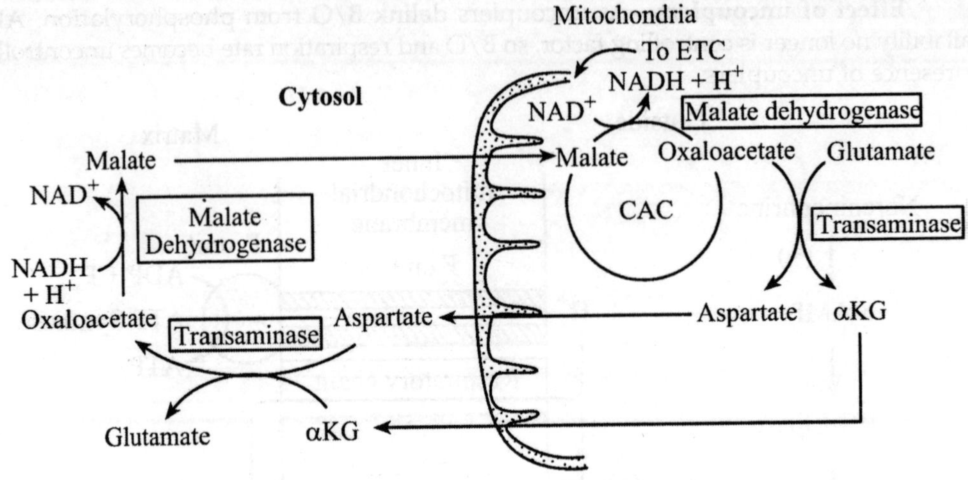

Fig. 6.10 : Malate-aspartate shuttle to transport cytosolic reducing equivalent to respiratory chain inside mitochondria

Glycerol phosphate shuttle—Glycerol phosphate shuttle also transport reducing equivalents (NADH+H) from cytosol, by a reaction mechanism as shown in the Fig. 6.11, but to a lesser extent, as compared to malate shuttle. This is because glycerol phosphate is re-oxidized to produce dihydroxy acetone phosphate and liberates $FADH_2$ (plomit) instead of $NADH+H^+$ **inside mitochondria** and so on ETC on IMM, only two ATP molecules are generated (Fig. 6.11).

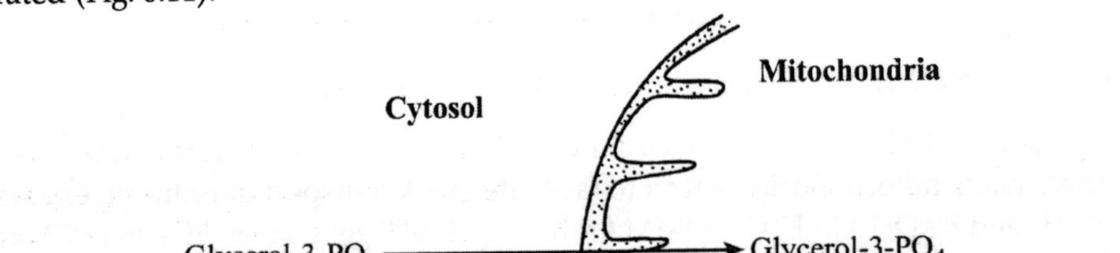

Fig. 6.11 : Mechanism of glycerol phosphate shuttle to transport reducing equivalents

The glycerol phosphate shuttle is active manly in tissues such as brain. This shuttle is not well established in most of the tissues of body, as the mitochondrial enzyme glycerol phosphate dehydrogenase is deficient in these tissues.

❑❑❑

7

Gluconeogenesis and Minor Metabolic Pathways of Glucose

Gluconeogenesis is the process, by which glucose is synthesized denovo in the body from non carbohydrate compounds. The major precursors for gluconeogenesis are glycogenic **amino acid, lactic acid, glycerol** and **propionate**.

Site of Gluconeogenesis —Liver and kidneys are the principal organs, where glucose and glycogen are synthesized denovo. The key enzymes of gluconeogenesis are present in the mitochondria, as well as in cytosol of these tissues.

Importance of Gluconeogenesis —Significance

1. Denovo synthesis of glucose occurs in the body primarily for the **maintenance of blood sugar level** in fasting conditions and in between meals, when hypoglycaemia develops. Hypoglycaemia may also occur due to less or non availability of carbohydrates in the diet (malnutrition).

2. A regular supply of glucose is necessary, especially for the tissues, such as CNS, brain and RBCs etc. as they are dependant solely on glucose as fuel.

3. Severe hypoglycaemia (low blood sugar level) if not corrected immediately, can result in fatal consequences as it causes dysfunctioning of brain and the subject may go into hypoglycaemic shock, that may result in death.

4. Gluconeogenesis scavenges (remove) and clears metabolic byproducts, such as lactic acid, glycerol and propionate from the tissues and blood.

5. In prolonged fasting conditions, gluconeogenesis must takes place to meet the **basal requirement for glucose** by body tissues, such as CNS, RBC etc., as well as to maintain citric acid cycle intermediates to prevent **ketosis** (Chapter 11).

Reactions of Gluconeogenesis

Gluconeogenesis is basically, but **only partial reversal** of glycolysis with certain modifications. Three reactions of glycolysis are **irreversible** due to the great energy barrier, as shown below:

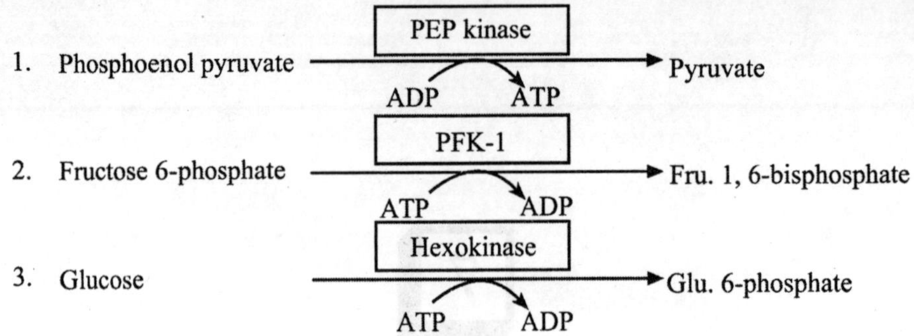

These irreversible reactions of glycolysis are circumvented by **alternative enzyme systems,** to achieve denovo synthesis of glucose in the body from non-carbohydrate substances.

Gluconeogenesis involves following discrete steps, that are catalyzed by key enzymes of the process as described here under:

1. Conversion of Pyruvate to Phosphoenol Pyruvate —This irreversible reaction of glycolysis is circumvented by the key enzymes of gluconeogenesis in two steps. In these key steps, the pyruvate is converted into high energy compound phosphoenol pyruvate as described here under:

a. **Pyruvate Carboxylase** is a biotin dependent **mitochondrial enzyme** which catalyzes the conversion of **pyruvate** (3C–compound) to **oxaloacetate** (4C–compound) by a CO_2 fixation reaction utilizing ATP as energy. This conversion is **prerequisite** of gluconeogenesis and takes place inside mitochondria. This reaction is also significant in providing **oxaloacetate** as the starter to establish citric acid cycle, in which finally all nutrients are oxidized completely.

b. **Phosphoenol Pyruvate Carboxykinase (PEP Kinase)** is a key enzyme of gluconeogenesis that catalyzes the conversion of oxaloacetate to form phosphoenol pyruvate. But this enzyme is present in the cytosol, while oxaloacetate formed inside mitochondria is impermeable through mitochondrial membrane. So oxaloacetate is first reduced by **malate dehydrogenase** to malate, which is permeable and comes out of the mitochondria into cytosol, where it is converted back into **oxalocetate** by oxidation (Fig. 7.1).

Oxaloacetate so formed in the cytosol is then oxidatively decarboxylated to produce **phosphoenol pyruvate** (PEP). This reaction is catalyzed by a specific enzyme **PEP kinase,** that utilizes GTP as energy and a molecule of CO_2 is released during the reaction as shown in the Fig. 7.1.

Phosphoenol pyruvate so produced then enters the reversible reactions of glycolysis and is converted in to fructose 1, 6-bisphosphate by cytosolic reversal of glycolysis (Fig. 7.1).

2. Conversion of Fructose 1, 6 bisphosphate to fru. 6-PO_4 —The second irreversible reaction of glycolysis is circumvented by the **enzyme fructose 1–6 bisphosphatase,** a key enzyme of gluconeogenesis, that converts fructose 1, 6 bisphosphate to fructose 6-PO_4 by hydrolyzing a phosphate group from C-1 position. This enzyme is active in liver and kidneys mostly. Fructose 6-PO_4 so produced is then isomerized to glucose 6-PO_4 catalyzed by a hexose isomerase enzyme as shown in the Fig. 7.1.

3. Conversion of glucose 6-PO$_4$ to free glucose —The third irreversible reaction of glycolysis is achieved by a specific enzyme called **glucose 6-phosphatase,** that hydrolyzes the phosphate group from glu. 6-PO$_4$. This enzyme is present only in liver, but it is absent in the muscles. Glucose so formed in liver can be released in blood to maintain BSL or can be converted to glycogen for the storage purposes (Fig. 7.1).

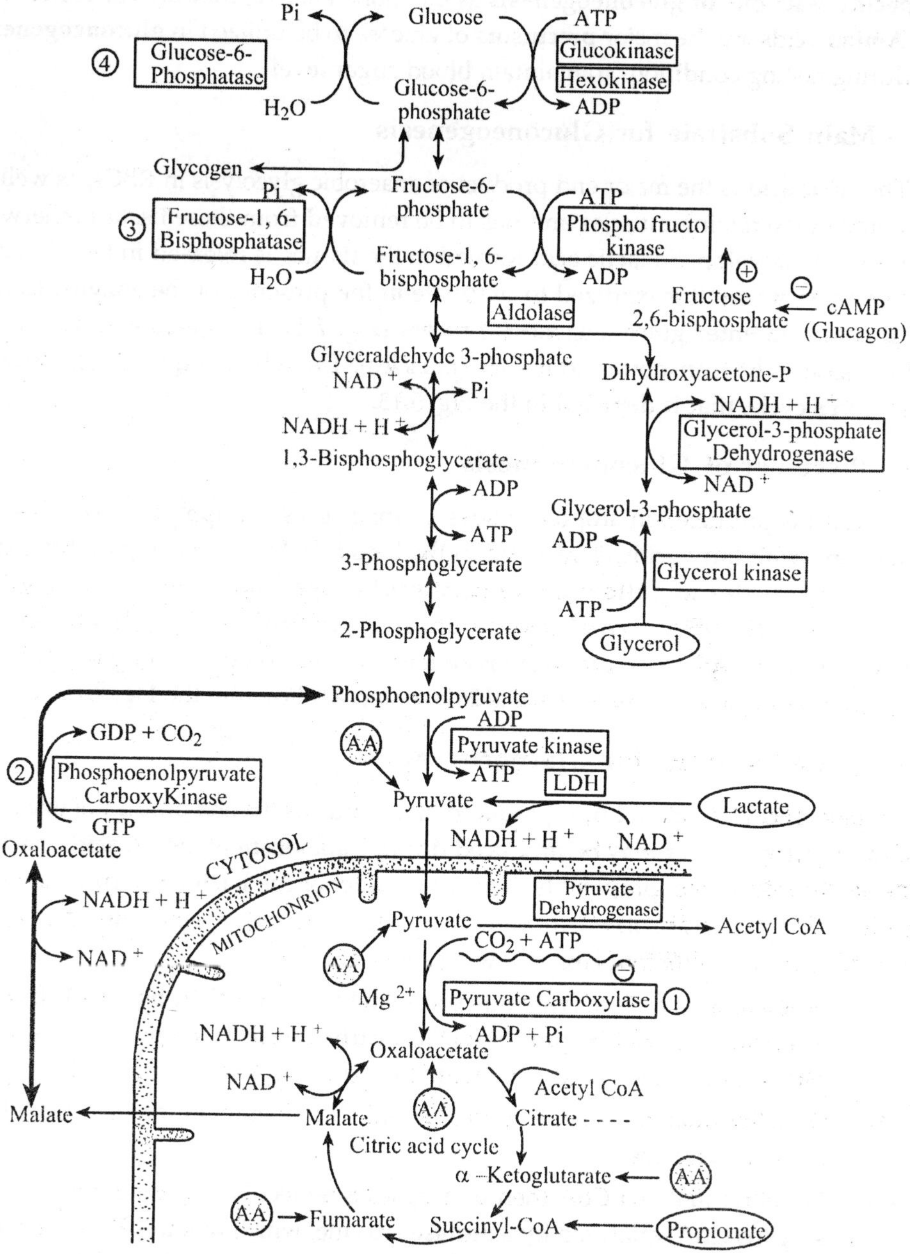

Fig. 7.1 : Key reactions of gluconeogenesis 1–4, along with depiction of the entry of lactate, amino acids (marked as AA) glycerol and propionate in the process

Gluconeogenesis from Amino Acids

Most of the amino acids (glycogenic) after transamination and deamination reactions are converted to α–keto acids, that are intermediates of citric acid cycle. So they enter at various points (Fig. 7.1) of CAC and are ultimately converted to malate or oxaloacetate. Then by the **specific reactions of gluconeogenesis** as mentioned above, they are converted into glucose. **Amino acids** are the major **precursors** of glucose to be utilized in **gluconeogenesis** process during fasting conditions to maintain blood sugar level.

Lactate - Main Substrate for Gluconeogenesis

The lactic acid is the major end product of anaerobic glycolysis in RBCs, as well as in the vigorously exercizing muscles and has to be removed from these tissues otherwise lactoacidosis may develop. So lactic acid formed in the tissues is released in blood and is taken up by liver, where it is oxidized to pyruvate in the presence of the enzyme **lactate dehydrogenase** and enter gluconeogenesis process (Fig. 7.1). The glucose so formed is released in blood to be transported to tissues (muscles). The whole sequence of events is known as Cori's cycle, that is depicted in the Fig. 5.13.

Glycerol–Precursor of Gluconeogenesis

Glycerol is produced in adipose tissues by the process of lipolysis (hydrolysis of TAG) and along with fatty acids, is released in the blood. In liver glycerol is activated to glycerol 3-PO_4 by the hepatic enzyme **glycerokinase** and is then oxidized to dihydroxyacetone phosphate, catalyzed by enzyme **glycerol 3-PO_4 dehydrogenase. Dihydroxyacetone phosphate** so produced is an intermediate compound of glycolysis and enters the **gluconeogenesis** process at the **aldolase** step as shown in the Fig. 7.1.

Propionate is a Substrate for Gluconeogenesis

Propionic acid (3C–atom FA) is produced in the body as the end product of oxidation of odd C–atom fatty acids or may be produced during catabolism of amino acids, such as isoleucine etc. In ruminants, propionate is major product of the bacterial fermentation of cellulose in GIT, that is absorbed in blood. The only fate of propionate is to enter gluconeogenesis process that involves following steps:

(1) Propionate is first activated as propionyl CoA and then by a CO_2 fixation reaction, catalyzed by **propionyl CoA carboxylase** in the presence of biotin and ATP, it is converted to D-methyl malonyl CoA (4C-compound).

(2) D-methyl malonyl CoA is then isomerized to L-methyl malonyl CoA by a racemase enzyme.

(3) L-methyl malonyl CoA then undergoes a molecular rearrangment to from succinyl CoA, catalyzed by a **mutase** enzyme, which requires **B_{12} conezymes** (Fig. 7.2). Succinyl CoA so formed is an inportant intermediate of CAC and is converted to malate, which then enters the gluconeogenesis process (Fig. 7.1).

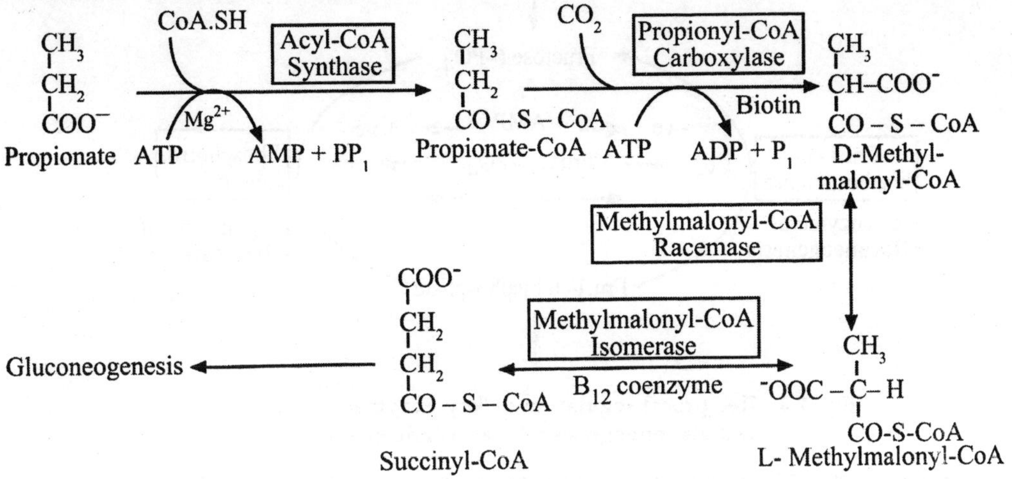

Fig. 7.2 : The activation and conversion of propionate into glucose by the gluconeogenesis process

Regulation of Gluconeogenesis

Gluconeogenesis, the denovo synthesis of glucose and glycolysis, the oxidative pathway of glucose are regulated **reciprocally;** that is, at a time only one process will proceed depending on blood sugar level status or immediate metabolic need of the cell. The reciprocal regulation of key enzymes of the two pathways is achieved by three mechanisms, that include—(1) Hormonal regulation of synthesis of key enzymes at the gene level, (2) Covalent modifications to activate/inactivate key enzymes and (3) Allosteric activation / inhibition of the key enzymes.

1. Regulation by Hormones—Hormones insulin and glucagon or epinephrine, depending on BSL status, **induce** and/or **suppress** synthesis of key enzymes of gluconeogenesis or glycolysis at the gene level. But this mechanism requires several hours and is referred to as a **long term effect** (details in the Chapter 8.).

2. Covalent Modifications—The key enzymes of gluconeogenesis and glycolysis are promptly and **reciprocally** regulated, depending on blood sugar level. This is achieved by reversible phosphorylation/ dephosphorylation of the key enzymes of the two pathways by c-AMP-protein kinase system, as illustrated in the Fig. 7.3.

3. Allosteric Effectors —The key enzymes of gluconeogenesis and glycolysis are allosteric enzyme and are reciprocally regulated by allosteric effectors as shown in the Fig. 7.4.

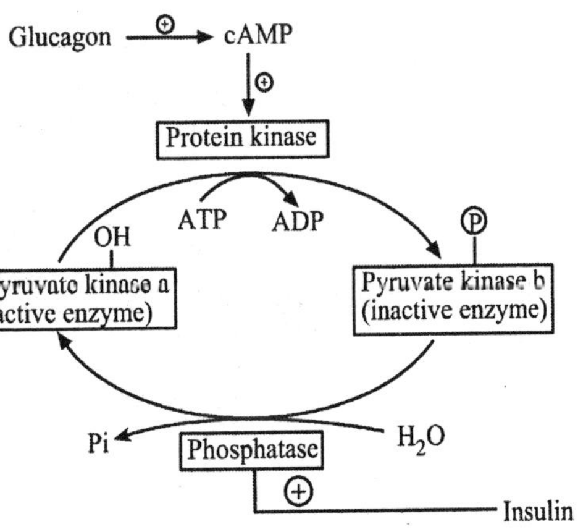

Fig. 7.3 : Activation/Inactivation of pyruvate kinase enzyme by phosphorylation/dephosphorylation

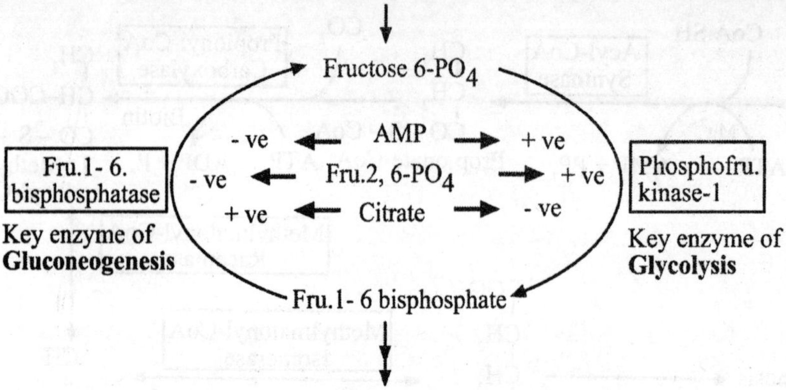

Fig. 7.4 : Reciprocal regulation of key enzymes of glycolysis and gluconeogenesis by allosteric effectors

Fructose 2, 6 Bisphosphate-An Allosteric effector—Fructose 2, 6-bisphosphate is a potent allosteric effector, that promptly and reciprocally regulates the two processes i.e. gluconeogenesis and glycolysis. When blood sugar level is high, the fructose 2, 6-bisphosphate synthesis increases and it allosterically activates the key enzyme **PFK-1** and thus glycolysis is stimulated. And at the same time it inhibits the **fru. 1, 6 bisphosphatase activity to decrease gluconeogenesis** process. This is important, as there is no need for glucose synthesis during hyperglycaemia.

The mechanism of action of fru.2, 6-bisphosphate to regulate the glycolysis and gluconeogenesis reciprocally, depending on BSL status is illustrated in the Fig. 7.5.

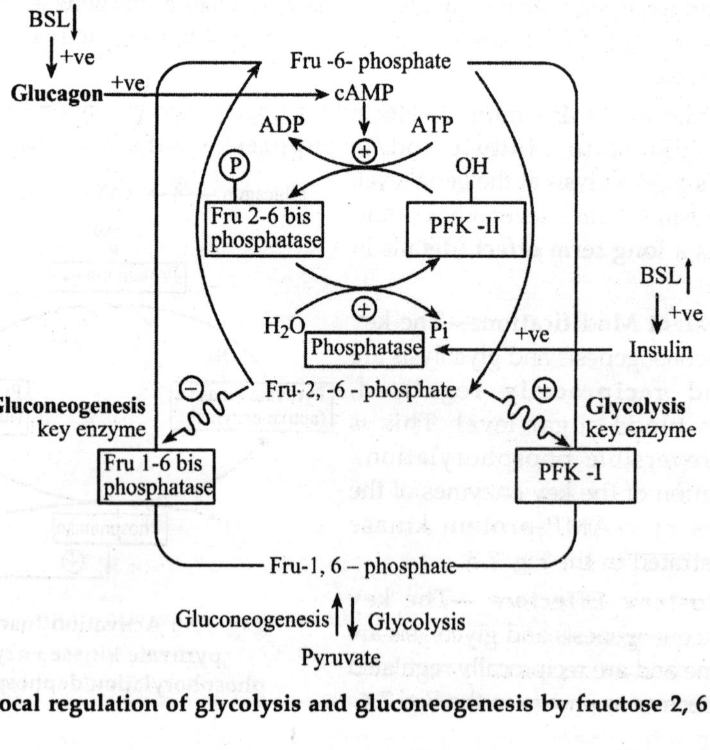

Fig. 7.5 : Reciprocal regulation of glycolysis and gluconeogenesis by fructose 2, 6 bisphosphate

HEXOSE MONO PHOSPHATE PATHWAY

Hexose mono phosphate shunt (HMP shunt), also known as **pentose phosphate pathway** is an alternative oxidative metabolic cycle of glucose. About 10% of glucose is metabolized by HMP shunt in various body tissues. It is a minor metabolic pathway and is **not for ATP generation,** but it has some specialized functions in our body that are as summarized here under:

HMP Shunt-a Multifunctional Pathway—Significance

1. HMP shunt generates reducing power—The reduced NADPH + H^+ produced in HMP shunt is utilized for the **reductive biosynthesis of lipids and steroids** such as—a) Fatty acid synthesis and lipogenesis and b) Cholesterol and steroid hormone synthesis.

2. Synthesis of pentose sugars —Ribose and deoxyribose sugars are synthesized in HMP shunt, which are required for the synthesis of nucleic acids RNA and DNA, as well as for the synthesis of other important nucleotides, such as ATP, coenzymes NAD, FMN, FAD etc.

3. To maintain sulphhydral buffer —In erythrocytes, HMP shunt pathway provides reduced NADPH + H^+ to keep glutathione in reduced state. The reduced glutathione detoxifies the oxy free radicals, which otherwise would cause hemolysis of RBCs. Thus HMP shunt pathway is necessary to mantain the fragility of RBCs.

Site of HMP Shunt —HMP shunt is actively going on in tissues, such as **liver, adipose tissues, adrenal gland, RBCs, ovaries, testes** and **lactating mammary glands,** that are actively involved in **reductive biosynthesis processes** as mentioned above.

HMP shunt is a multicyclic process and comprizes two phases

a. Oxidative phase

b. Non oxidative phase

Oxidative Phase —In oxidative phase, glucose 6–PO_4 is **oxidatively decarboxylated** into pentose sugars (unlike glycolysis), in two steps as described here under:

I Step—Three molecules of glucose 6-PO_4 are first dehydrogenated, catalyzed by specific enzyme **glucose 6-phosphate dehydrogenase.** This enzyme specifically requires NADP$^+$ coenzyme to remove 2 H^+ ions as NADPH + H^+, to produce **6-phosphogluconolactone,** which is then hydrolyzed to **6 phosphogluconate.**

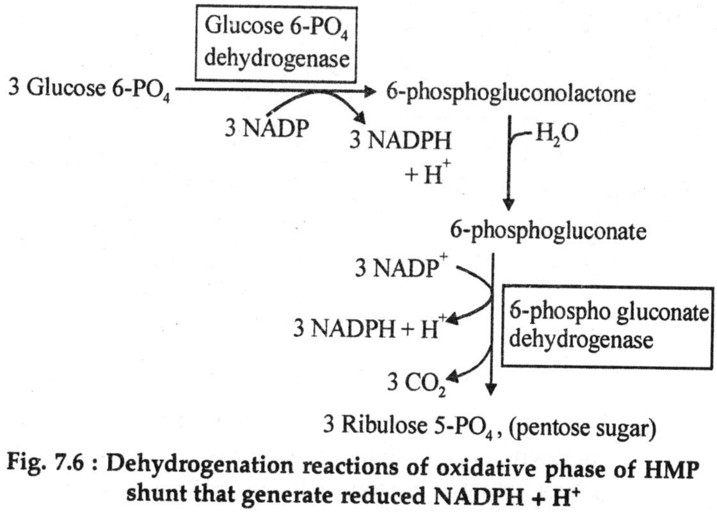

Fig. 7.6 : Dehydrogenation reactions of oxidative phase of HMP shunt that generate reduced NADPH + H^+

II Step —In a second dehydrogenation

step, 3 molecules of 6-phosphogluconate produced in above step are oxidized with simultaneous decarboxylation to produce 3 molecules of pentose sugar called **ribulose 5 phosphate** as shown in the Fig. 7.6.

Overall reaction of oxidative phase

$$3\ \text{Glucose 6-PO}_4 + 6\ \text{NADP}^+ \xrightarrow[\text{decarboxylation}]{\text{Oxidative}} 3\ \text{Ribulose 5-PO}_4 + 6\ \text{NADPH} + \text{H}^+ + 3\text{CO}_2$$

Pentose phosphates—Ribulose phosphate, produced in the oxidative phase of HMP shunt is isomerized to ribose 5-phosphate, that can be reduced to **deoxyribose 5-phosphate,** which are essential constituent of nucleic acids RNA and DNA respectively. So the end product of oxidative phase of HMP shunt, the pentose sugars ribose and deoxyribose are utilized for DNA and RNA synthesis and for the synthesis of other important nucleotides in the tissues. But if pentose sugars are not needed, the **unutilized** pentoses are rearranged to regenerate back glucose 6-PO_4 in the **next phase** of HMP shunt as given below:

Non Oxidative Phase—In this phase, three molecules of ribulose 5-PO_4 are first isomerized into one molecule of ribose 5-PO_4 and 2 molecules of xylulose 5-PO_4, catalyzed by **isomerase** and **epimerase** enzymes respectively (Fig. 7.7). These pentose sugars so produced are then rearranged, catalyzed by two specific enzyme systems called **transketolase** and **transaldolase** to regenerate back glucose 6-PO_4 as described in following steps:

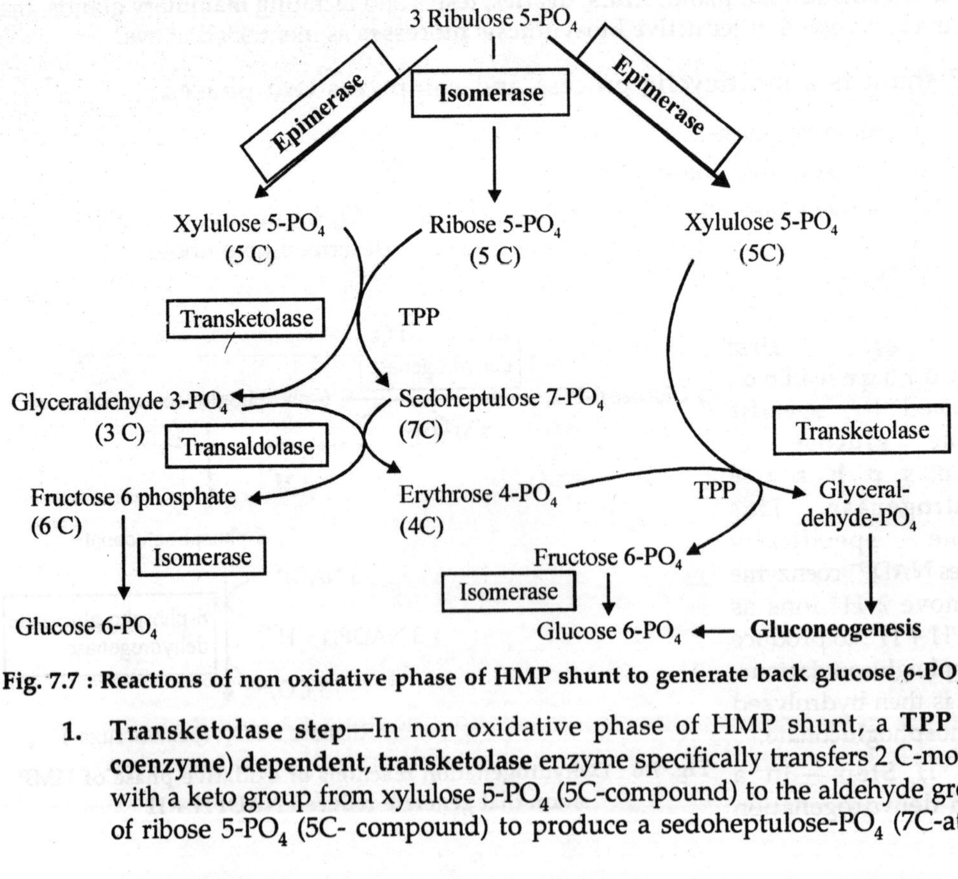

Fig. 7.7 : Reactions of non oxidative phase of HMP shunt to generate back glucose 6-PO_4

1. **Transketolase step**—In non oxidative phase of HMP shunt, a **TPP (B_1 coenzyme) dependent, transketolase** enzyme specifically transfers 2 C-moiety with a keto group from xylulose 5-PO_4 (5C-compound) to the aldehyde group of ribose 5-PO_4 (5C- compound) to produce a sedoheptulose-PO_4 (7C-atom

sugar) and glycerdehyde 3-PO_4 is left behind. The coenzyme TPP actively participates in this transfer reaction, where **nothing is lost or gained** but involves only transfer and simple rearrangement of groups between two pentose sugars, as shown in the Fig. 7.7.

2. **Transaldolase step**—The enzyme **transaldolase** then transfers back a 3 C–atom moiety from sedoheptulose-PO_4 to glyceraldehyde 3-PO_4 left as mentioned above to regenerate fructose 6-PO_4 and a tetrose monosaccharide erythrose 4-PO_4 is left behind (Fig. 7.7).
3. **Second transketolation step**—In a second transketolation step, the enzyme transketolase (TPP dependent) then catalyzes transfer of a 2 C-atom moiety from 3rd pentose sugar, xylulose 5-PO_4, on to the CHO group of erythrose-PO_4 left in above mentioned reaction to produce one more fructose 6-PO_4 molecule as depicted in the Fig. 7.7. The glyceraldehyde 3-PO_4 left behind then can enter gluconeogenesis process.

The reactions of non oxidative phase of HMP shunt in which the pentose sugars gets rearranged to generate back glucose 6-PO_4 are summarized in Fig. 7.7.

Overall Reactions of Non Oxidative Phase

In the non oxydative phase of HMP shunt, three molecules of pentose sugars are rearranged to regenerate 2 molecules of fructose 6-PO_4 and one glyceraldehyde 3-PO_4 molecule.

Significance of HMP Shunt

HMP shunt pathway of glucose oxidation is primarily for the generation of two important products i.e. pentose sugars and reduced NADPH + H^+. The significance and importance of HMP shunt pathway is described here under:

1. Production of Pentose sugars—In HMP shunt, pentose sugars ribose and deoxyribose phosphate are produced, which are required for the nucleotide and nucleic acid synthesis in our body. The pentose sugars are also utilized for the synthesis of high energy phosphate compounds, such as ATP, GTP, UTP, CTP and coenzymes NAD^+, $NADP^+$, FMN, FAD, Coenzyme A etc.

2. Skeletal muscles and some other tissues have only half HMP shunt operative, primarily for the synthesis of pentose sugar. These tissues possess enzymes of only **non oxidative phase** i.e. transketolase and transaldolase. As these reactions are freely reversible, so starting from fructose 6-PO_4 and glyceraldehyde 3-PO_4, by the reversal of non oxidative phase, pentoses can be synthesized in these tissues for the nucleotide synthesis and for RNA and DNA synthesis.

3. Generation of Reducing Power for Reductive Biosynthesis—

a. **Biosynthesis of fatty acids and lipogenesis**—HMP shunt pathway provides reduced NADPH + H^+, required essentially for the reductive biosynthesis of fatty acids and lipogenesis in liver and in adipose tissues. (Details in Chapter 11.)
b. **Cholesterol and steroid hormone biosynthesis**—Reduced NADPH + H^+ is required essentially for cholesterol biosynthesis, as well as for the synthesis of

adrenal corticosteroid hormones and male and female sex hormones in tissues, such as adrenal cortex, testes, ovaries etc. In these tissues the HMP shunt pathway of glucose oxidation is well established to provide reduced NADPH + H^+. (Chapter 12.)

4. Significant of HMP Shunt in Erythrocytes —HMP shunt is actively going on in RBCs and a direct relationship exists between RBC fragility and HMP shunt pathway. The HMP shunt in RBCs provides reduced NADPH + H^+, which is necessary for the maintenance of **sulphhydral buffer** as **glutathione.** The reduced glutathione detoxifies oxy free radicals, which are continuously being generated in RBCs by free radical chain reaction as given under:

$$O_2 \xrightarrow{e^-} O_2^- \xrightarrow[H^+]{e^-} OH^- \xrightarrow{e^-} H_2O_2$$

Oxy free radicals —The killer agents

The oxy free radicals, namely superoxide anion (O_2^-) hydroxyl ion (OH^-) and H_2O_2 are killer agents and can kill cells by peroxidation of lipid bilayer of the cell membranes.The free radicals are potentially injurious to tissues, as they cause single stranded breaks in DNA molecule, damage protein molecular structural integrity, as well as cause depolymerization of mucopolysaccharide components of tissues.

(1) Glutathione sulphhydral buffer—Glutathione, a tripeptide is present in rich amounts in RBCs. It has an active SH group and thus detoxify and remove oxy free radicals in RBCs as soon as they are generated as illustrated in the Fig.26.4, Chapter 26.

(2) HMP shunt maintain RBC integrity—Thus by detoxifying H_2O_2, catalyzed by the enzyme glutathione peroxidase, the cell membrane integrity of RBC's is preserved and hemolysis is prevented.

(3) Iron of hemoglobin is kept in **ferrous state** by active HMP shunt in RBC's and the reduced NADPH + H^+ provided by this pathway helps in converting methemoglobin (Fe^{3+}) back to normal hemoglobin with ferrous iron (Fe^{2+}), that is necessary to transport O_2.

(4) Sulphhydryl (SH) group at catalytic site —Reduced NADPH + H^+ is important in keeping active SH group at the catalytic site of key enzymes of glycolysis, for example **glyceraldehyde 3-PO_4 dehydrogenase.** The active SH group at catalytic site of this enzyme is necessary for the catalysis of key reaction of glycolysis in the RBCs.

Inherited Genetic Disorders of HMP Shunt

Glucose 6-PO_4 dehydrogenase defect is a genetic metabolic disorder of HMP shunt, prevalent in certain African tribes. In this congenital metabolic defect, because of mutation in genes, a key enzyme **glucose 6-PO_4 dehydrogenase** is deficient in RBCs. Due to this a block occur in HMP shunt pathway and NADPH + H^+ is not produced to keep glutathione in reduced state to detoxify oxy free radicals. This results in excessive hemolysis of RBCs, when subjects are exposed to certain oxidants, such as primaquin (an antimalarial drug) and results in **hemolytic anaemia** with severe jaundice.

This genetic defect is transmitted by X-linked recessive trait, by the heterozygous females (carriers of abnormal gene) and thus only males suffer from this congenital disorder. Several varients of Glu. 6-phosphate dehydrogenase defect have been identified.

This genetic disorder is prevalent in malaria infested population of African tribes. It is proposed that this abnormality in genes is **acquired by induction** in response to malarial endemicity in this geographical area. This genetic disorder is advantageous since it offers resistance to plasmodium malarial infection. This is because hemolysis of RBCs in this enzymes defect cause termination of life cycle of this parasite to kill them.

Favism—When the subjects suffering from Glu.6-PO_4 dehydrogenase deficiency consume a type of bean called **favabean,** excessive hemolysis of RBCs occurs resulting in hemolytic anaemia and the condition is referred to as **favism.** The favabean is rich in certain oxidants, which cannot be detoxified, as reduced NADPH + H^+ is not produced in these subjects.

URONIC ACID PATHWAY

Uronic acid pathway is another minor metabolic process of glucose oxidation in our body. In this pathway ATP is not generated, but it primarily is for the production of some specialized biologically important compounds, such as **glucuronic acid** and **pentose sugar** in human beings and **ascorbic acid** (vitamin C) in lower animals.

Significance of uronic acid pathway—In uronic acid pathway following physiologically important compounds are synthesized:

(1) Glucuronic Acid production—Glucuronic acid is a special hexose sugar acid in which the aldehyde group at anomeric C–atom 1 of glucose is protected by UDP, while the alcoholic group at 6 C-atom is oxidized to carboxylic group. The glucuronic acid produced in uronic acid pathway is an important precursor for the synthesis of several important compound, as well as it is utilized for conjugation reactions in liver as describe below:

(a) **Glycosamine glycans synthesis**—The glucuronic acid is utilized for the synthesis of glycosamine glycans, required for the production of brain lipids.

(b) **Mucopolysaccharide and chondroitin sulphate synthesis**—The glucuronic acid is the precursor for the synthesis of mucopolysaccharides and chondroitin sulphates in the body. These compounds are main ground cementing material of connective tissues, cartilages and bones etc. that provide strength to these tissues.

(c) **Conjugation reactions in liver** —Glucuronic acid is essentially required for the conjugation of bile pigment **bilirubin** as **mono** and **diglucuronide,** to make it water soluble for its excretion via bile into the feces.

(d) **Conjugation of steroid hormones and phenols** —Many steroid hormones and certain phenol are conjugated with glucuronic acid in liver, that is prerequisite for their exceretion.

(e) **Detoxification**—Certain harmful and unwanted chemicals and drugs need conjugation with glucuronic acid for their excretion.

(2) Pentose Sugar Production—Uronic acid is another pathway for the production of pentose sugars ribose and deoxyribose, that are required for nucleotide and DNA and RNA synthesis in body tissues.

(3) Ascorbic acid (Vitamin C)—Lower mammals, except guinea pig, primates and human beings, synthesize vitamin C from L–gulonolactone, that is an intermediate of uronic acid pathway.

Reactions of Uronic Acid Pathway—The main reactions involved in uronic acid pathway of glucose oxidation are summarized here under:

a) **Formation of UDP–Glucuronic Acid** —The aldehyde group at the anomeric C-atom of glucose is first protected by attachment with UDP nucleotide to produce UDP-glucose (reactions details in Chapter 5.). UDP-glucose so produced is then oxidized, catalyzed by enzyme **UDP-glucose dehydrogenase** to **UDP-glucuronic acid.** The **UDP-glucuronic acid** is an **active compound** and is utilized for various specialized functions in our body as summarized in the Fig. 7.8.

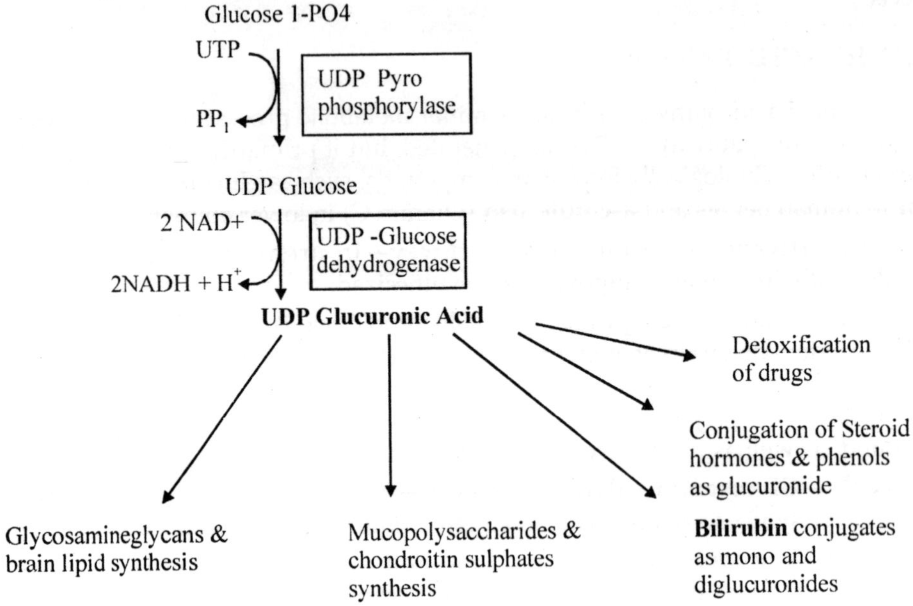

Fig. 7.8 : Synthesis, production and functions of glucuronic acid

b) **Synthesis of Pentose Sugars**—The major reactions involved in the synthesis of pentose sugars by uronic acid pathway are summarized in the Fig. 7.9.

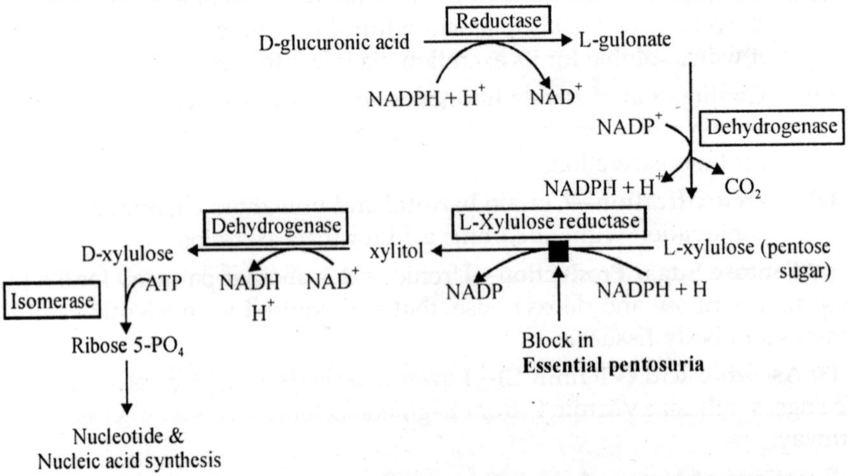

Fig. 7.9 : Production of pentose sugar in uronic acid pathway.
The block (■) shows genetic defect in essential pentosuria

3. Synthesis of Ascorbic Acid (Vitamin C) in Lower Animals—Lower mammals except primates, human beings and guinea pig can synthesize ascorbic acid (vitamin C) from the intermediate compound of uronic acid pathway called L-gulonate. The chemical structure of L-gulonate is very similar to ascorbic acid and thus it can easily be converted into vitamic C as described here.

L-Gulonate is first dehydrated to **L-gulonolactone,** which is then oxidized to 2- keto L-gulonolactone and then to ascorbic acid, catalyzed by the enzyme **L-gulonolactone oxidase** (lacking in human beings) as illustrated in the Fig. 7.10.

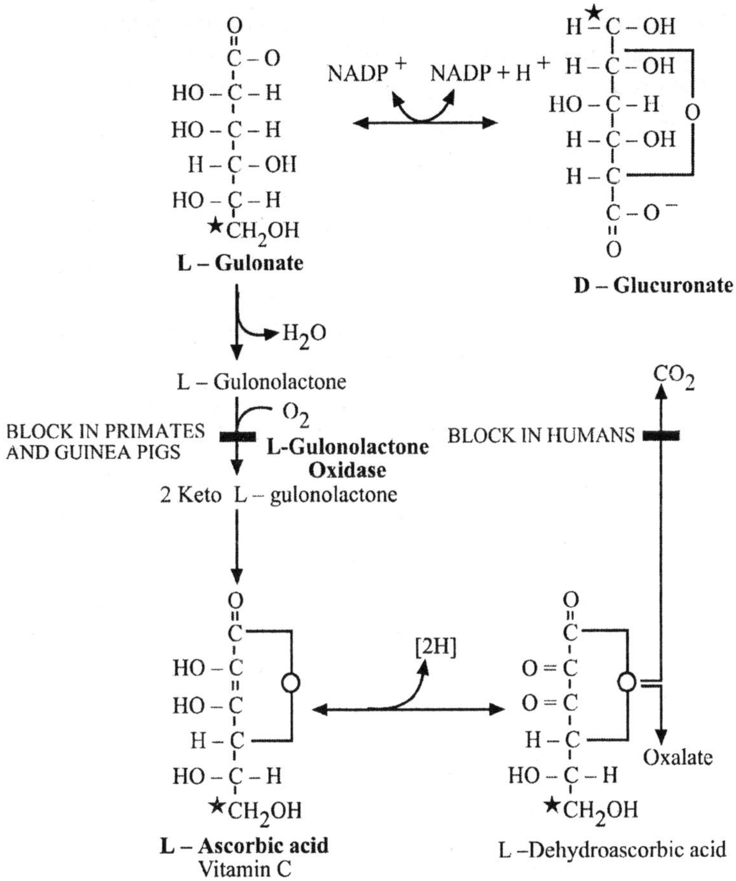

Fig. 7.10 : Synthesis of ascorbic acid (vitamin C) in lower animals

Higher primates and human beings cannot synthesize vitamin C from uronic acid pathway because they lack enzyme L-gulonolactone oxidase. The explanation proposed for this is, that during evolution process, the primates, guinea pig and humans have lost the gene which is responsible for the synthesis of enzyme **L-gulonolactone oxidase.**

This is a classic example of alterations in genetic constitution of animals **as induced** by change in environmental conditions. The primates (monkeys) had much vitamin C in their diet (fruits and vegetables are their staple diet), so they had no need to synthesize this vitamin and thus the gene encoding this key enzyme gradually disappeared during evolution process.

Essential Pentosuria — It is a rare genetic disorder, in which a block occurs in the uronic acid pathway, where L-xylulose cannot be converted into D-xylulose due to the lack of enzyme L-xylulose reductase (Fig. 7.9). As the L isomers of pentoses cannot be utilized by humans, they accumute in blood and are excreted in the urine. The essential pentosuria is otherwise a benign (mild) condition without any clinical adverse effects.

METABOLISM OF GALACTOSE

Galactose is a constituent of the disaccharide lactose (milk sugar) and is produced in the GIT by the hydrolysis of lactose, catalyzed by the enzyme **lactase.** Galactose so liberated is then absorbed along with glucose, by active transport mechanism from the small intestine (Chapter 4) and is brought to liver for it's further metabolism.

Importance of Galactose—Infants are fed only milk and thus galactose and glucose (digestive product of lactose), are the main carbohydrates provided by their diet. Galactose is very important for the infants, as it is required for the synthesis of brain glycolipids and mucopolysaccharides, that are essential for the brain development and skeletal growth.

The enzyme system which can synthesize galactose in liver is not developed in infant at an early age, so galactose from diet is necessary for the development of the child. That is why the nature has provided lactose as the main carbohydrate in milk to provide galactose to infant for its proper development and growth.

Conversion of Galactose to Glucose—Galactose is metabolized exclusively in the liver by its conversion to glucose as described here :

Activation of Galactose—Liver has a specific enzyme **galactokinase** that phosphorylates galactose to **galactose1-PO$_4$** utilizing ATP as energy. The galactose 1-PO$_4$ then undergoes an exchange reaction with UDP-glucose to produce UDP-galactose, in the presence of key enzyme **uridyltransferase.** UDP galactose so produced is an active compound and is utilized for the biosynthesis of various physiologically important compounds, such as glycolipids, mucopolysaccharides and lactose in the lactating mammary gland. (Fig. 7.11).

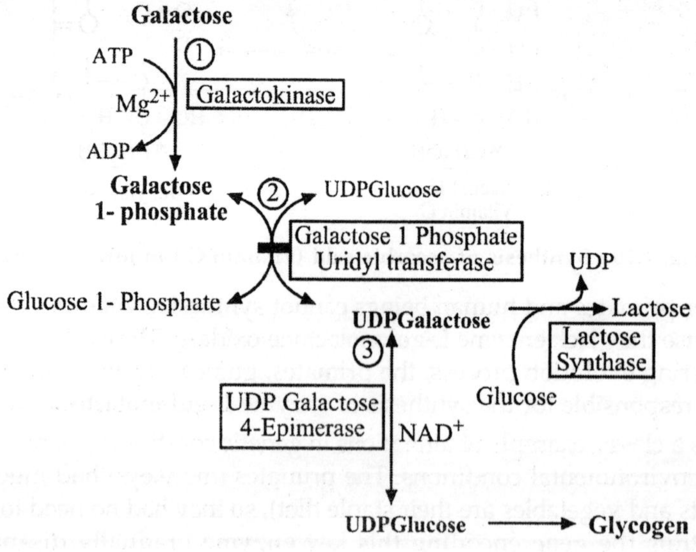

Fig. 7.11 : Metabolism of galactose in liver. Inborn error galactosemia is shown with a block (▬)

UDP-galactose can easily be converted into UDP-glucose in the presence of enzyme **epimerase** and thus enters the metabolic pathway of glucose. (Fig. 7.11).

Galactosemia—Genetic Disorder of Galactose Metabolism

Galactosemia is a serious inborn error of galactose metabolism with severe clinical manifestations, which can be life threatening.

Galactosemia is caused by abnormal genes resulting in deficiency of the key enzymes of galactose metabolism, namely **galactokinase** and / or **galactose –1-PO_4 uridyltransferase** (Fig. 7.12.). Because of this galactose cannot be converted into glucose, resulting in impairment of its utilization. Hence galactose 1-PO_4 accumulates in blood leading to metabolic dosorder **galactosemia** and is excreted in excess in urine causing **galactosuria.**

Clinical manifestations of galactosemia include—

(a) Excessive vomiting, aversion to milk and child fails to thrive.

(b) Hepatomegaly with severe jaundice.

(c) Hypoglycaemia and its adverse effects on brain.

(d) Mental retardation due to CNS damage.

(e) Renal damage resulting in aminoaciduria and albuminuria.

(f) Early cataract development in children results due to galactose accumulation and its reduction into dulcitol, which accumlates in the lens of the eye. This causes osmotic effects leading to cataract development in the child

(g) Death occurs at an early age.

Screening and Management of Galactosemia

Early detection of the condition is the most important factor to prevent irreversible damage to CNS. This severe metabolic disorder can be detected early by osazone test for galactose in urine.

If detected early, the irreversible clinical manifestations, such as CNS damage and consequences can be prevented by removing the child from mother's milk and by feeding lactose free formulae diet. This lactose free diet should be continued upto 6 years of the age of child, till the brain development is complete.

FRUCTOSE METABOLISM

Major dietary source of fructose is sucrose (cane sugar), the sugar we take as sweetner in our food and in beverages. Minor sources of fructose in our diet are fruits, honey and jaggery. On hydrolysis of sucrose by enzyme **sucrase** at the brush border of intestine, fructose is produced, which is absorbed from GIT by **facilitated diffusion** that requires a specific carrier but energy as ATP is not needed.

Utilization and Metabolism of Fructose

Fructose absorbed in portal blood from GIT comes to liver, where it is metabolized futher. In liver, fructose is phosphorylated by a specific enzyme **fructokinase** to form **fructose 1-PO_4**. The fructose 1-PO_4 is not the substrate for the usual enzymes of glycolysis, so it is metabolized promptly in liver by an alternative pathway utilizing different enzyme **aldolase B** (Fig. 7.12).

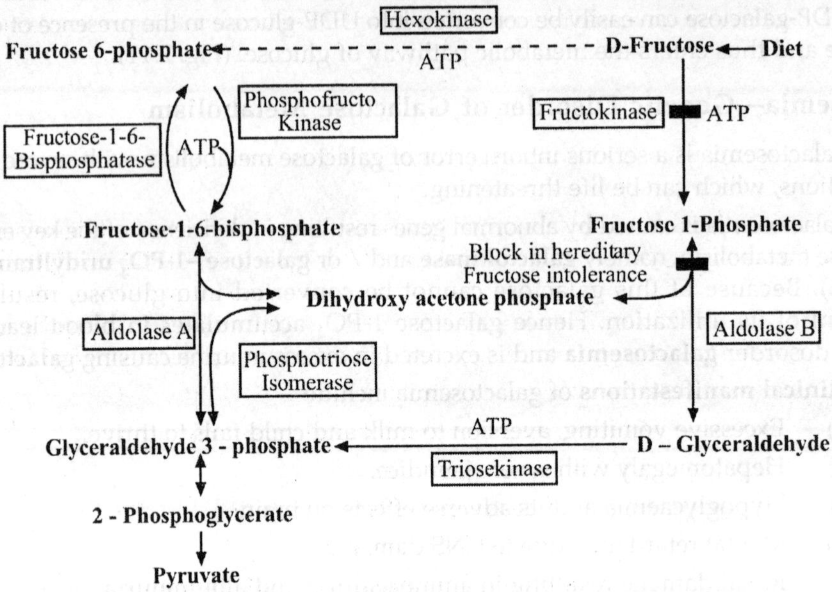

Fig. 7.12 : Metabolism of fructose by fructokinase and aldolase B enzymes

Aldolase B—It is a specific enzyme which split fructose-1-PO_4 into free glyceraldehyde (triose) and dihydroxyacetone phosphate. Free glyceraldehyde in liver is then activated by the enzyme **triosekinase** and ATP to glyceraldehyde 3-PO_4, which along with dihydroxyacetone-PO_4 can enter either glycolysis process or gluconeogenesis, depending on blood sugar status.

The key enzymes fructokinase and aldolase B are not dependent on hormone insulin. So diabetic patients can metabolize fructose rapidly. Fructose is metobolized more rapidly by normal persons also, as fructose metabolism by this route escapes bottlenecks of glycolysis at phosphofructokinase step.

Diabetic patient can utilize fructose faster than glucose because the rate limiting step of glycolysis, catalyzed by **phosphofructokinase-**I is bypassed and also because the enzymes of fructose metabolism, as well as its entry into tissues are not dependent on insulin.

Free fructose is present in rich amount in seminal plasma, where it is utilized as energy source necessary for the motility of sperms.

Inborn Errors of Fructose Metabolism

1. Congenital Fructose Intolerance—It is a serious genetic metabolic disorder of fructose metabolism, transmitted as autosomal recessive trait. It is caused by the deficiency of enzyme **aldolase B,** due to mutation in gene, because of which fructose 1-PO_4 cannot be metabolized and it accumulates in blood and is excreted in the urine.

Clinical manifestations include—Hepatomegaly with severe jaundice, infant fails to thrive and severe liver impairment may result in early death of the child.

Withdrawl of fructose and sucrose from diet of the child will immediately relieve the symptoms of this genetic disorder.

2. Fructosuria—This is a mild metabolic abnormality, caused by enzyme frutokinase defect. There are no serious clinical symptoms except the excretion of fructose in urine in this disorder.

□□□

8

Blood Sugar Homeostasis : Insulin, Glucagon and Diabetes Mellitus

Blood sugar level (BSL) regulation is the most important homeostatic mechanism of our body. BSL is normally regulated most efficiently and finely within the narrow range of physiological limits of 70-110 mg/dl of plasma. This is important because both hypoglycaemia (low BSL) as well as hyperglycaemia (high BSL) adversely affect functioning of the body organs, as described here under.

Hypoglycaemia—When blood sugar level tends to decrease, as observed in fasting conditions, initial symptoms observed are headache, nausea, vomiting etc. and if it is severe enough (<45 mg/dl), it adversely affects brain and CNS functions, resulting in dizziness, convulsions, seizures and person may go into coma and hypoglycaemic shock (as seen in insulin overdose) which can be life threatening. This is because brain and CNS are solely dependent on a regular supply of glucose for metabolism and energy production for their normal functioning.

Hyperglycaemia—Very high blood sugar level, as seen in uncontrolled, severe diabetes mellitus, results in osmotic pressure alterations and electrolyte imbalance between intra and extracellular fluids leading to osmotic diuresis, polyuria, dehydration, and other severe clinical complications, which can be fatal.

Normally blood sugar level is maintained efficiently by an equilibrium between various glucose producing (sources) and utilizing processes as illustrated in the Fig. 8.1.

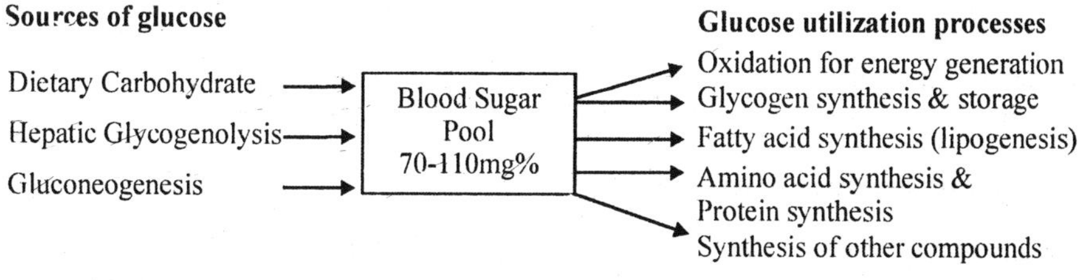

Fig. 8.1 : Blood glucose pool maintenance by a balance between sources and utilization of glucose

Factors Regulating Blood Sugar Level

Various factors are involved in maintenance of blood sugar level within normal limits. Important factors include organs such as liver, extrahepatic tissues, several metabolic and hormonal mechanisms etc. The various factors regulating BSL are described here.

Structural and functional integrity of gastrointestinal tract is necessary for proper digestion and absorption of food stuffs. Diarrhea, gastroenteritis and other GIT disorders impair digestion and absorption of nutrients including glucose.

Role of Liver in BSL Maintenance

Liver plays a central role in blood sugar regulation as it can add glucose to blood or increase its uptake depending on blood sugar level status. The salient features of the functions of liver in BSL maintenance are enumerated here.

1. Liver cells are freely permeable to glucose in comparision to extrahepatic tissues which depends on insulin for glucose uptake and thus transport of glucose through cell membranes in these tissues becomes **rate limiting** factor.

2. Hepatic glucokinase plays important role in BSL regulation after meals. Hepatic glucokinase activity is not affected by accumulation of glucose 6-phosphate in hepatic cells, thus liver is not subjected to any constraint in uptake of glucose form blood when BSL is elevated after meals. On the otherhand the extrahepatic enzyme **hexokinase** is inhibited by glucose 6 -phosphate by feed back mechanism and thus a restrain is exerted on extra glucose uptake by extra hepatic tissues from the blood in hyperglycaemic state.

3. Liver convert extra glucose into various compounds such as a) glycogen for storage purpose.b) Fatty acid synthesis and lipogenesis—Liver plays important role in removing extra glucose from blood after meals by converting it into fatty acid and c) synthesis of non essential amino acid to be utilized for protein synthesis.

4. Liver is net producer of glucose in hypoglycaemic (low BSL) state as seen during fasting conditions and also at normal BSL status during post absorptive states. This ensures a regular supply of glucose as fuel to tissues such as brain, CNS, RBC, etc. which solely depend on glucose for energy production porposes.

Kidneys also help in regulation of blood sugar to some extent. Gluconeogenesis can occur in renal tissues to produce glucose in hypoglycaemic states. However the primary role of kidneys is to excrete extra glucose in urine when BSL is very high (> 180 mg/dl). This is important to prevent adverse effects of severe hyperglycaemia as mentioned earlier.

Effect of Hormones on Blood Glucose Level

Hormones are most the important factors and play central role in maintenance of stable blood sugar status inspite of variations in dietary habits of an individual and in adverse environmental situations. Hormones involved in homeostasis of blood sugar can broadly be diffentiated into two categories a) hypoglycaemic hormones that lower BSL and b) hyperglycaemic hormones which increase blood sugar levels.

Insulin—Hypoglycaemic Hormone

Insulin plays a central role in regulation of blood sugar level. It is the only hormone that **decreases blood sugar** level in hyperglycaemic state as seen after meals. Insulin is secreted from the β cells of islets of Langerhan's of pancreas, in direct response to rise in post prandial BSL and in high blood sugar status.

After a carbohydrate rich diet, post prandial BSL tends to rise, which stimulates insulin secretion. Insulin lowers blood glucose by removing extra glucose from blood by **enhancing glucose uptake** by extra hepatic tissues, such as muscles and adipose tissues. Insulin stimulates all **glucose utilizing** metabolic cycles and simultaneously it lowers all **glucose producing** metabolic processes in the body (details lateron in this section).

The hypoglycaemic effects of insulin on various metabolic pathways are brought about by 1) Increasing glucose uptake by tissues, 2) by increasing key enzyme synthesis of glucose utilization processes at gene level and 3) Activation/inactivation of key enzymes of rate limiting steps by covalent modification of various metabolic cycles .

A brief account of hypoglycaemic effects of insulin along with its mechanism of action is summarized in the Table 8.1.

Table 8.1 : Hypoglycaemic effects of Insulin by its action on various metabolic processes

Carbohydrate metabolism—Effects of Insulin	
Glucose utilization processes	**Glucose production processes**
(a) ↑ Glucose uptake by tissues	↓ **Glycogenolysis** by ↓ c-AMP and inactivating key enzymes
(b) ↑ Glycolysis & HMP shunt by activating key enzymes	↓ **Gluconeogenesis** by suppressing synthesis of key enzymes at gene level.
(c) ↑ **Glycogenesis** in liver & muscles by activating key enzmes	
Lipid Metabolism—Effects of Insulin	
Convert extra glucose into fatty acid	↓ Lipolysis in adipose tissues by
↑ Fatty acid synthesis from acetyl CoA by	↓ c-AMP & inhibits hormone sensitive lipase
↑ Glycolysis and NADPH + H by ↑ HMP shunt by activating key enzymes	
↑ Lipogenesis & storage of fats in adipose tissues.	
Protein Metabolism—Effects of Insulin	
	↓ Protein catabolism
↑ Amino acid synthesis from components of CAC	
↑ Protein synthesis ↑ Growth and promotes development	

Hyperglycaemic hormones

1. Glucagon—The hormone glucagon is antagonist to insulin. It increases blood glucose level by its opposite effects to that of insulin on various metabolic cycles. Glucagon is secreted from the α -cells of islets of Langerhan's of the pancreas in response to hypoglycaemia (low BSL) as seen in fasting conditions or in between meals.

Liver is the target organ—Glucagon reaches to its target tissue liver and stimulates **hepatic glycogenolysis** to add glucose to the blood for immediate correction of hypoglycaemic state. Glucagon enhances hepatic glycogenolysis by activating the key enzyme phosphorylase through c-AMP-protein kinase activation cascade system as described in details in Chapter 6.

Glucagon also stimulates denovo synthesis of glucose from amino acid and lactic acid by activating key enzymes of **gluconeogenesis**. Thus by enhancing glucose producing processes (effect opposite to insulin) glucagon increases blood glucose to bring BSL to normal values.

Glucocorticoids

The glucocorticoids are steroid hormones secreted from the adrenal cortex and play important role in carbohydrate metabolism. The glucocorticoids **elevate blood glucose level** by stimulating **denovo synthesis** of **glucose** particularly from amino acids. Administration of these steroids causes tissue protein catabolism, increased hepatic uptake of amino acids and increased gluconeogenesis by acting directly **on genes** to enhance the synthesis of key enzyme of gluconeogenesis (details in Chapter 6.)

Glucocorticoids are the main hormones that regulate blood sugar level during prolonged fasting and starvation conditions by enhancing tissue protein catabolism to provide amino acids for glucose synthesis. This is essential to provide glucose to brain in order to survive and sustain life during starvation, even at the cost of tissue protein catabolism.

Glucose sparing action—In addition, glucocorticoids increase lipolysis in adipose tissue and free fatty acids so released are preferably oxidized in skeletal muscles etc. to spare glucose for **BSL** maintenance.

All these actions of glucocorticoids are opposite and **antagonistic** to the **insulin hormone.**

Epinephrine

The hormone epinephrine is secreted from the adrenal medulla in response to **stressful** and **adverse** conditions such as emotional stress, fear, anxiety, nervousness, hypoxia, hypoglycaemia etc. Epinephrine prepares the subject to counteract the challenging situations by its physiological stimulating effects on blood pressure, heart rate etc. Epinephrine **increases blood** sugar levels in stressful conditions to provide more glucose to brain (metabolism of brain increases during stress). Epinephrine increases BSL by stimulating hepatic glycogenolysis by activating phosphorylase enzyme via c-AMP protein kinase phosphorylation process (Chapter 6.). It also activates muscle glycogenolysis by activating myophosphorylase. However muscles lack the enzyme glucose 6–phosphatase, so muscle glycogenolysis produce lactate rather then free glucose.

Epidenphrine in addition to its hyperglycaemic effects, stimulates lipolysis in adipose tissues to provide fatty acids as fuel to extraheptic tissues thereby sparing glucose to be utilized by brain.

Growth Hormone and ACTH are secreted from anterior pituitary gland and **increase blood sugar** level by their indirect effects. ACTH stimulates glucocortoid secretion from adrenal cortex which increase BSL as mentioned above.

Growth hormone mobilizes lipids from adipose tissues by stimulating lipolysis via c-AMP hormone sensitive lipase activation. The free fatty acids so liberated are metabolized preferably in extra hepatic tissues, thus sparing glucose to raise BSL indirectly.

Chronic administration of growth hormone leads to persistant hyperglycaemia, which causes β–cell exhaustion and insulin insufficiency resulting in a diabetes like condition.

Thyroid Hormone

The thyroid hormones increase blood glucose level. Experimental evidence indicates that thyroxine has a diabetogenic effect. In humans, the blood sugar level increases in hyperthyroidism while it decreases in hypothyroidism. Exact mechanism by which thyroxine increases BSL is still not clear.

INSULIN

Insulin is an adiabetogenic (i.e. against diabetes) hormone, secreted from the β **cells** of **islets** of **Langerhan's of pancreas**. It is the major hormone that **regulates blood sugar level** and has profound **influence** on the **metabolism** of carbohydrate, lipids and proteins. Insulin is an **anabolic hormone** as it increases protein synthesis and promotes growth. This hormone has important role in promoting DNA replication and cell division and during fetal development it stimulates cell differentiation and fetal organogenesis. Thus insulin and insulin like factors are considered as important growth factors.

Historical Importance

The hormone insulin is known since ancient time as an active principle from pancreas, that prevented diabetes mellitus. The link between the islets of Langerhan's of pancreas and diabetes was suggested by De Mayer in 1909, but it was **Banting** and **Best** who proved this association in 1921, when they were able to extract insulin from pancreas of dogs, that had potent hypoglycaemic effects.

Insulin has a special place in the history of biochemistry and medicine, as it was the **first hormone** to be extracted in **pure form** from the pancreas of dog by **Banting** and **Best** (1921). Insulin is the **first protein hormone** to be chemically analyzed and the complete amino acid sequence of its two polypeptide chains was determined by **Sangers et al** by their special technique. Insulin is also the **first protein** to be **synthesized** commercially by recombinant DNA technology to be used in clinical medicine for the treatment of insulin dependent diabetes mellitus.

Chemical Structure—Insulin is a protein hormone having two polypeptides A and B chains as heterodimeric molecule. Human insulin (mol.wt 5737) contain **51 amino acids** arranged in two polypeptide chains. The A chain has 21 amino acids while the B chain has **30 amino acids**. The two chain are linked and held together by two **disulphide linkages** between 7th cys (cystein residue) of A and 7th cys of B chain and 20th cys of A and 19th of B chain respectively. A third intrachain disulphide bond is present in A chain between 6th and 11th cysteine residues as illustrated in the Fig. 8.2.

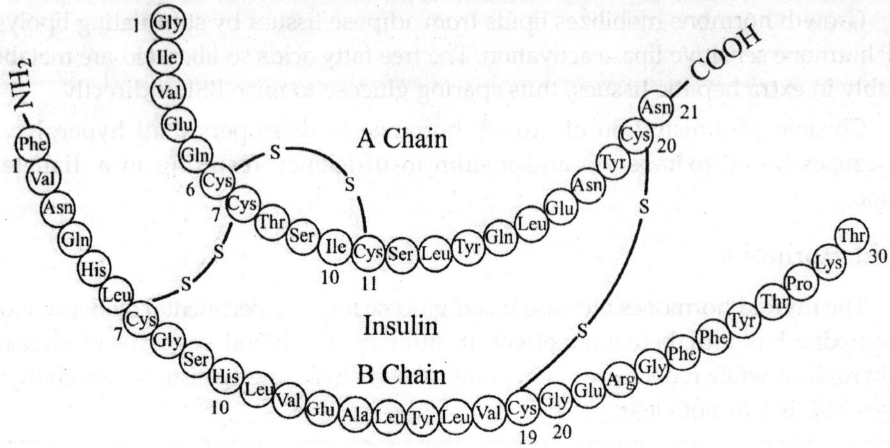

Fig. 8.2 : Chemical structure of insulin

Biological activity of insulin molecule resides in the regions that are highly conserved in most of the animal species including humans. These regions include the positions of : (1) the three disulphide linkages, (2) the hydrophobic residues at the carboxyl terminal region of B chain and (3) the amino and carboxyl terminal regions of A chain. Insulin looses its biological activity on reduction of the disulphide bonds.

Insulin from other animal species closely resembles in chemical structure to that of human insulin, with slight variation in amino acid sequence particularly at position 8, 9 and 10 of A chain. A very close similarity exists among human, porcine (pig) and bovine insulin.

Porcine insulin differs form human insulin, by only a single amino acid alteration i.e. an alanine is substituted in place of threonine at 30th position of B chain of human insulin. While the bovine insulin has additional modifications at 8th, 9th and 10th position of the A chain. These modifications donot affect the biological functions of insulin much and has least effect on its antigenicity. Thus porcine and bovine insulins were the standard therapy used for the treatment and control of type I diabetes mellitus (IDDM) in humans, until human insulin was produced commercially by recombinant DNA technology.

Biosynthesis of Insulin

Insulin is synthesized at the rough endoplasmic reticulum in the β cells of islets of Langerhan's of pancreas, as a **single large precursor polypeptide chain**. This precursor molecule has 107 (108) amino acids and is known as **preproinsulin** (mol. wt. 11,500). The preproinsulin has an extra leader sequence of hydrophobic 23 (24) amino acids (pre peptide) that directs the transfer of the precursor molecule from the ribosomes (site of synthesis) into the cisternae of the endoplasmic reticulum and then it is (pre peptide) cleaved (removed) off by specific proteolysis, leaving behind proinsulin.

Proinsulin—The proinsulin has 84 amino acids in a single polypeptide chain (mol wt 9000) and is biologically **inactive** precursor of insulin. The proisulin has an extra **C–peptide** chain with 33 amino acids, that masks the core biological activity of insulin as shown in the Fig. 8.3.

Significance—Insulin is synthesized as a single large polypeptide precursor proinsulin to provide proper conformation by folding, for the formation of correct disulphide bridges, necessary for the biological activity of insulin hormone. The major steps of synthesis of insulin and structure of proinsulin are shown in the Fig. 8.3.

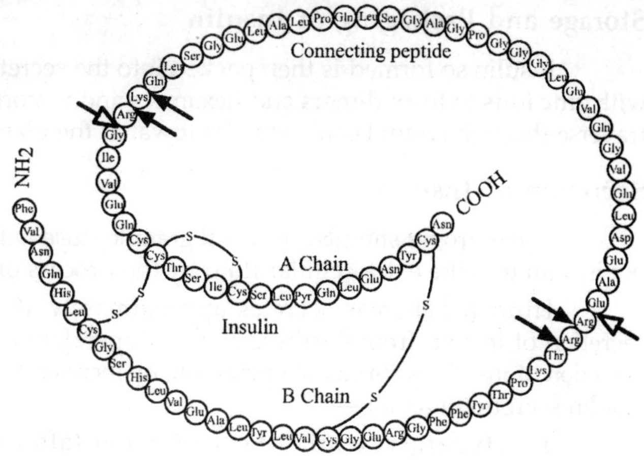

Fig. 8.3 : Diagrammatic representation of proinsulin structure depicting specific sites with basic amino acids where C-peptide is cleaved to produce active insulin

The complete amino acid sequence of insulin and proinsulin has been elucidated by Sanger's et al and is shown in the Fig. 8.3. It is evident from the structure that proinsulin has an extra C-peptide chain with 33 amino acids, that links the carboxy terminal of B chain (30th AA) of insulin to that of amino terminal (1st AA) end of the A chain.

Conversion of Proinsulin to Insulin

The proinsulin is transported to the golgi apparatus where it is converted into insulin by specific proteolysis to remove the extra C-peptide. A trypsin like protease enzyme specifically attacks the peptide bonds formed by basic amino acid **arg-arg** and **arg-lys,** present just before the carboxy terminal end of B chain and amino terminal of A chain respectively (Fig. 8.4). The hydrolysis of these bonds results in the cleavage of the extra C-peptide (33AAs) correctly, leaving behind **active insulin** in equimolecular quantities.

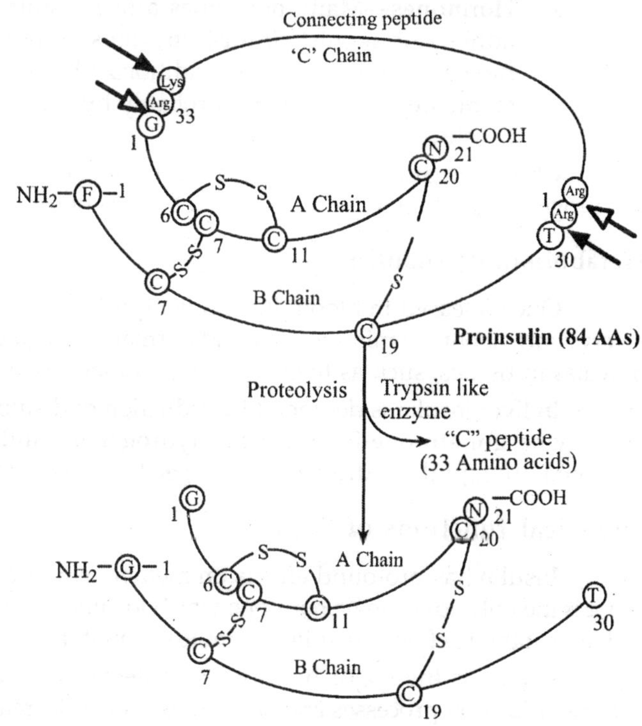

Fig. 8.4 : Synthesis of insulin as single chain precursors and its conversion to insulin by specific proteolysis. (AAs- amino acids)

Storage and Packaging of Insulin

Insulin so formed is then packed into the secretory β granules, where it combines with **zinc ions** to form dimers and hexamers and is stored. The β granules mature as they traverse through endoplasmic vesicles towards the plasma membrane of β cells.

Secretion of Insulin

Upon proper stimulation, the β granules fuse with the cell membrane and discharge the insulin into the extra cellular fluid by the process of exocytosis.

Human pancreas secretes approximately 40-50 units of insulin per day. The secretion of insulin from β cells is energy dependent and is **precisely regulated process.** Secretion of insulin is stimulated primarily by increase in blood sugar level. Factors affecting insulin secretion include—

1. **Hyperglycaemia**—Post prandial (after meals) rise in BSL is the major physiological stimulating factor for insulin secretion. β Cells have receptors on which glucose gets attached and activates c-AMP and Ca^{++} ions release which mediate the insulin secretory process. Along with hyperglyceamia, elevated plasma levels of **amino acids** and **ketone bodies** also increase insulin secretion from β cells.
2. **Hormones**—Many hormones affect insulin secretion. **Epinephrine** and α–adrenergic agonists **inhibit** insulin secretion even when blood sugar level is high as seen in stressful conditions. On the other hand β adrenergic agonists stimulate insulin release probably by increasing c-AMP concentration in the β–cells.

Chronic excessive levels of growth hormone, cortisol, placental lactogen, estrogens and progestins also increase insulin secretion, as observed in the later stage of pregnancy.

Metabolism of Insulin

Once released in blood, insulin is rapidly metabolized. Plasma half life of insulin is very short, about 3-5 minutes only and normally it is quickly degraded by specific protease enzymes in tissues, such as liver, kidney and placenta etc.

In liver insulin is degraded by reduction of disulphide bridges, that is catalyzed by the enzyme **glutathione-insulin transhydrogenase** and A and B polypeptide chains so separated are rapidly hydrolyzed by a specific protease enzyme into amino acids.

Biological functions of Insulin

Insulin has profound effects on metabolism of carbohydrates, lipids and proteins and it is the only adiabetogic hormone which lowers blood sugar level. It has overall anabolic and anticatabolic effects. Insulin promotes growth by stimulating protein synthesis.

Insulin is the major hormone that **decrease** blood sugar level by stimulating all glucose utilization processes and simultaneously it decreases glucose production processes. Major functions of insulin are described here.

Effects on Glucose Entry in Tissues

Insulin decreases BSL, when blood sugar level tend to rise, for instance after meals, by stimulating glucose uptake from blood into the tissue. The extrahepatic tissues, such as skeletal, cardiac and smooth muscles and adipose tissues are **impermeable to glucose** and are **dependent on insulin** to stimulate glucose entry inside the tissues. Insulin activates certain **glucose transporter systems** that facilitate glucose transport across the plasma membranes (Fig. 8.5) and glucose is taken inside for its metabolism. This transport of glucose is an **energy requiring active process.**

Effect on Glucose Utilization

Insulin increases the intracellular utilization of glucose by activating enzyme **hexokinase** which phosphorylates glucose to glucose 6 PO_4, that enters metabolic pathways. About 50% of glucose taken inside the tissues is oxidized for energy production, while 10% is converted into **glycogen** for the storage purpose. Rest 30 to 40% of the extra glucose, absorbed from GIT after digestion of food is utilized for fatty acid synthesis, lipogenesis and for synthesis of nonessential amino acids, needed for the protein biosynthesis.

Effects on Glucose Oxidation

Insulin stimulates **glycolysis** and **HMP shunt** pathways of glucose oxidation within seconds or minutes of its release from the β cells of pancreas. It activates the key enzymes of oxidative pathways directly by **covalent modifications** (Chapter 6). Also insulin **promotes (induces) synthesis** of these **enzymes** at **gene level** by stimulating transcription process. This takes some time and are referred to as long term effects of insulin.

Effects on Glycogen Synthesis

Insulin promotes synthesis of glycogen from glucose (glycogensis) by activating key enzyme, **glycogen synthase** and thus enhances glycogen storage in liver and muscles (Chapter 6).

Effects on Glucose Production

Insulin lowers BSL not only by promoting glucose utilization, but it simultaneously **inhibits** net glucose production and its release from the liver. Insulin **inhibits glycogenolysis** and **gluconeogenesis** rapidly within short time of its release by decreasing c-AMP concentration and activating enzyme **phosphatases** (Chapter 6.). Insulin thus inactivates key enzymes glycogen phosphorylase and fructose 1,6 bisphosphatase and thus reduces glucose production (Chapter 6). Insulin also suppresses the synthesis of key enzyme of gluconeogenesis at gene level (long term effect).

Effects on Lipid Metabolism

Insulin stimulates active **fatty acid synthesis** from glucose in liver and adipose tissues and thus favours **lipogenesis**. Insulin provides acetyl CoA by increasing glycolysis and reduced NADPH+H by stimutating HMP shunt needed for the fatty acid synthesis. Insulin activates **acetyl CoA carboxylase,** the key enzyme of fatty acid synthesis as well as it provides glycerol phosphate for the esterification of fatty acids to produce triglycerides, which are stored in adipose tissues. The adipose tissue is highly sensitive to the effects of insulin. to stimulate active lipogenesis.

Anti lipolytic effects—Insulin decreases mobilization of fatty acids from adipose tissue by **reducing lipolysis**. Insulin decreases c-AMP concentration and thus inhibits hormone sensitive lipase enzyme, thereby reducing lipolysis in adipose tissues.

Antiketogenic effect—Insulin decreases ketogenesis by inhibiting **HMG CoA synthase** enzyme. Insulin depresses ketogenesis also by promoting oxidation of acetyl CoA in the TCA cycle as well as by increasing its utilization for fatty acids synthesis. So less acetyl CoA is available for ketogenesis.

Effects on Protein Synthesis

Insulin promotes amino acids uptake by tissues and stimulates protein synthesis. It is an anabolic hormone and in general stimulates growth and protein synthesis by enhancing **transcription** and **gene expression**. It activates **RNA polymerase** and **initiation factors elf 4 A** of the transcription process to increase protein synthesis

Effect on DNA Replication

Insulin promotes DNA replication and cell multiplication in all mammalian cells. Insulin is an important growth factor and along with other insulin like growth factors (IGF-I and IGF-II), it is very important during pregnancy and fetal growth as it promotes cellular differentiation and fetal organogenesis.

A brief account of metabolic effects of Insulin is given in the Table 8.2.

Table 8.2 : Metabolic effects of Insulin

Metabolic processes	Net effect	Activation/Inactivation of key enzymes	Induction/Suppression of genes/key enzymes synthesis
Carbohydrate Metabolism	Activate glucose uptake	Activate glucose transporters	
1. Glucose uptake			
2. Glycolysis	Increased	↑ Phosphofructokinase ↑ Pyruvate kinase	↑ Glucokinase ↑ PFK-I ↑ Pyruvate kinase
3. Gluconeogenesis	Deceased	↓ c-AMP ↓ Fru. 1,6-Bisphosphatase	↓ Glu. 6 phosphatase ↓ Fru. 1,6 Bisphosphatase ↓ Pyruvate carboxylase ↓ PEP Carboxykinase
4. Glycogenesis	Increased	↑ Glycogen synthase	↑ Key enzyme synthesis
5. Glycogenolysis	Decreased	↓ c-AMP ↓ Glycogen phosphorylase	
6. HMP shunt	Increased	↑ Glu-6-PO_4 dehydrogenase	
Lipid metabolism			
7. Fatty acid synthesis	Increased	Provide acetyl CoA & NADPH +H^+	
8. Lipogenesis	Increased	↑ Acyl CoA carboxylase provide glycerol phosphate	
9. Lipolysis	Decreased	↓ c-AMP ↓ Hormone Senstive Lipase	
10. Ketogenesis	Decreased	↓ HMG CoA synthase	
Protein metabolism			
11. Protein synthesis	Increased	↑ RNA polymerase Activate Initiation factor eIF- 4A	↑ Gene expression ↑ Transcription

Mechanism of Insulin Action

Insulin being a protein hormone, it cannot permeate through plasma membrane of the cells of its target tissues. It brings about its wide range of biochemical and physiological actions inside the target cells by binding with **specific cell membranes receptors**. On binding with cell membrane receptors, insulin brings about its diversified actions inside the cells by inducing transmembrane signals which act on various levels of hormone action (Chapter 28).

Insulin receptors—Insulin receptor is a glycoprotein, consisting of 4 subunits in a tetrameric form of $\alpha_2\beta_2$ linked together by disulphide bridges (Fig. 8.5). The α subunits (mol. wt. 135 kDa) are entirely extracellular and have insulin binding domain. The β subunits (mol. wt. 95 kDa) are transmembrane proteins, exposed to cytoplasmic side as shown in the Fig. 8.5. The β subunit has **tyrosine kinase activity,** that is activated on binding of insulin on the α subunit and **transduce signals** inside the cells.

Signal transduction—When insulin binds on the specific domain of α subunit of the receptor, some conformational changes occur which activate the catalytic activity of tyrosine kinase of the β subunit. The activated tyrosine kinase then auto phosphorylates the tyrosine residues of β subunit proteins and also of some **insulin receptor substrates** (IRS 1-5) causing generation of transmembrane signals, which brings about diversified actions of insulin at various levels as illustrated in the Fig. 8.5.

Transduction of signals by insulin is brought about through a cascade of activation of several kinases. The activated tyrosine kinase autophosphorylates tyrosine residues of many insulin receptors substrates (IRS_{1-4}). Activation of IRS_1 is then involved in transduction of signals at DNA level to stimulate DNA replication, cell division and growth promoting effects of Insulin. While IRS_2 activation is involved in activation / inactivation by phosphorylation / dephosphorylation of key enzymes of various metabolic cycles (Chapter 6).

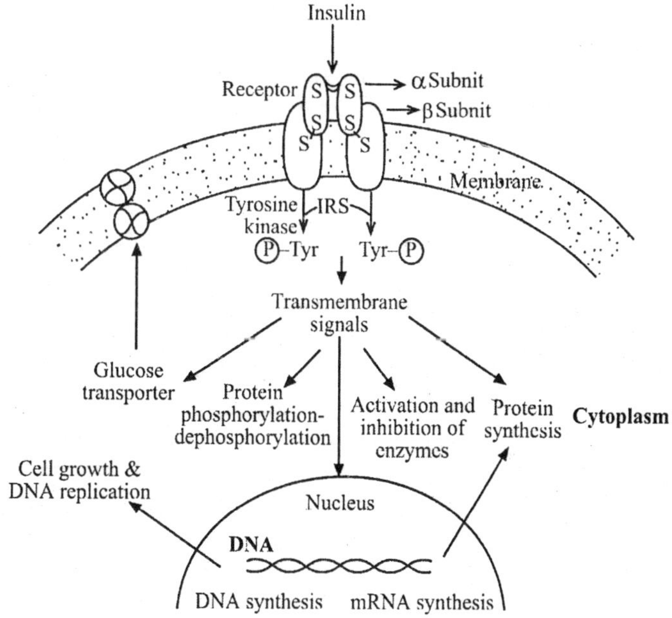

Fig. 8.5 : Diagrammatic representation of mechanism of insulin action via its binding on cell membrane receptor to transduce signals that modulates wide range of intracellular events

Activation of glucose transporters—The transduction of signals by tyrosine kinase of β subunit activate and translocate intracellular glucose transporters (carriers) towards plasma membrane, that bind with ECF glucose and facilitate its transport across plasma membrane to increase glucose uptake by the tissues. (Fig. 8.6).

There are normally 20,000 insulin receptors present on the plasma membrane of the target cell. The insulin receptors are constantly being synthesized and degraded. The half life of the insulin receptors is about 7-12 hours. When insulin binds to the receptor, several events occur which include—1) Some conformational changes occur in the receptors. 2) The receptors cross link and form microaggregates. 3) The insulin- receptor, microaggregates are internalized and. 4) Transduction of signals occurs to bring about effects of insulin inside the cells. 5) Insulin receptor complex is then degraded by lysozymes. Thus for some time after insulin action, the numbers of spare receptors on plasma membrane decreases, till the new receptors are synthesized.

Down Regulation—Insulin action inside the target cells depends on number of spare receptors on the cell membrane and their numbers is inversely proportional to the plasma insulin level. In condition of high plasma insulin level, as observed in obesity, acromegaly etc., the number of spare receptors are reduced very much and thus the tissues become less sensitive to the insulin action. This is known as down regulation.

The down regulation occurs due to the loss of spare receptors, as more insulin-receptor complexes are internalized and are degraded, when plasma insulin level is high. This explains the development of type II diabetes mellitus (NIDDM) in obese persons, as a result of down regulation and marked insensitivity to insulin action (insulin resistance) by the fat laden tissues.

GLUCAGON

Glucagon is a polypeptide hormone secreted from the α cells of islets of Langerhan's of pancreas. It is **antagonist** to **insulin action**, as it increases blood sugar level and has opposite effects on various metabolic processes to that of insulin. Glucagon is secreted from the α–cells directly in response to low blood sugar level (hypoglycaemia) and it **increases blood sugar level** by stimulating all glucose producing processes.

Chemical nature

Glucagon is **a protein hormone** and contains 29 amino acids in a **single polypeptide chain** (mol. wt. 3485). It is also synthesized as a large precursor molecule called proglucagon which on sequential proteolysis is converted into glucagon. Enteroglucagon, secreted from duodenal mucosa is similar to glucagon in its amino acid sequence and has similar physiogical and immunological properties to that of glucagon.

Glucagon secretion is stimulated by low blood sugar level, as observed in fasting conditions and is released in the blood. It circulates in plasma in free form (unbound to proteins) and thus its half life is very short, only 5 minutes. It is rapidly metabolised in liver by a specific enzyme, that hydrolyzes the first two amino acids from the amino terminal end and the hormone is inactivated.

Physiological Functions of Glucagon

Glucagon is anti insulin in nature and in general it favours the rapid mobilization of energy (lipids) and glucose from the storage depots, whereas insulin promotes energy storage by stimulating glycogenesis, lipogenesis and protein synthesis. Glucagon causes degradation of glycogen, fats and proteins and increases BSL to correct hypoglycaemic state. The main actions of glucagon are as described here.

Glucagon increases BSL—The main target organ for glucagon action is liver and it **increases hepatic glycogenolysis**. Glucagon binds to specific **receptors** on the plasma membrane of hepatic cells and activates membrane bound enzyme adenyl cyclase to produce c -AMP. The c-AMP so generated activates a protein kinase that stimulates the key enzyme **phosphorylase** of glycogenolysis (details in Chapter 6). The active phosphorylase a then causes **glycogenolysis** to liberate free glucose to be added to blood to elevate blood sugar level. Simultaneously the c-AMP-proteinkinase system decreases glycogenesis by inactivating glycogen synthase enzyme (details in the Chapter 6).

Glucagon increases gluconeogenesis—In liver glucagon enhances denovo synthesis of glucose from lactic acid and amino acids by activating key enzymes of gluconeogenesis through c-AMP-protein kinase activation. (Chapter 7, Fig. 7.5)

The net effect of glucagon on its target tissue liver is to increase glucose production, which is added to blood to increase blood sugar level and to correct hypoglycaemic state to prevent its adverse effects on tissues, particularly in brain.

Effects on lipolysis—Glucagon is a potent lipolytic hormone and stimulates lipolysis and mobilization of fatty acid from adipose tissues by activating **hormone sensitive lipase** through release of c-AMP. The free fatty acid level in plasma increases and they are preferably oxidized in extrahepatic and muscle tissues for energy production, thereby sparing glucose. The increased oxidation of fatty acids may lead to **ketogenesis.**

Effects on Gene expression—When BSL tend to decrease, as in fasting conditions, glucagon increases c-AMP concentration in the tissues by activating adenyl cyclase (detail in Chapter 28). The c-AMP so generated binds to certain **hormone sensitive elements on DNA** and induce transcription of genes to increase synthesis of key enzymes of gluconeogenesis, the phosphoenol pyruvate carboxykinase (PEPCK), fructose 1,6 bisphosphatase and glucose 6–phosphatase to enhance gluconeogenesis process. This takes some time and is referred to as long term affect of glucagons to increase BSL.

Insulin has opposite effects on these genes. It suppresses the synthesis of key enzymes of gluconeogenesis.

Glucose Tolerance

The ability of a person to metabolize give glucose load is referred to as **glucose tolerance.** Individual's response to utilize glucose is indicated by the nature of bloos glucose curve following administration of a test amount of glucose. Glucose tolerance decreases in all types of diabetes mellitus, in liver dysfunction, in some infections as well as in stressful conditions and in certain endocrinal disorder.

To assess the status of subject's ability to metabolize glucose properly, an oral glucose tolerance test (oral GTT) is done in a standardized manner.

Glucose Tolerance Test (GTT)

Normally glucose tolerance test is done to **diagnose** and **confirm diabetes mellitus** at an **early stage.** Glucose tolerance test is a well standardized test and is highly useful in the diagnosis of diabetes mellitus in doubtful cases.

However glucose tolerance test is rather an unphysiological test, as such large quantities of glucose normally does not enter systemic blood, so to evolve the carbohydrate tolerance ability of a person under physiological conditions, fasting and postprandial (after meals) blood sugar levels are analyzed.

Fasting and postprandial blood sugar levels—For this patient is asked to come after overnight fast and fasting blood and urine samples are collected next morning. The subject is then asked to take a heavy breakfast and blood and urine samples are collected at 1, 2 and 3 hours period. The blood samples are denoted as postprandial (after meals) samples and are analyzed for sugar levels. The peak value of blood sugar level is usually observed between 1-2 hours and it should be below 150 mg/dl in a normal healthy person indicating good glucose tolerance.

Oral Glucose Tolerance Test—Oral GTT is a well standardized test and is very useful to diagnose and confirm diabetes mellitus at an early stage in suspected cases.

Indications for performing are when patient comes with early signs and symptoms suggestive of diabetes mellitus (details in chapter 8), but his fasting (between 100-120 mg/dl) and two hours postprandial blood sugar levels (>150-170 mg/dl) are inconclusive. However GTT is not indicated in a persons with **confirmed diabetes mellitus,** where fasting BSL is >140 mg/dl and 2 hours postprandial values more than 200 mg/dl. Also GTT is not indicated in periodic checkups in the follow up cases of established diabetic patients.

Instructions to patients for GTT—As Indian diet is already carbohydrate rich diet, so patient is asked to come after overnight fast. He is instructed to discontinue drugs which are likely to influence blood sugar level at least two days prior to the test

Procedure—In morning patient is made to sit comfortably and fasting blood and urine samples are collected. The patient is then given 75g glucose (lg per kg body weight) in a glass of water, flavoured with some sharbet (to prevent vomiting) and is asked to drink it slowly. Blood and urine samples load are then and collected at 30 minutes interval for next 2½ hours, blood samples are analyzed for sugar concentration and urine samples are tested for the presence of sugar by Benedict's test. A graph is plotted with blood sugar values against time.

Interpretation of GTT—Glucose tolerance ability of the subject is then interpreted from the GTT curves obtained as illustrated in the Fig. 8.6.

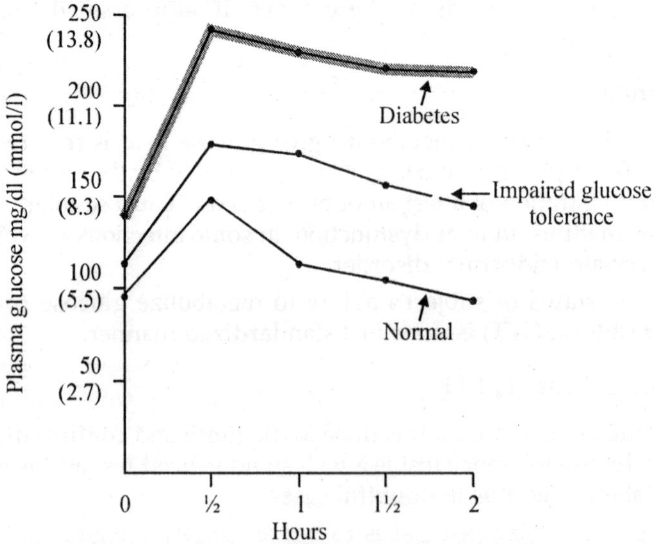

Fig. 8.6 : Oral glucose tolerance test

In normal healthy person, the fasting blood sugar level is around 70-110 mg/dl as seen from the Fig. 8.6. Following glucose load, the level rises and reaches to a peak value (<140 mg/dl) within 45 minutes to one hour and then starts declining and usually comes down to fasting levels by 2½ hours. This type of curve shows good glucose tolerance and is referred to as normal GTT (Fig. 8.6) Normal GTT curve is explained as follows: after giving glucose load, the blood sugar tend to rise to 130-140 mg/dl, that stimulates insulin secretion from j3-cells of islets of Langerhan's of pancreas, Insulin lowers blood sugar by facilitating glucose uptake and its utilization by tissues (Chapter 8) and thus in normal person BSL comes down to fasting level within 2½ hours. Glucose is not detected in any of the urine samples in normal person.

Impaired Glucose Tolerance—The subjects with impaired glucose tolerance have elevated blood glucose levels (>100-140 mg/dl) and 2 hours postprandial value between 140-200 mg/dl. Such persons are said to be with impaired glucose tolerance and need careful monitoring, as they are prone to develop diabetes mellitus within a year. These subjects should be followed up with periodical checkups and are advised to restrict refined carbohydrates in diet with more fibers and to take up regular physical exercise.

The 3rd GTT curve (Fig. 8.6) is that of a patient with **frank diabetes,** where fasting blood glucose level is more than 130-150 mg/dl and one hour peak value is very high (>200-240 mg/dl) and it remains high even after 2½ hours. Urine samples also **test positive** for the **presence of sugar.**

During pregnancy glucose tolerance is impaired sometimes and GYJ2 curve is slightly higher. GTI is also abnormal in case of gestational diabetes. Established diabetic patients should be monitored by periodical checkups with fasting and 2 hours postprandial blood glucose estimations (mini GTT) and with frequent sugar test in urine. For conducting G'JT in children, oral glucose load is given according to body weight (as 1.5 to 1.75 g/kg).

Intravenous GTT—It is done in certain case with suspected malabsorption. In this test, 25g (0.5 g/kg wt) glucose dissolved in sterile distilled water is administered intravenously very slowly (in 5-10 minutes) and blood samples are collected every 15 minutes interval for one hour. Peak blood sugar value is obtained with in 20-30 minutes (>200-250 g/dl) and comes down to 100 mg/dl in 45 to 60 minutes. Normally intravenous GTT is not conducted to avoid complications.

Corticosteroid Stressed GTT (CGTT)—Persons with latent diabetes shows almost normal glucose tolerance and a normal curve is obtained on conducting GTT but they may have some symptoms and complaints like frequent urination with polyuria, weakness, poor wound healing etc. In such cases, the corticosteroid stressed GTT is done to detect latent diabetes mellitus. For conducting CGTT, an oral dose of cortisone (100 mg) in two divided doses at 8 hours and 2 hours prior to performing oral GTT is given to decrease the glucose tolerance of the subject. Now when standard oral GTT is performed (after cortisone stress), a highly impaired abnormal diabetic like curve is obtained confirming **latent diabetes.** In normal healthy person CGTT curve is either normal or slightly higher.

The patients with latent diabetes (with abnormal CGTT) are prone to develop frank diabetes mellitus within a short time. Such person should be monitored regularly and are advised to change their life style, with regular physical exercise and dietary restrictions as mentioned above.

DIABETES MELLITUS

Diabetes Mellitus is a common metabolic disease caused by the insufficiency of hormone insulin or due to the tissue resistance against insulin action. The disease is characterized by high blood sugar concentration (hyperglycaemia) and glycosuria. Diabetes mellitus is known since ancient times. Charaka (Circo 400BC) described the disease with elaborate clinical description as **Madhumeha,** which can not be surpassed even now. Roman physicians coined the term diabetes mellitus from the Greek words diabetes meaning syphoning through urine and mellitus related to sugar.

Diabetes mellitus is a disease which **cannot be cured** completely and person has to live with it throughout his life. This disease can only be controlled and managed with dietary and life style changes and with hypoglycaemic drugs taken daily by the diabetic patient.

Diabetes is a major cause of mortility and morbidity and it is the most prevalent disease in the world. Approximately 10% of the world population and $1/5^{th}$ of the persons above 45 years of age suffer from diabetes mellitus. The disease is the major cause of mortility because of the serious clinical complications associated with diabetes, that can be life threatening.

Diabetes-A Silent Killer Disease

Diabetes is called a silent killer disease, because it can progress undetected and uncontrolled to serious clinical complications. This is because the initial clinical symptoms are not severe, and patient may not know that he has diabetes, unless he goes for blood test for some other medical problem. Thus if not detected early, uncontrolled diabetes with persistent hyperglycaemia can cause irreversible and severe clinical complications, that can be fatal. The clinical complications of diabetes mellitus affect cardiovascular system, eyes, kidneys and nervous system and may lead to heart attack, blindness, renal failure, stroke etc. resulting in death of the patient.

Types of Diabetes Mellitus

In modern clinical medicine, diabetes mellitus is classified into two major types:

1. Type I – Insulin dependent diabetes mellitus (IDDM)—It is severe disease, but comparatively rare. Only 5 to 10% of diabetic patients belong to this type and can only be managed by giving insulin injections daily.

2. Type II – Non insulin dependent diabetes mellitus (NIDDM)—This type is most common diabetes and about 90% diabetics have non insulin dependent diabetes mellitus. NIDDM patients are usually obese, above 40 year of the age and they have elevated plasma insulin levels. This down regulates insulin receptors causing tissue resistance against insulin action. This type of diabetes mellitus is also known as **maturity onset diabetes mellitus.**

The salient features of the two types of diabetes mellitus are summarized in the Table 8.3.

Table 8.3 : Comparison of salient features of two types of diabetes mellitus

Type I – IDDM	Type II- NIDDM
1. **Cause**—Lack of insulin due to β cell damage.	Marked tissue **insensitivity** to insulin action.
2. Rare (5 to 10%), but **severe** type	Common (90% of diabetes cases) but **mild** type
3. Develops at an early age and known as **juvenile** or **adolescent diabetes.**	**Maturity onset diabetes** occurs above 40-45 years of age. Obese persons more prone.
4. **Management** difficult, controlled by insulin injections daily, subcutaneously	Easily managed by diet & life style changes, regular exercise and by **oral hypoglycaemic drugs.**
5. **Ketosis** is frequently associated	Usually non ketotic if properly controlled
6. Predisposition - heriditary	Heriditary
7. Pathological or autoimmune destruction of β cells.	Caused by resistance or antibodies against insulin

Metabolic Abnormalities

Insulin deficiency cause major changes in metabolism of carbohydrates, lipids and proteins resulting in various metabolic abnormalities as described here.

Hyperglycaemia is the primary manifestation of diabetes mellitus. Insulin insufficiency or lack of its action results in elevation of blood sugar levels due to—1) decreased entry of glucose into the cells, 2) decreased utilization of glucose for metabolic purposes and 3) increased production of glucose due to increased rate of glycogenolysis and gluconeogenesis, as a result of unopposed action of diabetogenic hormones, such as glucagon, glucocorticoids, growth hormone, epinephrine etc.

Glycosuria occurs when blood glucose level is very high and exceeds the renal threshold level i.e. 180 mg/ dl, at which the renal tubular reabsorption capacity is saturated and the extra glucose is excreted in the urine. The condition is known as **glucosuria.**

Hyperlipidemia—In deficiency of insulin, glucose is metabolically not available (though present in excess in blood) for energy production, so fats are mobilized from the adipose tissues and fatty acids are oxidized in excess to meet energy demands of the body. The net effect **is hyperlipidemia** i.e. increased plasma levels of free fatty acid, increased **triglyceride** levels and **hypercholesterolemia.**

Ketogenesis—Increased fatty acid oxidation results in accumulation of acetyl CoA, because it cannot be completely oxidized in TCA cycle, where availability of oxaloacetate becomes **rate limiting**. Increased rate of gluconeogenesis in diabetes mellitus results in **depletion** of **intermediates of TCA cycle** and oxaloacetate production is also reduced from pyruvate (decreased glycolysis). Therefore the excess of acetyl CoA, produced by fatty acid oxidation is diverted towards ketone bodies and cholesterol production leading to **ketosis** and **hypercholesterolemia,** causing adverse effects with severe clinical manifestation (details along with the clinical complications of diabetes).

Protein catabolism increases in diabetes mellitus mainly to provide amino acids for elevated rate of gluconeogenesis process. Increased tissue protein breakdown results in muscle wasting. The metabolic derangements due to insulin deficiency or due to the lack of its action in diabetes mellitus are summarized in Fig. 8.7.

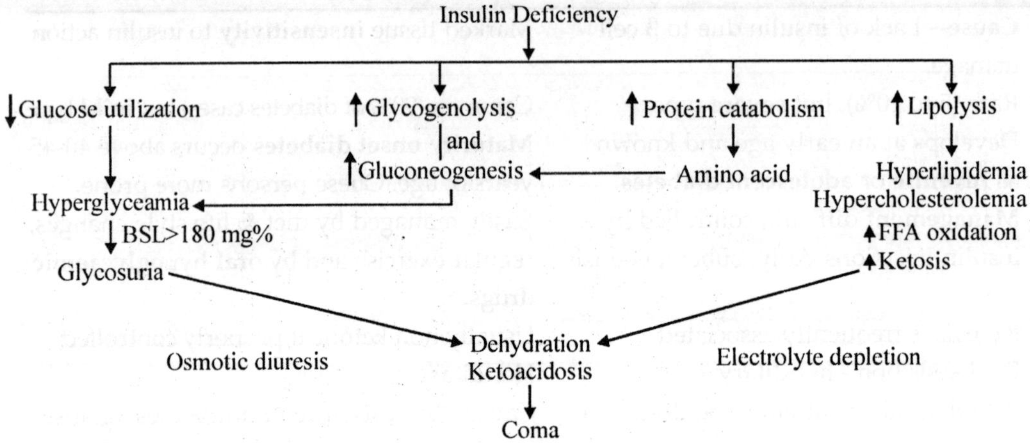

Fig. 8.7 : Metabolic alterations occuring due to insufficiency of insulin action in Diabetes mellitus

Clinical Manifestations

Initial clinical signs and symptoms of diabetes mellitus include—

- Polyuria with increased frequency of urination
- Sudden weight loss inspite of consumption of adequate diet.
- Polyphagia and polydypsia are common.
- Weakness, tiredness, bodyache and backaches are frequent complaints.
- Recurrent skin infections such as boils, cellulites, abscesses and poor wound healing.

These symptoms can be explained as—When blood sugar level exceeds renal threshold, extra glucose is excreted in urine causing **glycosuria** that is accompanied with osmotic diuresis and obligatory water loss in urine resulting in **polyuria.** This results in hyperosmolality of plasma and thirst centres of hypothalamus are stimulated leading to **polydypsia.**

Due to insulin deficiency glucose is metabolically unavailable, so fats are mobilized from adipose tissue and tissue proteins are excessively catabolized to provide amino acids for gluconeogenesis resulting in **weight loss.** To compensate excessive loss of glucose in urine, the patient has tendency to take more food i.e. **polyphagia.**

The recurrent skin infection and poor wound healing in diabetic patients is probably because, high ECF glucose provide good nutrition to the bacteria for their multiplication.

Diagnosis of Diabetes Mellitus

Diabetes is diagnosed by following test—1) Fasting and post prandial blood sugar analysis 2) Urine sugar test is done to detect the presence of glucose and 3) diabetes is confirmed by glucose tolerance test (details given before in this section).

Early detection and **management** of diabetes mellitus is necessary to prevent clinical complications, that can be life threatening. Persistant hyperglycaemia in uncontrolled and chronic diabetes causes **irreversible damage** to the body tissues, resulting in severe clinical complications.

Clinical Complications of Diabetes Mellitus

Uncontrolled hyperglycaemia in diabetic patient is very harmful and results in severe clinical complications, that are described here. The metabolic derangement in the diabetic patients may lead to following adverse consequences:

1. Hyperlipedimia and **hypercholesterolemia** in diabetes with elevated **LDL cholesterol** cause **vascular complication.**

2. Atherosclerosis in medium blood vessels—Atherosclerosis in blood vessels develops due to the deposition of cholesterol (LDL), triglycerides etc. forming atherosclerotic plaques in the intima of the arterial walls. This results in hardening of walls and narrowing of lumen of the blood vessels which lead to elevation of blood pressure and **hypertension.** Atherosclerosis and consequent intravascular thrombosis may block the medium sized blood vessels causing vascular complication with severe clinical consequences, such as :

- **Coronary artery blockage** leads to **myocardial infarction** and cardiovascular disease which may result in heart attack.
- **Cerebral shock** may result due to atherosclesrosis and thrombosis in cerebral artery blocking blood supply to some parts of the brain, which may lead to **paralytic attack.**
- **Diabetic foot**—A block in peripheral blood vessels causes impaired blood supply to the lower limbs resulting in ischaemia which may lead **to gangrene** (diabetic foot) and the leg has to be **amputed** to save the life of the patient.

3. Microangiopathy : **Microaneurysm**—Persistant hyperglycaemia in uncontrolled diabetes mellitus adversely damage small blood vessels and capillaries causing **microangiopathy**. The **endothelial** and **mural** (cement) cells of connective tissue of the walls of blood capillaries are damaged due to **glycosylation of proteins** (details later in the section) and by **deposition of amyloglycans** resulting in loss of cementing abilities of the mural cells. Due to this, the capillary walls become fragile and blood exudes into the surrounding tissue i.e. hemorrhages causing **microaneurysms** which leads to diabetic complications in retina of eye, kidneys and CNS as described below.

Retionopathy—Haemorrhage in the retina of eye severely affects vision and may result in blindness.

Nephropathy and **nephrosclerosis**—This results due to **glycosylation** of basement membrane proteins of renal tissues resulting in impairment of renal function. This results in **microalbuminuria** and may lead to renal failure.

Neuropathy—Peripheral neuropathy with paresthesia is very common clinical complication of diabetes. Hyperglycaemia leads to reduction of glucose to **sorbitol** and its accumulation in the **schwann cells** of nervous tissue is believed to be the main cause of neuropathy (Polyol mechanism)

Cataract—Early cataract development of eye lens in diabetics is also due to accumulation of sorbitol. Eye lens is freely permeable to glucose, where it gets reduced to sorbitol which is impermeable and thus accumulates in the lens. Sorbitol is hygroscopic and causes disturbances of water and electrolyte balance in eye lens and inhibits Na^+/K^+ ATPase resulting in lens becoming opaque and cataract develops.

Diabetic Ketoacidosis

Ketosis is more common in uncontrolled, type I insulin dependent diabetes mellitus (IDDM) and in chronic diabetic state. Ketosis is a serious clinical complication of diabetes mellitus and cause severe adverse effects that can be life threatening. The adverse consequences of diabetic ketoacidosis are—1) Acidosis with lowering of ECF and ICF pH, due to the loss of alkali reserve, 2) Electrolyte depletion and **dehydration,** 3) **Hyperventillation** and **kussmaul breathing** to compensate acidosis, 4) **acetone breath** with characteristic fruity odour and 5) Diabetic ketoacidsis if not treated and managed promptly, can lead to fatal consequences. In severe conditions patient goes into coma and may result in death (details in Chapter 35).

Management of Diabetes Mellitus

Early diagnosis and management of diabetes is essential to prevent irreversible clinical complications. Prompt and proper control of blood sugar level of diabetic patient is the most important aspect to manage diabetese, to prevent clinical complications, so that patient can lead almost symptom free healthy life. Approximately 50% of mild, type II NIDDM diabetes in early stage can be controlled by strict dietary and life style changes.

Dietary Management—A diabetic patient is advised to consume low carbohydrate (refined), low fat but with enough **good quality proteins** and **fiber rich** diet. Less amount of saturated fats but more intake **of mono** and **polyunsaturated fatty acids** in a proper ratio (Chapter 12 & 13) is good for diabetics to lower blood cholesterol level (low LDL) to prevent risk of vascular clinical complications.

Regular exersize and weight control is important for diabetic patients as **sedentary (inactive) life style** aggravates clinical complications. Daily exercise, as advised by doctor is beneficial in lowering blood sugar and cholesterol level to prevent risk of heart diseases.

Hypoglycaemic drugs—Oral hypoglycaemic drugs when taken daily (as advised by physician) control hyperglycaemia in type II NIDDM patients. Commonly used oral hypoglycaemic drugs are sulphonylurea and biguanide drugs. The sulphonylurea drugs such as **tolbutamide** and **glibenclamide** stimulate insulin secretion. New generation potent hypoglycaemic drugs, with least side effects are now available commercially to manage diabetes in clinical practice.

Type I – IDDM is managed by Insulin injections—Pure synthetic human insulin is now available with least **immunoantigenic effects** and can safely be used clinically to control hyperglycaemia in IDDM patients. Insulin injections are given **subcutaneously** before meals to control post prandial rise in BSL and also to control blood sugar level for 24 hours. The insulin injections are available in different units (doses) and should be taken as prescribed by a doctor.

Biochemical Indices in Monitoring Diabetes

Periodical monitoring of diabetic patient to check blood sugar control is the most important aspect of prognosis to prevent irreversible clinical complications. The important biochemical and clinical indices in monitoring diabetic patient include:

1. Fasting and **postprandial** blood sugar levels and **urine sugar** should be assessed regularly (every 3 months).

2. Glycosylated hemoglobin—In recent years glycosylated hemoglobin levels in blood is considered as a better parameter of long term monitoring of BSL in diabetic patients. Glycosylated Hb estimation gives more reliable and accurate assessment of BSL control. This is because glycated Hb level remains unchanged for long period inspite of variations in dietary intake of sweets and carbohydrates by diabetes patient. Hyperglycaemia in diabetics cause glycosylation of body proteins including hemoglobin by a non enzymatic process which may alter biological properties of proteins of body as mentioned earlier in clinical complications of diabetes.

Glycosylation of proteins takes place by non enzymatic reaction between carbonyl group of glucose forming first a schiff's base with the N-terminal amino group of proteins to produce an aldimine which then reverts to ketimine linkage. The glycosylation is an irreversible process and once glycated, the glycosylated Hb remains unchanged in the intact RBCs for at least 6 to 8 weeks till the erythrocytes complete their life span and are destroyed. Thus glycated Hb level remains unchanged inspite of daily alteration in dietary carbohydrate consumption.

Normally the level of glycosylated Hb is about 3-5%. But in diabetic patients the glycated Hb levels are raised to 6-15% or more depending on BSL status and remain elevated for 8 to 10 weeks. Thus its estimation is a better index in long term monitoring of diabetic control.

3. Serum lipid profile—Total blood cholesterol, LDL, HDL and triglyceride levels in plasma should be analyzed periodically to assess diabetic control as their elevated levels are **high risk factors** for the development of vascular clinical complications and heart diseases.

4. Assessment of renal functions—The kidney functions of diabetic patients should be assessed periodically by estimation of blood urea and serum creatinine, at least every six months. **Microablumiuria** should be checked periodically. It is very significant clinically as it gives the earliest indication of renal function impairment in diabetic patients. (Chapter 36.)

5. Eye fundus examination—The eye fundus should be examined periodically in diabetics to assess any damage to the retina of eye to screen out microaneurysms (retinopathy) and haemorrhages in the eye tissues, that can cause blindness.

HYPERINSULINISM

Insulin secreting tumors or overdose of insulin injection may result in **hypoglycaemic shock** (BSL less than 40-50 mg/dl) which can be life threatening. In hypoglycaemia, if not corrected promptly, the patients may go into coma with fatal consequences. Prolonged hypoglycaemia (severe) may cause irreversible brain and CNS damage and even death.

Management is done by immediate intravenous glucose administration. The differentiation between **hypoglycaemic coma** from **hyperglycaemic shock** (BSL>700-1200 mg/dl) is important since **clinical symptoms are similar but treatment is exactly opposite**. Severe hyperglycaemia can occur in uncontrolled diabetes mellitus or on the face of glucose load in severe diabetes. The differential diagnosis is done mainly by blood sugar status.

❑❑❑

9

LIPIDS OF PHYSIOLOGICAL IMPORTANCE

Lipids are a heterogeneous group of compounds, that are characterized by their insolubility in water but they are soluble in organic solvents such as alcohol, ether, chloroform etc. Chemically lipids are esters of higher fatty acids with certain alcohols.

DIETARY LIPIDS—IMPORTANCE

Fats and oils are the main cooking medium of our food, that makes it tasty and palatable and thus are important constituents of our diet. They are of high calorific value (9 Cal /gm) and are the main dietary source of fat soluble vitamins A, D, E and K, as well as of essential fatty acids, that are required essentially in our diet to maintain good health and growth.

FUNCTIONS OF LIPIDS IN BODY

1) Lipids are structural components of biomembranes, present as lipid bilayer, that is responsible for maintaining their characteristic semipermeability and structural integrity.
2) Lipids are present abundantly in white and grey matter of the brain and in myelin sheath of nerves and act as insulators.
3) Lipids can be stored as high energy reserves in adipose tissues to an unlimited extent as reserve energy source.
4) Fatty acids are preferably oxidized for energy production purposes in most of the body tissues.
5) Lipids provide cushioning effect (pads of adipose tissue) to protect internal organs of the body against thermal and mechanical trauma.

CLASSIFICATION OF LIPIDS

Lipids are classified according to their chemical nature into three major groups as— a) Simple lipids, b) Compound lipids and c) Derived lipids. A brief account of major classes and subclasses with important examples is given in the Fig 9.1.

Simple Lipids—Simple lipids are esters of fatty acids with various alcohols. Important examples are given here under—

(i) **Triglycerides**—The triglycerides are neutral fats and oils, which are chemically triesters of glycerol with higher fatty acids.

(ii) **Waxes**—Waxes are esters of fatty acids with some higher monohydric alcohols.

Compound Lipids—In addition to esters of fatty acids with certain alcohols, compound lipids contain some other groups or radicals. Examples are as given below—

1. **Phospholipids**—Phospholipids contain phosphate and usually a nitrogenous base in addition to fatty acids and alcohol.
2. **Glycolipids**—Glycolipids contain carbohydrate unit in addition to fatty acid and alcohol.
3. **Lipoproteins**—Lipids are transported via blood to tissues in association with proteins as lipoprotein complexes.

Derived Lipids—Derived lipids are the substances obtained during hydrolysis of lipids, for example fatty acids, glycerol, cholesterol etc. Some of them still retain the properties of lipids such as they are insoluble in water but are soluble in fat solvents.

Lipids

- Simple Lipids
  - Triglycerides e.g.–Fats & Oils
  - Waxes
- Compound Lipids
  - Phospholipids
    - Lecithin
    - Cephalin
    - Sphingomyelins
  - Lipoproteins
    - VLDL
    - LDL
    - HDL
    - Chylomicrons
  - Glycolipids
    - Cerebrosides
    - Gangliosides
- Derived Lipids
  - Fatty Acids
  - Glycerol
  - Cholesterol

Fig. 9.1 : Classification of lipids in three major classes with important examples

FATTY ACIDS

Fatty acids occur in nature mainly as esters with certain alcohols, as simple and compound lipids. Chemically the fatty acids are aliphatic straight chain derivatives of monocarboxylic acids and their C-chain range in length from C_4 to C_{24}. Fatty acids occurring in natural fats are of two types.

(i) Saturated fatty acids and

(ii) Unsaturated fatty acids.

Nomenclature

Fatty acids are named after the corresponding hydrocarbon chain. The carbon atoms are numbered from the carboxylic group (- COOH) as α, β, γ and so on.

Saturated Fatty Acids

The naturally occurring saturated fatty acids in lipids are mostly with even number carbon atoms and have the general formula as CnH_2nO_2. A brief account of natural saturated fatty acids, with even C-atoms along with their chemical formulas and sources is given in the Table 9.1.

Table 9.1 : Saturated fatty acids occurring in natural fats and oils

S. No.	Fatty Acid	No. of C - atoms	Formula	Sources
1	Butyric acid	4	$CH_3(CH_2)_2COOH$	Butter, Ghee
2	Caproic acid	6	$CH_3(CH_2)_4COOH$	Butter, Vegetable oils
3	Caprylic acid	8	$CH_3(CH_2)_6COOH$	Butter, Vegetable oils
4	Capric acid	10	$CH_3(CH_2)_8COOH$	Butter, Vegetable oils
5	Lauric acid	12	$CH_3(CH_2)_{10}COOH$	Coconut oil
6	Myristic acid	14	$CH_3(CH_2)_{12}COOH$	Coconut oil
7	Palmitic acid	16	$CH_3(CH_2)_{14}COOH$	Most vegetable oils & animal fats
8	Stearic acid	18	$CH_3(CH_2)_{16}COOH$	Most vegetable oils & animal fats
9	Arachidic acid	20	$CH_3(CH_2)_{18}COOH$	Peanut oil

Unsaturated Fatty Acids

The unsaturated fatty acids have one or more than one double bonds in their hydrocarbon chains. Unsaturated fatty acid with one double bond is named as monounsaturated fatty acid, while polyunsaturated fatty acids contain two or more than two double bonds.

Monounsaturated fatty acid with 16 C- atoms is known as palmitoleic acid. It is the most abundant fatty acid in nature and is found to occur in almost all natural fats and oils. Its chemical formula is as given under—

$$[CH_3 - (CH_2)_5 - CH = CH - (CH_2)_7\ COOH]$$

Polyunsaturated fatty acids (PUFA)—The polyunsaturated fatty acids contain more than one double bond in their hydrocarbon chains. They are also called as **essential fatty**

acids, as they cannot be synthesized in our body and have to be taken in diet to maintain normal health and growth. The insufficient intake of PUFA in diet may result in deficiency disease conditions. Three polyunsaturated fatty acids found to occur commonly in natural fats and oils are—a) Linoleic acid, b) Linolenic acid and c) Arachidonic acid. The chemical nature, important sources and characteristics of these polyunsaturated fatty acids are summarized in the Table 9.2.

Table 9.2 : Polyunsaturated fatty acids with their chemical formulas and characterstics

S. No.	Unsaturated fatty acid	Number of C-atoms	Number of double bonds/ location (Δ)	Sources
1	Linoleic acid	18	Two / ω_6 series $C_{18}\,\Delta_{9,\,12}$	Peanut oil, soybean oil, corn & cottonseed oil.
2.	Linolenic acid	18	Three / ω_3 series $C_{18}\,\Delta_{9,\,12,\,15}$	Linseed oil, soybean oil & fish liver oil
3	Arachidonic acid	20	Four / ω_6 series $C_{20}\,\Delta_{5,\,8,\,11,\,14}$	Peanut oil

Chemical Characteristics of PUFA

1. Unsaturated fatty acids exhibit geometric isomerism and naturally occurring PUFA in natural fats and oils mostly have Cis-configuration.

2. Unsaturated fatty acids are also designated as ω_3 or ω_6 series fatty acids. For instance – linolenic acid is ω_3 fatty acid, in which the double bond is present on 3rd C-atom starting from- CH_3 end while linoleic and arachidonic acid are ω_6 fatty acids with double bond present on 6th C-atom starting from -CH_3 end.

Importance of Linolenic acid—The linolenic acid is the most important PUFA in our diet, since it is the precursor from which arachidonic acid can be synthesized in our body. The linolenic acid is found to occur in certain oils and fish liver oil (Table 9.2).

It is the most efficient PUFA to lower blood cholesterol level and thus is considered to be protective against cardiovascular diseases. Low incidence of hypercholesterolemia and heart diseases observed in fish eating population is believed to be due to the ω_3 linolenic acid in their diet.

Important Functions of Essential Fatty Acids (PUFA)

1. Polyunsaturated fatty acids are **essential constituent** of **lipid bilayer** of cell membranes and mitochondrial membranes and maintain the structural integrity of biomembranes.
2. PUFA are **essential constituents** of **phospholipids** and maintain the functional **integrity** of **mucosal linings** of germinal epithelium of reproductive organs. Deficiency of PUFA in diet of experimental animals results in infertility.
3. PUFA is involved in the **metabolism of cholesterol**, as HDL specifically scavenges cholesterol esterified with PUFA from the tissues and transport it back to liver for its oxidation to bile acids and excretion via bile into GIT and in the feces. Thus PUFA lowers the blood cholesterol level and should be taken in diet to prevent cardiovascular diseases and hypertension (Chapter 12.).

4. Arachidonic acid is the precursor of potent physiological compounds such as **prostaglandins** and **prostacyclins** which virtually occur in all mammalian tissues. They function as local hormones and act as potent smooth muscle contractors, vasodialators and inhibit platelet aggregation. The prostaglandins are also used as pharmacological agents (drugs) in certain disease conditions.
5. **Thromboxanes** are derived from PUFA (20 – C polyenoic fatty acids) and play important role in platelet aggregation during blood coagulation and cause vasoconstriction at the site of injury to prevent blood loss.
6. PUFA is required essentially for phospholipid and lipoprotein VLDL synthesis and plays central role in distribution of lipids from liver to the tissues. Thus they prevent accumulation of fats in liver, which otherwise may lead to the development of fatty liver (Lipotropic factor).

Deficiency of PUFA—Clinical Aspects

Deficiency of PUFA in diet may cause impairment of normal health and results in severe clinical manifestations. Some important deficiency manifestations of PUFA are listed below—

A) Fatty liver resulting in cirrhosis of liver and jaundice.
B) Skin lesions, dermatitis, eczema etc.
C) Degenerative changes in tissues.
D) Damage to epithelial and mucosal lining of reproductive tract resulting in sterility in male and infertility in female experimental animals.

The evidence of the role of essential fatty acid (EFA) in epidermal proliferation was provided by animal experiments, where deficiency of EFA resulted in severe scaling of skin and loss of hairs. Omega 6-fatty acid deficiency is the major cause of skin symptoms. The omega 3-fatty acid present in fish liver oil decreases plasma lipoprotein levels, especially VLDL and LDL cholesterol and thus prevents risk of cardiovascular diseases.

Eicosanoids—Prostaglandins

Prostaglandins and structurally related prostacyclins, thromboxanes and leukotrienes are all eicosanoids because they contain 20 C-atoms (Greek eikas = 20). They are potent physiologically active compounds and act as local hormones. They are synthesized from common precursor arachidonic acid by cyclization (a pentane ring is introduced) in the center of the fatty acid molecule with certain other modifications. The eicosanoids have several diversified physiological functions in the body.

The prostaglandins stimulate inflammation, modulate synaptic transmission and are potent smooth muscle contractors. Drug aspirin decrease inflammation, pain and fever by inhibiting synthesis of prostaglandins, prostacyclins and thromboxanes in the body. Aspirin also decreases intravascular thrombus formation by inhibiting thromboxane synthesis, there by decreasing platelet aggregation.

Glycerol

Glycerol is derived from hydrolysis of neutral fats and oils. It is a trihydroxy alcohol with chemical structure as given below—

Commercially glycerol is a by product of soap industry and is also known as glycerine. It is used widely in the preparation of cosmetics. In body, glycerol is produced by lipolysis, i.e. mobilization of fats from adipose tissue and is released in blood. Glycerol from blood is taken up by liver and is converted into glucose by the process of gluconeogenesis (Chapter 7).

$CH_2 - OH$
|
$CH - OH$
|
$CH_2 - OH$

Glycerol

Cholesterol

Cholesterol is widely distributed in animal kingdom. It is an important steroid present in all cells of our body but particularly it is present in rich amounts in brain, nerves, liver and other tissues, in free as well as esterified with fatty acids.

Chemically it belongs to steroid compounds, which have a complex ring structure called cyclopentano perhydrophenanthrene nucleus or steroid ring structure as shown in the Fig. 9.2.

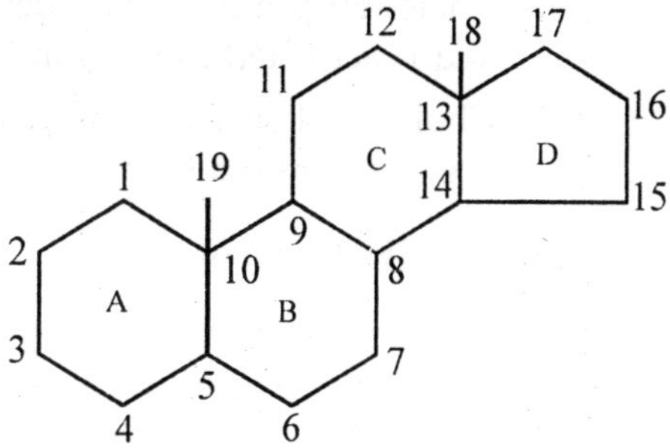

Fig. 9.2 : Steroid ring structure

The steroid ring system contains three cyclohexane rings desingated as A, B and C and one cyclopentane ring D. Two methyl groups are present on C-atom 10 and 13 and the steroid ring contains a total of 19 C-atoms as shown in the Fig. 9.2.

Cholesterol—Chemical Structure

Cholesterol has a total of 27 carbon atoms with a steroid ring and a side chain of C-8 hydrocarbons present at $C_{17.}$ The cholesterol molecule has a double bond in ring B between C_5 and C_6 and has one – OH group at ring A on C_3, thus it is designated as 3–hydroxy 5, 6-cholestene (Fig. 9.3).

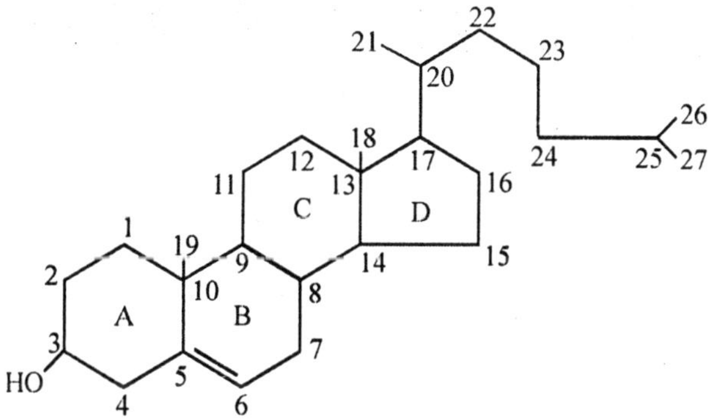

Fig. 9.3 : Chemical structure of cholesterol

Importance of Cholesterol—Biological Fuctions

Cholesterol is an important component of our body fluids and tissues and performs several important physiological functions, that are summarized here under—

1. Cholesterol is the important constituent of all biomembranes of the body tissues.

2. It is the precursor of all steroid hormones synthesized in adrenal cortex such as glucocorticoids, mineralocorticoids etc.
3. Cholesterol is the precursor of male sex hormones (testosterone) as well as of female sex hormones such as progesterone, estrogens etc.
4. Cholesterol in liver is oxidized to synthesize bile acids and bile salts, which are secreted in bile into GIT. Bile salts play important role in emulsification and digestion of fats and their absorption from GIT as micelles.
5. Cholesterol derivative 7 – dehydrocholesterol is the precursor (provitamin – D) and is activated as vitamin D in skin in the presence of sunlight.
6. Cholesterol serves as important insulator in myelin sheath of nerves and prevents leakage of electric nerve impulses.

Properties of Cholesterol

1. Cholesterol is whitish yellow crystalline solid, insoluble in water, but is soluble in organic solvents such as ether, chloroform etc.
2. The –OH group on 3rd C-atom of cholesterol (A ring) forms ester with higher fatty acids in our body, preferably with PUFA, that is obtained from β C-atom of phospholipid lecithin. The reaction is catalyzed by the enzyme lecithin-cholesterol – acyl transferase (LCAT), present on HDL fraction of plasma lipoproteins (details in the Chapter 13.)

Blood cholesterol level

Normal blood cholesterol level is maintained within physiological limits of 150 to 220 mg/dl.Maintenance of blood cholesterol level within normal range is of prime importance, since hypercholesterolemia is a high risk factor implicated in the development of atherosclerosis and cardiovascular diseases.

Hypercholesterolemia and Clinical Manifestations

Hypercholesterolemia (high blood cholesterol, >250 mg%) develops secondary to diseases like diabetes mellitus, hypothyroidism, nephrotic syndrome, obstructive jaundice etc, while the primary hypercholesterolemia occurs in certain inherited genetic diseases (Chapter 13.).

High blood cholesterol level is major risk factor for the development of atherosclerosis, that is characterised by the deposition of cholesterol, TAG etc. in the walls of blood vessels and results in cardiovascular diseases and hypertension (high blood pressure).

Cholesterol is transported in blood as lipoprotein fractions. LDL is cholesterol rich lipoprotein and has affinity (tendency) to get deposited in the arterial walls and is thus considered as **bad cholesterol** lipoprotein. While HDL is considered as **good cholesterol** lipoprotein, as it scavenges (removes) cholesterol from the tissues and blood and bring it back to liver for oxidation and excretion via bile in GIT, thereby lowering blood cholesterol level. Thus HDL is protective and reduces the risk of heart diseases.

SIMPLE LIPIDS

The simple lipids include mainly neutral fats, oils and waxes and are chemically esters of fatty acids with certain alcohols.

A. Neutral Fats and Oils

Neutral fats are triacylglycerols (TAG) or triglycerides and are chemically esters of trihydric alcohol glycerol, with 3 molecules of fatty acids. The triglycerides are synthesized by esterification of glycerol with fatty acids as shown in the reaction given below—

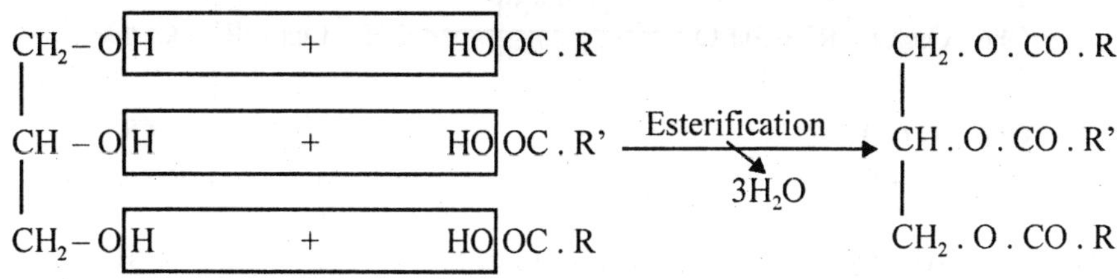

Glycerol **Fatty Acids** **Triglyceride**

Characteristics of Triglycerides

(1) Triacylglycerols (TAG) are the major high energy reserve storage form of lipids present in adipose tissues of our body. Mixed triglycerides, in which three different fatty acid are esterified with 3-OH groups of glycerol, are of common occurrence in natural fats and oils. In general 1st and 3rd OH group of glycerol is esterified with saturated C_{16} or C_{18} fatty acid (palmatic or stearic acid), while 2nd OH is preferably esterified with unsaturated fatty acid (PUFA). Examples are vegetable oils such as peanut oil, linseed oil, soybean oil, corn oil and fish liver oil etc.

(2) Vegetable oils are generally liquid at room temperature and have higher proportions of unsaturated fatty acids, whereas animal fats are solid and have more saturated short, medium and long chain fatty acids, for example pig fat. Fats containing medium chain fatty acids are soft in nature, for examples butter, ghee, coconut oil etc.

(3) Vegetable oils rich in PUFA are good for health and should preferably be used as cooking medium, as they lower blood cholesterol level and prevent risk of heart diseases.

Properties of Simple Lipids

I) Physical Properties—

i) Neutral fats are colourless, odourless and tasteless in nature.

ii) **Solubility**—Simple fats are insoluble in water and soluble in non polar solvents such as ether, alcohol, chloroform etc.

iii) **Specific gravity**—The specific gravity of lipids is less than one and therefore when added in water they float on the surface of water.

iv) **Emulsification**—Emulsion of fats in water can be made by shaking oil drops vigorously in water in the presence of certain emulsifying agents such as soaps, bile salts etc.

Emulsification of dietary fat in GIT is necessary for their digestion and absorption. Bile salts and bile acids secreted in bile from the liver, lower surface tension of water and thus emulsify fats to facilitate their digestion and absorption from the GIT.

II) Chemical Properties

i) **Hydrolysis**—Triglycerides are hydrolyzed by boiling with dilute acid or by the action of enzyme lipase, into three molecules of fatty acids and one molecule of glycerol as shown in the reaction given below—

$$\begin{array}{l} CH_2-O.CO-R \\ | \\ CH-O.CO-R' + 3H_2O \\ | \\ CH_2-O.CO-R \\ \textbf{Triglyceride} \end{array} \xrightarrow[\text{hydrolysis}]{\text{HCl/boil}} \begin{array}{l} CH_2-OH + R-COOH \\ | \\ CH-OH + R'-COOH \\ | \\ CH_2-OH + R-COOH \\ \textbf{Glycerol} \quad \textbf{Fatty acids} \end{array}$$

ii) **Saponification**—Hydrolysis of triglycerides with alkalies such as NaOH or KOH produce sodium or potassium salts of fatty acids, which are called soaps and are used as detergents. The process is known as saponification and the reaction is as given under—

$$\begin{array}{l} CH_2-O.CO-R \\ | \\ CH-O.CO-R + 3\ Na\ OH \\ | \\ CH_2-O.CO-R \\ \textbf{Triglyceride} \end{array} \longrightarrow \begin{array}{l} CH_2\ OH \\ | \\ CHOH + 3\ R\text{–}COONa \ (\text{Soaps}) \\ | \\ CH_2\ OH \\ \textbf{Glycerol} \end{array}$$

Saponification number is the number of mg of KOH required to completely saponify 1gm of oil or fat.

iii) **Hydrogenation**—Unsaturated fats and oils, on addition reaction with hydrogen become solidified as vanaspati ghee. For example dalda etc are manufactured by hydrogenation of unsaturated double bonds of PUFAs of vegetable oils.

iv) **Halogenation and Iodine Number**—The halogenation is also an addition reaction in which halogen such as I_2 is added on the double bonds of unsaturated fatty acids of vegetable oils.

Iodine number is thus index of degree of unsaturation and is directly proportional to the content of PUFA of the oil. **Iodine number** is defined as number of gms of I_2 taken up by 100 gm of fat / oil.

v) **Rancidity**—Natural fats and oils have a tendency to become rancid and develop unpleasant smell and taste on standing and exposure to O_2. Rancidity is of two types as described below—

a) **Hydrolytic rancidity**—It is caused by partial hydrolysis of triglyceride (fat / oil) molecule, due to the presence of traces of lipase enzyme in the natural fats and oils.

b) **Oxidative rancidity**—This occurs as a result of oxidation of the polyunsaturated fatty acids (PUFA) by atmospheric oxygen. Rancidity is catalyzed by certain metals such as copper etc. Oxygen attacks the double bonds of unsaturated fatty acids, causing peroxidation and decomposition of fatty acid molecule. The peroxidation results in the splitting of fatty acid molecule into aldehyde and ketones, which polymerize and impart bad smell and taste to the fat. The reaction is as given below:

Peroxidation of Unsaturated Fatty Acid—The auto oxidation of PUFA takes place by highly reactive O_2 as given under—

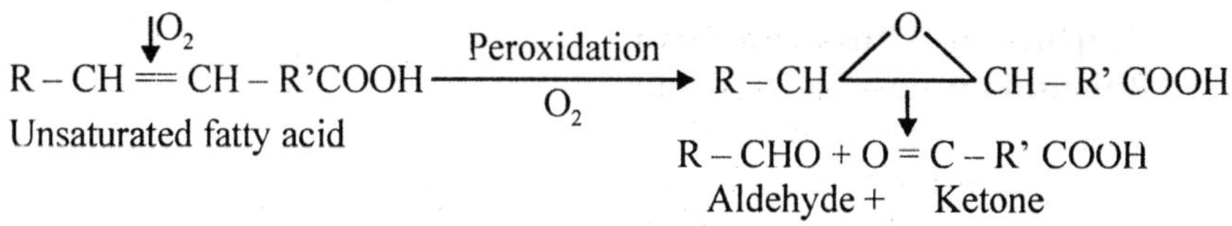

Free Radical Chain Reaction

Peroxidation of unsaturated fatty acids of lipids of biomembranes of body may cause free radical chain reactions resulting in the production of harmful free radicals. The free radicals are potentially injurious to body organs, as they cause irreversible damage to tissue proteins, DNA and lipid bilayer of cell membranes etc. (Chapter 25, 26.)

Free Radicals—Clinical Aspects

Free radicals, due to their potentially damaging actions are implicated in the development of some deadly diseases like heart diseases, neurological diseases, cancer etc and are also involved in the ageing process.

Antioxidants—Role in Removing Free Radicals

Certain antioxidants are found to occur in nature associated with natural oils such as vitamin E (α - tocopherol) and vitamin C that can prevent peroxidation of unsaturated fatty acid of vegetable fats and oils. In body vitamin E accumulates in lipid bilayer of biomembranes and protects peroxidation of PUFA. The phenolic group of vitamin E reacts with peroxyl free radical of PUFA and detoxifies it. This protects the structural integrity of lipid biomemberanes of the body tissues. α–Tocopherol thus acts as the most effective naturally occurring antioxidant in our body (Details in Chapter 25).

COMPOUND LIPIDS

The compound lipids are esters of fatty acids with certain alcohols and contain some other groups in addition to the fatty acids and alcohol. Examples are as described below—

I) Phospholipids—Phospholipids are esters of fatty acids and contain a nitrogen base and phosphoric acid residue in addition to the fatty acid and the alcohol such as glycerol or sphingosine.

Examples of phospholipids are—

a) **Lecithin—Phosphatidyl Choline—Lecithin** contains a nitrogenous base choline and phosphoric acid, esterified with 3rd OH group of glycerol, in addition two fatty acids esterified with glycerol as shown below—

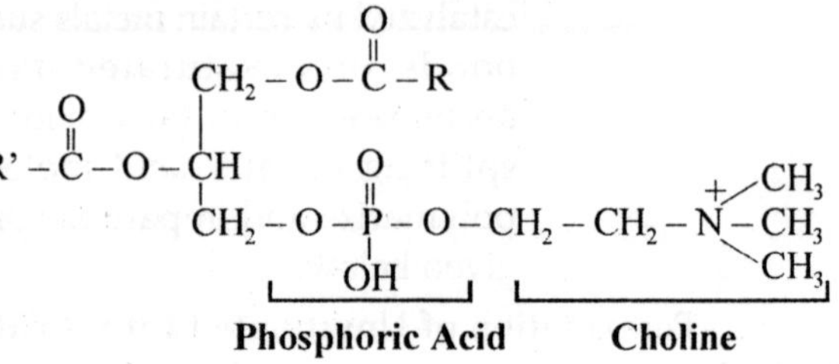

b) **Cephalin—Phosphatidyl Ethanolamine**—The cephalin has a nitrogenous base ethanolamine instead of choline of lecithin, as shown below—

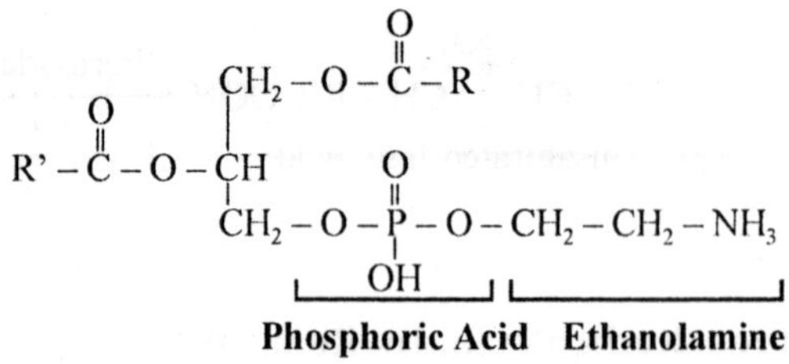

c) **Plasmalogens**—The plasmalogens occur in brain and muscle and constitute as much as 10% of phospholipids of these tissues. Chemically they resemble cephalin, except that fatty acid is linked by an ether, rather than ester linkage at C-1 position, as an unsaturated long chain aldehyde.

d) **Sphingomyelins**—Sphingomyelins have an unsaturated C_{18} amino alcohol called sphingosine instead of glycerol, which is esterified with a fatty acid and contain a phosphoric acid residue and a nitrogen base as shown under—

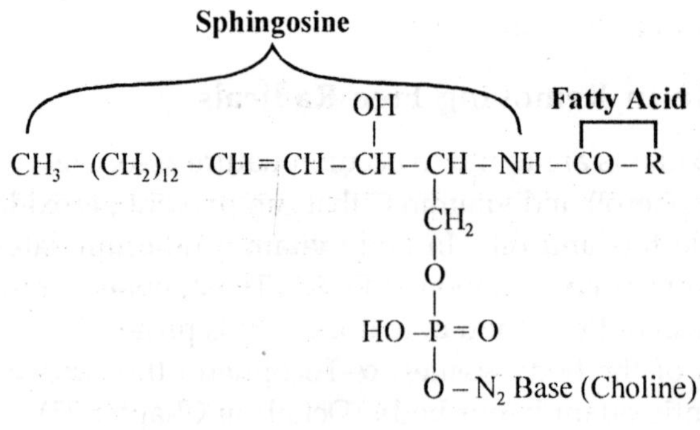

Sphingomyelines are found in large quantities in brain and nervous tissues.

Functions of Phospholipids

1. Phospholipids are the essential components of biomembranes and maintain their structural integrity.
2. Phospholipids are required essentially for the synthesis of lipoprotein complexes for their transport via blood to tissues.
3. Lecithin is widely distributed in brain and nervous tissues.

4. Lecithin can provide choline, to synthesize acetyl choline, a neurotransmitter.
5. Dipalmityl lecithin acts as a 'lung surfactant' that reduces the surface tension and thus prevent adherence of the lung alveoli. Absence of dipalmityl lecithin causes collapse of the alveoli leading to 'respiratory distress syndrome'.
6. Plasmalogens are present in phospholipids of brain, nerve tissue, muscle and mitochondria.
7. Sphingomyelins make up a large part of myelin sheath of nerves.

Glycolipids

Compound lipids containing a carbohydrate moiety, in addition to fatty acid and amino alcohol sphingosine are termed as glycolipids. They do not contain phosphoric acid and glycerol. The glycolipids are of two types—

a) **Cerebrosides**—Cerebrosides contain sphingosine, a higher fatty acid (24 C-atoms) called lignoceric acid or its derivative and a carbohydrate moiety galactose. These are present in rich amounts in white matter of brain. Examples of cerebrosides are kerasin, nervon, cerebron and oxynervon, which contain characteristic derivatives of C_{24} fatty acids.

   Cerebrosides accumulates in excess in the reticuloendothelial cells of spleen, liver, lymph nodes etc. in **gaucher's disease.** It is a rare hereditary disease of lipid metabolism caused by genetic abnormality.

b) **Gangliosides**—Gangliosides are most complex glycolipids and are composed of N–acetylneuraminic acid (sialic acid), sphingosine, fatty acid and 3 molecules of hexose sugars such as glucose or galactose.

   Gangliosides are found in the grey matter of brain. Large amount of gangliosides accumulates in the brain in cases of **Tay Sach's** and **Niemann Pick's disease,** which are the hereditary genetic diseases of lipid metabolism.

Lipoproteins

Lipoproteins are lipids with protein as a prosthetic group. Lipids are insoluble in water, so they are transported in the plasma by associating the nonpolar lipids (triglycerides and cholesterol esters) with amphipathic lipids (phospholipids and cholesterol) and protein part, called **apoproteins** to make them water soluble lipoprotein complexes.

Apoproteins associated with lipoproteins in blood serve as recognition site for cell surface receptors, so that lipoproteins are taken inside tissues for further metabolism. Specific apoproteins associated with lipoproteins also serve as activators for the enzymes involved in lipoprotein metabolism (details in chapter 13).

A typical lipoprotein comprizes of a lipid core of nonpolar lipids, surrounded by a surface layer of phospholipids and cholesterol with the protein moiety attached to this surface layer as apoprotein.

There are four major classes of lipoproteins transported in blood circulation to various tissues. These are as given under—

(1) Chylomicrons

(2) Very low denisity lipoproteins (VLDL)

(3) Low density lipoprdeins (LDL)

(4) High density liporoteins (HDL)

Triglyceride is the predominant lipid in chylomicrons and VLDL while cholesterol and the phospholipids are main constituents of LDL and HDL respectively.

Sulpholipids

Sulphated derivatives of glycolipids are known as sulpholipids. These are abundantly found in white matter of brain. The sulphogalactoceremide, known as sulphatides are present in high amount in myelin sheath of nerves.

❑❑❑

10

LIPIDS : DIGESTION AND ABSORPTION

DIETARY LIPIDS

Major dietary lipids are triacylglycerol (TAG) along with small amounts of cholesterol, mostly as esters and phospholipids. Average quantity of lipid intake is approximately 50-60 g in normal diet per day. Dietary fats and oils also bring essential fatty acids (PUFA) and fat soluble vitamins A, D, E and K along with them, which are absorbed along with digested products of lipids in gastrointestinal tract (GIT).

Dietary lipids are insoluble in water, so their digestion and absorption in gastro intestinal tract needs **specialized mechanisms.** These are —

1. **Emulsification** of lipids to very fine (small) droplets is necessary to make them soluble in aqueous medium, so as to provide access for the action of the digestive enzymes **lipases.**
2. **Micelle formation** is prerequisite to facilitate absorption of products of fat digestion (insoluble in water) by intestinal mucosa.

Digestion of Lipids in Mouth and Stomach

Digestion of lipids in mouth and stomach is very limited, inspite of the presence of lingual (salivary) and gastric **lipases.** However **in stomach** the lipids are partially emulsified by relatively high temperature and peristaltic movements and some digestion does occur here, as described below—

Digestion of lipids with **short and medium chain fatty acid** is achieved to some extent in **stomach** by lingual lipase, which comes along with food from the mouth, because it has low optimum pH (2.5 to 6) and is active in the acidic pH of gastric chyme. This enzyme acts specifically on short and medium chain TAG, present in milk fat and hydrolyzes them into fatty acids. Thus the action of lingual lipase is more significant in newborn infants. Since food remains in stomach for about 2-4 hours, about 30% of short and medium chain triglycerides are digested in stomach. The short and medium chain fatty acids so released are absorbed directly into portal blood, as they are moderately water soluble. However the bulk of dietary lipids from stomach passes into duodenum for further digestion.

Digestion in Intestine

Most of the digestion of dietary lipids occurs in small intestine and involves following stages—

(a) Emulsification of lipids—In duodenum the acidic chyme from stomach mixes with bile and the alkaline pH of bile first neutralizes acidic gastric chyme, to provide favorable pH for the action of pancreatic lipase. The **bile salts and bile acids present in the bile** are strong surfactants and lower surface tension of water to facilitate emulsification of fats. **Peristaltic movements** of **intestine** further facilitates distribution of lipids as very fine droplets in the aqueous medium (emulsion). Emulsification increases surface area of fine droplets of lipids to provide access for the action of **lipolytic** enzymes.

(b) Lipid digestion by pancreatic lipase—Pancreatic lipase is the major enzyme that digest dietary triacylglycerol (TAG). For the action of pancreatic lipase on lipid emulsion, a specific protein **colipase** is necessary to anchor and stabilize lipase on TAG droplets. Colipase is secreted by pancreas and its binding to TAG molecules at the oil water interface is obligatory for the action of pancreatic lipase. Partially digested phospholipids also facilitate lipase binding on TAG molecules.

The pancreatic lipase is specific for hydrolyzing the primary ester bonds at position 1 and 3 C-atom of TAG, resulting in the release of two fatty acid molecules and one molecule of 2- monoacylglycerol as shown in the Fig. 10.1.

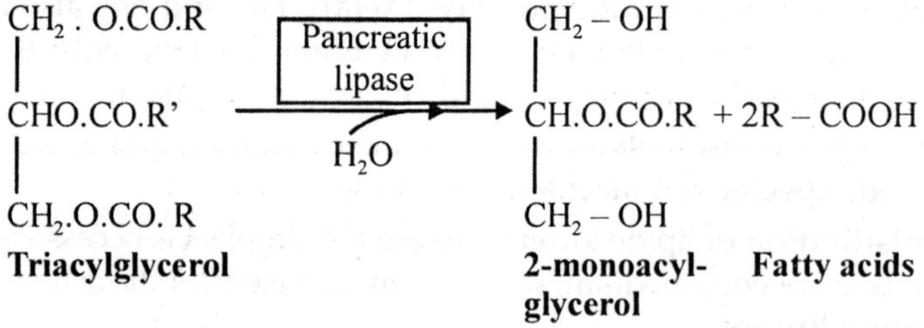

Fig. 10.1 : Hydrolysis of triacylglycerol by pancreatic lipase

Further hydrolysis of 2-monoglyceride does not occur much in GIT. However it can be achieved to some extent by an **isomerase** enzyme which shifts acyl group from 2nd to 1st position, which then can be hydrolyzed by pancreatic lipase.

Other Lipolytic Enzymes of Pancreatic Secretion

1. Pancreatic cholesterylesterase—It hydrolyzes dietary cholesterylesters to cholesterol and fatty acid.

2. Pancreatic phospholipase A_2—The action of phospholipase A_2 on phospholipid lecithin produces lysolecithin and a fatty acid molecule.

Absorption of Lipids—Mechanism

The glycerol and short chain fatty acids (chain length < 8-12 C) are fairly soluble in water and are directly absorbed into portal blood and taken to liver. All other products of fat digestion are insoluble in water and are absorbed as **micelles.** Former theories of mechanism of lipid absorption given by Fraser and Verzar have now been replaced by recent Bergstrom's micelle formation hypothesis, that comprises the salient features of both the previous lipolytic and partition theories of Frazer and Verzar.

According to Bergstrom's hypothesis, the TAG are partially hydrolyzed to fatty acids and 2-monoglyceride, which are then absorbed as **mixed micelles.** Micelles are vey small particles formed by a core of insoluble lipid droplet around the periphery of which, amphipathic compounds like bile salts, bile acids and phospholipids arrange themselves in such a way, that their hydrophobic groups are oriented towards central core lipids (soluble in fat droplet) and hydrophilic portion towards the surrounding aqueous medium as illustrated in the Fig. 10.2. By micelles formation, the bile salts exert a solubilizing effect on insoluble lipid digestion products in aqueous medium and thus they can easily be absorbed by the intestinal mucosa.

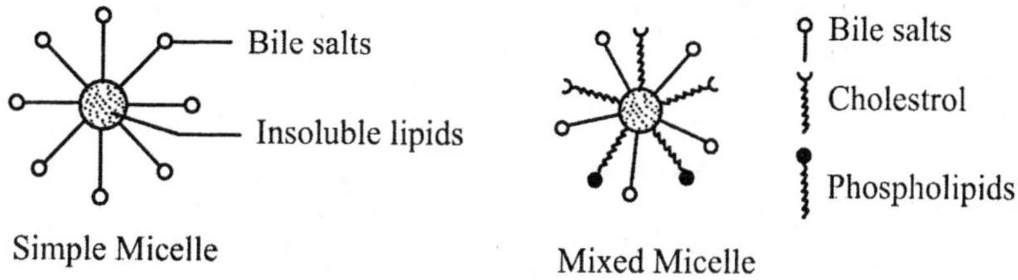

Fig. 10.2 : Diagrammatic representation of mechanism of micelle formation

Absorption of Lipid Digestion Products

In this micelles form, the products of digestion of dietary lipids along with fat soluble vitamins are absorbed by the microvillous surface of mucosa of **jejunum**. The fatty acids, 2-monoglyceride and other digestion products of lipids then diffuse into mucosal cells, while bile acids and salts are left behind, which later on are absorbed in ileum, to be returned to liver for resecretion in bile (enterohepatic circulation).

Resterification of Fatty Acids

Inside mucosal cell, the fatty acids are reesterified to reproduce TAG. For this, first the fatty acids are activated to acyl CoA, which then form ester on position 1 and 3 C-atom of 2- monoacylglycerol to produce triacylglycerol as shown in the Fig. 10.3.

Chylomicron Formation

The insoluble TAG so formed along with cholesterol esters are incorporated into a lipoprotein complex called **chylomicron,** which is then surrounded by **apoproteins A, B-48** and amphipathic phospholipid coat to solubilize it in aqueous medium. The chyle(milky fluid) from the intestinal mucosal cells loaded with chylomicrons is then secreted through lacteals into thoracic duct, which opens into the systemic blood. The serum may appear milky for sometime after a fat rich diet, due to the presence of TAG rich chylomicrons. Normally it clears within few hours by the uptake of TAG from chylomicrons by adipose tissues under the action of the enzyme **lipoprotein lipase** (details in Chapter 13.)

A brief account of lipid digestion, absorption as mixed micelles, resterification in intestinal mucosa and their secretion and transport as lipoprotein chylomicrons is illustrated in the Fig. 10.3.

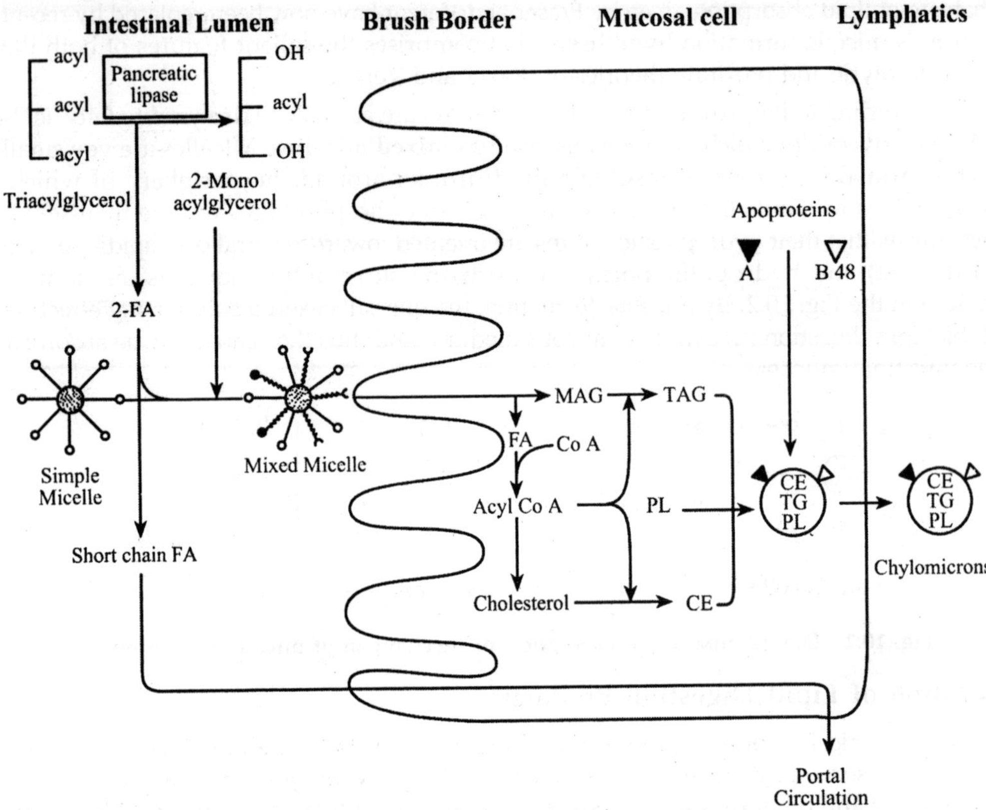

Fig. 10.3 : Diagrammatic representation of digestion and absorption of lipids by intestinal mucosa (MAG-monoacylglycerol, CE-cholesterol ester, PL-phospholipids TG-triglycerides, FA-Fatty acid)

Abnormalities in Digestion of Lipids—Clinical Aspects

Any defect in digestion and absorption of lipids would result in **steatorrhea** which is characterized by excessive fat excretion in feces. Abnormalities in fat digestion and absorption may impair digestion and absorption of other components of diet, particularly minerals which form insoluble soaps with fatty acids and are excreted in feces.

Conditions associated with abnormalities of fat digestion and absorption include—

1. Chronic diseases of pancreas such as pancreatitis.
2. Bile duct obstruction leading to defective fat absorption from GIT.
3. Intestinal diseases such as sprue, coelic disease, crohn's disease etc.
4. Surgical removal of pancreas or parts of intestine.

In the above mentioned conditions, there is lack or deficiency of either pancreatic lipolytic enzymes or bile acids and bile salts required essentially for the emulsification and digestion of dietary lipids. Due to this fats are not digested or absorbed properly in GIT and are excreted in feces causing steatorrhea and its adverse clinical complications.

11

LIPID METABOLISM

Lipids are essential constituents of cell structure, and serve indispensable functions in our body which include—

(a) Lipids are most concentrated and unlimited storage form of energy as **reserve source** in adipose tissues in form triacylglycerol (TAG).

(b) They are **structural component** of **bio-membranes** of tissues as **lipid bilayer** containing phospholipids and cholesterol, and regulate cell permeability.

(c) **Insulators**—Lipids are important in providing insulation and protection to body organs against mechanical and thermal shocks, as well as they prevent electric nerve impulse leakage in brain and myelin sheath of nerves.

Due to the hydrophobic and nonpolar nature, lipids are different from other compounds of body in their functions, as well specialized mechanisms are needed for their digestion and absorption in GIT, their transport to body tissues and in their metabolism.

Triacylglycerol—The Body Reserve Source of Energy

Triacylglycerol(TAG) is the most abundant lipid of body and is stored in **adipose tissues** as reserve body fuel in appreciable amounts, in contrast to carbohydrates, which can be stored in very limited amount as glycogen in liver(7%) and in muscles to the extent of 1% only.

Adipose tissue TAG reserves are sufficient to meet energy requirement of body for several weeks, in case of food deprivation in humans, as well as in animals for example during hibernation and in migratory birds etc.

Mobilization of TAG from Adipose Tissue—Lipolysis

When the need arises, fats are mobilized from adipose tissues, for instance during fasting conditions and hypoglycemic states. Under the influence of various hormones like epinephrine, growth hormone, glucocorticoids, glucagon, TSH, ACTH etc, stored TAG are hydrolyzed to free fatty acids and glycerol which are released in blood. The hormones mentioned above, activate a **hormone sensitive lipase** enzyme via **c-AMP-proteinkinase activation** system as shown in the Fig. 11.1.

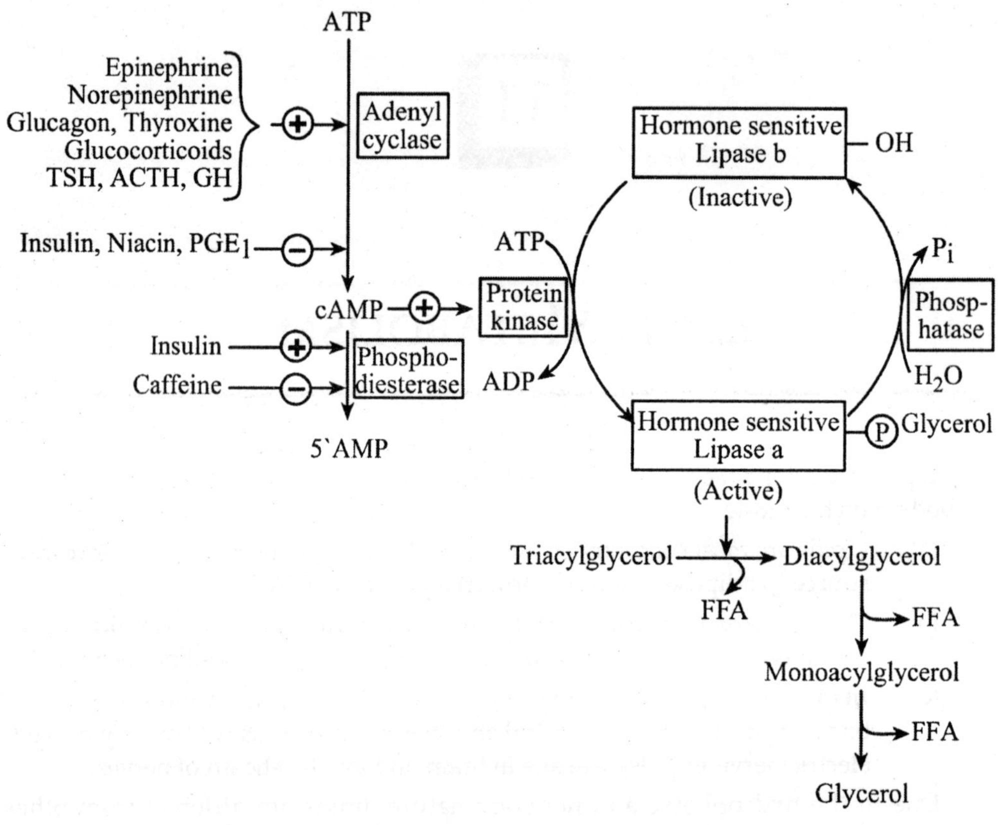

Fig. 11.1—Regulation of lipolysis in adipose tissues via c-AMP and hormones

Fate of fatty acids and glycerol

The products of lipolysis of TAG hydrolysis in adipose tissues i.e. free fatty acid (FFA) and glycerol are released in blood. The only fate of glycerol is to be taken up by liver for gluconeogenesis process. (Chapter 7, Fig 7.1), while fatty acids are **preferably** taken up by skeletal muscles, heart muscles and other tissues for oxidation purpose to liberate energy. Thus they spare glucose(to correct hypoglycaemia), to be utilized by brain, RBC etc. which solely depends on glucose as fuel.

FATTY ACID OXIDATION—β OXIDATION

The fatty acids in body are oxidized by a special pathway called **β oxidation,** that is established inside **mitochondria** in the **matrix.** The β oxidation pathway was proposed by scientist Knoop and according to this, oxidation of fatty acids(FA) occurs at β C-atom, resulting in successive removal of 2C-atom units as **acetyl CoA.** The acetyl CoA molecules so liberated then enter **citric acid cycle (CAC),** which is actively going on close by in the mitochondrial matrix for complete oxidation to liberate energy.

β-oxidation—It takes place at β carbon atom of long chain FA with even number C-atoms e.g. **palmitic acid**(16 C-atom, FA), to remove at a time, a two C-atom compound, acetyl CoA, leaving behind, **acyl CoA** with two C-atoms short (14 C-atom FA). This acyl CoA then reenter next β oxidation cycles, resulting in successive removal of acetyl CoA molecules, until the fatty acid molecule is oxidized completely to acetyl CoA as shown below.

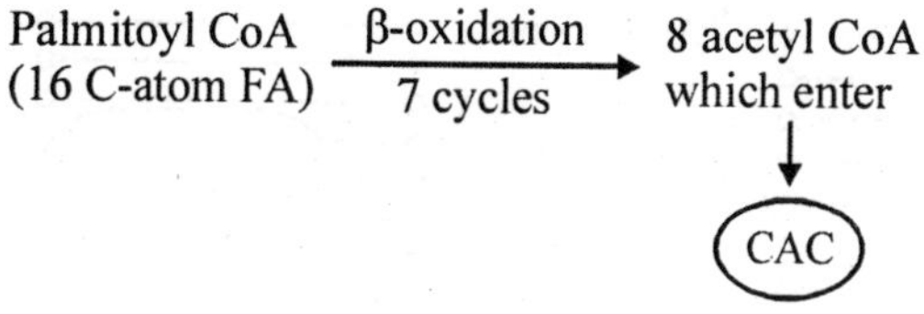

Fig. 11.1—Overview of β oxidation of fatty acids to liberate acetyl CoA

β Oxidation—Mechanism

The β-oxidation of fatty acids takes place in most of the tissues of the body, such as skeletal muscles, heart muscles, liver etc. except brain and RBCs, which depend solely on glucose as fuel. Reactions of β-oxidation of FA are described in following steps—

(a) Activation of fatty acid—Before FA can enter β-oxidation, its activation as acyl CoA is prerequisite. Activation of FA takes place in cytoplasm by enzyme **thiokinase,** coenzyme A (CoA)and ATP as energy source . The fatty acyl CoA has a high energy thioester bond and is ready to enter a metabolic process (Fig. 11.3).

(b) Transport of fatty acyl CoA into mitochondria—As β-oxidation of fatty acids is established inside mitochondria in the matrix, acyl CoA has to be transported across mitochondrial membrane. Short chain acyl CoA (up to 10 C-atoms) can cross the mitochondrial membrane freely, but higher fatty acyl CoA cannot cross inner mitochondrial membrane and thus a special transport system is required , known as carnitine carrier system.

Carnitine Carrier System

Carnitine (β OH γ-trimethylammonium butyrate) is widely distributed in body and it specifically binds fatty acids to form **acylcarnitine,** which is permeable and thus facilitates fatty acid transport across mitochondrial membrane.An enzyme **carnitine palmitoyl transferase-I** transfers fatty acid from acyl CoA to the OH group of carnitine on the cytosolic side of outer mitochondrial membrane and the complex **acylcarnitine** so formed permeates across the membrane, as shown in the Fig. 11.2.

At the inner mitochondrial membrane, a special carrier protein, **translocase,** transports acyl carnitine inside mitochondria, where a membrane bound enzyme **transferase -II** catalyzes transfer of acyl group back to coenzyme A (Fig. 11.2).

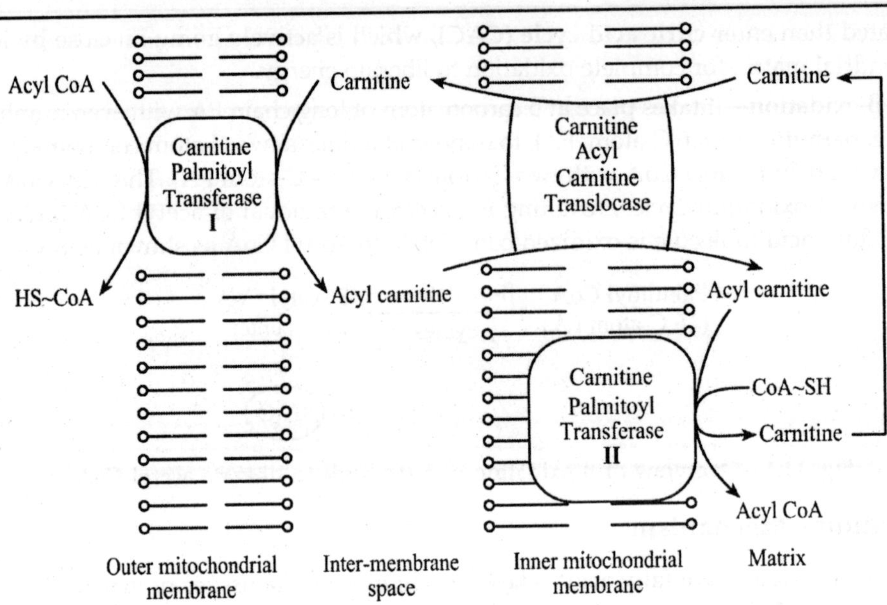

Fig. 11.2 : Diagrammatic representation of transport of Fatty acid by carnitine carrier system through mitochondrial membrane

β-Oxidation Cycle—Fatty acid β-oxidation cycle involves four discrete steps as describe here under—

Step I—Oxidation by FAD linked dehydrogenase

In the mitochondrial matrix, the fatty acyl CoA is oxidized at β C-atom by a specific dehydrogenase enzyme (FAD dependent), by removing two reducing equivalents as $FADH_2$ from α and β C-atoms of long chain fatty acid to introduce a double bond, producing α-β unsaturated acyl CoA (Fig. 11.3).

Step II—Hydration

In the next step a hydration reaction occurs, in which elements of water (H_2O) are added at the double bond to produce β-hydroxy acyl CoA and the reaction is catalyzed by enzyme **enoyl CoA hydratase** (Fig. 11.3).

Step III—2nd Dehydrogenation

This step involves a second dehydrogenation of β–OH acyl CoA to remove $2H^+$ ions by NAD^+ coenzyme in the presence of enzyme **β OH acyl CoA dehydrogenase** to produce a β- keto acyl CoA (Fig. 11.3).

Step IV—Cleavage of 2-C-atom unit

In the final reaction of β- oxidation cycle , a thiolytic cleavage at the β C-atom of β-keto acyl CoA takes place. It is catalyzed by **enzyme thiolase** to liberate 2 C-atom compound acetyl CoA, leaving behind a 2 C-atom short acyl CoA, which reenter the next β - oxidation cycle (Fig. 11.3). This process of β oxidation continues till fatty acid is completely oxidized to acetyl CoA molecules.

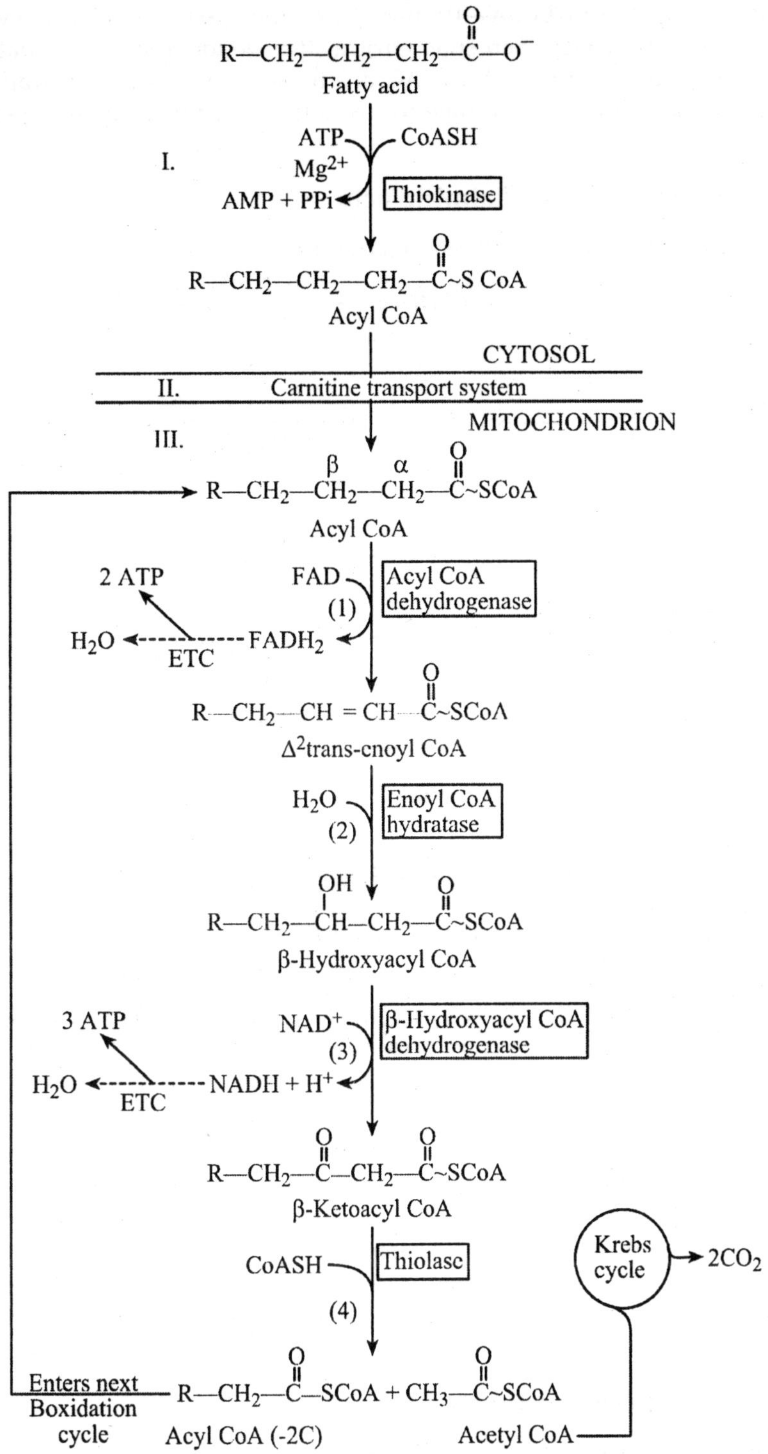

Fig. 11. 3 : Reaction of β- oxidation of fatty acids to produce acetyl CoA

Oxidation of palmitoyl CoA—Palmitoyl CoA undergoes 7 β oxidation cycles to yield 8 acetyl CoA molecules, which then enter citric acid cycle for complete oxidation to CO_2. The reducing equivalents removed in CAC are transferred to electron transport chain (ETC) on inner mitochondrial membrane for oxidative phosphorylation to generate ATP (Chapter 6).

Energetics of β-oxidation of Palmitic acid

1. **In One β- oxidation cycle**—ATP generation at steps

I step—dehydrogenase ⟶ 1 $FADH_2$ $\xrightarrow[\text{Phosphorylation}]{\text{ETC/BO}}$ 2 ATP

III step—dehydrogenase ⟶ 1 NADH + H^+ $\xrightarrow[\text{Phosphorylation}]{\text{ETC / BO}}$ 3 ATP

Total = 5 ATP

2. 1 palmitoyl CoA undergoes 7 β oxidation cycles, So — 7 x 5 =35 ATP are generated

3. 1 palmitoyl CoA produce 8 acetyl CoA by β oxidation which enter CAC to produce (12 ATP/1 CAC) — 8 x 12 CAC = 96 ATP are generated

Total = 131 ATP

4. 2 high energy phosphate bonds as ⟶ PPi are used during FA activation — (-) 2 ATP = -2 ATP

129 ATP

So complete oxidation of palmitic acid yield 129 ATP molecules/mol.

Oxidation of Odd Carbon Chain Fatty Acids

Normally dietary fats and our body lipids contain even C-atom chain fatty acids which are oxidized by β oxidation. However, some times odd C-number fatty acid may come through diet, that are also oxidized by β oxidation, which proceed in same manner as described for even number fatty acid, except at the last and final β oxidation cycle, a 3 C-atom compound, **propionyl CoA** is left behind.

Fate of propionyl CoA—In the liver, the propionyl CoA is converted into glucose by the gluconeogenesis process as described in detail in Chapter 7. Propionyl CoA first undergoes a **carboxylation**(CO_2 fixation) reaction, catalyzed by the enzyme **propionyl CoA carboxylase** in presence of ATP, and **biotin coenzyme.** By this CO_2 fixation the 3 C- propionyl CoA is converted to 4 C- compound **methyl malonyl CoA,** which then is isomerized to **succinyl CoA** in the presence of B_{12} coenzymes and enter gluconeogenesis pathway (Fig. 7.2 Chapter 7).

Oxidation of Unsaturated Fatty Acids

The **polyunsaturated fatty acid (PUFA)** are also oxidized by β oxidation pathway, with successive removal of acetyl CoA molecules as in case of saturated fatty acids, till a double bond in the molecule is reached. Then two additional enzymes are needed for further oxidation as decribed below—

(a) **Isomerase enzyme**—It isomerizes the cis configuration of FA to **trans isomer** and then elements of water are added on the double bond at β C-atom to produce a D-isomer of β hydroxy acyl CoA.

(b) **β OH acyl CoA epimerase**—This enzyme converts the D-isomer to L form of β OH acyl CoA, which then undergoes next steps of β oxidation cycle (III step and IV steps). In this alternative oxidation process, the I step of β–oxidation (the FAD dependant dehydrogenation) is bypassed and because of this, there is less ATP generation in the β oxidation cycle at the position of double bonds in fatty acid molecule.

α -Oxidation of Fatty Acids

Though β-oxidation is the most predominant pathway of fatty acid oxidation, however **α -oxidation** i.e. oxidation at α C-atom and removal of one C-atom unit at a time, has been detected in brain. The α - oxidation does require activation of fatty acid to acyl CoA, but ATP is not generated in the process. The α- oxidation is particularly important in removal of methyl groups present on β C- atom of plant fatty acids, such as phytanic acid. This is necessary as the methyl group block the β oxidation of these fatty acids in our body.

Refsum's disease—It is a rare genetic metabolic disorder of fatty acid metabolism causing severe **neurological damage.** In this genetic metabolic defect, patient lacks enzymes of α -oxidation, which removes CH_3 groups of phytanic acid. The phytanic acid occurs in green plant foodstuffs as well as in milk lipids. This phytanic acid cannot undergo β-oxidation, until the CH_3 group at β C -atom(which block β oxidation) is first removed by α-oxidation. In refsum's disease due congenital lack of enzyme **α- hydroxylase of α-oxidation,** phytanic acid accumulates in excess in blood and cause severe neurological damage and certain skeletal abnormalities.

ω-Oxidation

It is a minor pathway occuring in certain tissue, in which medium and long chain fatty acids are oxidized. In ω-oxidation the methyl end (ω C-atom, CH_3 group) of fatty acid is first oxidized to carboxylic group. For this additional enzymes are needed. The **ω–hydroxylase** with cofactor cytochrome p450 first hydroxylate the ω CH_3 group to primary alcoholic group (CH_2OH), which subsequently is oxidized to a carboxylic group (COOH), thus producing a dicarboxylic fatty acid. This fatty acid with COOH groups at both the ends then undergoes β-oxidation from both the sides, usually to produce adipic (C_6) acid in the end, which is then excreted in urine.

KETOGENESIS

Ketogenesis is a process in which there is excessive production of **ketone bodies** which include **acetoacetate, β–hydroxybutyrate** and **acetone.** Ketogenesis occurs under certain metabolic conditions, in which there is **excessive but incomplete** fatty acid oxidation.

Ketone Bodies are Interconvertible

The three compounds, acetoacetate, β-hydroxybutyrate and acetone are collectively known as ketone bodies, due to the presence of a keto group in their molecule. Acetoacetate and β-hydroxy butyrate are interconvertible by a mitochondrial enzyme **β-OH butyrate dehydrogenase,** while acetone is produced by **spontaneous decarboxylation of acetoacetate,** which is an irreversible reaction as shown in the Fig. 11.4.

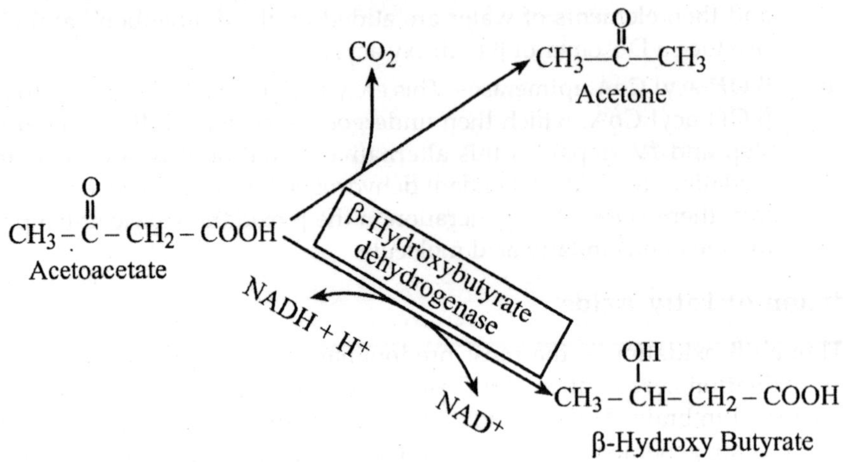

Fig. 11.4 : Interconversion of ketone bodies, with their chemical structures

Normally total concentration of ketone bodies in blood does not exceed 0.2 m.mol/L. The blood level of ketone bodies is highly elevated in **ketosis.** This clinical condition ketosis is characterized by **ketonemia** and excessive excretion of ketone bodies in urine, called **ketonuria.**

Liver is Site of Ketogenesis

Production (synthesis) of ketone bodies occurs mainly in **liver** in **mitochondrial matrix,** in certain metabolic conditions associated with high rate of fatty acid oxidation.

Acetyl CoA is the precursor—Acetyl CoA produced by β-oxidation of fatty acids is the precursor for ketone bodies formation. The reactions of ketogenesis are as given under—

1. Two molecules of acetyl CoA condense by a reversal of **thiolase** reaction of β-oxidation, whenever there is excessive accumulation of acetyl CoA, to produce acetoacetyl CoA.
2. In next step, acetoacetyl CoA condenses with yet another molecule of acetyl CoA to produce **β–hydroxy β–methyl glutaryl CoA (HMG CoA),** in the presence of enzyme **HMG CoA synthase** as shown in the Fig. 11.5.
3. HMG CoA so produced is then cleaved to produce **acetoacetate** (ketone body) and acetyl CoA, by a cleavage enzyme called **HMG CoA lyase** (Fig. 11.5).

Metabolism of Ketone Bodies

Liver is unable to oxidize and utilize ketone bodies as it lacks the enzyme that can **activate** acetoacetate, which is prerequisite for its oxidation. This accounts for net production of ketone bodies by liver and their release in blood. The ketone bodies are then transported to **extra hepatic tissues,** where they can be oxidized for energy production purposes to some extent. The ketone bodies are preferably oxidized in certain tissues such as **skeletal muscles,**

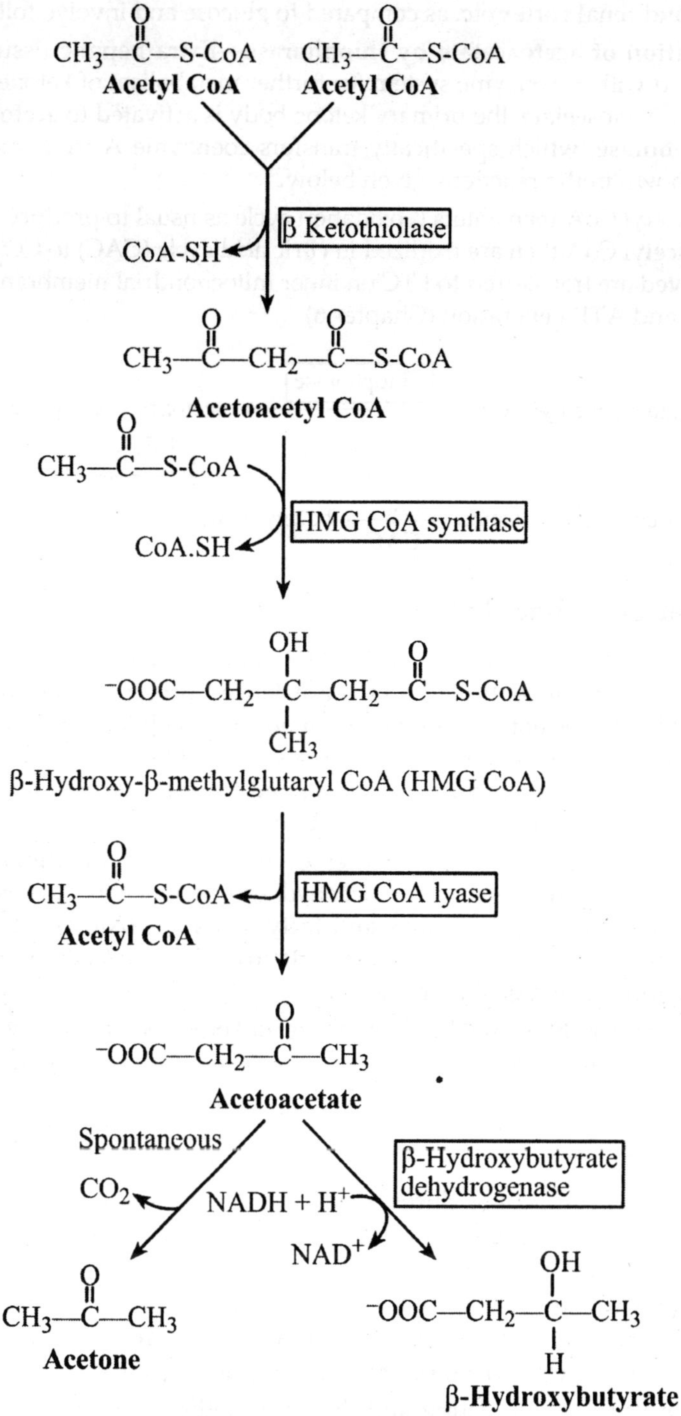

Fig. 11.4 : Pathway of ketogenesis in liver

cardiac muscles and **renal cortex** etc. as compared to glucose and involve following steps—

1. Activation of acetoacetate by thiophorase—Extra hepatic tissues mentioned above are equipped with the enzyme system for further metabolism of ketone bodies, but to a limited capacity. Acetoacetate, the primary ketone body is activated to **acetoacetyl CoA** by an enzyme **thiophorase**, which specifically transfers coenzyme A from succinyl CoA to acetoacetate as shown in the reactions given below.

2. Acetoacetyl CoA then enters β-oxidation cycle as usual to produce 2 molecules of acetyl CoA. The acetyl CoA then are oxidized in citric acid cycle (CAC) to CO_2 and reducing equivalents removed are transferred to ETC on inner mitochondrial membrane for oxidative phosphorylation and ATP generation (Chapter 6).

Acetoacetate + succinyl CoA —[Thiophorase]→ acetoacetyl CoA + succinate

acetoacetyl CoA + succinate ↓ [Thiolase]

For complete oxidation ←(Enter **CAC**)— 2 acetyl CoA

Overproduction of Ketone Bodies

Normally production of ketone bodies by liver and their utilization (oxidation) in extra hepatic tissues is well controlled and is a balanced process. Thus concentration of ketone bodies in blood does not exceed 1mg/dl in normal conditions and their excretion in urine is in traces only. However under certain metabolic conditions, which are accompanied by high rate of fatty acid oxidation such as **prolonged fasting, starvation** and in **diabetes mellitus etc.,** there is **overproduction** of **ketone bodies,** leading to **ketosis.**

The overproduction of ketone bodies occurs in above mentioned conditions, primarily due to metabolic unavailability of carbohydrates. So for energy generation purposes fats are mobilized from adipose tissues and fatty acids are oxidized excessively but incompletely to acetyl CoA, resulting in accumulation of acetyl CoA, which is diverted towards excessive production of ketone bodies.

Ketosis—It is condition in which blood level of ketone bodies is very high (>70 mg/dl), known as **ketonemia**, which results in excessive excretion of ketone bodies in urine, called **Ketonuria**.

Causes of Ketosis

1. A minimum amount of glucose oxidation in body is **essential** for accomplishing complete oxidation of fatty acid, as they burn in the common furnace i.e. citric acid cycle(CAC).
2. When glucose is metabolically unavailable, citric acid cycle is not well established. The **starter** of CAC i.e. **oxaloacetate** is mainly derived from pyruvate, which is produced from glycolysis of glucose.
3. During prolonged fasting and diabetes mellitus intermediates of CAC are diverted towards gluconeogenesis, resulting in further less availability of oxaloacetate and other CAC intermediates. These compounds are necessary to establish CAC, in which acetyl CoA can be oxidized completely.

4. **Ketosis** results due to increased rate of fatty acid oxidation, producing excessive amount of acetyl CoA, which cannot be completely oxidized in CAC, so is diverted towards ketone bodies formation.

Regulation of ketogenesis is achieved normally by insulin—glucagon ratio, which regulates the rate of lipolysis in adipose tissues. The ketogenesis would occur only when there is excessive mobilization of fatty acids from adipose tissues (lipolysis), which occur when glucagon—insulin ratio increases, or in case of increased levels of other diabetogenic hormones. The activity of glucagon and other diabetogenic hormone increases in diabetes mellitus, prolonged fasting, starvation etc. which are associated with ketosis.

Adverse Effects of Ketosis—Clinical Aspects

1. **ketoacidosis**—The ketone bodies, acetoacetate and β–hydroxybutyrate are fairly strong acids and their excessive accumulation in blood and tissues result is lowering of pH causing **metabolic acidosis.**

2. The plasma bicarbonate level is depleted, since it is used up for **buffering** ketone bodies and thus the **carbonic acid—bicarbonate ratio** increases. This stimulates chemoreceptors in hypothalamus, causing **hyperventilation,** and clinical state is known as **Kussmaul respiratory syndrome,** i.e. a typical **acidotic breathing.**

3. **Acetone,** the third ketone body is volatile and is eliminated out by lungs during expiration, resulting in a typical **acetone breath** with **fruity smell.**

4. **Osmotic diuresis**—Increased urine volume induced by **ketonuria** and **associated** vomiting may lead to **dehydration.**

5. **Electrolyte loss**—The ketone bodies are excreted in urine as Na^+ or K^+ salts leading to loss of important electrolytes.

6. **Hypokalemia, dehydration** and **acidosis** together contribute to **lethal effects** of ketosis, in which patient may go into **coma** and death may result.

Management of ketoacidosis—Administration of bicarbonate, and proper maintenance of electrolyte and fluid balance by monitoring, are the major aspects of ketoacidosis management to save the life of patient.

BIOSYNTHESIS OF FATTY ACIDS

Living organisms, including human beings synthesize major amount of saturated and monounsaturated **long chain fatty acids** with **even number** C-atoms in the body. The **denovo** FA synthesis takes place by sequential condensation of two C-atom moiety, **acetyl CoA** to produce long chain fatty acid molecules with C-atoms 16, 18, 20 etc.

Lipogenesis

The fatty acids so synthesized, as well as dietary fatty acid absorbed from GIT are then esterified with glycerol phosphate to produce triglyceride or triacylglycerol (TAG). The triglycerides are major lipids (neutral fat), which are stored in adipose tissues as reserve energy source and the process is known as lipogenesis.

Factors Stimulating Lipogenesis

1. **Welfed state** and carbohydrate rich diet **stimulates** active fatty acid synthesis and lipogenesis. After digestion and absorption, extra glucose (>40%) is converted to fatty acids and lipids and are stored in adipose tissues as reserve energy source.

2. Insulin hormone is the major factor that stimulates active fatty acid synthesis and lipogenesis. Insulin secretion is stimulated after a carbohydrate rich diet as post prandial blood sugar level (BSL) tends to increase. Insulin is major hormone that lowers BSL, by increasing uptake of glucose from blood to tissues for its utilization and extra glucose is converted into fatty acids.

3. Raw materials for fatty acid synthesis—

- **Acetyl CoA** is the main precursor for FA synthesis and is provided by oxidation (**glycolysis)** of **glucose,** that is stimulated by hormone insulin.
- **Reducing equivalents NADPH+H$^+$** are necessary for **reductive biosynthesis** of fatty acids, and are provided by **HMP shunt pathway** of glucose oxidation; which is also stimulated by insulin. This pathway is well established in the tissues where active fatty acid synthesis(lipogenesis) is going on.

4. Insulin activates the key enzymes of FA synthesis.

Denovo Fatty Acid Synthesis—Salient Features

1. Tissues involved in lipogenesis—Denovo synthesis of fatty acid predominantly takes place in liver, adipose tissues, kidneys, lungs and lactating mammary gland.

2. The enzymes of active fatty acid synthesis are located in **cytoplasmic fraction** of cells as **multienzyme** complex known as **fatty acid synthase complex.**

3. Procurement of acetyl CoA—Acetyl CoA (precursor of FA synthesis) is provided by glucose (glycolysis) that form pyruvate; which is then converted into acetyl CoA **inside mitochondria.** However denovo fatty acid synthesis takes place in the cytoplasm, therefore a specialized system is required to transport acetyl CoA out of the mitochondria, since mitochondrial membrane is impermeable to it.

ATP-Citrate Lyase System—For the transfer of acetyl CoA from the mitochondria, it first combines with oxaloacetate to form citrate a member of CAC, which is permeable through the mitochondrial membrane and comes out. In the cytoplasm, a specific enzyme **ATP- citrate lyase** then cleaves citrate into acetyl CoA, which enter FA synthesis process and oxaloacetate which via malate shuttle goes back into the mitochondria, and enters CAC as illustrated in the Fig. 11.5.

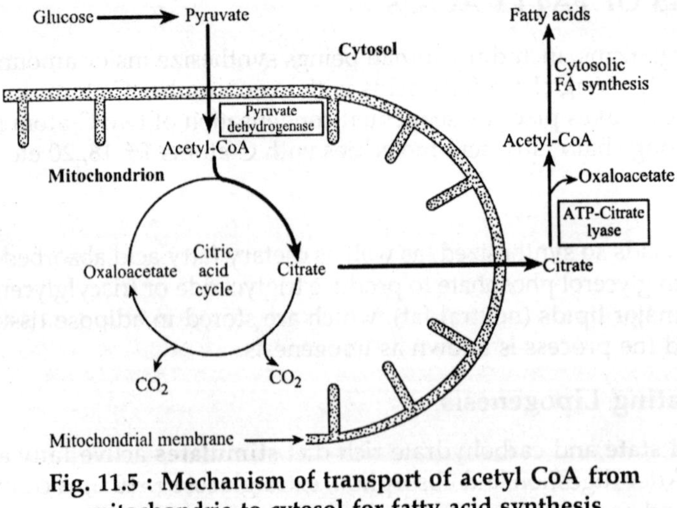

Fig. 11.5 : Mechanism of transport of acetyl CoA from mitochondria to cytosol for fatty acid synthesis

Process of Denovo Fatty Acid Synthesis

The **extra mitochondrial** cytoplasmic fatty acid synthesis system consists of a **multi enzyme complex.** This fatty acid multienzyme system works as a **dimer,** each with one polypeptide chain (monomer) containing **seven enzyme** activities necessary for FA synthesis, that is synthesized from a single mRNA transcribed by fusion of several genes as shown in the Fig. 11.6.

Fatty acid synthase dimer—Each monomeric polypeptide with seven catalytic units lies in **head to tail configuration** in the dimer, in such a manner that thiol(-SH) group of a special **acyl carrier protein (ACP)** of one monomer is in close proximity and facing to –SH group of cysteine residue of **ketoacyl synthase** of other monomeric polypeptide chain as shown in the Fig. 11.6. Since both the thiol groups participate simultaneously in fatty acid synthesis, only a **dimer multienzyme complex** is active.

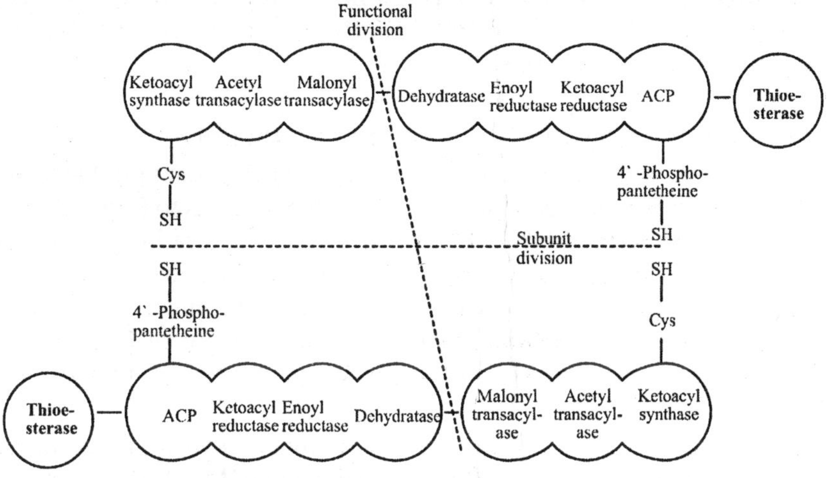

Fig. 11.6 : Fatty acid synthase multienzyme complex as dimer with two identical monomeric polypeptide chains with seven catalytic units and ACP-SH groups lying head to tail configuration

Domains of Fatty Acid Synthase Multienzyme

The fatty acid synthase multienzyme has two main domains, which are included in functional unit of enzymes and reside on half of the two facing monomers of the fatty acid synthase and therefore only a dimeric form of the enzyme is active. The two domains of fatty acid synthase are—

a) **Condensing unit**—The first domain of fatty acid multienzyme consists of the sites for initial substrate binding and condensing enzyme activity. This domain includes the enzymes acetyl transferase, malonyl transferase and β–ketoacyl synthase as shown in the Fig. 11.6.

b) **Reductive multienzyme unit**—The second domain of fatty acid synthase is the reductive unit and resides on the facing ACP - SH of the 2nd half of monomeric unit of the multienzyme. The reductive domain has three enzyme units called ketoacylreductase, hydratase and enoyl reductase (Fig. 11.6).

Denovo fatty acid synthesis involves successive condensation of 2 C-compound acetyl CoA to increase the chain length of fatty acid to be synthesized. The whole reaction sequence occurs while the intermediates are bound to ACP unit of multienzyme. The denovo fatty acid synthesis occurs in following stages.

Initiation of Fatty Acid Synthesis

1. Malonyl CoA synthesis—The initial step in fatty acid biosynthesis is a **CO_2 fixation** reaction which requires **coenzyme biotin** to increase carbon chain length of a 2 C-atom moiety acetyl CoA. The reaction is catalyzed by a **key** enzyme **acetyl CoA carboxylase,** which is also a multienzyme with biotin as coenzyme. This enzyme is activated by insulin and is a key regulatory allosteric enzyme. The reaction and regulatory modulators are shown in the Fig. 11.7.

2. Substrate binding on multienzyme dimer—Malonyl CoA (3 C-atom compound) is then transferred to -SH group of ACP of fatty acid synthase complex, by **malonyl transacylase** unit of multienzyme complex.

A second acetyl CoA combines with -SH group of cysteine residue of facing 2nd monomer, as catalyzed by the enzyme subunit **acetyl transacylase.**

3. Condensation—In the next step the condensing enzyme ketoacyl synthase catalyzes the condensation between the acetyl and malonyl units to form a 4 carbon β–keto acyl compound, called **acetoacetyl ACP** with simultaneous decarboxylation of malonate. **Decarboxylation** of malonate provides energy for the condensation to form 4 C-atom fatty acid. In this condensation reaction the acetyl group attacks the methylene group of malonyl residue with simultaneous **decarboxylation,** catalyzed by enzyme **3 keto acyl synthase,** forming a β–ketoacyl(acetoacetyl) ACP enzyme. This frees the SH group of cys. residue of Ist monomer (Fig. 11.7).

The fatty acid synthesis is basically a **reductive biosynthesis** requiring specially **NADPH+H^+** as reducing equivalents, mainly provided by HMP shunt pathway of glucose oxidation, that is actively going on in these tissues.

4. Reductive stage—The 2nd monomer with ACP domain of synthase multienzyme complex is the **reductive** domain and carries on I reduction step, in which acetoacetyl ACP formed as mentioned above, is reduced by reducing equivalents provided by NADPH + H^+ in the presence of enzyme β **ketoacyl reductase** to produce β hydroxy acyl ACP (β OH butyryl ACP). The β OH acyl ACP is then dehydrated by a **dehydratase** enzyme subunit of this domain to form a 2, 3 unsaturated β enoyl acyl ACP, which undergoes 2nd reduction step utilizing second NADPH + H^+ to form a saturated four C-atom fatty acid called butyryl ACP catalyzed by enoyl reductase subunit as illustrated in the Fig. 11.7.

5. Increase in chain length—The butyryl group from ACP is then transferred to the SH group of condensing unit of facing monomer and enters the next synthesis cycle to increase the C-chain length from 4 to 6. For this it condenses with a second malonyl unit, catalyzed by condensing enzyme to produce a 6 C-atom, 2-keto acyl ACP, which then enter the reduction reaction steps of next fatty acid synthesis cycle as described above.

This sequence of the reactions of fatty acid synthesis cycle are repeated 6 times to successively increase carbon chain length to 6, 8, 10 and finally to C-16 fatty acid to form the **palmitoyl S-ACP multienzyme complex.**

6. Cleavage of Palmitate—The enzyme **thioesterase** (7th enzyme) of the multienzyme complex finally deacylase (cleave) palmitoyl -ACP complex to liberate palmitic acid (16 C-chain fatty acid) and frees the **synthase multienzyme** complex to restart **synthesis cycle** process as illustrated in the Fig. 11.7.

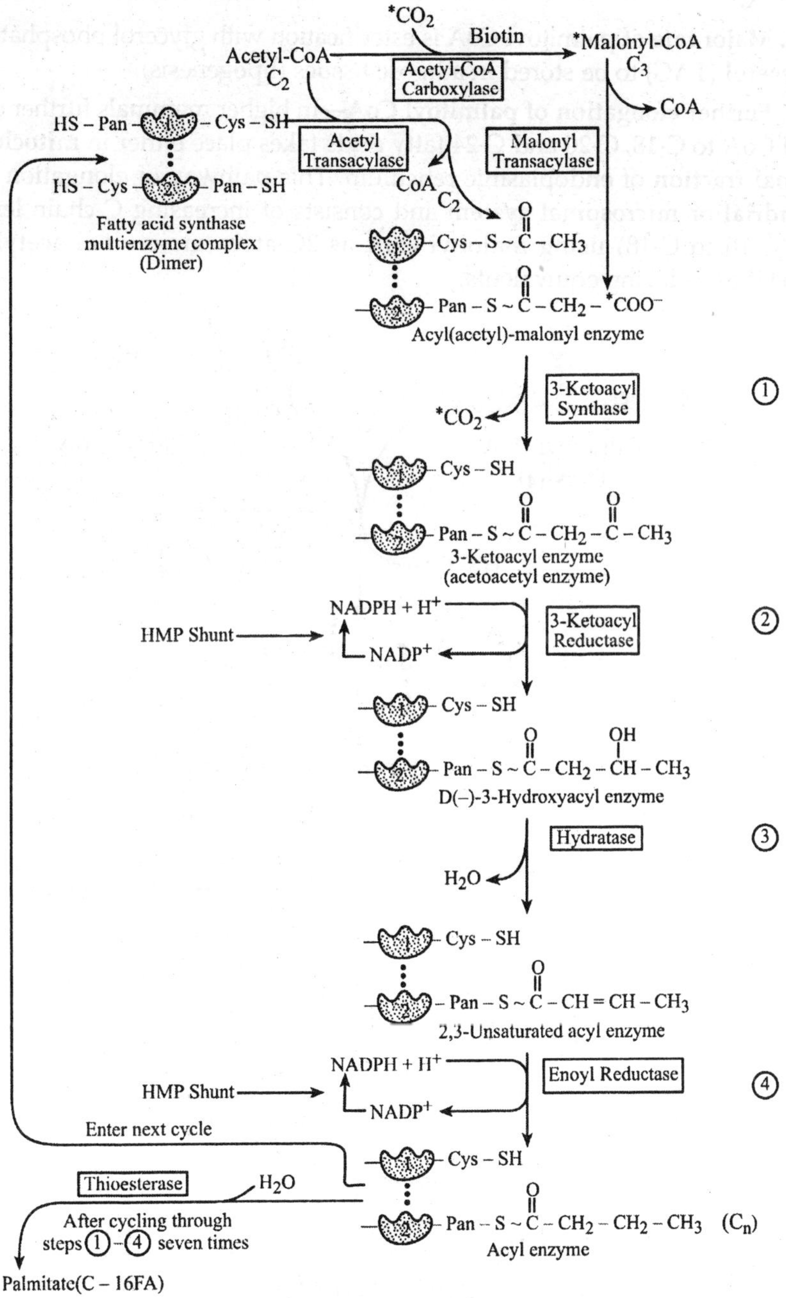

Fig. 11.7 : Denovo biosynthesis of fatty acids by sequential additions of malonyl residues to increase C-chain length by 2 C-atoms in one cycle with steps 1-4

Fate of Palmitate

1. The free palmitate must be activated to palmitoyl CoA before it can enter any other metabolic pathways by acyl CoA synthase enzyme in the presence of ATP and coenzyme -A.

2. Major fate of palmitoyl CoA is esterification with glycerol phosphate to produce **triacylglycerol** (TAG) to be stored in adipose tissues (lipogenesis).

3. Further elongation of palmitoyl CoA—In higher mammals further elongation of palmitoyl CoA to C-18, C-20 and C-24 fatty acids takes place either in **mitochondria** or in **microsomal fraction** of endoplasmic reticulum. This pathway of elongation is known as **mitochondrial** or **microsomal system** and consists of increasing C-chain length by two carbons (C-16 to C-18) using malonyl CoA as 2C-atom moiety i.e. acetyl donor and NADPH+H^+ as reducing equivalents.

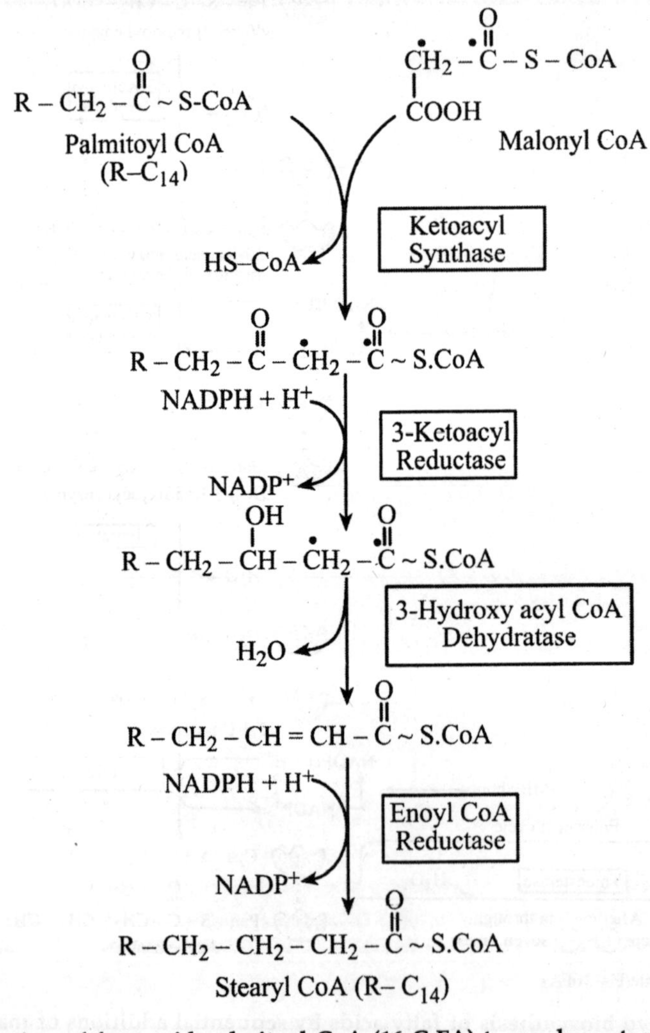

Fig. 11.8 : Elongation of fatty acid palmitoyl CoA (16 C-FA) to produce stearyl CoA (18 C-FA)

- This process is catalyzed by **mitochondrial** or **microsomal elongase** enzyme complex and involves enzymes ketoacyl synthase, β ketoacyl CoA reductase, dehydratase and 2nd reductase as shown in the Fig. 11.8.
- **Elongation of C-chain length**—For instance conversion of palmitoyl CoA(C-16) to stearate (C-18) is summarized in the Fig. 11.8.
- In fasting and diabetes mellitus, the elongation of fatty acids is abolished
- In **brain,** elongation of stearyl CoA to C-22 and C-24 fatty acids increases rapidly during myelination process for the synthesis of **sphingolipids,** as the C-24 fatty acid lignoceric acid or its derivative is essential components of compound lipids of CNS.

Regulation of Lipogenesis

1. Nutritional state is major factor which control lipogenesis. Welfed state and carbohydrate rich diet stimulates fatty acid synthesis and lipogenesis, while fasting and hypoglycaemic state decrease it.

2. Hormonal regulation—Hormone insulin in welfed state increase lipogenesis while diabetogenic hormones during fasting conditions inhibit it via activating c-AMP protein kinase system which inhibit the key regulatory enzyme of fatty acid synthesis i.e. **acetyl CoA carboxylase**

3. Allosteric regulation—The rate limiting step regulating lipogenesis is the key enzyme **acetyl CoA carboxylase** (initial step), which is **allosterically activated by citrate** concentration which increases in welfed state.

4. Feedback inhibition by long chain FFA—Acetyl CoA carboxylase enzyme is **inhibited** by increased plasma levels of long chain FFA by feed back inhibition.

5. Longterm regulation at gene level—Insulin induces synthesis of acetyl CoA carboxylase directly at the gene level in welfed state, while glucagon and epinephrine **suppress genes** responsible for its synthesis during fasting and hypoglycemic state.

SYNTHESIS OF TRIACYLGLYCEROL(TAG)

Liver and **adipose tissues** are major sites for TAG synthesis, however other tissues can also synthesize them, but to a lesser extent.

TAG synthesis in liver—Triacylglycerol are synthesized in liver mainly for the esterification of dietary fatty acids and fatty acid synthesized denovo from the extra dietary glucose. TAG so produce are then incorporated and secreted as lipoprotein complex **VLDL** for their transportation to extrahepatic tissues for further metabolism, or for the storage in adipose tissues.

In adipose tissues TAG synthesis occurs primarily for the storage as reserve fuel source, to be utilized later on when needed. TAG synthesis takes place by esterification of fatty acyl CoA with **glycerol phosphate.**

Activation of fatty acids and glycerol is necessary prior to esterification and lipogenesis process and is achieved by following mechanisms.

1. Fatty acids activation—Activation of fatty acids is catalyzed by **thiokinase** enzyme, which requires ATP and coenzyme A, as given below.

2. Activation of glycerol

(a) **In liver** glycerol is activated by a specific enzyme **glycerokinase** utilizing ATP to produce glycerol phosphate as given under—

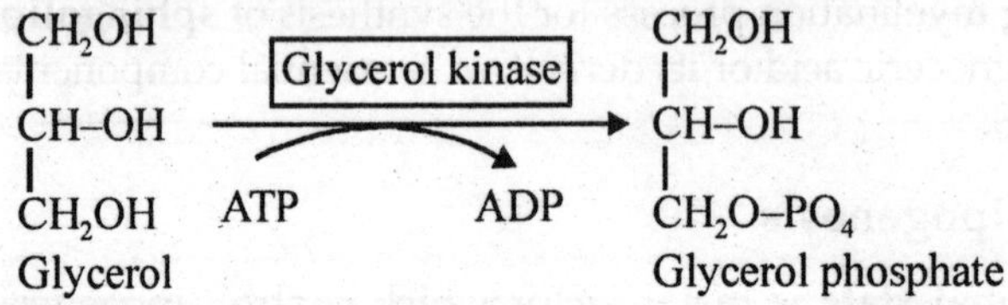

(b) In adipose tissues enzyme glycerol kinase is **deficient** and thus glucose oxidation (glycolysis) serves as the major source of dihydroxyacetone phosphate(DHAP), which is then reduced to **glycerol phosphate** by a dehydrogenase enzyme. The glycerol phosphate is utilized for esterification of fatty acids to synthesize triglycerides in adipose tissues.

TAG Synthesis—Esterification

The fatty acyl CoA transfers the acyl molecule to the hydroxyl group of glycerol phosphate at position Ist in the presence of enzyme **acyltransferase** to produce first a lysophosphatidic acid and then another acyl group is added by ester bond at 2nd position to form **phosphatidic acid.** This phosphatidic acid is the **precursor** for synthesis of **phospholipids** also, for example **lecithin, cephalins** etc. as shown in the Fig. 11.9.

For TAG synthesis a **phosphatase** enzyme hydrolyzes (cleaves) phosphate group from phosphatidic acid to produce **diacylglycerol,** which is then esterified with third acyl group to form TAG as shown in the Fig. 11.9.

Mixed Triglycerides

TAG usually contain different fatty acids—The three fatty acyl groups in naturally occurring triglycerides are not same. Usually a saturated fatty acid is present on C-1 of glycerol and an unsaturated (PUFA) on 2nd C-atom of glycerol. A monounsaturated (MUFA) or a saturated acyl group is esterified on 3rd C-atom in naturally occurring TAG.

PHOSPHOLIPID METABOLISM

Phospholipids are compound or conjugated lipids with **phosphate** as an essential constituent and include **lecithin, cephalin, phosphotidyl inositol, plasmalogens, cardiolipin and sphingomyelins.**

Specific functions of phospholipids—Phospholipids are essential structural constituent of lipid bilayer of biomembranes and regulate the cell permeability. The phospholipids are the major components to form lipids of brain, myelin sheath of nerves and acts as lung surfactant to prevent adherance of alveoli in lungs. The phospholipids are also necessary for blood clotting process.

1. Synthesis of lecithin and cephalin—For the synthesis of lecithin and cephalins, choline and ethanolamine first get phosphorylated and then combine with high energy

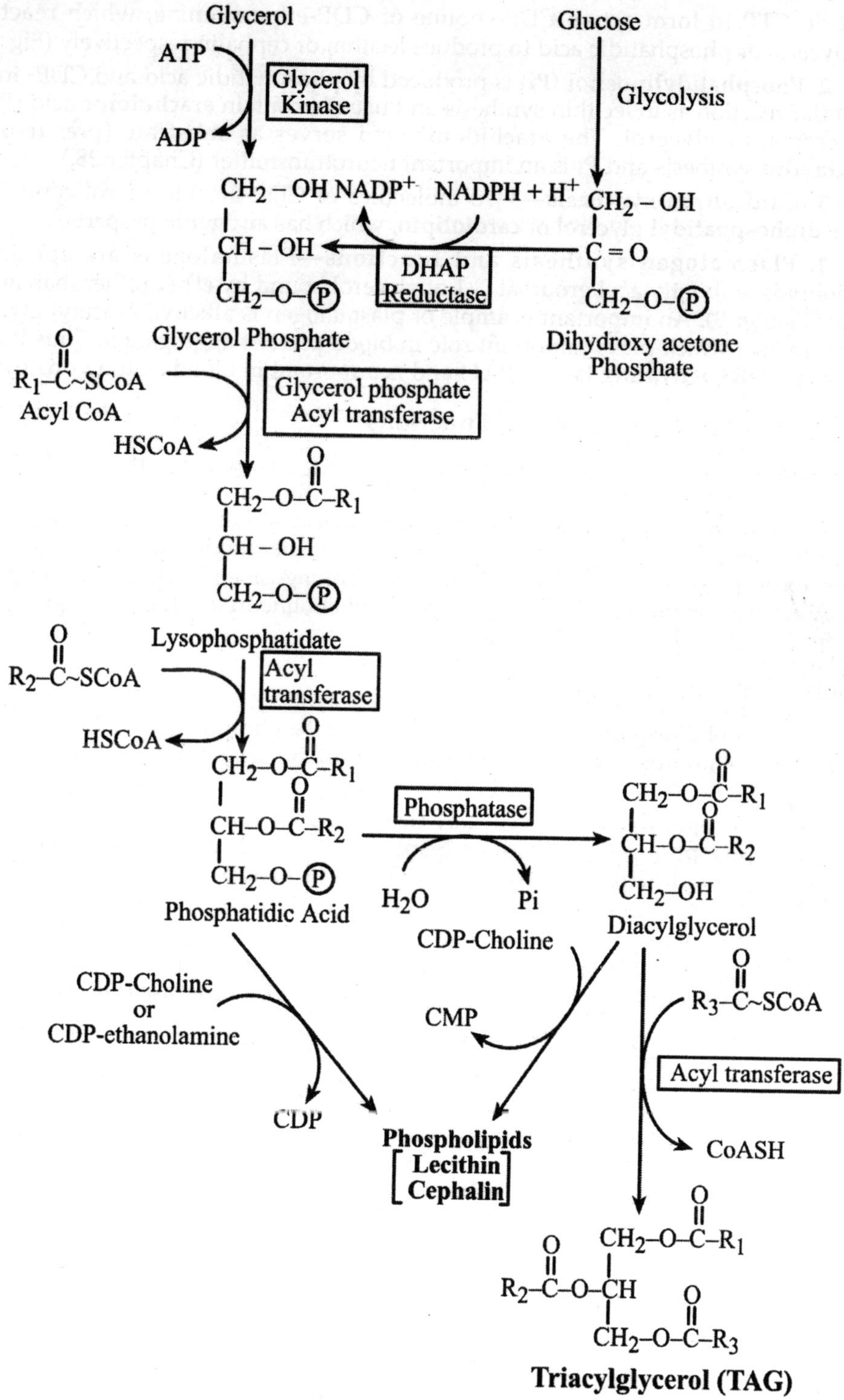

Fig. 11.9 : Syntheis of TAG and phospholipids by esterification

nucleotide **CTP** to form active CDP-choline or CDP-ethanolamine, which react with diacylglycerol or phosphatidic acid to produce lecithin or cephalin respectively (Fig. 11.9).

2. Phosphatidylinositol (Pi) is produced by phosphatidic acid and CDP- inositol by a similar reaction as in lecithin synthesis and usually contain **arachidonic** acid (PUFA) on 2nd carbon of glycerol. The arachidonic acid serves as substrate (precursor) for **prostaglandin** synthesis and Pi is an important neurotransmitter (Chapter 28.)

3. Cardiolipin synthesis—Two molecules of CDP-diacylglycerol condense to produce **diphosphatidyl** glycerol or **cardiolipin,** which has **antigenic** properties.

4. Plasmalogen synthesis and functions—Plasmalogens are specialized phospholipids with fatty acyl group at C-1 of glycerol bound by **ether** rather than an ester linkage (Chapter 9). An important example of plasmalogen is alkenyl, 2–acetyl glycerol 3 phosphocholine, which plays important role in blood platelet aggregation. Thus it is also known as **platelet activating factor (PAF)** and is important in blood clotting process.

Sphingomyelins : Synthesis and Functions

Sphingomyelins are phospholipids, containing a complex amino alcohol called, **sphingosine** instead of glycerol as their molecular constituent. Sphingosine is synthesized from 16 C-atom fatty acid palmitoyl CoA and amino acid serine, which condense together to produce a condensation complex. This complex on further modification by a sequence of reactions form sphingosine. The hydroxal group of sphingosine is then esterified by a fatty acyl CoA to form **ceramide,** which reacts with CDP-choline to produce sphingomyelin as shown in the Fig. 11.11.

Catabolism of Phospholipids

A group of **phospholipase** enzymes hydrolyze the phospholipids and degrade them to respective acyl groups, glycerol, choline and phosphate. Important phospholipases and their mechanism of action is described below—

(a) **Phospholipase A** specifically cleaves the fatty acyl group at C-1 position of phospholipid to produce lysolecithin, which is further acted upon by **phospholipase B** to remove 2nd acyl group from C-2 position. The phospholipids can hydrolyze both acyl groups of a phospholipid molecule.

(b) **Phospholipase A_2** present in pancreatic juice, **snake venom** and **bee-venom** specifically hydrolyze acyl group from **C-2** position of lecithin to produce **lysolecithin** which can be reacylated by acyl transferase (Fig. 11.10).

(c) **Phospholipase C and D** attack ester bond at position C-3 to cleave choline phosphate and free choline respectively as illustrated in the Fig. 11.10.

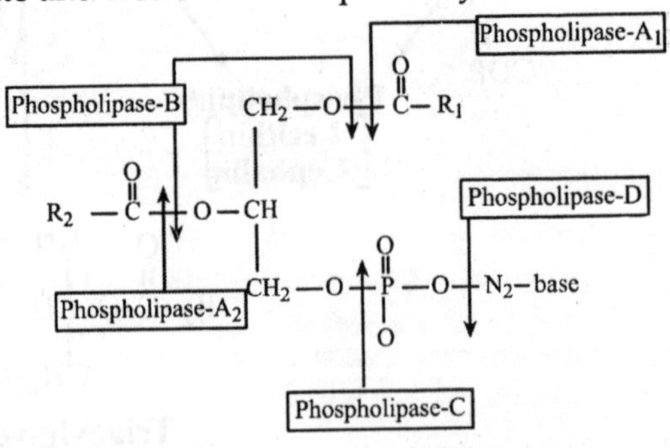

Fig. 11.10 : Sites of hydrolytic attacks by different phospholipases on phospholipids to hydrolyze them

Glycolipid metabolism—Glycolipids are produced by combination of **ceramide** with one or two sugar(galactose) units. The simplest glycolipids are **cerebrosides** formed as shown in the Fig. 11.11.

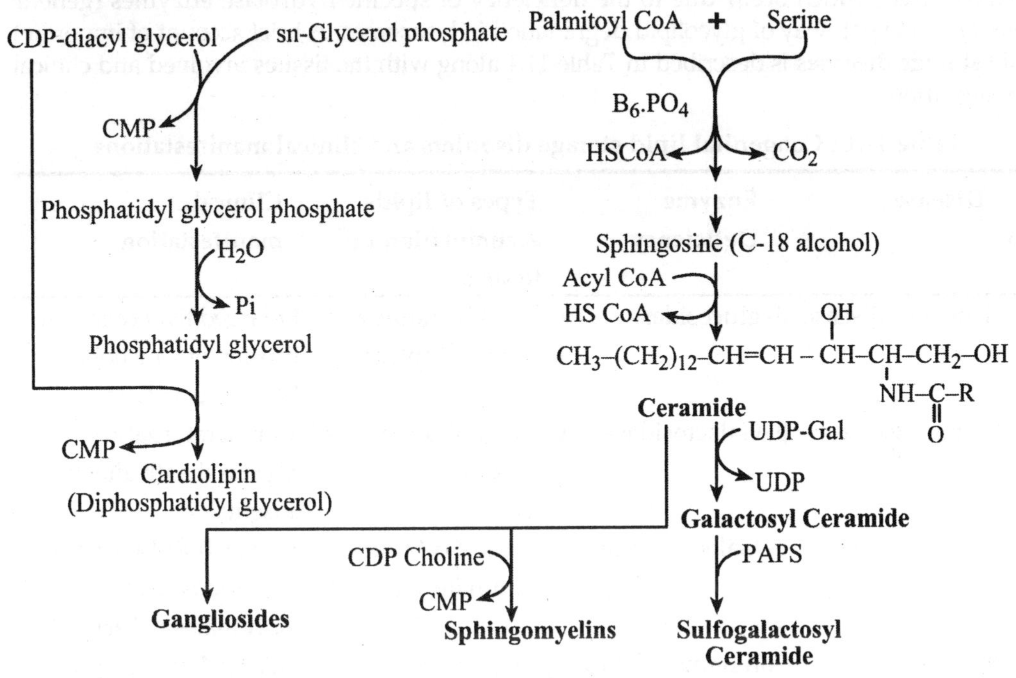

Fig. 11.11 : Biosynthesis of ceramide, sphingomyelins and galactosylceramide (cerebrosides) & its sulpho derivates (PAPS is active sulphate group donar)

Gangliosides—The gangliosides are highly complex lipids and are synthesized from ceramides by the stepwise addition of sugar units from UDP-glucose, UDP-galactose and **sialic acid** i.e. N, acetylneuraminic acid (NANA).

Biomedical importance of compound lipids —

1. **Essential constituent of biomembranes**—Phospholipids, sphingolipids and glycolipids are amphipathic compound and are ideally suitable as main constituent of lipid bilayer of **biomembranes.**
2. **Lung surfactants**—Phospholipids such as lecithin, cephalins are natural lung surfactants and prevent alveoli from collapsing duc to adhcrance. Lecithin is synthesized shortly before parturition in full term infants. **Deficiency** of **Lung surfactants** in pre-term newborns results in **respiratory distress syndrome** which can be life threating.
3. **Phosphotidylinositol** provide II messenger for actions certain hormone after their binding on cell membrane receptors inside target cell (Chapter 28.)
4. **Plasmalogens** as alkenyl phospholipids are platelet-activating factor (PAF) and are important in blood clotting process.
5. **Glycosphingolipids** attached at outer side of plasma membrane are considered to be important for **intracellular communication,** for cell to cell contact and are also important constituents of ABO blood group substances.

Glycolipid Storage Diseases—Clinical Aspects

A group of congenital inherited **glycolipid storage diseases** (10-12 numbers) have been reported, which occur due to the deficiency of specific hydrolase enzymes (genetic defects), in the pathway of glycolipid degradation in lysozomes. A brief account of important **lipid storage diseases** is described in Table 11.1 along with the tissues involved and clinical manifestations.

Table 11.1 : Congenital lipid storage disorders and clinical manifestations

S. No.	Disease	Enzyme Deficiency	Types of lipids accumulation in tissues	Clinical manifestation
1.	Gaucher's disease	β- glucosidase	Glycosyl ceramides in liver and spleen	Enlarged liver & spleen Erosion of long bones Mental retardation
2.	Krabbe's disease	β -Galactosidase	Galactosyl ceramides in tissues	Mental retardation Myelin sheath almost absent
3.	Fabry's disease	α- Galactosidase	Glycoacyl ceramides accumulate in tissues	Skin rash,Kidney failure X-linked recessive trait (only males suffer)
4.	Niemann Pick's disease	Sphingomyelinase	Sphingomyelins accumulate in liver and spleen	Enlarged liver & spleen Mental retardation Death in early life
5.	Tay-Sach's disease	Hexosaminidase A	Gangliosides accumulation in tissues	Mental retardation blindness, muscular weakness. Early death
6.	Faber's disease	Ceramidase	Sphingo ceramides accumulate in tissues	Hoarseness, dermatitis Skeletal deformities Mental retardation Death in early life

□□□

12

CHOLESTEROL METABOLISM

Cholesterol is exclusively an **animal sterol** found in almost all body tissues and in blood. Cholesterol in body is present either in free form or esterified with long chain fatty acids as cholesteryl ester, and thus it is classified as derived lipid and it exhibits **characteristic properties** of lipids. Chemically it is a sterol with steroid ring structure known as cyclopentano perhydrophenanthrene ring. (Fig. 12.1).

Functions of cholesterol—It performs several important functions in the body which are enumerated here under—

1. Cholesterol being amphipathic in nature, it forms essential constituent of lipid bilayer of biomembranes and plays important role in regulating cell membrane permeability.
2. Cholesterol is **precursor** of all steroid hormones synthesized in body (a) Adrenocorticoid steroid hormones such as cortisol, cortisone, corticosterone etc., (b) Male sex hormones e.g. testosterone, and (c) Female sex hormones such as estrogens, progesterone etc.
3. **Vitamin D synthesis—7 dehydrocholesterol** (an intermediate in cholesterol biosynthesis) is precursor for vitamin D synthesis in the skin, in presence of sunlight.
4. **Synthesis of bile acids and bile salts**—Cholesterol is oxidized and metabolized in liver to produce **bile acids** and **bile salts,** which are necessary for fat digestion and absorption in the GIT.
5. **As insulator**—Cholesterol is present in rich amount in the lipids of brain, CNS and myelin sheath of nerves and acts as an insulator and prevents leakage of nerve impulses.

Total body cholesterol is about 140 g, which is distributed as—approximately 30g in brain and nerves, 30 g in muscles, 30 g in adipose tissues and rest in liver (10g), skin (10g), blood(10g) and is found to occur in almost all cell of the body.

A part of cholesterol in the body is obtained from diet (animal food stuffs) but most of it is synthesized in body mainly in liver. However it can also be synthesized in almost all tissues of body.

BIOSYNTHESIS OF CHOLESTEROL

Site of synthesis—Most of the cells (except RBC) can synthesize cholesterol, but largely **liver and intestine** are the major sites for cholesterol biosynthesis, followed by skin, adrenal cortex and reproductive organs etc. These tissues are actively involved in steroid hormone synthesis, where cholesterol is needed as precursor.

The enzymes involved in cholesterol biosynthesis occur partly in **cytosol** and rest in **microsomal fraction** of reticuloendothelial system of the tissues.

Raw materials (precursor) required for the cholesterol synthesis include —

(a) **Acetyl CoA,** that form all the 27 C-atoms of steroid ring of cholesterol and is provided mainly from glycolysis of glucose. **ATP citrate lyase** in cytoplasm liberate acetyl CoA from citrate in cytosol for cholesterol synsthesis, as in case of fatty acid biosynthesis (details in the Chapter 11).

(b) **Reducing equivalents as NADPH+H$^+$** are necessary for cholesterol biosynthesis and are provided by HMP shunt pathway of glucose oxidation. This pathway is actively going on in all tissues, which synthesize cholesterol and other steroid compounds.

Cholesterol biosynthesis occur in two phases—

(a) Ist phase involves formation of **isoprenoid units**(5 C- hydrocarbons) from acetyl CoA.

(b) In 2nd phase the condensation of six 5 C-atom isoprenoid units takes place to produce a long chain 30 C-hydrocarbon, called **squalene** that cyclizes to form a **steroid ring structure.**

Ist Phase—Synthesis of Isoprenoid units (5 Carbon compound)—This involves the following steps—

(a) **Synthesis of HMG CoA**—Two molecules of acetyl CoA condense in the presence of enzyme **thiokinase** to form acetoacetyl CoA (4 C-compound), which combines with another molecule of acetyl CoA to produce **HMG CoA** (6 C- compound). The reactions are similar to ketone bodies synthesis (Fig. 11.5 Chapter 11), but site is different, since ketone bodies are produced in liver, inside mitochondria, while cholesterol synthesis takes place in cytoplasm (two pools of HMG CoA exists).

(b) **Formation of mevalonate**—Cytosolic HMG CoA is then reduced to mevalonate catalyzed by a key regulatory enzyme **HMG CoA reductase.** This is **a rate limiting step** in cholesterol biosynthesis and requires 2 molecules of **NADPH+H$^+$** to provide reducing equivalents as shown in the Fig. 12.1.

(c) **Activation of Mevalonate—Energy as 3 ATP molecules** is then utilized to activate mevalonate to **3-phosphomevalonate 5-pyrophosphate** in 2 steps, catalyzed by specific **kinase enzymes** (Fig. 12.1).

(d) **Formation of Isopentenyl Pyrophosphate**—In the presence of a specific **decarboxylase** enzyme, phosphomevalonate pyrophosphate(6C) is decarboxylated to an active 5 C-hydrocarbon **isoprenoid unit,** known as **isopentenyl pyrophosphate(isopentenyl pp)** as shown in the Fig. 12.2.

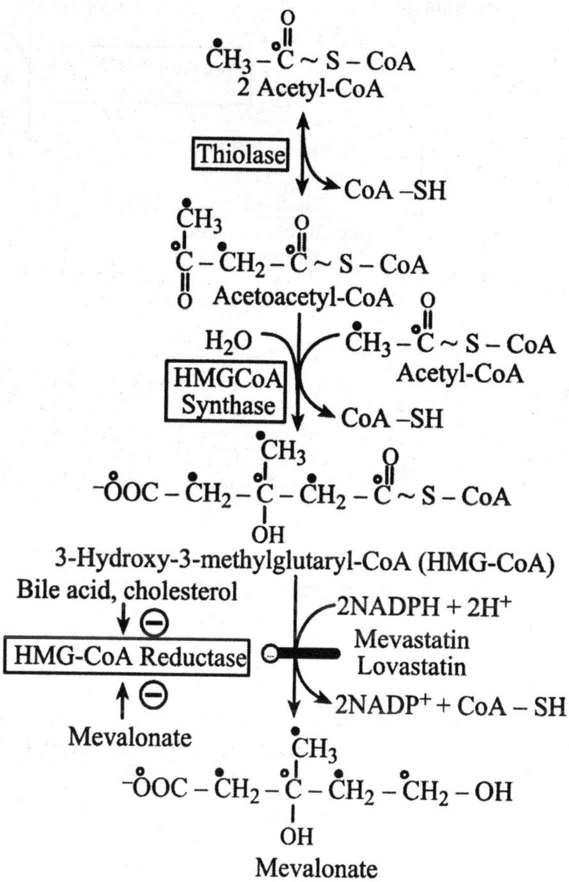

Fig. 12.1 : Synthesis of HMG CoA and its reduction to mevalonate (6C-Unit) from acetyl CoA

IInd **Phase**—Synthesis of Squalene (30C-HC)—A long chain hydrocarbon squalene (30 C-chain) is synthesized by the condensation of six isoprenoid units, which involves following steps—

1. Isomerization of isopentenyl diphosphate—First a molecule of isopentenyl pp is isomerized catalyzed by an **isomerase** enzyme, by shifting a double bond to form **dimethylallyl diphosphate**(5C). This then **condenses** with another molecule of isopentenyl diphosphate (5C) to produce a **10 C- compound called geranyl diphosphate** (10C) as shown in the Fig. 12.2 in the presence of **cis-prenyl transferase enzyme.** Hydrolysis of ppi group provides energy to form covalent bond during the condensation reaction.

2. Formation of Farnesyl diphosphate (15 C-compound)—Geranyl diphosphate (10 C) so formed, then condenses with another molecule of isopentenyl diphosphate (5C) in presence of enzyme cis-prenyl transferase to produce a 15 C- chain intermediate **farnesyl diphosphate** (15 C-compound).

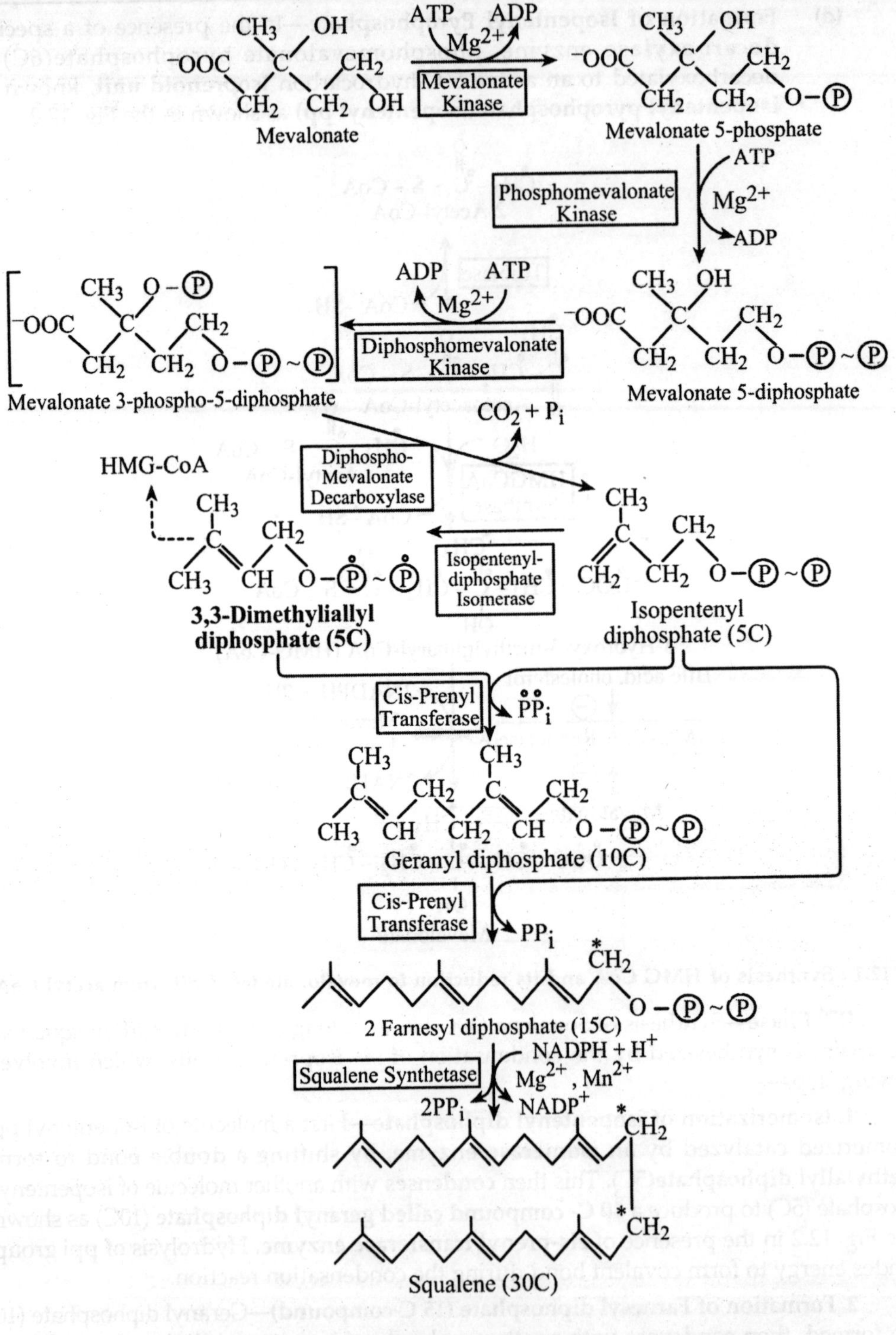

Fig. 12.2 : Formation of squalene (30C-hydrocarbon chain) from six isoprenoid (5C-compound) units

3. Synthesis of Squalene (30 C-compound)—In the next step, two molecules farnesyl diphosphate fuse together, at the diphosphate end(head to head position) by a **reductive condensation** reaction. This involves first an elimination of phosphate group, followed by reduction utilizing $NADPH^{+}+H^{+}$ with further elimination of remaining ppi group. The pyrophosphate group so removed is hydrolyzed to provide energy for the condensation of two farnesyl residues to produce a 30 C- chain compound called squalene.

4. Synthesis of Lanosterol from Squalene—Squalene then undergoes a special cyclization reaction to produce a I[st] tetracyclic steroid compound **lanosterol** (30 C-sterol). During this cyclization, squalene is first converted to **squalene epioxide** by a mixed function oxidase enzyme utilizing O_2 and $NADPH+H^{+}$ (Fig. 12.3), which then cyclizes to form lanosterol as catalyzed by an **cyclase** enzyme.

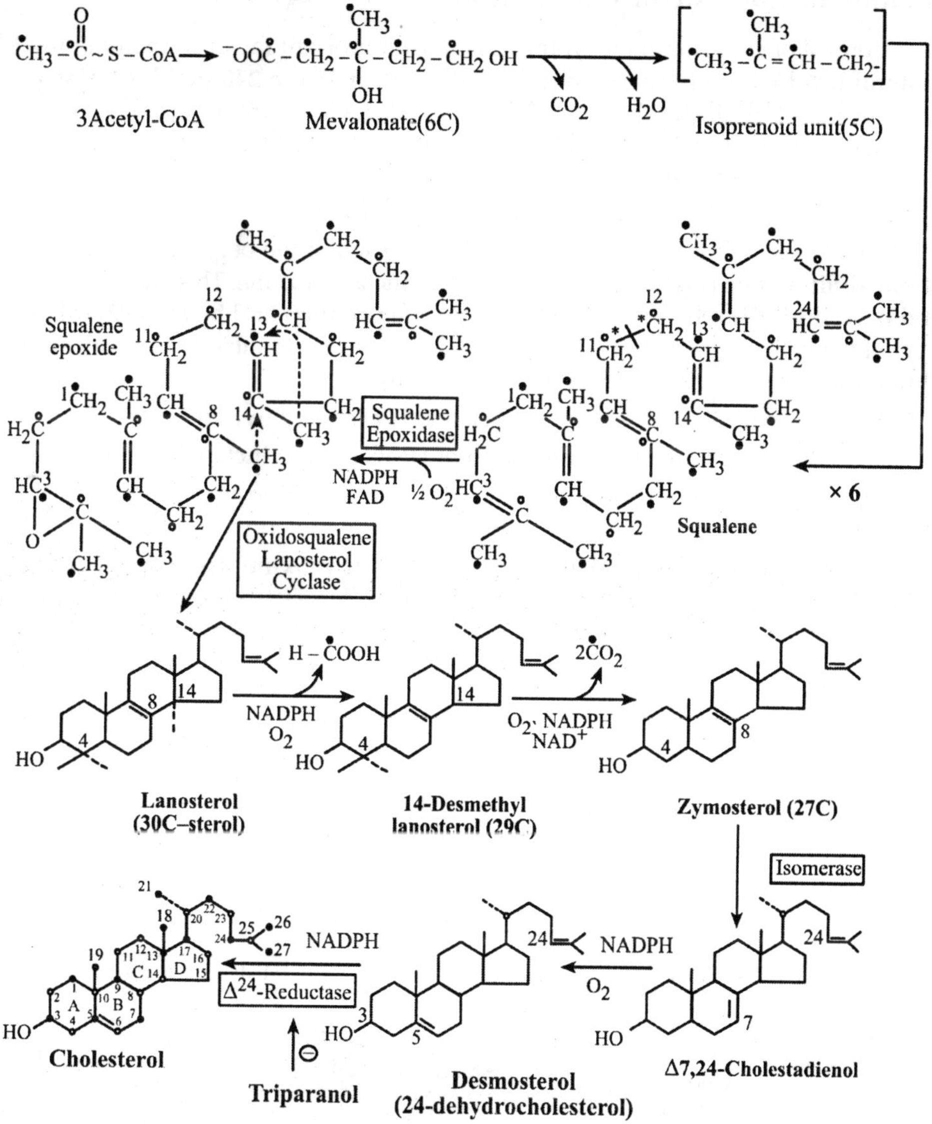

Fig. 12.3 : Overview of cholesterol biosynthesis from acetyl CoA and conversion of lanosterol to form cholesterol

5. Conversion of lanosterol to cholesterol—The 3 extra methyl groups of lanosterol on C-atoms 4 and 14 of steroid ring (Fig. 12.3) are then removed by oxidation and successive decarboxylation reactions to produce a 27 C - steroid known as **zymosterol.** This steroid is similar to cholesterol, but has two extra double bonds, one in B ring at 8-9 position and second in the side chain at C 24-25 position. (Fig. 12.3). To convert zymosterol into cholesterol, first the double bond in B ring of zymosterol is shifted from C 8-9 position to C 5-6 position by isomerization(Fig. 12.3) to form **desmosterol**. Finally **cholesterol is produced** from desmosterol by reduction of double bond of the side chain at position C 24-25 by the enzyme **Δ^{24} reductase** (Fig. 12.3).

Regulation of Cholesterol Biosynthesis

Control of cholesterol biosynthesis is most important factor in maintaining **blood cholesterol level** within normal physiological limits of **150 to 240 mg/dl of plasma**. Any increase in blood cholesterol level is a major risk factor causing **atherosclerosis** and cardiovascular diseases. Cholesterol biosynthesis is regulated as described below—

HMG-CoA Reductase—A key regulatory enzyme

Regulation of cholesterol biosynthesis is exerted at the beginning of the pathway, at the **rate limiting** step catalyzed by **HMG CoA reductase** enzyme. This enzyme is subjected to **feed back inhibition** by **mevalonate** (immediate product) and by **cholesterol,** the final product of the pathway (Fig. 12.4). The HMG CoA reductase is subjected to both **short term** and **long term** regulations by following mechanisms.

1. Short term regulation—It is quick and is achieved in two ways—

(a) Feed back **allosteric inhibition** of HMG CoA reductase by mevalonate.

(b) **Covalent modification** by phosphorylation/dephosphorylation of HMG CoA reductase by c-AMP and **hormones** as shown in the Fig. 12.4.

(c) **Hormonal regulation**—Insulin and thyroxine are known to increase cholesterol biosynthesis by activating HMG CoA reductase enzyme by dephosphorylation (covalent modification), while glucagon and glucocorticoids inactivate it via c-AMP protein kinase system as illustrated in the Fig. 12.4.

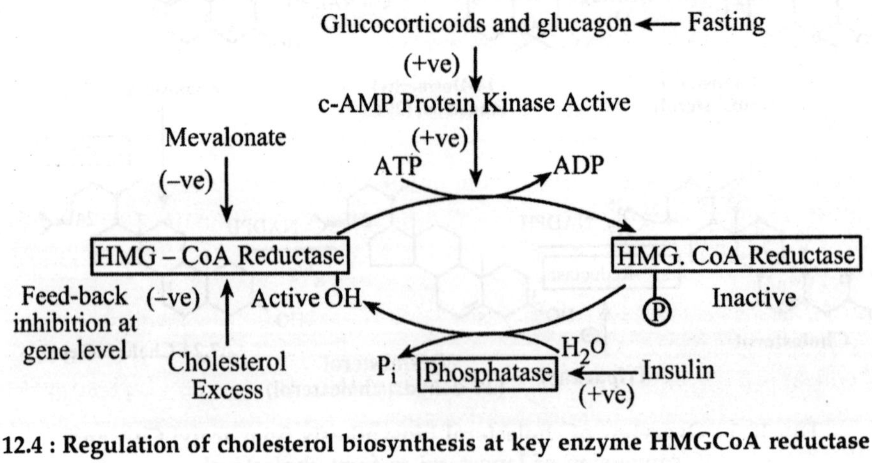

Fig. 12.4 : Regulation of cholesterol biosynthesis at key enzyme HMGCoA reductase step

2. Long term regulation at gene level—In case of increased body cholesterol pool and hypercholesterolemia, a metabolite or cholesterol itself feed back inhibits and **suppresses** transcription of gene, responsible for HMG CoA reductase synthesis, thereby decreasing cholesterol biosynthesis (details in Chapter 22).

Body cholesterol pool—The concentration of body and blood cholesterol pool is regulated by various factors as describe here—

(a) A balance normally exists between rate of cholesterol biosynthesis and its rate of metabolism (oxidation to bile acids) and excretion to maintain body cholesterol pool.

(b) Amount of cholesterol taken in diet and fasting conditions would accordingly influence cholesterol biosynthesis to regulate normal blood cholesterol level.

(c) **Cholesterol uptake by LDL receptors**—Cholesterol biosynthesis is inhibited in tissues as a result of cholesterol uptake by LDL receptors. The influx of cholesterol and high cholesterol contents of the cells would **suppress** transcription of gene responsible for HMG CoA reductase synthesis (feed back inhibition at gene level).

(d) **LDL-receptor synthesis**—The synthesis of LDL-receptors is also a highly regulated process. It has been observed that the number of LDL-receptors on cell surface is regulated by cellular need for cholesterol. Thus influx of cholesterol **down regulates** the LDL-receptors synthesis by feed back inhibition at gene level to **suppress** transcription process.

CHOLESTEROL TRANSPORT

Cholesterol is transported by blood to various tissues as **lipoproteins fractions** of plasma. The highest amount of cholesterol is found in LDL-fraction, mostly as esterified form, followed by VLDL and HDL fractions.

Absorption from GIT—Dietary cholesterol and cholesterol esters after hydrolysis in GIT are absorbed as **micelles** by intestinal mucosa along with other lipids and incorporated into **chylomicrons,** after reesterification with long chain fatty acids. Chylomicrons are then released via lacteals into thoracic duct and then into systemic circulation.

Lipoprotein lipase (LpL)—Lipoprotein lipase enzyme on cell membranes of extra hepatic tissues, acts on chylomicrons and hydrolyze triglycerides (dietary neutral fat) and fatty acids released are taken up by tissues, leaving behind **chylomicron remnants,** containing mainly cholesterol esters and phospholipids.

Chylomicron remnants are taken up by liver—The **receptors** on hepatic cell membrane facilitates uptake of both cholesterol containing chylomicron remnants and LDL fraction mediated by respective apo proteins. In liver cholesterol esters are hydrolyzed to free cholesterol.

VLDL synthesis in Liver—Liver then incorporates dietary as well as endogenously synthesized (in liver itself) TAG, cholesterol esters, phospholipids, long chain fatty acids etc along with specific **apoproteins** to produce lipoprotein **VLDL.** The VLDL so synthesized then is transported via blood to peripheral tissues for further metabolism. (Details in the Chapter 13).

From plasma VLDL- TAG are taken up by adipose tissues—TAG from plasma VLDL fraction are taken up by adipose tissue for storage as reserve energy source by the action of lipoprotein lipase enzyme (LpL) and VLDL is converted to **IDL (VLDL remnant).** IDL is later on converted to **LDL fraction,** rich in cholesterol by the addition/deletion of certain **apoproteins** (Chapter 13).

LDL cholesterol metabolism—The LDL particles are cholesterol rich (75% of total plasma cholesterol) and contains only **apoprotein B-100.** Most of the **LDL particles** are formed from VLDL and IDL in the plasma as mentioned above, but a small portion is also released directly from the liver. Major role of LDL is to transport cholesterol from liver to peripheral tissues for its utilization for various purposes.

LDL receptors—The LDL receptors present on the plasma membrane of peripheral tissues are specific for the uptake of cholesterol rich LDL particles inside the tissue for further metabolism. The LDL receptors are localized in special regions called **coated pits** which contain a protein **clathrin.** The LDL-receptors recognize apo B-100 of LDL particle and bind them. The LDL receptor-LDL complex is then internalized by **endocytosis** and the vesicles fuse with lysozomes, which hydrolyze them and cholesterol is released inside the cells. Uptake of cholesterol rich LDL is a **well regulated process** and is controlled by **LDL receptor** synthesis and their concentration (number) on the cell membrane, which is achieved by feed back inhibition at gene level.

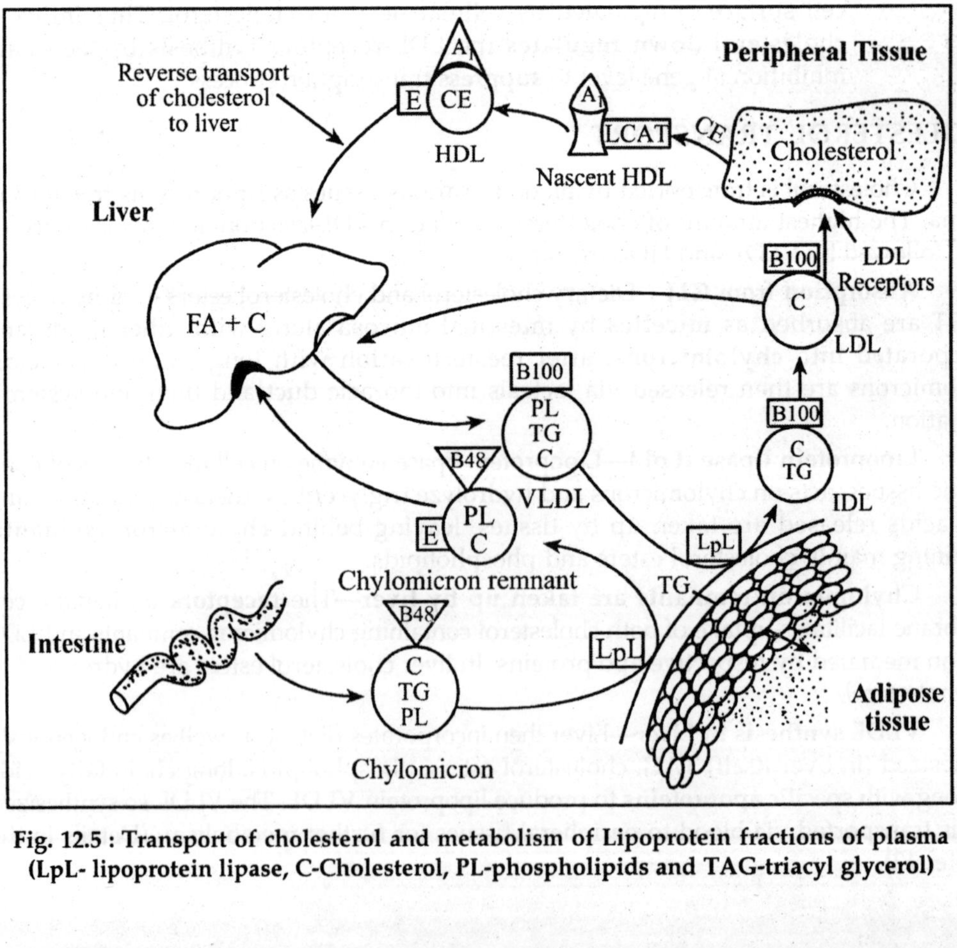

Fig. 12.5 : Transport of cholesterol and metabolism of Lipoprotein fractions of plasma (LpL- lipoprotein lipase, C-Cholesterol, PL-phospholipids and TAG-triacyl glycerol)

Fate of cholesterol in tissues—

The free cholesterol released into the peripheral tissues has following fate—

(a) It is incorporated into **cell membrane lipids** to regulate its fluidity and permeability.

(b) Cholesterol is utilized for **steroid hormones synthesis** specially in adrenal cortex, ovaries and testes.

(c) **Storage after reesterification in cell**—The reesterification of cholesterol in the tissues is catalyzed by the enzyme ACAT(acyl cholesterol-acyl transferase), which is specific to catalyze esterification of cholesterol with **monounsaturated fatty acid (MUFA)**, e.g. oleic acid. The esterefied cholesterol with MUFA is stored in the tissues to some extent for its use lateron when needed.

(d) **Expulsion of extra cholesterol out of cell**—The extra cholesterol not needed in peripheral tissues any more, is expelled out and extracted by **HDL** fraction of plasma lipoproteins, for its reverse transport as described here under—

Reverse Cholesterol Transport—Role of HDL

From peripheral tissues, the extra cholesterol is transported back to the liver for its final oxidation to bile acids and excretion via bile into GIT. **HDL fraction of plasma with apo A-1 and enzyme LCAT**(lecithin-cholesterol acyl transferase) has great role in the reverse transport of cholesterol to liver for its oxidation and excretion. HDL cholesterol is **antiatherosclerotic** in nature and is protective against heart diseases, and thus is termed as **good cholesterol**.

Role of PUFA—Polyunsaturated fatty acids are required for the reesterification of cholesterol, which is **prerequisite** for its extraction out of peripheral tissues. This is because the enzyme LCAT is specific in esterification of cholesterol with **PUFA** only, which is transferred from 2nd C-atom of lecithin. The cholestrol esterified with PUFA is then taken inside HDL_2 fraction for the reverse transport and oxidation to liver. Thus PUFA has important role in **lowering body cholesterol pool** and decreasing **blood cholesterol level.**

HDL_2—The high density lipoprotein (HDL_2) is the main transporter of cholesterol from peripheral tissues to liver which is the only route for its oxidation into bile acids and its excretion. The intestine cells and liver synthesize phospholipids (lecithin with PUFA at 2nd C-atom) and apo A needed for HDL particles synthesis. The HDL newly synthesized is secreted in plasma as **nascent HDL,** which is **discoidal** in shape. The free cholesterol extracted from peripheral tissue is taken up by the nascent HDL after its esterification with PUFA. The apo A-1 of HDL activates the enzyme LCAT(lecithin -cholesterol acyl transferase) present in plasma, which then bind to the HDL disc and catalyzes the esterification reaction (details in Chapter 13).

The cholesterylester with PUFA then moves into the interior of nascent HDL disc. This process continues and ultimately HDL_2 **spheres,** with lots of cholesterol esters are formed. **Mature** HDL_2 then are taken up by liver by apo A-1 mediated receptors of hepatic cells. Inside liver, the cholesterol ester, phospholipid and TAG are released, for further metabolism.

Metabolism and Excretion of Cholesterol

The only route through which cholesterol can be metabolized, is its oxidation in liver to **bile acids,** and its **excretion in bile** into GIT. Approximately lg of cholesterol is excreted per day, half of it as bile acids and bile salts, while the rest as neutral sterols in the feces.

Enterohepatic circulation—Much of bile acids, bile salts and cholesterol secreted in bile in GIT are reabsorbed into portal circulation, taken up by liver and is resecreted into the bile. The process is known as **enterohepatic circulation.** This is important as bile acids, bile salts as well as cholesterol is needed for micelles formation to facilitate absorption of lipids in GIT. Thus only a part of bile acids, bile salts and sterol are excreted in feces normally (lg/day). **Coprostanol** is the main sterol in feces, which is produced by bacterial action on cholesterol in lower intestine.

Oxidation of Cholesterol to Bile Acids

The **primary bile acids,** that is **cholic acid** (main) and **chenodeoxy cholic acid,** are synthesized in the liver from cholesterol oxidation and involves following reactions —

1. 7 α-hydroxylation—The first step in cholesterol oxidation to synthesize bile acids, is addition of a **hydroxyl group** at 7th position in ring B of cholesterol molecule by a **microsomal 7-α hydroxylase enzyme.** It is a **monooxygenase enzyme** that requires cofactors NADPH+H$^+$, vitamin C and **cytochrome p450** to reduce 1 atom of O_2 to water, while the other is added at 7th C-atom of B-ring as hydroxyl group (Fig. 12.6). The 7α hydroxylase is a key regulatory enzyme and the reaction is a **rate limiting step** in the synthesis of bile acids.

Role of vitamin C—Vitamin C deficiency impairs bile acid synthesis at 7α-hydroxylation step, resulting in **hypercholesterolemia** and **atherosclerosis** in **scorbutic** guinea pigs.

2. Sub pathway of bile acids biosynthesis—The two primary bile acids, **cholic** acid and **chenodeoxy cholic** acid differ in an additional **OH group** at **12th position in "C" ring** of steroid molecule of **cholic acid.** So bile acid biosynthesis pathways is divided early, into two sub pathways. Apart from this difference, both sub pathways involve similar reactions, such as (a) hydroxylation and (b) oxidation of side chain to shorten it from 8 C-chain to 5 C-chain as illustrated in the Fig. 12.6.

3. Conjugation of bile acids—In liver **peroxysomes,** the primary bile acids are conjugated with glycine and taurine(oxidation/decarboxylation product of cysteine) in the ratio of 3 : 1 to produce glycocholic and taurocholic acid and glyco and taurochenodeoxy cholic acid respectively. These are then secreted in bile as sodium or potassium salts i.e. the **bile salts** into **GIT** (Fig. 12.6).

Secondary bile acids—These are produced in GIT by bacterial fermentation of primary bile acids, which cause **de-conjugation** (removal of glycine and taurine) and **de-hydroxylation** at 7th position(removal of 7 α OH group). This results in the formation of secondary bile acids i.e. **deoxycholic** acid from **cholic acid** and **lithocholic** acid from chenodeoxycholic acid as shown in the Fig. 12.6.

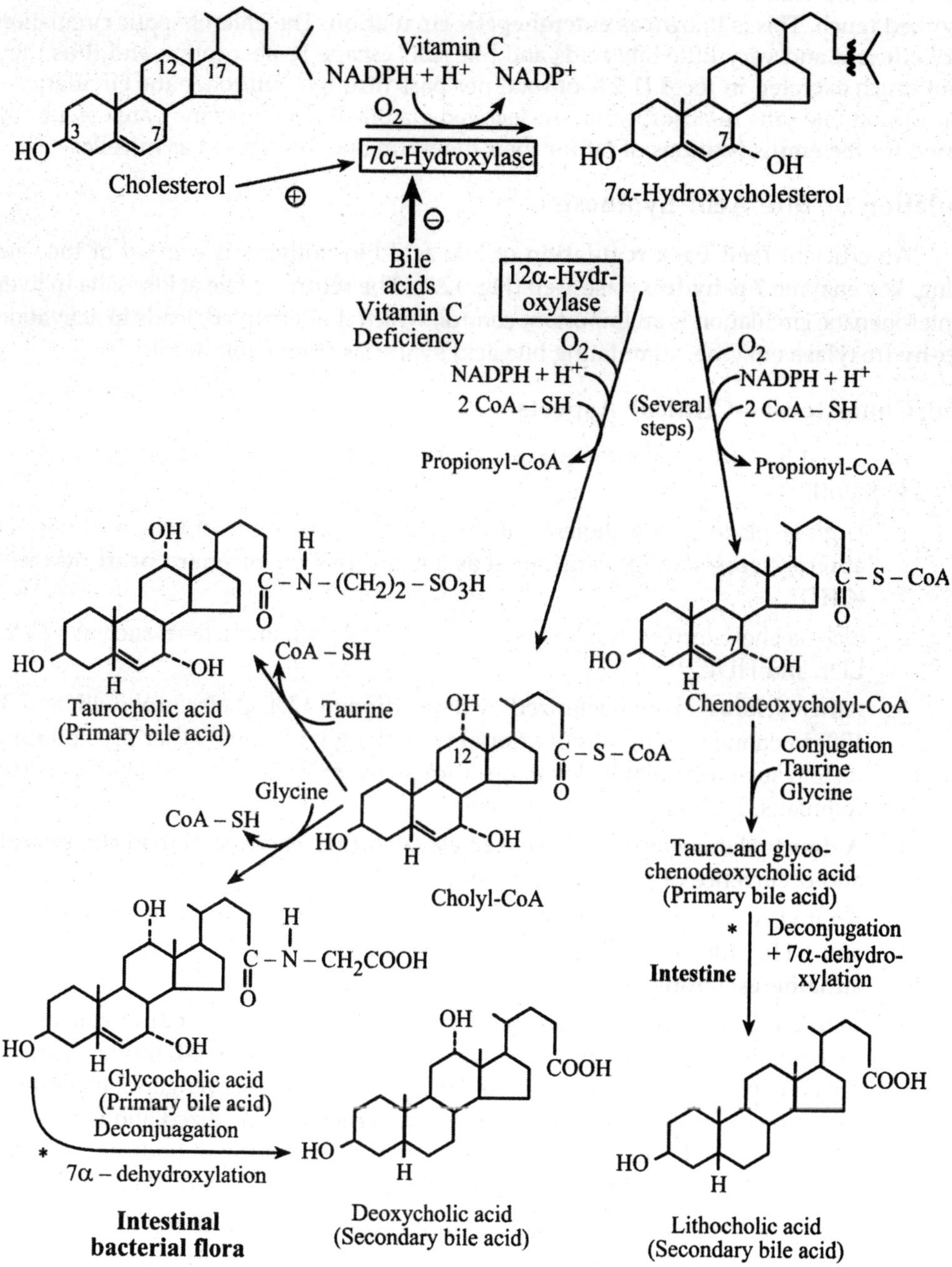

Fig. 12.6 : Oxidation of cholesterol to bile acids, their conjugation and secretion in the bile and production of secondary bile acids in GIT by the action of bacterial flora

Enterohepatic circulation—An appreciable amounts (98%) of primary and secondary bile acids are **absorbed** almost exclusively in the ileum (small intestine) and via portal blood are transported to liver, except lithocholic acid, which is insoluble and not reabsorbed much. This is known as **enterohepatic circulation.** The enterohepatic circulation is very efficient and very little bile acids and bile salts escape re-absorption and thus they are not much excreted in feces (1-2% of total per pass through enterohepatic circulation). Bile acids and bile salts re-absorbed are re-secreted via bile into GIT, as they are essentially required for the emulsifications of fat for their digestion and absorption as micelles.

Regulation of Bile Acid Synthesis

An efficient **feed back regulation** of bile acid biosynthesis is exerted at the rate limiting, key enzyme **7 α-hydroxylase step** (Fig. 12.7). The return of bile acids/salts to liver via enterohepatic circulation is an important control, which if interrupted, leads to activation of **7 α-hydroxylase** enzyme, stimulating bile acid synthesis from cholesterol.

Blood Cholesterol—Clinical Aspects

In normal healthy person the plasma cholesterol level is maintained in the **range of 150-240 mg/dl.**

1. High levels of serum cholesterol are strongly correlated with the incidence of **atherosclerosis** and is considered as a major risk for **coronary heart diseases (CHD).**
2. Plasma cholesterol is transported as various lipoprotein fractions such as VLDL, LDL and HDL.
3. **Atherosclerosis** is characterized by **deposition** of LDL cholesterol (with apo B-100) in connective tissues of arterial wall along with some other precipitating factors such as prolong elevated serum levels of VLDL, IDL and chylomicron remnants, TAG etc.
4. A direct relationship exists between elevated LDL cholesterol (bad cholesterol) and occurrence of atherosclerosis and CHD.
5. An inverse relationship exists between HDL_2 plasma level (good cholesterol) and CHD. Elevated HDL-2 plasma levels is protective against CHD as it is **antiatherosclerotic.**
6. **LDL cholesterol-HDL cholesterol ratio**—Most predictive parameter as the precipitating factor for atherosclerosis and CHD is **LDL—HDL ratio** in plasma. This is explained in terms of role of HDL_2 acting as the scavenger of cholesterol from tissues for its reverse transport to liver where it is oxidized and is excreted via bile into GIT and in feces. This results in the lowering of body and plasma cholesterol pool and reducing the risk of atherosclerosis and CHD.
7. **Hypercholesterolemia—Increased plasma cholesterol > 240 mg/dl.**—Elevated plasma total cholesterol and LDL cholesterol levels are major precipitation factors leading to atherosclerosis and is considered as **high risk factor** for coronary heart diseases (CHD).

Hypercholesterolemia—High blood cholesterol level is termed as hypercholesterolemia and is caused by several factors. The hypercholesterolemia is broadly divided in two types as described below—

1. Primary hypercholesterolemia—It is mainly caused by **genetic defects** and is inherited as **familial hypercholesterolemia.** In this, due to **genetic defect, LDL receptors** on cell membranes of peripheral tissues are not synthesized. Because of the lack or insufficiency of LDL receptors on peripheral tissues, LDL cholesterol will not be taken up by the tissues for its further metabolism from plasma and results in elevation of plasma cholesterol (mostly LDL cholesterol increases) to very high levels.

2. Hypercholesterolemia —Secondary to certain diseases

Cholesterol levels in blood increases in association with many clinical diseases as listed below—

(a) **Diabetes mellitus**—It is usually associated with hypercholesterolemia. This is because of excessive but incomplete fatty acid oxidation, resulting in **accumulation** of acetyl CoA, which is channeled towards **cholesterol synthesis.** (Details in the Chapter 11).

(b) **Obstructive jaundice**—It is associated with elevated plasma cholesterol levels, as its excretion via bile is impaired due to obstruction in biliary ducts.

(c) **Hypothyroidism**—Elevated levels of cholesterol (>300mg/dl) are associated with hypothyroidism because of decreased number of HDL receptors on hepatic cells, resulting in reduction in cholesterol uptake by liver for its oxidation and excretion.

(d) **Nephrotic syndrome**—In this condition serum cholesterol level is very high, (>400 mg/dl), probably because albumin is lost in urine, while globulin level rises resulting in elevated lipoprotein levels along with rise in plasma cholesterol.

Atherosclerosis—Increased blood cholesterol level, particularly elevation in LDL cholesterol and decrease in HDL fraction and simultaneous hyperlipemia for prolonged period will result in deposition of LDL cholesterol and TAG in subintimal regions of walls of arteries. Aorta, coronary arteries and cerebral vessels are mainly affected by atherosclerosis.

A part of LDL particles at **atherosclerotic site** are degraded by nonspecific uptake by macrophages. Later on the macrophages become overloaded with cholesterol esters and TAG and become **foam cells,** which is the **hallmark** of **atherosclerotic plaques** formation in the intima of blood vessels. This hardens them and results in narrowing of lumen of the arteries, leading to the **hypertension.** Any blood clot or thrombus would finally block the narrowed lumen of artery leading to **ischemia** of tissues involved. Finally insufficiency of O_2 would cause death of tissue involved. For instance myocardial infarction (MI) is caused by atherosclerosis and blockage of the coronary artery supplying blood to heart tissue resulting in ischemia. The lack of O_2 supply to heart muscles finally causes heart attack.

Measures to Lower Blood Cholesterol Level

1. Dietary measures—Dietary changes are highly beneficial in lowering moderately elevated blood cholesterol levels. The goal should be to lower total cholesterol, which ideally should be less than <200 mg/dl. LDL cholesterol should be less than 130 mg/dl, while HDL cholesterol **more** than 35 mg/dl is good for the prevention of atherosclerosis. To achieve this following dietary changes are necessary—

(a) **Low fat – low cholesterol diet** should be taken.

(b) **Unrefined, unpolished, whole grain cereals** should be taken and sugar should be restricted.

(c) **Proper ratio of PUFA-MUFA in diet**—Less saturated fats and **more unsaturated** lipids in diet should be taken to lower blood cholesterol levels. Oils and fats with a proper ratio of PUFA, ω_3 fatty acid and mono unsaturated fatty acid (MUFA) are best for lowering cholesterol level. They should be taken along with plenty of antioxidants. (Details in the Chapter 27).

(d) **Dietary fiber**—Both insoluble and soluble dietary fibers present in soyabean, guar, beans, legumes, raw salad etc lower blood cholesterol level (Chapter 27).

2. Life style changes —

(a) **Regular but moderate exercise** is a must to maintain body weight and to lower blood lipid levels including cholesterol. Regular exercise lower LDL cholesterol where as HDL is raised in plasma to protect from risk of atherosclerosis and CHD.

(b) Smoking, excessive alcohol intake and emotional stress. etc all increase lipolysis and fat mobilization from adipose tissues, thereby increasing FFA and cholesterol blood levels. So it is advised to avoid smoking and excessive alcohol intake and to reduce stress level to prevent risk of cardio vascular diseases.

3. Hypocholesterolemic drugs—When the above measures fail to lower plasma cholesterol levels, hypercholesterolemia can be treated by drugs. Some important drugs which lower cholesterol level are listed below—

(a) **Sequestrial agents and drugs**—The plant sterol **sitosterol, galactomammans** (soluble dietary fiber) and drug **cholestestyramine, resins etc** bind **bile acids** and **sterols** in GIT and prevent their re-absorption, thus interrupting enterohepatic circulation. So more bile acids and sterols would be excreted in feces. This simulates **more oxidation** of cholesterol to bile acid synthesis and thus results in lowering of blood cholesterol level.

(b) **Drugs inhibiting cholesterol biosynthesis**—Several drugs such as **mevastatin** and **lovastatin,** reduce cholesterol biosynthesis in body, at the **key regulatory step** of cholesterol biosynthesis by inhibiting **HMG CoA reductase enzyme** (Fig. 12.4).

(c) When both TAG and cholesterol are elevated in the blood, the lipid lowering drugs, such as **clofibrate** and **gemfibrozil** are used. These drugs activate enzyme **lipoprotein lipase** thus increasing clearing of VLDL and chylomicrons from blood. But these drugs have many side effects.

(d) **Nicotinic acid**—Niacin (3 mg/day) is beneficial in lowering blood cholesterol level and so B complex should be taken.

(e) **Vitamin C** is beneficial in reducing blood cholesterol levels and thus its intake should be increased in the diet.

□□□

13

LIPOPROTEIN METABOLISM : TRANSPORT OF LIPIDS

The total plasma lipid level is approximately 700-1000 mg/dl and comprises mainly triglycerides, cholesterol, free fatty acids and phospholipids. Lipids are insoluble in water, and thus they are transported in the blood as **lipoprotein complexes** to various tissues for their metabolism or storage purposes. Lipoproteins are complex lipids in conjugation with proteins, to make them soluble in aqueous medium of blood for their transport.

COMPOSITION OF LIPOPROTEIN

Lipoproteins are composed of a central **lipid core** containing triacylglycerides (TAG), cholesterol esters and long chain fatty acids, surrounded by a **coat layer** of amphipathic compounds such as **phospholipids, cholesterol, apoproteins** etc. The lipophilic groups of phospholipids(PL) and cholesterol are arranged in such a way that they are toward central core insoluble lipids while the polar groups are oriented towards surface of lipoprotein particles so that lipoprotein particles, becomes soluble in the aqueous medium of plasma for their transport. Apoproteins attached on the surface of lipoprotein particles further facilitate their transport in aqueous medium of the blood (Fig. 13.1).

Major types of lipoproteins—The lipoprotein fractions of the plasma are categorized according to their **density** and **apoprotein** contents into **four major types**. Pure fat is less dense than water, and it is observed that, as the proportion of lipid to protein content in lipoprotein particle increases, the density decreases. The plasma lipoproteins are separated, depending on their density and their rate of sedimentation by ultracetrifugation into **4 major groups.** These are as described here—

1. Chylomicrons—The dietary lipids after digestion and absorption from GIT are incorporated in a lipoprotein complex, called chylomicron in the intestinal mucosal cells (details in chapter 10). Chylomicrons contain highest amount of TAG (99%) with lowest protein contents (only1%). Their particle size is largest but with lowest density amongst the lipoproteins of the plasma, as is evident from contents of the Table 13.1.

2. Very low density lipoproteins (VLDL) —These are synthesized in the liver and are primarily responsible for the transport of endogenously synthesized TAG, cholesterol as well as dietary lipids absorbed from GIT. The VLDL particle lipid core consists of mainly TAG, long chain fatty acids, cholesterol esters and is surrounded by a coat of phospholipids and apoprotein B-100 as summarized in the Table 13.1.

3. Low density lipoproteins (LDL)—LDL is formed from VLDL in the plasma, after TAG is removed and taken up by extrahepatic tissues by the action of lipoprotein lipase(LpL). LDL is rich in cholesterol and transport it to the peripheral tissues for its utilization.

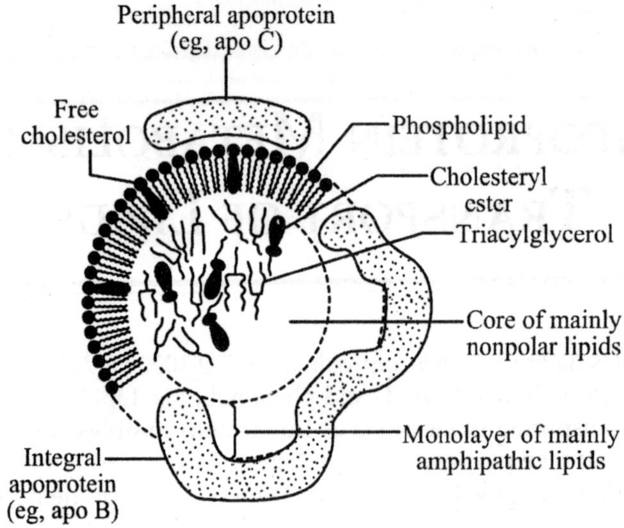

Fig. 13.1 : Generalized structure of a plasma lipoprotein

4. High density lipoproteins (HDL)—HDL has more protein contents and thus have highest density as compared to other lipoprotein fractions of the plasma. HDL is mainly responsible for the **reverse transport of cholesterol** i.e. from peripheral tissues to liver for its **oxidation to bile acids** and its excretion in gastrointestinal tract (GIT). Three types of HDL fractions have been identified in the plasma and are designated as HDL_1, HDL_2 and HDL_3 respectively.

The TAG is predominant lipid in chylomicrons and VLDL, while cholesterol is in highest amount in the LDL, while phospholipid contents are more in HDL fraction of lipoproteins. A brief account of composition, apoprotein contents, electrophoretic mobility and density of various lipoprotein fractions is summarized in the Table 13.1.

Apoproteins—The protein part of lipoproteins is called **apo-protein**, which is a characteristic feature of the lipoprotein particles. The apoproteins are very important for solubilizing the lipid components, for their transport in the blood to various tissues. In addition, the apoproteins have several other important functions as listed below—

(a) Apoproteins have major **regulatory role** in lipoprotein metabolism.

(b) **Apo C-II** is essentially present in chylomicrons and VLDL and **activates** enzyme **lipoprotein lipase,** which clears and removes TAG from these lipoprotein fractions to be taken up by extra hepatic tissues for **further metabolism** or for storage purposes.

(c) **Apo A-I** of **HDL_2** activates enzyme lecithin - cholesterol acyl transferese **(LCAT)** necessary for esterfication and extraction of cholesterol from the tissues, for its reverse transport to liver. This is the main route by which cholesterol is oxidized to bile acids and excreted in GIT.

(d) Apoproteins of lipoproteins are also important as **ligands** for recognition and binding of lipoprotein to the specific receptors present on plasma membranes of target tissues. For instance apo B-100 of LDL is necessary for the recognition and its binding by **LDL receptor** on the peripheral tissues, for the uptake of LDL for its utilization and further metabolism.

Table 13.1 : Characteristic features of lipoproteins of plasma

Characteristic features	Chylomicrons	VLDL	IDL	LDL	HDL
Density g/L	<0.96	0.96-1.006	1.006-1.019	1.019-1.063	1.063-1.21
Electrophoretic mobility (band)	Origin	Pre-β	Broad-β	β	α
Diameter (nm)	500	30-90	30	20-25	10-15
Composition (%)Protein	2	10	16	20	40
Lipid (Total)	98	90-93	89	80	60
Lipid component (%)TAG	88	55	30	12	8
Phospholipid	8	20	22	28	47
Cholesterol	4	24	32	59	40
Free fatty acid	-	1	-	1	1
Apoprotein	A,B-48	B-100	B-100	B-100	A-I
Components	C II, E	C II, E	E	-	C II, E

METABOLISM OF LIPOPROTEINS

Chylomicrons

Dietary lipids after digestion and absorption by intestinal mucosa are incorporated into **nascent chylomicrons,** which consist of a lipid core comprising mainly TAG, cholesterol ester and is surrounded by a coat of phospholipids with apoproteins **A** and **B-48.** The chylomicrons are then secreted via lacteals into the thoracic duct and then to systemic blood. In plasma, they combine with apo C II and apo E (derived from HDL) to mature as chylomicrons, as illustrated in the Fig. 13.2.

Site of chylomicron metabolism—The chylomicrons are metabolized mainly in adipose tissues and skeletal muscles. Half life of chylomicrons is very short, about 1 hour only and they are rapidly metabolized and cleared from the blood within 1-2 hours of their release from GIT. TAG of chylomicrons are hydrolyzed by enzyme **lipoprotein lipase** (LpL) present at the endothelial layer of blood capillaries of the tissues. Free fatty acids so released are transported inside the extrahepatic tissues for oxidation purposes or are reesterified and stored as TAG in the adipose tissues.

Cofactors of Lipoprotein Lipase (LpL)

Both phospholipids and apo C II are required as cofactors for LpL activity. Apo C-II has a specitic site which binds phospholipid of **lipoprotein particle,** through which it get attached on the enzyme LpL. The apo C II activates the enzyme LpL, which hydrolyzes TAG, while lipoproteins remains attached to the enzyme lipoprotein lipase on endothelium of blood capillaries of the tissues. The free fatty acids released are transported inside the tissues for further metabolism.

Metabolism in adipose tissue—Insulin hormone **enhances gene expression** and **induces** synthesis of lipoprotein lipase enzyme in adipocytes, which clears the bulk of TAG from chylomicron to be taken up for reesterification and **storage** purpose. After removal of bulk of TAG, the chylomicron are converted to **chylomicron remnants** as shown in the Fig. 13.2.

Liver takes up chylomicron remnants—chylomicron remnants are taken up by the liver by a **apo E specific receptor mediated** endocytosis and inside hepatocytes the cholesterylesters and remaining TAG are hydrolyzed as shown in the Fig. 13.2.

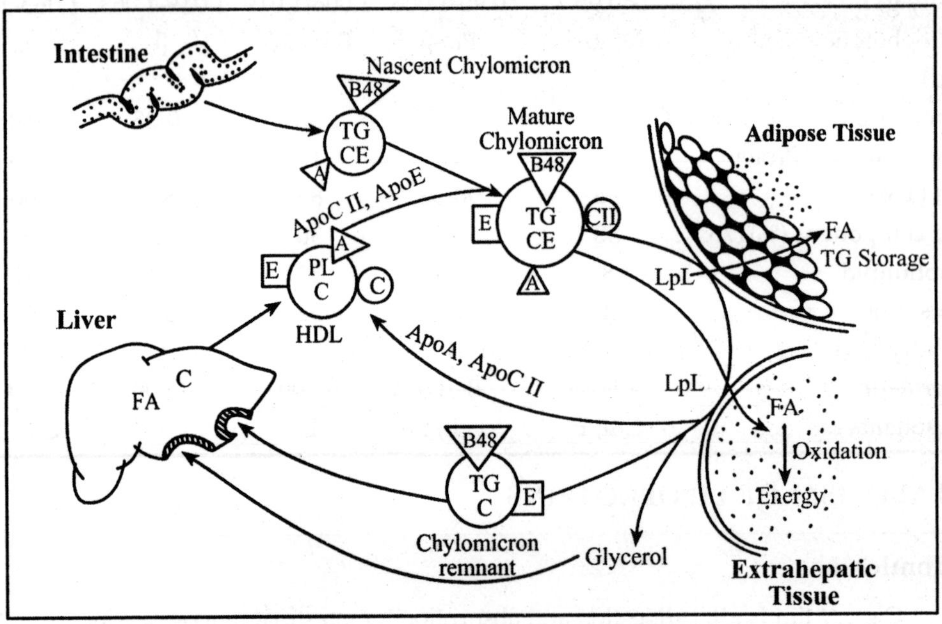

Fig. 13.2 : Metabolism of chylomicrons (TG-triglycerides, PL-phospholipids, CE-Cholesterol ester, C-Cholesterol, FA-Fatty acid)

VLDL Metabolism

Liver synthesizes **very low density lipoproteins** (VLDL), which are the major transport carriers for endogenous TAG synthesized by **hepatic lipogenesis** process during welfed state. Liver incorporates TAG and cholesterylester as the core lipids surrounded by layer of phospholipids and cholesterol (amphipathic lipids) and **apoprotein B-100.** The VLDL so synthesized are secreted in plasma where they acquire apoproteins **C II** and **apo E** (derived from HDL) and get matured.

The half life of VLDL in plasma is very short (1-3 hours only) and are rapidly metabolized when they reach to extrahepatic tissues, such as adipose tissue, muscles and heart. Apo C II activates enzyme LpL of endothelium of blood capillaries of the tissues and bulk of TAG of VLDL particles is rapidly hydrolyzed. The fatty acids so liberated are transported inside adipose tissue where they are resesterified and stored, while in muscles they are oxidized to generate energy.

Formation of Intermediate Density Lipoproteins (IDL)

After removal of TAG, the VLDL is changed into **VLDL remnants** also known as IDL (intermediate density lipoproteins). Some amount of IDL is taken back into liver, where apo E is a ligand for its binding on receptors for the uptake of IDL by hepatic cells, while major fraction of IDL further looses TAG and apo E and is converted into LDL as shown in the Fig. 13.3. Thus the main function of VLDL is to transport endogenous and dietary TAG from liver to the extrahepatic tissues for their metabolism and after removal of TAG, it is converted to **low density lipoprotein (LDL),** which now mainly contain cholesterol.

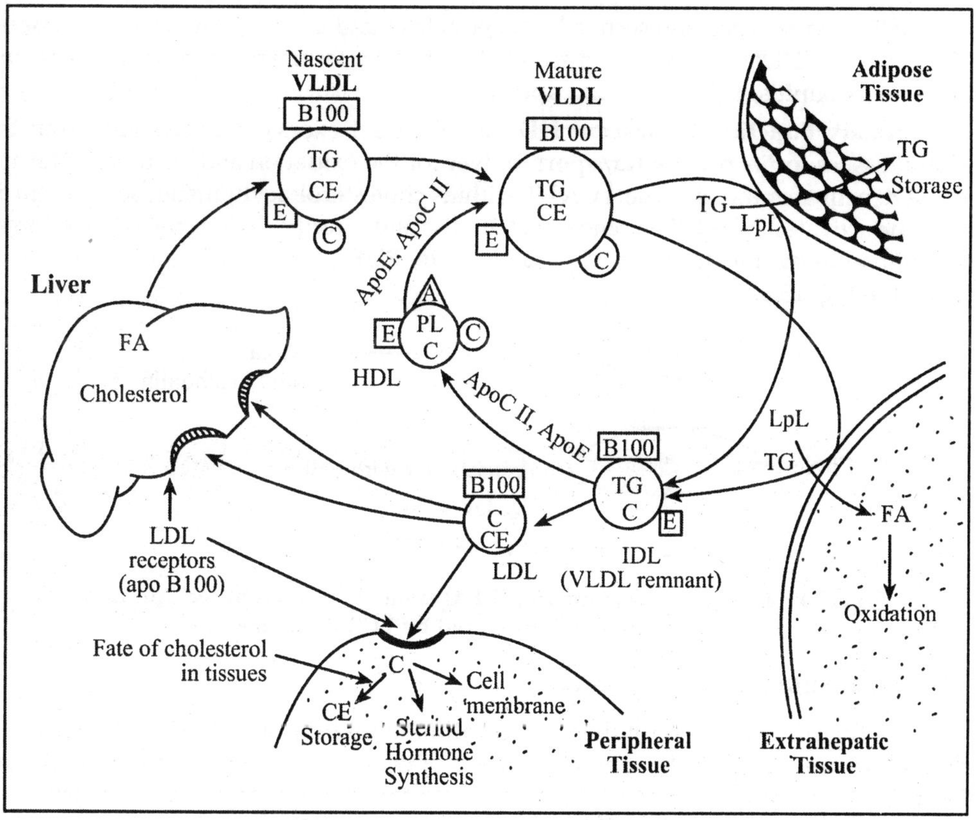

Fig. 13.3 : Metabolism of VLDL and production of IDL and LDL. Metabolic fates of IDL and LDL

LDL Metabolism

The LDL particles are rich in cholesterol esters and contain only **apo B-100** and are produced mainly from IDL after it looses apo C II and apo E. Out of the total plasma cholesterol, around 75% is present in LDL fraction of lipoproteins. Half life of LDL in blood is about 2 days and they are metabolized slowly via **LDL receptors** of peripheral tissues. The **LDL receptors** are present on all cells of the body but are most abundant on hepatic cells and adrenal cortex.

LDL uptake by extra hepatic tissues—Approximately 30% of LDL is taken up and are degraded in **extrahepatic tissues.** These tissues contain specific **LDL receptors** on the cell membranes located in specialized regions called **coated pits,** which contain a specific protein **clathrin.** These LDL receptors recognize apo B-100 of LDL and binds the LDL particles to form receptor- LDL complex which are then internalized by the process of endocytosis. Inside the cells the vesicles then fuse with lysosomes and are degraded. The cholesterol esters are hydrolyzed and are utilized for further metabolism in the tissues. (details in the chapter 12).

HDL Metabolism

HDL is synthesized and secreted from both liver and intestine. However the **nascent** (newly secreted) HDL synthesized and secreted from the intestine is **discoidal** in shape and contain phospholipids, cholesterol and **apo A-I.**

Primary Function of nascent HDL is to extract excess of cholesterol from the peripheral tissues for its **reverse transport to liver** for the oxidation and excretion. The apo A-I of HDL activates the enzyme LCAT(Lecithin- cholesterol acyltransferase) present in plasma. The activated LCAT then binds to HDL disc and esterify cholesterol extracted from peripheral tissues by transferring acyl group, usually PUFA from 2nd C-atom of lecithin as shown in the Fig. 13.4.

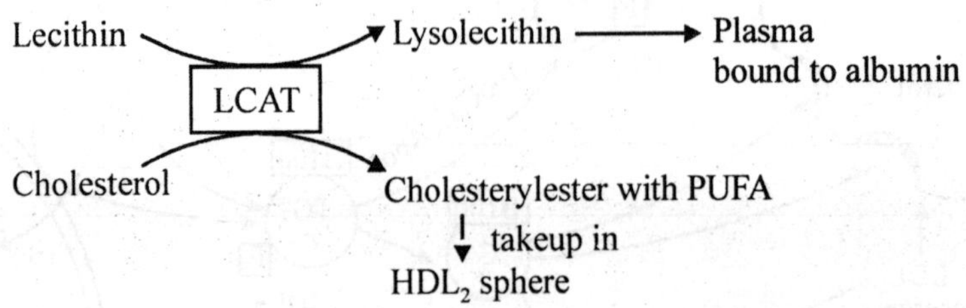

Fig. 13.4 : Transfer of acyl group (PUFA) from 2nd C-atom of lecithin to esterify cholesterol, catalyzed by LCAT enzyme

Reverse Transport of Cholesterol

The esterified cholesterol then moves into the interior of HDL disc. The reaction continues and ultimately HDL_2 spheres with lots of cholesterol esters are formed. **Mature** HDL_2 spheres are then transported to liver, where apo A-I is recognized by hepatic cell membrane receptors and binds HDL_2 to form receptor- HDL_2 complex which are then internalized in the liver. Hepatic lipase then hydrolyzes HDL phospholipids, while cholesterol esters are oxidized to bile acids and are secreted in bile. (details in chapter 12). The events of HDL metabolism are illustrated in the Fig. 13.5.

Role of Liver in Lipid Metabolism

Liver plays central and unique role in lipid metabolism. Major functions of liver are listed below —

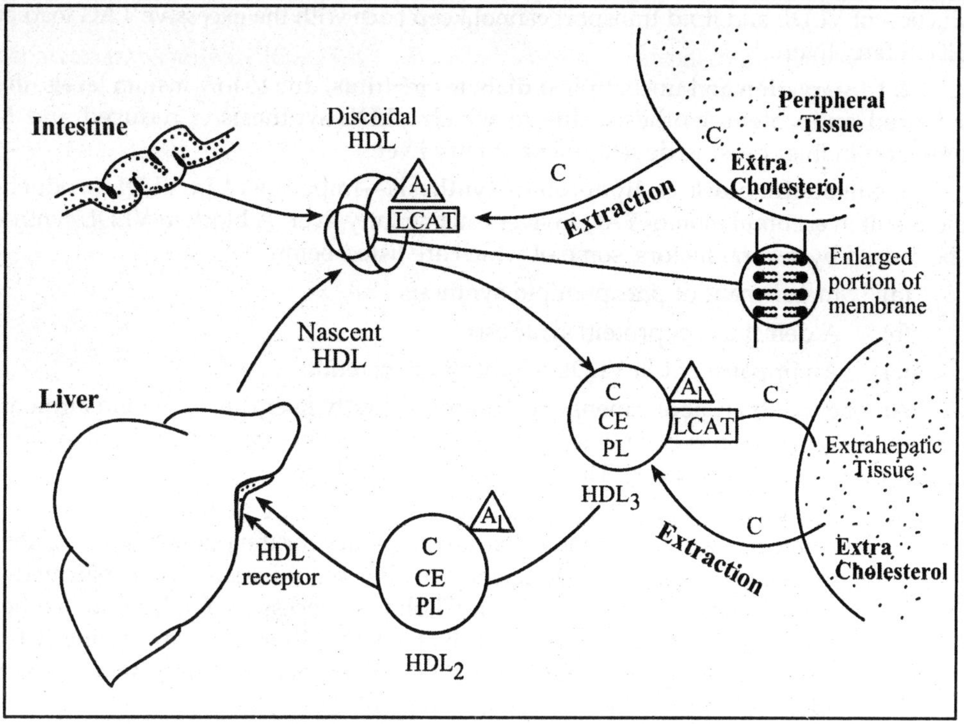

Fig. 13.5 : Metabolism of HDL and its role in reverse transport of cholesterol (LCAT—Lectithin cholesterol acyltransferase)

1. Synthesis and secretion of bile acids and bile salts in bile, which are necessary for fat digestion and absorption in GIT.

2. Liver has active enzyme systems for denovo synthesis of fatty acid and for active **lipogenesis** i.e. synthesis of TAG, phospholipids and cholesterol etc.

3. Liver plays integral role in the synthesis of lipoproteins for the transport of lipids to various tissues for their metabolism.

4. Liver oxidizes fatty acid and converts acetyl CoA into **ketone bodies** (ketogenesis).

Clinical Aspects—Fatty Liver

Any imbalance in the rates of triglyceride synthesis (lipogenesis) and their transport from liver as VLDL would result in the development of **fatty liver.** In this clinical condition, **lipids,** mainly as TAG accumulate in the hepatic tissues in excess and is regarded as pathological condition. This is because when it becomes chronic, it results in clinical manifestations. Fat laden hepatic cells undergo **fibrotic changes** which progress to **cirrhosis** of **liver,** with severe **liver function impairment** leading to **jaundice,** which can be life threatening.

Causes of Fatty Liver

1. Excessive fat mobilization from adipose tissues or excessive **dietary intake** of lipids elevate FFA plasma levels, resulting in increased uptake of fatty acids by the hepatocytes. This stimulates lipogenesis and accumulation of excessive TAG in liver. The production of VLDL and lipid transport cannot keep pace with the excessive TAG load and results in fatty liver.

2. In starvation and uncontrolled **diabetes mellitus,** due to low insulin levels, there is impaired apoprotein synthesis, due to which VLDL synthesis is reduced and fats accumulates in liver causing development of fatty liver.

3. Metabolic block in lipoprotein synthesis—Impairment in VLDL production would result in accumulation of TAG in liver causing fatty liver. A block in VLDL synthesis can be caused by several factors, some of which are listed below —

(a) Impairment of phospholipid synthesis

(b) A defect in apoprotein synthesis

(c) An impairment in VLDL synthesis or secretion

Among the three, above mentioned causes, the fatty liver due to the impairment of phospholipid synthesis has been extensively studied. A defect in phospholipid synthesis can occur due to various reasons. These are summarized here —

(a) **Deficiency of lipotropic factors**—The lipotropic factors such as choline, betaine, inositol, methionine etc.are necessary for the phospholipid synthesis in hepatocytes. Their deficiency in diet would result in impairment of phospholipid synthesis in liver. Choline is necessary for lecithin synthesis and can be produced in humans by **trimethylation** of ethanolamine or from betaine by **transmethylation** reaction, provided **methionine** as SAM (S-adenosyl methionine) is available in plenty (Chapter 16).

(b) **Deficiency of essential fatty acids (PUFA)**—Polyunsatured fatty acids are required for the phospholipid synthesis, since OH group at 2nd C of phospholipid is usually esterified with PUFA.

4. Defect in apoprotein synthesis—Toxic chemicals such as puromycin, carbon tetrachloride, chloroform, ethionine (α amino γ mercaptobutyrate) and phosphorus, lead and arsenic poisoning etc. inhibit protein synthesis, as well as they block apoprotein synthesis in the liver. This results in impairment of VLDL synthesis or its secretion resulting in development of fatty liver.

5. Free radical injury—Deficiency of natural antioxidants, such as vitamin E and selenium is found to be associated with fatty liver, because of excessive necrosis of hepatic cells. This probably involves free radical formation which causes peroxidation of lipid bilayer of hepatic cells. The free radical injury to hepatic cells results in excessive necrosis of hepatic tissue and fatty liver develops. Lipid peroxidation and free radical chain reaction is normally prevented by the presence of plenty of antioxidants in diet (Chapter 25).

Lipotropic Factors

Lipotropic factors are the substances, deficiency of which results in impairment of VLDL synthesis leading to the accumulation of TAG in liver, causing fatty liver. The important lipotropic factors include choline, betaine, methionine and inositol etc. Choline and inositol are constituents of phospholipids and thus are necessary for VLDL synthesis and distribution of lipids from liver to the tissues through blood circulation .

Other compound which are also considered as lipotropic factors include the substances which are concerned with **transmethylation** reactions. These are required for the synthesis of choline and carnitine etc. which are necessary for phospholipid synthesis or for the oxidation of fatty acids to lower free fatty acid plasma levels. Methionine is the most potent methylating agent and its deficiency results in development of fatty liver.

Vitamins B complex members such as folic acid, B_{12}, B_6, pantothenate etc are included in lipotropic factors as their deficiency causes fatty liver. They are important for VLDL synthesis as potent coenzymes, needed for the synthesis of phospholipids and apoproteins.

Polyunsaturated fatty acids are also included in lipotropic factors as they are required for the synthesis of phospholipids, deficiency of which result in development of fatty liver.

Alcoholism also causes fatty liver—Excessive intake of alcohol can also lead to lipid accumulation in liver which ultimately results in **cirrhosis** of liver with **fatal consequences**. Prolonged chronic consumption of alcohol causes increased TAG synthesis in liver, which is stimulated by accumulation of fatty acids. This is because of decreased oxidation of fatty acids and reduced citric acid cycle activity in cases of chronic alcoholism.

These events can be explained by the fact that ethyl alcohol is metabolized in liver by oxidation to acetaldehyde in the presence of **alcohol dehydrogenase** enzyme which shift **NAD^+/$NADH+H^+$** ratio to reduced side. Therefore oxidized NAD^+ is less available for citric acid cycle and other metabolic processes, such as biological oxidation on electron transport chain. This causes accumulation of acetyl CoA which is diverted towards other metabolic process as enumerated below—

1. Increased fatty acid synthesis and enhancement of **lipogenesis**, causing hyperlipemia and TAG accumulation in liver resulting in fatty liver.
2. Increased cholesterol synthesis and hypercholesterolemia is associated with alcoholism.
3. Increased uric acid level in plasma is due to accumulation of lactic acid($NADH+H^+$ accumulation reduces pyruvate to lactate) and lactoacidosis which impairs the uric acid excretion by kidneys (Chapter 18).

Lipoprotein Abnormalities—Clinical Aspects

Primary lipoprotein disorders are genetic, inherited lipoprotein metabolic abnormalities due to the defect at one or other stage, in the course of lipoprotein **synthesis,** their **transport** or **degradation.** The lipoprotein metabolic abnormalities can be grouped as hypo and hyperlipoproteinemias. The major abnormalities along with the clinical manifestations are summarized in the Table 13.2 & 13.3.

Table 13.2 : Important Lipoprotein disorders—Primary hypolipoproteinemias

Name	Congenital defect	Clinical aspects/ Biochemical abnormalities
1. Abetalipoproteinemia	- rare, inherited disorder - impaired Apo B synthesis	- Severe condition - Fat & fat soluble vitamin are not absorbed from GIT

contd. ...

... contd.

Name	Congenital defect	Clinical aspects/ Biochemical abnormalities
	- VLDL & chylomicrons are not synthesized	- Serum levels of TAG, PL & cholesterol are extremely low, - retardation of growth - Death in early life.
2. Familial hypolipoproteinemia	Apo B containing lipoproteins synthesis is defective	- LDL 10-50% of normal level - Chylomicron level normal - Mild condition.
3. Hypo α-lipoproteinemia Tangier's disease Fish eye disease	- Apo A-1 deficiency, low or near absence of HDL reverse transport of cholesterol is impaired.	- Cholesteryl esters accumulate in reticuloendothelial system of tissues - adenoids are common, yellow orange tonsil - **Atherosclerosis** in elderly persons leading to CHD

Table 13.3 : Primary hyperlipoproteinemias

Name	Defect - Abnormal Genes	Clinical aspects/ Biochemical abnormalities
Type I Familial hyperlipoproteinemia	Lipoprotein lipase deficiency	Increased levels of VLDL, chylomicrons LDL and HDL in plasma -Atherosclerosis & CHD
Type II Familial hyper cholesterolemia	LDL receptor defect	Increased levels of LDL and total cholesterol in plasma - Atherosclerosis & CHD
Type III Hyperlipemia Broad β band disease	Abnormality in Apo E, so remnants are not cleared by liver	Increased levels of Chylomicron & VLDL-remnants in blood. Hyper-cholesterolemia, xanthomas - atherosclerosis in coronary artery leading to cardiac diseases.

contd. ...

... contd.

Name	Defect - Abnormal Genes	Clinical aspects/ Biochemical abnormalities
Type IV Hypertriglyceridemia	Overproduction of TAG. increased **VLDL** & TAG levels in plasma	- Atherosclerosis and CHD - associated with obesity, Diabetes mellitus & alcoholism
Type V Hyperlipemia	Defect in apo C-II & LpL deficiency	Increased TAG, chylomicrons, LDL & HDL in plasma - Atherosclerosis,risk for CHD

❑❑❑

14

PROTEINS : DIGESTION AND ABSORPTION

Digestion involves degradation of macromolecules of food stuffs by hydrolysis, catalyzed by specific digestive enzymes into smaller molecules, which can be absorbed by intestinal mucosa. **Proteolytic hydrolase enzymes** of gastrointestinal secretions catalyze hydrolysis (digestion) of complex native protein molecules, by breaking **peptide bonds** to librate smaller amino acid molecules, which are absorbed by intestinal mucosa.

Protein digestion begins in stomach

Gastric juice secreted by gastric glands is a clear, pale yellow fluid with an highly acidic pH (pH 1-2). It mainly consists of following components—1). Hydrochloric acid (0.2-0.5%), 2). Protein digestive enzymes pepsinogen, rennin and 3). Mucous. All these components play great role, mainly in the protein digestion in stomach and thus their secretion and functions are described in this section.

Secretion of HCl

Hydrochloric acid is secreted from parietal (oxyntic) cells of gastric mucosa by an **active process**, similar to that of **chloride shift mechanism** of RBCs (Chapter 33 Fig. 33.8). In gastric glands the enzyme **carbonic anhydrase** catalyze formation of H_2CO_3 (carbonic acid), which then ionizes into H^+ and HCO_3 ions. The H^+ ions so produced are secreted in the lumen of stomach by an active process driven by a

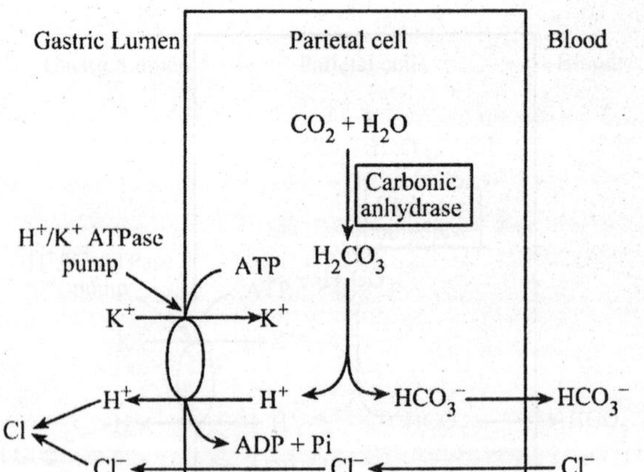

Fig. 14.1 : Mechanism of secretion of HCL by gastric glands

membrane bound **H^+/K^+- ATPase** pump as shown in the Fig. 14.1. Chloride ions from the plasma are then exchanged with HCO_3^- ions from the mucosal cells and are secreted into the lumen of stomach along with H^+ to form HCl as illustrated in the Fig. 14.1.

Functions of HCl—HCl of gastric juice plays important role, particularly in the digestion of proteins and has many important functions, as enumerated here.

1. Hydrochloric acid is an antiseptic and kills bacteria, which come along with food in GIT.
2. HCl denaturates complex native protein by destroying tertiary and secondary structures of protein molecule, thereby causing unfolding of polypeptide chain, to make it more accessible to the action of proteolytic enzymes.
3. HCl activates precursor zymogen, the pepsinogen into active **enzyme pepsin** for the digestion of proteins.
4. It provides optimum pH for the proteolytic action of enzyme pepsin to hydrolyze peptide bonds during digestion of proteins in the stomach.

Pepsin Starts Protein Digestion

Protein digestion begins in stomach by the hydrolase enzyme pepsin. Chief cells of stomach glands synthesize it as an inactive zymogen called pepsinogen, which is activated to pepsin by H^+ ions. The HCl cleaves (split) a protective (extra) polypeptide chain of 44 amino acids of pepsinogen, to expose the active sites of pepsin molecule. The active pepsin then autocatalyzes further activation of pepsinogen as shown in the Fig. 14.2.

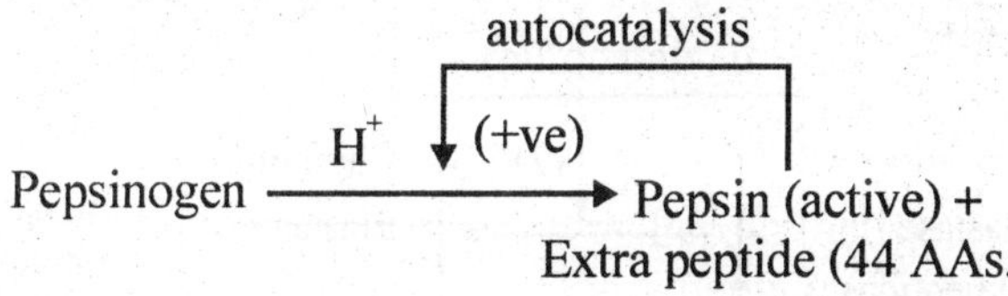

Fig. 2 : Activation of pepsinogen to pepsin by HCl

Specificity of Pepsin

(a) Pepsin is an **endopeptidase** and hydrolyzes peptide bonds within protein molecule to degrade it into smaller proteoses and peptone molecules (intermediate digestive products).

(b) Pepsin is specific in hydrolyzing peptide bonds of protein molecule formed by aromatic amino acids, such as phenylalanine, tyrosine, tryptophan etc or peptide bonds between dicarboxylic amino acid, glutamate.

Rennin—Rennin enzyme is important in digestion of milk in newborns and infants and thus is present in gastric juice of infants only (absent in adults). Rennin prevents rapid passage of milk from stomach to intestine, by curdling it. In presence of Ca^{++} ions rennin converts casein (milk protein) into paracasein, which is precipitated as white precipitate of calcium paracaseinate. This is then digested by pepsin to smaller intermediary protein molecules proteoses and peptones.

- Rennin is used in commercial preparation of cheese in industries.

Digestion of Proteins in Intestine

The acidic chyme containing partially digested food stuffs from stomach comes in duodenum, where the acidic pH is neutralized by alkaline contents of the bile and pancreatic juice. This provide optimum pH for the action of pancreatic enzymes for the further digestion of food in intestine.

Proteolytic Enzymes of Pancreatic Secretion

The proteolytic digestive enzymes of pancreatic secretion are synthesized as zymogens called **trypsinogen, chymotrypsinogen, proelastase** and **procarboxypeptidases.** These zymogens are activated after their secretion into duodenum as described here.

1. Activation of zymogens is initiated by an enzyme secreted by intestinal mucosa, known as **enteropeptidase** or **enterokinin.** This enzyme hydrolyzes a peptide bond between basic amino acids such as lysine, from the NH_2 terminal end of trypsinogen, and split (cleave) off an hexapeptide (6 amino acid peptide) to expose active **trypsin enzyme.**

2. Activated trypsin so liberated then autocatalyzes activation of its own zymogen molecules as well as it activates other proteolytic zymogens of pancreatic secretion, to produce active enzymes namely, **chymotrypsin, elastase** and **carboxypeptidase** respectively as shown in the Fig. 14.3.

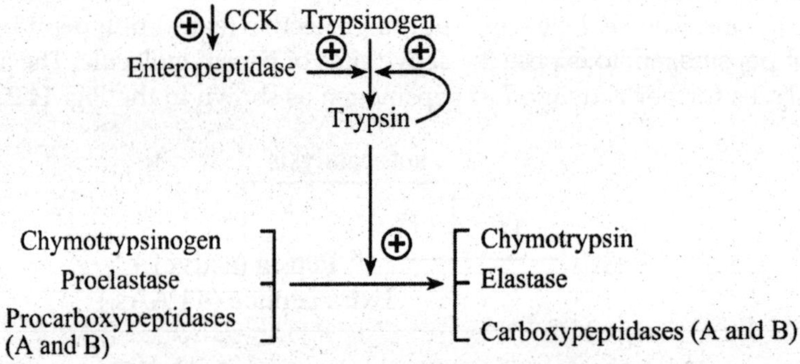

Fig. 14.3 : Mechanism of activation of pancreatic zymogens to active enzymes

Specificities of proteolytic enzymes—mechanism of action

1. Trypsin, chymotrypsin and elastase are all **endopeptidases** and contain an amino acid serine at their catalytic sites, required essentially for the digestion of proteins. Thus they are referred to as **serine proteases.**
2. Trypsin, chymotrypsin and elastase act on protein digested product from the stomach such as proteoses, peptones and polypeptides to hydrolyze them into smaller peptides.
3. **Trypsin** is specific in hydrolyzing peptide bonds formed by **basic amino acids** only, such as arginine and lysine etc.
4. **Chymotrypsin** hydrolyzes peptide bonds between uncharged, **neutral amino acids,** such as aromatic amino acids, phenylalanine, tyrosine, tryptophan etc.
5. **Elastase** specifically hydrolyzes peptide bonds formed by small amino acid residues, such as glycine, alanine, serine etc. A brief account of specificities of proteolytic enzymes is summarized in the Table 14.1.

6. **Carboxypeptidase**—It is an **exopeptidase** and hydrolyzes peptide bond from carboxyl end of the polypeptides to cleave amino acids successively, one by one.

Intestinal Mucosal Secretions

Intestinal secretions consists of two important proteolytic enzymes. These are—

1. **Aminopeptidase**—It is also an **exopeptidase** and hydrolyzes terminal peptide bonds from the amino end of the polypeptide and oligopeptide chains, to successively remove **free amino acids.**
2. **Dipeptidases**—A variety of dipeptidases of varying specificity are secreted by intestinal mucosa into GIT, which hydrolyze various dipeptides into free amino acids. Some of the dipeptides are absorbed directly into mucosal cells, where certain dipeptidases are active in the cell itself and hydrolyze dipeptides into free amino acids.

Table 14.1 : Mechanism of action and specificities of proteolytic enzymes

	GIT secretions & enzymes	Substrate	End products of digestion	Hydrolyzes peptide bonds formed by AAs
1.	Gastric secretion			
	Pepsin	Native proteins	Proteose, peptones, polypeptides	**Endopeptidase** Aromatic amino acids, phe, tyr, trypt & dicarboxylic acids glutamate etc.
	Rennin in Infants	Casein of milk Ca^{++} ions necessary	Coagulation of milk calcium paracasein.	
2.	Pancreatic secretion			
	Trypsin	Proteose, peptone polypeptides.	Smaller polypeptides	Endopeptidases Basic amino acids Arg, lysine etc
	Chymotrypsin	Large polypeptides milk proteins	Small polypeptides Coagulate milk	Uncharged aromatic amino acids phe, tyr, trypt & val, leu etc.
	Elastase	Large polypeptides	Smaller peptides Dipeptides	Small amino acids e.g. gly, ala, ser.

Absorption of Amino Acids—End products of protein digestion

The proteolytic enzymes of GIT secretions are very potent and digestion of proteins is highly efficient in GIT. The proteolytic enzymes completely hydrolyze proteins into **free amino acids.** So very little or no undigested proteins are lost in feces. L- amino acids are rapidly absorbed from intestine by an **active transport, energy dependent process** into the portal blood. D-isomer of amino acids are not utilized in our body for metabolic purpose and they may diffuse passively into mucosal cells. A variety of amino acid specific carriers, along with Na^+ ions are required for amino acid absorption, some of which are linked to Na^+/ K^+ ATPase pump—

Mechanism of Amino Acids Absorption

1. Na^+ dependent active transport—It is an energy dependent process and requires B_6 PO_4 (PLP) as a cofactor for the transport of amino acids. α-Amino acids are transported through intestinal brush border by **multiple specific carriers** which have specific sites to bind amino acids, as well as Na^+, similar to that of glucose absorption process (Fig. 4.3, Chapter 4). At least six specific carriers exist on intestinal brush border. Some important carries system found to exist are—(a) For neutral amino acid, (b) For aromatic amino acid phenylalanine and methionine and (c) For imino acid such as proline, hydroxy proline etc.

2. Na^+ Independent carrier transporters—These specialize in transport of neutral and lipophilic amino acids. Important examples are phenylalanine and leucine carriers and cationic amino acid carriers for lysine, arginine etc.

γ -Glutamyl Cycle—Meister Cycle

To explain the mechanism of amino acid transport across biomembranes in tissues, such as intestine and kidneys, another transport system, known as **γ.glutamyl** or **Meister cycle** has been proposed. This transport system involves a tripeptide called **glutathione** (γ glutamyl- cysteinyl-glycine). It is an energy dependent process and require **3 ATP molecules** for the transport of one amino acid, as compared to Na^+ dependent carrier transport system mentioned above, which requires only one ATP molecule . Thus physiological importance of γ-glutamyl cycle is still not very significant. However this cycle may be involved in rapid transport of certain amino acids such as cysteine and glutamate for the synthesis of glutathione. Details of the cycle are as illustrated in Fig. 14.4.

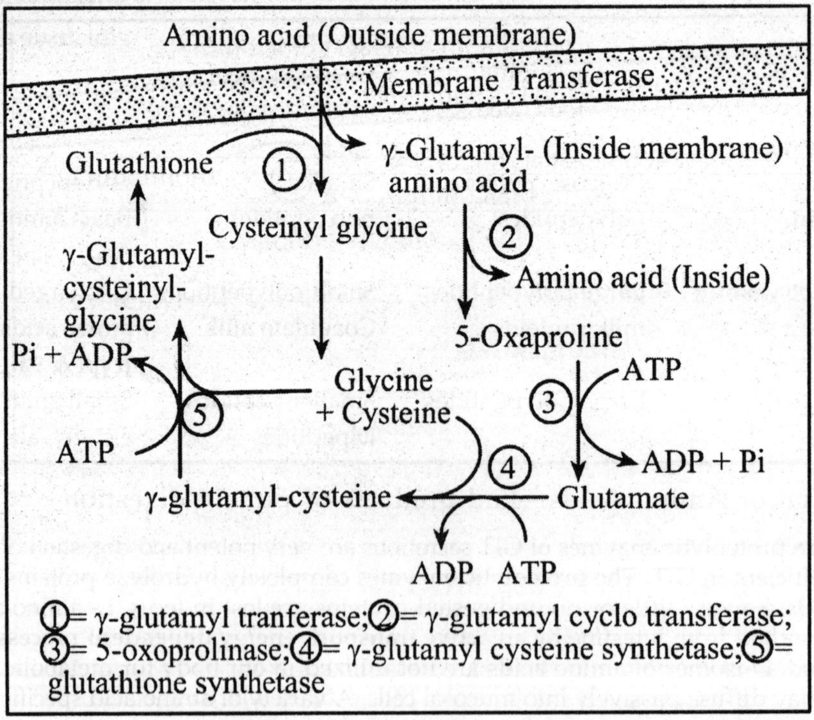

Fig. 14.4 : γ-Glutamyl cycle (Meister cycle) involved in the transport of amino acids across biomembranes

Biomedical and Clinical Aspects

- Cooking of food stuffs denaturates proteins. This facilitates digestion of proteins by proteolytic enzymes. For example **pepsin** gets easy access to the peptide bonds of unfolded and uncoiled polypeptide chain of denatured proteins.
- **Absorption of intact proteins** is observed in infants and newborns. This is very **significant** for transfer of maternal **immunoglobulins** by breast feeding to the offsprings to provide them immunity and protection against infections.
- **Food allergies**—Absorption of intact protein normally does not occur in adult intestine. However, certain individuals do absorb intact protein and is the major cause of **antibody** formation, resulting in **food allergy.** The uptake of intact protein molecule probably occurs by the process of **endocytosis** or **pinocytosis,** in intestinal mucosal cell.
- **Pancreatitis cystic fibrosis** or surgical removal of pancreas result in deficiency of pancreatic enzymes, which impair protein and lipid digestion and they are lost in feces.
- **Hartnup's disease**—It is an inherited genetic defect in which there is insufficiency or lack of carrier system to absorb neutral amino acid, specially **tryptophan** by the intestine and renal tubules. This defect results in severe deficiency of **tryptophan.** As tryptophan is an essential amino acid and is also a precursor for the synthesis of niacin coenzymes in the body, the disorder is associated with deficiency state of niacin. Thus typical symptoms of **"pellagra"** are observed in the affected individuals of Hartnup's disease. (Chapter 16).

❑❑❑

15

PROTEIN METABOLISM

Proteins are major constituents of living organism as - **Structural components** of the body tissues and **Functional components** of living cells, such as enzymes, hormones, carriers, receptors etc that make complex life processes possible.

Chemically proteins are polymers of α-amino acids, linked by **peptide bonds,** in a **specific sequence** to produce a broad spectrum of a variety of proteins, with unique functions to make life possible. All type of proteins are synthesized in body from **20 naturally occurring** α amino acids, just by changing their **number** and **sequence** in the polypeptide chain i.e. protein molecule.

Body Proteins are in Dynamic State

An appreciable amount of body proteins, approximately 400 g (1-2% of total) are degraded daily and are replenished by resynthesis.

This Protein turnover is a key **physiological** process to **revive** and to keep body tissues **healthy** and **active**. Body protein turnover takes place at varying rates, as evident by their half life. This is controlled and regulated according to wear and tear of the tissues, by a small regulatory factor called **'ubiquitin'** which **tags** a protein and **mark** it for its degradation. Rapid turnover proteins includes- plasma proteins, enzymes, muscle proteins, uterine tissue proteins etc, their half life being in hours to few days only. While structural proteins such as **collagen** have long half life, often in weeks or months and thus are degraded and resynthesized slowly.

Protein metabolism is metabolism of amino acids—Basically protein metabolism is the metabolism of amino acids, as body proteins are built up of α amino acids.

Two unique features of protein metabolism are:

1. Proteins and amino acids unlike lipids and carbohydrates, **cannot be stored** in body, so a daily regular supply of good quality proteins in our diet is necessary to meet the demand for **essential amino acids** for **protein biosynthesis** activity going on in the body. Excess of dietary amino acids (unutilized) are **degraded** (not stored), and catabolized rapidly or converted into glucose or ketone bodies.

2. **Nitrogenous moiety (NH_2 group) of proteins is toxic**—Unlike carbohydrates and lipids, which are completely oxidized to CO_2 and H_2O, the NH_2 group of proteins cannot be oxidized in animals and if it accumulates in free form, is highly toxic to body tissues. Animals excrete it, by first converting it into **less toxic compounds**. The mode of excretion of nitrogen waste products is different in different animals.

   Mode of nitrogen (NH_2) excretion—Different animals have developed different mode of N_2 waste product excretion as described here under—

   (a) **Ammonotelic animals** such as fishes excrete NH_3 directly in to the surrounding water through their **gills** and are known as ammonotelic animals.

   (b) **Uricotelic animals**—Reptiles and birds convert NH_2 moiety of proteins into **uric acid** (less harmful) and excrete it as **semi solid urine** to conserve water as they live in conditions of scarcity of water (e.g. in deserts) or migratory birds have to conserve body water for long periods.

   (c) **Ureotelic animals**—Higher primates (mammals) including humans convert NH_2 group into urea (end product of protein metabolism), which is very soluble in water and can easily be excreted in the urine.

Nitrogen Balance

A healthy adult who is maintaining his body weight constant remains in a perfect N_2 balance i.e. in an equilibrium between the amount of N_2 intake and to that of the N_2 excreted out of the body.

Positive N_2 Balance is when N_2 intake in diet is more than N_2 excreted and is observed during pregnancy, growth period and recovery period after serious illness etc.

Negative N_2 Balance is when dietary N_2 intake is less than N_2 excreted out of the body. It occurs in old age, disease states associated with excessive tissue catabolism, starvation and cancers etc.

Fate and Functions of Amino Acids

Apart from their important role to synthesize various proteins, necessary for carrying out life processes, amino acids serve many other important specialized functions in the body. These can be summarized as given under—

1. Many physiologically active nitrogenous compounds are synthesized from amino acids, such as hormones, neurotransmitters, purine and pyrimidine nucleotides, DNA, RNA, heme containing compounds, pigments etc. All these compound are synthesized in our body from 20 naturally occurring α amino acids.

2. During fasting and hypoglycaemic conditions, amino acids are main raw material for the denovo glucose synthesis to maintain blood sugar level (BSL).

   A brief account of sources, fate and functions of α amino acids along with plasma amino acid pool is depicted in Fig. 15.1.

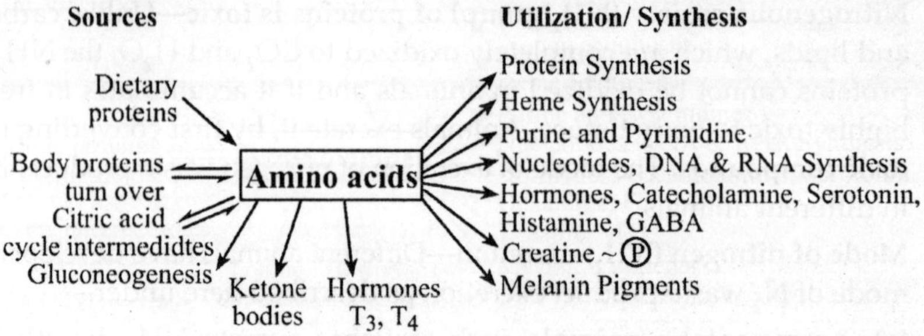

Fig. 15.1 : Fate and functions of α amino acids

Some other specialized biochemical functions of α-amino acids include —

1. Methionine serves as an important methylating agent to synthesize **choline** required for the synthesis of phospholipids and acetylcholine (neurotransmitter) and for the synthesis of d TMP (deoxythymidylate) for DNA synthesis etc.
2. **Conjugation and detoxification in liver**—Glycine and taurine amino acid are conjugated with **bile acids,** to form **bile salts** glycocholate and taurocholate, which are necessary for the digestion and absorption of fats in GIT.
3. Suphur containing amino acid cysteine conjugates and detoxify many drugs and aromatic halogens to less harmful compounds which are then excreted.
4. In muscles, creatine phosphate (a high energy phosphate) is synthesized from amino acids glycine, glutamine and methionine, and serves as storage form for rapid production of ATP during muscle contraction.
5. **Thyroid hormone synthesis**—Oxidized Iodide(I^+) is utilized for iodination of tyrosine residues of thyroglobulin to synthesize T_3 and T_4.
6. **Melanin pigment synthesis**—Tyrosine amino acid is the precursor for the synthesis of melanin, which imparts normal pigmentation to skin, hair, iris of eye etc. to protect us from the damaging effects of ultra violet radiation.
7. **Glutathione Synthesis**—Glutathione is a tripeptide with an active sulphhydral group required to detoxify oxy free radicals in erythrocytes and thus prevents their adverse damaging effects.
8. **Catecholamine synthesis**—Phenylalanine and tyrosine are precursors for the synthesis of hormones dopamine, epinephrine and norepinephrine.
9. Tryptophan amino acid is a precursor for the synthesis of vitamin niacin (B complex) and serotonin hormone in the body.
10. Arginine and methionine are involved in polyamine synthesis, which are important growth factors of the body.

Details of specialized functions are given along with normal metabolism of individual amino acids in this chapter.

AMINO ACID METABOLISM—GENERAL ASPECTS

Daily protein turnover liberates amino acids, some of which and the excess of dietary amino acids, which remain unutilized are rapidly degraded in the body. The NH_2 group of amino acids is converted into a nontoxic compound **urea**, as the end product of amino acid metabolism—

Urea is less harmful as compared to ammonia (NH_3), which is highly toxic to body tissues particularly to brain and CNS and causes **ammonia intoxication** with severe clinical consequences. Urea is the major nitrogenous waste product of our body. It is highly soluble in water and thus can easily be excreted in urine. Approximately **30 g of** urea is excreted in urine per day as main nitrogenous waste product of amino acid metabolism.

α-Amino Group is Removed by Transamination

α-Amino groups of amino acids is usually not deaminated as **free ammonia** (NH_3) in the tissues during amino acid catabolism, as it is highly toxic. So in our body the α amino group of amino acids is first channeled and concentrated into amino acid **glutamate** by **transamination reactions,** while the C-skeleton of amino acid left behind (α-keto acids) have a variety of fates as shown in the Fig. 15.2. The amino group as glutamate is then transported to liver, where it is converted into less harmful **end product urea**. The general aspects of α-amino acid metabolism is summarized in Fig. 15.2.

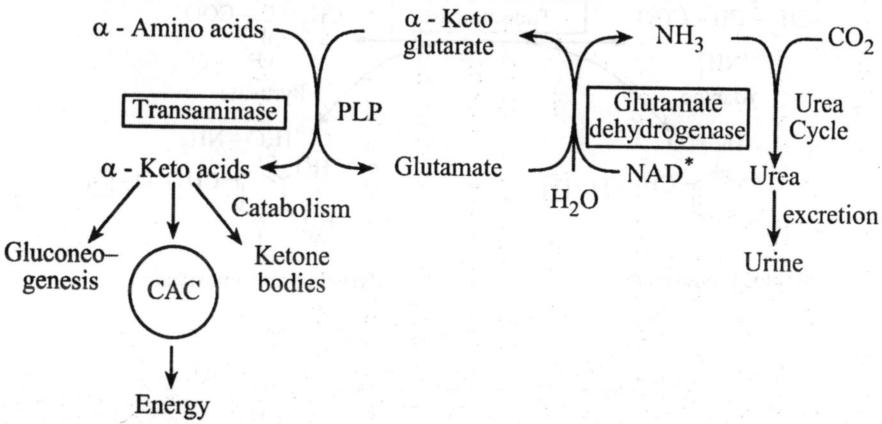

Fig. 15.2 : Overview of metabolism of α-amino acids and fate of N_2 moiety and C-skeleton left of amino acids, after transmination

Amino Acid Metabolism—Urea Synthesis

The amino acid catabolism and urea synthesis is divided into following stages —

1. Transamination
2. Oxidative deamination
3. Ammonia transport and
4. Reactions of urea cycle

Transamination

As the name suggest, in transamination reaction the α amino group of an amino acid is transferred to an α-keto acid, in the presence of specific **transaminase enzymes** which require **pyridoxal phosphate** (PLP) or B_6 PO_4 as co-enzyme.

Salient Features of Transamination Reaction

- All α-amino acid undergo transamination reaction except lysine, threonine and proline.

- Transamination reactions are **freely reversible,** so they can interconvert a pair of α amino acid and a pair of α keto acids. Thus transamination can function both in amino acid degradation (catabolism) as well as for the **biosynthesis** of **nonessential amino** acids in the body.
- Pyridoxal phophate (PLP) actively participates in the transfer of NH_2 group during transamination reaction but reverts back to coenzyme after completion of the reaction.

Mechanism of transamination—The CHO group of **pyridoxal phosphate coenzyme** first form a **Schiff's base** with α-amino group of amino acid to produce an **aldiamine,** which is then broken to leave behind an α keto acid and **pyridoxamine phosphate** is formed. The amino group from pyridoxamine phosphate is then transferred to another α keto acid to produce a new amino acid and pyridoxal phosphate (PLP) is released back to continue the transamination reaction as illustrated in the Fig. 15.3.

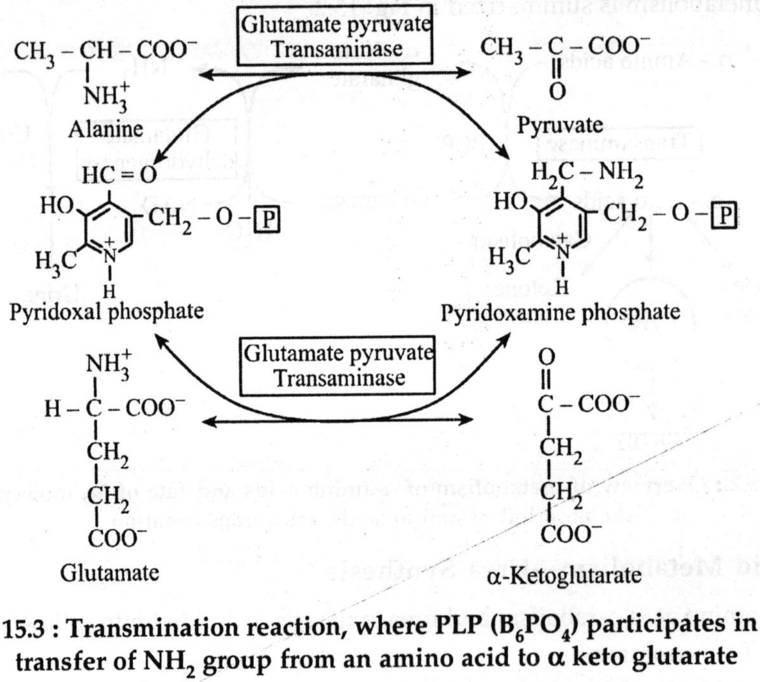

Fig. 15.3 : Transmination reaction, where PLP (B_6PO_4) participates in the transfer of NH_2 group from an amino acid to α keto glutarate

Two Important Transaminases

Alanine pyruvate transaminase (SGPT) and glutamate oxaloacetate (SGOT) transaminases, now known as alanine aminotransferase and aspartate aminotransferase respectively, are present in most of the mammalian tissues. They can transfer α-amino group of most of the amino acids to concentrate it in amino acids, alanine or glutamate. This is because the above mentioned transaminases are specific for only one pair of amino acid and α-keto acid, and non specific for the other pair. As alanine can also donate NH_2 group to α ketoglutarate to form glutamate, ultimately NH_2 group of all amino acids is concentrated as glutamate.

Clinical Significance—The serum levels of these two transaminases is clinically important in the **diagnosis** and **prognosis** of certain diseases. Serum alanine aminotransferase (SGPT or ALT) is significantly elevated in **liver diseases,** while serum aspartate transaminase (SGOT or AST) is raised in myocardial infarction and **heart disease** and thus their determination in serum is important in differential diagnosis of diseases.

Oxidative Deamination of Glutamate

α-Glutamate dehydrogenase enzyme, **present in liver,** plays central role in the metabolism of NH_2 moiety of amino acids. It is a potent enzyme present in hepatic cells and catalyzes oxidative deamination of glutamate in liver to liberate ammonia (NH_3) which immediately enters urea cycle for its conversion into urea, as shown in the Fig. 15.4.

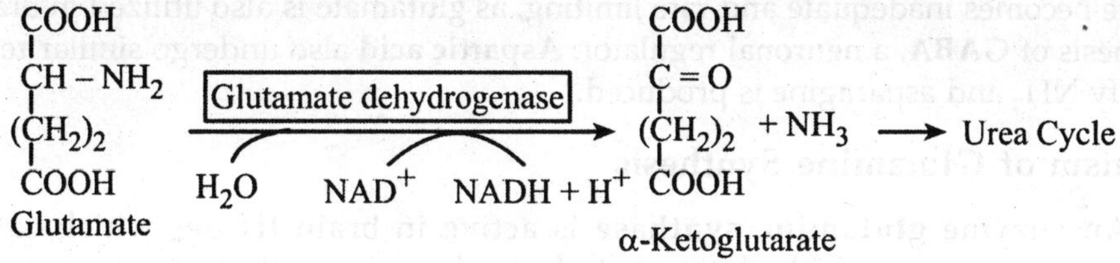

Fig. 15.4 : Oxidative deamination of glutamate in liver

α-Amino acid oxidases are autooxidizable, flavoprotein containing enzymes (with limited activity in tissues), can also oxidatively deaminate amino acids in tissues such as liver and kidneys. The reduced flavine coenzyme produced during deamination are oxidized by molecular O_2 and H_2O_2 produced is degraded by enzyme **catalase** into water and O_2. Details of reactions are shown in the Fig. 15.5.

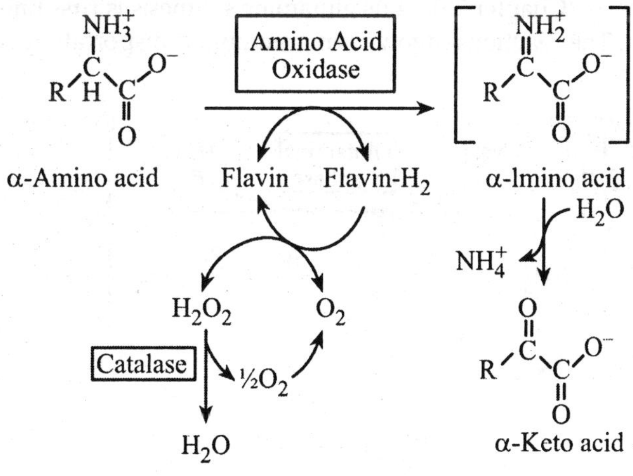

Fig. 15.5 : Oxidative deamination of amino acids by α-amino acid oxidases

Ammonia Intoxication is Life Threatening

Ammonia is constantly being produced by bacterial fermentation of food in GIT and is absorbed in portal blood and brought to liver. Healthy liver promptly removes NH_3 liberated by oxidative deamination of glutamate, as well as NH_3 brought by portal blood from GIT, for urea synthesis. So normally free ammonia is not present much in systemic circulation. This is essential, since even traces of ammonia are toxic to tissues, particularly to **brain and CNS.** In **liver diseases** such as **cirrhosis of liver,** the NH_3 level in blood increases and cause **ammonia intoxication.** The clinical symptoms of ammonia intoxication include tremors, slurred speech, blurring of vision and in severe cases, it leads to **coma** (hepatic coma) and even death.

Ammonia Transport

Although brain tissue can synthesize urea to some extent, this mechanism of NH_3 removal is insignificant. Major mechanism to capture and detoxify ammonia in brain tissue is **glutamine synthesis,** but that is also to a limited extent. This is because availability of the glutamate becomes inadequate and rate limiting, as glutamate is also utilized in brain for the synthesis of **GABA,** a neuronal regulator. **Aspartic acid** also undergo similar reaction to detoxify NH_3 and asparagine is produced.

Mechanism of Glutamine Synthesis

An enzyme **glutamine synthase** is active in brain tissues which catalyze condensation of ammonia with glutamate to form glutamine with an amide group. ATP provides energy for the covalent bond formation as shown in the Fig. 15.6.

The glutamine (and asparagines) so formed in brain is then transported to liver, where it is deaminated by specific enzyme **glutaminase** (asparaginase) and ammonia so liberated is immediately converted into urea by the reactions of urea cycle.

In **kidneys** the glutamine is deaminated by renal enzyme glutaminase to ammonia which is excreted as ammonium ion in exchange with sodium ion to eliminate and excrete the acid load of the body (Chapter 35). Thus glutamine synthesis is **first line of detoxification** of ammonia in brain. The reactions involved in ammonia transportation are summarized in the Fig. 15.6.

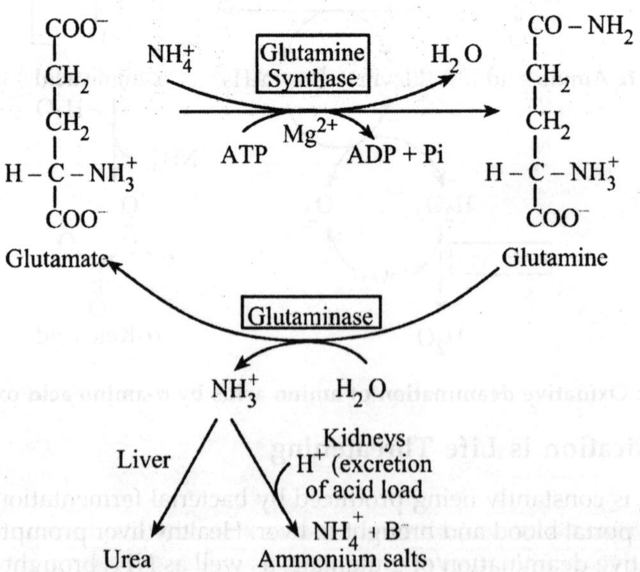

Fig. 15.6 : Detoxification of ammonia in brain and NH_3 transported as glutamine to liver and kidneys

UREA CYCLE

Urea is the end product of protein and amino acid metabolism and accounts for 80-90% nitrogenous waste products excreted in the urine. Urea is synthesized mainly in the liver in a five step cyclic process known as urea cycle. The reactions of urea cycle were first proposed by Hans Kreb and Hanseleit (1932) which, later on were modified by Ratner and Kohen as **modern urea cycle** as described here.

Urea has two amino groups, one derived from NH_3 liberated by oxidative deamination of glutamate (described above) in liver and the other from aspartic acid. Chemically the two amino groups in urea molecule are linked to carbonyl group by covalent bonds as shown in the Fig. 15.7. The carbonyl group of urea is derived from CO_2, produced during metabolic reactions.

Reactions of Urea Cycle

Urea synthesis is a five step cyclic process and five specific enzymes catalyze the steps of the urea synthesis. The first two enzymes catalyzing the initial two steps of urea synthesis are mitochondrial while the rest of the enzymes are present in the cytosol of hepatic cells. Important enzymes and the reactions of the urea synthesis are described here.

1. **Carbamoyl phosphate synthesis**—Urea synthesis begins with the synthesis of a compound, carbamoyl phosphate (NH_2-CO-PO_4) by forming a covalent bond between NH_3 and CO_2 in the presence of enzyme **carbamoyl phosphate synthase** and the reaction occurs inside mitochondria of hepatic cells.

   —Two ATP molecules provide energy for the condensation of CO_2 and NH_3, which is provided by oxidative deamination of glutamate as described earlier in the section.

   —The enzyme **carbamoyl phosphate synthase** is a key regulatory enzyme and requires **N- acetyl glutamate** as an enzyme activator. This reaction is irreversible and is a rate limiting step in the process of urea synthesis.

2. **Formation of citrulline**—Ornithine amino acid is the starter of urea cycle and this is the major function of this amino acid in the body. Ornithine is a dibasic amino acid and has an extra amino group (in addition to α amino group), on which **urea molecule** is built as illustrated in the Fig. 15.7. Ornithine is regenerated back after the completion of urea cycle.

   —**Ornithine transcarbamoylase** is also an hepatic mitochondirial enzyme, which catalyzes transfer of carbamoyl group (NH_2-CO) from carbamoyl phosphate on to the 2nd NH_2 group of ornithine to produce **citrulline** as shown in the Fig. 15.7. The citrulline so formed is then transported out of the mitochondria through a transporter system, as the rest of reactions of urea cycle take place in the cytosol.

3. **Argininosuccinate Formation**—In the hepatic cytosol, the amido group of citrulline then condenses with aspartic acid through its NH_2 group, (which provides 2nd NH_2 group of urea), in the presence of enzyme **argininosuccinate synthase** to form argininosuccinic acid. One ATP molecule is utilized for the covalent bond formation between the carbonyl group of amido group of citrulline with NH_2 group of aspartate to build a urea molecule as illustrated in the Fig. 15.7.

4. **Cleavage of Argininosuccinate**—The argininosuccinate is then hydrolytically cleaved by the enzyme **argininosuccinase** in a specific manner, to retain the NH_2 group of asparate to form **arginine** and fumarate is released as shown in Fig. 15.7. The fumarate so released enters citric acid cycle and is converted into oxaloacetate. Transamination of oxaloacetate by SGOT enzyme regenerates aspartate to continue the reaction of urea cycle as shown in the Fig. 15.7. The arginine formed in the reaction is the immediate precursor of urea.

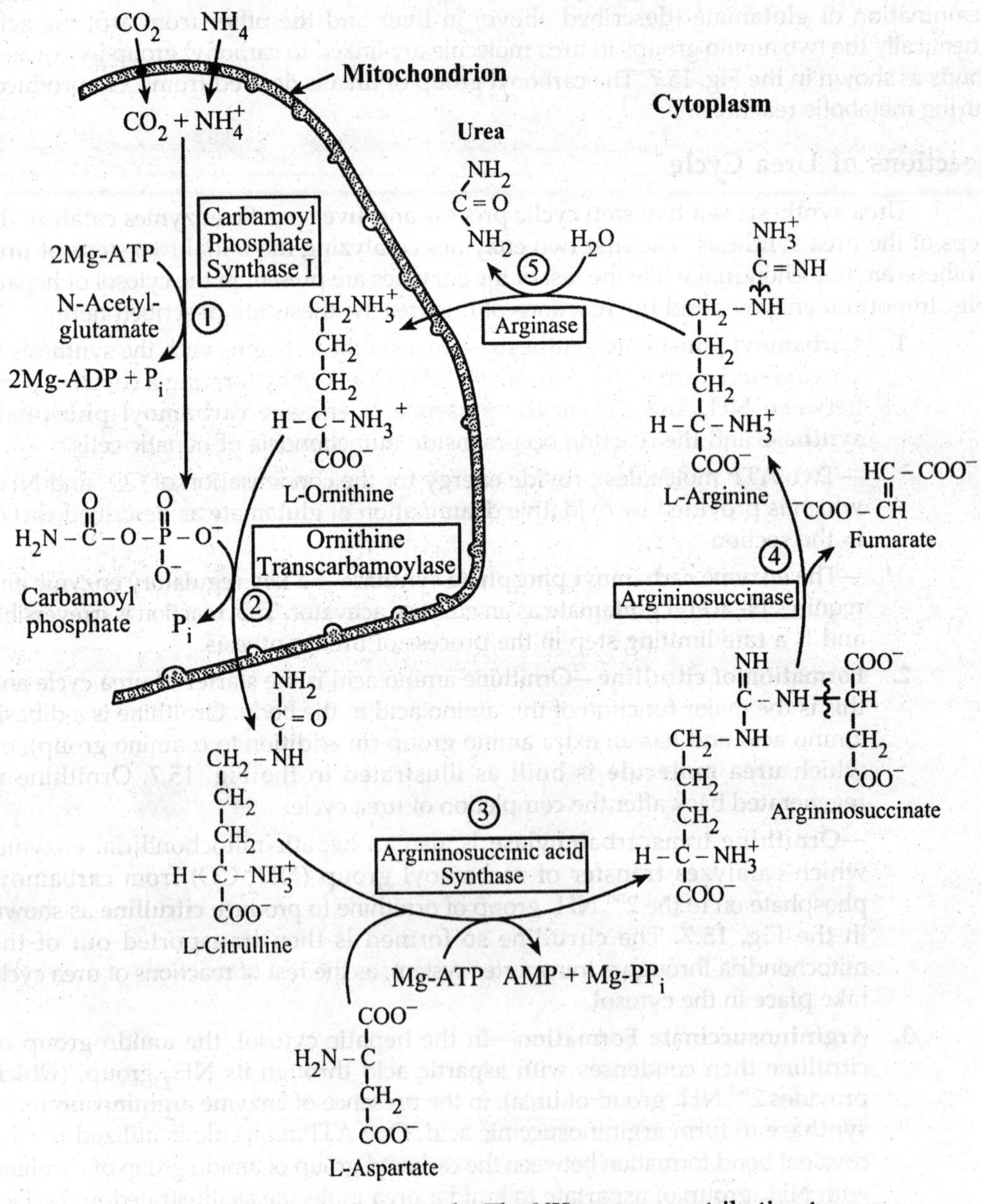

Fig. 15.7 : Reactions of urea cycle. The NH_2 groups contributing to urea synthesis are highlighted

5. **Formation of urea**—The 5th and final reaction of urea cycle is the hydrolytic cleavage of guanido group of **arginine** to liberate **urea** and the reaction is catalyzed by the enzyme **arginase** as shown in the Fig. 15.7. During the reaction ornithine is regenerated which reenter liver mitochondria to continue the process of urea cycle. Ornithine and lysine are competitive inhibitors of the enzyme arginase.

Regulation of Urea Cycle

The first reaction catalyzed by hepatic mitochondrial enzyme **carbamoyl synthase** is a **rate limiting step** in urea synthesis. This enzyme is allosterically activated by **N-acetyl glutamate (NAG).** High protein diet increase NAG by increasing its synthesis from glutamate (more amounts available) and acetyl CoA. So its concentration in the liver increases which results in increase in urea synthesis.

- Excess of dietary proteins or amino acids also induce gene expression to increase synthesis of individual enzymes of urea cycle, as unutilized amino acids are degraded rapidly and more ammonia has to be converted into urea.
- In starvation conditions, blood sugar level has to be maintained by increasing tissue protein catabolism, to provide more amino acids for the process of **gluconeogensis.** Thus during starvation and prolonged fasting conditions, the levels of enzymes of urea cycle increase 10 to 20 folds to cope up the increased production of ammonia for urea synthesis.

Metabolic Disorders of Urea Cycle

Congenital inherited genetic disorders of urea cycle due to the lack or insufficiency of each of the enzymes of urea cycle exists, with severe to mild clinical manifestations. A brief account of metabolic disorders associated with urea synthesis along with clinical manifestations are listed in the Table 15.1.

Since urea synthesis convert toxic ammonia to less harmful, easily excretable compound urea, all metabolic defects of urea cycle results in clinical manifestations of ammonia intoxication. The condition is more severe in abnormalities associated with enzyme defects involving I and II reactions of urea cycle.

Clinical symptoms common to all urea cycle disorders include vomiting, aversion to high protein foods, irritability, lethargy, intermittent ataxia and mental retardation due to irreversible CNS lesions, caused by ammonia intoxication.

Management—Significant improvement is observed in the metabolic disorders of urea synthesis on low protein intake and much of the irreversible damage to brain can be prevented by restricting proteins in diet.

Table 15.1 : Congential metabolic disorders of urea cycle

Metabolic disorder	Enzyme insufficiency/ defect	Biochemical alteration with clinical manifestation
1. Hyperammonia type I	Carbamoyl phosphate synthase I	Severe ammonia intoxication Mental retardation
2. Hyperammonia type II	Ornithine transcarba-moylase	Both mother & infant have ammonia intoxication

contd. ...

... contd.

Metabolic disorder	Enzyme insufficiency	Biochemical alteration with clinical manifestation
		↑Glutamine level in blood & CSF. ↓GABA in brain resulting in CNS damage & ataxia
3. **Citrullinemia** (rare)	Argininosuccinate synthase	Less severe, mild disorder as Citrulline and argininosuccinate serves as alternative carrier of ammonia and excreted in urine.
4. Argininosuccinateurea	Argininosuccinase	↑Argininosuccinate in blood Friable brittle tufted hair (trichorhexis nodosa)
5. Hyperarginemia	Arginase	Mild condition↑Arginine in blood, CSF. ↑urinary arginine, lysine and cystiene excretion.

Blood Urea—Clinical Aspects

Normal blood urea level is 15-40 mg/dl. High protein diets marginally increase blood urea level, as excess of urea formed is readily excreted in urine.

Biomedical Importance—Blood urea level estimation along with serum creatinine level is frequently used as screening test for assessing **kidney function.** High blood urea level is associated mainly with inadequate renal functions. (Chapter 36). The conditions in which blood urea levels are elevated can be broadly divided into 3 categories.

(a) **Pre renal conditions**—Blood urea level is elevated apparently in chronic fever and dehydration due to **hemoconcentration.**

(b) **Renal disorders and diseases**—Due to renal insufficiency in diseases such as glomerulonephritis, nephrotic syndrome, nephritis etc blood urea level is highly elevated.

Uremia—The term **"uremia"** is referred to a clinical syndrome observed in **acute renal failure,** where blood urea level along with creatinine is excessively high, as nitrogenous waste products are not excreted and accumulate in blood. In excess, urea also become toxic to brain and CNS and results in **azotemia** causing CNS damage, CNS dysfunction, convulsions, ataxia and patient may go into coma and may results in death due to renal failure.

(c) **Post renal causes**—Any obstruction in urinary tract results in decreased GFR with corresponding increase in blood urea levels. Obstruction in urine flow can be caused by stones in urinary tract, due to stricture of urethra, enlarged prostate, malignant tumors of bladder and may result in acute renal failure.

❑❑❑

16

Metabolism of Individual Amino Acids

Metabolism of individual amino acids has great biochemical as well as biomedical importance and is dealt here with special emphasis on following points.

1. Fate of C-skeleton of amino acid left after the removal of NH_2 group by transamination reaction.
2. Inherited **congenital** metabolic abnormalities associated with normal metabolism of amino acids.
3. Specialized physiologically potent compounds synthesized from various amino acids.
4. Dietary importance, whether amino acid is essential or non essential.

GLYCINE

Glycine is the smallest member of the aliphatic class of amino acids, with a single α carbon atom. It is an nonessential amino acid and can easily be synthesized in the body. It is **glycogenic** in nature and have great functions in the body, in addition to its role in biosynthesis of proteins. The important functions of glycine are descried here.

1. Synthesis of Heme—Glycine is very important for the synthesis of heme and heme containing proteins such as **hemoglobin, myoglobin, cytochromes** etc in the body to transport O_2 for biological oxidation (details in the Chapter 34).

2. Synthesis of purine nucleotides and RNA and DNA synthesis—Whole molecule of glycine is incorporated in the purine ring at position 4, 5 and 7 during purine nucleotide synthesis, as shown in the Fig. 16.1 (Chapter 18).

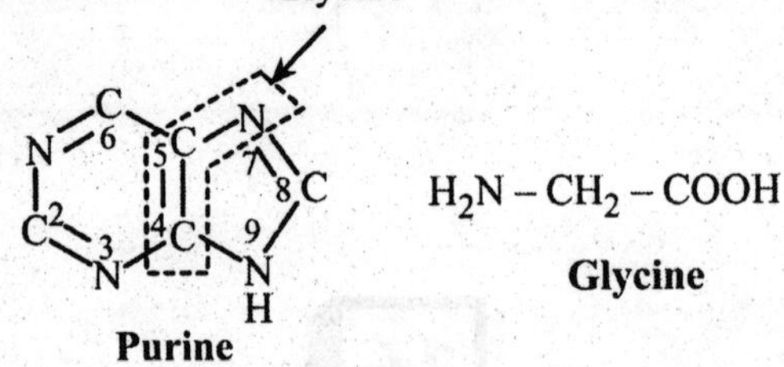

Fig . 16.1 : Incorporation of glycine molecule in purine ring, during nucleotide synthesis

3. Synthesis of creatine phosphate—Creatine phosphate is an high energy phosphate compound present in substantial amounts in tissues, such as cardic muscles, skeletal muscles and brain etc as **storage form** of energy for ready availability of ATP(energy) during muscle contraction. Creatine phosphate is synthesized from **glycine** as described here under.

Amido group or arginine is first transferred to NH_2 group of glycine, in the presence of enzyme **amido transferase** to produce **guanidoacetate,** which is then methylated by active methionine(SAM) to produce creatine. The creatine is then converted into high energy compound, creatine ~ PO_4 by ATP in the presence of the enzyme **creatine phosphokinase(CPK)** as shown in the Fig. 16.2.

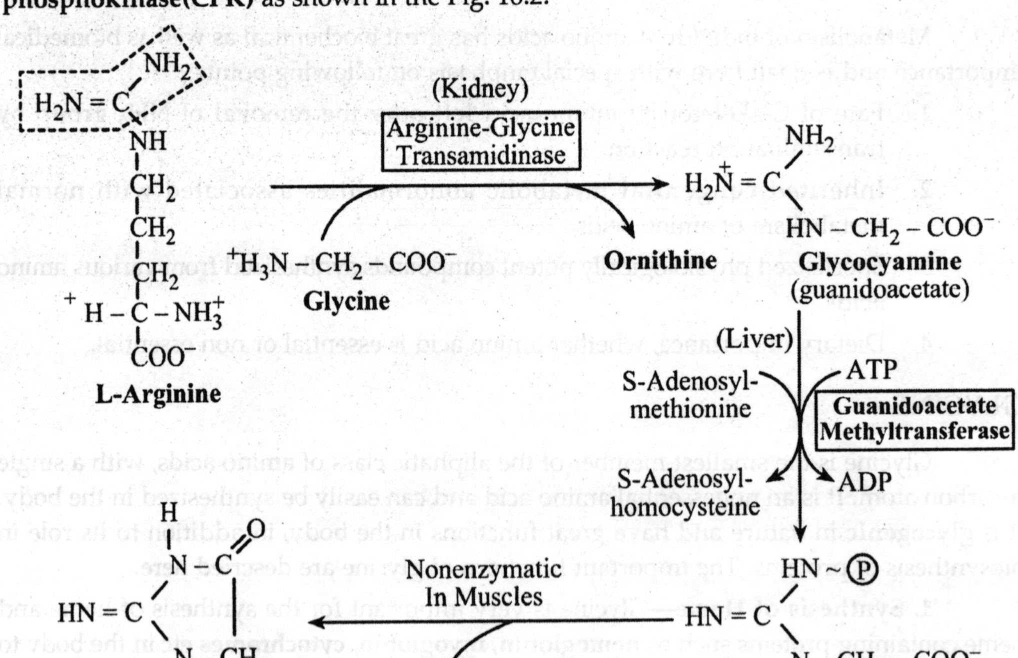

Fig. 16.2 : Synthesis of creatine phosphate from glycine

Biomedical Importance of CPK Isoenzymes

a. CPK Isoenzymes—Creatine phosphokinase(CPK) is a dimer with two subunits called B and M and it exists as **3 isoenzymes – CPK BB** present in brain, **CPK MM** present in skeletal muscles and **CPK MB** present in the cardic muscles.

b. CPK MB is elevated in serum in myocardial infarction (MI) to very high levels, 48 hours before the heart attack and thus is helpful in early diagnosis of heart diseases (MI), so that patient can be managed to save his life.

c. CPK Nac (total enzyme level) is elevated in muscle dystrophies and muscle wasting diseases.

Creatinine is an anhydride of creatine, formed by non-enzymatic reaction in muscles and is released in serum as a nitrogenous waste product to be excreted in urine.

Creatinine Coefficient—Creatine turnover to creatinine takes place at a constant rate and amount excreted depends on muscle mass of an individual and is knows as creatinine coefficient.

4. Glutathione synthesis—Glutathione is a tripeptide, synthesized from the amino acids glycine, cysteine and glutamate linked together by peptide bonds. It has an **active sulph hydral** group and plays important role in the detoxification of H_2O_2 formed in tissues, particularly in RBCs. Glutathione thus maintains the sulphhydral buffer and protects the RBCs from hemolysis, due to damage caused by oxy-free radicals and H_2O_2.(details in Chapter 7).

5. Conjugation to form bile salts—Glycine conjugates with bile acids in liver to form glyocholate, that with Na^+ ions produce bile salt and are secreted via bile into intestine. The bile salts play important role in fat digestion and absorption in GIT (Chapter 10).

6. Detoxification of aromatic compounds—Glycine is important in detoxification of many aromatic compounds. For example, glycine conjugates with benzoic acid (used as food preservative) in liver to form **hippuric** acid, which is less harmful and is excreted in urine as normal waste product (Chapter 36).

7. Synthesis and cleavage of glycine—The **glycine synthase** complex of liver mitochondria catalyzes both glycine synthesizing and glycine catabolic pathways, as it is a **reversible reaction.** The glycine synthase is a multienzyme complex and requires pyridoxal phosphate (PLP), lipoic acid, THF and FAD as coenzymes. **Tetrahydrofolate** (THF) participates in the reaction as a donor or acceptor of I C – moiety (CH_2 group) in glycine synthesis or degradation respectively as shown in the Fig. 16.3.

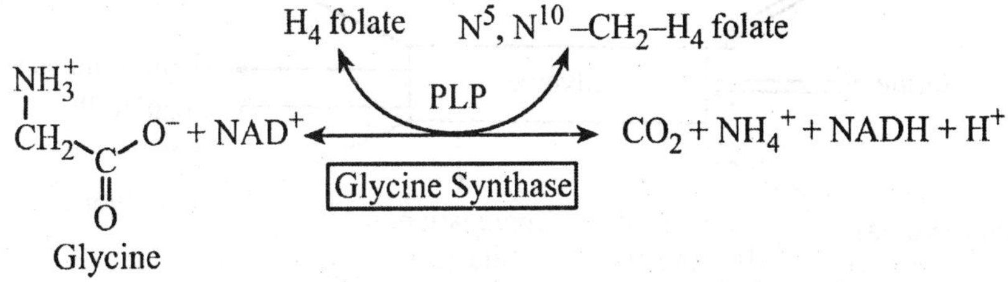

Fig. 16.3 Glycine synthase complex catalyzes both synthesis and degradation of glycine

7. Glycine—Source of IC- moiety— Glycine and serine amino acids are interconvertible and are catabolized to provide IC- moiety to be taken up by tetrahydrofolate coenzyme for various biosynthetic purposes as shown in the Fig. 16.4.

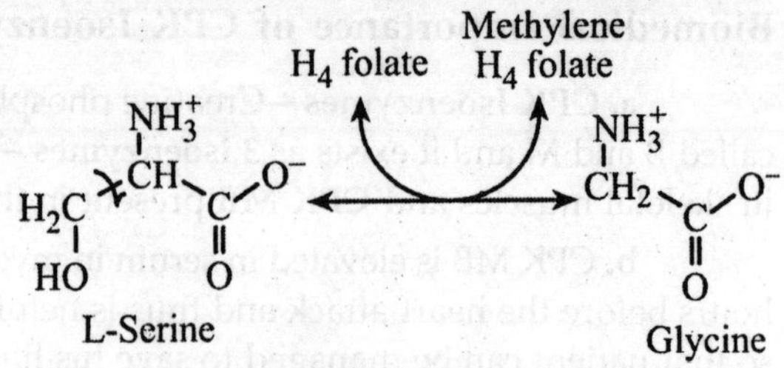

Fig. 16.4 : Freely reversible serine hydroxymethyl transferase reaction

Glycine undergoes Transamination

To some extent, glycine is also catabolized by transamination to form glyoxaldehyde, in the presence of **glycine transaminase** enzyme. Glyoxaldehyde so formed is then decarboxylated to formaldehyde, which is also a **source of IC-moiety** as illustrated in the Fig. 16.5.

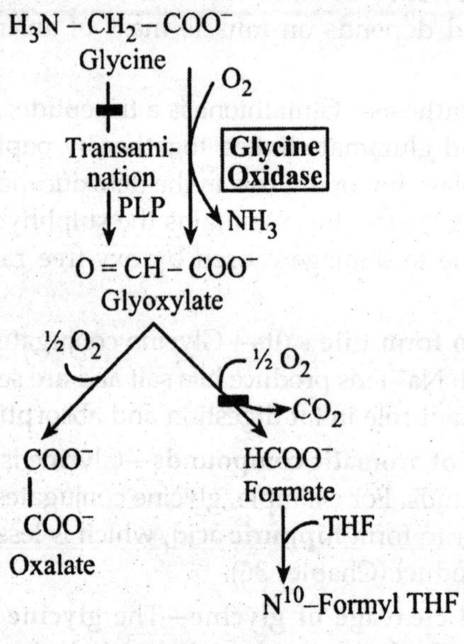

Fig. 16.5 : Catabolism of glycine. Blocks (■) indicate inborn errors of glycine metabolism

A brief account of functions of glycine are summarized in the Fig. 16.6.

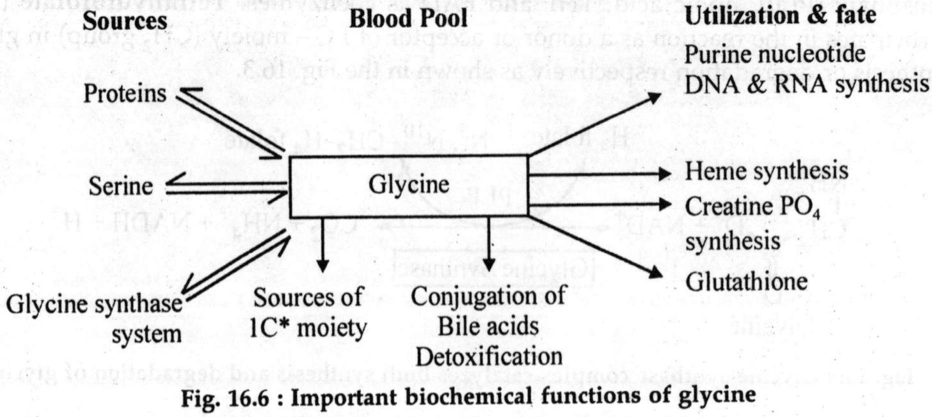

Fig. 16.6 : Important biochemical functions of glycine

Inborn Errors of Glycine Metabolism

1. Primary hyperoxalouria—This congenital metabolic disorder is very severe and results due to a genetic **defect** in **glycine transamination,** coupled with impairment of **decarboxylation** of glyoxaldehyde to formaldehyde. Due to this block, glycine is directly oxidized into glyoxaldehyde and further into **oxalic acid,** that accumulates in excess in blood and is excreted in the urine (Fig. 16.5).

Clinical manfestations—Excess of oxalic acid and urinary oxalates cause renal stone formation along with deposition of calcium oxalates in various soft tissues. Clinical manifestations of this metabolic disorder includes—Progressive bilateral calcium oxalate **urolithiasis, nephrocalcinosis** and **recurrent urinary tract infections.** Death may result due to renal failure at an early age.

2. Glycinuria—This genetic abnormality is due to **renal tubular reabsorption defect,** in which excess of glycine is excreted in urine (0.6 to lg/day), along with a tendency to form renal stones. However this is a mild condition and is compatible with life.

SULPHUR CONTAINING AMINO ACIDS

There are three sulphur containing amino acids, **methionine, cysteine** and **cystine** occurring in the natural proteins. Among them, only **methionine is an essential amino acid** while cysteine and cystine can be synthesized easily in body from the methionine, during its normal course of metabolism. All S- containing amino acids are **glycogenic** in nature, as after deamination and desulphuration they are converted to pyruvate, that enters gluconeogenesis process.

Importance of methionine—In addition to its function in various body protein synthesis, methionine serves as important **methylating agent** required for the synthesis of many physiologically important compounds as summarized in the Table 16.1. To act as a methylating agent, methionine is first activated to S-adenosyl methionine (SAM) by ATP in the presence of **methionine adenosyl transferase enzyme** as shown in the Fig. 16.7.

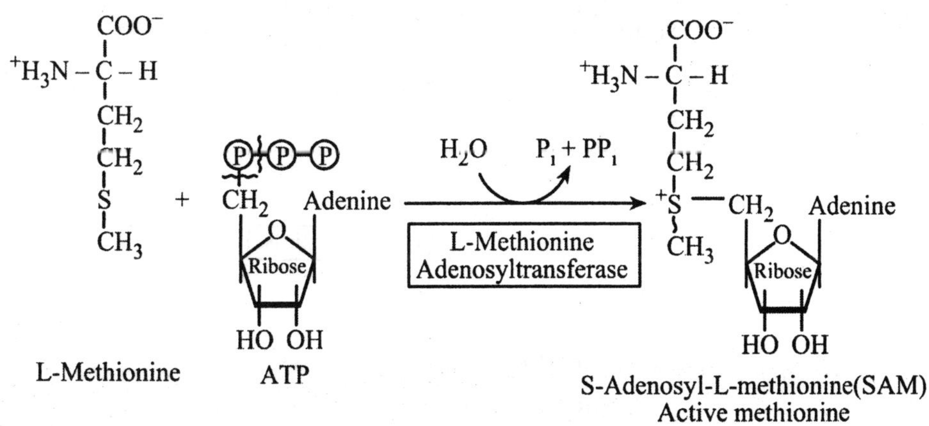

Fig. 16.7 : Activation of methionine to S-adenosyl methionine (SAM)

Table 16.1 : Biosynthesis of important compounds by transmethylation with active methionine

Methyl group acceptor	Compounds synthesized	Physiological functions
1. Guanidoacetate	Creatine - phosphate	High energy storage form of ATP for muscle contraction.
2. Norepinephrine	Epinephrine	Hormone
3. Ethanolamine	Choline	Synthesis of acetyl choline (neurotransmitter) & lecithin (phospholipids)
4. 5OH tryptamine (serotonin)	Melatonin	Hypothalamic hormone
5. Carnosine	Anserine	Muscle compounds.

2. Synthesis of cysteine and cystine—After transmethylation reactions, methionine is converted to **homocysteine** (Fig. 16.8), that is a precursor for the synthesis of cysteine and cystine amino acids in our body as described in the following reaction.

(a) Homocysteine condenses with serine to form **cystathionine** in the presence of enzyme **cystathionine synthase** which requires PLP (B_6PO_4) as coenzyme.

(b) **Cystathioninase** enzyme (also PLP dependent) then cleaves cystathionine in a specific manner to produce cysteine amino acid and homoserine left behind is deaminated to α keto butyrate (Fig. 16.8). In this specific cleavage reaction, sulphur is transferred from homocysteine to the OH group of serine to synthsize cysteins as shown in the Fig. 16.8.

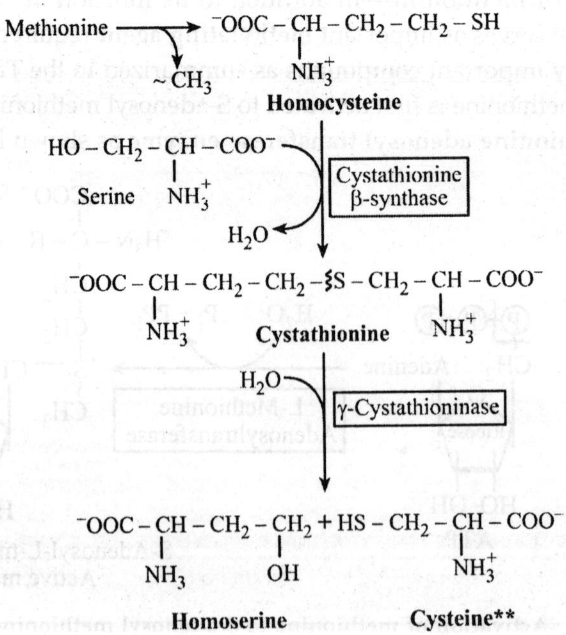

Fig. 16.8 : Synthesis of cysteine from methionine

3. Synthesis of polyamines—Polyamines are physiologically potent compounds having many amino groups. Examples are **putrescine, spermine** and **spermidine.** The polyamines are potent growth factors and play important role in cell multiplication. These are synthesized in body from **ornithine** and **methionine** amino acids after decarboxylation (Fig. 16.23). Details of their functions and biosynthesis will be discussed later on in this section.

Functions and metabolism of cysteine and cystine

Cysteine and cystine are interconvertible by an NAD dependent, **cystine reductase enzyme.** These S-containing amino acids though nonessential.(can be synthesized from methionine) but serve many important biological functions, that are summarized below.

(a) Glutathione synthesis—Glutathione is a tripeptide of 3 amino acids, glycine – cysteine – glutamate with an active **sulphhydral group,** provided by cysteine. This compound is important for the detoxification of H_2O_2 and other oxy free radicals formed in RBCs to prevent hemolysis. (Details in Chapter 7).

(b) Synthesis of coenzyme A—Cysteine is required for the synthesis of **coenzyme** A as it is an essential constituent of the coenzyme (Chapter 25). The coenzyme A has an **active SH group** provided by cysteine residue and acts as an acyl carrier in all metabolic reactions of fatty acids (details in Chapter 11).

(c) Provide SH group at catalytic site of enzymes—Cysteine residue of enzyme protein molecule is important for providing SH group at the catalytic site of key enzymes, such as **glyceraldehyde 3-PO_4 dehydrogenase.** The SH group at catalytic site is necessary for catalysis of the reaction (glycolysis).

(d) Conjugation in liver—Cysteine is decarboxylated to taurine, which conjugates with bile acids to produce bile salts (taurocholate), necessary for fat digestion and absorption in GIT.

In liver, cysteine conjugates with many drugs, toxins and harmful aromatic compounds to form **mercapto derivatives,** that are less harmful and are excreted out of the body.

(e) Provide disulphide linkages—Many protein molecules gain strength by formation of disulphide (-S-S-) linkage between 2 cysteine residues of the facing two folds of polypeptide chain in their secondary and tertiary structure. The disulphide linkages are strong and not broken by denaturating agents and thus provide strength to the protein molecule.

A brief account of enumerable functions of cysteine and cystine are summarized in Fig. 16.9.

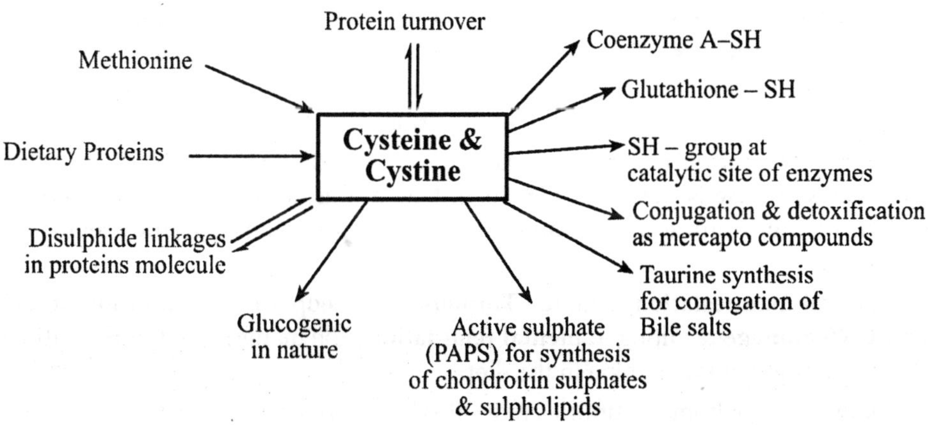

Fig. 16.9 : Biological functions of cysteine and cystine

Interconversion of Cysteine and Cystine

The enzyme **cysteine dehydrogenase oxidizes** and removes two H^+ ions that are taken up by NAD coenzyme from the SH groups of two molecules of cysteine amino acid, which get linked together by disulphide linkage to produce cystine amino acid. This is the basis of disulphide bonds formation in protein molecule between two cysteine residues. The reaction is reversible, so cysteine and cystine are interconvertible as shown in the Fig. 16.10.

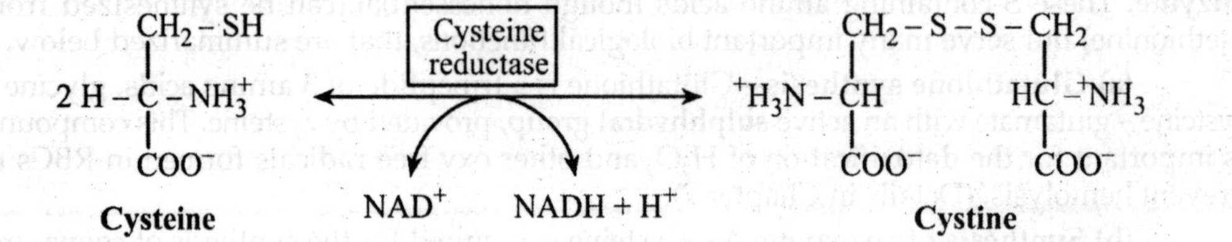

Fig. 16.10 : Formation of disulphide linkage and interconversion of cystine and cysteine

Catabolism of Cystine and Cysteine

Cysteine is catabolized by two pathways as given under—

1. Direct oxidation of sulphur of cysteine amino acid takes place by a dioxygenase enzyme to form cysteine sulphinate. This is then desulphurated to alanine and sulphinate, produced is oxidised to sulphate. The sulphate so formed can be excreted in urine, or some amount is activated to **active sulphate.**

Active sulphate is phosphoadenosyl phospho 5- sulphate (PAPS) and is formed by the activation of sulphate (desulphination of cysteine sulphinate) by ATP. The PAPS so produced is utilized for the synthesis of chondroitin sulphate (mucopolysaccharide), sulphoglycolipids etc. as well as it is utilized for the detoxification of harmful compounds in the liver.

2. Transamination and desulphuration of cysteine produce pyruvate, which is the common end product of both the pathways and finally enter **gluconeogenesis** process or citric acid cycle for oxidation as shown in the Fig. 16.11.

Inborn errors of metabolism of S-containing amino acids—

1. Homocystinurias—These are a group of inherited metabolic defects of methonine catabolism, that are characterized by excretion of excessive amounts (>300 mg/day) of homocystine and S-adenosyl methionine in the urine.

Classic Homocystinuria Type I is due to a genetic defect in which the enzyme **cystathionine synthase** is deficient (Fig. 16.8). Due to this, a block occurs in the normal catabolism of methionine and homocysteine to synthesize cysteine, resulting in homocystinuria.

Clinical manifestation include - Thrombosis, osteoporosis, dislocation of lens of eyes and **CNS damage** resulting in **mental retardation.** The management of the patients is done by providing low methionine in the diet.

Other types of homocystinurias have also been reported, but they are rare and are due to defect in the remethylation cycle.

2. Cystinuria—Cystinuria along with lysinuria is an inherited **renal tubular reabsorption defect,** characterized by excessive urinary excretion of cysteine (30 times more than normal) and other basic amino acids—lysine, arginine and ornithine. In this genetic disorder a common renal carrier system is defective leading to the excretion of all the four amino acids mentioned above. As cysteine is insoluble amino acid, there is a tendency of formation of renal calculi, otherwise the conditions is mild in nature.

3. Cystinosis—It is a rare inherited **lysozomal enzyme** abnormality, characterized by impaired cysteine transport mechanism. Because of this, cysteine amino acid accumulates in plasma and as it is sparingly soluble, its crystals get deposited in the soft tissues and organs, particularly in reticuloendothelial system. There is a generalized aminoaciduria (due to renal reabsorption defect), with severe renal function impairment. Death results at an early age due to renal failure.

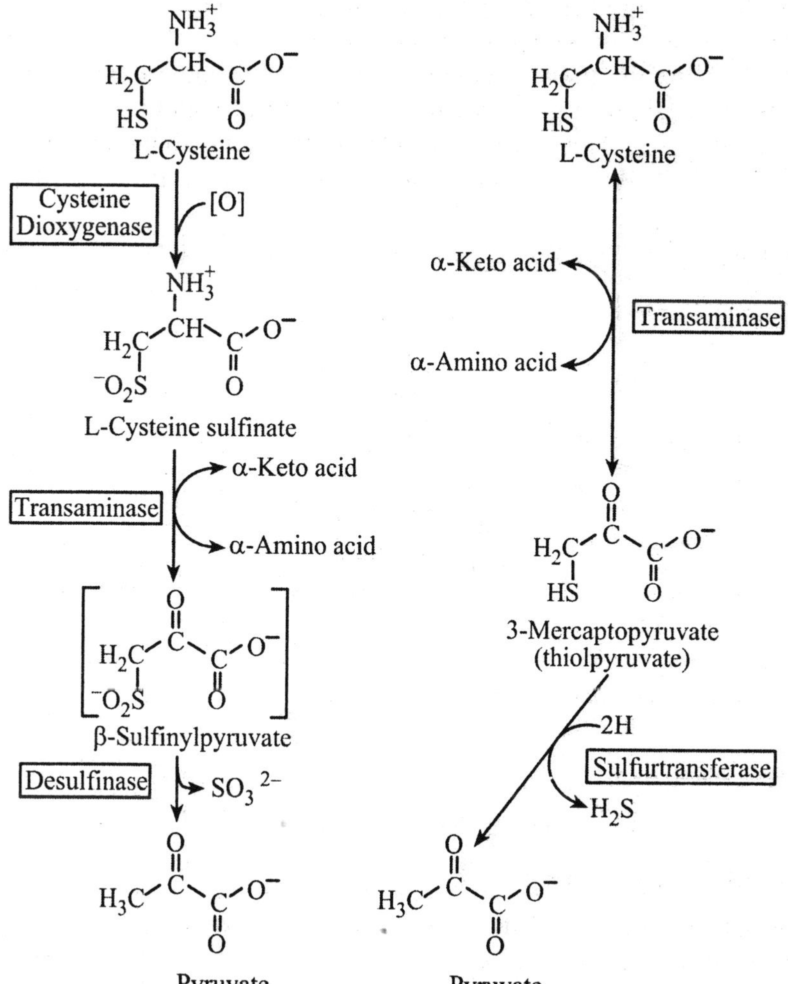

Fig. 16.11 : Metabolism of cysteine to pyruvate by transamination and transulphuration

PHENYLALANINE AND TYROSINE

Phenylalanine and tyrosine are structurally related aromatic amino acids and thus their metabolism will be dealt here together. Phenylalanine is an **essential amino acid,** while tyrosine can easily be synthesized in the body from phenylalanine, by addition of an hydroxyl group at para position of phenyl ring as illustrated in the Fig. 16.13.

Specialized Functions

Apart from their important role in the synthesis of **structural** as well as **functional proteins** in body, these amino acids have many important specialized biochemical functions. These amino acid are precursors for the synthesis of physiologically potent compounds, that include—1) Catecholamine hormones, 2) Thyroid hormones T_3 and T_4, and 3) Pigment melanin.

Phenylalanine and tyrosine after transamination are normally catabolized into fumarate and acetoacetate, which enter gluconeogenesis and ketogenesis pathways respectively. Thus they are **partly glycogenic** and **partly ketogenic** amino acids. **Several inborn errors** of metabolism of these amino acids are **prevalent** with serious clinical manifestations, such as CNS damage and mental retardation with high mortility rate at an early age. These will be discussed along with normal metabolic pathway of these amino acids.

The major functions and fate of phenylalanine and tyrosine amino acids are summarized in the Fig. 16.12.

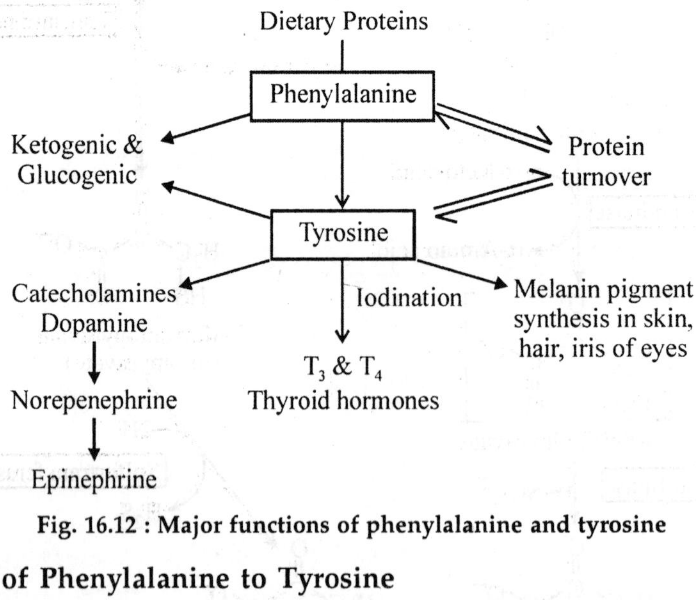

Fig. 16.12 : Major functions of phenylalanine and tyrosine

Conversion of Phenylalanine to Tyrosine

Phenylalanine is catabolized after its conversion to tyrosine, which is the precursor for the synthesis of various biologically important compounds as shown in the Fig. 16.12.

Phenylalanine is hydroxylated at the para position of benzene ring by the enzyme **phenylalanine hydroxylase,** which requires a **biopterine coenzyme,** that is similar to folate coenzymes. This enzyme is an hepatic **monooxygenase,** that incorporates one atom of O_2 at para position of phenyl ring to hydroxylate it, while the other atom of O_2 is reduced to water, by H^+ ions provided by tetrahydrobiopterine as shown in the Fig. 16.13. The reduced biopterine coenzyme is regenerated by H^+ ions provided by reduced NADPH coenzyme to continue the cycle.

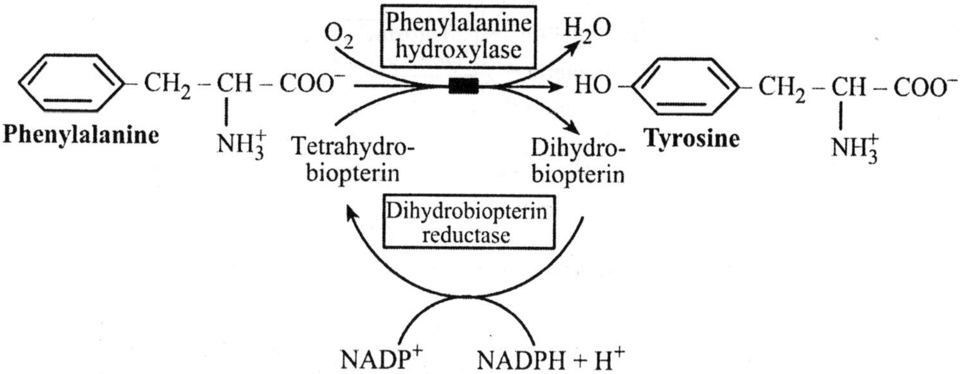

Fig. 16.13 : Conversion of phenylalanine into tyrosine and the genetic block in metabolic abnormality phenylketonuria(PKU) depicted by ▬

Specialized Functions of Tyrosine

1. Biosynthesis of catecholamines—Tyrosine is the precursor for the synthesis of catecholamine hormones, synthesized in adrenal medulla, sympathetic ganglions, autonomous nervous system and CNS. They are named as catecholamines because of the presence of a catechol ring (3, 4 dihydroxy benzene) in their chemical structure (Fig. 16.14).

Catecholamines are **potent biological amines** and act as hormones with a variety of physiological and biochemical functions. Catecholamines include **dopamine, norepinephrine** and **epinephrine.**

Catecholamine are synthesized from tyrosine by **hydroxylation** and **decarboxylation** reactions. **Decarboxylation** of amino acids produces cartain physiologically important **primary amines** which are active as **hormones, neurotransmitters, neuronal modulators etc.** in the body. The reactions involved in the biosynthsis of catecholamines are as given here.

1. Hydroxylation of tyrosine—Tyrosine hydroxylase enzyme catalyzes addition of a 2nd hydroxyl group at 3rd position of phenyl ring of tyrosine to produce **3, 4 dihydroxy phenylalanine** (DOPA). This hydroxylase enzyme is similar to phenylalanine hydroxylase and requires tetrahydrobiopterine and reduced NADPH coenzymes (Fig. 16.14).

2. Decarboxylation of DOPA—In the next step DOPA is decarboxylated (COOH group of the side chain) by **DOPA decarboxylase** – a pyridoxal phosphate (PLP) dependent enzyme to produce first catecholamine, the **dopamine.**

3. Dopamine is further hydroxylated at β carbon of the side chain to produce **norepinephrine,** catalyzed by enzyme **dopamine hydroxylase.** This enzyme requires **Cu^{++} ions** to activate molecular O_2 to nascent oxygen (O^-), to add an hydroxyl group at β carbon of the side chain, while other atom of O_2 is reduced to water by **vitamin C** as illustrated in the Fig. 16.14.

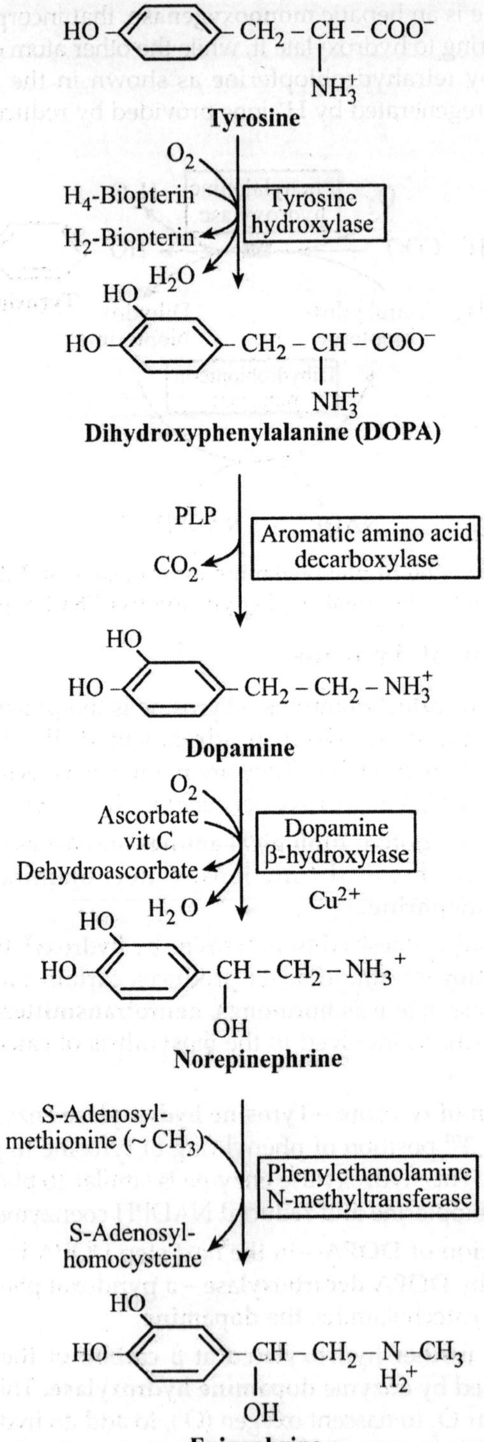

Fig. 16.14 : Synthesis of catecholamines from tyrosine

4. Methylation at nitrogen of amine group—Norepinephrine in the next step is methylated at nitrogen of amine group by active S-adenosylmethionine(SAM) to produce **epinephrine** by a transmethylation reaction (Fig. 16.14).

Biological Functions of Catecholamines

1. Dopamine—It is mainly produced in brain and CNS and acts as an important neuronal modulator and neurotransmitter in extrapyrimidal and straital tracts. Dopamine is also an natural inhibitor of **prolactin hormone** secretion.

Parkinson's disease—This neurological disorder is prevalent in old age (1% in > 60 years of age) and is characterized by **loss** of control of **voluntary** movements of skeletal muscles and facial muscles with tremors in extremities etc.

Biochemical hallmark of **Parkinsonism** is decreased production of dopamine in brain (low dopamine levels), that results in neurological defects.

Treatment—As dopamine cannot cross the blood-brain barrier, the drug of choice in Parkinson's disease is L -dopa or levodopa, which reaches to brain and is converted into dopamine by decarboxylation and the neulogical symptoms are alleviated.

Epinephrine and Norepinephrine

Norepinephrine is an important neurotransmitter while epinephrine has wide range of physiological and biochemical functions.

Epinephrine is major hormone secreted as a result of sympathetic stimulation from adrenal medulla during stressful conditions. The hormone prepares the subject to counteract adverse and challenging situations, such as emotional stress, anxiety, nervousness, fear etc. (fight or flight phenomenon). Epinephrine is a vasoconstrictor and increases blood pressure, and thus blood supply to all the organs including heart and brain is increased.

Biochemical functions—Epinephrine increases blood sugar level (BSL) by increasing glycogenolysis and gluconeogenesis processes in the liver. The hormone stimulates lipolysis and fat mobilization from adipose tissue and FFA released in blood are preferably oxidized in extrahepatic tissues, thus sparing glucose to be utilized by brain. (details in Chapter 7).

2. Thyroid hormone synthesis—Tyrosine amino acid is the precursor for the synthesis of thyroid hormone T_3 and T_4, that are synthesized in thyroid gland by **iodination** of tyrosinyl residues of thyroglobulin. The details of biosynthesis and functions of thyroid hormones are described in the Chapter 29.

3. Melanin pigment synthesis —Melanin is a brown black pigment, that imparts an uniform colour to the skin, hair, iris of eye etc and provides protection from the damaging effects of ultraviolet rays of sunlight. Melanin synthesis takes place in **melanosomes** of **melanocytes** of skin, hair and iris of eye, from amino acid tyrosine. Only one enzyme is required at the first two steps of biosynthesis of melanin, known as **tyrosinase,** which is a copper containing oxygenase. The rest of the steps are non enzymatic and occurs spontaneously as shown in the Fig. 16.15.

Ist Step in the biosynthesis of melanin is hydroxylation of tyrosine at 3rd position of phenyl ring, catalyzed by enzyme **tyrosinase** to produce **DOPA.**

IInd Step—Dopa is then again acted upon by **tyrosinase** enzyme, that brings about **rearrangement** in the phenyl ring to convert DOPA to **dopaquinone** as shown in the Fig. 16.15. Dopaquinone so formed is then converted to **indolequinone** through a series of reactions including decarboxylation, oxidation and cyclization of side chain as illustrated in the Fig. 16.15. Indolequinone is then polymerized to form melanin pigment.

Abnormalities of Melanin Synthesis

Freckles (black moles) on skin are caused by a collection of melanocytes with hyper pigmentation. While **Leukoderma** (white patches on skin) is due to localized absence (genetic inherited disorder) or degeneration of melanocytes, resulting in white patches on skin, where melanin pigment is not synthesized. **Greying of hair** is also due to the disappearance of melanocytes from the roots of the hairs due to the aging process.

Albinism—Albinism (Greek - albino means white) is an inborn metabolic disorder transmitted as autosomal recessive trait with a prevalence rate of 1 in 20,000 births. It is an inherited genetic disorder due to a block in pigment melanin synthesis. Albinism includes a group of clinical syndromes characterized by **nonpigmentation** of skin, hair and iris of the eyes. At least ten types of albinism are known, all are characterized by decreased or absence of melanin pigmentation in skin, hair and iris of eyes. Some of them are discussed below—

1. Tyrosinase Negative Albinism—In this disorder, **albinos** lack enzyme **tysosinase** as a result of mutation in genes in melanocytes of skin, hair and iris of eyes. Because of this a block occurs in the synthesis of pigment melanin, resulting in white discolouration of skin, light pale yellow or tan coloured hair and pale yellow iris of eyes and the affected subjects are called albinos.

Clinical Manifestations

(i) Skin of albinos is highly sensitive to sunlight with frequent sunburns, rashes, eczema etc. due to the lack of melanin pigmentation, that normally protects our skin from the damaging effects of ultra violet radiations.

(ii) UV radiations damages DNA of skin, and may cause mutations, resulting in **skin cancers.**

(iii) **Photophobia occurs** due to nonpigmentation of iris of eyes and albinos are unable to open eyes in the bright sunlight.

2. Tyrosinase – positive albinism —These albinos have some (little) visible pigment on their skin and have yellow to light tan hairs. The malonocytes of the hair bulbs may synthesize some pigment, but the amount is insufficient. Clinical manifestations are mild in these cases. The defect lies in the transport of tyrosine amino acid into the melanocytes.

Catabolism of Phenylanine and Tyrosine

As phenylalanine in body is converted to tyrosine, metabolism of these amino acids are interrelated and shall be considered together, along with the major inborn congenital abnormalities of their metabolism. The reactions of metabolism of these amino acids are as described here—

1. Tyrosine undergoes **transamination** in the presence of enzyme **tyrosine transaminase**, to form para hydroxy phenylpyruvate. Pyridoxal phosphate(PLP) coenzyme is required in the reaction.

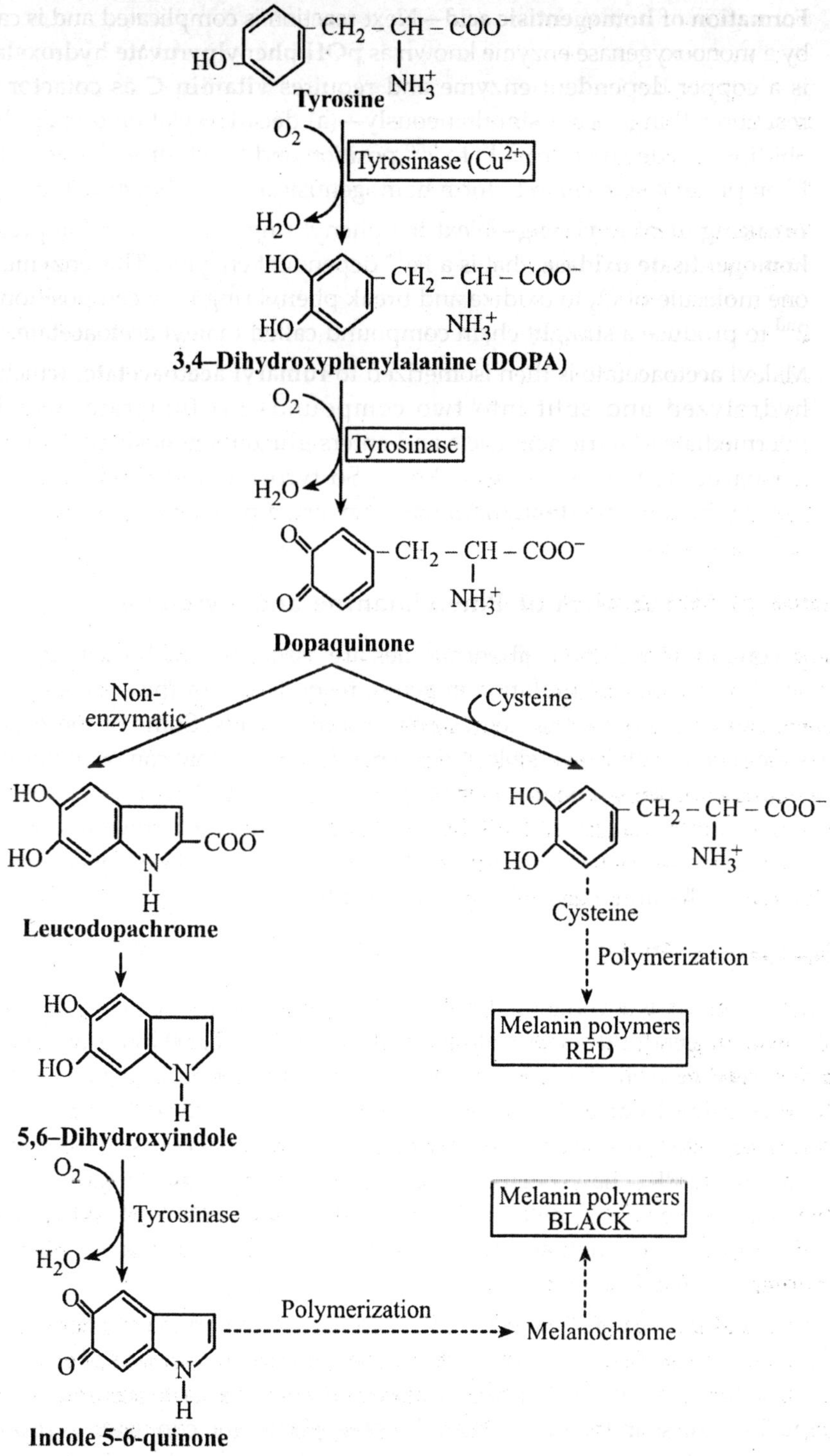

Fig. 16.15 : Biosynthesis of melanin pigments from amino acid tyrosine

2. **Formation of homogentisic acid**—Next reaction is complicated and is catalyzed by a monooxygenase enzyme known as **pOH phenylpyruvate hydroxylase,** that is a copper dependent enzyme and requires **vitamin C** as cofactor. In this reaction 3 things occur simultaneously—(a) decarboxylation of side chain, (b) Shifting of side chain from Ist to 2nd position and (c) Hydroxylation at position Ist, in place of side chain to form homogentisic acid as shown in the Fig. 16.16.
3. **Breaking of phenyl ring**—Next the phenyl ring is broken in the presence of **homogentisate oxidase,** that is a Fe^{++} dependent enzyme. This enzyme utilizes one molecule of O_2 to oxidize and break phenyl ring between position 1st and 2nd to produce a straight chain compound called maleyl acetoacetate.
4. Maleyl acetoacetate is then isomerized to **fumaryl acetoacetate**, which is then hydrolyzed and split into two compounds (a) fumarate, which is an intermediate of citric acid cycle and enters **gluconeogenesis** pathway and (b) acetoacetate, that is a ketone body. So tyrosine and phenylalanine after transamination and **degradation** are converted partly into glucose and partly to ketone bodies.

Inborn Errors of Metabolism of Phenylalanine and Tyrosine

Many congenital metabolic abnormalities are associated with phenylalanine and tyrosine metabolism because of mutation in genes, responsible for the synthesis of almost each and every enzyme required to degrade these amino acids. Some of these metabolic abnormalities are severe with irreversible clinical manifestations, that can be life threatening. These abnormalities are known since ancient times and are challenging to physicians for the early detection and management (suitable treatment) to prevent irreversible damage to CNS and brain, that causes mental retardation of the affected child. The key enzyme defects and respective metabolic abnormalities are described here.

1. Phenylketonuria (PKU)

Phenylketonuria is a congenital metabolic disorder of amino acid phenylalanine. This is most prevalent genetic disorder with an incidence of 1 in 10,000 births and is inherited as **autosomal recessive** trait. The disorder is caused by mutation in gene due to which enzyme **phenylalanine hydroxylase** is not synthesized. Because of this, phenylalanine cannot be converted into tyrosine and a block occurs in the normal course of its metabolism. Many other variants of PKU have been discovered in recent years that cause classic inborn errors of phenylalanine metabolic defects. The other important variants of PKU are due to—(a) defect of biopterin reductase enzyme, (PKU type II & III) and (b) defect in dihydroxybiopterin synthase enzyme.

Biochemical abnormalities—Due to the **block** in conversion of phenylalanine to tyrosine, phenylalanine and its alternative metabolites (abnormal), such as phenyl pyruvate (α keto derivative hence name PKU), phenyl lactate and phenyl acetate accumulate in blood and are excreted in excess in the urine. Phenyl acetate glutamine conjugate is produced in liver and is excreted in urine, which imparts a typical **mousy odour** to the urine. The formation of alternative metabolites of phenylalanine are shown in the Fig. 16.17.

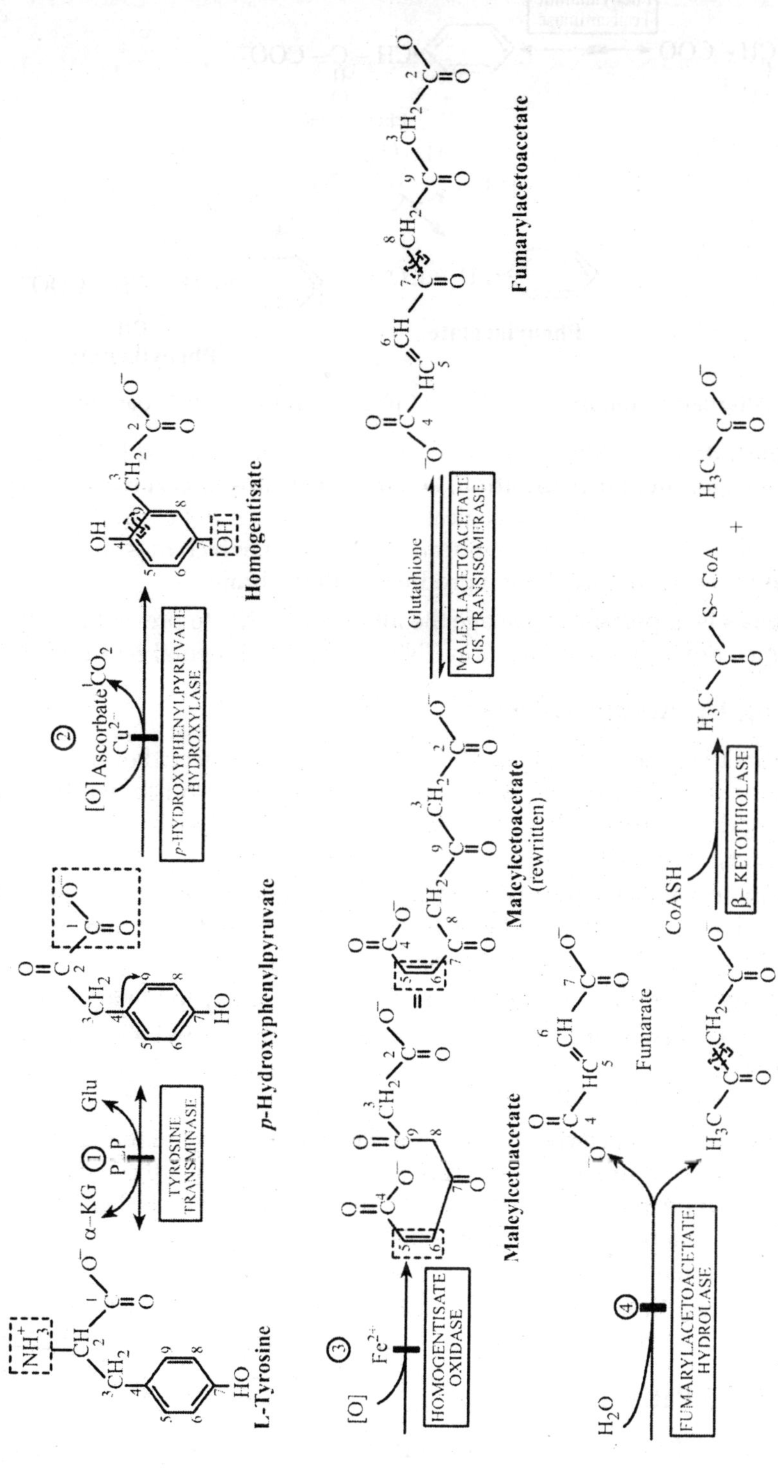

Fig. 16.16 : Metabolism of phenylalanine and tyrosine along with inherited metabolic abnormalities, with respective enzyme defects (block shown by ▬)

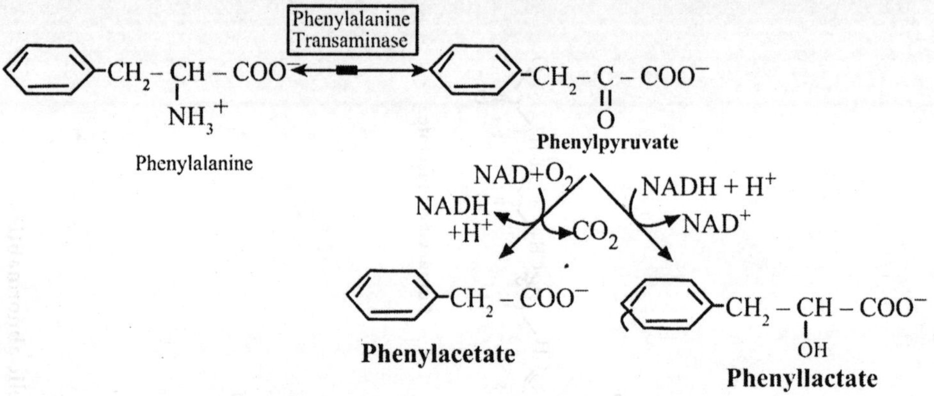

Fig. 16.17 : Alternative abnormal products of phenylalanine in PKU patients

Clinical manifestations—Major clinical consequences of unmanaged PKU include—irreversible CNS damage, **mental retardation, seizures, psychosis, eczema** and typical mousy odour of urine. Exact cause of neurological lesions of PKU is not clear. It can be due to elevated levels of phenylalanine and its abnormal metabolites, which may interfere with GABA and **serotonin** (neuronal modulators) production in the brain.

Early diagnosis is important to manage the infant with PKU to prevent irreversible damage to CNS, which otherwise may result in CNS damage and mental retardation.

Prenatal Screening by Amniocentesis

In recent years the genetic metabolic abnormalities can be detected even before the birth by amniocentesis. It is a technique in which amniotic fluid is withdrawn by a long needle from uterus and cultured to detect the enzyme defect. This enables the doctor to manage the abnormality at an early stage.

Gene therapy to replace abnormal gene by normal gene in intrauterine fetus has now became possible by the recombinant DNA technology and child born maybe normal (Chapter 23) , without much clinical complications.

Post natal diagnosis—PKU is detected in newborn babies by following tests—

Determination of increased plasma levels of phenylalanine can be carried out by chromatographic techniques. Normal level of phenylalanine in serum is 1-2 mg/dl that may be elevated in PKU to as high as 20-65 mg/dl.

Guthrie's Test—It is a bacterial (Bacillus subtilis) bioassay technique to assess the level of phenylalanine in the serum. In this method, bacteria are grown in the culture media with standard amounts of phenylalanine, and the size of bacterial colony is then compared with bacteria grown in serum of the patient. This will give the approximate levels of elevated phenylalanine in PKU patients.

Ferric chloride test in urine—As urine of PKU patient contains high amounts of phenyl ketones (> 500 to 3000 mg/day), it will give a blue green colour with ferric chloride.

Management of PKU Patient

Early detection of PKU in the newborn infants is necessary to prevent extensive CNS damage in the affected child. Baby should be weaned away from mother's milk and instead is given artificial milk substitutes with very low phenylalanine (10-20 mg/kg body weight/day), but sufficient tyrosine in diet to meet the requirement necessary for growth of the child. This low phenylalanine diet is continued till 5-6 years of the age of child to allow for proper brain development.

2. Alkaptonuria

Alkaptonuria is an inherited metabolic disorder of phenylalanine and tyrosine amino acids. It is a mild condition with no serious clinical manifestation, but the disorder has great historical importance in biomedical sciences.

Historical importance—Alkaptonuria was known since ancient times, and is one of the foremost metabolic defects that gave the idea about inborn errors of metabolism to the scientists. They were curious to know the cause about the fact that in some families, the nappies of the infant or their urine became brown black on exposure to air in the affected subjects.

Sir Garrod in 1909 was first to coin the term genetic **inherited inborn errors of metabolic disorders,** by observing that these abnormalities had certain common features, such as—a) their familial occurrence and b) the localized prevalence in the offspring of consanguineous marriages (marriages within family). Sir Garrod then proposed **one gene - one enzyme** hypothesis to explain the cause (defect) of the metabolic diseases, which are now referred to as molecular diseases.

Molecular Diseases

According to the one gene- one enzyme hypothesis, the mutations (altered genetic code on DNA) in genes results in lack or insufficiency of a key enzyme synthesis, thereby a block occurs in the normal course of metabolism of a particular amino acid or a metabolite. This sometimes results in severe clinical manifestations, which can be even fatal. These genetic metabolic abnormalities are now recognized as **molecular diseases** and are main cause of inborn errors of metabolic disorders.

Alkaptonuria—Characteristics

Alkaptonuria is an autosomal recesive trait with incidence of 1 in 2 million births. It is a mild genetic defect caused by the deficiency of enzyme **homogentisic acid oxidase** (Fig. 16.16), resulting in a block in tyrosine metabolism, where homogentisic acid cannot be further catabolized. This results in accumulation of homogentisic acid in the blood and its excessive excretion in urine.

Characteristic Sign and Symptoms

1. Urine of the patients become brown-black on exposures to air, because of the oxidation of homogentisic acid to benzoquinone acetate that polymerize to form a brown black pigment called **alkapton** as shown in the Fig. 16.18.

2. Ochronosis—Later on in this disorder, the excess of homogentisic acid gets deposited as brown black **alkapton** pigment in the connective tissues, cartilages of nose, pinna of ear, sclera, intervertebral disc (causing back ach) and in the bone joints causing mild type of **arthritis** and the condition is called as **ochronosis**. By the 3rd or 4th decade of life, the black pigment is visible through thin skin of nose and pinna of the ear.

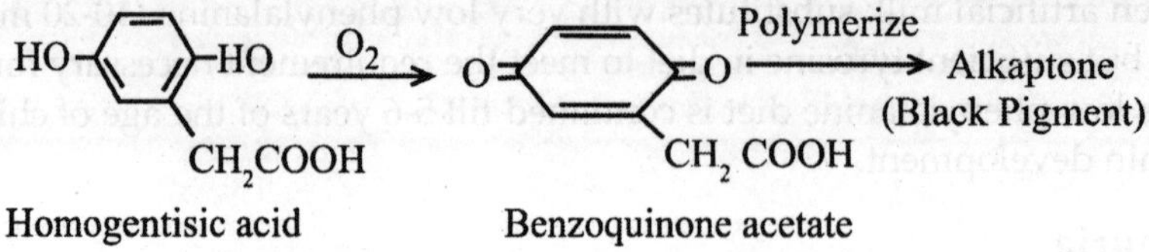

Fig. 16.18 : Oxidation of homogentisate to benzoquinone acetate (alkaptone)

3. Tyrosemia type I (Tyrosinosis)

It is a severe clinical metabolic disorder due to the congenital lack of enzyme **fumaryl acetoacetate hydrolase**, because of which a block occur in the normal course of phenylalanine and tyrosine metabolism (Fig. 16.16).

Biochemical alterations—Plasma levels of phenylalanine and tyrosine, along with methionine are elevated due to the impairment of transport system and enzyme inhibition by the accumulation of abnormal metabolites.

In acute tyrosinosis, the infant exhibits diarrhea, vomiting, "cabbage like" odour and the child fails to thrive. If not treated properly, it results in liver impairment, severe jaundice and death within 6-8 months of the age of the child. While in chronic condition, mild symptoms are observed and death occurs within 10 years of the age of the child. To manage the patient, a diet low in phenylalanine, tyrosine and methionine should be given.

4. Tyrosinemia type II (Richner Hanhart syndrome)

In this inborn error of phenylalanine and tyrosine metabolic disorder, there is a congenital lack of enzyme **tyrosine transaminase** (Fig. 16.16). Clinical manifestations include eye and skin lesions and moderate mental retardation. Plasma level of tyrosine is elevated. Alternative metabolites like pOH phenyl pyrvate, pOH phenyl lactate and pOH phenyl acetate, N-acetyl tyrosine and tyramine are excreted in the urine in excess.

5. Neonatal Tyrosenemia type III

It is due to the lack or insufficiency of enzyme pOH phenyl pyruvate hydroxylase (Fig. 16.16) that causes increased excretion of tyrosine along with alternative metabolites (mentioned in tyrosenemia II) in the urine. This condition is mainly observed in premature infants.

TRYPTOPHAN

Tryptophan is an **essential amino acid,** with an aromatic heterocyclic **indole ring** (Fig. 16.19). After transamination, the indole ring of tryptophan is degraded via **kynurenine** pathway, finally to yield acetyl CoA and alanine. Tryptophan thus is both **glycogenic** and **ketogenic** in nature.

Specialized functions of tryptophan

In addition to its essential requirement for various body protein synthesis, tryptophan is precursor for the biosynthesis of many physiologically potent compounds such as —

1. **Serotonin**—It is a vasocontrictor, neurotransmitter and important neuronal modulator.

2. **Melatonin** is a hormone synthesized from pineal gland.
3. **Niacin**—Tryptophan to some extent is normally metabolized via kynurenine pathway to vitamin B member niacin and its coenzymes NAD and NADP as shown in the Fig. 16.19.
4. Tryptoptan is a **source of IC* moiety** to be utilized for the biosynthesis of purine and pyrimidine nucleotides needed for DNA replication and RNA synthesis.

Tryptophan Metabolism—Kynurenine Pathway

Tryptophan is metabolized via a specific pathway called kynurenine- anthranilate pathway, in which the amino acid is degraded into amphibolic intermediates, the acetoacetate(a ketonc body) and alanine- a glycogenic amino acid . The kynurenine pathway in the body also serves as an alternative metabolic process, by which trytophan is converted (to some extent) into vitamin B member **niacin**, from which its coenzymes NAD and NADP are synthesized. The reactions of this pathway are as described here.

1. **Tryptophan oxygenase** enzyme, also known as **tryptophan pyrrolase**, catalyzes the cleavage of the indole ring of tryptophan by oxidation, forming an open chain compound called **formylkynurenine.** The enzyme tryptophan oxygenase is an inducible hepatic enzyme, containing a heme like iron porphyrin as a cofactor and utilizes both atoms of O_2 for the cleavage of the indole ring as shown in the Fig. 16.19.
2. In the next step, the formyl group is removed from fromylkynurenine by a hydrolytic cleavage, catalyzed by the enzyme **kynurenine formylase** to produce kynurenine. The formyl group so removed is taken up by the conenzyme tetrahydrofolate (THF) and is utilized for various biosynthetic pathways. (Chapter 25).
3. Kynurenine so produced, either undergoes transamination reaction, or it is converted into 3 hydroxy kynurenine and then to **3 OH anthranilate** by removal of alanine molecule from the side chain as illustrated in the Fig. 16.19. The reaction is catalyzed by enzyme – **kynureninase** that requires pyridoxal phosphate (PLP) as coenzyme.

   B_6 deficiency—Vitamin B_6 conenzyme (PLP) is necessary for the above mentioned reaction and thus in deficiency of vitamine B_6, the pathway is blocked and an alternative metabolite **xanthurenic** acid is produced and excreted in the urine. This can be taken as an indication of B_6 deficiency state. Pyridoxal (B_6) deficiency also impairs the synthesis of vitamin **niacin** and its coenzyme NAD and NADP from trypotophan and thus it is manifested in the development of **pellagra** like symptoms.
4. In the kynurenine pathway, the 3 OH anthranilate produced as mentioned above, is then degraded by further oxidation reactions, finally to produce acetoacetate (ketone body). Thus tryptophan is partly ketogenic in nature, while alanine removed in the above mentioned reaction is glycogenic in nature.
5. **Niacin synthesis**—Some amount of 3OH anthranilate (about 3%) in body is diverted towords the synthesis of vitamin B member niacin and coenzyme NAD and NADP formation. In this alternative pathway the aromatic ring of 3OH anthranilate is first broken and then recyclized to incorporate the N_2 of amino group of the side chain into the ring to form a pyridine compound, **niacin** as illustrated in the Fig. 16.19.

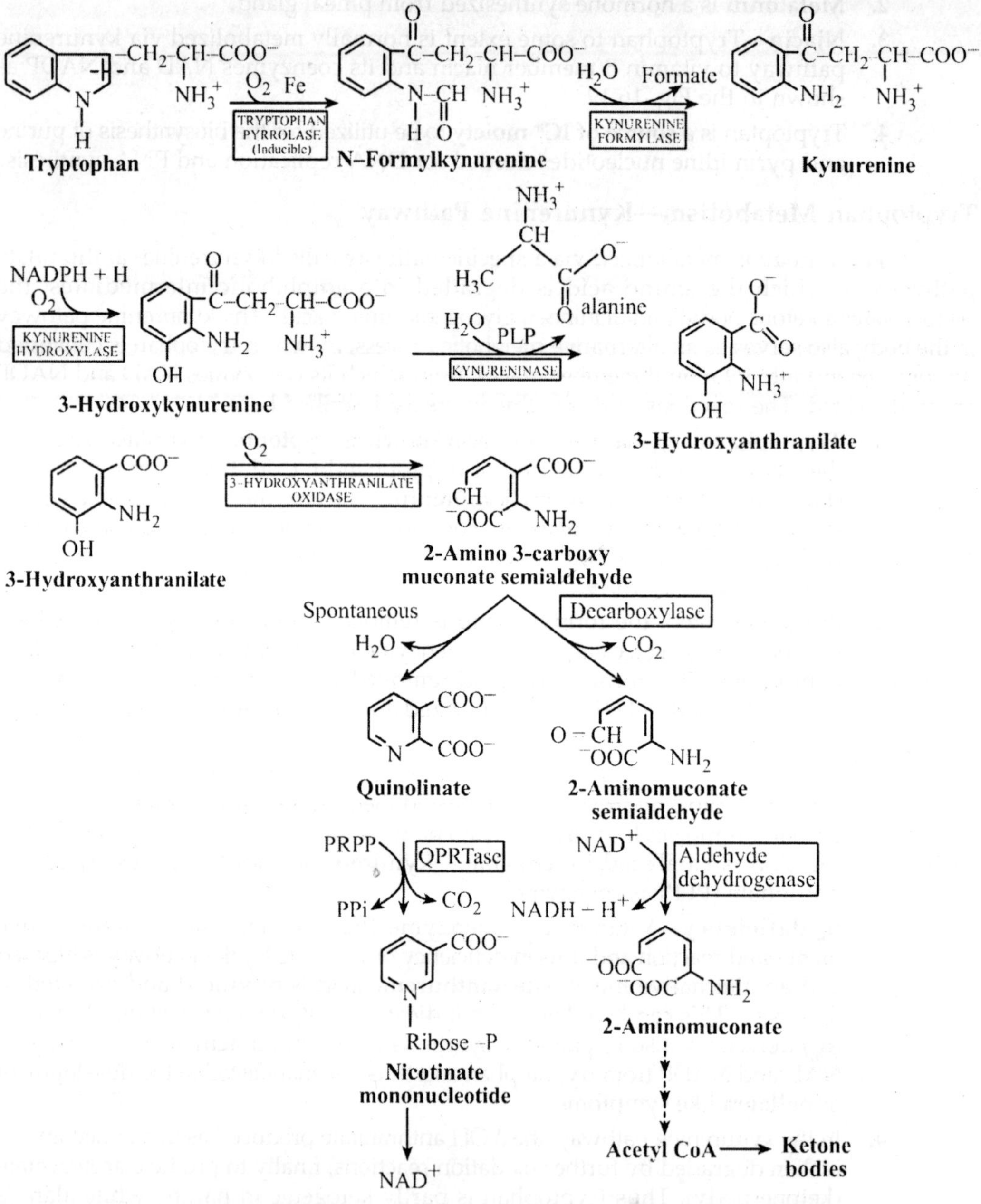

Fig. 16.19 : Kynurenine pathway of tryptophan metabolism and niacin conenzyme synthesis

Serotonin Biosynthesis

One of a very important specialized functions of tryptophan is to synthesize **serotonin** in the tissues such as brain, GIT, mast cells and platelets. Serotonin is a potent primary amine and has wide variety of physiological function in the body.

Serotonin is synthesized from tryptophan by first hydroxylation at 5th position of the indole ring to from 5-hydroxytryptophan. This hydroxylation reaction is catalyzed by the enzyme **tryptophan hydroxylase,** which is similar to the enzyme phenylalanine hydroxylase and requires the conenzyme biopterine as described in the conversion of phenylalanine to tyrosine. It is also proposed that phenylalanine hydroxylase enzyme also hydroxylates trypotophan in the tissues.

Next the 5-hydroxytryptophan so produced is converted into serotonin by decarboxylation of the side chain, catalyzed by the enzyme **decarboxylase**, that requires pyridoxal phosphate as coenzyme. The reactions are given in the Fig. 16.20.

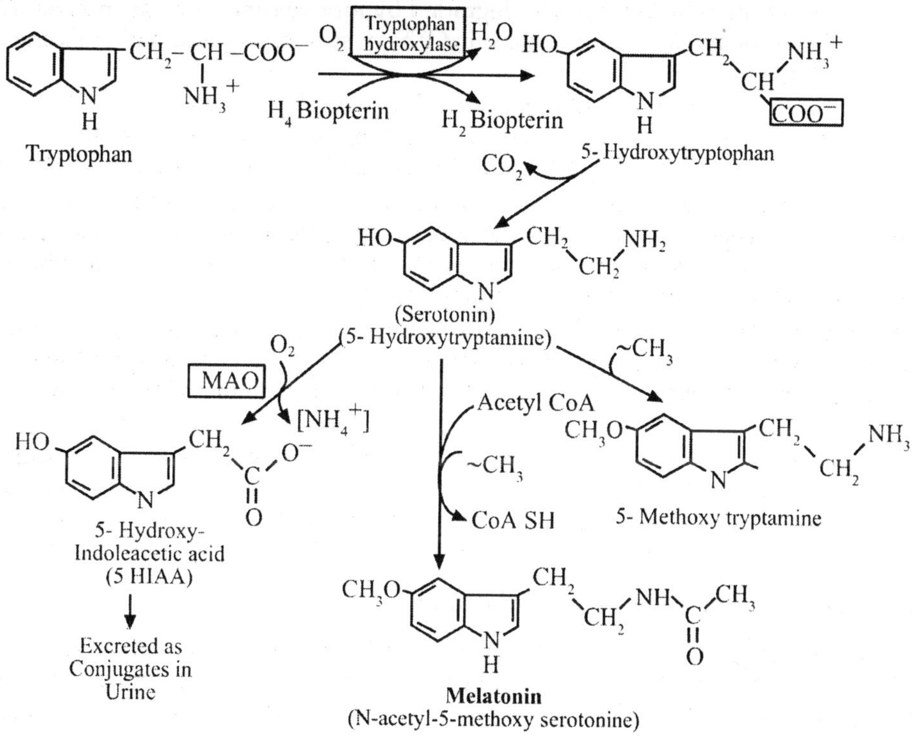

Fig. 16.20 : Synthesis of serotonin and melatonin from tryptophan and their metabolism

Metabolism of serotonin—The serotonin in tissues, such as brain is stored in bound form with certain proteins. The free serotonin is slowly released from the bound storage form and acts as a physiologically potent hormone or neuronal modulator.

The free serotonin is degraded and inactivated by the enzyme **monoamine oxidase** (MAO) which oxidatively deaminate the side chain to produce **5 hydroxy indole acetic acid** (5OH-IAA) as shown in the Fig. 16.20. The 5OH-IAA is then excreted in the urine as a normal constituent.

Function of serotonin—Serotonin is a potent physiologically active amine and has diversified biological functions in various tissues as listed here.

1. Serotonin is a potent vasoconstrictor and thus affects blood pressure.
2. It is a potent smooth muscle contractor, synthesized in GIT and plays an important role to regulate motility of gastrointestinal tract.

3. As vasoconstrictor it causes smooth muscle contraction in bronchioles and arterioles.
4. Serotonin is an important neurotransmitter as well as neuronal modulator in CNS and regulate cerebral activity.
5. It induces sleep, control behavioral pattern and regulates body temperature.
6. In brain, free serotonin level determines the normal neuronal status. Serotonin level is reduced in brain during **depression**, while elevated serotonin levels cause **agitation** and **excitation** of the subject.
   - Tranquilizer drugs such as reserpine acts via rapid release of bound (storage) serotonin, which is rapidly degraded by monoamine oxidase enzyme. Due to this serotonin levels are reduced in CNS for sometime, resulting in calming down effects.
   - On the other hand antidepressant drugs, such as iproniazid, inhibit enzyme MAO, thereby prevents serotonin degradation. So the serotonin level increases in CNS resulting in mood elevation and patient comes out of depression.

Argentaffinoma—It is a carcinoid tumor of serotonin producing cells in the abdominal cavity. These cells are known as argentaffin cells, as they get stained with silver salts. The localized malignant tumors of these cells are called argentaffinomas, which cause excessive production of serotonin in abdominal cavity.

Clinical Manifestations—The excessive production of serotonin and its effects on smooth muscle contraction and as vasoconstrictor results in the clinical symptoms such as red flushing of face, hypertension, intermittent diarrhea, respiratory distress and excessive sweating etc.

Biochemical abnormalities include excessive excretion of 5OH indole acetate in urine, as high as 500 mg/day (normal is < 5 mg/day). The patients exhibit pellagra like symptoms due to the decreased production of niacin from tryptophan, most of which (> 60%) is diverted towards excessive production of serotonin in argentaffinoma.

Biosynthesis and Functions of Melatonin

Melatonin is a potent compound synthesized in the pineal gland from serotonin by N-acetylation and O-methylation with active methionine as shown in the Fig. 16.18. Melatonin plays important role in sleep-wake cycle and is believed to be intimately involved in diurnal variation and biological rhythms.

Melatonin also regulates MSH and ACTH secretion from pituitary gland and thus control melanin synthesis and its uniform distribution on skin, hair etc.

Inborn Error of Tryptophan Metabolism

Hartnup's Disease

It is an inherited congenital defect in the transport of tryptophan and other neutral aromatic amino acids across biomembranes. This genetic abnormality impairs the absorption of tryptophan from the intestine as well as its reabsorption by renal tubules due to a carrier defect and causes servere clinical symptoms.

Clinical Manifestations

In Hartnup's disease, there is a generalized neutral amino aciduria. In this condition tryptophan is not absorbed from GIT properly leading to reduction in synthesis of niacin in the body. This is manifested in pellagra like skin rashes, cerebral ataxia and mental deterioration.

BRANCHED CHAIN AMINO ACIDS

Branched chain aliphatic amino acids **valine, leucine** and **isoleucine,** are all **essential** amino acids. Valine is a glucogenic amino acid, leucine is ketogenic, while isoleucine is partly glucogenic and partly ketogenic in nature. **Metabolism** of all the three branched chain amino acids is very similar and initial reactions are catalyzed by same enzyme systems, therefore their metabolism is considered together. A brief account of catabolism of branched chain amino acids is described in the following steps.

(a) All the three branched chain amino acids are transaminated by **same transaminase** enzyme to produce respective branched chain α keto acids (Fig. 16.21) in liver, kidneys, muscles etc.

(b) These branched chain α keto acids, then enter mitochondria, where they are oxidatively decarboxylated by a multiunit **decarboxylase** enzyme to produce branched chain α keto acyl CoA with one C-atom short than the original amino acids as shown in the Fig. 16.21.

(c) These branched chain α ketoacyl CoA of valine, leucine and isoleucine are then catabolized in several steps by different pathways to succinyl CoA from valine, acetoacetate from leucine and propionyl CoA and acetyl CoA from isoleucine, as described in the Fig. 16.21.

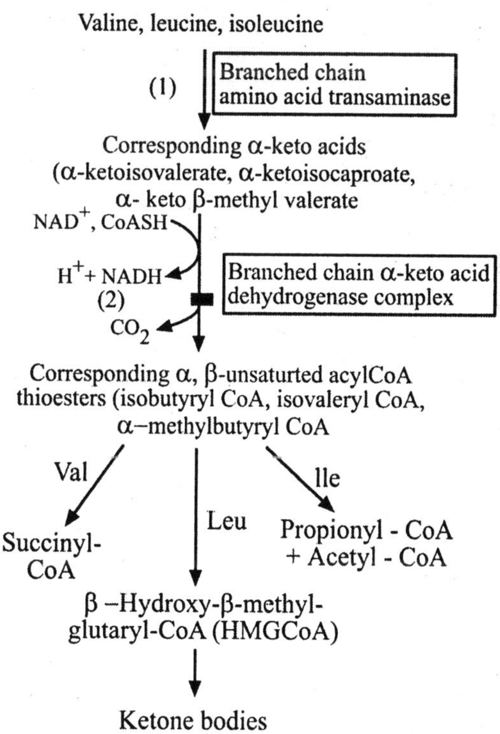

Fig. 16.21 : Catabolism of branched chain amino acid along with the inborn errors of metabolism shown by block (▬)

Metabolic Disorders of Branched Chain Amino Acids

1. Maple syrup urine disease—This is a rare congenital disorder due to the deficiency of enzyme **decarboxylase,** that impairs the oxidative decarboxylation of branched chain α keto acids (reaction-2 Fig. 16.21).

The disorder is very severe with fatal consequences, if not managed early. This metabolic disorders is characterized by a most striking feature i.e. the odour of urine resembles that of **maple syrup** or **burnt sugar** and hence the name maple syrup urine disease. This typical odour of urine is due to the increased excretion of branched chain amino acids and their α keto acids. Other varients of this disorder has also been reported but they are with mild symptoms.

Clinical manifestation occurs just after the birth of infant and include convulsion, CNS damage which may result in mental retardation in the surviving children. Accumulation of branched chain α ketoacids causes **acidosis,** that may lead to coma and death within first year of the age of child.

Management—Early detection of the disorder by the typical smell of urine is highly beneficial in the management of the infant and irreversible damage to CNS can be prevented. The infant should be weaned away from mother's milk and should be provided diets low in branch chain amino acids to prevent CNS damage and other clinical manifestations.

2. Isovaleric acidemia—This metabolic disorder occurs due to a congenital defect in synthesis of enzyme isovaleryl CoA dehydrogenase (Fig. 16.21), that oxidizes and metabolizes isovaleric acid, an intermediate of leucine metabolism. The block results in excessive accumulation of isovaleric acid and its excretion in urine. Clinical symptoms include typical **"cheesy" odour** of the breath and body fluids, excessive vomiting, severe acidosis and coma which can be fatal.

METABOLISM OF HISTIDINE

Histidine has a heterocyclic imidazole ring with an alanine like side chain. It is considered as a semi essential amino acid, since it can be synthesized to a limited extent in body, but the amount is insufficient during growth and have to be supplied in the diet. Histidine is **glycogenic** in nature, since during its metabolism the imidazole ring is broken to produce α keto glutarate which enter glyconeogenesis pathway.

Functions of Histidine—In addition to its specific role in various body protein synthesis, histidine performs several important functions in the body as listed below—

1. It is a source of IC moiety (formimino group) to be taken up by tetrahydro folate(THF) coenzyme, and is utilized in various biosynthetic processes. (Chapter 25)
2. **Histamine formation**—Histidine is decarboxylated to produce histamine, a **potent biological amine** with wide range of physiological and biochemical functions (described later in the section).
3. **Formation of anserine and carnosine**—These are unusual dipeptides found to occur normally in muscles. These are produced by the condensation of histidine with β alanine by a peptide bond. Functions of these histidine compounds in muscles are still not clear.

Metabolism of Histidine

Catabolism of histidine takes place by specific enzyme system and is described here in following steps—

1. **Histidase**—This enzyme specifically oxidizes and deaminates the side chain of histidine to produce a compound called **urocanic acid** with an unsaturated side chain as shown in the Fig. 16.22.
2. **Urocanase**—The enzyme uroconase hydratate urocanic acid in a unique way by the addition of elements of water at the double bond in the side chain with formation of a compound called 4 imidazolone 5 propionate. The imidazolone ring of this compound is then broken to produce an open chain compound know as **formimino glutamate (Figlu)** and the reaction is catalyzed by enzyme hydrolase as shown in the Fig. 16.22.

Figlu is source of IC moiety—Formimino glutamate produced on breaking of imidazolone ring of histidine serves as an important source of IC* moiety. The formimino group is subsequently transferred to tetrahydro folic acid (THF) in the presence of an enzyme THF-formimino transferase and glutamate liberated is deaminated to α ketoglutarate, which enters gluconeogenesis pathway.

In folic acid deficiency further metabolism of figlu is impaired and it is excreted in urine in excess. This is utilized as diagnostic test **(Figlu test)** to detect and diffentiate folic acid deficiency anaemia. (Chapter 25).

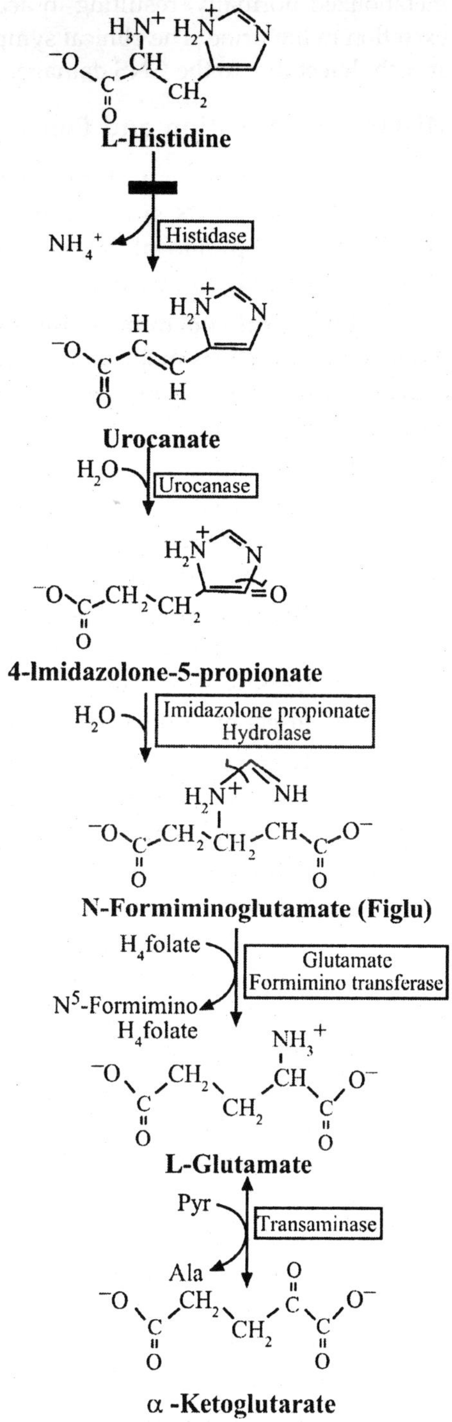

Fig. 16.22 : Catabolism of histidine to α-ketoglutarate

Inborn Errors of Histidine Metabolism

Histidinemia is an metabolic disorder of histidine amino acid, that is inherited as autosomal recessive trait. The congenital defect is due to the lack of the enzyme **histidase,** because of which histidine amino acid cannot be metabolized normally, resulting in accumulation of histidine in blood and its increased excretion in the urine. The clinical symptoms include mental retardation and characteristic speech defect due to the CNS damage.

Histamine Formation and Functions

Histamine is a potent primary amine produced from the histidine by decarboxylation of its side chain. The reaction is catalyzed by enzyme **histidine decarboxylase,** a pyridoxal PO_4 independent enzyme. Histamine is mostly produced by platelets, mast cells and basophils.

Biological functions of histamine—Histamine has a wide range of diversified physiological functions in the body. It acts through three different types of receptors on the target tissues, which may explain its diversified effects in the body. The important effects of histamine are described here.

1. Histamine is a very potent vasoactive amine. It causes smooth muscle contraction and increases vascular permeability. H_1 receptors are present on blood vessels, that specifically bind histamine and are mainly responsible for the vasoactive effects of histamine.
2. **Triple response**—Histamine is released from the injured tissues, mostly by the mast cell and is responsible the for "triple response" i.e. itching, swelling and redness of skin, which develops when an injury is caused by sharp knife or by insect bite on the skin.
3. **Hypersensitive reactions : Allergic reactions**—Histamine is released in excessive amounts in response to sensitization by several **allergens.** The major effects of histamine in **allergic conditions** are fall of blood pressure, asthamatic bronchiospasms leading to respiratory distress etc.
4. **Gastric secretion**—H_2 receptors are present on gastric mucosa on which histamine binds and are responsible for stimulating HCl secretion by gastric glands. This action is mediated through release of c-AMP. **H_2 antagonists are** widely used in the treatment of hyperacidity and peptic ulcers.
5. **Effects on CNS**—H_3 receptors are presynaptic, autoreceptors controlling synthesis and release of histamine in the brain. The histaminergic neurons or mast cells produce histamine within brain. It is considered as waking amine, that increases cerebral sensitivity to excitary inputs.
6. **Anaphylatic Shock—Certain allergens** such as penicillin stimulate production of IgE antibodies which get fixed on mast cells. When next dose of penicillin is injected, it interacts with IgE and stimulates degranulation of mast cells causing explosive liberation of histamine and slow reacting substances (SRS). This leads to peripheral vasodilation with severe fall in blood pressure and even peripheral vascular collapse causing anaphylactic shock, which can be life threatening.

Histamine is inactivated by enzyme monoamine oxidase, that oxidatively deaminate it into imidazole acetic acid, which is excreted in the urine.

Biomedical importance

1. Histamine is used as a stimulant for gastric secretion during gastric analysis for the assessment of gastric functions.
2. Antihistamine drugs are mainly used to block H_1 and H_2 receptors to control allergic and anaphylactic reactions.
3. Antihistaminic drugs are widely used in anticough syrups and in allergic asthamatic conditions.

PROLINE METABOLISM

Proline is a nonessential glycogenic amino acid. It is found in large amounts in collagen fiber of connective tissue proteins. During post translational modification of procollagen (precursor), the proline and lysine residues are hydroxylated to produce collagen (Chapter 21). Hydroxy proline plays important role in forming cross linkages along with hydroxylysine, conferring a fine network of fibrous proteins of connective tissue in which ground cementing substances like chondroitin suphates are laid down. This provides rigidity and strength to the fibrous proteins collagen, elastin and keratin of connective tissues.

Metabolism

Proline is catabolized by dehydrogenation and breaking of heterocyclic ring to open chain compound, glutamic semialdehide which is oxidized to finally glutamic acid as shown in the Fig. 16.23. The glutamic acid is then deaminated to α ketoglutarate, that enters gluconeogenesis process to synthesize glucose.

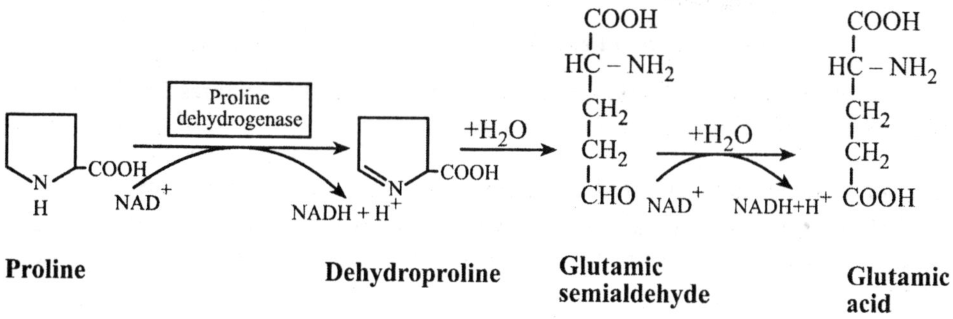

Fig. 16.23 : Reactions of catabolism of proline to form glutamate

ARGININE METABOLISM

Arginine is an highly basic α amino acid with an additional amide group at δ amino group. Arginine is formed mainly in liver during **urea cycle,** where it is cleaved by enzyme **arginase** into urea and ornithine (Chapter 15). It is synthesized in other tissues also, but to a very limited extent.

Arginine is considered as a semi essential amino acid since during growth its requirement is increased and should be supplied in diet. It plays important role in the synthesis of **creatine phosphate,** the high energy compound, which provides readily available energy (ATP) during muscle contraction (Fig. 16.2).

Nitric oxide synthesis—Arginine is a substrate for the production of nitric oxide (NO) in tissues. Nitric oxide is considered as a **wonder compound**, as it possesses more potential biological functions than any other substance in the body. **The diversified functions** of nitric oxide includes—i) It dilates blood vessels, relaxes smooth muscles and is constantly being produced and released by endothelial cells to regulate blood pressure, ii) Nitric oxide inhibits platelet aggregation and thus prevents intravascular thrombus formation and iii) Nitric oxide is an important neurotransmitter, active in CNS and brain.

Synthesis of Nitric Oxide

Nitric oxide is synthesized from **arginine** in the presence of a cytosolic enzyme **nitric oxide synthase** (NOS), that oxidize and split off terminal N_2 of amido group of arginine as nitric oxide. The enzyme NOS is a dioxygenase and uses one O_2 molecule and reduced NADPH. During the reaction one atom of O_2 is incorporated into NO, while the other oxygen atom is utilized to convert rest of the molecule into citrulline formation, as shown in the Fig. 16.24.

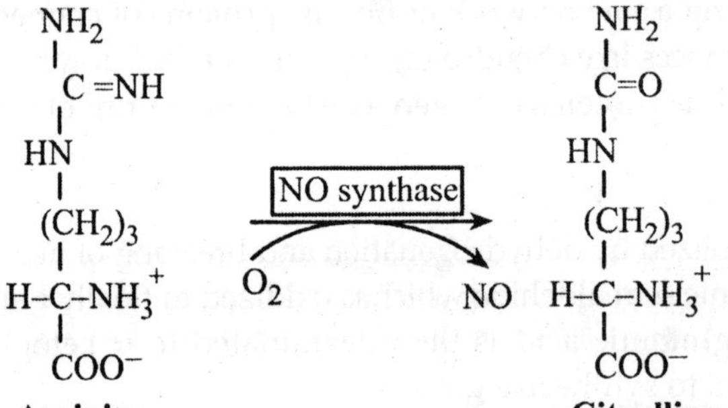

Fig. 16.24 : Synthesis of nitric oxide from arginine

Nitric oxide synthase (NOS)—It is a complex enzyme, that requires 5 redox cofactors namely NADPH, FAD, FMN, heme and biopterine during the synthesis of nitric oxide. The nitric oxide has a very short half life (3-4 seconds only) as it is scavenged by oxy free radicals, such as super oxide, or is inactivated by hemoglobin and other hemoproteins.

Nitric oxide is also produced from the nitrites used as vasodilator drugs in cardiovascular diseases, such as nitroglycerine and amyl nitrates, during their degradation in the body.

Functions of Nitric Oxide

The role of nitric oxide as therapeutic agent in form of **nitroglycerine** and **amylnitrate** used in the treatment of **angina pectoris** (heart diseases) has been known since 1980s. However only recently it has been realized that, it is produced in **vivo constantly** and has great role as smooth muscle relaxant and potent vasodilator, thereby is involved in regulation of blood pressure.

Three isoenzymes of nitric oxide synthase (NOS) are found to occur in the body tissues showing wide range of functions of nitric oxide at these sites as described below—

1. Endothelial NOS is found to occur in the endothelial cells, platelets, endocardium and myocardium. In these tissues the nitric oxide is constantly being produced and causes **arterial relaxation** and thus normal **blood pressure** is maintained.

- Nitric oxide causes vascular dilatation by relaxing smooth muscles via activating G-protein, guanylcyclase- cyclic GMP kinase system. Nitric oxide acts as an inhibitor of platelet aggregation and thus prevents intravascular thrombosis to regulate blood flow in the circulatory system.

Acetylcholine is a smooth muscle relaxant and vasodialator, but it will not cause these effects, if endothelial cells are striped away from underlying smooth muscles of the blood vessels. This finding indicates that acetylcholine initially interacts with endothelial cells of small blood vessels and via Ca^{++} and phosphotidylinositol causes release of an **endothelium derived relaxing factor (EDRE)** which was found to be none other than **nitric oxide.**

2. Neuronal NOS—It is the second isoenzyme of NOS found to occur in the central and peripheral neurons. **Nitrogenic neurons** (which produce NO) are seen in cerebellum and its production and release is Ca^{++} dependent. Nitric oxide is a powerful **neurotransmitter** and chemical messenger that relaxes **sphincters** and smooth muscles in urinary and gastrointestinal tracts.

3. Macrophage NOS—Nitric oxide mediates the bactericidal action of macrophages and is Ca^{++} indenpend in its release at these sites.

In **septic shock**, the bacterial lipopolysaccharides released in blood induces NOS synthesis in endothelial and smooth muscle cells, producing explosive and uncontrolled amounts of nitric oxide. This causes a profound vasodilatation resulting in severe fall in blood pressure and patient goes into septic shock, that can be fatal.

Inhibitors of NOS—Chemical inhibitors of NOS enzyme (e.g. N-mono methylarginine) may be beneficial in the management of septic shock in the affected subjects.

Chemical inhibitors of nitric oxide synthase when administered to animals and humans lead to **vasoconstriction** and marked elevation of blood pressure. This shows importance of **NO** as a major regulator of blood pressure.

Polyamines

Arginine and ornithine in addition to their prime role in urea synthesis serve as precursors for specific ubiquitous compounds known as **polyamines.** These are called as polyamines because they contain **many amino** groups in their molecular structure. Important polyamine are **spermidine** and **spermine** and humans synthesize about 0.5 m.mol of spermine per day. The molecular structure and synthesis of polyamines is described here under—

Polyamine Synthesis

The polyamines are synthesized by condensation of putrescine (decarboxylated ornithine) and a portion of methionine from decarboxylated SAM (Fig. 16.23) to produce **spermidine,** that again condenses with another decarboxylated SAM to produce **spermine** as shown in Fig. 16.25. The key enzyme of polyamine synthesis is **ornithine decarboxylase (ODC),** that decarboxylate ornithines to form putrescine, the precursor of polyamines.

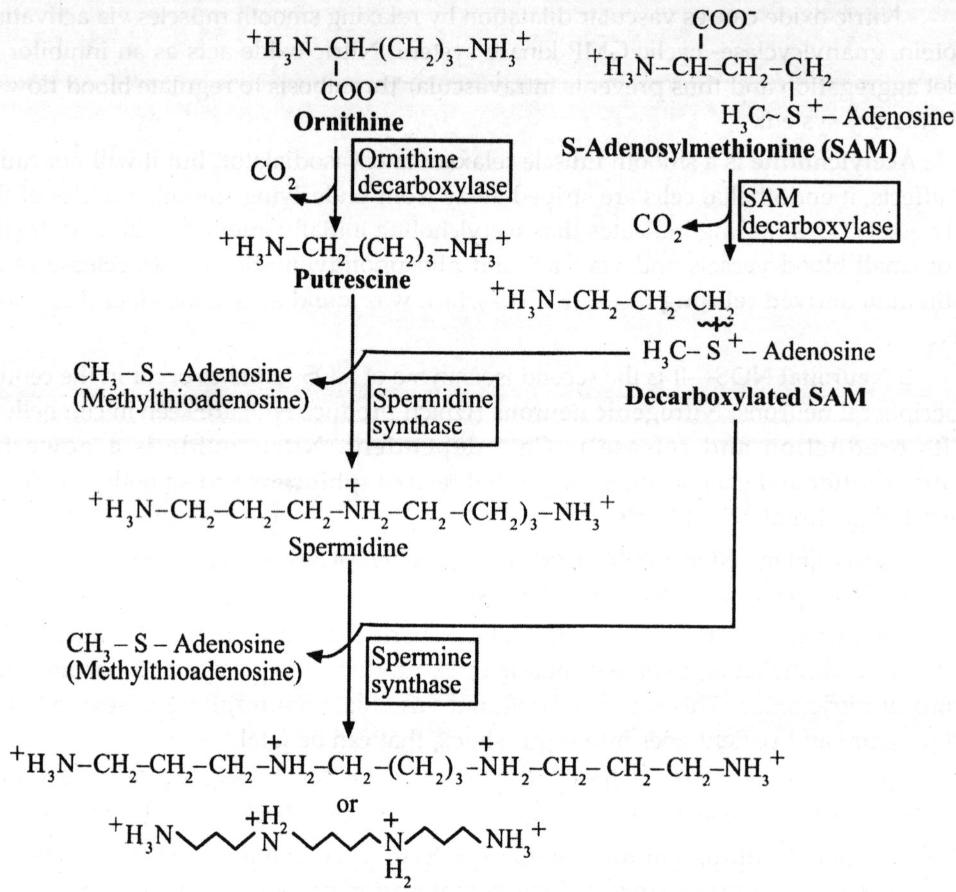

Fig. 16.25 : Biosynthesis of spermidine and spermine by the condensation of decarboxylated ornithine (putrescine) and decarboxylated S- adenosyl methionine (decarboxylated SAM)

Biochemical Functions Polyamines

Humans synthesize about 0.5 m.mol of spermine per day. Polyamines are highly basic in nature, since they contain multiple positively charged amino groups (Fig. 16.25). Because of the basic nature, the polyamines readily associate with nucleic acids DNA and RNA and thus have significant and diversified physiological functions, as listed below —

1. Polyamines are associated with **cell proliferation** and cell growth. They are reported to be important **growth factors** for cultured mammalian and bacterial cells in vitro.
2. They play significant role in **stabilization** of intact cell, biomembranes and subcellular organelles.
3. Polyamines are involved in DNA and RNA biosynthesis by stimulating the synthesis process and thus act as growth factors.
4. They are associated with DNA stabilization and packaging of DNA in bacterophages.

5. **Role in protein synthesis**—Recently it has been reported that polyamines are required for initiation of protein synthesis. The **hypusine residue** derived from spermidine is essential constituent of **initiation factor** elf 4D requirel essentially to initiate transcription process (Chapter 20). A correlationship is observed between rate of spermidine synthesis and rate of cellular growth, that shows significant role of polyamines in **growth process** by stimulating protein synthesis. Polyamines inhibit certain protein kinases.

Regulation of Key Enzymes of Polyamine Synthesis

The **ornithine decarboxylase** and **S- adenosyl methionine decarboxylase** are the key regulatory enzymes in polyamine biosynthesis process. The half life of ornithine decarboxylase in humans is only 10 minutes (shortest amongst the enzymes of body) and its activity responds rapidly to several stimuli. For example a 10 to 200 fold increase in its activity was observed by administration of growth hormone, corticosteroids, testosterone etc. to cultured mammalian cells. Same effect (increase) was observed in the activity of SAM decarboxylase.

Biomedical Importance

Ornithine and SAM decarboxylases are of clinical interest with respect to their potential use for **enzyme directed chemotherapy** to check cancerous growth. Difluoro methylornithine (DFMO), a **"suicide inhibitor"** of Ornithin decarboxylase (ODC), has been used successfully in isolating mutant cell lines that overproduce ODC as well as to inhibit rapid **cancerous cell multiplication** by inhibiting enzyme ODC.

DFMO is also useful drug against pneumocystitis carnii parasite infection, common in AIDS patient and in African sleeping sickness and Indian kalaazar. In these parasites DFMO inhibits polyamine synthesis by inhibiting ODC (half life in bacteria is more, in many hours) and so parasite cannot multiply and are killed. The drug will not harm humans as ODC half life in our body is in minutes only and is constantly being resynthesized rapidly.

Clinical Significance of Polyamines

Polyamine concentration increases in cancer tissues and in malignancies. The regulation of ornithine decarboxylase enzyme is of potential interest during chemotherapy in the treatment of cancer.

Tumor markers—The excretion of polyamines is elevated in almost all types of cancers, such as leukemia, carcinoma of lungs, bladder, kidney etc and their determination is beneficial in early detection of certain cancers.

Putresciene is an important tumor marker for cancerous growth and rapid cell proliferation, while spermidine is considered diagnostically important in diseases associated with excessive cellular destruction.

❑❑❑

17

NUCLEIC ACIDS

The nucleic acids are **biopolymers** synthesized by the condensation of large number of purine and pyrimidine nucleotides, that are the building blocks of nucleic acids. The nucleic acids are of two types called **deoxyribonucleic acid (DNA)** and **ribonucleic acid (RNA)** and are the main genetic material of the living organisms. The nucleic acid DNA is the principal storage form and transmitter of genetic information characteristic of a particular species to its progeny during reproduction.

DNA STORE GENETIC INFORMATION

Genes are chemically a DNA molecule or part of DNA molecule. DNA is the **chemical basis** (molecular) **of heredity** and is organized into genes on chromosome. All the genetic information characteristic of a specie is present on DNA molecule in code language called **genetic code,** which reside in the specific sequence of nucleotides on its polynucleotide chains. The genes govern the synthesis of RNAs as well as they direct the synthesis of various types of proteins with unique and specific biological functions that are absolutely necessary to carry on life processes. The genetic information from DNA **flows** through RNAs to specific structural and functional proteins, which ultimately determine the type of cell, its function, tissues formed or a specie produced, as illustrated in the Fig. 17.1.

Functions and Comparison of Nucleic Acids

DNA is organized into genes, arranged linearly on chromosomes in nucleus and has all the **genetic information** characteristic of a specie in code form known as genetic code

m – RNA transfers the genetic code from DNA (gene) to the site of protein synthesis i.e. the **ribosomes** (r – RNA) in cytoplasm.

t – RNA translates the genetic code on m - RNA into a specific sequence of amino acids to synthesize specific proteins, as directed by genes (DNA).

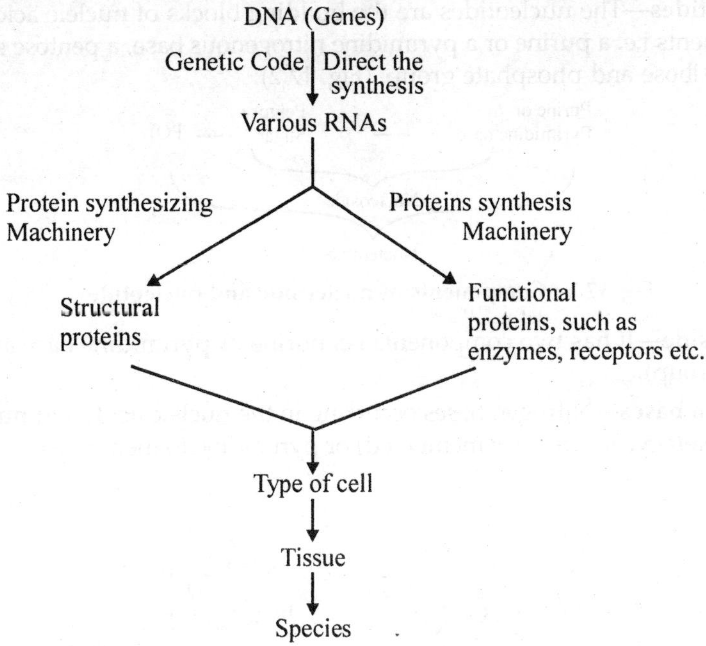

A brief account of the salient features and comparison of DNA and RNA is given in the Table 17.1.

Table 17.1 : Comparison of DNA and RNA

	DNA	RNAs
1.	Molecular basis of heredity, organized as genes on chromosomes.	Transfer genetic information from gene (DNA) into types of proteins to be synthesized for a specific cell for its functions.
2.	Transmits hereditary characters from parent to offspring during reproduction.	
3.	Chemically DNA is double helix formed by 2 polynucleotide chains.	Single polynucleotide chain.
4.	Contains 4 types of purines & pyrimidine nucleotides of A,G,C and T.	Contain 4 types of nucleotides of A,G,C and U (in place of thymine).
5.	Contains deoxyribose sugar as component of nucleotides.	Contain ribose sugar as component of nucleotides.

Chemical Structure of Nucleic Acids

Polymeric nucleic acids are formed by the condensation of large number of monomeric units called **nucleotides.** The four nucleotides of purine and pyrimidine **bases** are linked by diphosphoester linkages to form a long polynucleotide chain.

Nucleotides—The nucleotides are the building blocks of nucleic acid and consist of three components i.e. a purine or a pyrimidine nitrogenous base, a pentose sugar such as ribose or deoxyribose and phosphate group. (Fig. 17.2).

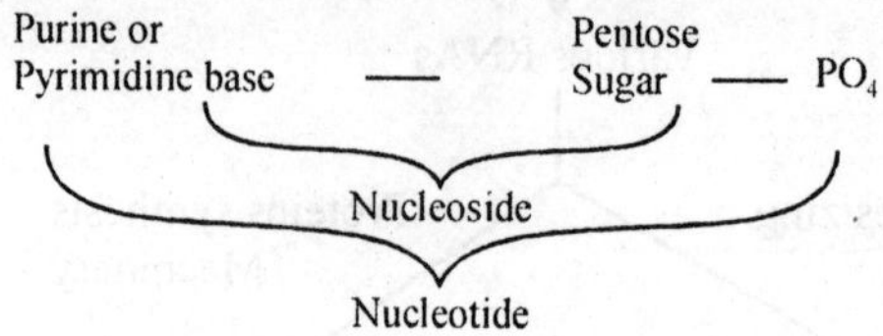

Fig. 17.2 : Components of nucleoside and nucleotide

Nucleoside—It has two components i.e. purine or pyrimidine base and a pentose sugar (no PO_4 group).

Nitrogen bases—Nitrogen bases occurring in the nucleic acids and nucleotides are derivatives of heterocyclic purine (9 membered) or pyrimidine (6 membered) rings as shown in the Fig. 17.3.

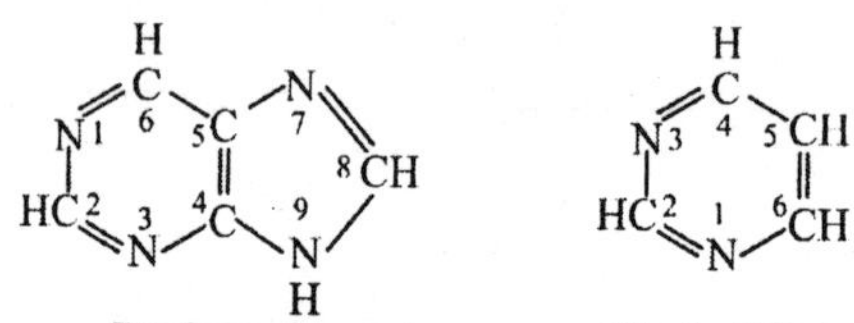

Purine **Pyrimidine**

Fig. 17.3

Purine bases—Two purine bases are found to occur in almost all types of nucleotides and nucleic acids. These are called **adenine** and **guanine** and are derivatives produced by substitution of groups on purine ring as shown in the Fig. 17.4.

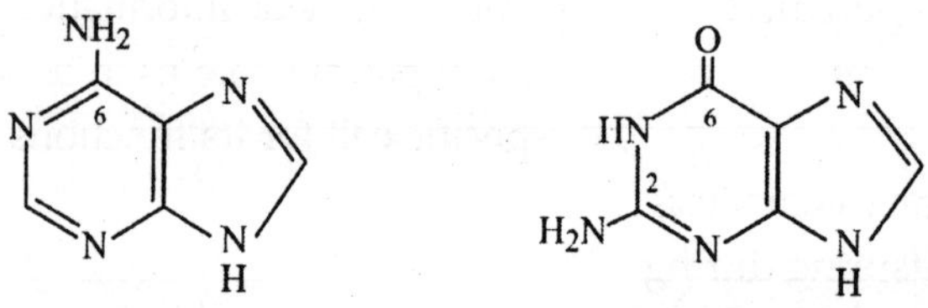

Adenine (6-aminopurine)

Guanine (2-amino-6 oxypurine)

Fig. 17.4

Pyrimidine bases—Three pyrimidine bases i.e. **cytosine, uracil** and **thymine** are the components of **nucleic acids,** and are formed by the substitution of groups on pyrimidine ring as shown in the Fig. 17.5.

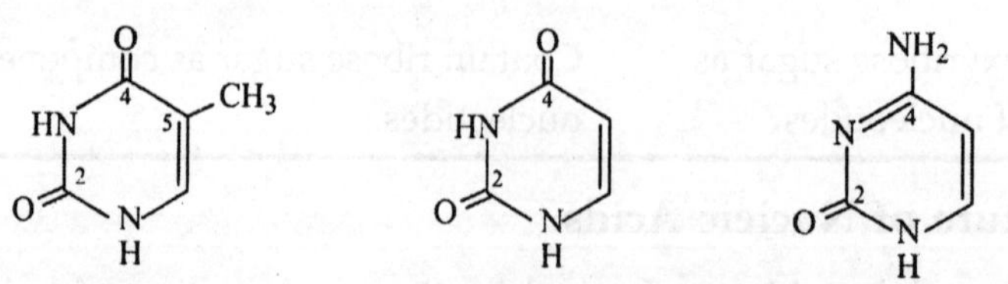

Thymine **(2,4 dioxy, 5-methyl pyrimidine)**

Uracil **(2,4 dioxy pyrimidine)**

Cytosine **(2-oxy 4-amino pyrimidine)**

Fig. 17.5

Nucleoside—It is formed by linking purine or pyrimidine nitrogenous bases with a pentose sugar, either ribose or deoxy ribose by a covalent N- glycosidic bond to N-9 of purine or to N-1 of a pyrimidine base (Fig. 17.3). The deoxyribose has one oxygen atom less at C- atom 2 of ribose sugar, as shown in the Fig. 17.6.

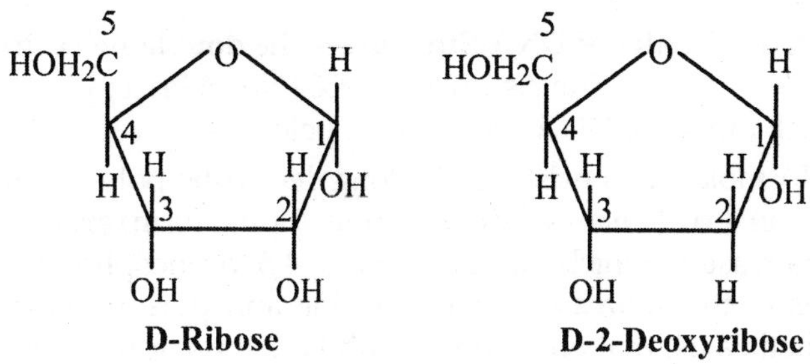

Fig. 17.6

Depending on the type of pentose sugar present, the nucleotide and nucleic acid are named as ribonucleotides or deoxyribonucleotides and ribonucleic acid (RNA) or deoxyribonucleic acid (DNA) respectively.

Structure of DNA—DNA is Polymer of Four Deoxynucleotides

Deoxyribonucleic acid (DNA) has two long polynuclotide chains formed by the condensation of many molecules of 4 purine and pyrimidine deoxyribonucleotides i.e. of adenine, guanine, cytosine and thymine linked together by 3′ – 5′ diposphoester linkage, as shown in the Fig. 17.7.

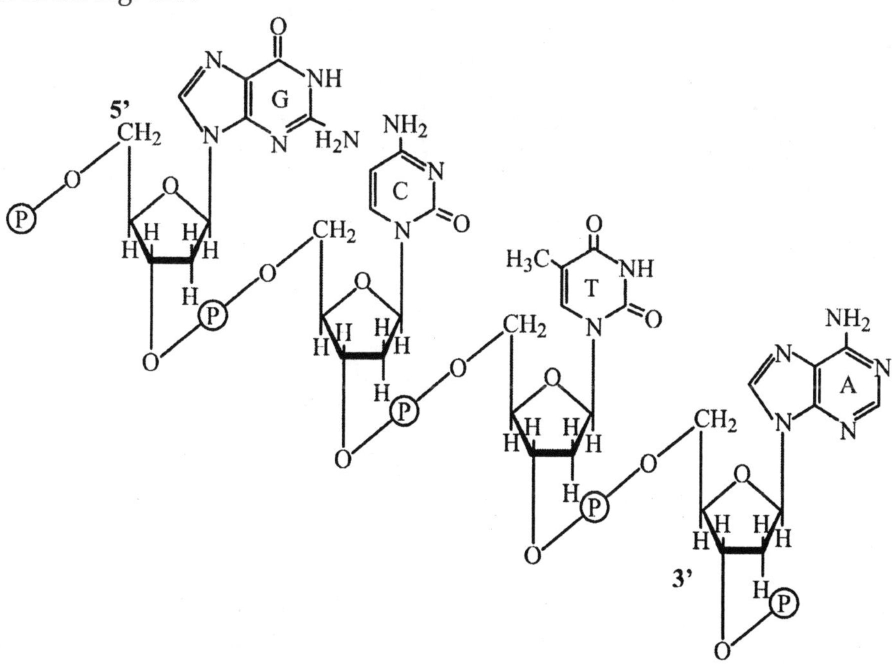

Fig. 17.7 : Condensation of deoxynucleotides by 3′ – 5′ diphosphoester linkage in a DNA polynucleotide chain (A-Adenine, T-Thymine, C-Cytosine, G-Guanine)

Genetic Code—In each polynucleotide chain of DNA, approximately 1 million nucleotides may be present, that are linked by 3′ – 5′ diphosphoester bonds in a highly conserved specific sequences. In these conserved specific nucleotide sequences on DNA strands, the whole genetic information, characteristic of species resides in a code form, known as genetic code.

Watson – Crick Model of DNA Structure—The **double helix** structure of DNA was proposed by James Watson and Francis Crick in 1953 (Nobel Prize, 1962). The salient features of Watson and Crick model of DNA are as given below—

1. The DNA molecule consists of two long deoxyribo polynucleotide chains twisted around one another in a **right handed double helix** (Spiral form) around an imaginary axis. It is like a spiral staircase in which the sugar and 3′ – 5′ diphosphoester linkages comprise the two hand rails of the staircase and the purine and pyrimidine bases jutting inside perpendicular to helix axis, that are bonded with H_2 bonds, forms the steps (Fig. 17.8).

2. Base pairing rule—The two long polynucleotide chains of the double helix spiral DNA molecule are held together by hydrogen bonds between N_2 bases of two polynucleotide chains. The base pairing rule is strictly maintained, according to which an adenine base will always be bonded with thymine by two H_2 bonds and cytosine is bonded with guanine by three H_2 bonds. These bases are known as **complementary base pair** as adenine have affinity to bind with thymine only while cytosine always binds with guanine respectively.

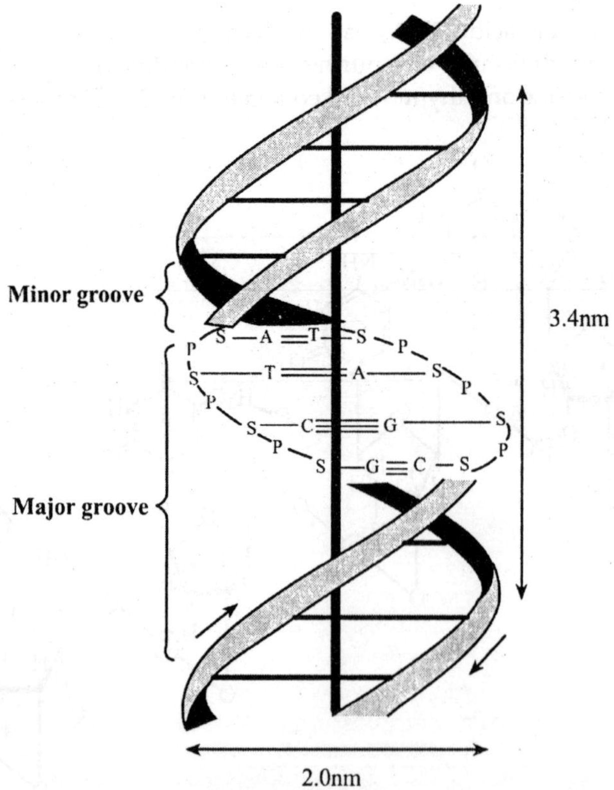

Fig. 17.8 : The Watson and Crick Model of the double helical structure of DNA along with salient features

3. Antiparallel polynucleotide strands—The two polynucleotide chains of DNA, though have same nucleotide sequence (genetic code) but is running in **opposite direction** or **antiparallel** to each other. The nucleotide sequence of one strand runs in 5′ to 3′ diphosphoester linkage direction while the other is in 3′ to 5′ direction. Thus the nucleotide sequence on the two DNA strands are **complementary**, meaning that adenine nucleotide will always face thymine of the other strand to form base pair and cytosine will always be in front of the guanine to pair by H_2 bonds as shown in the Fig. 17.8.

4. The double helix DNA molecule has a pitch of 34 A° (3.4 nm) per turn and has 10 base pairs in a single turn. There is a regular repetition of a minor groove (0.6 nm) and a major groove (1.2 nm) as shown in the Fig. 17.8.

Different forms of DNA—Double helix DNA may exist at least in six different forms designated as A to E and Z. The 'B' form of DNA double helix described by Watson and Crick is most common form of DNA occurring in a living cell under physiological conditions.

Prokaryotes DNA—In prokaryotic cells e.g. bacteria with no distinct nucleus or chromosomes, the double helix DNA is loosely packed as circular DNA or other forms.

Super Coiling and Organization of DNA Molecule

The two polynucleotide strands of double helix DNA having more than one million base pairs are of very long size (length). So in eukaryotes the double helix DNA is **super coiled** and packed along with some basic proteins, such as **histones** to form **nucleosomes** and exists as nucleoproteins.

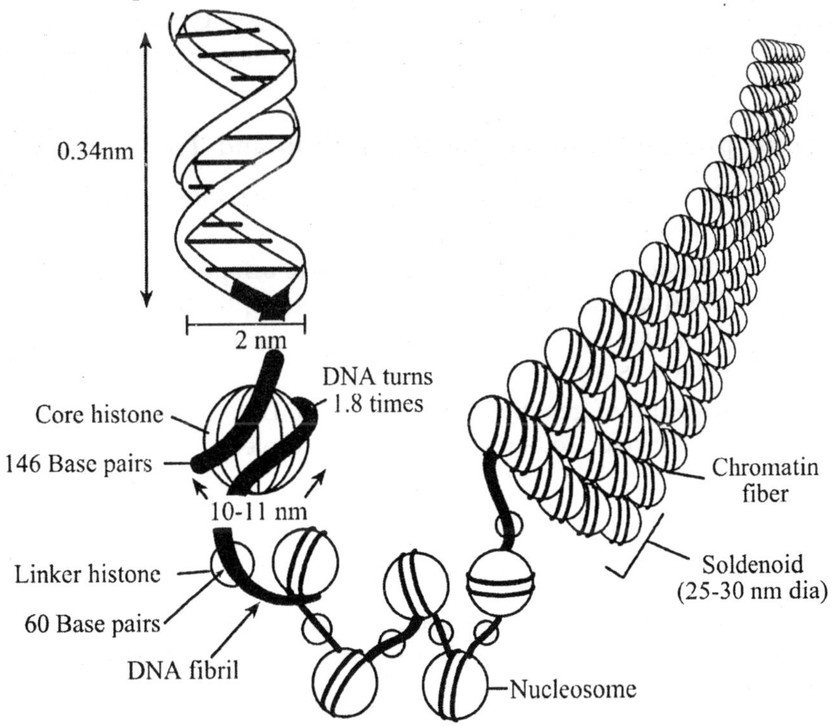

Fig. 17.9 : DNA coils organization to form chromosome

Nucleosomes—Under electron microscope the nucleosomes appear like **beads on a string of DNA fibril.** The beads are nucleosomes and each nucleosome core particle consists of a double stranded DNA, coiled around a complex of eight histone (octamers) molecules in approximately **1.8 turns of a left handed supercoil** as shown in the Fig. 17.9.

DNA Fibrils—DNA fibrils are thus formed by supercoiling of double helix DNA around **histone octamers,** at specific distance (55 bp is **linker DNA)** forming nucleosome (beads) on DNA string.

Chromatin—Six such DNA fibrils are then further packaged into chromatin fibers which are then organized into chromosomes as illustrated in the Fig. 17.9.

Separation of two DNA strands—Denaturation

The two strands of double helix DNA are held together by H_2 bonds between A and T and G and C bases. By heating or change in pH, the H_2 bonds between the base pairs are broken, resulting in separation of two strands of DNA. This is called as melting or denaturation of DNA molecule.

Annealing—Renaturation of DNA strands require base pair matching

When temperature is lowered (cooling), the two strands of DNA get bonded again in natural DNA double helix. This is because the nucleotide sequence of 2 strands is complimentary and all base pair match as the sequence is running in opposite direction (antiparallel nucleotide sequence). Thus the two separated DNA strands have great affinity to bind again, as base A of one DNA strand faces T and G faces C of another strand through out, to facilitate base pairing back by H_2 bonds. This is called annealing (renaturation) of DNA molecule.

DNA Strands provide Template for Replication and Transcription

The genetic code present in the specific nucleotide sequence on DNA strands has genetic information characteristic of species. It primarily serves two purposes—

1. **Transmission of heredity characteristics to offsprings**—During cell division each DNA strand acts as template and is replicated to produce a 'xerox' (identical) copy of DNA molecule to transfer information inherited by daughter cells to the offsprings during reproduction.
2. **Transcription of genetic code**—One of the DNA strand act as a template to transcribe genetic code regarding type of proteins to be synthesized on a m-RNA strand and thus govern all types of protein synthesis and functions of a living cell.

Ribonucleic Acid (RNA) Structure and Functions

RNAs mostly have single polynucleotide chain formed by the polymerization of 4 ribonucleotides of A, G, C and uracil (in place of thymine of DNA) ; linked together by 3' – 5' phosphodiester bonds.

Types of RNA—Three major types of RNA are found to occur in living cells, which are as listed here under—

			Cellular composition
1) Messenger -RNA	–	m–RNA	05 – 10 %
2) Ribosomal-RNA	–	r–RNA	50 – 80 %
3) Transfer-RNA	–	t–RNA	10 – 20 %

Messenger RNA (m–RNA)

As the name suggests, the m–RNA takes message from the DNA (genes) as to the type of protein to be synthesized, to the ribosomes. The genetic information regarding type of protein with specific function to be synthesized is present on the DNA molecule as **genetic code,** which is transcribed as such into a long polynucleotide strand called m–RNA in the nucleus. The m – RNA after processing comes out of the nucleus via endoplasmic reticulum and gets attached to the ribosomes, the protein synthesizing machinery (details in Chapter 21).

Ribosomal RNA (r–RNA)

The r–RNA is a large single stranded RNA linked to certain basic proteins to form **ribosomes.** The ribosomes are spherical dense particles of approximately 100 A° attached to endoplasmic reticulum, giving it a rough appearance as seen by electron microscope. Ribosomes are the major sites of protein biosynthesis in the cell.

The eukaryotic ribosome has a sedimentation rate of 80 S unit which is further separated into 2 subunits, i.e. a small 40 S subunit and a larger 60 S subunit, while the prokaryotic (micro organisms) ribosomes are smaller in size and have different RNA sequences.

The prokaryotic ribosomes are smaller and have a sedimentation rate of 70 S unit which are further separated into a smaller 30 S subunit and a bigger 50 S subunit.

Transfer RNA (t–RNA)

The t–RNA is small, special RNA containing approximately 70 – 80 nucleotides with regular intrastrand folding and base pairing to give it a **clover leaf** like 3 dimensional structure with some specific features and functions as shown in the Fig. 17.10.

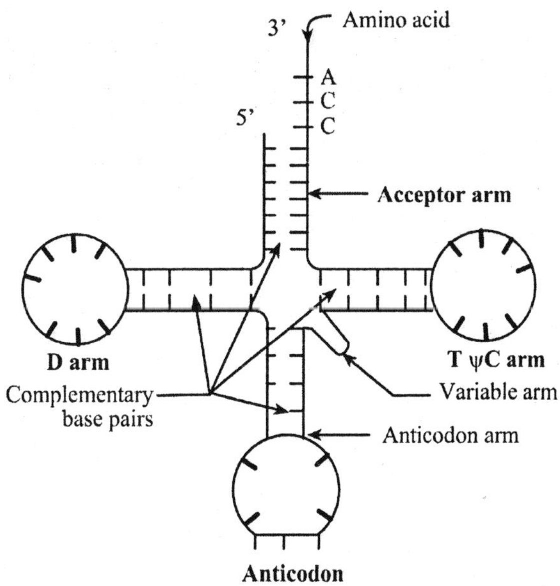

Fig. 17.10 : Characteristic features of t-RNA molecule

Characteristic Features of t-RNA

The t-RNA consists of—

1) Acceptor arm where specific amino acid is charged.
2) DHU loop or 'D' arm for the recognition of specific enzyme to charge specific amino acid at acceptor arm.

3) Thymidine pseudouracil cytosine (T ψ C) loop for the recognition and binding at specific site (A) on ribosome.
4) Anticodon loop which has 3 nucleotides (Triplet) complementary to recognize "genetic codon" on m–RNA by base pairing rule.

The t–RNA plays an important role in protein biosynthesis process. It acts as an adaptor intermediate molecule, which facilitates the **translation** of genetic code, as directed by genes (DNA) on messenger RNA, into a specific sequence of amino acids to synthesize a specific protein molecule (details in the Chapter 21).

❑❑❑

18

NUCLEOTIDE METABOLISM

Nucleotides are the building blocks of nucleic acids, DNA and RNA. Nucleic acids are polymers formed by the condensation of many molecules of purine and pyrimidine nucleotides linked together by 3′, 5′ diphosphoester linkage. Nucleotides are made up of a nitrogenous base purine or pyrimidine, sugar ribose or deoxyribose and a phosphate group. Almost all living organisms **synthesize nucleotides denovo** as required for DNA and RNA synthesis.

Dietary purine and pyrimidine nucleotides and nucleic acids are **non essential** as they are not utilized and not incorporated directly in the biosynthesis of tissue nucleic acids. Although humans consume nucleoprotein rich diet, the dietary purine and pyrimidine nucleotides after digestion and absorption are degraded and catabolized and are excreted out in the urine.

Nucleotide Biosynthesis—Salient Features

Purine and pyrimidine nucleotides are synthesized in most of the tissues of body **denovo** from **amphibolic intermediates** and are incorporated into nucleic acids and other biologically important nucleotides such as ATP, coenzyme A, NAD, FAD etc.

Denovo nucleotide biosynthesis is a **precisely regulated process** and occurs at a controlled rate, as required by the cell and a proper supply in appropriate amounts is ensured during growth and cell division etc. The rate of biosynthesis is regulated by **intracellular mechanisms,** that are able to sense the pool size and demand for nucleotides and accordingly control the synthesis by **efficient feedback system.**

Salvage pathway—In addition to denovo nucleotide biosynthesis, the **salvage pathway** is operative in certain tissues to utilize free purine and pyrimidine bases and incorporate them into **nucleotides** for DNA and RNA synthesis. This pathway salvage and removes the breakdown products left of nucleic acid catabolism in tissues, that are reutilized for nucleic acid synthesis. The salvage pathway is active in tissues such as erythrocytes, brain etc. where the denovo synthesis is not established.

Denovo Biosynthesis of Purine Nucleotide

Liver is the major site of denovo nucleotide biosynthesis, though it occurs in most of the tissues of body. The major pathway of purine biosynthesis is termed as **denovo** since the purine ring is build up from small amphibolic compounds. During denovo synthesis, a free purine base is not produced, rather the purine ring is **assembled** on **phosphoribosyl pyrophosphate (PRPP)** backbone, to directly **form nucleotides.** The different amphibolic compounds contributing to the various atoms of **purine ring** are shown in Fig. 18.1.

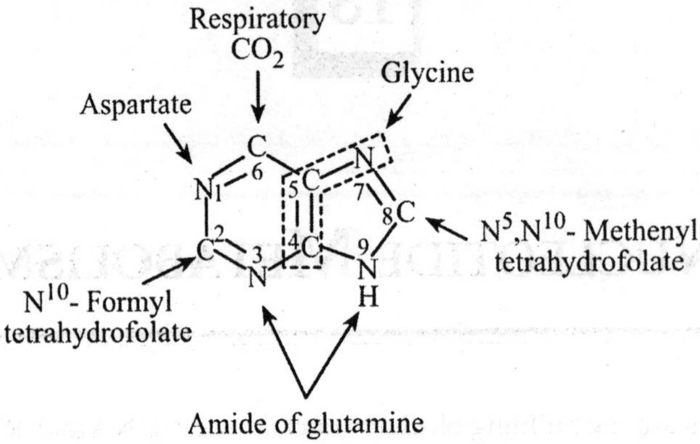

Fig. 18.1 : The assembly of purine ring from different amphibolic intermediates during denovo biosynthesis of nucleotides

Reactions of denovo Purine Nucleotide Synthesis

The parent or Ist nucleotide formed in denovo biosynthesis is inosine monophosphate (IMP) from which adenine monophosphate (AMP) and guanine monophosphate (GMP) are produced. There are in total 10 steps in the denovo purine nucleotide biosynthesis and in these steps, one by one the various carbon and N_2 atoms of purine ring are assembled on as PRPP molecule as shown in Fig. 18.2.

The enzymes catalyzing the denovo synthesis exist as **multienzyme complex** in **eukaryotic** cells to enhance the efficiency of biosynthetic pathway. The steps of denovo biosynthesis of nucleotides are as given under:

Ist Step : Phosphoribosyl pyrophosphate (PRPP) synthesis

1. PRPP is synthesized from ribose 5 phosphate, derived from HMP shunt or uronic acid pathways of glucose metabolism. The enzyme catalyzing the reaction is **PRPP synthase,** which is the most important **key regulatory** enzyme and the step is **rate limiting** in the nucleotide biosynthesis pathway.

The **multienzyme** of purine nucleotide synthesis is produced from a single m-RNA, transcribed by gene fusion. This multienzyme complex has **three main catalytic activities** as described below and shown in Fig. 18.2.

2. First committed step in IMP synthesis is the addition of N_9 (nine) atom of purine ring on PRPP molecule from amido nitrogen of **glutamine.** The reaction is catalyzed by enzyme **PRPP glutamyl amidotransferase** with the release of pyrophosphate from PRPP, energy of which is utilized to power the covalent bond formation and NH_2 group is added on the OH group of ribose 5 PO_4 (Fig. 18.2). The enzyme is allosteric and its activity is **controlled** by **tight feedback mechanism** by IMP, AMP and GMP (Fig. 18.2).

3. Phosphoribosyl amine formed in above step then condenses with **glycine** amino acid (forming C-atom 4, 5 and N_7 of purine ring) in presence of **glycinamide ribotide synthase** enzyme and ATP as energy source to form **glycinamide ribosyl 5-PO_4** or glycinamide ribotide **(GAR).**

4. N^{10} formyl tetrahydrofolate then donate C_8 of purine ring on NH_2 group (N_7) of glycinamide ribotide to produce **formyl glycinamide ribotide (FGAR)**, in the presence of enzyme **GAR transformylase** (Fig. 18.2).

5. In next step, glutamine transfers the 2nd amido N_2 to FGAR (to form N_3 of purine) to produce **formyl glycinamidine ribotide** (FGAM)

6. Cyclization of imidazole ring of purine—Next the imidazole ring is formed by the cyclization of FGAM, in presence of ATP to produce **amino imidazole ribotide** as shown in the Fig. 18.2.

7. Carbondioxide fixation—Seventh step is a CO_2 fixation reaction, without using biotin or ATP, and incorporates CO_2 directly to form C_6 of purine, that is added on C_5 as shown in the Fig. 18.2.

8. Aspartate then transfers its NH_2 group to condense with carboxyl (C_6 atom) group to add N_1 of developing purine ring to **form 5-aminoimidazole 4-carboximide ribotide.**

9. Transfer of a formyl group from N^{10} formyl THF (C_2 of developing purine ring) on carboxyamido group of the compound formed in above reaction, occurs in presence of **formyl transferase** enzyme to produce 5-aminoimidazole 4-formyl carboxamine ribotide.

10. The final reaction catalyzed by **cyclohydrolase** enzyme leads to ring closure to form 2nd ring of purine and **inosine monophosphate(IMP),** the first purine nucleotide is produced as described in detail in Fig. 18.2.

Inhibitors of Purine Nucleotide Synthesis : Biomedical Importance

Denovo synthesis of purine nucleotide, as well as DNA replication and RNA synthesis, **essentially** require **tetrohydrofolic acid** coenzyme (THF) to build up purine ring as emphasized in the biosynthesis pathway. Deficiency of folic acid or impairment of its activation to THF would inhibit purine nucleotide biosyntheis and would severely affect rapidly multiplying tissues such as **erythropoiesis** and **fetal development** in pregnancy.

- Folic acid antagonists methotrexate and **aminopterine** inhibit enzyme folic acid reductase (which form THF) and thus would inhibit purine nucleotide biosynthesis and DNA replication. These folic acid antagonists thus are used as **anticancer drugs** (chemotherapeutics) to check certain **cancerous growths** (Details in Chapter 25).

- Azaserine is a glutamine antagonist and inhibits transfer of amido nitrogen of glutamine in reaction 2 and 5 of purine nucleotide biosynthesis and thus block purine biosynthesis. Azaserine is used as anticancer drug, as it checks rapidly growing tissues by blocking DNA replication and cell division.

- 6 Mercaptopurine inhibits the conversion of IMP to AMP and GMP (Fig. 18.3) and is also used as anticancer drug in the treatment of certain cancers (Chapter 39).

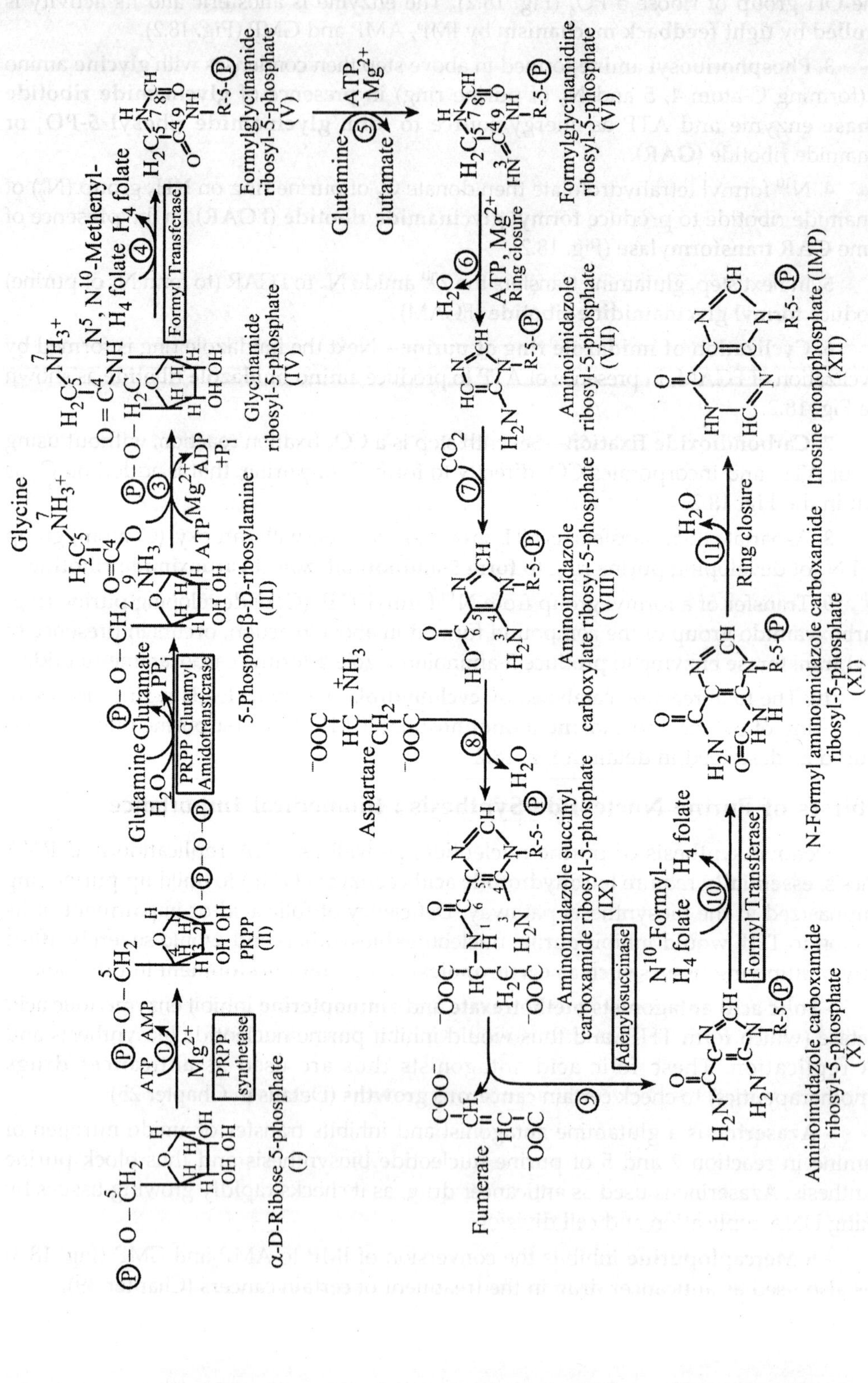

Fig. 18.2 : Major steps involved in denovo biosynthesis of purine nucleotides

Formation of AMP and GMP from IMP

Conversion of IMP to GMP and AMP occurs in two separate pathways as described below—

Formation of AMP—Addition of an amino group from aspartate at C_6 carboxyl group of IMP takes place in a two step reaction—1) Condensation of aspartic acid at carboxyl group of C_6 of IMP require GTP as energy source and enzyme **adenosuccinyl synthase** (Fig. 18.3).

2) Release of fumarate by hydrolyzing adenosyl succinate formed in the above step, takes place in the presence of enzyme adenosylsuccinase and NH_2 is added at C_6 of purine ring of form AMP, as shown the Fig. 18.3.

Formation of GMP from IMP

1. **Oxidation of IMP** is catalyzed by IMP dehydrogenase (NAD dependent) and is converted into xanthosine monophosphate (XMP) by addition of a carbonyl group at C_2 of purine ring.

2. Transamidation of 6 oxy group of XMP with amido group of glutamine proceed next, in presence of **transamidase** enzyme and ATP to produce GMP as described in the Fig. 18.3.

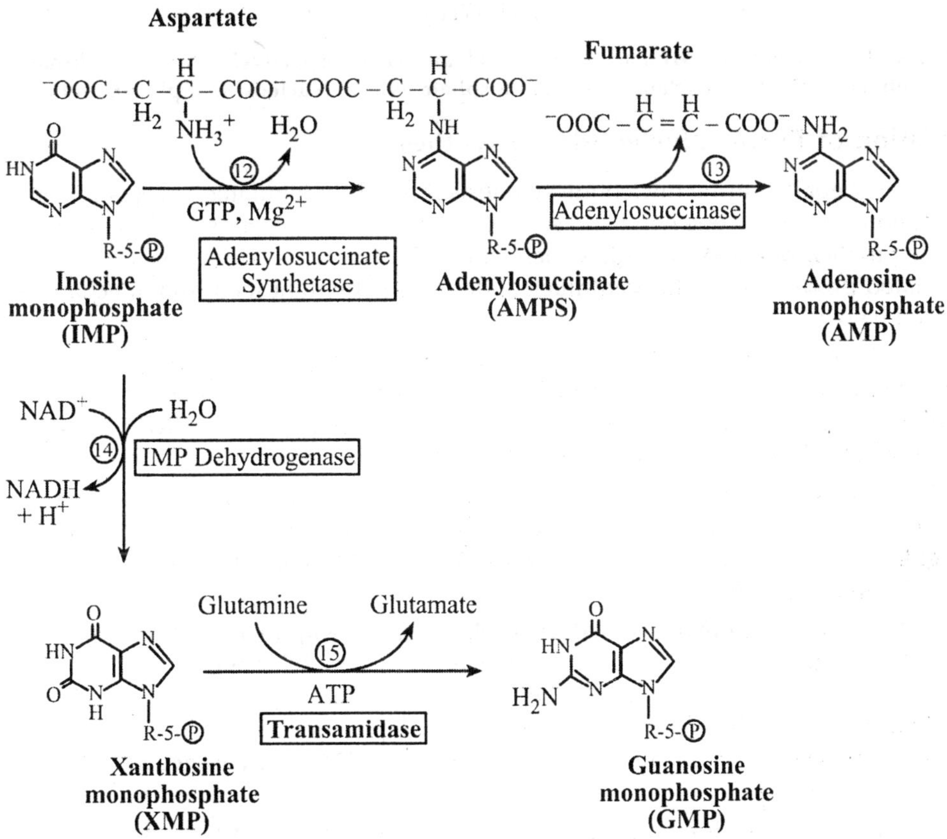

Fig. 18.3 : Conversion of IMP to AMP and GMP

Formation of Di and Triphosphates

The mononucleotides AMP and GMP are then converted to high energy di and triphosphates by phosphoryl transfer from ATP, as catalyzed by specific **nucleoside kinase** enzymes respectively (Fig. 18.4).

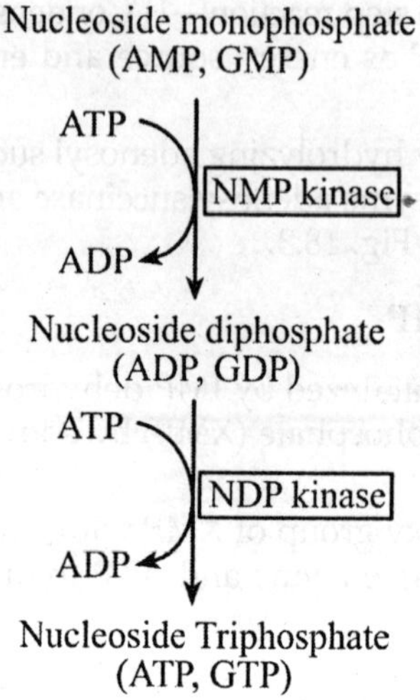

Fig. 4 : Conversion of AMP and GMP to ATP and GTP respectively by specific kinase enzymes (NMP—Nucleoside monophosphate; NDP—Nucleoside diphosphate)

Regulation of Purine Nucleotide Biosynthesis

The hepatic purine nucleotide synthesis is a tightly regulated process to conserve energy and nutrients on one hand and secondly to ensure the proper supply of nucleotides, according to their demand for wide variety of cellular functions. Thus a balanced and well coordinated production of different nucleotides occurs by tight feed back mechanisms as described below—

1. Concentration of PRPP is the major factor regulating rates of denovo purine nucleotide biosynthesis. The rate of **PRPP synthesis** depends on availability of ribose 5-phosphate and activity of enzyme **PRPP synthase.** The enzyme **PRPP synthase** is allosterically **feedback regulated** by both AMP and GMP concentration (Fig. 18.5).

2. PRPP glutamyl transferase—This enzyme catalyzes the first committed step in the purine biosynthesis and is under tight feedback inhibition by AMP and GMP. However the regulation of purine biosynthesis is more effective at PRPP synthase level than at PRPP glutamyl tranferase step.

3. Conversion of IMP to AMP and GMP—Another important regulation mechanism of purine biosynthesis is at the stage of conversion of IMP to AMP and GMP. The control is imparted in two ways—(1) AMP and GMP when are available in adequate amount to meet the cellular demands, they feedback inhibit their own production from IMP. AMP inhibits enzyme **adenosyl succinate synthase** while GMP inhibit **IMP dehydrogenase** by feedback mechanism. (Fig. 18.5A).

(2) Cross regulation occurs at the conversion of IMP to AMP as it requires GTP as energy source and conversion of IMP enroute XMP to GMP require **ATP,** thus this serves as cross regulation to decrease synthesis of one purine nucleotide when there is deficiency of other nucleotide for the balanced production of purine nucleotides as illustrated in the Fig. 18.5B.

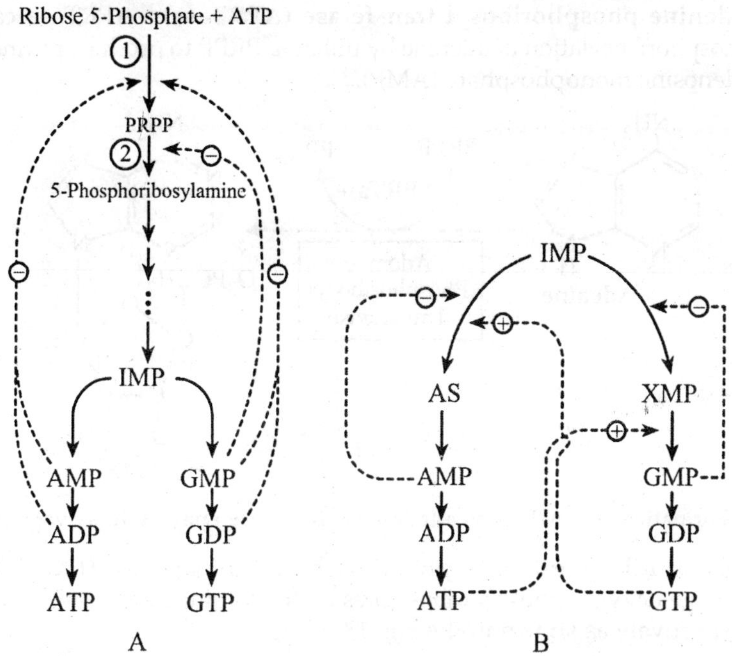

Fig. 18.5 : (A) Regulation of purine nucleotide denovo biosynthesis at various levels by negative feed back mechanisms (B) Cross regulation of AMP and GMP synthesis by GTP and ATP respectively

Salvage Pathway of Purine Nucleotide Synthesis

Free purine bases formed as a result of normal turnover of tissue nucleic acids can be reutilized in certain tissue for conversion into nucleotides and nucleic acid synthesis by **salvage pathway.**

Salient features—

- Salvage pathway is of special importance in tissues such as brain, RBC, leukocytes etc in which the denovo nucleotide biosynthesis is not operative. Liver is the major tissue to provide purine bases and nucleosides for purine nucleotide salvage pathway in these tissues.
- The **key reactant** in the salvage pathway is also **PRPP,** as in the case of denovo biosynthesis, hence the two pathways are closely interrelated and are controlled by the availability of PRPP and **PRPP synthase enzyme activity.**
- Nucleosides can also be salvaged to nucleotides by phosphorylation reaction in presence of kinase enzymes.
- In salvage pathway very less amount of energy is required in form of ATP as compared to denovo biosynthesis, thus it is very **economical**. The salvage pathway is **very quick** to provide nucleotides for rapid DNA synthesis, as it involves only two steps.

Two different enzymes salvage purines—The purine bases, adenine and guanine are salvaged by two different enzymes to their respective nucleotides. The key enzymes of salvage pathway and the reactions catalyzed are described here under—

1. **Adenine phosphoribosyl transferase (APRTase)**—APRTase catalyzes the phosphoribosylation of adenine by utilizing PRPP to produce purine nucleotide, adenosine monophosphate (AMP).

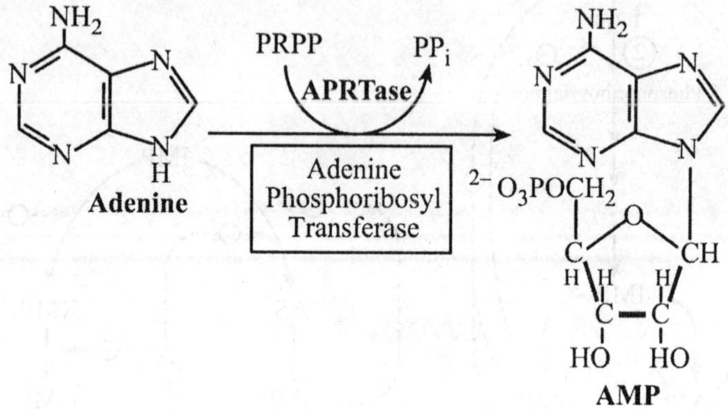

Fig. 18.6 : Formation of AMP from adenine by APRTase enzyme in salvage pathway

2. **Hypoxanthine guanine phosphoribosyl transferase (HGPRTase)**—This salvage enzyme phosphoribosylates hypoxanthine or guanine to IMP or GMP respectively as shown in the Fig. 18.7.
3. **Salvation of nucleosides**—Another salvage mechanism involves conversion of nucleosides directly to purine nucleotides by **phosphorylation** utilizing ATP in presence of specific kinase enzymes.

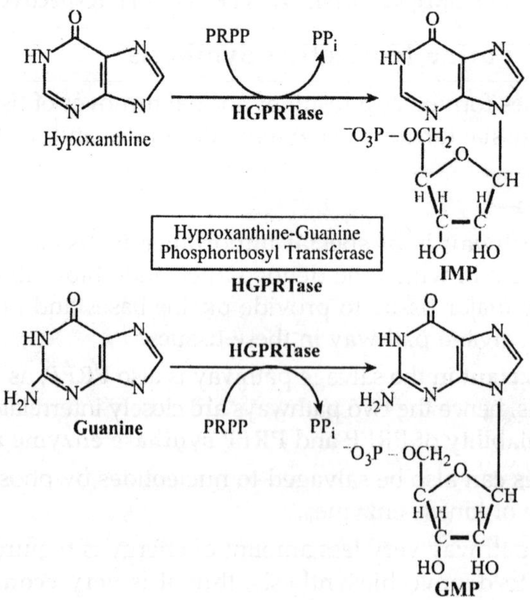

Fig. 18.7 : Formation of GMP and IMP by HGPRTase salvage pathway

Availability of PRPP is the most important regulatory factor for purine nucleotide biosynthesis. Quantatively the salvage pathway consumes more PRPP and quickly forms **purine nucleotide** (only two steps are required) for nucleic acid synthesis in tissues.

Conversion of Ribonucleotide to Deoxyribonucleotides

Reduction of ribonucleotide diphosphate at 2nd carbon of ribose moiety is catalyzed by **ribonucleotide reductase** complex and results in the formation of **deoxyribonucleotide diphosphates** (dNDP), that are required for DNA synthesis.

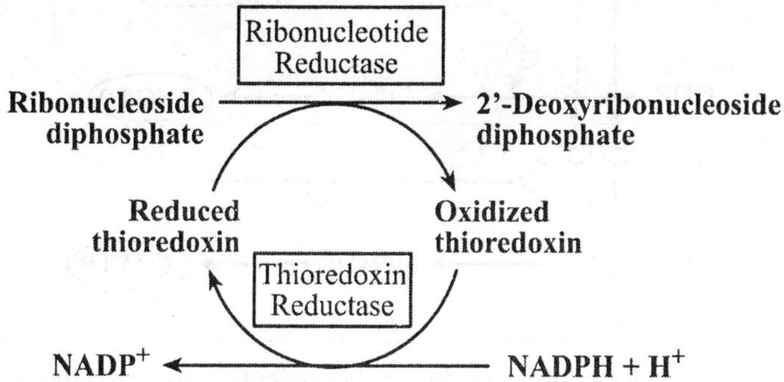

Fig. 18.8 : Reduction of ribonucleotide to deoxyribonucleotides by thioredoxin system

Reductase enzymes is a multisubunit (two β_1 and two β_2 subunits) and requires **thioredoxin,** a protein cofactor with two cysteine residues with sulphhydral groups (SH) to donate reducing equivalents for the reduction of ribonucleotide. Finally $NADPH^+$+H dependent **thioredoxin reductase** coverts the oxidized thioredoxin to reduced state to complete the cycle, as described in the Fig. 18.8.

Regulation of deoxyribonucleode production

A balanced and well coordinated production of deoxycibonucleotide occurs by following mechanisms—

Salient features

The enzyme **ribonucleotide reductase** is active only when cell division is to occur where deoxyribonucleotide triphosphates are needed for DNA replication.

The reduction reaction is subjected to complex feedback regulatory mechanism to ensure a balanced production of deoxyribonucleotides of A, G, T and C as high energy triphosphates, required for DNA replication.

The regulation of purine and pyrimidine deoxyribonucleotide formation is well coordinated to provide proper amounts of deoxyribonucleotides of A, G, C and T and a mole to mole parallelism of purine and pyrimidine deoxyribonuleotide formation exists. This is achieved as described below—

Several sites of cross regulation characterize fine coordinated control of purine and pyrimidine deoxyribonucleotide triphosphate production as shown in the Fig. 18.9.

Availability of PRPP is common regulatory factor in the production of purine as well as pyrimidine nucleotide formation and **PRPP synthase enzyme** is subjected to a tight control by **multiple feedback loops mechanism** of regulation by both purine and pyrimidine nucleotides.

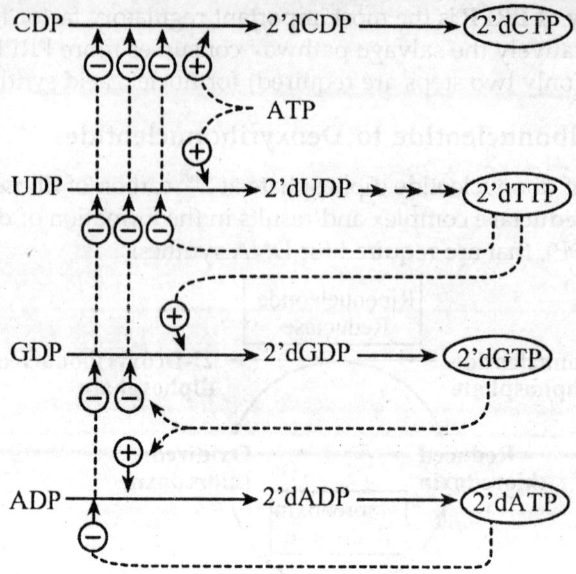

Fig. 18.9 : Regulation of deoxyribonucleotide production by tight multiple feed back control

PURINE CATABOLISM

The end product of nucleic acid and purine nucleotide catabolism in humans is **uric acid.** Both endogenous (nucleic acid turnover) and exogenous (dietary purines absorbed from GIT) **purines** are **degraded** to **uric acid mainly in liver.**

- Mammals other than higher primates and humans further convert uric acid to **allantion** in the presence of enzyme **uricase** (absent in humans). The allantoin is very soluble in water and can easily be excreted in urine.
- Birds and reptiles convert NH_2 group of amino acids to **uric acid** as the **end product** of **all protein metabolism.** Uric acid is sparingly soluble in water and thus is excreted as **semisolid** urine to conserve water in these species.

Reactions of purine catabolism—Humans convert the major **purine nucleosides,** adenosine and guanosine to uric acid. The sequence of reactions are given below and summarized in the Fig. 18.10.

Reactions involved in Catabolism of Purine Nucleotides to Form Uric Acid

1. Adenosine is **first deaminated** to inosine by enzyme **adenosine deaminase.**
2. Phosphorylatic cleavage of inosine and guanosine nucleosides then split off ribose phosphate in the presence of specific **purine nucleoside phosphorylase** enzyme to release free purine bases hypoxanthine and guanine respectively.
3. Hypoxanthine (from inosine) so liberated is oxidized to **xanthine** by the enzyme **Xanthine oxidase.**
4. Guanine purine base is deaminated in the presence of **guanase enzyme** to xanthine (2, 6 dioxypurine).

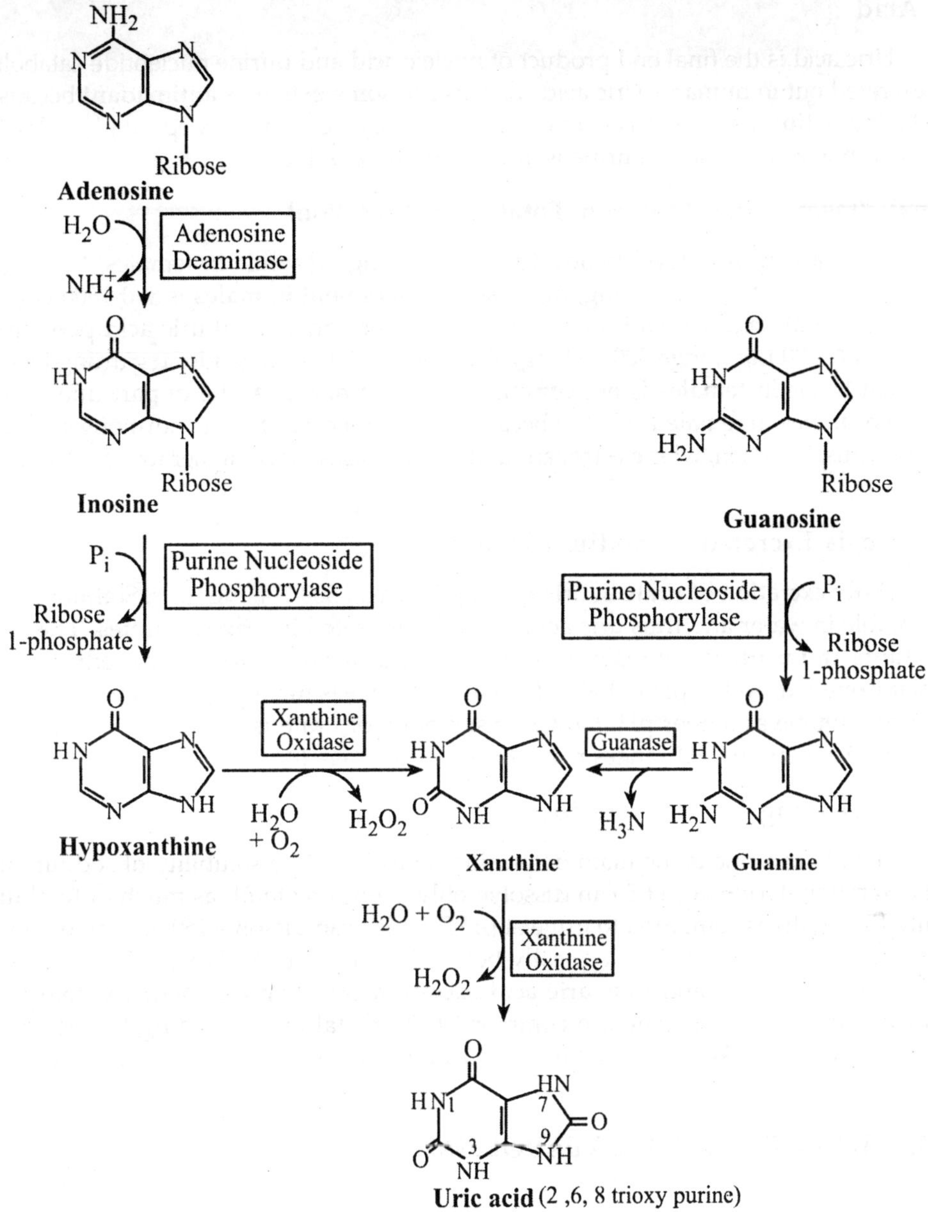

Fig. 18.10 : Catabolism of purine nucleosides to uric acid

5. Xanthine so produced from both adenosine via hypoxanthine and from guanine is finally oxidized to uric acid (2, 6, 8 trioxypurine) by the enzyme xanthine oxidase (Fig. 16.10). This enzyme contains molybdenum and iron and requires FAD as coenzyme. During oxidation by xanthine oxidase H_2O_2 is produced which is detoxified by the enzyme **catalase** to H_2O and O_2 as illustrated in the Fig. 18.10.

Uric Acid

Uric acid is the final end product of nucleic acid and purine nucleotide catabolism, to be excreted out in humans. Uric acid can serve to some extent as **antioxidant** because of its **reducing action** and itself is converted to allantion. Uric acid is sparingly soluble in water and thus its excretion in urine is affected by several factors.

Normal Plasma Uric Acid and Total Uric Acid Pool

Plasma uric acid level in adult male is 3-7 mg/dl while in females it is slightly lower and ranges between 2.5-5 mg/dl. **Total uric acid pool in males** is 800-1600 mg with an average of 1200 mg, which is just double to that of female total uric acid pool, that is approximately 600 mg (range 500-700 mg). **Cause** of variation in gender regarding low uric acid serum levels in females is not completely known but it can be in part due to more excretion of urates in female probably because of the effects of female hormone estrogens. Postmenopausal women, where estrogen production ceases, have higher uric acid level (2.5 to 6 mg/dl).

Uric Acid is Excreted as Sodium Urates

Daily excretion of urate in urine ranges between 500-700 mg/day. Sodium urate is more soluble in water and thus uric acid is mainly excreted in urine as **urates. The pH of body fluids and urine** is the main determining factor for the solubility of sodium urates and their excretion. At the pH of blood (7.4) uric acid exists mainly as sodium urate, which is far more soluble at higher pH. However solubility of urate is reduced at low pH as in case of urine, where the pH ranges from 4.8 to 7.6 (average 6).

Factors Affecting Urate Excretion

The pH of urine is the main determinant factor for the solubility of sodium urate and its excretion. Urine at pH 5 can dissolve only about one tenth as much of total urate (i.e. only 15 mg/dl) as compared to urine at pH 7, which can dissolve 150-200 mg of urates/ dl. Normally the pH of urine is typically below 5.8 and at this pH very less amount of sodium urate is soluble, and thus **uric acid excretion** is **very problematic. Urinary tract crystallization** of sodium urate are common at the distal and collecting tubules where acidification of urine takes place and most of the renal stones formed are composed of uric acid and urate crystals.

DISORDERS OF PURINE CATABOLISM

Gout

- **Hyperuricemia** is when serum uric acid level is more than 7 mg/dl, where it exceeds the solubility limits of sodium urates. The plasma and extracellular fluid become supersaturated with respect to urate and favours crystallization of sodium urates in soft tissues and bone joints where they get deposited, in form **of 'tophi'.** The needle shaped urate crystals deposited, stimulate phagocytosis resulting an **inflammatory** reaction and **acute arthritis** (typically monoarticulate type) develops which may progress to **chronic arthritis.** The condition is called as **Gout.**

- Under polarized light microscope, needle shaped birefrigent sodium urate crystals can be seen in the joint fluid, that can be taken as diagnostic sign of **clinical condition gout.**
- Interstitial deposition of urate crystals in renal parenchyma leads to **urolithiasis** i.e renal calculi formation. This may results in renal failure.

Urate Solubilizing Factors in Plasma

However precipitation of sodium urate crystals as tophi deposition may not occur even at high plasma urate pool (as high as 2000-4000 mg) in some individuals, perhaps because of the presence of certain solubilizing factors in their plasma, such as proteins and mucopolysaccharides, that prevent precipitation of urates. There are individual variations for the presence of factors that prevent the precipitation of sodium urates in the body tissues and in bone joints.

Temperature affects tophi formation—Solubility of uric acid and urates is reduced at low temperature. At 30° C uric acid is deposited as urate crystals in the cooler areas of the body, such as distal joints of foot, big toes etc.

Hyperuricemia—Gout

Gout comprises a group of clinical states associated with hyperuricemia alone or combined disorders. Gout is often associated with high living, excessive meat eating and alcohol consumption. Gout **mostly affect males** over 30 years of age as they have high plasma uric acid pool as compared to the females. Gout is mainly of two types—Primary and secondary gout.

Primary Gout

It is a genetic metabolic disorder due to **over production of uric acid resulting from uncontrolled increased** synthesis of purine nucleotides. The primary gout is inherited and shows familial incidence.

Causes of primary gout—The primary gout is caused by the mutations in the genes encoding for the key enzymes of purine nucleotide synthesis as described below—

1. Enzyme **phosphoribosyl amidotransferase** is active but not sensitive to feedback inhibition.
2. **PRPP synthase** enzyme is resistant to feedback inhibition by normal allosteric control mechanisms, resulting in increased production of PRPP. This abnormality is inherited as X-linked recessive trait.
3. **Deficiency of enzymes of salvage pathway**—Deficiency of enzymes of salvage pathway would result in increased availability of PRPP, causing increased denovo synthesis and overproduction of purine nucleotide and uric acid. For instance—
   (a) **Lesch Nyhan syndrome**—This inborn error of purine nucleotide metabolism is a rare but severe, X-linked inherited disorder, caused due to the deficiency of **HGPRTase**, a key enzyme of purine nucleotide **salvage pathway.** This results in **accumulation of PRPP,** that is diverted towards excessive **denovo**

purine nucleotide biosynthesis, which are not needed and thus are catabolized to uric acid in excess.

Clinical manifestation—Main clinical symptoms of Lesch Nyhan syndrome includes CNS lesions resulting in **bizarre behavior, self mutilation** and **mental retardation.** Early manifestations are hyperuricemia causing urinary tract stone formation and nephrolithiasis. Gout develops later in life. A **mild type** of this abnormality has also been reported, caused due to **partial** HGPRTase deficiency.

4. **Von Gierke's glycogen storage disease** due to the congenital lack of enzyme glucose 6-phosphatase (Chapter 5) is accompanied by hyperuricemia. This is because more glucose 6-PO_4 is diverted to HMP shunt path resulting in more availability of ribose 5-PO_4 and PRPP leading to overproduction of uric acid with adverse manifestations of hyperuricemia.

Secondary Gout

Hyperuricemia occurs secondary to many other disorders and diseases causing secondary gout. These include—

1. **Increased purine catabolism**—This is associated with increased nucleoprotein turnover rates seen in—a) rapidly growing malignant cancers such as **leukemia, lymphomas, psoriasis, polycythemia** etc; b) Increased **tissue degradation** as seen in **trauma** or during cancer treatment by chemotherapy, radiation therapy etc.
2. **Decreased renal excretion of uric acid and urates** secondary to a) Renal failure in kidney diseases, b) Hypertension and **preeclamptic toxemia** of pregnancy and c) **Acidosis** such as **lactoacidosis** and **ketoacidosis**. In acidosis conditions, pH of urine decreases resulting in decreased urate solubility thereby decreasing their excretion.

In all these above mentioned conditions, uric acid level in plasma increases which may lead to secondary gout and other adverse effects of hyperuricemia.

Management of Gout

1. Primary Gout is characterized by uncontrolled overproduction of purine nucleotide and uric acid and is best managed by **reducing rate** of uric acid production by the compound **allopurinol.** It is an structural analogue of hypoxanthine and thus when administered **competitively inhibits** the key enzyme **xanthine oxidase,** thereby decreasing oxidation of hypoxanthine and xanthine into uric acid (Fig. 18.10). Hypoxanthine and xanthine are more soluble in water and can be easily excreted in urine with least adverse effects.

2. Uricosuric drugs—Certain drugs such as **probenecid** and **salicylates** (aspirin) decrease the reabsorption of uric acid from the renal tubules, thereby increase renal excretion of urates and thus are beneficial in reducing serum uric acid levels.

3. Antiinflammatory drugs such as **colchicine** is very useful in arresting the arthritis in gout. Other useful antiinflammatory drugs are phenylbutazone, iodomethacin, corticosteroids etc which relieves arthritis accompanying gout.

Hypouricemia

Hypouricemia is associated with deficiency of enzyme **xanthine oxidase** due to either a genetic defect or in case of severe liver damage. In severe conditions, xanthinuria occurs with xanthine lithiasis (renal stones).

Immuno Deficiency Diseases

These are associated with genetic congenital deficiencies of enzymes of purine catabolism and result in severe immuno deficiency. These include—

1. **Adenosine deaminase deficiency**—This genetic disorder causes severe combined **immuno deficiency** involving both thymus derived lymphocytes (T-cells) and bone marrow derived lymphocytes (B-cells). In this condition both T lymphocytes and B lymphocytes are very less in number in blood and are non-functional.
2. **Purine nucleoside phosphorylase deficiency** is associated with severe T-lymphocytes deficiency with severe immuno supression. B lymphocytes may be normal in this genetic abnormality

Both these immunodeficiency diseases are inherited as autosomal recessive trait. The T and B lymphocyte **dysfunction** appears to be due to the lack of key enzymes **adenosine deaminase** and **purine nucleotide phosphorylase** of purine nucleotide catabolism (Fig. 18.10) in these cells. This result in accumulation of dATP and dGTP, which feedback inhibit enzyme **ribonucleotide reductase,** thereby blocking the synthesis of deoxyribonucleotides, particularly of dCTP synthesis. This results in a **imbalance in availability** of **deoxyribonucleotides,** and DNA replication and cell division is impaired causing T and B lymphocytes deficiency resulting in severe immuno suppression state in the affected subjects.

PYRIMIDINE NUCLEOTIDE BIOSYNTHESIS

The denovo biosynthesis of pyrimidine as **free base** takes place first and then they condense and get attached with ribose phosphate to produce nucleotides. This is the major difference from denovo synthesis of purine nucleotides, which are synthesized on ribose phosphate moiety directly as **nucleotides.**

Reactions of Denovo Pyrimidine Synthesis

The denovo pyrimidine synthesis involves following major steps—

1. Formation of carbamoyl phosphate—Synthesis of pyrimidine commences with the formation of **carbamoyl phosphate,** catalyzed by cytosolic **carbamoyl phosphate synthase II,** unlike urea synthesis in which carbamoyl phosphate is synthesized by mitochondrial enzyme (Chapter 15). The reaction requires glutamine as NH_2 group donor, CO_2 and ATP as energy source, as shown in the Fig. 18.11.

2. Aspartate transcarbamoylase then catalyzes condensation of carbamoyl phosphate with aspartic acid to produce a condensation product, that cyclize to form **pyrimidine ring** by ring closure.

3. Ring closure of the condensation product, with the loss of elements of water is catalyzed by **dihydrooratase** enzyme, and first pyrimidine, **dihydroorotic acid** is formed as shown in the Fig. 18.11.

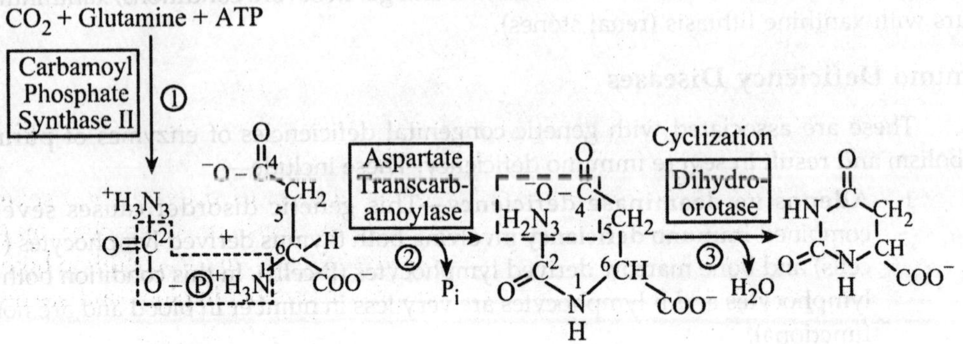

Carbamoyl phosphate (CAP) **Aspartic acid** **Carbamoyl aspartic acid** **Dihydroorotic acid (DHOA)**

Fig. 18.11 : Denovo synthesis of a pyrimidine derivative dihydrooratate

4. Conversion of dihydroorotic acid to OMP—Dihydroorotic acid is dehydrogenated (from C_5 and C_6 of ring) to orotic acid by a mitochondrial **dehydrogenase enzyme.** The orotic acid produced is then attached with ribosyl phosphate by a β N-glycosidic bond to N_3 of pyrimidine ring to produce **first** nucleotide, the **orotidine monophosphate (OMP)** in the presence of **orotate phosphoribosyl transferase** enyme as shown in the Fig. 18.12. Here PRPP is the donor of ribosyl phosphate moiety.

5. Formation of Uridine monophosphate (UMP) takes place by decarboxylation of OMP in the presence of enzyme **OMP decarboxylase.**

6. Synthesis of cytosine triphosphate—UMP is first converted to high energy phosphate UTP by respective kinases utilizing ATP as the energy source (Fig. 18.12). The UTP is then converted to CTP by transfer of NH_2 group from the amido group of glutamine in presence of **CTP synthase** enzyme and ATP as energy source as shown in the Fig. 18.12.

7. Synthesis of deoxy thymine monophosphate—Reduction of ribose moiety of UDP in the presence of **ribonucleotide reductase** produces deoxy UDP. The dUDP so formed is first converted to dUMP and **then** is **methylated** in the presence of **thymidylate synthase enzyme,** where N^5, N^{10} methylene tetrahydrofolate donate the methyl group. Folic acid coenzyme (THF) is essentially required in purine and pyrimidine nucleotide biosynthesis and DNA replication.

Anticancer drug methotrexate competitively inhibits enzyme **dihydrofolate reductase** and THF cannot be produced (details given along with purine nucleotide biosynthesis) and formation of deoxy thymidine monophosphate is blocked. The dTMP and dTTP along with other deoxy ribonucleotides are needed in excess for rapidly multipling cancerous tissue. Methotrexate block the production of purine and pyrimidine nucleotides needed for DNA replication and thus cancerous growth is checked and the drug is used widely as **anticancer drug.**

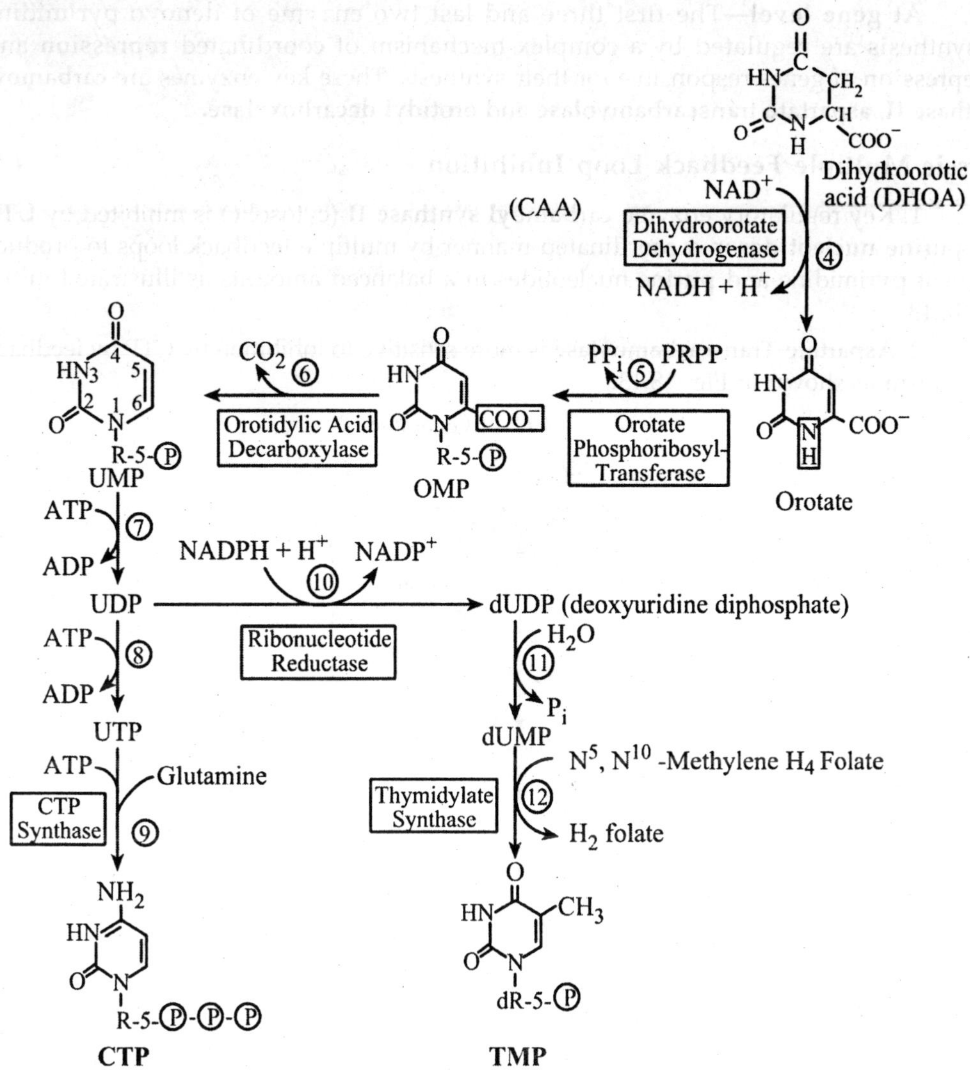

Fig. 18.12 : Conversion of oratate to pyrimidine nucleotides UTP, CTP and TMP and their deoxyribosyl derivatives

Pyrimidine Nucleotide Salvage Mechanism

Mammalian cells though lack efficient mechanism to salvage free pyrimidine base, but can salvage **pyrimidine nucleosides** to their respective nucleotides in presence of ATP and the respective kinases for nucleic acid synthesis.

Regulation of Pyrimidine Nucleotide Biosynthesis

Pyrimidine nucleotide biosynthesis is regulated at the levels of both **gene expression** and at the enzyme level by **classic feedback allsoteric inhibition** by the final products CTP, UTP and TDP.

At gene level—The first three and last two enzyme of denovo pyrimidine biosynthesis are regulated by a complex mechanism of coordinated **repression** and **derepression** of genes responsible for their synthesis. These key enzymes are **carbamoyl synthase II, aspartate transcarbamyolase** and **orotidyl decarboxylase.**

Classic Multiple Feedback Loop Inhibition

1. Key regulatory enzyme **carbamoyl synthase II** (cytosolic) is inhibited by **UTP** and **purine nucleotides** in a coordinated manner by multiple feedback loops to produce different pyrimidine and purine nucleotides in a balanced amounts as illustrated in the Fig. 18.13.

2. **Aspartate Transcarbamoylase** is more sensitive to inhibition by CTP by feedback mechanism as shown in Fig. 18.13.

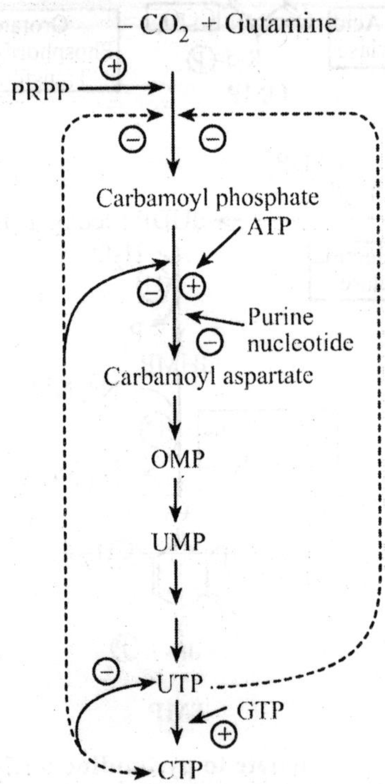

Fig. 18.13 : A Classic multiple loop feed back regulation of pyrimidine nucleotide synthesis

CATABOLISM OF PYRIMIDINES

The catabolism of pyrimidine ring takes place mainly in the liver and produce highly soluble end products such as β alanine and β amino isobutyrate, that can be easily excreted in the urine. This is in contrast to purine nucleoside catabolism, which produces insoluble uric acid, excretion of which is difficult and depends on many factors such as pH, temperature, renal efficiency etc (as described earlier).

The degradation of pyrimidines involves conversion of **cytosine** by **oxidative deamination** to **uracil** which is then reduced to **dihydrouracil.** The pyrimidine ring is then broken to liberate CO_2, NH_3 and β alanine (Fig. 18.14). Thymine is similarly reduced to dihydrothymine and then the ring is broken to liberate CO_2, NH_3 and β amino isobutyrate as shown in the Fig. 18.14.

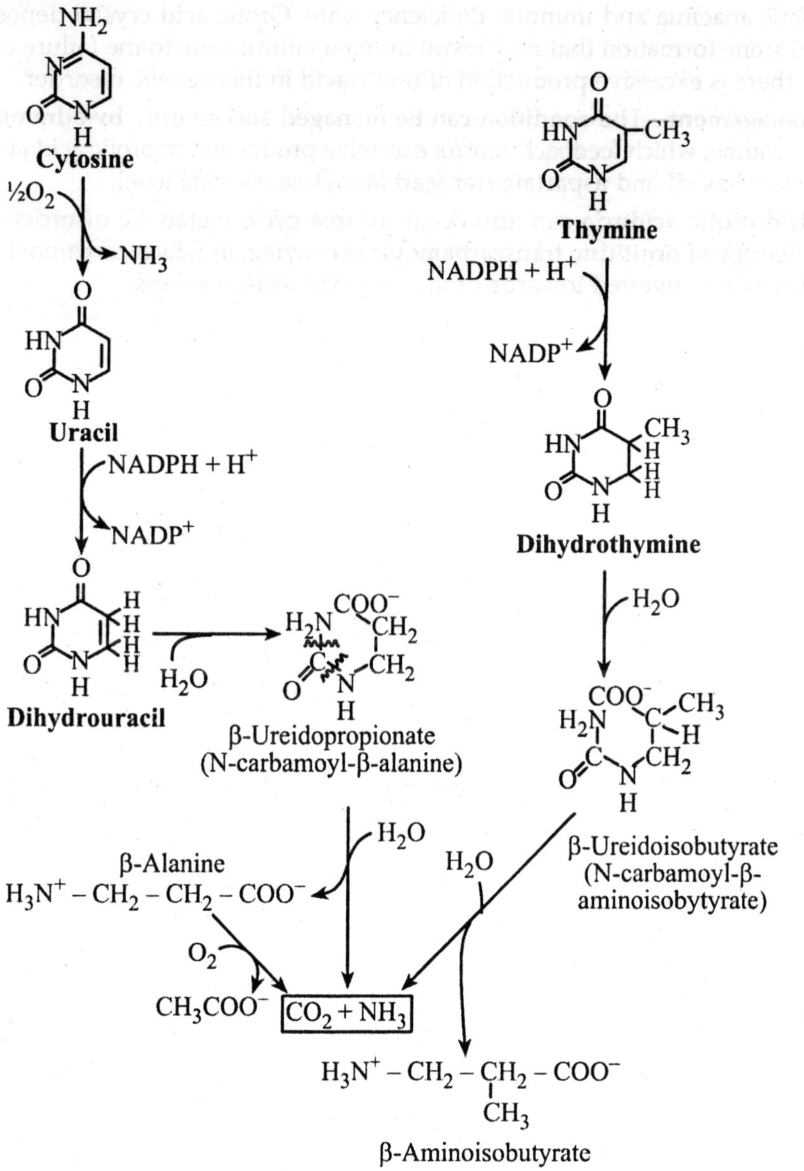

Fig. 18.14 : Degradation of pyrimidines to CO_2 and NH_3 and β-amino acids

Disorders of Pyrimidine Metabolism

1. Orotic acidura—This **genetic** disorder is severe and is caused by the congenital lack of enzymes **oratate phosphoribosyl transferase** (OPRTase) or **OMP decarboxylase** (Fig. 18.12) or both, resulting in the metabolic disorder **orotic aciduria.** This is an autosomal recessive trait and is characterized by accumulation of orotic acid in body and **orotic acid crystalluria.**

Clinical manifestations includes—Growth retardation, failure to thrive, megaloblastic anaemia and immuno deficiency state. Orotic acid crystal deposition may cause renal stone formation that may result in renal failure. Due to the failure of feedback inhibition, there is excessive production of orotic acid in this genetic disorder.

Management—The condition can be managed successfully by administration of uridine or cytidine, which feedback control excessive production of orotic acid at carbamoyl phosphate synthase II and aspartate transcarbamoylase enzyme level.

Mild orotic aciduria can also occur in urea cycle metabolic disorder such as in genetic deficiency of **ornithine transcarbamoylase** enzyme, in which carbamoyl phosphate accumulates and is diverted towards orotic acid synthesis in excess.

❑❑❑

19

DNA REPLICATION

DNA has all the genetic information characteristic of a specie, in form of **genetic code,** that resides in the specific sequence of nucleotides present on its two polynucleotide strands.

- During cell division, each daughter cell gets an exact copy of whole genome of the mother cell. When cells divide, the DNA strands are replicated into identical copy of whole genetic code, and each daughter cell receives an exact (xerox) copy of DNA with complete genetic information from the parent cell.
- **Primary function of DNA replication** is thus to provide and transmit complete genetic information from the parent to progeny (offsprings) during reproduction and from parent cell to daughter cell during cell division.

HIGH FIDELITY IS MAINTAINED IN DNA REPLICATION

Replication of DNA is carried out with **high fidelity,** that is essential for **genetic stability** of the species and also for its **survival.** Thus DNA replication is very complex process involving many **regulatory factors** and several **verification signals** to ensure high fidelity during replication.

Salient Features of DNA Replication

1. **DNA Replication is Semi Conservative**—Double helix DNA has two polynucleotide strands with complementary and antiparallel nucleotide sequences. Thus they have same genetic code but the nucleotide sequence is running in opposite direction i.e. from 3′ to 5′ and from 5′ to 3′ diphosphoester linkages in the two strands of DNA. Both strands of DNA undergo **simultaneous replication** to synthesize two **identical** daughter DNA double helix molecules, each having one **original** strand from the parent DNA and other **newly** synthesized strand.

One out of these two identical copies of DNA so produced will go to two daughter cells during the cell division. This type of replication is known as **semi conservative,** since the **half** (one strand) of the **original DNA(parent)** is passed on and is conserved in the daughter DNA molecule, that is going to the daughter cell during the cell division as shown in the Fig. 19.1.

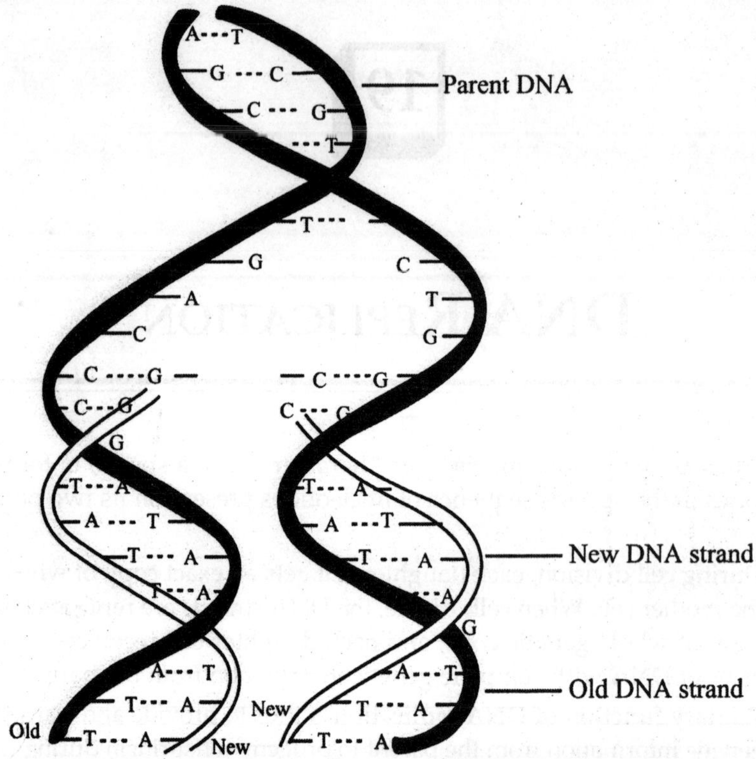

Fig. 19.1 : Semiconservative DNA replication (A new complementary strand is synthesized on the old DNA template)

2. DNA replication occurs in"S" phase of cell division cycle—Mitosis in cell division occurs in the "M" phase of the cell division cycle, followed by "G" phase (gap) and then "S" phase (synthesis phase) during which chromosomal DNA is replicated. Cells divide (after "S" phase) in the G_2 phase.

3. DNA synthesis is bidirectional—Replication of DNA takes place on both the DNA strands **simultaneously but in opposite direction** in a replicating bubble and requires formation of a replicating fork. This is because the enzyme **DNA polymerase** can synthesize new DNA only in 5′ to 3′ direction, and since the two strands of DNA molecule are antiparallel with nucleotide sequence in opposite direction, the new DNA synthesis occurs in opposite direction on the two DNA strands and thus it is **bidirectional** (Fig. 19.2).

4. DNA Polymerase is the main enzyme involved in DNA synthesis. During DNA replication, DNA polymerase catalyzes synthesis of a complementary DNA strand on the original DNA template (parent) by polymerizing deoxyribonucleotides, sequentially according to **base pair rule** by forming **5' to 3' diphosphoester linkages.**

Initiation of DNA Replication

The initiation of DNA replication starts (originate) at some **specific regions** on DNA molecule, known as **origin sites ("ori")** of replication. In prokaryotes with circular DNA, there is **single origin site,** whereas in eukaryotes, **multiple origin sites** occur forming **multiple replicons** or **replication bubbles. The ori**(origin sites) usually consist of **short consensus AT base pair rich nucleotide sequences.**

The ori sites are recognized by specific protein factors known as **dna-A,** that get bound to ori near AT rich region on DNA double helix. The AT base pair rich regions near ori have less number of hydrogen bonds, so are broken easily to unwind DNA double helix. By unwinding, the two DNA strands separate to provide access to the nucleotide sequence for enzyme **DNA polymerase** to start replication process.

Formation of Replication Bubble (Replicon)

The two complementary strands of double helix DNA at ori site get separated by an unwindase enzyme called **helicase** that starts unwinding at AT rich region onwards (Fig. 19.2). The two separated strands are then stabilized (to prevent premature reannealing) by the **single strand binding proteins (SSB)** and a replication bubble is produced as shown in the Fig. 19.2.

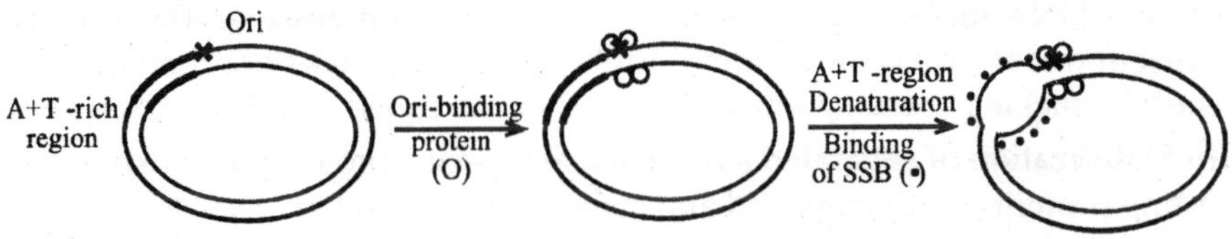

Fig. 19.2 : Formation of replicating bubble on prokaryotic DNA

Multiple Replicon Formation in Eukaryotes

There are multiple origin sites on long eukaryotic double helix DNA and thus multiple replicating bubbles are formed during mammalian DNA replication process. This is necessary to **ensure quick** and **rapid DNA** replication within the time limit during cell division. Thus during eukaryotic DNA replication, **multiple replicons** are formed on long DNA molecule, which synthesize new DNA strands at various sites **simultaneously,** so that whole DNA is replicated quickly. Later on these nascent DNA fractions at multiple replicons are ligated (joined) together to form a **complete DNA molecule** as depicted in the Fig. 19.3.

Formation of Replication Fork—Replisomes

DNA replication requires a number of protein factors and enzymes for several protein - protein and protein - DNA interactions to initiate DNA synthesis and to maintain **high fidelity** during replication process and collectively they are known as **replisomes.** For the formation of a replication fork, four protein (enzyme) components are required that work in following sequence—

(a) Unwinding of double helix DNA—DNA helicase enzyme (dna-B) starts the unwinding of DNA double helix by breaking H_2 bonds of AT rich segments adjacent to ori site, after dna A binds on them. As the enzyme helicase moves further on DNA molecule the unwinding progress and the two strands get separated to form a replication bubble.

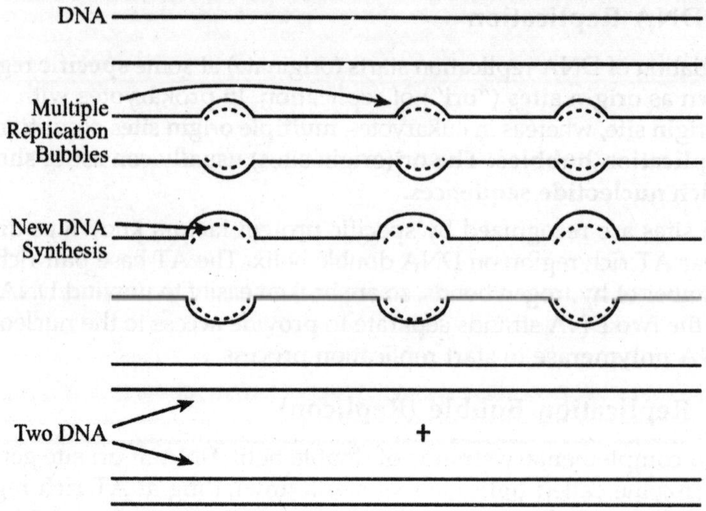

Fig. 19.3 : Multiple replicons formation on eukaroytic DNA during replication process

Topoisomerases (gyrases)—The unwinding of DNA creates topological tension in the double helix DNA molecule. This is **relieved** by **topoisomerase** (gyrases) enzymes by cutting a **nick** at one strand of unwinding double helix. The **nicks** are then quick resealed (ligated) and the nicking - sealing enzymes together are called as topoisomerases.

(b) Stabilization of separated stands—Some specific protein factors known as **single strand binding proteins** (SSB) then bind and stabilize the separated DNA strands, to prevent premature reannealing and a replication fork is formed.

(c) Initiation of DNA synthesis : DNA polymerase complex—DNA polymerase is the main enzyme that catalyzes the polymerization of deoxyribonucleotides by forming 5' to 3' diphosphoester linkage to synthesize a new complementary DNA strand on the parent DNA template according to base pairing rule. In prokaryotes there are three DNA polymerases namely, DNAP I, II and III, while in eukaryotes there are 5 DNA polymerase enzymes with distinct functions and characteristics and collectively are known as DNA polymerase complex.

A brief account of various **DNA polymerases** of bacteria (E.coli) and mammalian enzymes involved in DNA replication with their specific actions is given in Table 19.1.

Table 19.1 : A comparison of prokaryotic and eukaryotic DNA polymerases along with their specific actions

Prokaryotic DNA polymerase	Mammalian DNA polymerase	Functions
I	α	DNA synthesis on lagging strand and gap filling
II	ε	DNA proof reading and repair
-	β	DNA repair
-	γ	Mitochondrial DNA synthesis
III	δ	Processive leading strand DNA synthesis and proof reading.

In eukaryotes the DNA polymerase α is a tetramer with 4 subunits and is the primary enzyme responsible for the DNA replication. The mammalian DNA polymerase is capable of polymerizing about 100 nucleotides per second, at a rate, ten fold less than the rate at which prokaryotic DNA polymerase polymerizes, which is around 1000 nucleotides/second.

(d) RNA Primers : RNA Primase enzyme—The initiation of DNA synthesis requires **priming** by a **short** length RNA primers (10 to 100 nucleotides long). The RNA primers are synthesized by the enzyme **RNA primase,** as a complementary sequence to that of the nucleotide sequence on DNA strand at the origin site (ori) strictly according to the **base pairing rule.**

DNA Synthesis is Bidirectional—DNA replication is **bidirectional** and **asymmetric** in a replication fork and proceeds in opposite direction.

At the replication fork, the DNA polymerase enzyme initiates synthesis (replication) of nascent DNA strands on both the sides of replication fork, but in opposite direction and in different manner. This is because **DNA polymerase enzymes** can synthesizes new DNA **only in one direction i.e. 5′ to 3′ phosphodiesterase linkage.** As the two strands of double helix DNA are antiparallel in which the nucleotide sequence is running in opposite direction (Fig. 19.4), thus the DNA synthesis on the two strands in a **replication fork occurs in opposite direction** and is **asymmetrical** and they are differentiated as (a) Leading strand and (b) Lagging strand.

(a) **Leading strand with 5′-3′ orientation**—On this strand of the replication fork, the new DNA synthesis is **continuous** (Fig. 19.4) and is catalyzed by DNA polymerase δ on a single RNA primer.

(b) **Lagging strand with 3′-5′ orientation**—On the lagging strand the nascent DNA is synthesized in opposite direction to that of leading strand. Here the replication is **discontinuous** and occurs in small pieces (80 to 100 nucleotide long) on many RNA primers. These small nascent DNA fractions synthesized on RNA primers are called as **okazaki pieces** (Fig. 19.4).

The various enzymes involved in the bidirectional DNA replication at the replication fork along with **okazaki pieces** on **lagging strand** are shown in the Fig. 19.4.

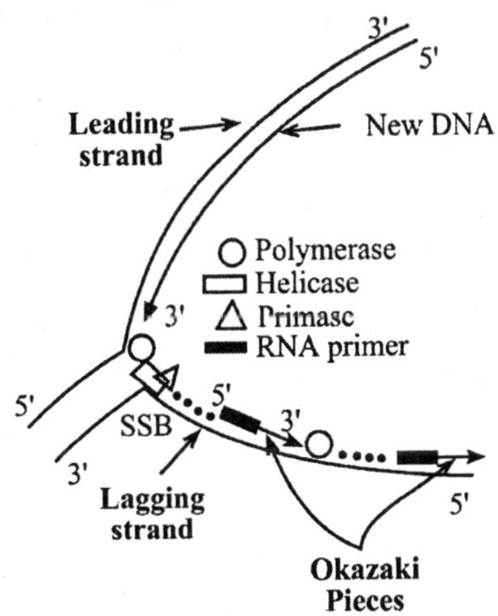

Fig. 19.4 : Replication fork with bi-directional and asymmetrical DNA replication on the leading strand in continuous manner and lagging strand with okazaki pieces

DNA Synthesis—Elongation Process

The synthesis of new DNA strand takes place in 5′-3′ direction and is catalyzed by DNA polymerase enzyme. The four deoxyribonucleotide triphosphates i.e. dATP, dGTP, dCTP and dTTP, are required for the DNA replication. The sequence in which these nucleotides are added to be polymerized in the nascent DNA strand, is directed by the template DNA (parent) strand strictly according to **base pair rule.** According to this rule, the adenine nucleotide on template DNA strand will direct incorporation of thymine nucleotide and guanine nucleotide would direct addition of cytosine nucleotide respectively, so that a complementary copy of the parent DNA with same genetic code is produced.

- **For the elongation** of new DNA strand, the incoming deoxyribonucleotides are added in a specific sequence as determined by the template DNA, to the **3′ end of** growing nascent DNA chain, with splitting off the pyrophosphate group of high energy nucleotide triphosphates. The pyrophosphate is then hydrolyzed to provide energy for the condensation of nucleotides by 5′-3′ phosphodiester linkage in a stepwise manner.
- **On lagging strand,** short segments of growing DNA (up to 250 nucleotides) attached to RNA primers, known as **okazaki fragments** are formed in same manner as above, but in opposite direction. After many okazaki fragments are synthesized, they are joined together, after removal of RNA primers to produce a continuous DNA strand.
- **For joining of okazaki pieces,** the RNA primers are first removed by exonuclease action, and then the gap left is filled up by addition of correct nucleotide with proper base pairing by the DNA polymerase enzyme. Then the fragments are sealed by the enzyme **DNA ligase** to form a continuous DNA strand on lagging strand also ultimately.
- When polymerization is complete in the replicating fork, the RNA primers are removed by exonuclease activity and the gap left is filled by correct deoxy nucleotide addition by DNA polymerase. The DNA pieces of multiple replicons are then ligated by **DNA ligase enzyme.** After replication at one site is complete, the core basic protein histones are attached.

DNA replication thus occurs on both the strands simultaneously, but in short pieces (about 100 - 250 nucleotide long) in replication forks as shown in the Fig. 19.4.

The entire mammalian genome replicates in nine hours, which is an average period required during cell divisions for the reproduction process.

Mismatch Repair and Proof Reading

During DNA replication the **maintenance** of **high fidelity** is the **most important** aspect to **conserve** the genetic information (**genetic code**) characteristic of a specie. Besides its 5′-3′ polymerization catalysis, the DNA polymerase III (δ DNA polymerase in mammals) also has **proof reading capability.** It checks the incoming nucleotide to be polymerized strictly according to the base pair match as on template DNA strand, so that only **correct nucleotide** is added in the growing DNA strand. Further, it proof read and scan the entire nascent DNA after replication process and any mismatched (mistake) is immediately **edited** by removing wrongly added nucleotide. This double monitoring does not permit errors of mispairing during replication of DNA and thus high fidelity in DNA replication is maintained.

Mismatch Base Repair—Excision Mechanism

During DNA replication, maintenance of high fidelity is achieved by full proof mechanism to correct any mistake, that has occured. The DNA polymerase δ scan the entire nascent DNA replicated and detect any wrong base incorporated. The 3' - 5' exonuclease unit of DNA polymerase does **the editing** to remove the mismatched base and the correct base is then added by the nucleotide excision mechanism as shown in the Fig. 19.5. The open end of the nick is then sealed by DNA ligase (Fig. 19.5).

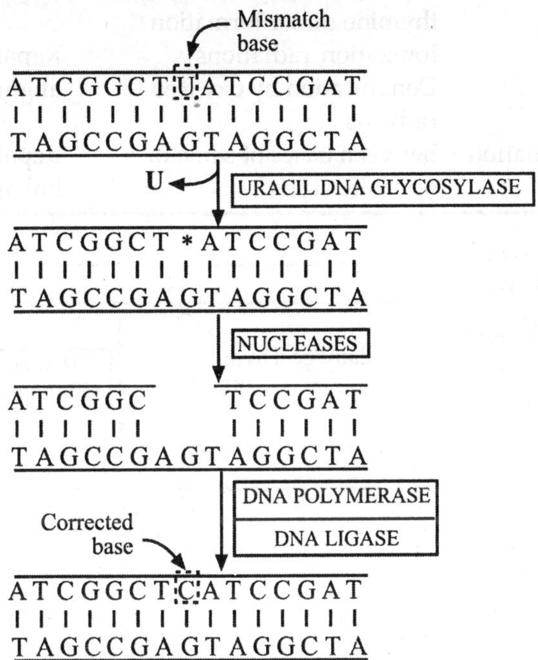

Fig. 19.5 : Nucleotide excision mechanism to repair the mismatch base during DNA replication

DNA Damage and Repair

The maintenance of integrity of the genetic code on DNA molecule (genes) is of **utmost importance** for the survival of an organism and for the stability of a specie. The DNA of our body however is continuously being subjected to damage by many environmental **physical** and **chemical** factors. DNA damage may also take place due to the **replication error**, though extremely rare. Any alterations in the original nucleotide sequence of DNA (genes) are known as **mutations (details in the Chapter 21).**

Mutations—Serious consequences result if damaged DNA is not repaired immediately, as it leads to **mutations** i.e. alteration (abnormality) in the genetic code. Mutations in DNA (genes) results in **congenital genetic inborn errors** or heridītary diseases which may cause severe clinical manifestations that can be fatal.

Fortunately living organism is endowed with an highly efficient DNA repair system to correct the damage done to DNA. The various causes and types of DNA damage along with repair mechanisms in our body are summarized in the Table 19.2.

Table. 19.2 : Types of DNA damage and repair mechanisms

Mutations	Causes	Mechanism of repair
1. Single base alteration	Deamination of adenine to hypoxanthine. Alkylation of base, insertion or deletation of nucleotide	Mismatched repair by DNA polymerase III (δ DNAP in eukaryotes)
2. Two base alterations	UV induced thymine - thymine dimer formation	Nucleotide excision repair
3. Double stranded chain breaks	Ionization radiations Denaturation by oxy free radicals	Repair by unwinding alignment and ligation
4. Cross linkage formation	Between bases of same or opposite strand	Repair by breaking cross linkages

Ultraviolet Radiation DNA Damage and Repair

Exposure to UV light may result in the formation of **incorrect thymine – thymine dimer** formation instead of normal **A-T base pairing.** These thymine dimers **block** the **DNA replication** process, which cannot be allowed to ensure proper growth and health of an individual. For this we have an efficient mechanism to repair the damage as soon as the thymine dimers are formed.

This repair mechanism is known as **nucleotide excision repair.** In this the damaged region of DNA with thymine dimers (up to 30 base pairs) is removed by the **endonuclease** catalytical activity of DNA polymerase and then the correct nucleotide bases are added by β- DNA polymerase. Then the open ends of the nick are joined by DNA ligase enzyme. Nucleotide excision repair mechanism is illustrated in the Fig. 19.6.

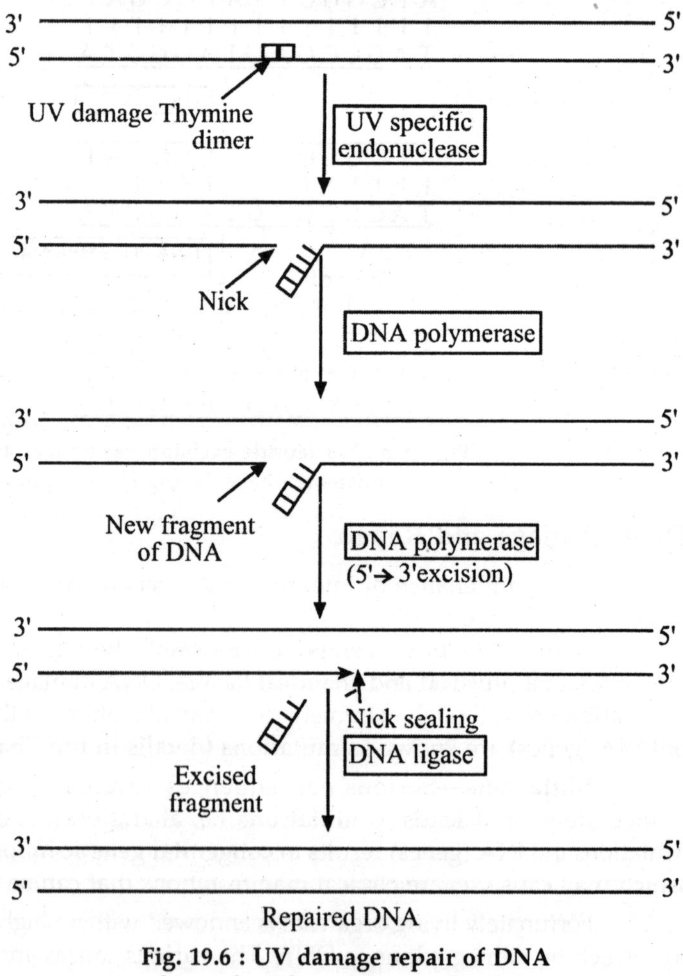

Fig. 19.6 : UV damage repair of DNA by nucleotide excision mechanism

Abnormalities of DNA Repair Mechanism : Genetic Disorder

Xeroderma pigmentosa

It is a rare autosomal recessive inherited disorder, due to the defect in DNA repair mechanism. In **this** genetic disorder there is an absence (lack) of enzyme **UV specific endonuclease,** which is involved in the repair mechanism (Fig. 19.6) of U.V. damaged DNA. The **clinical manifestions** are high sensitivity to UV light, blisters and rashes on the skin, **hyper pigmentation** with subsequent development of multiple **skin cancers** with fatal consequences.

Ataxia - telangiectasia

It is an autosomal recessive disorder in humans with severe clinical manifestations, such as **cerebral ataxia** and **lymphoreticular neoplasm (blood cancer).** These patients have increased **sensitivity** to **DNA damage** by x-rays, which cannot be repaired due to the genetic defect in DNA repair enzymes.

Fanconi's anaemia with increased incidences of cancers is also an autosomal recessive trait in which genetic defect is in a failure to repair cross linking DNA damage.

❑❑❑

20

RNA Synthesis and Functions

Ribonucleic acids are primarily involved in all types of protein synthesis in a living organism, necessary to carryon life processes. RNAs are usually made up of a single polynucleotide chain formed by the polymerization of four types of ribonucleotides of adenine, guanine, cytosine and uracil (in place of thymine of DNA), joined together by 3′-5′ diphosphoester linkages.

RNA exists in four major types

In eukaryotic cells, four major kinds of RNAs are found to occur namely—Ribosomal RNA (r-RNA), messenger RNA (m-RNA), transfer RNA (t-RNA) and small nuclear RNA (sn RNA). These various types of RNAs have different three dimensional molecular structures and perform diversified functions. These will be described along with their specific role in the process of protein biosynthesis.

RNAs are synthesized on DNA template

All types of RNAs mentioned above are synthesized, as directed by certain nucleotide sequences on DNA templates. The entire DNA molecule is not expressed in RNA synthesis. Rather, **some specific selected regions of DNA** act as template, on which a complementary copy of specific nucleotide sequence of DNA strand is copied to synthesize different RNAs. Whereas in **DNA replication process,** during the cell division, the entire genome(whole DNA molecule) is copied, which is a **must** so as to **transfer entire genetic information** characteristic of a specie to the daughter cells, during cell division and to the offsprings during reproduction.

TRANSCRIPTION

Transcription is a process by which the genetic information present as **genetic code,** in the specific sequence of nucleotides in certain regions of DNA molecule (genes) is transferred into a **complementary copy** to synthesize the **m-RNA.** In other words by **transcription** process the genetic code stored in specific nucleotide sequence of the genes on DNA is **expressed** (transcribed) into m-RNA, that carries the message as to the type of protein to be synthesized in the cytoplasm.

The process of transcription is best characterized in prokaryotes and is quite similar to that of mammalian cells as well. However the **regulatory signals, factors** facilitating the transcription and **processing** of **RNA primary transcript** are highly complicated in eukaryotes and shall be described along with the primary prokaryotic RNA transcription process.

Transcription is directed by genes

Genes, the **functional units** of **DNA** and the **molecular basis** of **heredity,** are the **coding regions** on DNA double helix, that are transcribed into a complementary copy of m-RNA. This m-RNA carries the message as genetic code, to be translated into a specific sequence of amino acids to synthesize a protein molecule.

Template and antitemplate strands of DNA

One of the strand (polynucleotide chain) of DNA double helix acts as a **template** to direct the synthesis of m-RNA and is known as **coding strand** or **sense strand** and has a nucleotide sequence in 3′ to 5′ diphosphoester direction. The other strand of DNA, that has antiparallel nucleotide sequence i.e. in 5′ to 3′ direction, is called as **antitemplate** (non-coding) or **antisense strand** (Fig. 20.1).

During transcription process the template strand of DNA is copied into a complementary strand of m-RNA, which is like a "xerox" copy, with the only difference that the nucleotide sequence is **antiparallel** i.e. running in **opposite direction (5′ to 3′)** and has **uracil nucleotide in place of thymine of DNA** as illustrated in the Fig. 20.1.

Base Pairing Rule Guide Transcription

DNA dependent RNA polymerase (RNAP) catalyzes the transcription of genetic code on DNA template strand into a sequence of ribonucleotides to synthesize m-RNA, strictly according to the base pairing rule. According to this, the thymidine(T) nucleotide on DNA template directs the incorporation of adenine (A) nucleotide while guanidine (G) nucleotide will direct the addition of cytidine (C) nucleotide and vise versa, during the process of transcription , except uracil (U) nucleotide is incorporated instead of thymidine (T) into the growing m-RNA as illustrated in the Fig. 20.1.

DNA [5' ——— A T G C A T G G C A ——— 3' Non-coding strand
      3' ——— T A C G T A C C G T ——— 5' Template strand

RNAP ↓ **Transcription**

5' ——— A U G C A U G G C A ——— 3' m- RNA

Fig. 20.1 : Polymerization of ribonucleotides by RNA polymerase to synthesize m-RNA according to base pairing rule as directed by template DNA strand

Thus during trancription a complementary copy of DNA template as m-RNA is synthesized with same **genetice code**, only the nucleotide sequence is antiparallel, that is running in opposite direction (5' to 3' direction) and it is an **exact xerox copy** of the **antisense strand of DNA,** except uracil is in place of T of DNA (Fig. 20.2).

DNA Dependent RNA Polymerase

Transcription of coding region of DNA template strand is catalyzed by the enzyme **DNA dependent RNA polymerase (RNAP).** This enzyme is specific in catalyzing polymerization of ribonucleotides by forming 5' to 3' diphosphoester linkage, into a sequence, **complementary** to the genetic code on template strand of DNA, according to base pairing rule.

Prokaryotic DNA Dependent RNA Polymerase

Prokaryotic RNA polymerase of E-coli bacteria is a **multisubunit enzyme** (mol.wt. 350 kDa) with an **active core complex** consisting of two small identical α subunits and two larger, unidentical β and β′ subunits. In this core RNAP complex, β subunit is the ribonucleotide polymerization catalyzing unit (Fig. 20.2).

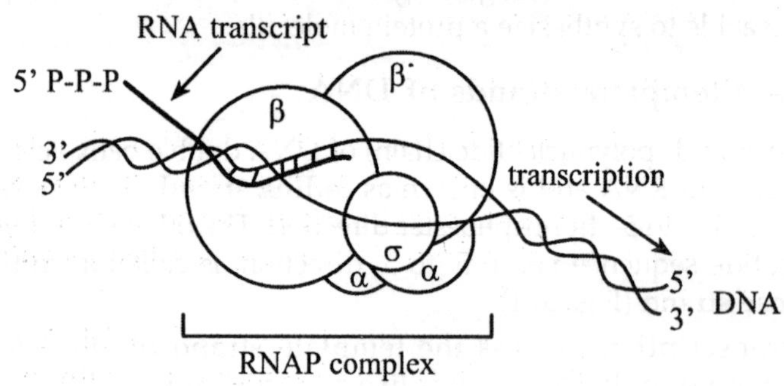

Fig. 20.2 : Diagrammatic representation of core RNA polymerase complex along with sigma (σ) factor

RNA polymerase is a **metalloenzyme** containg two zinc ions and requires a **sigma factor**(σ), which is a specific protein factor necessary for the **recognition** of **promoter site** on DNA template. The sigma factor helps RNAP to get attached **tightly** to a specific consensus nucleotide sequence on promoter region of DNA template, just before the transcription start, as shown in the Fig. 20.3.

Mammalian cell contains **three** different DNA dependent **RNA polymerases.** Eukaryotic **RNAP II** is responsible for the transcription of genes into m-RNA strand, while **RNAP I** catalyzes synthesis of ribosomal RNA and **RNAP III** is responsible for the synthesis of t-RNAs.

Promoter Region

It is a specific region on DNA template **just before** the **start of coding region** (gene), on which the RNAP binds after recognition with the help of sigma factor. Prokaryotic promoter site is relatively simple as compared to the eukaryotic site, that shall be described later on.

Bacterial promoter is approximately 40 nucleotide (40 bp) in length, that is sufficient in size for the attachment of enzyme RNA polymerase. Promoter region has two short (conserved) consensus nucleotide sequences as described below—

(1) Upstream 35 bp from the **start of transcription** of gene, there is a specific sequence of 8 nucleotides as TGTTGACA -3 ′ to which RNAP binds to form a **closed complex,** after sigma factor recognizes it and facilitates tight binding of RNAP .

(2) More near to the **transcription start,** about 10 base pair upstream, another consensus short sequence of **AT-rich 6 nucleotides** as **TATAAT** is present, known as **pribnow box** or **TATA box** as shown in Fig. 20.3. This TATA box has less number of H_2 bonds (low melting point) and thus facilitates easy unwinding of DNA double helix to provide access to the nucleotide sequence to be transcribed by RNA polymerase.

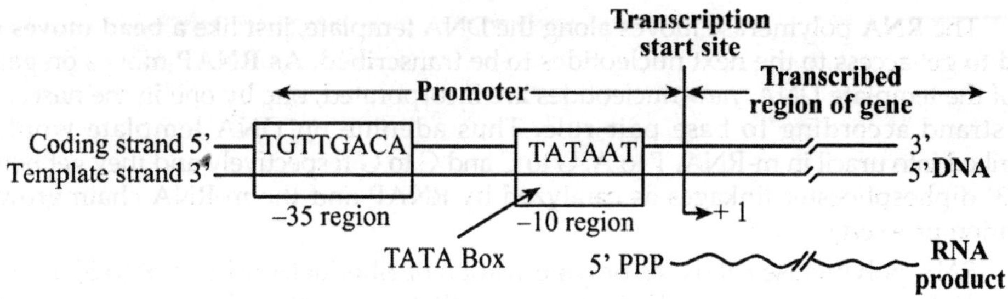

Fig. 20.3 : Promoter region with TATA box on DNA template in prokaryotes

RNA transcription process on DNA template involves initiation, elongation and termination steps that are described here.

1. Initiation of Transcription

To initiate the transcription of genetic code of DNA template into m-RNA strand, the RNA polymerase with the help of sigma factor recognize the **"promoter region"** just near the transcription start and gets attached (bind) on DNA template near TATA box.

The **correct binding** of RNAP on promoter site, just before the **transcription start** i.e. the first nucleotide of **first triplet** or **codon** of the coding region (i.e. gene) of DNA, is the most important aspect in the transcription process, since there is no punctuation between genetic codons or triplets on DNA strand. Sigma factor plays vital role in correct binding of RNAP on the promoter site.

Unwinding of DNA double helix—After proper binding of RNAP on promoter site, the sigma factor get detached and unwinding of DNA double helix start at the 'TATA box' onwards. **RNAP** can catalyze both unwinding of DNA by its **leading "unwindage" activity** and **lagging** rewinding of DNA double helix during the transcription process, to form a **transcription bubble** as shown in the Fig. 20.4.

Transcription Start

The first nucleotide to be polymerized at the transcription start is always a **purine** nucleotide and the first initiation codon on m-RNA near its 5' cap is **mostly** 'AUG', that code for methionine amino acid. This ensures the correct site of transcription start, that is the beginning of coding region (gene) of DNA. The first nucleotide to be incorporated according to base pair rule, gets attached to the β subunit of RNAP, which becomes the 5' end of m-RNA to be synthesized.

Polymerization Require High Energy Ribonuleotides

High energy ribonucleotides as ATP, UTP, GTP and CTP are utilized in the transcription process for polymerization, as catalyzed by RNA polymerase. The pyrophosphate group (PPi) removed is hydrolyzed to provide energy for 5′-3′ **diphosphoester bond formation** between the incoming ribonucleotides. The ribonucleotides are polymerized in a sequence on the nascent m-RNA strand as directed by the DNA template strand strictly according to base pairing rule.

2. Elongation of m-RNA Strand

The RNA polymerase moves along the DNA template, just like a **bead moves on a thread** to get access to the next nucleotides to be transcribed. **As RNAP** moves on genetic code of the template DNA, new nucleotides are incorporated, one by one in the nascent m-RNA strand according to base pair rule. Thus adenine on DNA template would be transcribed into uracil in m-RNA, T to A, G to C and C to G respectively and they get bonded by 5′-3′ diphosphoester linkages as catalyzed by RNAP and the m-RNA chain grows as elongation proceeds.

RNA polymerase catalyzes polymerization of ribonucleotides in 5′ to 3′ direction, **antiparallel** to the sequence on DNA template, that has nucleotide sequence in 3′ to 5′ direction. The elongation complex (Fig. 20.4) containing RNAP and nascent m-RNA attached to β subunit of enzyme, progress along the DNA template and new ribonucleotides are incorporated as directed by genetic code of genes by base pairing and m-RNA chain elongates till it reaches to the **termination site.**

3. Termination of Transcription

When the **coding region (gene)** of DNA template is completely transcribed into m-RNA, specific **termination signal** on termination site stops further transcription. In prokaryotes, the **termination signal** on DNA template is recognized by a specific protein, known as the **rho** (ρ) **factor.** This rho factor is an ATP dependent RNA-DNA helicase which **disrupt** further transcription, by dissociating the nascent m- RNA from the DNA template and RNAP enzyme also gets separated from DNA as illustrated in the Fig. 20.5.

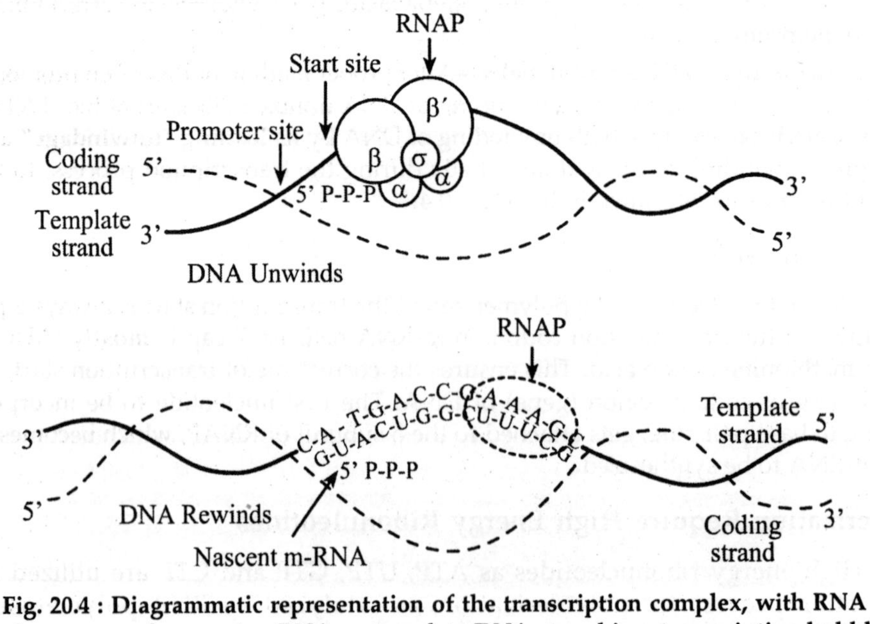

Fig. 20.4 : Diagrammatic representation of the transcription complex, with RNA polymerase and nascent m-RNA on template DNA strand in a transcription bubble

Termination Signals

Rho dependent transcription termination signals in E-coli bacteria appear to have a conserved **consensus** nucleotide sequence of about **40 bp long.** It contains a **hyphenated** or **palindrome** sequence. The palindrome means a word read same in forward or backward direction, for example 'madam'. Here at the **termination site,** there is a specific hyphenated, **interrupted invert repeat** nucleotide sequence followed by a AT bp chain (Fig. 20.5). As the transcription proceeds through this termination sequence, the growing nascent m-RNA strand gets folded into a **hairpin loop** because of the intramolecular hydrogen bonds formation between (facing) enough complementary bases, due to the interrupted invert repeat nucleotide sequences as shown in the Fig. 20.5. At this **hairpin** loop, the **rho (ρ) termination factor** stops further transcription and dissociates the RNAP from the DNA template and releases nascent **transcript** (m-RNA). Eukaryotic termination signals however are not well defined.

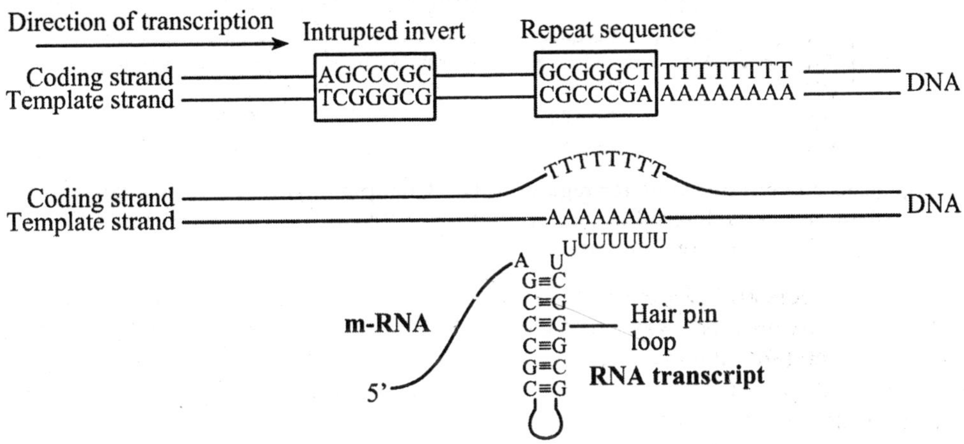

Fig. 20.5 : Prokaryotic transcript with hairpin loop at the termination region on DNA template containing interrupted invert repeat termination signal

Eukaryotic Transcription Signals are Complex

The **high fidelity** and **frequency** of transcription process in mammalian cells is controlled by many specific protein factors bound to certain DNA regions. These are as described here under—

1. Initiation at promoter region needs correct binding of RNAP II—For the correct binding of RNAP on promoter site, two types of consensus sequence elements are present on DNA template, that are recognized by various factors facilitating the binding of RNAP just before the transcription start as described below—

(a) **Hogness box**—It usually has a consensus nucleotide sequence rich in AT base pairs as **'TATAAAAG'**,resembling to that of prokaryotic **TATA** box and defines the exact transcription start point. This box has affinity to bind a 30 kDa specific **TATA binding protein (TBP),** which in turn binds several protein factors called **TBP associated factors (TAFs).** The whole complex is referred to as **TF II D** and its binding on hogness (TATA) box is stated to be the start of the formation of **initiation complex** with correct binding of RNAP-II on DNA template at promoter site as shown in the Fig. 20.6

(b) **Eukaryotic initiation transcription complex formation** is highly complicated as compared to prokaryotes, where only sigma factor is required to recognize the TATA box on promoter region for RNAP binding. However to maintain high fidelity during transcription, in eukaryotes, a number of protein factors (around ten in numbers), such as A, B, D, E, F and H are required that facilitate correct binding of RNA polymerase II on TATA box, just near the transcription start.

2. GC and CAAT boxes—The GC and CAAT boxes are situated at some distance from the TATA box on DNA template and are responsible for the determination of **frequency of transcription** process (Fig. 20.6). Each of the two GC boxes bind a specific protein **SPI** and the CAAT box binds a specific **CTP** protein to determine the frequency of transcription process as illustrated in the Fig. 20.6.

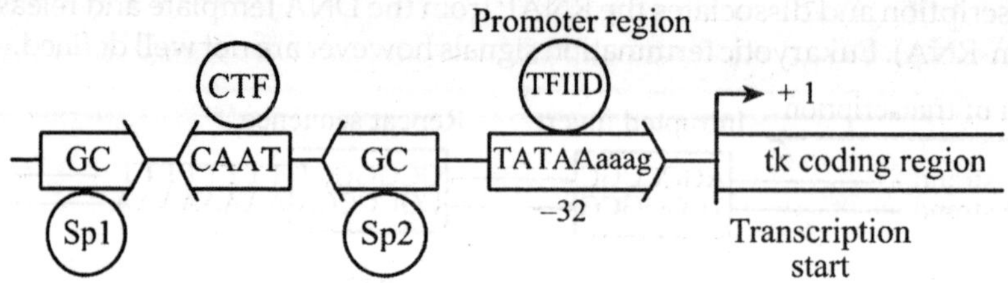

Fig. 20.6 : Eukaryotic promoter region on DNA template with various signals and protein factors to determine correct binding of RNAP II on TATA box and to control the rate of frequency of transcription

3. Enhancers and Repressors—A third class of transcription signals on eukaryotic DNA template has **hormone sensitive elements (HSE),** that **coordinate gene expression** with that of the **metabolic need** of the cell through hormones and II messengers. These HSE on DNA are called **enhancers,** that **increase rate** of **transcription** (gene expression) and **repressors** which decrease transcription process.

The hormone sensitive elements are located both **upstream** and **downstream** from the transcription start site, may be 100 or 1000 bp away on the same chromosomes and still can exert their **inducer** or **suppression effects,** when hormone receptor complex or II messenger (c-AMP) bind on them (detail in Chapter 22).

Processing of Primary Transcript (m-RNA)

In prokaryotes, the **primary transcript** (m-RNA) begins to start protein synthesis (translation) even before completion of the transcription process, as there is no compartmentalization of nucleus from the cytoplasm. However, addition of 5' cap with certain other modifications are prerequisite to start translation process.(described later in the section)

Eukaryotic Primary Transcript Processing

In eukaryotes, the **primary transcript** is very large and is known as **heterogenous nuclear RNA (hnRNA)** that undergoes **extensive processing to form m-RNA** in the nucleus. The nuclear primary transcript has 50-70% extra **non-coding regions** called '**introns'** interspersed between the **coding regions** known as the '**exons'**, probably because many exons (1, 2, 3......) are the genes that code for enzymes 1, 2, 3...... of a metabolic cycle. In fact these introns interrupt the coding regions of the structural genes encoding the enzymes 1,2,3.........of a metabolic cycle. The intron RNA sequences are cleaved out of the primary transcript and the exons (genes 1,2,3) of the transcript are appropriately spliced together in the nucleus during the processing of hnRNA to produce a m-RNA. The mechanism of the removal of introns requires a special structure called **spliceosomes.**

Spliceosome is involved in converting primary transcript (hnRNA) into m-RNA and it cleaves out introns and splice together exons as described here. The spliceosome consists of specific **ribozymes** that are chemically RNA containing enzymes and catalyze correct splicing and excision of exones from the introns (extra non coding regions) during processing of hnRNA to form a m-RNA strand.

Mechanism—The ribozymes specifically cut at 5' terminal end of the intron from the end of the exon I and a special RNA structure called **lariat** is formed as shown in the Fig. 20.7, where the 5' end of the intron loops to join its 3' end. The ribozymes then apply a second cut at the 3' end of intron and begining of exon 2 to remove the intron portion between exon l and 2 as **RNA lariat**. After removal of the intron as lariat, the two exons are ligated by ligase enzyme to form m-RNA as illustrated in the Fig. 20.7.

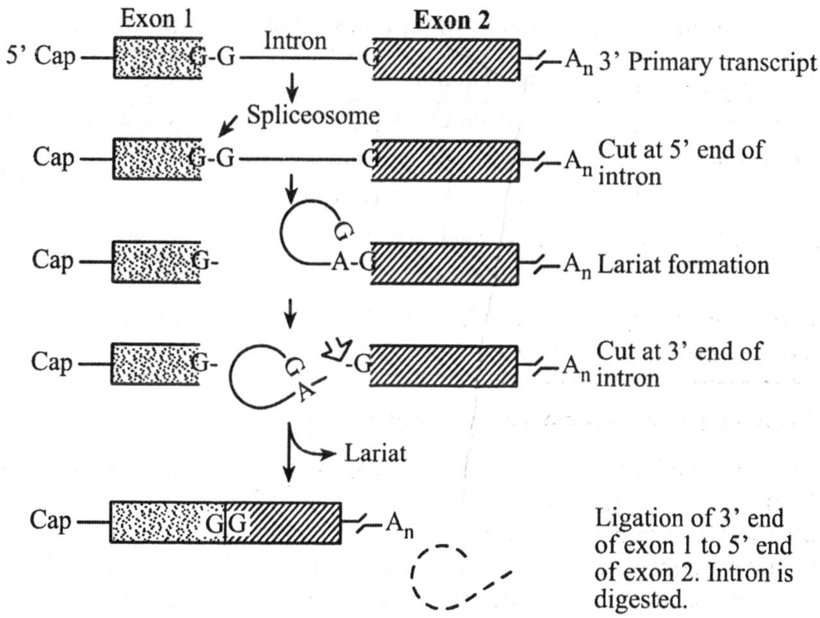

Fig. 20.7 : Conversion of hnRNA (primary transcript) into m-RNA by spliceosome, that remove introns as RNA lariat and ligate together exons (genes)

Post Transcriptional Modification of m-RNA

RNAs are extensively modified after transcription in several ways, such as 5′ **capping** and 3′ terminal additions of adenine nucleotides on m-RNA. In addition extensive base modifications in the nucleotides of m-RNA takes place after its transcription.

m-RNA Modification and Processing

Addition of **5′ cap**—A 7-methylguanosine cap is added at 5′ terminal end of the newly transcribed m-RNA precursor in the nucleus before it is transported to cytoplasm. This **5′ cap** is required for the **initiation** of protein synthesizing ribosome complex formation, at the correct site i.e. at the start of coding region on m-RNA. The 5′ cap also **protects** the premature degradation of m-RNA (hydrolysis) by exonucleases from 5′ terminal end.

Addition of poly (A) tail at 3′ end—A poly (A) tail is added at 3′ terminal during post transcriptional processing of m-RNA. The exact function of this poly(A) tail at 3′ end of m-RNA is not very clear, but it probably is required to protect the 3′ end from **exonucleases** attack.

Inhibitors of Transcription—Clinical Importance

The synthesis of RNA (transcription) on DNA template is inhibited by several antibiotics and toxins. Certain **antibiotics** specifically inhibit **prokaryotic transcription** i.e. in bacteria and other microorganisms only, but will not affect mammalian transcription process. Thus they can safely be used as drugs to kill bacteria and cure infections and diseases. Important antibiotics used in clinical medicine for the treatment of some dreaded diseases are given in the Table 20.1 along with their mode of action.

Table 20.1 : Inhibitors of transcription process along with their clinical uses

S.No.	Antibiotic/Inhibitors	Site of action	Clinical use
1.	Actinomycin D from **streptomyces**	Binds on DNA template and block RNAP movement	Anticancerous drug
2.	Rifampicin (synthetic derivative of rifamyce)	Binds to β subunit of RNAP of bacteria, which is inactivated	Tuberculosis & Leprosy drugs
3.	Alfa amanitin toxin from mushroom	Inactivates RNAP II (eukaryotic enzyme)	Poisonous to humans

Reverse Transcription—Biomedical Importance

Some of the viruses known as **retroviruses** have RNA **as genetic material** instead of DNA. These viruses infect host and utilize its DNA for multiplication by reverse transcription and the enzyme catalyzing the process is called **reverse transcriptase.** This process was first described by Howard Tamin and David Baltimore independently in 1970 (both got noble prize in 1975) and the mechanism involved in the process is described below—

Mechanism of Action of Reverse Transcriptase

The RNA genome of the retrovirus can direct the synthesis of first a single stranded DNA on its RNA template as catalyzed by enzyme **RNA dependent DNA polymerase** or **"reverse transcriptase"**. The nascent, (newly synthesized) DNA strand on viral RNA template strand forms a RNA-DNA hybrid, as this nascent DNA is complementary copy (c-DNA) to the viral RNA.

Then a specific enzyme RNAase H degrades the RNA strand and the c-DNA strand left behind in turn serves as a template to produce a complementary double stranded DNA molecule (proviral DNA) containing the genetic information originally present on RNA genome of the **retrovirus.**

Human Immunodeficiency Virus (HIV)

HIV is a good example of retrovirus, that on infecting humans causes the deadly immunodeficiency disease commonly known as **AIDS.** The RNA genome of this retrovirus (HIV) after entry into the host (human) CD_4 cell, acts as a template on which enzyme reverse transcriptase (RT) first synthesize a complementary DNA copy(c-DNA). The viral RNA is then removed and degraded by RNAase H. Next this c-DNA (containing viral genome) synthesize double strand DNA (proviral DNA) which migrates into the nucleus of the host cell and gets integrated into host (human) genome, catalyzed by the enzyme viral integrase. As this proviral DNA is replicating form, it multiplies in the dividing host cells and large population of viral RNA (HIV) are produced that cause AIDS. (details in the Chapter 38).

Oncogenic Viruses – Retroviruses

Certain retroviruses can cause cancers in mammals and are called oncogenic viruses. This is one of the probable etiological factor of **carcinogenesis** by integration of retroviral genes into host DNA in humans, that activates certain oncogenes. (details in Chapter 39).

Mechanism of Oncogene Viruses in Carcinogenesis

When the Rous sarcoma retrovirus infects humans, its circular DNA is incorporated into host DNA and may act as **oncogenic** agents. The powerful drive for **multiplication** by this viral **genome overrules** the **regulatory checks** of normal human cell divisions, resulting in uncontrolled cell multiplication causing cancerous growth (tumor). This is known as **transformation** of normal process of cell division by **oncogenic virus** into a rapidly multiplying malignant tumor (details in the Chapter 39). In addition to Rous sarcoma virus more than 20 oncogenic retroviruses have been discovered that are capable of activating oncogenes and cause malignancies in animals and humans.

Gene Therapy—Role of retrovirus

Certain inborn metabolic diseases and inherited congenital disorders can be treated by **gene replacement therapy** to **correct** the basic genetic abnormality. Theoretically the gene therapy involves replacement of a **mutant** (abnormal) gene with the **normal gene**. One of the **major route** by which normal gene can be introduced into humans, to replace the abnormal gene is **via retrovirus.** The mechanism is given in brief here under—

Retroviruses can take normal human gene inserts up to approximately 9 k bp and act as **vectors** which are then introduced into the actively growing (multiplying) target cells taken out from the bone marrow or hepatocytes and then are reintroduced into the human body. The introduced normal gene would begin to direct the expression of its protein products (that were deficient in the genetic disorder) and thus would correct the deficiency in the host to alleviate the symptoms of the genetic disorder (Chapter 23). Clinical trials of gene therapy that have been successfully completed or going on include – Adenosine deaminase defect, cystic fibrosis, familial hypercholesterolemia and various types of cancers.

❑❑❑

21

PROTEIN BIOSYNTHESIS

The genetic information regarding the type of protein to be synthesized is **transcribed** from the coding region (gene) of DNA, into a specific sequence of nucleotide on m-RNA strand in form of **genetic code**. This genetic code is then **translated** into a specific sequence of amino acids in which they condense by peptide bonds to synthesize a specific protein.

Messenger RNA (m-RNA)—It is appropriately called as messenger RNA, since it carries the message from genes in the nucleus, as to the type of protein to be synthesized, to **ribosomes,** the **protein synthesizing machinery** in the cell cytoplasm. The m-RNA strand has the genetic information in form of **genetic code**, that resides in the specific sequence of **triplets** or **codons** in its ribonucleotide chain.

Ribosomes – the protein synthesizing machinery

The ribosomes are the centers of protein biosynthesis in the cell cytoplasm, attached to the endoplasmic reticulum. Ribosomes are made up of ribosomal RNA (80%) with some basic proteins and are in form of **dense spherical particles** of 100 **A°** size, attached to the endoplasmic reticulum to give it a **rough appearance** as seen by the electron microscope.

Prokaryotic and Mammalian Ribosomes

According to their sedimentation rate and the size of their two subunits, the prokaryotic and eukaryotic ribosomes are **differentiated** as **70S** and **80S ribosomes** respectively. They can be further separated into two subunits, one small and other larger as shown in the Fig. 21.1.

Translation of Genetic Code—Role of t-RNA

The process of deciphering of genetic code of m-RNA strand into a specific sequence of amino acids to synthesize proteins is known as **translation**. For this, an **adapter**

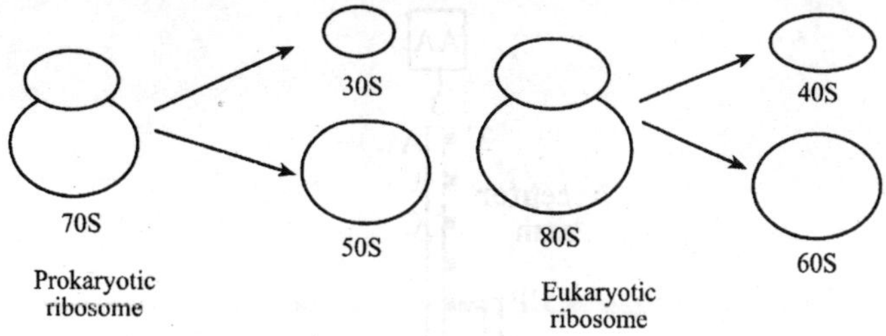

Fig. 21.1 : Comparison of prokaryotic and eukaryotic ribosomes

intermediate compound **t-RNA** is required. This adapter molecule is a small special RNA molecule known as **transfer RNA** or **t-RNA** and plays very important role during protein synthesis. It facilitates the translation of genetic code into a specific sequence of amino acid to synthesize a protein molecule. This adapter t-RNA molecule recognizes the **triplet** or **codon** on m-RNA by its **anticodon loop** on one hand and on the other, it brings an amino acid, as specified by that codon, to be added in a particular sequence during protein synthesis.

For 20 naturally occurring amino acids present in proteins of the body, there exist at least **20 specific t-RNA** molecules, one for each amino acid and play significant role in the process of protein biosynthesis.

t-RNA Structure—Salient features

1. All 20 t-RNA molecules specific for 20 amino acids have extraordinary similar three dimensional structure, resembling to that of a **clover leaf** (Fig. 21.2). This clover leaf like structure of t-RNAs is formed, because the nucleotide sequence of t-RNA strands allows extensive folding in a similar way, due to intramolecular complementary base pairing by the formation of hydrogen bonds.
2. All t-RNA molecules consist of **four main arms** with loops at the end of three arms as shown in the Fig. 21.2.
3. **The acceptor arm**—It is at the 3′ terminal end of t-RNA polynucleotide chain with an **ACC** nucleotide sequence, on which a specific amino acid is attached (charged) by ester bond.
4. **The anticodon loop**—It is at the lower end of the clover leaf structure of all t-RNAs and consists of a **specific** and **antiparallel** (complementary) **sequence** of three nucleotide to that of a **codon on m-RNA**, known as **"anticodon".** The anticodon loop recognizes the codon on m-RNA by its complementary base pair bonding.
5. **The DHU arm**—It is also known as 'D' arm as it contains a dihydrouracil nucleotide. This arm along with the variable arm (Fig. 21.2) is most important for recognition by specific **amino acyl t-RNA synthase** enzyme for proper charging of specific amino acid, for which it has the anticodon triplet.
6. **The TψC arm**—This arm of t-RNA molecules has a pseudouracil (ψ) nucleotide and facilitates proper binding of t-RNA on ribosomes.

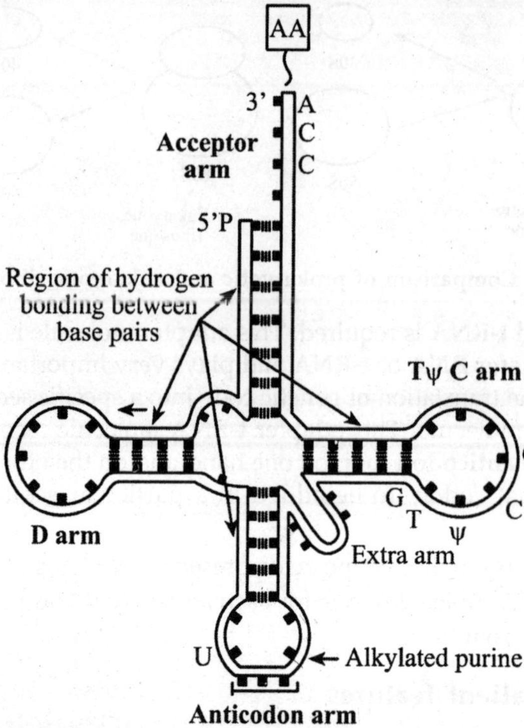

Fig. 21.2 : Typical clover leaf structure of t-RNAs along with characteristic features

Role of t-RNA in Translation

The t-RNA plays an highly significant role in translation of genetic code on m-RNA strand into a specific sequence of amino acids, that is the most important aspect of the protein biosynthesis. This intermediate adapter molecule (t-RNA) recognizes the "triplet" (codon) on m-RNA by its anticodon loop with base pairing and add the specific amino acid for which this genetic codon is meant on m-RNA, at proper position in the specific sequence of the growing polypeptide chain.

Genetic Code

The genetic information characteristic of a specie resides in a **code language,** in the specific sequence of four deoxyribonucleotide of A,T,C and G arranged on the polynucleotide strand of DNA and is known as **genetic code.**

Triplet—Genetic code constitutes various **triplets** with a **specific, non overlapping sequence** of **three purine** or **pyrimidine nucleotides** known as **codons.** The triplet acts as a code word for a specific amino acid.

Brief History

The fact that the code word of a particular amino acid resides on a specific sequence of three nucleotides (triplet) on DNA, was first discovered by Nirenberg and Mattheai (Noble prize, 1969) during their unique experiments in vitro. They used synthetic polyuridine m-

RNA in vitro experiments and observed that it is translated into a polypeptide chain containing only phenylalanine. So accordingly they suggested that UUU triplet should be the code word for phenylalanine.

In their next experiments, they used synthetic polynucleotide m-RNA strand with alternate cytidine and uridine nucleotides. This m-RNA directed the synthesis of a polypeptide chain with alternate amino acids, serine and leucine. This **sequence is possible only** when the **code word for an amino acid resides on a triplet** as is evident from the Fig. 21.3.

m-RNA - CUC UCU CUC UCU

In vitro ↓ Translation

polypetide- serine- leucine-serine- leucine

Fig. 21.3 : In vitro experiments with synthetic m-RNA (with alternate CUCUCU), that was translated in a polypeptide with alternate sequence of AAs

The whole genetic code for all the 20 amino acids later on was deciphered by similar experiments using synthetic m-RNAs of known sequences by scientist Khurana and his associates (Nobel prize 1968) and is summarized in the Table 21.1.

As there are four nucleotides in nucleic acids containing A,G, C and U (T in DNA), the possible specific sequences for triples would be **4^3=64 codons** are possible and in actual fact all of them have been found to exist in the living organisms. Out of these, 61 codons or triplets code for 20 naturally occurring amino acid that constitute various proteins.

Three codons (UAA, UAG, UGA) do not code for any amino acid and are called **nonsense codons** and they act as **stop** or **chain termination codons**. AUG and sometimes GUG act as chain initiation or first codon to be transcribed in m-RNA, during most of the protein synthesis going on in the living organisms.

Table 21.1. The Genetic code (codon assingnments in messenger RNA).[1] for specific amino acids

First Nucleotide	Second Nucleotide				Third Nucleotide
5' End	U	C	A	G	3' End
U	Phe	Ser	Tyr	Cys	U
	Phe	Ser	Tyr	Cys	C
	Leu	Ser	Term	Term[2]	A
	Leu	Ser	Term	Trp	G
C	Leu	Pro	His	Arg	U
	Leu	Pro	His	Arg	C
	Leu	Pro	Gin	Arg	A
	Leu	Pro	Gin	Arg	G
A	Ile	Thr	Asn	Ser	U
	Ile	Thr	Asn	Ser	C
	Ile[2]	Thr	Lys	Arg[2]	A
	Met	Thr	Lys	Arg[2]	G
G	Val	Ala	Asp	Gly	U
	Val	Ala	Asp	Gly	C
	Val	Ala	Glu	Gly	A
	Val	Ala	Glu	Gly	G

Salient features of genetic code

1. **Universality**—The genetic code with 64 codons (Table 21.1) is of universal occurrence in almost all living organisms with few exceptions e.g. mitochondrial genetic code.

2. Genetic code is **non-overlapping,** continuous and **without any puctuation** on m-RNA, therefore the **starting point is extremely important** in the commencement of the translation process.

3. Genetic code is unambiguous and a particular triplet is highly specific and will always code for only **one particular amino acid.**

4. Degeneracy in genetic code—As there are 61 triplets (codons) that code for 20 natural amino acids, there are more than one codons (triplets) for most of the amino acids, (Table 21.1). This is known as **degeneracy in genetic code** i.e. multiple codons decode the same amino acid. Some amino acids are encoded by several codons e.g. serine has six different codons, while other amino acids, such as methionine and tryptophan have a single codon, as is evident from the Table 21.1.

First two nucleotides in a triplet are important

It has been observed that in the multiple codons that code for same amino acid, the **first two nucleotides** are **same**, and are necessary for the correct decoding in **translation** process. In general the third nucleotide of codon is less important and can be different, as is evident from the Table 21.1.

Recognition of codon by "anticodon"

The recognition of specific codon on the m-RNA is achieved by the **anticodon loop** of t-RNA that is necessary for the translation of genetic code into specific sequence of amino acids during biosynthesis of proteins. The codon in m-RNA is recognized by the anticodon of t-RNA according to **base pairing** rule, since it has a **complementary nucleotide sequence** to that of codon.

Wobble phenomenon—According to wobble hypothesis, as put forth by Crick, a single t-RNA, specific for an amino acid with only one **anticodon** can recognize **more than one codons** for that amino acid on m-RNA by what is known as "wobble phenomenon". During this wobbling, the base pairing of first two nucleotides of codon by anticodon loop of t-RNA is **strict,** since the first two nucleotides of codon—anticodon are **essentially complementary**. While the 3rd nucleotide (different & mismatched) binding is loose or does not occur, and is recognized by just "wobbling" of anticodon on the codon of m-RNA.

This explains the degenerancy in genetic code in terms of molecular basis. Thus the multiple codons for a single amino acid can be recognized by one specific t-RNA with single **anticodon** by **wobbling,** to overcome the degeneracy in the genetic code.

TRANSLATION—PROTEIN BIOSYNTHESIS

The deciphering of genetic code of m-RNA into a specific sequence of amino acids to be polymerized to produce a polypeptide chain (protein) is known as translation process. It is achieved in following steps.

(1) Activation of amino acids.

(2) Proper charging of amino acid on specific t-RNA.

(3) Protein synthesis proper in 3 steps—initiation, elongation and termination.

(4) Post translational modifications.

Activation and Charging of Amino Acid on t-RNA

As mentioned earlier, for each 20 naturally occurring amino acids, there are **20 (at least)** specific t-RNA molecules with specific anticodon loop. But by itself the t-RNA cannot recognize its own amino acid. So for the recognition and proper charging of amino acid on t-RNA, specific enzymes known as **amino acyl t-RNA synthase** are required to facilitate attachment of correct amino acid on the t-RNA molecule.

There are 20 **amino acyl t-RNA synthase** enzymes present in the body, one for each of the 20 t-RNA molecules. These enzymes recognize the specific t-RNAs by their D and variable arms and at the same time charge them with the appropriate amino acid on the ACC- 3′ terminal acceptor arm (Fig. 21.2). These enzymes have **two active sites,** one for the recognition and binding of an amino acid and other for recognition of t-RNA with **anticodon** specific for that particular amino acid.

Activation of Amino Acids and Proper Attachment on t-RNA

Amino acid is first activated by ATP and then get attached, after recognition on the enzyme aminoacyl t-RNA synthase (E) to form an intermediate **aminoacyl-enzyme complex.** This complex then recognizes a t-RNA with specific anticodon meant for this particular amino acid and facilitates the charging of amino acid on the acceptor arm of t-RNA by an ester bond at 3' CCA nucleotide as shown in the Fig. 21.4. Two high energy phosphate bonds by the hydrolysis of pyrophosphate group of ATP are consumed during the activation of amino acid.

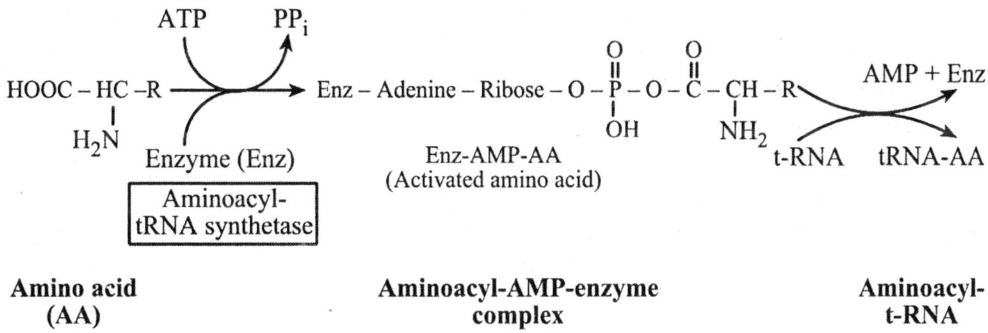

Fig. 21.4 : Activation and charging of amino acid on its specific t-RNA, as catalyzed by aminoacyl t-RNA synthase enzyme

Salient Features of Protein Synthesis

The ribosomes in cell cytoplasm are main protein synthesizing machinery on which the m-RNA with specific genetic code, as transcribed from DNA (genes), comes and gets attached.

The translation of genetic code on m-RNA commences from its 5′ terminal end near the G^m TP cap, to 3′ direction and is facilitated by t-RNA. The condensation of amino acid occurs one by one, in specific sequence during translation by peptide bonds formation and proceeds from NH_2 terminal end to the carboxyl terminus of polypeptide chain.

The actual process of translation of genetic code (that is universal) is very similar in prokaryotes and eukaryotes but the regulatory factors and signals are more complex in mammalian protein synthesis.

PROTEIN BIOSYNTHESIS—TRANSLATION

The process of protein biosynthesis is divided in 3 phases i.e. initiation, elongation and termination and are as described here under—

Initiation involves Several Protein Factors

As there is no punctuation between triplets (codons) on m-RNA strand, the correct attachment of m-RNA and assembly of ribosome machinery on its **first codon** is the most important aspect to maintain high **fidelity of translation process.** To ensure that no mistake occurs, in most of protein synthesis the first codon on m-RNA, near its 5′ cap is usually **AUG,** that code for **methionine** amino acid. Thus the initiation process usually starts with binding of **methionyl t-RNA complex** on first codon AUG on m-RNA with the formation (assembly) of **ribosome-m-RNA initiation complex.** This is achieved in following steps and requires **initiation factors** and GTP as energy source.

1. The recognition and attachment of 40S ribosome subunit (30S in prokaryotes) along with methionyl t-RNA on m-RNA is facilitated **by initiation factor 2 (If-2)** bound to GTP that provides energy for the formation of **initiation complex** as shown in the Fig. 21.5 step 1.

2. The recognition and correct attachment of 40S subunit with methionyl t-RNA on the first initiation codon (AUG) near 5′ cap of m-RNA also requires **initiation factor 1A (IF 1A)** and **initiation factor IF-3** (Fig. 21.5 step II).

3. Formation of 40S initiation complex on m-RNA strand—The proper binding of 40S ribosome on the first initiation codon AUG on m-RNA strand is facilitated by its 5′ cap. For this a **5′ cap recognition factor 4F** (comprising 4E, 4G and 4A) binds to the m-RNA near its 5′ cap, while factors **4A and 4B bind to m-RNA strand** to prevent secondary coiling of m-RNA as illustrated in the Fig. 21.5 Step III.

This whole complex (m-RNA with 4F near its cap and 4A and 4B) is then **scanned by 40S t-RNA-IF2 complex.** The **anticodon** loop of methionyl t-RNA recognizes the first codon (AUG) by **base pairing** and binds with it by forming H_2 bonds on m-RNA near 5′ cap. The whole **40S t-RNA-IF2** complex gets properly attached on m-RNA, which is facilitated by factors IF-1A, IF-2 and IF-3 as shown in Fig. 21.5.

4. Formation of 80S initiation complex—The binding of 60S subunit to the 40S initiation complex, with the hydrolysis of GTP as energy source, results in release of initiation factors IF A1, 2, 3 and other factors and a rapid association of 60S with 40S to form 80S ribosome along with m-RNA strand near its 5′ cap occurs. (Fig. 21.5 step IV.)

A and P sites on 80S ribosome—The newly assembled 80S ribosome have two specific sites, called **P site** for the attachment of growing peptide chain and **A site** for the incoming new amino acyl t-RNA according to the genetic code on m-RNA.

Formation of mammalian 80S ribosome complex involves at least **10 protein initiation factors,** for proper binding of m-RNA to the ribosome for correct deciphering of genetic code, while prokaryotes need only 3 initiation factors IF-1, IF-2 and IF-3 and is more simpler as compared to eukaryotes.

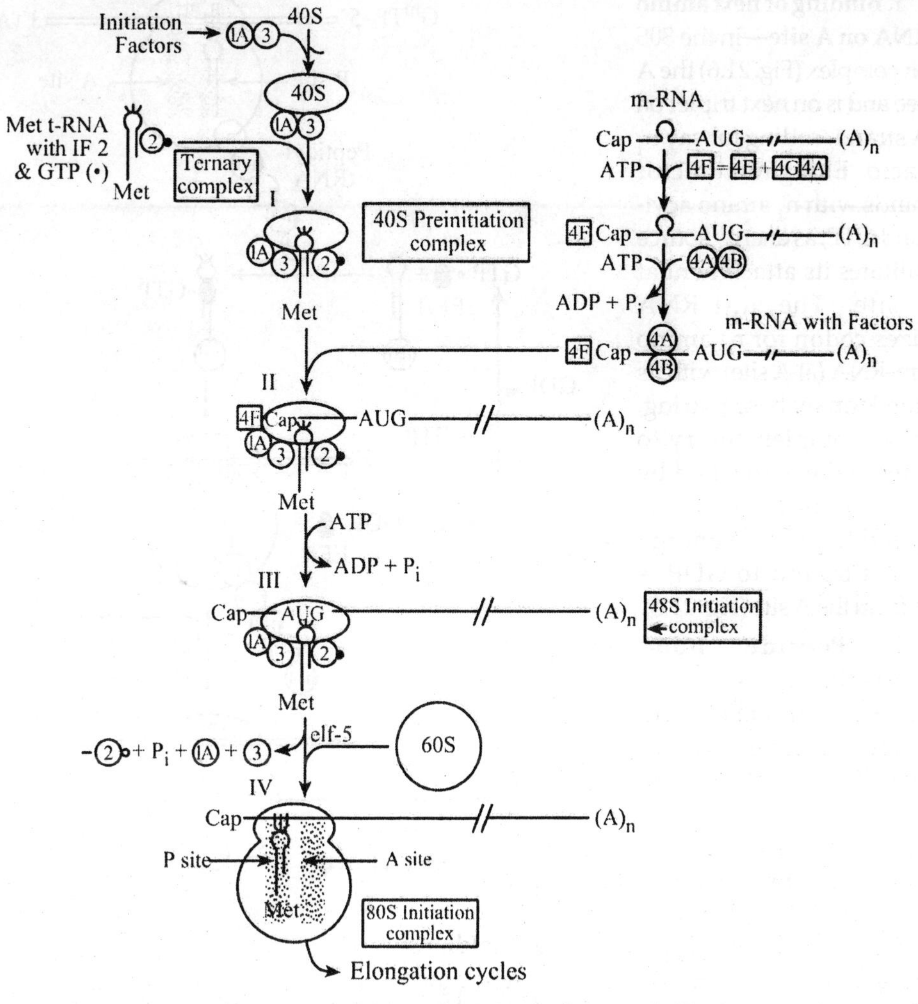

(Fig. 21.5 : Diagrammatic illustration of formation of 80S ribosome initiation complex on m-RNA strand near its 5' cap as facilitated by various initiation factors. (Modified, courtesy Harper)

Regulation of Rate of Translation in Eukaryotes

The factors 4F (comprising 4E, 4G, 4A) bound on initiation complex at the 5′ cap of m-RNA controls the rate of the translation process. Insulin and other growth factors activate 4E-BPI (of 4F) by phosphorylation and the active 4E then recognize and bind on 5′ cap of m-RNA. This effect on activation of 4E (4F) shows anabolic and growth promoting functions of insulin in posttranscriptional increase of the rate of protein synthesis.

Elongation of Nascent Polypeptide

After 80S initiation complex is formed, it is ready for the translation of successive codons (triplet) on m-RNA strand into a sequence of amino acids to synthesize a specific protein and elongation cycle commences. Elongation process requires **elongation factors,** ATP and amino acyl t-RNAs. Elongation cycle comprizes following steps—

1. Binding of next amino acyl t-RNA on A site—In the 80S initiation complex (Fig. 21.6) the A site is free and is on next triplet on m-RNA strand, coding for say n_1 amino acid. Elongation factor EF1-α binds with n_1 amino acyl-t-RNA and GTP, as energy source and facilitates its attachment at the A site. The n_1-t-RNA recognizes codon for n_1 amino acid on m-RNA (at A site) with its anticodon loop by base pairing, as both are complementary to each other and get attached by hydrogen bonds. GTP is hydrolyzed to provide energy and EF-1 α bound to GDP is released from the A site (Fig. 21.6).

2. Peptide bond formation—In next step, the amino group of n_1 amino acid on A site carries out a nucleophillic attack on the carboxyl group of methionine amino acid (first amino acid) attached on P site to form a peptide bond. The reaction is catalyzed by enzyme **peptidyl transferase**, a component of 60S subunit of ribosome. The methionine (I[st] amino acid) leaves P site as it gets bonded by peptide bond to n_1 amino acid on A site as illustrated in the Fig. 21.6, thereby releasing its t-RNA, which is then discharged from the P-site.

3. Translocation of ribosome—Next the enzyme **peptidyl transferase** transfers the growing peptide chain with t-RNA from A site to P site with **simultaneous translocation** of

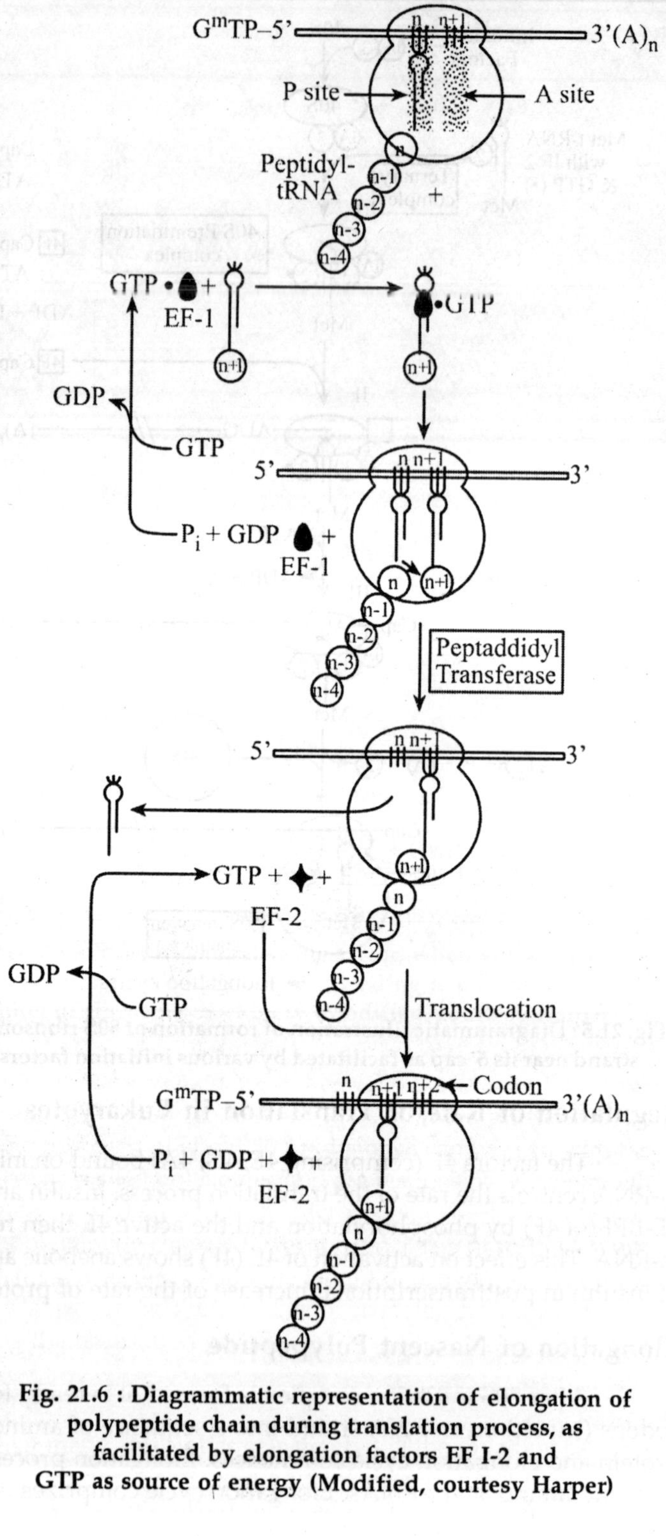

Fig. 21.6 : Diagrammatic representation of elongation of polypeptide chain during translation process, as facilitated by elongation factors EF I-2 and GTP as source of energy (Modified, courtesy Harper)

the ribosome on m-RNA to bring the next triplet (codon for n_2 amino acid) on the A site. Elongation factor 2 with GTP is responsible for this **translocation,** where hydrolysis of GTP provide energy for shifting of the growing peptide chain from A site to P site and simultaneous translocation of ribosome to bring the next codon on m-RNA for n_2 amino acid on the A site as illustrated in the Fig. 21.6.

- Now A site of the ribosome is free and is on next codon for n_2 amino acid on m-RNA. The elongation factor EFI brings the n_2 amino acyl-t-RNA complex at A site to begin next elongation cycle.
- Elongation of nascent polypeptide chain takes place as the **ribosome moves** on m-RNA strand to next codons, one by one to bring them on A site and amino acids are successively added in the growing polypeptide chain in a specific sequence as directed by genetic code on m-RNA. The ribosome move from 5′ to 3′ direction on m-RNA like a bead moves on thread. As elongation cycles are repeated, the nascent polypeptide chain grows.
- **Energy requirement for one peptide bond formation**—Two high energy phosphate bonds as pyrophosphate are utilized for the activation of amino acid and two GTP are hydrolyzed in one elongation cycle. So total of 4 high energy bonds are used in one peptide bond formation during protein synthesis. Translation process occurs rapidly and a mammalian ribosome can form about 400 peptide bonds per minute during protein biosynthesis.

Termination of Translation Process

The termination of polypeptide synthesis occurs when the genetic code on the m-RNA is completely translated into a specific polypeptide chain and a **nonsense** or a **terminating codon** of m-RNA appears on the A site of ribosome. There is no t-RNA to recognize this termination codon. A **releasing factor recognizes** this termination signal on A site (Fig. 21.7) and along with hydrolysis of GTP, in the presence of enzyme peptidyl transferase, **dissociates** the nascent polypeptide chain from the ribosome, which also separates into **40S** and **60S** subunits as illustrated in the Fig. 21.7.

Post-translational Processing

Most of the nascent polypeptide chains synthesized by the translation undergo extensive post-translational modifications before they become fully functional protein molecules. These modification include—

1. Proteolytic cleavage of N-terminal methionine (I^{st} amino acid at the initiation complex), since most of the time it is not needed in the protein molecule.
2. Activation of precursors to active protein molecule e.g. preproinsulin to proinsulin and then to **active insulin** (Chapter 8) by specific proteolysis.
3. Activation of zymogens to active enzymes (e.g. pepsinogen, trypsinogen etc to active pepsin, trypsin) by specific cleavage of extra small polypeptides by specific proteolysis.
4. Procollagen is converted to collagen by hydroxylation of proline and lysine amino acids which then crosslink to form fine mesh between three collagen polypeptide strands to impart greater strength to the collagen fibers.
5. Covalent modifications by acetylation, phosphorylation, glycosylation etc are common modifications before a protein becomes functional.

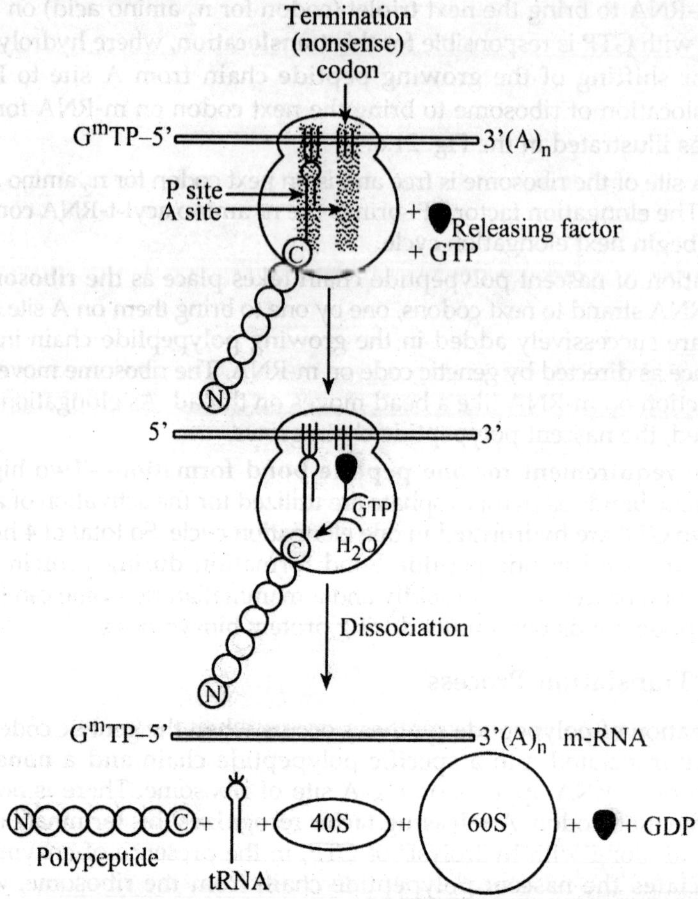

Fig. 21.7 : Termination of protein synthesis by the releasing factor, when a nonsense termination codon of m-RNA appears on the A site of ribosome, leading to release of nascent transcript (mRNA) and dissociation of ribosome machinery (Modified, courtesy Harper)

Polyribosomes

Many ribosomes can translate the genetic code on m-RNA strand **simultaneously.** A single m-RNA strand with 4 or 5 ribosomes, at about **80 bp distance** can be seen with the help of an electron microscope. These are called **polysomes** and are responsible for the rough endoplasmic reticulum appearance viewed by electron microscope. The first ribosome assembly with the **longest** nascent polypeptide chain is **farthest** from 5′ cap of m-RNA while last one with short polypeptide chain is near the 5′ cap. The length of growing polypeptide chain on the ribosome machinery depends on the number of triplets, the ribosome has travelled on m-RNA strand during the translation process.

Inhibitors of Protein Synthesis

The discovery of the fact that many **antibiotics selectively** kill the invading bacteria by inhibiting protein synthesis in microorganisms only and would not affect the protein synthesis in the mammalian cells, has revolutionized the medical practice. Currently the modern medical practice depends heavily on the use of antibiotics to check infections and cure several diseases, many of which previously were difficult to be cured and resulted in fatal consequence, for instance tuberculosis, typhoid etc.

Antibiotics inhibit protein synthesis in bacteria because the prokaryotic ribosomes are smaller in size (70S) and comprise different types of RNAs than the mammalian 80S ribosomes. Many effective antibiotics were found to interact specifically to inhibit protein synthesis in 70S ribosomes of bacteria only and would not affect the mammalian protein synthesis in 80S ribosomes. The most useful antibiotics, by inhibiting protein synthesis in 70S ribosomes of microorganism, arrest their multiplication and growth, causing death of the bacteria. Since these antibiotics do not affect mammalian protein synthesis, they are not toxic to human beings and thus can safely be used as drugs to check infections and cure diseases.

Some important antibiotics used as drugs along with their mechanism of action to cure specific diseases are listed in the Table 21.2.

Table 21.2 : Important antibiotic drugs with their mechanism of action to cure disease states

S. No.	Antibiotic	Inhibit prokaryotic protein synthesis by	Used as drugs in Infection/ diseases
1.	Tetracyclin	Inhibit t-RNA binding to ribosomes	Broad spectrum bactericidal agent
2.	Chloramphenicol/ Chloromycitin	Inhibit binding of 30S ribsome to m-RNA	Typhoid
3.	Streptomycin	Compete with tyrosinyl t-RNA to bind at m-RNA causing premature release of polypeptide	Broad spectrum anti-biotic & bactericidal agent
4.	Erythromycin	Translocation of ribosomes is inhibited by inactivating enzyme peptidyl transferase.	Upper respiratory tract infections.

Some antibiotics inhibit protein synthesis in both prokaryotic and mammalian ribosomes and they are toxic to humans. However these antibiotics have been used in research projects as experimental compounds.

Experimental antibiotics are helpful in study of mechanism of various stages of protein synthesis and regulation of the gene expression as described here under—

1. **Puromycin**—It blocks protein synthesis in both 70S and 80S ribosomes. Its structure is similar to tyrosinyl t-RNA and thus competes for ribosome binding at A site. This results in premature release of polypeptide from the ribosomes.
2. **Actinomycin D and mitomycin** inhibit protein synthesis both in prokaryotic as well as in mammalian ribosomes. These drugs inhibit DNA replication and transcription and thus are used as **anticancer drugs** in the treatment of certain cancers.
3. Amantin toxin from a kind of mushroom is highly poisonous in humans as it inhibits RNA polymerase II in eukarayotes and blocks transcription of m-RNA.

4. Diphtheria toxin inactivates elongation factor 2 and inhibits protein synthesis in 80S ribosomes in humans leading to fatal consequences.

Mutations in DNA Cause Genetic Diseases

Any alteration or change in the nucleotide sequence of genetic code of DNA (genes) results in **mutations.** There are many environmental and chemical factors which can cause mutations in normal genome and are collectively known as **mutagens.** Mutations on DNA are reflected in impairment (defective) of DNA replication, transcription and finally in translation process, resulting in either insufficient synthesis of specific protein or in abnormal protein synthesis which is nonfunctional. This results in inborn genetic disorders that are transmitted to the offsprings (heriditary).

Types of mutations—There are two major types of mutation known as point mutations and frame shift mutations. These are described here under.

1. **Point mutations**—The point mutations occur by a **single base alteration** that may be due to transition or transversion as described below—
   (a) **Transition**—In this case a purine is replaced by another purine base or a pyrimidine is replaced by another pyrimidine base in the genetic code on DNA.
   (b) **Transversions**—These are due to the change of a purine base into a pyrimidine base or vice versa.
2. **Frame shift mutations**—These occur by either **insertion** or **deletion** of a base pair from the genetic code on DNA, thereby disrupting the whole genetic code and results in **altered reading frame** of the sequence of nucleotides on DNA (genes).

Point mutation manifestations—Change in genetic code by a single base alteration (point mutation) may result in missense consequences, when they are translated into amino acids during the protein biosynthesis. Abnormal hemoglobin synthesis illustrates well the consequences of single base point mutation occurring in the genes encoding α and β globin polypeptide chains. These point mutations result in the substitution of single amino acid by another to synthesize an abnormal hemoglobin with following consequences.

1. Acceptable missense mutations—A mutation of a single base change in the genetic code in a "triplet" (codon) at the 3rd **nucleotide** may be **acceptable** regarding the **function of a protein,** with no severe clinical manifestations. For example **Hemoglobin Hikari** (HbH), seen in 2 famililes of Japanese people, transport O_2 normally and individual has healthy life. This type of mutation and its consequence is shown in Table 21.3.

2. Partially acceptable missense mutation—This single base mutation is best characterized in **hemoglobin S** of the congenital disorder **sickle cell anaemia.** In this genetic disease, the mutation occurs in the β chain gene encoding hemoglobin in one codon at 2nd nucleotide, resulting in replacement (incorporation) of valine at 6th position of β chain (Table 21.3) in place of glutamate residue causing production of an abnormal hemoglobin S (HbS). This HbS transport O_2 normally and is partially acceptable, but on exposure to certain **conditions, sickling of RBC** occurs resulting in excessive hemolysis causing sickle cell anaemia (Chapter 33).

3. Unacceptable missense mutation—This mutation (single base change) occurs in the I^{st} nucleotide of a **triplet** in gene encoding α chain of hemoglobin resulting in unacceptable missense consequence. This mutation results in the replacement of histidine at 58th position of α chain by tyrosine residue causing abnormal met hemoglobin (HbM) synthesis, in which ferrous ion is oxidized to ferric iron and it is unable to transport O_2 resulting in death of the infant.

The single base point mutations resulting in abnormal hemoglobin synthesis with single amino acid substitutions are summarized in the Table 21.3.

Table 21.3 : Point mutations causing Hemoglobinopathies

Mutations	Change in hemoglobin molecule	Amino acid change	Mutation in codon
Acceptable missense	HbA-β chain	61 lysine	AAA or AAG
	↓	↓	AAA → AAC; AAG → AAU
	Hb hikari β chain	Aspargine	AAU AAC
Partially acceptable missencse	HbA-β chain	6 glutamate	GAA or GAG
	↓	↓	↓ ↓
	HbS	Valine	GUA GUC
Unacceptable missense	HbA-α chain	58 histidine	CAU or CAC
	↓	↓	↓ ↓
	HbM (fe^{+++})	tyrosine	UAU UAC

Frame Shift Mutations

These mutations result from deletion or insertion of a nucleotide in DNA genetic code and cause completely altered reading frame of the sequence of codons on m-RNA strand. This will result in either a block in protein synthesis or an abnormal protein may be synthesized which is biologically nonfunctional.

A deletion or insertion of a single nucleotide in the gene would result in the shifting of reading frame in m-RNA, so that a **garbled** protein, with completely irrelevant amino acid sequence would be produced. This is because the ribosome machinery does not recognize the altered frame as there is no punctuation between the codons.

Frame shift mutations may produce a termination codon in the middle of the coding region (gene), that would cause premature chain termination, e.g. in case of β - thalassemia.

Mutations usually result from faulty DNA repair mechanisms. Statistically, out of every 10^6 cell division, one mutation occurs. A mutation in the germ cell will be transmitted to the offspring and is the major cause for most of the congenital genetic disease.

Carcinogenic Mutations

The mutations sometimes may alter the regulatory control on cell division, resulting in unrestrained cell proliferation leading to a cancerous growth. Any substance (mutagen) causing mutation that cause activation of oncogenes results in malignant tumors (details in the Chapter 39).

❑❑❑

22

REGULATION OF GENE EXPRESSION

The whole genetic information characteristic of a specie is present in every cell of higher mammals and the **entire genome** is practically identical in each somatic cell on 23 pairs of chromosomes. However expression of genes is different in different tissues, and usually only a fraction of **entire genome** is expressed in a cell at any given moment. Rest of the genes remain dormant (repressed) for long periods, some times may be for whole life. To explain this regulation of gene expression, for instance, the genes responsible for insulin synthesis is present in nuclei (genome) of all the cells of body, but is expressed only in specialized β cells of islets of langarhan's of pancreas and not by cells of other organs like kidney or liver.

Regulation of gene expression is extremely important during growth, development and tissue differentiation and organogenesis during fetal growth. This gene expression control is absolutely essential during ontogenesis, cell differentiation and differentiation of cellular components.

Gene Expression is Responsive to Environmental Changes

In order to adapt itself and to survive in the ever changing environmental conditions, the genetic expression of an organism has to be regulated and should be responsive to external signals. Also in order to conserve energy and nutrients, the gene expression of an organism is regulated according to the immediate metabolic need of the cell.

In broad terms there are only two types of regulation of gene expression. These are—

1. **Positive regulation** is when gene expression is quantitatively increased by a specific regulatory signal or **inducer.**
2. **Negative regulation** is when gene expression is decreased by the presence of a regulatory element or **repressor.**

Constitutive and Inducible Genes

A **constitutive gene** is that which is expressed at a reasonably **constant rate,** as the expression is constantly needed all the time to synthesize certain functional proteins and enzyme for proper functioning of the cell. These genes are not much subjected to regulation. For example the enzymes of citric acid cycle have to be produced all the time, at a constant rate for the oxidation of nutrients and to provide energy to carry on the life processes.

Inducible Genes—The inducible genes are those, whose expression is increased in response to the presence of a specific **inducer or activator** and result in a positive response. Usually an inducible gene has a low basal rate of transcription. However some genes in contrast have higher basal rate of transcription and their expression has to be inhibited (-ve response) or **suppressed** by a **repressor molecule.**

Regulation of Gene Expression in Prokaryotes

The regulation of gene expression is best characterized in prokaryotes and it provides models for the study of mammalian gene expression. To illustrate the control of gene expression in prokaryotes in response to the external signals, some salient features should be considered as described here under:

Operon

The concept of **operon** was proposed by Jacob and Monod (1961) and according to this the genes encoding for a **set of enzymes** involved in a particular metabolic cycle are called **structural genes** and are present adjacent to each other on DNA, so that their expression can be **regulated simultaneously** by an **operator gene** present nearby on the promoter region. The structural genes along with the operator is called **OPERON** (Fig. 22.1).

One Cistron – One subunit Concept

Many functional enzymes exist as dimers or tetramers, or as multienzyme complex and catalyze key regulatory steps of a metabolic cycle. The **smallest unit** of **genetic information** on DNA encoding one subunit of a multienzyme is known as **cistron** and accordingly **one cistron-one subunit** concept was proposed in place of one gene-one enzyme hypothesis.

A polycistronic complex on DNA (Chromosome) thus transcribes a m-RNA with genetic code for simultaneous synthesis of many subunits of a multienzyme complex to produce a **functional enzyme.** As the polycistronic genetic code lies in vicinity of each other, their expression can be regulated by a single operator (regulatory) gene (at promoter site) simultaneously in response to an external signal.

Lac Operon Model

Jacob and Monod in 1961 (nobel prize 1965) introduced **operon concept** to explain the regulation of prokaryotic gene expression in E-coli bacteria to metabolize lactose and named it as **Lac operon.** According to this model, Lac operon comprises **three structural genes** encoding for three enzymes required for lactose metabolism. These enzymes are **β–galactosidase** (gene Lac z) to hydrolyse lactose, **permease** enzyme (gene Lac Y) for the transport of lactose inside the bacterial cell, and **transacetylase** (gene Lac A), the function of which is not clear. Near to these three structural genes (Lac Z, Y and A) on DNA, an **operator** gene is present on promoter region as shown in Fig. 22.1A a and together they constitute **Lac operon.** A third regulatory gene (inhibitor I) is present at some distance away from the lac operon on the same chromosome (Fig. 22.1A).

Repression of Lac Operon

The regulatory gene (I) is a constitutive gene in E coli and is expressed at a constant rate to transcribe m-RNA for the synthesis of a specific protein factor called **repressor** molecule. Active form of this repressor is **tetrameric** (4 subunits) and has affinity to bind to **operator** at promoter region (Fig. 22.1A), and after binding on promoter, it **inhibits expression of structural genes** (repression). This is because the operator gene lies between structural genes and promoter region, where the **RNA polymerase** binds to start transcription process. The **binding** of **repressor on operator blocks** the RNA polymerase to move on DNA to get access to structural genes for the transcription and thus **gene expression is repressed** (inhibited).

Thus **Lac operon** remains dormant (no expression) normally as E coli bacteria grow mainly on glucose as fuel source and enzymes for lactose metabolism are not needed. When addition of glucose was stopped and instead lactose is added to the culture media, for some time the bacteria were starved as glucose is exhausted and c-AMP concentration was increased. The c-AMP along with CAP factor then bind on promoter region of lac operon, that facilitates the binding of RNA polymerase enzyme at the promoter (Fig. 22.1B).

Derepression of Lac Operon

The presence of the **inducer** (lactose or its metabolite), which has affinity to bind with the repressor subunits and the tetrameric repressor molecule is degraded and now it cannot bind to the operator, which become free and **derepressed.** With this block being removed, RNA polymerase can move from the promoter site on DNA and gets access to the coding region of the structural genes Z, Y and A and transcription begins. The m-RNA so produced start synthesizing enzymes required for the lactose metabolism as illustrated in the Fig. 221B.

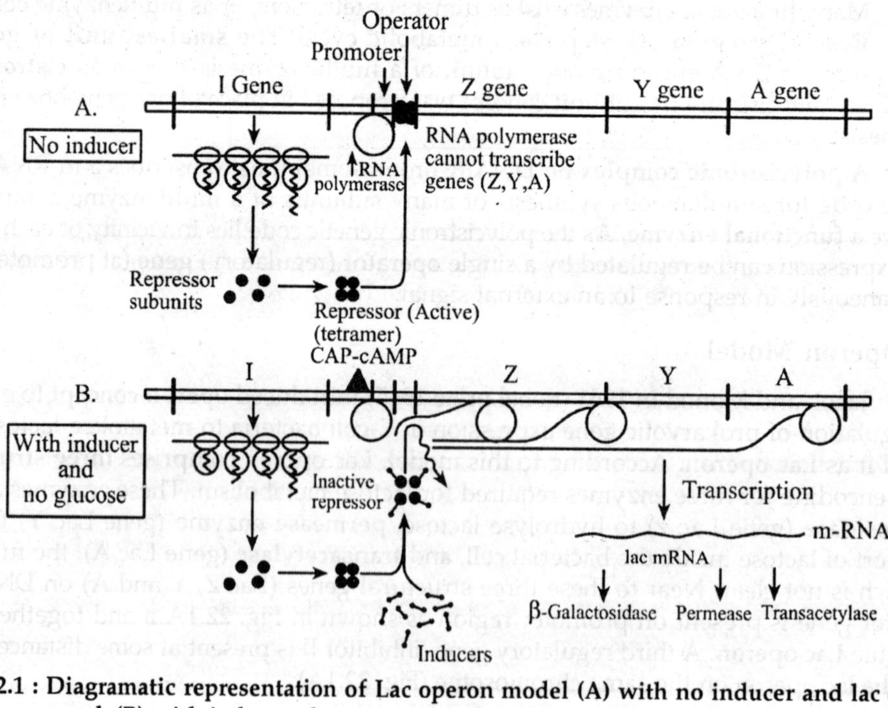

Fig. 22.1 : Diagramatic representation of Lac operon model (A) with no inducer and lac operon is repressed. (B) with inducer, derepression of lac operon occurs and transcription starts

Thus inducer lactose has positive effect and enhance gene expression of lac operon to enable *E-coli* to start lactose metabolism in order to survive in the changed environmental conditions.

Regulation of Gene Expression in Eukaryotes

Regulation of eukaryotic gene expression is very complex and the mammalian cell employs a variety of mechanisms at various stages, such as transcription, processing of primary transcript, post transcriptional modification of m-RNA, its transport to cytoplasm etc. Some important gene expression regulatory mechanisms are described here.

Gene expression is highly selective in mammalian cells and differ from tissues to tissues.

Enhancers and Repressor DNA Elements

Eukaryotic gene expression is increased or decreased by certain DNA elements (regions) called **enhancers** and **repressors** and their interaction with certain specific protein factors. These enhancers and repressor DNA elements can exert their influence on gene expression even when present thousands of base pair distance from the promoter region, where the transcription commences.

The enhancer and repressor DNA elements are responsive to external signals (inducers), hormones and II messengers and coordinate the gene expression to the metabolic need of the cell (Details in chapter 20). Several specific motifs mediate the DNA-element protein interactions, and the specific binding of **hormone receptor complex**, or binding of II messengers such as c-AMP, CAP etc at the correct regions of DNA to regulate gene expression. These specialized motifs are—

1. Helix turn helix motif facilitates binding of homeobox proteins.
2. Zinc finger motifs are specific for binding of steroid hormone-receptor complex on hormone sensitive elements (HSE) of DNA.
3. Leucine zipper motif for binding of CRE etc.

Repression of Gene Expression in Eukaryotes

In eukaryotes the synthesis of many key regulatory enzymes and other functional proteins is controlled at **gene level** by regulating gene expression according to the need of the cell. For example, the synthesis of key regulatory enzymes **ALA synthase (Heme synthesis)** and **HMG CoA reductase** of **cholesterol biosynthesis** is controlled by negative feed back regulation of gene expression as described here—

The gene encoding for ALA synthase is a **constitutive** gene and is expressed at reasonably constant rate as it is necessary for hemoglobin synthesis and remains **derepressed** normally. The regulatory gene (heme – R) transcribes synthesis of an **aporepressor** (inactive) which cannot bind on operator at promoter site (Fig. 22.2A). Thus operator remains derepressed and will allow the RNAP II to move on DNA to continue transcription of structural genes for the synthesis of **ALA synthase** and other enzymes required for heme synthesis.

But in the case of excess of heme accumulation, more than is required for heme containing enzyme production in liver, the heme feed back inhibits its own biosynthesis at gene level. In excess, heme reaches to nucleus and acts as a **corepressor,** as it has affinity to bind on aporepressor and convert it to an **active repressor.** The active repressor binds to **operator gene** at the promoter region and blocks the movement of RNAP II and thus stops transcription of structural genes responsible for the synthesis of key enzyme ALA synthase needed for heme biosynthesis as shown in the Fig. 22.2B. This is an example of **negative effect** on **gene expression** by repressor activated molecule by a feedback mechanism.

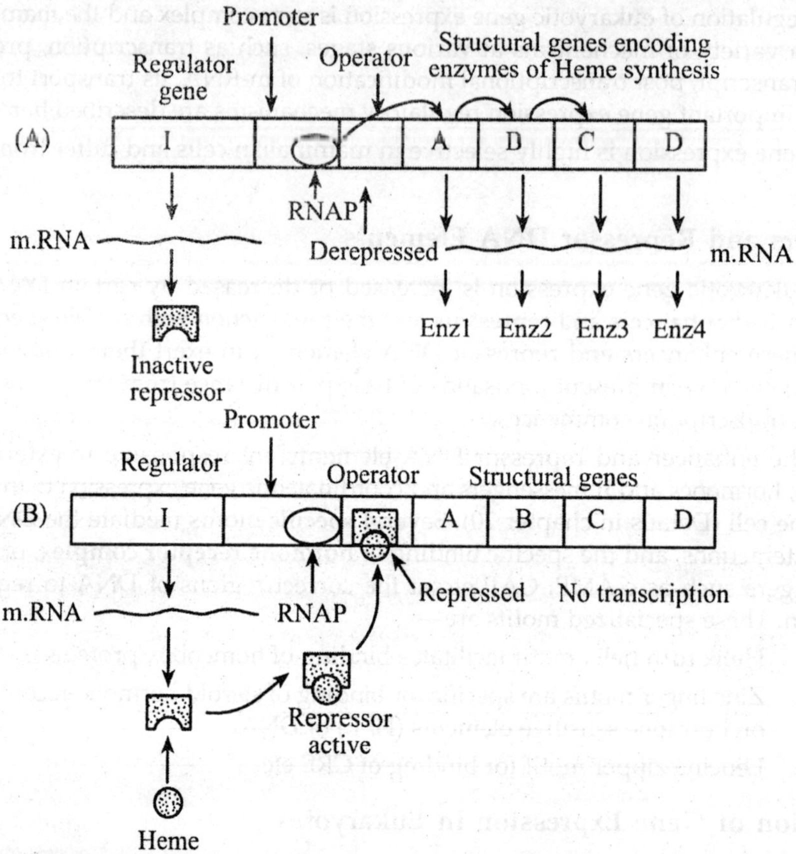

Fig. 22.2 : Repression of genes encoding the ALA synthase enzyme by excess of heme (corepressor) by feed back inhibition (B)

Gene Amplification

Eukaryotic gene expression is increased several folds during development by the process known as **gene amplification.** For instance during development in the **metozoans,** there is sudden increased demand for ribosomal nucleic acid, t-RNA and m-RNA for increased protein requirment. To meet this abrupt increase demand of RNAs and proteins, the number of genes from one gene increases several folds by gene amplification. For example at the time of oncogenesis in fruitfly (Dorsophila), the amplification of preexisting genes coding for eggshell proteins (chorionic proteins) occurs several fold to provide for increased protein synthesis. This amplification of gene is probably achieved by repeated intiations at the ori, resulting in multiple replicating DNA strands in a big replicating bubble, which is illustrated in the Fig. 22.3, as seen by the electron microscope.

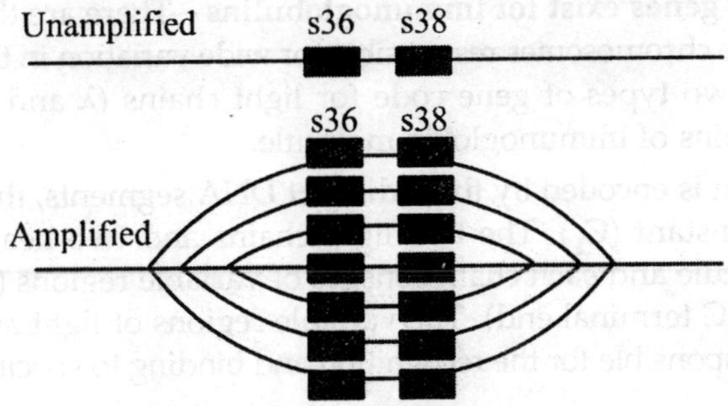

Fig. 22.3 : Diagrammatic representation of gene amplification by repeated initiations at the ori in DNA replication bubble

Gene amplification in response to certain **drugs** have been observed in human also. Recent evidence shows that eukaryotic gene amplification occurs in response to anticancer drugs, for instance **methotrexate.** This drug acts by **inhibiting** enzyme **dihydrofolate reductase** enzyme needed essentially in rapidly multiplying cancerous tumors, for activation of H_2 folate to **tetrahydrofolate** (THF). Active THF is required for rapid synthesis of deoxy nucleotides needed for enhanced rate of DNA replication in cancerous tumor.

It has been observed that prolonged use of drug methotrexate resulted in development of resistance against the drug by the malignant cells. The cancerous cells develop resistance and restart multiplying rapidly by increasing **number of genes** encoding for H_2 folate reductase enzyme to several folds by gene amplification and the anticancer drug becomes ineffective.

Oncogene amplification—Some oncogenes are also observed to be amplified several folds during cancerous growth in certain fast growing malignant tumors.

Gene Switching Phenomenon

Gene switching is an important phenomenon observed, for instance during hemoglobin synthesis. In gene **switching phenomenon** one gene is **switched off** during development and another closely related gene takes over its function. For example during embryonic development, hemoglobin has **two zeta and two eta** polypeptide chains initially. Then during 6 months of intrauterine life, this is changed to **fetal hemoglobin** with two α and two γ chains. After birth HbF is replaced by HbA with α_2, β_2 terameric form. All these changes occur by gene switching mechanism and is necessary for the proper transport and sufficient supply of O_2 to the growing infant (details in Chapter 33).

Selective Gene Rearrangement

Production of wide range of antibodies (approximately 10 billions) by our immune system in response to different antigen stimulation is explained by formation of **functional gene combination** by selective gene rearrangement. These **active genes** so formed are responsible for the synthesis of **immunoglobins** with a specific hypervariable region for recognition and binding to the specific antigens to destroy them.

Three types of genes exist for immunoglobulins—There are three unlinked types of genes occuring on the chromosomes responsible for wide variation in the immunoglobulin molecular structure. Two types of gene code for light chains (λ and κ) and one family encoding for heavy chains of immunoglobin molecule.

Each light chain is encoded by three distinct DNA segments, the variable (V_L), the joining (J_L) and the constant (C_L). The two light chains and two heavy chains form an immunoglobulin molecule and each chain consists of variable regions (at N terminal end) and a constant region (C terminal end). The variable regions of light and heavy chains of immunoglobulin are responsible for the recognition and binding to specific antigen (Chapter 32) to destroy it.

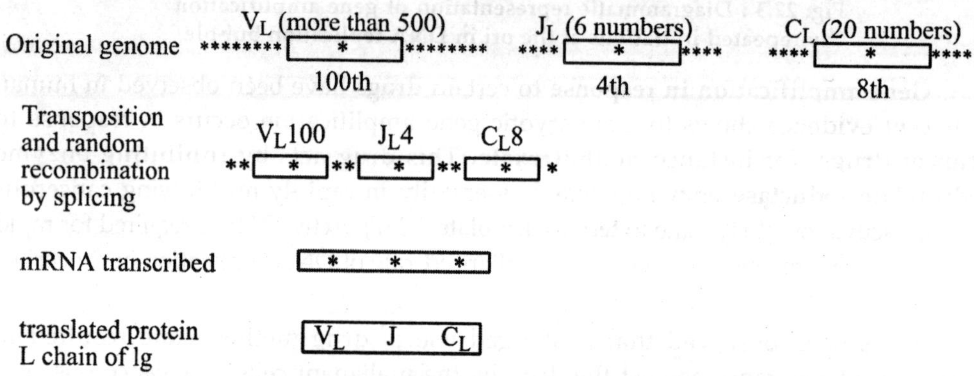

Fig. 22.4 : Selective gene rearrangement in response to stimulation by specific antigen to synthesize a specific immnoglobulin (Ig)

A mammalian genome contains about 500 V_L DNA segments, around 6 J_L DNA segment and 20 C_L regions on same chromosome for specific light chain synthesis. During course of differentiation of B lymphocytes, one of V_L segment (out of 500) is brought closer to J_L and C_L segment by **gene rearrangement** and are responsible for the synthesis of a wide range of immunoglobulin. So by innumerable combinations of V_L (out of 500), J_L (out of 6) and C_L (out of 20), the immune system of our body can produce billions of different antigen specific antibodies to bind and destroy different antigens. (Fig. 22.4). This is the basis of our body's powerful immune system against infection and diseases.

❑❑❑

23

RECOMBINATION DNA TECHNOLOGY

Genetic engineering and recombinant DNA technology is the most advanced practice in the field of genetics and it has revolutionized the application of molecular biology in the medical sciences, with a great impact on current clinical practice.

Genetic engineering and recombinant DNA technology is the manipulation of a DNA segment to produce a **chimeric** or **recombinant DNA** by incorporating a particular genetic information (genetic code) from DNA of one organism into the DNA from different organism.

Recombinant DNA technology in nutshell involves the incorporation of a human DNA segment with a specific genetic information, by artificial means in laboratory, into a prokaryotic circular DNA to produce a **chimeric DNA**. This recombinant DNA is then allowed to multiply (cloning) in the bacteria to generate a large population of human DNA (gene), for the use in medical practice or for other scientific purposes.

Biomedical Importance

1. Synthesis of proteins for medicinal use—With the advancement of recombinant DNA technology, now it is possible to synthesize artificially in laboratory, large quantities of a variety of human proteins. For instance hormones such as insulin, growth hormone, interleukin-2, antihemophillic globulins, interferon etc are produced commercially with least contamination to enable their use, safely as therapeutic agents in various diseases.

2. Vaccine production—A variety of vaccines with least contamination are produced by recombinant DNA technology in large amounts, for the prevention of killer diseases such as hepatitis B.

3. Commercial production of specific proteins (antibodies) required in ELISA technology for the diagnosis of HIV infections, AIDS and other dreadful diseases.

4. Production of probes and markers for early screening of inherited congenital diseases, for their treatment and early management to prevent irreversible damage to the body organs, particularly to CNS and brain.

5. Tumor markers and probes are now available by recombinant DNA technology for early diagnosis of cancers and oncogene activation. This is important to control cancerous growth at an early stage by proper treatment to prevent clinical complications in order to save life.

6. Gene therapy—Inborn errors of metabolism and genetic diseases can be detected by prenatal screening of amniotic fluid (amniocentesis). The affected cases can be treated now by introducing a normal gene for the genetic abnormality, with the advent of DNA recombinant technology. The gene therapy is going to become a standard treatment procedure in coming years in the mangement of genetic diseases before birth.

7. Application in forensic medicine by DNA finger printing—The recombinant DNA technology has great application in forensic medicine and medicolegal cases. With this technology it is now possible to match DNA of a suspect with that of DNA from the samples of blood, hair, tissue etc. obtained from the site of crime to identify the culprit. Also the disputes of parenthood of children can be settled by DNA matching in medicolegal cases.

8. Application in industries—Several enzymes (proteins) can be produced commercially by recombinant DNA, for use in various industries for the production of cheese, food additives, detergents etc.

9. Agriculture applications—This technology has proved quite fruitful for higher crop yields of better qualities. By recombinant DNA, plants are being developed to survive in extreme draught conditions and with more resistance to infections and common diseases.

10. Transgenesis—By the genetic engineering, it is now possible to alter the germ cell of animals with better characterstics to be passed on to the offspring. A certain percentage of the altered germ cells injected into the fertilized ovum (e.g. of mouse) is incorporated in the genome of the newborn animal and the process is called **transgenesis.** These transgenic animals are proving to be useful in biotechnology research to study the gene expression during development, to correct genetic abnormality, to produce better animal breed with increased milk production or with other better qualities etc.

Process of DNA Recombination

The technique of recombinant DNA involves the following key steps—

1. Isolation of DNA segment (gene) to be cloned.
2. Incorporation of isolated DNA segment into an appropriate **vector** to produce **chimeric DNA.**
3. Induction of chimeric DNA into the bacteria for cloning to produce a large population of identical DNA copies.

Restriction Endonucleases

Isolation of a specific DNA (gene) segment requires certain endonuclease enzymes that can cut DNA strand at specific regions. These enzymes are called restriction endonucleases and are referred to as molecular scissors. The restriction endonucleases (RE) are sequence specific and cut DNA from any source, only at the specific nucleotide sequences, like a pair of scissors cut a cloth as desired. More than 400 different sequence specific REs are now available commercially to be used in genetic engineering. These enzymes are named after the bacterial species from which they are extracted, by adding a roman letter specifying the order of their discovery e.g. EcoRI is the first RE isolated from Ecoli bacteria.

Sequence specifications—Restriction enzymes recognize and cut (cleave) a DNA double helix specific sequence, which is generally 4-7 bp long and with **palindrome arrangment** (meaning a word is read same from both the direction e.g. madam). These specific nucleotide sequences on DNA double helix are in form of **inverted repeat** i.e. same sequence in 5′ to 3′ direction on the both strands of DNA as shown in Fig. 23.1.

Table 23.1 : Specificity of restriction enzyme

Enzyme	Source of enzyme	Specific sequence identified cut by the enzyme
Eco RI	Escherichia coli RY 13	G↑AATT C C TTAA↓G
B gl II	Bacillus globigii	A↑GATC T T CTAG↓A
Hind III	Hemophilus influenzae Rd	A↑AGCT T T TCGA↓A
Taq I	Thermus aquaticus	T↑CG A A GC↓T
Hpa I	Hemophilus parainfluenza	GTT↑AAC CAA↓TTG

The EcoRI recognizes and cuts DNA double helix at these specific nucleotide sequences in a peculiar way to leave behind two DNA pieces with **sticky ends** as shown in Table 23.1. Some REs cut DNA at specific sequence to leave behind two DNA pieces with blunt ends but with specific sequences. Human DNA with specific gene (coding) segment may have such specific sequences on both side of a gene and thus can be cut by specific restriction endonuclease, to obtain a specific human gene as illustrated in the Fig. 23.1. The DNA segment so cut is then isolated by **electrophoresis** and **Southern blot technique** (details will be described later in the chapter).

Selection of Vector

The main technique for recombinant or **chimeric** DNA production is to introduce the cut piece of human DNA fragment (containing specific gene) with the RE as described above into a **carrier** known as **vector**. Commonly used vectors are **plasmids** or **bacteriophages**.

Plasmids—These are circular double helix DNA seen in bacteria, where their function is to confer resistance against natural **antibiotics** to the host bacteria. The plasmids have many properties that make them highly suitable for use as vectors in recombinant DNA technology. These properties are—(1) Plasmids can easily be exchanged between different microorganisms and bacteria, (2)They **replicate** and multiply **independently** from bacterial DNA and (3) **Plasmids** can easily accept a small fragment of foreign DNA upto 6-10 kbp (1 kbp length = 1000 base pairs).

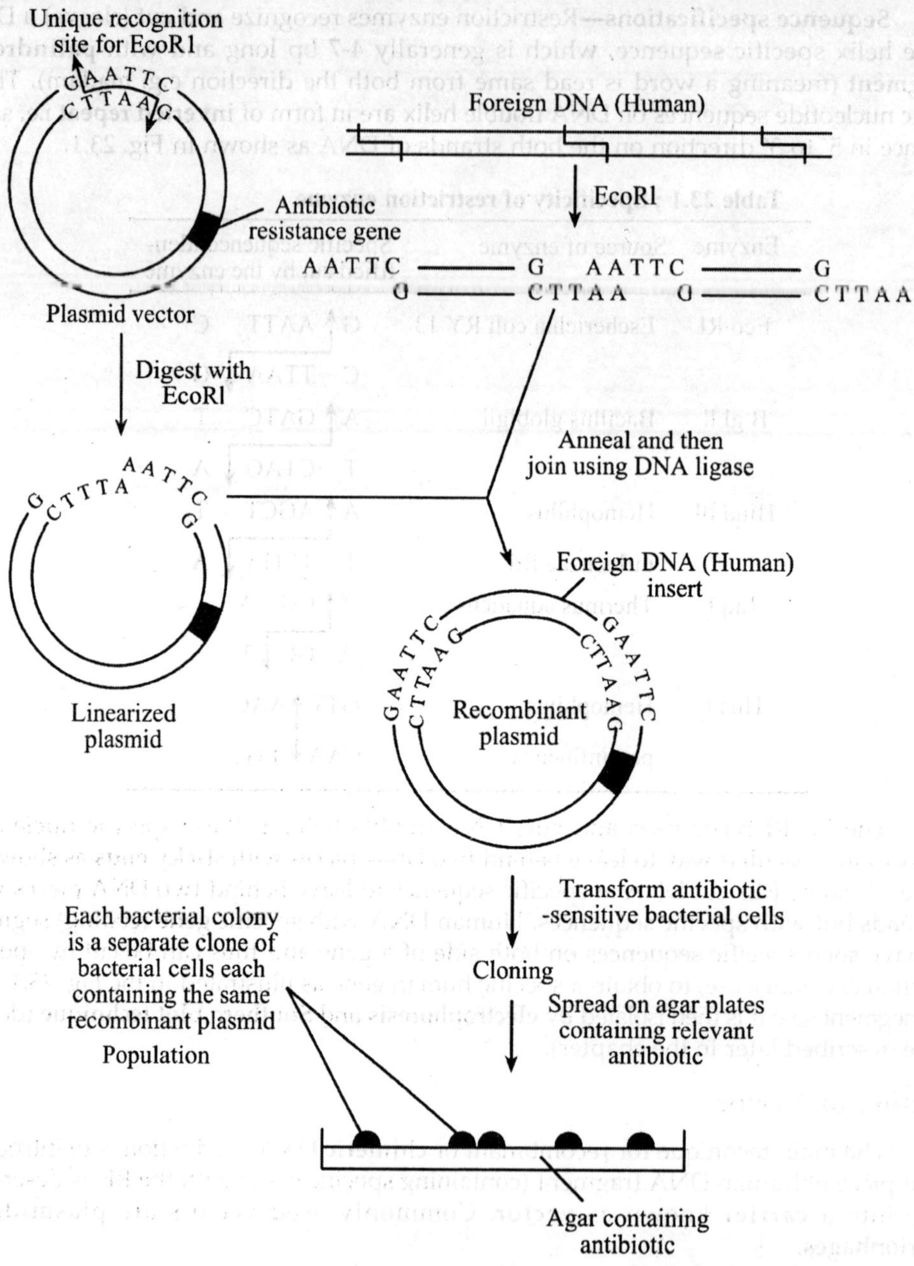

Fig. 23.1 : Major steps involved in DNA Recombination and in DNA cloning using a plasmid vector

Bacteriophages—These usually have linear DNA double helix and can accept large fragments of DNA from other organism (10-20 kbp long). **A phage DNA** for example can be used as **vector** as they have 40% extra genome, that can be replaced by the human DNA fragment to produce a chimeric DNA.

Preparation of Chimeric DNA —

The procedure for production of a chimeric DNA is illustrated in Fig. 23.1. The key step in recombinant DNA technology is that both human DNA and the vector circular duplex DNA (plasmids) are acted upon by **same restriction enzyme** (e.g. EcoRI), that recognizes specific palindrome nucleotide sequences on both types of DNA. Thus both DNAs are cut at specific sites leaving behind similar sticky ends with **complementary base sequences** as shown in the Fig. 23.1.

Insertion of human DNA fragment into vector DNA—The vector DNA and human DNA cut pieces with complementary nucleotide sequences at the sticky ends are then incubated together at proper temperature, so that **annealing** takes place between the complementary sequences of sticky ends of human and vector DNA fragments to form a hybrid DNA. The **DNA ligase** enzyme is then allowed to act on **hybrid DNA molecule** to seal the sticky ends of two DNA molecules by forming diphosphoester linkage between vector and inserted human DNA fragment to produce the **chimeric DNA** as shown in Fig. 23.1.

Cloning of Chimeric DNA

Cloning is the production of large population of identical DNA molecules or cells (organisms) from a common ancestor molecule or cell respectively in vitro. By cloning in laboratory, it is possible to produce a large number of identical DNA from the chimeric or recombinant DNA produced as described above for the use in medical practice.

For amplification of chimeric DNA by cloning technique, the vector carrying human DNA fragment (gene) is introduced into bacteria. The bacteria are then allowed to grow in proper conditions in laboratory. As bacteria grow, the vector with chimeric DNA containing human gene also multiply to generate large population of recombinant DNA, since it can replicate independently.

Transfection—It is a process by which the plasmid with chimeric DNA is introduced into E coli bacteria by incubating them in a hypertonic medium containing calcium for a few minutes. By this the calcium channels on bacterial surface opens and the vector permeate through them into the host bacteria. The bacteria are then allowed to grow in proper culture media containing amphicillin and tetracyclin, so that all other wild bacterias are destroyed by the antibiotics and only E coli containing recombinant DNA (in plasmid) will grow.

The bacterial colonies with plasmid are then selected with proper care and grown in culture media containing a sublethal dose of bacterial protein synthesis **inhibitor** to lower bacterial growth, so as to allow large increase in plasmid (with chimeric DNA) replication.

The large number of plasmid DNA so produced are then separated and incubated with same restriction enzyme EcoRI then will cleave inserted human DNA fragments from the plasmid DNA. In this way large copies of human genes are generated by cloning method, to be used in medical and clinical practice.

Commercial synthesis of human proteins is now possible if the E coli bacteria carrying the plasmid inserted with recombinant human DNA (e.g. gene encoding insulin) is allowed to grow in good culture medium. In the E coli bacteria the desired human protein (Insulin) will also be synthesized along with bacterial proteins in the culture. The insulin hormone so produced is then isolated for medicinal use.

Cloning—Amplification of Chimeric DNA

The term cloning in the recent scenario has two broad meanings.

1. **Molecular cloning**—When a DNA fragment (gene) of higher organism (humans) is introduced into bacterial DNA to produce chimeric DNA and is cloned into a large population of identical DNA insert (human gene), it is known as molecular cloning or cloning of gene. These genes so produced are then introduced into a **gene library** for their later use in medical practice.
2. **Cloning of living organism**—It is done by manipulating a somatic cell from an animal by genetic engineering technique, in such a way that it grows into an exact duplicate of that animal and is known as cloning of an animal. Classic example of animal cloning is production of **Dolly** the sheep clone by manipulating a somatic cell from the udder of a sheep, grown into an identical clone in vitro.

DNA Finger Printing and RFLP

The nucleotide sequences of the **intron** portion (noncoding regions) of DNA molecule, that do not code for any protein or characteristic, is different from person to person and is called **polymorphism.** But these sequences are characteristic of **a particular person,** and may match to his parents. These regions of DNA serves as **molecular finger prints** and the DNA samples of a person would match exactly like the finger prints of an individual.

Restriction fragment length polymorphism (RFLP)—This **polymorphism** in the non coding region (introns) of DNA is due to the presence of **variable numbers of tandem repeats** and are known as **hypervariable regions,** that varies from person to person but is characteristic of an individual or of his parents.

These hypervariable regions are located at scattered sites on non coding regions of DNA and are repeated sequentially and are known as **Tandem repeats.** These are specifically **recognized** by certain **restriction enzyme** and when they are allowed to act on DNA of an individual, **characteristic** length of DNA fragments (cut pieces) are obtained. These DNA fragments will **match exactly** with the DNA of samples obtained from the site of **crime** (blood, hair, tissue or even semen stains) of the suspect that are treated with same RE. This would facilitate the identification of the culprit, when his DNA is matched with that of the DNA samples taken from the site of the crime. This technique is called **DNA finger printing** and has **great** application in **forensic medicine** and in **medicolegal cases.**

Polymerase chain reaction (PCR)

The technique **polymerase chain reaction** is utilized in biotechnology for **amplification** of a **DNA** molecule **in vitro** to produce **millions of identical copies** of DNA within few hours. PCR is like a **'Xerox machine'** (photocopier) for gene copying with great speed. The process is very simple and is free from tedious procedures of isolation of a desired DNA, or from any contamination.

Procedure of PCR involves following key steps—

1. **DNA or gene** to be copied is isolated by **Southern blot technique** and the double helix DNA is separated into two strands by heating at 95°C for 15 seconds to 2 minutes.

2. **Annealing of primers**—Two small fragments of 20-30 nucleotide of complementary base sequence of flanking region (at each end of target DNA strand) as primers are added and annealed by cooling to 50° C (Fig. 23.2), which hybridize with the complementary regions of two ends of DNA strands as shown in the Fig. 23.2.

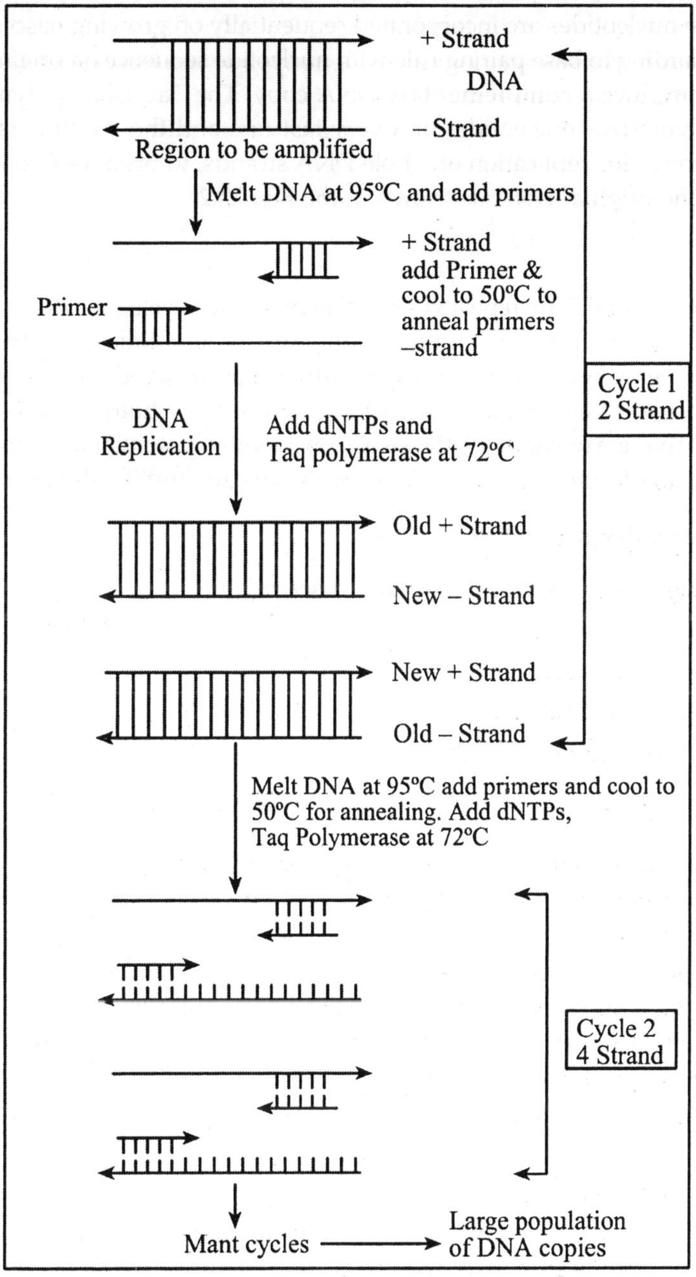

Fig. 23.2 : Diagrammatic representation of polymerase chain reaction (PCR) to produce large population of DNA copies by automation technique

3. **A DNA Taq polymerase enzyme** from bacteria **Thermus acquaticus** (from hot springs) is used for DNA synthesis because it does not gets denatured at high temperature of the reaction medium. This enzyme replicates and synthesizes new DNA strand at 72° C on **DNA primers** annealed at the two original DNA strands (Fig. 23.2) by utilizing high energy nucleotides ATP, CTP, GTP and TTP. The nucleotides are incorporated sequentially on growing nascent DNA strand according to base pairing rule with nucleotide sequence on original DNA strand to produce a **complementary** DNA copy. The Taq DNA polymerase enzyme polymerizes nucleotides at a very fast rate and the the time taken is only 30 seconds for replication of whole DNA strands, to produce two identical copies of the original DNA as shown in the Fig. 23.2.

PCR Cycles

The step 1, 2 and 3 are repeated several time. In each cycle the DNA strand produced are doubled. Thus 20 such cycles would provide for 1 million times amplification of DNA. These cycles are being repeated by **automated instrument** called **tempcycler.** In this way millions of copies of DNA or gene are produced within few hours by this technique at a very fast rate. **Polymerase chain reaction** has considerable potential in medical and clinical applications due to its speed, accuracy and **specificity** in amplification process.

Clinical Applications of PCR

(a) Diagnosis of bacterial and viral diseases.

(b) Medicolegal cases

(c) Detection of genetic disorders and uses in gene therapy.

(d) Detection of cancers

(e) Rapid diagnosis of AIDS

Techniques Used in Isolation of Desired DNA Fragments

To separate and isolate a desired DNA fragment for recombinant DNA technology, DNA probing, transfer and hybridization techniques are used. Some important techniques are describe in brief here—

Southern Blot Technique—The method was developed by the scientist Southern in 1975, (hence the name) and it facilitates detection of a particular DNA segment in the whole human genome, and its isolation for the use in recombinant DNA technology. The key steps of the Southern blot technique is described in brief here.

Procedure—Major Steps

(1) DNA is first isolated from the tissue and it is then fragmented by one or more restriction enzymes.

(2) The cut pieces of DNA are then **separated** on agarose gel by electrophoresis and then treated with mild alkali to denaturate double helix DNA to single stranded DNA.

(3) The single stranded DNA fragments are then blotted over a **nitrocellulose** membrane, by pressing the membrane over the agarose gel electrophoretogram. An exact replica of electrophoretic pattern of the separated DNA fragments in the gel is reproduced on the nitrocellulose membrane.

(4) DNA fragments are then fixed on membrane by baking at 80° C. The desired DNA fragment is then isolated by exposing the membrane to a radioactive labelled complementary DNA strand (c-DNA) probe (from gene library), that will hybridize to only the **complementary** DNA fragment on the membrane.

(5) After thorough washing to remove the excess of radioactive probe, the membrane is exposed to X-ray plate in dark place for a few days. The radiation from the fixed probe in hybridization with DNA segment needed, will form specific band on X-ray plate. The process is known as autoradiography and it facilitates the detection of a desired DNA segment on the nitrocellulose membrane, which can then be isolated by cutting the band for the use in DNA recombinant technology.

Northern Blot Technique—This technique is used for the detection and isolation of a specific RNA (instead of DNA) molecule and is named as Northern blot, just as a simili to Southern blot (which was discovered first) in what is known as laboratory jargons.

The procedure of Northern blot is similar to that used in Southern blot technique and the detection probes used are also same i.e. a radioactive complementary DNA strand (c-DNA). Thus c-DNA would form a RNA – DNA hybrid as they have complementary nucleotide sequence due to base pairing. This facilitates the identification of a specific RNA molecule separated from the tissues for its isolation and use in vitro experiments.

Western Blot Techniques—This technique was developed for the identification and isolation of a specific **protein molecule**. The procedure is similar to the above mentioned techniques, except that heating is avoided in fixing of the membrane as it would denaturate protein. The probes used for detection of a desired protein on nitrocellulose membrane is a radioactive **specific antibody,** that will bind only with the specific protein (antigen) to be separated. The excess of antibody is then thoroughly washed away and the membrane is radiographed as mentioned in the above techniques.

Mapping of Human Genome

Human genome project comprizes collaborative research efforts by different scientists world over to map the sequence of genes located on the 23 pairs of human chromosomes. It began in 1990 and uptill now only a small percentage of whole genome is identified. It is expected that within a decade, the entire sequence of human genome would be discovered.

Gene Library

Gene libraries have been setup world over where a collection of genetic information as specific DNA fragments (genes) are stored and compiled. In these libraries two types of collections are done to be used in genetic engineering and recombinant DNA technology for medical and clinical purposes as mentioned earlier. These are —

(a) **Genome library** containing human DNA fragments with specific genetic information or genes.

(b) **Complementary DNA** (c-DNA) strands are useful as DNA probes. These are copies of DNA sequences and are prepared from m-RNA molecules, that serves as a template to prepare a complementary copy of DNA as catalyzed by enzyme **reverse transcriptase.**

DNA probes—The gene libraries consists of thousands to millions of DNA fragments. A group of DNA single strand fragments labelled with radioisotopes are needed to search the DNA libraries for desired DNA fragment or gene. These probes contain complementary nucleotide sequence to that of specific human genes in the library. The DNA and RNA probes are also helpful in Southern and Northern blot techniques used in DNA recombinant technology. DNA probes are also very important for the detection of many genetic diseases, cancers and metabolic disorders for their early management and treatment.

❑❑❑

24

MINERAL METABOLISM

In addition to the major components of diet i.e. carbohydrates, lipids, proteins and vitamins, minerals are necessary for the growth and development of body, as well as to maintain normal health. Minerals play important functions that are absolutely essential for the existence of the living organisms. For instance calcium is essential for bone and teeth formation and neuromuscular irritability, while iron, as component of hemoglobin, myoglobin and cytochrome enzyme system, is necessary for the life to transport oxygen for biological oxidation, to generate energy to carry on life processes.

Essential minerals are grouped in two major categories—

(a) **Major elements**—There are seven major minerals, whose daily requirement is more than 100 mg and include calcium, phosphate, magnesium, sodium, potassium, chloride and sulphur.

(b) **Trace elements**—Certain minerals are required in minute amounts (< 100 mg/day) to carry out some essential vital functions of the body and are known as **trace elements or microminerals.** In **slight excess,** these minerals produce **toxicity** in the body with severe clinical manifestations. Essential trace minerals include iron, copper, iodine, zinc, fluoride, manganese, cobalt, molybdenum, selenium and chromium.

CALCIUM

Calcium is the most abundant mineral in the body. The total calcium contents of body is about 1 to 1.5 kg, 99% of which is present in bones and teeth, and rest 1% in blood and extracellular fluid of the body.

Biochemical Functions

Calcium is necessary for several vital functions in body that are absolutely essential for life and are as listed below—

1. **Bone and teeth formation**—Major function of calcium along with phosphate is the **mineralization** of bones and teeth to provide strength and make them strong, as well as to give firmness to skeleton that provides proper shape to the body.
2. **Neuromuscular excitability and muscle contraction**—Calcium mediates excitation and contraction of skeletal, cardiac as well as smooth muscle fibers. Ca^{++} ions are released in muscle on stimulation and its binding with the **troponin C**, activates enzymes **myokinase** and **ATPase** to provide energy, as well as it triggers excitation and muscle contraction by coupling of actin and myosin filaments.
3. **Nerve impulse transmission**—Calcium is necessary for the transmission of nerve impulse from the pre-synaptic to post synaptic region.
4. **Blood clotting factor**—Calcium is essential blood clotting factor IV, required in **cascade of activation** of clotting factors and for the conversion of prothrombin to thrombin to facilitate blood coagulation. (Chapter 32).
5. **Secretion of hormones and neurotransmitters**—Calcium facilitates release of certain hormones e.g. insulin, PTH, calcitonin etc from the endocrine glands.
6. **Second messenger for hormones**—Certain hormones exert their action inside the cell, by mediation of Ca^{++} ions release that acts as IInd messenger along with c-AMP, for example epinephrine, ADH etc.
7. **Activation of enzymes**—Ca^{++} ions directly activate many enzymes such as pancreatic lipase, ATPase, myokinase etc.
8. **Secretory processes**—Ca^{++} ions regulate and stimulate exocrine release of intestinal secretions, as well as secretory processes such as endocytosis, exocytosis etc
9. **Calcium** is important for **cell membrane integrity** and facilitates transport of substances across plasma membranes of cells.
10. **Motility of cilia**—Calcium is necessary for the ciliary movement of micro organisms.
11. **Cell communication**—Ca^{++} maintain cell to cell contact and adherance of cells in tissues. Calcium channels also facilitate cell communication.
12. **Calmodulin**—A calcium binding regulatory protein called calmodulin activates many enzymes e.g. adenyl cyclase, muscle phosphorylase etc.
13. **Action on myocardium**—Ca^{++} ions prolong systole in myocardium. One of the main adverse effects of hypercalcaemia is cardiac arrest in systole.

Sources and RDA of Calcium

Milk and milk products are rich sources of calcium, while vegetables, cabbage, cereals and fish are good to moderate sources of calcium in diet.

RDA for Adults is 0.8 g/day, for growing children it is 0.8 to 1.2 g/day, while during pregnancy and lactation, 1.5 g/day is recommended.

Absorption—Calcium is absorbed mainly in the duodenum. Active vitamin D as **calcitriol is necessary** for calcium absorption from GIT. Calcitriol stimulates **gene expression** in intestinal mucosal cell to synthesize a **carrier protein** called, **calcium binding protein,** that facilitates calcium absorption against concentration gradient from the intestine.

Other factors affecting calcium absorption are:

- Acidic pH, lactose and amino acids lysine and arginine **favours calcium absorption** from the intestine.
- **Factors inhibiting calcium absorption** include—High contents of phytate, oxalates and phosphates etc. in diet, that form insoluble salts with calcium and prevent its absorption.
- improper fat digestion and absorption, as in case of **steatorrhea,** the fatty acids form insoluble calcium soaps and thus prevent calcium absorption from intestine.

Serum calcium level

Most of the calcium in blood is present in plasma (as RBCs are devoid of Ca^{++}) and normal level ranges between **9-11 mg per deciliter of plasma.** Calcium in plasma is present mainly in 3 forms—

(a) **Ionized calcium Ca^{++}**- 5 mg/dl. The most active form of calcium is ionized Ca^{++} in the plasma.

(b) **Protein bound Ca^{++}** - 4 mg/dl. Binding of calcium with plasma proteins prevents the precipitation of calcium salts and calcification of soft tissues is prevented.

(c) **Complexed with anions** – 1 mg/dl. Calcium gets chelated with anions, such as citrate etc. and remains as complex form in plasma.

The three forms of calcium are **interconvertible** and remains in **an equilibrium state** in the plasma.

HOMEOSTASIS—SERUM CALCIUM REGULATION

Regulation of serum calcium level within physiological limits of 9-11 mg/dl is the most **efficient** and **minutely regulated homeostasis mechanism,** as both low, as well as high serum calcium levels adversely affect the body organs and can be even **life threatening.** A living organism **has very little tolerance** for any significant deviation from the normal range of calcium level in plasma, as described below—

Hypocalcaemia—Low serum calcium level (<8.5 mg /dl) results in neuromuscular hyper irritability, tremors etc. and if serum calcium level is severely low (<7 mg/dl), it results in **tetany,** in which muscles goes into spasmic contracted state leading to fatal consequences.

Hypercalcaemia—(high serum calcium level) similarly causes severe adverse effects, such as muscle paralysis, cardiac arrest, coma and even death.

Factors Regulating Serum Calcium Level

Three hormones, calcitrol (activated vitamin D), parathyroid hormone (PTH) and calcitonin play central role in maintaining serum calcium level within narrow range of 9 to 11 mg/dl, that is absolutely essential to maintain normal functioning of the body organs.

Calcitriol

It is the active form of vitamin D and is synthesized in the body by specific hydroxylation reactions i.e. **α 25-hydroxylation** of calciferol (Vitamin D) in the liver and **α–1 hydroxylation** in the kidneys (Chapter 25). Calcitriol acts like a hormone and plays predominant role in calcium and phosphate metabolism. Its action is mainly at the three target tissues as given below—

Action on intestine—Calcitriol stimulates gene expression for the synthesis of a carrier protein, known as **calcium binding protein,** necessary for calcium absorption from GIT (details in the Chapter 25).

Action on bones—Calcitriol **stimulates mineralization** of bone to make bones strong, as well as it coordinates remodelling of bones along with parathormone to keep them healthy and strong.

Action on kidney—Calcitriol decreases excretion of both calcium and phosphate in urine by increasing their reabsorption by renal tubules, as they are important minerals needed for mineralization of bones.

Parathyroid Hormone (PTH)

PTH is the main hormone that **increases serum calcium level.** PTH is secreted from 2 pairs of parathyroid glands (embedded in thyroid tissue) directly in response to **low serum calcium level.** Hypocalcaemia can be due to dietary deficiencies of calcium or vitamin D or due to impaired absorption of calcium or both. Secretion of PTH (as stimulated by low serum calcium) **restores** normal ECF calcium level by acting **directly** on bones and kidneys and indirectly on intestine as described below. (details in Chapter 29).

Effects on bone—PTH enhances **demineralization** of bone and **bone resorption** including both inorganic as well as organic matter dissolution, leading to mobilization of calcium from the bones to blood. Thus PTH restores normal serum calcium level and thus prevents adverse effects of hypocalcaemia, even at the cost of bone resorption and demineralization.

Effects on intestine—PTH **indirectly** increases absorption of calcium and phosphate from intestine through increasing calcitriol synthesis in kidneys by stimulating renal **α1-hydroxylase** enzyme.

Effects on kidneys—PTH has very **rapid effect** on reducing excretion of calcium from kidneys and thus **immediate correction of hypocalcaemia** is achieved. PTH at the same time increases excretion of phosphate in urine by reducing its reabsorption. This is an important effect of PTH to prevent supersaturation of calcium and phosphate in plasma, which otherwise would lead to calcium phosphate precipitation and its deposition and calcification of soft tissues.

Calcitonin

It is a polypeptide hormone containing 32 amino acids and is secreted by **parafollicular cells** embedded in the thyroid tissue. The biochemical action of calcitonin is still not very clear. However it is proposed that the actions of calcitonin are **antagonist** and **opposite** to PTH on calcium metabolism, as given below—

- Calcitonin **lower serum calcium level**
- It increases **(promotes) mineralization** of bones by stimulating osteoblastic activity.
- Calcitonin **reduces bone resorption** and demineralization of bones.
- It increases excretion of calcium and phosphate in the urine.

A brief account of the factors regulating calcium homeostasis and their effects are summarized in the Table 24.1.

Table 24.1 : Comparison of 3 major factors affecting calcium metabolism

Effects on	Vitamin D	PTH	Calcitonin
Serum calcium Level	↑	↑↑↑	↓
Calcium absorption from GIT	↑↑	↑ Via calcitrol synthesis	—
Bone mineralization	↑↑	–	↑
Bone resorption/ demineralization	–	↑↑	↓
Renal reabsorption of calcium	↑	↑↑	↓
Renal reabsorption of phosphate	↑	↓↓	↓

Bones as Reservoir of Calcium

Calcium in bones is in a **dynamic state** and **calcium hydroxyapatite** serves as reservoir of calcium in the body. Whenever plasma calcium level tend to decrease, PTH mobilizes calcium from the bones to restore normal calcium levels. Bones are constantly being remodelled to keep them healthy and strong. Calcitriol and PTH coordinate bone remodelling process by their potent action on **osteoblast** and **osteoclastic** activity.

Calcium : Phosphate Ratio in Blood

A reciprocal relationship exists between serum levels of calcium and phosphate. The ionic product of Ca^{+2} x PO_4^{-3} level is kept at 40 in serum in the adults, which is just **below** the **solubility product of calcium phosphate**. This is essential to prevent deposition of calcium phosphate and calcification of soft tissues, that causes formation of calculi. In children where active skeletal growth, bone formation and mineralization is needed, Ca^{++} and PO_4^{-3} ratio is higher and kept at **50. Serum proteins** indirectly influence serum calcium level. In hypoalbuminemia (as seen in nephrosis, protein malnutrition etc.), the total serum calcium decreases, as half of the serum calcium is protein bound, but ionized Ca^{++} level is kept normal, that is necessary for the proper functioning of body tissues.

Disease States Associated with Alteration in Serum Calcium

As PTH is the predominant factor involved in regulation of serum calcium level within 9-11 mg/dl, alterations in serum calcium level are mainly associated with abnormalities of parathyroid hormone secretions.

Hypocalcemia—Decrease in serum calcium level is mostly associated with hypoparathyroidism, that may occur due to accidental surgical removal of parathyroid glands, or due to autoimmune destruction of the glands.

Hypocalcemia below **8mg/dl** results in neuromuscular hyperirritability, tremors etc and if it is severe enough (below 7 mg%) it causes **tetany**. The tetany is a clinic syndrome characterized by severe spasmatic muscular contractions, such as **carpopedal spasms**, facial muscle spasms, **bronchio** and **laryngio spasms** leading to **respiratory arrest, cyanosis** and may results **in death.**

Hypercalcemia—Increased serum calcium level (more than 11 mg/dl) is associated with hyperparathyroidism, which can be due to parathyroid adenoma or due to PTH secreting gland tumors. The excessive production of PTH causes demineralization of bones and **bone resorption** is drastically increased resulting in **osteoporosis,** the bone density is reduced and bones become brittle and easily by fracturable. Calciuria results leading to the development of **urinary calculi (stones)** and **nephrocalcinosis** etc. Ectopic calcification may be seen in renal tissues, pancreas, arterial wall and muscles. Death may result due to renal failure. Marked elevated plasma calcium level may result is muscle paralysis and cardiac arrest, that can be fatal.

Biochemical hallmarks of hyperparathyroidism are elevated serum ionized calcium and decreased phosphate levels.

PHOSPHORUS

Total body phosphate is about 1 kg, most of which (80%) is in bones and teeth as calcium phosphate, 10% in muscles, while rest is distributed in all cells of the body.

Biochemical Functions

(a) **Formation of bones and teeth**—Along with calcium, phosphate is necessary for the mineralization of bones and teeth to make them strong and healthy.

(b) **High energy phosphate compounds**—Phosphate plays central role in generation and utilization of high energy phosphate compounds, such as ATP, CTP, GTP, UTP and creatine phosphate etc. ATP is the **energy currency** of the body, needed for carrying out life process.

(c) **Synthesis of nucleic acids**—Phosphate plays an important role in DNA and RNA synthesis, where nucleotides are linked together by **phosphodiester linkage that** forms the back bone of nucleic acid structure.

(d) **Synthesis of physiological important compounds**—Phosphate is essential constituent of compounds, such as phospholipids, phosphoproteins etc. that are potent physiological compounds.

(e) **Physiologically important nucleosides and nucleotides**—Phosphate is essential for synthesis of coenzymes NAD^+, $NADP^+$, FMN, FAD,, UDP etc and for the synthesis of c-AMP as second messenger for hormone action.

(f) **Activation of nutrients**—Low energy phosphate bonds is required for the activation of monosaccharides, such as glu-6 PO_4, fru-1, 2 bisphosphate, glyceraldehyde 3-PO_4 etc so that they can enter metabolic cycles for oxidation purposes.

(g) **Activation of enzymes**—Covalent modification of key enzymes by **phosphorylation and / or dephosphorylation** is essential for the reciprocal regulation of metabolic cycles according to the need of the cell. (Chapter 5).

(h) **Phosphate buffer system**—The phosphate buffer of blood plays important role in **acid base balance** maintenance and proper functioning of the body organs.

Sources and RDA

Milk and dairy products are good sources of phosphate, while cereals, nuts and meat etc. are moderate sources of phosphates in our diet.

RDA – 500 mg/day for adults.

Serum Phosphate Level

Normal serum phosphate level is 3-4 mg/dl in adults. Because of the active bone and teeth formation, plasma phosphate level in children is slightly higher i.e. 5-6 mg/dl. Serum phosphate level is regulated by its excretion through urine by hormones as described in serum calcium homeostasis. Usually 500 mg of phosphate is excreted in urine per day. **Abnormalities associated** with serum phosphate level are —

1. Low serum phosphate levels are associated with **hyperparathyroidism** and **rickets** (vitamin D deficiency disease).

2. Renal diseases—In severe kidney diseases, serum phosphate content is **elevated** leading to acid base balance disturbances and **acidosis.**

MAGNESIUM

Total body magnesium is approximately 20 gm, 70% of which is present in bones and teeth along with calcium and phosphate as mineral constituents, while rest 30% is distributed in blood and tissues (muscles).

Biochemical Functions

1. Mg^{++} ions are **cofactors** of several enzymes that catalyze phosphorylation of compounds by utilizing ATP, which are key steps of major metabolic cycles. Examples are **hexokinase, phosphofructokinase**(key enzymes of glycolysis) etc.
2. Mg^{++} is an **activator of enzyme adenyl cyclase** which synthesize c-AMP (II messenger for hormone action) as well as for the activation of **c-AMP-protein kinase** system, that is involved in regulation of glycogenolysis and gluconeogenesis to maintain blood sugar level (BSL).
3. Mg^{++} ions are necessary for **neuromuscular excitability** along with Ca^{++} ions.

Sources—Cereals, beans, green leafy vegetables and fish are good sources of magnesium.

RDA—300 to 350 mg/day for adults.

Absorption—Magnesium is absorbed by intestinal mucosal cells through a specific carrier protein, by facilitated diffusion mechanism.

Serum level—Normal serum Mg level is 2-3 mg/dl, and is present in form of ionized Mg^{++}(60%), bound to proteins (30%) and complexed with other anions (10%).

Deficiency states of magnesium is associated with the conditions, such as malnutrition, chronic alcoholism, chronic diaorrhea and liver cirrhosis etc.

Mg^{++} deficiency and low serum Mg level causes **neuromuscular hyperirritability,** tremors, convulsions, spasmic contractions etc. The signs and symptoms of low Mg are similar to **hypocalcaemia** but would be relieved only by giving magnesium supplements.

SODIUM

Sodium is the **chief cation of the extracellular fluid (ECF).** About 50% of body sodium is present in bones, 40% in ECF and rest (10%) is distributed in tissues.

Biochemical Function

1. **Regulation of plasma volume**—Sodium along with chloride maintains **osmotic pressure** and **fluid balance** of the body and blood and thus regulates blood pressure.
2. **Maintence of acid-base balance**—Sodium along with chloride and bicarbonate ions, regulates acid base balance of the body.
3. **Normal neuromuscular irritability**—Sodium is necessary for the regulation of normal neuromuscular irritability and muscle contraction.
4. **Maintenance of Na^+/K^+ ATPase** activity and **Na^+/K^+ pump** is very important to regulate osmotic pressure, electrolyte balance and fluid balance between ECF and ICF, as well as for the regulation of cell membrane permeability.
5. **Intestinal absorption of sugars and amino acids**—The active transport system (carrier SGLT-1) responsible for absorption of glucose from the intestine requires Na^+ and is linked with Na^+/K^+ pump (Chapter 4).

Normal diet contain 5-10g sodium, that is taken as common salt (NaCl) and same amount is excreted daily in urine to maintain Na^+ body pool. Normal plasma level of Na^+ is 136-145m.Eq./L, while in cells (ICF) its level is 35m.Eq./L.

Regulation of Sodium level—As sodium ECF level is primarily concerned with **osmotic pressure regulation** and **electrolyte** and **fluid balance between ECF and ICF,** its plasma level maintenance is of prime importance. Increased plasma sodium level is associated with high blood pressure and is a major risk for heart diseases.

At the same time it is an important electrolyte of the body that maintains fluid balance, so its excretion is regulated at the distal convoluted tubules in kidney by mineralocorticoid hormones.

Aldosterone is the major mineralocorticoid hormone secreted by adrenal cortex, that is mainly involved in plasma sodium level maintenance. During urine formation, as much as 800g of Na/day is filtered in glomerular filtrate. 99% of this is reabsorbed by renal tubules by an **active process** and its retention in body is controlled by the hormone aldosterone, according to the need of the body. (Chapter 30).

Clinical Aspects

1. In diseases associated with **edema**, along with water, sodium retention in body is increased causing ECF, ICF fluid disturbances.
2. **Diuretic drugs** increase sodium excretion, along with which water is also eliminated out in urine. This helps in excretion of extra sodium and water out of the body.
3. In hypertension sodium restricted diet is advised so as to avoid risk of congestive heart diseases (CHD).

Deficiency Disease—Clinical manifestation

a) **Hyponatremia**—Normally sodium deficiency does not occur on normal diets. However sodium levels may decrease in excessive vomiting, diaorrhea and in burn cases.

In **Addison's disease** (adrenal insufficiency) and in chronic renal diseases, hyponatremia may result with associated low blood pressure and blood circulation insufficiency.

b) **Hypernatremia**—High plasma sodium levels associated with following conditions—
   a) **Cushing's syndrome**—It is a clinical syndrome caused by hyperactivity of adrenal gland resulting in high plasma sodium levels.
   b) Prolonged corticosteroid therapy results in high sodium levels.
   c) In pregnancy, steroid hormone and placental hormones cause sodium retention along with water causing edema and hypertension.
   d) In dehydration blood volume decreases with apparent increase in plasma sodium level.

The clinical symptoms of high sodium levels are **edema** and **hypertension** in the susceptible individuals. Plasma levels of sodium are determined mostly by flamephotometry (Chapter 37).

POTASSIUM

Potassium is the **principal intracellular cation** and is involved in regulation of intra cellular fluid (ICF) osmotic pressure, ICF volume, electrolyte balance and acid-base balance of the body.

Excracellular potassium is equally important factor and plays significant role in the body as listed below—

a) **Skeletal and cardiac muscle excitation**—The K^+ dependent depolarization wave stimulates muscle contraction and contraction of cardiac muscles.
b) **Transmission of nerve impulse**—Potassium is equally important during conduction of nerve impulse, which depends on Na^+ influx and K^+ efflux by Na^+/K^+ pump which triggers a depolarization wave.
c) **Resting membrane potential**—It is maintained by difference in K^+ concentration between ICF and ECF with the help of Na^+/K^+ Pump.

Daily Requirement and Sources

Good sources of potassium include banana, apple, orange, pineapple, potato, beans, almonds etc. Tender coconut water is rich in potassium. Daily requirement is 3-4 g/day, that is easily available in our diet.

Plasma potassium level is 3.5 to 5 m.Eq/L and is estimated directly by flamephotometry (Chapter 37). As K^+ is an intracellular cation, hemolysis of RBCs should be avoided during estimation process.

Regulation of Potassium Level

Aldosterone is the main hormone that regulates plasma K^+ level. The hormone stimulates sodium reabsorption from the renal tubules in exchange of K^+ which is excreted in urine (obligatory loss of K^+ in urine).

Acid-base regulation depends mainly on sodium retention and potassium excretion in urine (Chapter 35). In acidosis K^+ is expelled from cells to ECF and thus plasma levels of K^+ are elevated, while in alkalosis a shift of K^+ from ECF to ICF occurs.

Disturbance in plasma potassium levels—Maintenance of plasma K^+ level within narrow range of 3.5 - 5 m. Eq/L is very important, as even **minor increase** can be **life threatening**.

Hyperkalemia—High plasma K^+ levels are seen in the following conditions —

(a) Renal failure results in impairment of regulation of electrolytes and water balance leading to high plasma potassium levels.

(b) Adrenal insufficiency—Aldosterone is mainly responsible for K^+ plasma level regulation and in adrenal insufficiency, (Addison's disease) plasma potassium level increases.

(c) Diabetic acidosis is associated with increased plasma and ECF K^+ levels.

(d) Severe dehydration results in apparent increase in K^+ levels.

Clinical manifestations of high K^+ level include, mental depression, confusion, muscular weakness, paralysis, small bowel ulcers, bradycadia and peripheral vascular collapse leading finally to **cardic arrest** with ECG abnormalilities like elevated T waves, with fatal consequences.

Management—Intravenous glucose and insulin in physiological doses is given, which stimulate glycogen synthesis and shifting of ECF K^+ inside cells. It has been observed that a trapping of 0.3 m.mol/ of K^+ /g of glycogen storage occurs inside cells, so plasma K^+ level is decreased.

Hypokalemia—Low plasma potassium levels are seen in conditions associated with following cases—

1. Postoperative prolonged I.V. fluid administration without potassium.
2. Metabolic alkalosis.
3. Cushing's syndrome.
4. Prolonged cortisone therapy.
5. Prolonged diaorrhea and vomitting.
6. Prolonged use of diuretic drugs.

Clinical manifestations of low plasma K^+ include

1. Muscular weakness and in severe cases it causes paralysis.
2. Tachycardia, cardiomegaly, cardiac arrest with ECG abnormalities as flatten waves, inverted T wave with low ST segment due to AV block, causing death.
3. Mental confusion.

Management—Administration of orange juice and inclusion of potassium in I.V. fluid therapy in physiological doses **is must** in hypokalemia to save the life of patient.

CHLORIDE

Chloride ion is equally important along with sodium as extracellular anion to maintain osmotic pressure, acid base balance and plasma volume of the blood. As common salt NaCl, its intake, excretion and functions parallel that of sodium.

Electrolyte balance of the body is closely interrelated to homeostasis of Na^+, K^+ and Cl^- plasma levels.

Biochemical Functions

1. **Along with Na^+ and K^+,** chloride is closely involved in acid-base balance, fluid balance and electrolyte balance of the body to regulate osmotic pressure of ECF and ICF compartments.
2. **Gastric juice and HCl secretion**—Chloride is necessary for secretion of HCl in gastric juice (Chapter 14).
3. **Chloride shift**—During transport of CO_2 and protons by hemoglobin and in carbonic acid buffering, chloride shift is an important phenomenon to maintain electric neutrality. (Chapter 33).

Normal Plasma Chloride level is 96 to 106 m Eq/L. Cerebrospinal fluid (CSF) chloride levels are slightly higher, about 120 m Eq/L. Since CSF protein contents are low as compared to plasma proteins, so elevated Cl^- ions in CSF are necessary for donnan equilibrium maintenance, across nerve membranes.

Excretion of Cl^- is through urine and sweat and is parallel to Na^+ as NaCl is the main inorganic component of urine and sweat.

Hypochloremia—Decreased plasma Cl^- levels are seen in following cases —

(a) Excessive vomiting with gastric HCl loss, that is accompanied by compensatory increase in bicarbonate resulting in **hypochloremic alkalosis.**

(b) Respiratory alkalosis—Hyperventilation is associated with increased HCO^-_3 levels and a compensatory decrease in Cl^- levels.

(c) In Addison's disease along with Na^+, Cl^- is also excreted out in excess in the urine.

(d) Excessive sweating—With Na^+, chloride is also lost during excessive perspiration.

Hyperchloremia—Elevated serum chloride levels are observed in conditions associated with following disorders—(a) Dehydration, (b) Respiratory acidosis, (c) Cushing's syndrome, (d) Severe diaorrhea, and (e) Renal acidosis etc. With loss of HCO^-_3 in acidosis, there is a compensatory retention of chloride in the plasma.

SULPHUR

Mostly organic sulphur is present in foodstuffs, mainly as sulphur containing amino acids of proteins and is absorbed from intestine along with amino acids and is utilized by body

Biochemical Functions

1. **Sulphur containing amino acids**—Methionine, cysteine and cystine are essential constituent of body proteins and necessary for formation of disulphide bonds.

   (a) Disulphide bridges between 2 cysteine residues, that maintain structural, conformation and functional integrity of proteins and are necessary to maintain secondary structure of proteins, such as insulin and immunoglobulin etc.

(b) Kerasin protein is rich in disulphide bridges to give more strength to hair, nail, hoof, horn etc.

2. **Enzyme catalytic site**—SH group of cysteine at the active site is essential for catalytic action of several enzyme. The coenzymes of thiamine, biotin, lipoic acid etc. needed for catalysis, all have active SH group needed for the activation of the substrate.
3. **Sulphhydral buffers—Glutathione,** a tripeptide present in erythrocytes has reduced SH group to detoxify H_2O_2 and thus prevents hemolysis of RBCs. (Chapter 26).
4. **Sulphomucopolysaccharides**—Chondroitin sulphate is an important ground cementing substance of connective tissues of bones, cartilages, walls of blood vesseles etc.
5. **Glycolipids** have sulphated mucopoly accharides and are present in rich amounts in brain and CNS.
6. **Conjugation and detoxification**—Cystiene with sulphhydral group is involved in conjugation and detoxification of many drugs and harmful aromatic halogens, phenols, indoles etc. forming **mercapto compounds** that are less harmful and are excreted out of the body.
   (a) Bile salts are produced by conjugation of bile acids with **taurine**, a sulphur containing compound formed by decarboxylation and oxidation of cysteine.
   (b) **Active sulphate** as phosphoadenosine phosphosulphate (PAPS) is important for the synthesis of glycosaminoglycans and other compounds. (Chapter 16).
7. **S-Adenosylmethionine (SAM)** is important in all **transmethylation** reactions and synthesis of nucleotides, nucleic acids DNA and hormone epinephrine etc.
8. **Coenzyme A** has an active SH group and acts as a carrier for fatty acyl group necessary for their metabolic reactions.

RDA for sulphur is not specifically recommended. However good quantity of proteins in diet will meet the body needs for sulphur.

Metabolism—In body sulphur of various organic compounds is ultimately oxidized to sulphates in liver during their metabolism and is excreted in the urine. The urinary sulphur is in following forms—

(a) Inorganic Sulphates—Total sulphates in urine is about 1g /day and 80% of urinary sulphate is as inorganic sulphates.

(b) Organic sulphate or ethereal sulphate (10%) as phenol and indoxylsulphate of tryptophan and tyrosine residues (formed in GIT by bacteria and absorbed to be detoxified in liver) are excreted in the urine.

(c) Neutral unoxidized sulphur—This portion constitutes 10% of total urinary sulphur and is excreted as incompletely oxidized amino acids or thiocyanates, urochromes etc.

TRACE ELEMENTS

IRON

Iron is a trace element, required by body in minute amount, but has great functions in the body and plays predominant role in oxygen carriage and biological oxidation to generate **energy** to carryon **life processes.**

Biochemical Functions

1. **Hemoglobin,** the major iron containing protein of the blood (75% of total body iron) and **myoglobin** of muscles (5% of total) are major O_2 transport and storage proteins of the body.
2. **Cytochromes, cytochrome oxidase** (heme iron containing enzymes) and certain non-heme iron-sulfur proteins are components of **electron transport chain**(ETC) necessary for biological oxidation to **generate ATP.**
3. **Many enzymes** like **peroxidases, catalase, xanthine oxidase, aconitase, tryptophan pyrrolase** etc. are the iron metalloenzymes which are key enzymes of important metabolic processes.

Total body iron is approximately 4 to 4.5 g, 75% of which is present as hemoglobin in erythrocytes, 5% as myoglobin in muscles and rest is distributed in liver and other tissues as ferritin and hemosiderin (storage forms), transferrin (transport form) and as various enzymes such as cytochromes, catalase, peroxidase etc.

Sources and RDA—Rich sources of iron are meat, liver, kidney etc. Good sources include green leafy vegetables, pulses, cereals, apple, jaggery and iron cooking pots.

RDA for adult male is approximately 10 mg/day. Women during fertile age require more iron to compensate the loss of iron in menstrual excretion and thus RDA for female is 15 mg/day.

During pregnancy 20mg/day are recommended as iron requirement increases in last trimester of pregnancy for the placental transfer of iron needed for fetal blood formation.

Absorption, Transport and Storage

Iron is absorbed mostly in stomach and duodenum. Only 10% of dietary iron is normally absorbed. Many factors affect iron absorption from GIT, that are enumerated below.

Factors facilitating iron absorption are—(a) Acidic pH of gastric juice favous iron absorption as it solubilizes iron salts and releases iron from bound to proteins, (b) Ascorbic acid (Vitamin C.) reduces ferric iron into ferrous form, which can cross mucosal cell membrane and thus is readily absorbed.

Factors inhibiting iron absorption are—Food stuff containing phytates, oxalates etc. such as green leafy vegetables, impairs iron absorption. The phytates and oxalates form insoluble iron salts and thus prevents its absorption. Rich amount of phosphates present in foodstuffs also forms insoluble iron salts and impairs its absorption from the intestine.

Mucosal block controls iron absorption—Iron is a **one way element,** meaning that very less amount of iron is excreted out of the body (<1 mg/day) and the body iron is efficiently **reutilized** and **recycled** for **heme synthesis.** Iron in excess **is toxic,** so very less amount of iron (<1 mg/day) is absorbed from GIT normally. This is achieved by regulation of iron absorption at the **intestinal mucosal level itself.** This is an **unique feature** of iron metabolism.

Mechanism—Iron in excess in body is highly toxic, **thus it is absorbed from GIT** only according to the need of body. The iron absorption is regulated by controlling the amount of a **carrier protein apoferritin** in the intestinal mucosal cells, that is required to bind absorbed iron to form **ferritin.** The ferritin so formed is stored in mucosal cell itself. If iron is needed by the body, then only the iron from mucosal ferritin stores is released to be takenup by a transporter protein known as **transferrin** into the blood, otherwise the mucosal cells containg ferritin stores, are **desquamated** and excreted in feces as illustrated in the Fig. 24.1.

Transferrin is the transporter form of iron in blood—Transferrin is a glycoprotein in nature and binds $2Fe^{+++}$ ions per molecule. For this ferrous iron from the ferritin have to be oxidized to ferric form, that is facilitated by **Cu^{++} containing ceruloplasmin** present in the blood, as shown in the reaction.

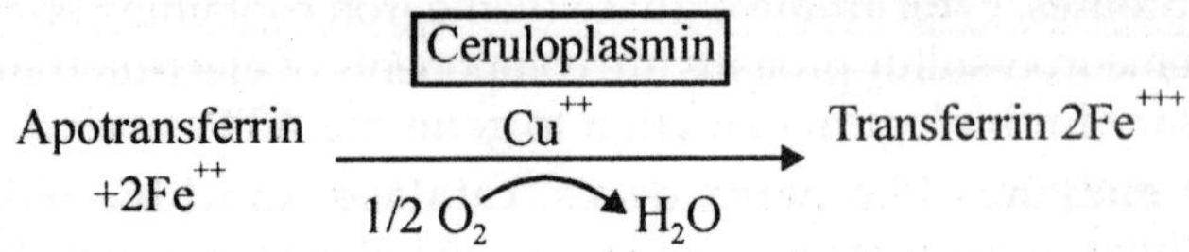

Ceruloplasmin is a **ferrodoxidase** enzyme containing Cu^{++} ions, which oxidizes ferrous iron to ferric state, that is takenup by transferrin for its transport to the tissues.

Plasma transferrin—Normal plasma transferrin level is 250 mg. / dl. The total iron binding capacity (TIBC) of plasma is 400 μg /dl and it is provided by transferrin. Usually only one third of this capacity is saturated with iron and thus protein bound iron (PBI) level is about 100 μg/dl of plasma. The plasma transferrin transports iron to various tissues for the synthesis of heme and other iron contaning compounds.

Storage of iron—There are two main storage form of iron, present in various tissues of the body as listed below—

a) **Ferritin** is the main storage form of iron, present in liver, spleen and bone marrow (GIT ferritin is temporary storage form). It is a high mol. weight (5, 00,000) protein, that can bind 4000 iron atoms/mol and is very **soluble in water.**

b) **Hemosiderin** is another storage form of iron, that is water **insoluble** and is present as **golden brown granules** in reticuloendothelial system of spleen and liver. It binds about 25% of total body iron for the storage purpose.

A brief account of iron metabolism and its distribution and recyclization in body is illustrated in the Fig. 24.1.

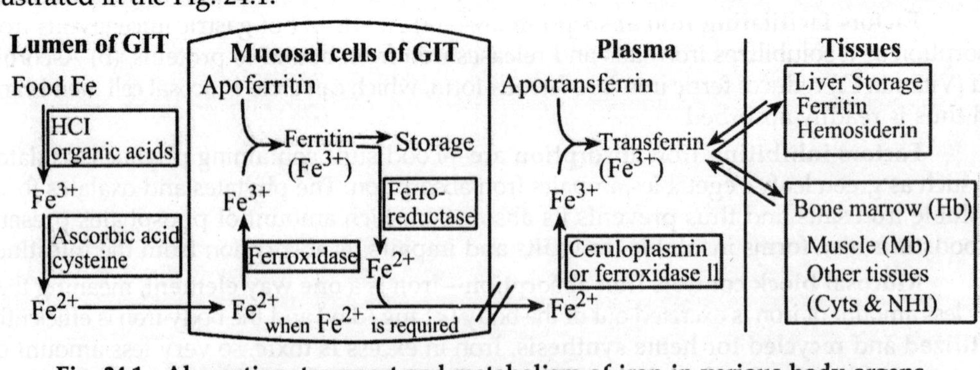

Fig. 24.1 : Absorption, transport and metabolism of iron in various body organs. GIT—Gastrointestinal tract, Hb—Hemoglobin, Mb—Myoglobin, CYCs—Cytochromes, NHI—Non heme iron

Iron Deficiency State

Iron deficiency anaemia—Iron deficiency results in **microcytic hypochromic anemia,** which is the most common clinical disorder all over the world. About 30% of world population is anaemic, while in India iron deficiency anemia is as high as 70%.

Iron deficiency mainly occurs in following conditions.

1. When iron **requirement increases,** for example during pregnancy and in growing children etc. with poor dietary intakes due to low socioeconomic status.

2. **Acute blood (iron) loss**—Excessive bleeding due to accidents and post operative bleeding etc. results in loss of iron from the body.

3. **Chronic bleeding disorders such as—**

(a) Hook worm disease

(b) Menstrual cycle disorders in females (excess bleeding)

(c) Bleeding piles, ulcers etc

(d) Nephrotic syndrome, with severe proteinuria and **hematuria.**

4. **Dietary deficiencies**—Indian diet is rich in phytates, oxalates, phosphates etc. that impairs iron absorption. So when iron requirement is high and dietary supply is insufficient with defective absorption, it may result in iron deficiency state.

Clinical Manifestations

Iron deficiency anaemia is characterized by low hemoglobin levels (<10-12 gm %) resulting in insufficiency of oxygen and impaired metabolism, causing apathy (dullness). sluggishness, loss of appetite, loss of memory etc. Apathy i.e. loss of interest in the surroundings is hall mark symptom of anaemias.

Iron Toxicity

Excess accumulation of iron in body may result in toxicity and severe damage to tissues leading to disease states. Excesive accumulation of iron in body may result in cases with—

(a) **Repeated blood transfusions** needed in patients suffering from hemophilia, aplastic anaemias, thalassemias etc. leading to **iron overload,** as very little iron is excreted out of the body, since it is mostly bound to proteins.

(b) **Bantu tribe of Africa** is prone to iron excess in body because their staple diet is corn, that is low in phosphates, which they cook in iron pots, so excess of iron is absorbed in GIT and is deposited in tissues, causing **hemosiderosis.**

(c) A **rare inherited defect** in the regulation of iron absorption from GIT has also been reported

Hemosiderosis—Excess of iron in body is deposited as insoluble golden brown hemosiderin granules in liver, spleen, pancreas etc. causing damage to tissues resulting in dysfunction and condition is called as hemosiderosis.

Hemochromatosis—When iron contents in body is high (total body iron > 25-30g) with high plasma transferrin levels, excessive iron is deposited as hemosiderin in most of the tissues of body, particularly in liver, pancreas, skin etc. leading to **bronze pigmentation** of skin (hemochromatosis) **bronzed diabetes** (pancreatic tissue damage) and **cirrhosis** of liver causing severe **jaundice.**

COPPER

Copper is a trace element required by body in small amounts, but it plays very important role in iron transport and as an essential cofactor of key enzymes of biological oxidation and other metabolic precesses.

Total body copper is approximately 100 mg, that is distributed in various organs and blood in following forms—

Plasma copper—Normal concentration of copper in plasma is 100-200 µg/dl, 95% of which is tightly bound to protein as **ceruloplasmin,** while a small fraction (5%) is loosely held to albumin, that is the transport form of iron.

Plasma ceruloplasmin—Normal plasma ceruloplasmin level is 25-50 mg/dl. Ceruloplasmin is a blue coloured glycoprotein present in the blood and contain 0.34% copper (6-8 atoms of copper per mole), half as cupric ion and half as cuprous ions. Ceruloplasmin is not a transporter of copper, rather it plays important role as an enzyme in iron transport. Ceruloplasmin acts as **ferroxidase,** and oxidizes ferrous iron to ferric state, so that it can be taken up by transferrin for its transport to various tissues for heme synthesis.

Storage of copper—Copper is stored in different forms in various organs, for instance **cuprothionine** in liver, **cerebrocuperiene** in brain, **erythrocuperiene** in bone marrow and RBCs etc.

Sources and RDA of copper—Liver, meat, egg, cereals, nuts and leafy vegetables are good sources of copper. RDA is 2-3 mg/day for adults

Absorption—About 10% of dietary copper is absorbed normally, mainly in duodenum, mediated by a copper binding carrier protein. Phytate, zinc and molybdenum in diet impair copper absorption.

Biochemical Role

Copper is essential constituent of several enzymes of the body. Important examples are listed below—

1. Oxidases such as **cytochrome oxidase, catalase, monoamine oxidase** and **superoxide dismutase** etc. contain copper as an essential constituent and thus are involved in biological oxidation to generate nascent oxygen.
2. **Synthesis of Heme**—Copper is an essential constituent of key enzyme of heme biosynthesis i.e. **ALA synthase** and thus is necessary for hemoglobin synthesis.
3. **The enzyme lysyl oxidase** contain Cu ions that is required for the oxidation of 4 lysine residues together to form **desmosine** (allyllysine) of **elastin fibers** that form crosslinkages, necessary for **elasticity** and **stability** of elastin of connective tissue of the blood vessels.
4. **Uptake and transport of iron**—Copper plays important role in iron absorption from GIT and transport of iron to tissues for heme synthesis as illustrated in the Fig. 24.1.
5. **Melanin synthesis**—Copper is a cofactor of enzyme **tyrosinase** of melanocytes of skin, hair etc. required for the synthesis of pigment melanin. Melanin imparts normal colouration to skin, hair etc. and protects them from the ultraviolet radiation damage (Chapter 16).

Deficiency States

1. Severe copper deficiency, secondary to malnutrition results in **microcytic hypochromic anaemia** (associated with iron deficiency), as copper is necessary for iron uptake and heme synthesis.

2. The enzyme cytochrome oxidase is copper dependent and therefore in deficiency of copper, O_2 and biological oxidation in tissues is reduced, affecting CNS, particularly motor neurons resulting in **neonatal ataxia** in newborns.

3. In copper deficiency, **elasticity** and **stability of elastin fiber** is not maintained properly resulting in weakenig and **increased fragility of** the walls of **blood vessels,** causing **aneurysms** (oozing of blood in surrounding tissue), **fatal rupture of aorta** and **fibrosis of myocardium** leading to heart diseases. The defect is due to impaired cross linkage formation of elastin fibers of the walls of blood vessels as described above.

4. **Menke's kinky hair syndrome** occurs due to defective intestinal absorption of **copper, resulting in anaemia, hypopigmentation, graying of hair** and sometimes depending on period of copper deficiency, as alternate white patches on the hair, giving it a **flag** like appearance. The defect is due to impaired melanin synthesis, as enzyme tyrosinase requires copper for its activity. (Chapter 16).

Copper Toxicity

Wilson's disease—It is a rare **genetic** abnormal metabolic disorder of copper, in which **ceruloplasmin** (protein bound copper) level in plasma is **drastically decreased**. Due to this **free copper gets deposited** in excess in liver and brain, that is highly toxic to these tissues and causes **hepatolenticular degeneration.** The clinical manifestations are cirrhosis of liver, severe jaundice, CNS damage with neurologic symptoms, kidney damage and renal failure. A typical **green blue coloured** ring develops around cornea of eye due to the copper deposits.

Cause of Wilson's disease—The disorder is caused by genetic abnormality, due to which the synthesis of ceruloplasmin (which binds copper in plasma) is reduced drastically, so free copper gets deposited in the tissues causing copper toxicity.

Treatment—Administration of zinc salts, that bind copper and prevents its absorption from the GIT, or drug **penicillamine,** which chelates copper, have proved beneficial in reducing copper toxicity. Hence they are used for the treatment of Wilson's disease to prevent irreversible damage to liver and brain.

IODINE

Iodine is necessary in diet, mainly for the synthesis of **thyroid** hormones, **thyroxine** (T_4) and **triiodothyronine** (T_3) and thus its metabolism is closely associated with thyroid hormones. **Total body iodine** is about 20-30 mg, most of which(80%) is present in the thyroid gland, in bound form as **iodothyroglobulin.**

Plasma Iodine—Normal level is 4-10 μg/dl, most of which is as protein bound iodine (PBI) and represents circulating thyroid hormones T_4 and T_3.

Daily Requirement and Sources

Daily requirement of iodine is 150-200 μg/day and is mainly obtained from **drinking water.** In most parts of the world, iodine is a **scarce element** in soil, so very less amount of iodine is present in vegetables. As iodine is constantly being eroded from the soil, mountain regions (high attitude) generally are devoid of iodine and are called **goiterous belt.** (Chapter 29).

Sources—Sea weeds and sea foods are commercial sources of iodine while **iodization of commercial salts** have resulted in increased availability of I_2 in our diet.

Absorption—Iodine is mainly absorbed from the small intestine. Generally only 30% of dietary iodine is absorbed. The presence of **thiocyanates** in foodstuffs such as cabbage, tapioca etc. prevent iodine uptake by intestinal mucosa and are called **goiterogenous compounds.**

Biochemical functions—Most important biochemical role of iodine is the synthesis of thyroid hormones. The thyroid hormones, T_3 and T_4 are necessary for the **normal functioning** of all cell of the body and are required as growth factors for normal fetal development, along with growth hormone, insulin and other growth factors. (details in chapter 29).

Disease states due to the dietary deficiency of iodine, that results in abnormalities in thyroid hormone synthesis, causing **simple goiter** and other thyroid diseases are discussed in detail in Chapter 29 along with thyroid hormones.

FLUORIDE

Fluoride is a trace element required in small amounts and plays important role in bone and teeth formation to make them strong and healthy, but in **excess** it is very harmful and toxic to the body tissues.

Biochemical Functions

1. Fluoride in small amount is required for **bone** and **teeth formation** and is beneficial in **keeping teeth and bones strong and healthy.**
2. **Exchange of Fluoride with calcium hydroxyapatite**—The exchange of fluoride with calcium hydroxyapatite of teeth and bones, hardens and makes them strong. This is highly beneficial in **preventing dental caries.** Fluoride first enters the hydration shell over the surface of apatite crystal and then it undergoes slow exchange with calcium and gets incorported into crystal lattice, forming **fluoroapatite,** that hardens tooth enamel. This makes it more resistant to bacterial attacks and acid action, thus preventing dental caries.
3. **Prevention of dental caries**—Fluoride prevents development of dental caries by forming a **protective fluoroapatite layer** on the enamel of the teeth and thus prevents tooth decay by bacteria. In the pits and fissures of teeth, bacterial fermentation of the left over food particles produce **acids,** that dissolve enamel and dentine of the teeth to expose pulp to bacterial infection causing **dental caries** and **inflammation of the tooth.** Flouride protects the tooth decay by protecting enamel from the acid action.
4. **Fluoride exchange in bone minerals**—Fluoride exchange with calcium hydroxyapatite in bones also occurs gradually along with the age and this hardens the bones. This is very benificial and **prevents osteoporosis in old age** by making bones resistant to resorption and demineralization.

Dietary sources and RDA—Main source of fluoride in our diet is **drinking water**. One part per million (ppm) of fluoride in drinking water is sufficient to supply the daily requirement of fluoride. RDA is 1-2 mg F^-/day.

Fluoride Deficiency

Dental caries—It is well established that low fluoride contents of drinking water (<0.5 ppm) is associated with development of dental caries in children. Considering the beneficial effects of fluoride in the prevention of dental carriers, fluorination of drinking water in some parts of USA, where fluoride contents are less than 0.5 ppm, have been tried. Also fluoride containing **toothpastes** are now available commercially, which are considered protective against dental caries.

Fluoride Toxicity

The beneficial effects of fluoride are masked by **deleterious** and **harmful** effects of slight excess of fluoride contents in drinking water. Fluoride levels more than 2 ppm in water results in loss of **appetite** and **chronic gastroenteritis.**

Dental Fluorosis—Fluoride levels more than 5 ppm in drinking water cause **fluoride toxicity** known as **fluorosis.** It is characterized by **mottling** of enamel, **stratification** and **discolouration of teeth** with yellow brown patches and roughening of the surface of teeth. The condition is know as **dental fluorosis.**

Skeletal Fluorosis—Fluoride levels more than **20 ppm** in water is highly toxic and if consumed, the flouride accumulates in bones causing **skeletal fluorosis.** This is a severe clinical condition and is associated with **alternative** areas of **bone resorption (osteoporosis)** and **osteosclerosis** (hypercalcificaton). Excess of fluoride particularly increases density of bones of spine, limbs and pelvis, causing **skeletal fluorosis** resulting in **crippling** of individuals with **hump back, stiff joints** and **brittle bones,** and this severe condition is referred to as **genu-valgum.**

Fluoride toxicity is common in many parts of the world and is very prevalent in northen parts of India where fluoride contents of water is more than 10ppm. In these parts defluorination of drinking water should be done, and toothpaste without fluoride should be used.

ZINC

Total body zinc content is about 2gm, distributed in appreciable amounts mainly in muscles, brain, pancreas, prostate gland, WBC, nails and hairs etc.

Biochemical Functions

1. **Cofactor of many enzymes**—Zinc is an essential constituent of several enzymes, such as carbonic anhydrase, alcohol and lactate dehydrogenases, alkaline phosphatase, DNA and RNA polymerases and retienine reductase etc.
2. **Transcription and protein biosynthesis**—Zinc is required for transcription process and protein synthesis as cofactor of enzyme RNA and DNA polymerase, so it is important for growth and normal health. (Chapter 20).

3. **Regulation of gene expression**—As zinc finger motif that mediates binding of regulatory protein factor on hormone sensitive element of DNA, it regulates transcription of genes.
4. **Storage and secretion of insulin**—β cells of islets of pancreas have rich contents of zinc that plays important role in storage and secretion of insulin hormone (Chapter 8).
5. **Taste and smell sensation**—Zinc containing protein **gusten** in saliva and nasal secretions is important in taste and smell acuity.
6. **Reproduction**—Zinc is necessary for proper reproduction and spermatogenesis

Deficiency of zinc is manifested in impaired growth, poor wound healing, skin lesions such as dermatitis, alopecia, hyperkeratinization etc. **Hypogonadism** and reproduction failure is observed in zinc deficiency. **Psychaitric** disorders like depression, dementia etc. occurs because zinc binds with amyloid proteins to form a plaque in Alzheimer disease. Zinc deficiency also result in diminished taste and smell acuity.

Zinc deficiency mainly results secondary to condtions like acrodermatitis, chronic alcoholism etc. Acrodermatitis enteropathica is a genetic zinc absorption disorder and is characterized by inflammation around mouth, nose, fingers, diarrhea, alopecia, hypogonadism etc.

Zinc toxicity—It is often observed in welders due to inhalation of zinc oxide fumes. The zinc toxicity is manifestated mainly in gastrointestinal irritation, nausea, vomiting etc. Many rat poisons contain zinc compounds which may result in accidental poisoning resulting in fatal consequences. Chronic zinc toxicity may result in gastric ulcers, pulmonary fibrosis, pancreatitis, anemia etc.

MANGANESE

Manganese is a cofactor of several enzymes of the body and is present in nuclei, complexed with nucleic acids. Total body manganese is 15mg, present mainly in liver and kidneys.

Source—It is widely distributed in foodstuffs, rich sources being nuts and tea leaves.

RDA – 5mg/day

Biochemical Functions

1. Manganese is an essential cofactor of important enzymes like hexokinase, pyruvate carboxylase, isocitrate dehydrogenase and superoxide dismutase.
2. **Synthesis of chondroitin sulphate** and other **glycoprotein**—Mn is an integral part of enzyme **glycosyltransferase** required for the synthesis of mucopolysaccharides and glycoproteins and thus help in laying down organic matrix of bones and cartilages.
3. **Structural integrity of nucleic acids**—As cofactor of enzyme RNA polymerases, manganese is required for RNA synthesis and it helps in maintenance of RNA structural integrity.
4. **Cholesterol biosynthesis**—Manganese helps in condensation of mevalonate into squalene formation, so it is required for cholesterol biosynthesis and steroidogenesis in the body. (Chapter 12).

Deficiency disease in humans is not known but in animals, Mn deficiency is manifested in impairment of growth, bone deformities and in severe cases sterility and decreased carbohydrate tolerance.

Manganese toxicity—Manganese ore miners inhale poisonous mangnese fumes and develop psychotic symptoms called **manganese madness** and Parkinson's disease.

MOLYBDENUM

Biochemical function—Molybdenum is an essential constituent of enzyme **xanthine oxidase** (purine catabolism), **aldehyde oxidase** and **sulfite oxidase**. Enzyme sulfite oxidase is necessary for the oxidation of sulfite derived from sulphur constaining amino acids, cysteine and cystine, to form inorganic sulfate for its excretion.

Molybdenum has been shown to be an essential element in nutrition for the activity of various enzymes. Normal growth is impaired in animals grown on molybdenum deficient diets.

SELENIUM

Selenium in small amounts is necessary in diet as it has important biological functions in body.

Biochemical Role

(a) Most of selenium in blood is present as constituent of **glutathione peroxidase** enzyme that protects the cells (RBC) against damage caused by H_2O_2 and oxy free radicals.

(b) Selenium acts **synergistically** as **antioxidant along with vitamin** E and protects tissues and cell membranes against lipid peroxidation by superoxide anion and thus maintains structural integrity and semipermeability of the biomembranes.

(c) Selenium along with vitamin E prevents **hepatic necrosis, cirrhosis** of liver and muscle dystrophy by scavenging oxyfree radicals.

(d) In the cells, selenium exists complexed with sulphur containing amino acids as **selenocysteine** and **selenomethionine** and may be incorporated as such during enzyme (protein) biosynthesis.

(e) Selenium is an essential constituent of enzyme **5-deiodinase,** that converts thyroxin (T_4) to triiodothyronine, which is prerequisite for its action in the target tissues. Selenium deficiency impairs the activity of 5-deiodinase enzyme leading to **hypothyrodism.**

(f) **Anticancerous activity**—Selenium being a natural antioxidant against free radical damage to tissue nucleic acids, it is considered to have anticancer activity and prevent risk of certain cancers.

(g) Se binds cadmium, mercury and other heavy metals and thus protect body against their toxic effects in case of heavy metal poisoning.

Selenium deficiency states—Certain parts of the world (e.g. Keshan in China) have soil deficient in selenium and **prevelance** of **Keshan's disease** is common, that is characterized by multifocal myocardial necrosis, cardiomegaly and cardiac arrythemias.

Selenium is known to cure this disease. In some other parts of the world Se deficiency in soil is shown to cause liver necrosis, cirrhosis, cardiomyopathy and muscular dystrophy. Selenium prevents these tissue lesions by its antioxidant property.

Selenium toxicity—High Se contents in soil in some regions of world or megadose of selenium, results in selenium toxicity, in which hair loss, dermatitis, hyperirritability and garlic (odour) breath are the main symptoms.

COBALT

As an essential constituent of vitamin B_{12}, cobalt is taken in body as an important nutrient. B_{12} vitamin contains 4% cobalt by weight. Cobalt is shown to stimulate hormone erythropoietin which promote erythropoiesis.

Chronic excess amount of cobalt is toxic, as it cause polycythemia.

CHROMIUM

Chromium as trivalent and hexavalent ions is biologically active form. It stimulates biotin synthesis by intestinal bacteria and is important in our body as **glucose tolerance factor.**

Biochemical Role

1. Chromium in small amount is beneficial in improving glucose tolerance of an individual. Trivalent chromium bound with niacin containing organic compound is a **glucose tolerance factor** present in our body.
2. This glucose tolerance factor (containing trivalent Cr) binds with insulin and **potentiates** its **secretion** from β cell of islets of Langerhan's.
3. Efficiency of binding of insulin on its specific receptors of target tissue cell memberanes is also facilitated by chromium and thus it increases glucose uptake and its metabolism by tissues.

Sources and RDA—Daily requirement is 20-50 μg for an adult. Good sources are brewers yeast, liver, meat and cereals. Cooking food in stainless steel containers improves chromium intake. Total body chromium contents are about 6 mg.

Deficiency of chromium results in following symptoms—

(a) Impaired glucose tolerance is observed in Cr deficiency. Chromium supplementations improve blood sugar levels status in mild diabetes (NIDDM).

(b) Impaired growth, infertility and low sperm count are observed in chromium deficiencies.

Chromium toxicity—It is an occupational hazard, for instance persons working in tanning industries and causes liver and kidney damage.

□□□

25

VITAMINS

Vitamins are certain **organic** compounds required **essentially in diet** in **small amounts** for some **specific biological functions** in the body to maintain normal **health** and **optimum growth** of an organism. Vitamins are activated as coenzymes or hormone like compounds and perform some important physiological functions in our body.

Vitamins are accessory food factors—It has become clear right from the beginning of 20th century, that natural food stuffs contain vitamins as the accessory food factors, apart from minerals, carbohydrates, proteins and fats, that are necessary for good health and growth of the animals. Deficiency of any one of these vitamins in the diet would result in characteristic **deficiency disease.**

Classification of vitamins—Vitamins are classified into two major groups based on their solubility in water or fat solvents.

1. **Water soluble vitamins** include B complex members and vitamin C.
2. **Fat soluble vitamins** are vitamin A, D, E and K, which are soluble in fats and oils as well as in the fat solvents.

WATER SOLUBLE VITAMINS—B COMPLEX

Vitamin B complex includes a group of vitamins which are not related to each other chemically but are grouped together, because their major functions in body is to serve as important **coenzymes**. These coenzymes are necessary for specific enzyme systems that catalyze the reactions for the production and utilization of energy to carryout life processes, as well as for proper functioning of the cells, so as to maintain **good health** and **growth** of an individual.

Multiple Deficiency States and Diseases

It has been observed since long that deficiency of a single vitamin B complex member is rare, as they are widely distributed in nature, as well as they are synthesized in intestine by the bacterial flora. Poor diets and malnutrition often result in multiple deficiency disease state. However definite characteristic deficiency symptoms of a particular vitamin B complex member are well recognized, mostly reported by experimental studies in animals.

Regular Dietary Supply is Essential

Vitamin B complex members are water soluble, thus they are easily excreted out and are not stored much in the body (except B_{12} in liver). Thus they must be **provided regularly in diet** to maintain good health. Indians has a variety of food stuffs in their diet, that ensures regular supply of B complex members to the individuals.

Vitamin B Members Function as Coenzymes

Vitamin B complex members in body are activated by specialized reactions, as important **coenzymes** that are the cofactors, necessary for most of the enzyme catalyzed reactions of major metabolic pathways of the living organisms. Deficiency of vitamin B-complex members would result in impairment of various metabolic cycles leading to cellular dysfunction and thus normal health and growth cannot be maintained. A brief account of coenzymes of important B complex members, their biochemical role and deficiency diseases are summarized in the Table 25.1.

Table 25.1 : Biochemical role and deficiency states of B complex vitamins

B-complex vitamins	Name	Coenzymes	Reactions catalyzed/ biochemical role	Deficiency state
B_1	Thiamin	TPP	Oxidative decarboxylation & Transketolase reaction	Beri-Beri, peripheral neuritis
B_2	Riboflavin	FMN, FAD	Dehydrogenases H^+ transfer/oxidation	Cheilosis Glossitis
B_3	Niacin	NAD^+, $NADP^+$	Dehydrogenases H^+ transfer/oxidation	Pellagra
B_6	Pyridoxine	B_6 PO_4 or PLP	Transamination Heme synthesis Amino acid metabolism	Neuritis Neuropathy Microcytic anaemia
-	Pantothenate	Coenzyme A	Acyl carrier, Fatty acid metabolism	Neuritis Neuropathy Microcytic anaemia
-	Biotin	Biotin	CO_2 fixation. Cocarboxylase	Dermatitis Depression Hallucination
-	Folic acid	THF or FH_4	1C moiety transfer DNA & RNA Synthesis	Megaloblastic anemia
B_{12}	Cyanocobal-amine	Cobamides	Biosynthesis of DNA myelin sheath integrity	Pernicious anemia Neurological lesions.

THIAMIN (VITAMIN B_1)

Thiamin is also known as **aneurine** (as it cures neuritis) or **anti beri-beri factor.** Vitamin B_1 in the body is activated as specific coenzyme called **thiamin pyrophosphate (TPP),** that acts as a cofactor of key enzymes of **carbohydrate metabolism.**

Chemistry—Chemically thiamin consists of a substituted pyrimidine ring joined by a methylene bridge to a thiazole ring as shown in the Fig. 25.1.

Thiamin is activated as TPP—In the presence of enzyme **thiamin pyrophosphate transferase,** ATP donates pyrophosphate group to thiamin, that forms an ester with alcoholic group of the side chain of thiazole ring to produce coenzyme thiamin pyrophosphate (TPP) as shown in the Fig. 25.1.

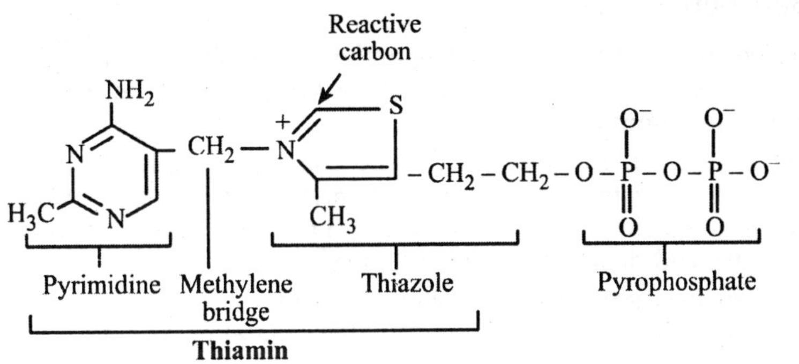

Fig. 25.1 : Chemical structure of thiamin and thiamin pyrophosphate (TPP) coenzyme

Biochemical Functions

Thiamin pyrophosphate is an active coenzyme of enzyme catalyzed reactions of glucose metabolism, such as glycolysis, citric acid cycle and HMP shunt pathways. The TPP coenzyme specifically transfers an activated 2C- atom moiety with an aldehyde group in the reactions given below (details in Chapter 5).

1. **Oxidative decarboxylation** of pyruvate (end product of glycolysis) to from acetyl CoA.
2. **Oxidative decarboxylation** of α ketoglutarate to succinyl CoA (Citric acid cycle).
3. **Transketolase** reaction of HMP shunt pathway, where TPP transfers an active 2C moiety with an aldehyde group. (Chapter 7).

Dietary sources—Cereals, pulses, oil seeds, nuts and yeast are good sources of the thiamin. Thiamin is mostly concentrated in the outer layer of grains, so bran and unpolished rice should be taken in diet. Animal sources include meat, liver, egg yolk and milk

RDA—Daily requirement of thiamin increases with increase in carbohydrates in diet, as thiamin is required for glucose metabolism. Recommended daily allowance is therefore 0.5 mg/1000 calories/day.

Biochemical Alterations in Thiamin Deficiency

In thiamin deficient subjects, TPP dependent reactions of glucose metabolism are impaired resulting in the accumulation of pyruvate and lactate in the muscles and other tissues. Impaired glucose metabolism results in reduced amount of ATP generation in muscles and nerves leading to **peripheral neuritis, muscle weakness** and **muscle wasting** etc, that are characteristic symptoms of B_1 deficiency disease.

α-Keto carboxyl derivatives of branched chain amino acids leucine, isoleucine and valine may also accumulate in thiamin deficiency causing neuronal damage as TPP is required as coenzyme in their metabolism (details in the Chapter 16).

Beriberi—Thiamin deficiency disease

Beriberi is caused by carbohydrate rich, low thiamin diets such as polished rice or food stuffs made from highly refined wheat flour and sugar as staple diet. Early symptoms of beriberi include **anorexia** (loss of appetite) weakness, exhaustion, numbness of legs, frequent pinprick sensation, tiredness, muscle cramps etc, finally leading to **peripheral neuritis.**

Types of Beri-Beri

1. Dry beriberi—In this condition neurological manifestations are mainly observed with muscle weakness, tiredness, **difficulty in walking, muscle degeneration** that finally progress to **peripheral neuritis.**

2. Wet beriberi—It is accompanied by **neurological symptoms** along with **retention** of water and **edema** of legs that may progress into **muscle wasting** and **cardiac involvement.** Clinical complications include breathlessness, palpitation, increased systolic blood pressure, weakening of heart, and death due to heart failure.

3. Infantile beriberi—It affects infants of vitamin B_1 deficient mothers. Symptoms include irritability, vomiting, convulsions, cardiac dilatation and sudden death due to cardiac failure.

4. Wernicke Korsakoff Syndrome : Cerebral beriberi—Severe deficiency of B_1 associated with chronic alcoholism and malnutrition results in **encephalopathy** characterized by apathy, loss of memory, rhythmical to and fro eye ball movements etc. and the condition is referred to as Wernicke Korsakoff syndrome.

Thiamin deficiency may also occur in conditions listed below, due to—

a. Excess of alcohol intake increases thiamin demands and thus chronic alcoholism results in B_1 deficiency state.

b. Raw fish and certain sea foods contain enzyme thiaminase, that destroys thiamin resulting in vitamin B_1 deficiency disease.

c. Thiamin antagonists such as **pyrithiamine** is present in certain plants e.g. ferns. If these are eaten by animals (horse, cattles etc) they may develop thiamin deficiency state.

RIBOFLAVIN (VITAMIN B_2)

Riboflavin was the first B complex vitamin to be isolated in the pure form. This vitamin is synthesized by all green plants and microorganisms. Warberg (Nobel prize, 1931) was the pioneer scientist who isolated yellow flavoprotein enzyme and showed that it is essential for the **cellular respiration. Warburg's yellow enzyme** was then shown to contain a flavin cofactor called riboflavin.

Chemical Nature

Riboflavin consists of a heterocyclic 3 rings structure called as isoalloxazine ring attached to ribitol (sugar alcohol) as shown in the Fig. 25.2. Riboflavin is a yellow fluorescent pigment, that is heat stable but decomposes when exposed to light.

Coenzymes of riboflavin—Riboflavin in body is activated by ATP to coenzyme **flavin mononucleotide (FMN)** and further by addition of another adenine nucleotide from ATP to **flavin dinucleotide (FAD)** coenzymes as shown in the Fig. 25.2.

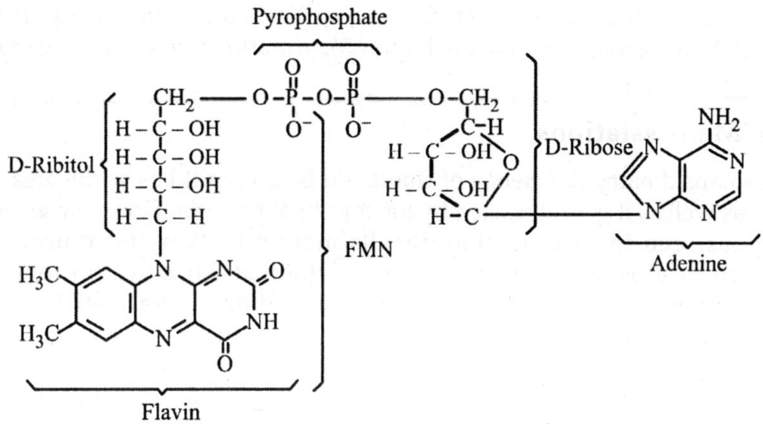

Fig. 25.2 : Structures of riboflavin, FMN and FAD coenzymes

Biochemical Functions

The flavin coenzymes, FAD and FMN participate in many important oxidation reduction reactions of the body and are prosthetic groups of flavoprotein enzymes. Many of these flavoprotein enzymes also have some metals as essential cofactors and are called **metalloflavoproteins.**

FMN Dependent Oxidoreductases

There are two important FMN containing enzymes, as described below—

1. Amino acid oxidases—These are autooxidizable flavoproteins that oxidatively deaminate L-amino acids as illustrated in the Fig. 15.4. Hydrogen peroxide produced in the reaction is degraded by enzyme **catalase.**

2. NADH dehydrogenase—On respiratory chain the NADH dehydrogenase (complex I) contains FMN coenzyme that transports electrons from reduced NADH to coenzyme Q (Chapter 8) for biological oxidation and ATP generation.

FAD dependent enzymes

There are several FAD dependent **oxidoreductase** enzymes present in the body. Some important FAD dependent enzymes are listed here along with the reactions they catalyze.

1. Succinate $\xrightarrow[\text{FAD} \rightarrow \text{FADH}_2]{\text{Succinate dehydrogenase}}$ Fumerate (Citric acid cycle).

2. Fatty acyl CoA $\xrightarrow[\text{FAD} \rightarrow \text{FADH}_2]{\text{Fatty acyl dehydrogenase}}$ α, β unsaturated fatty acyl CoA (β Oxidation)

3. Hypoxanthine —[Xanthine oxidase; FAD → $FADH_2$]→ Xanthine (Uric acid formation)

Other important oxidoreductases having FAD coenzyme are pyruvate and α-Ketoglutrate dehydrogenase complex (Chapter 5), glutathione reductase, glycine cleavage enzyme etc.(Chapter 16).

Deficiency Manifestations

In humans dietary deficiency of riboflavin is rare, as it has widespread occurrence in the nature, as well as it is synthesized by the intestinal bacteria. However severe deficiency of riboflavin, as seen in malnutrition, results mainly in skin and mucous membrane symptoms such as **glossitis** (magenta tongue), **cheilosis** (fissures on tongue), angular stomatitis (inflammation at the corners of mouth), **seborrhic dermatitis**, circumcorneal **vascularization** and **photophobia.**

Dietary sources and RDA—Rich sources of riboflavin are liver, yeast, egg, whole milk etc., while plant food stuffs are moderate sources and include whole grains (cereals), legumes, green leafy vegetables and tuber roots.

RDA – 1.5 mg/day for adults.

NIACIN

Niacin or nicotinic acid is a vitamin B complex member and is also known as **pellagra preventing factor**. Niacinamide (Nicotinamide), an acid amide of niacin present in natural foodstuffs, is also a source of vitamin in the diet.

Chemical Nature

Niacin is a 3- carboxylic derivative of pyridine while niacinamide is an acid amide of niacin as shown in the Fig. 25.3.

Coenzymes of Niacin are NAD^+ and $NADP^+$

NAD^+ (Niacin adenine dinucleotide) and $NADP^+$ (Niacin adenine dinucleotide phosphate) coenzymes are synthesized in the body by first condensation of niacin with ribose phosphate and then with AMP to produce NAD^+, which is then phosphorylated by ATP to $NADP^+$. The structures of NAD^+ and $NADP^+$ are shown in the Fig. 25.3.

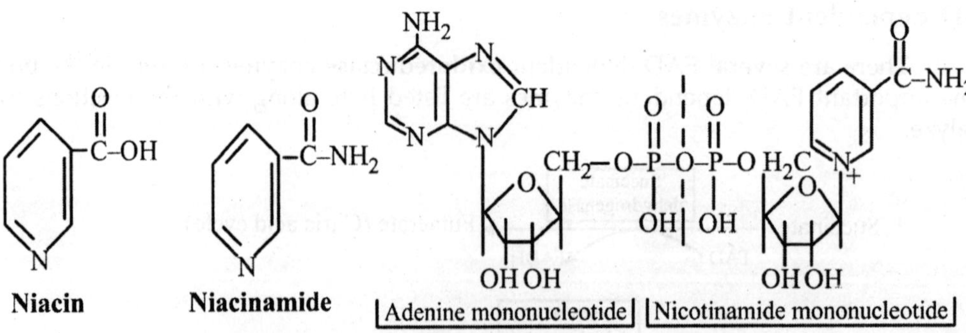

Nicotinamide Adenine dinucleotide (NAD^+)

Fig. 25.3 : Structures of niacin, niacinamide and NAD^+ ($NADP^+$ has one extra PO_4 group)

Biochemical Functions

NAD and NADP are coenzymes, required by oxidoreductase enzymes that are key components of most of the metabolic pathways of carbohydrates, fats and amino acid metabolism. Generally **NAD^+ linked dehydrogenases** initiate oxidation by removing H^+ ions from metabolites undergoing catabolism e.g. in glycolysis, citric acid cycle etc. and transfer them to the respiratory chain (ETC) for biological oxidative phosphorylation to trap energy of oxidation as ATP.

Whereas $NADP^+$ linked reductase (dehydrogenase) are mostly involved in the **reductive biosynthesis** of fatty acids, cholesterol, steroid hormones, amino acids etc and bring reducing equivalents (NADPH + H^+) from alternative glucose metabolic process, called HMP shunt pathway (Chapter 7).

Deficiency disease—Pellagra

Niacin deficiency results in a condition called **pellagra** (meaning rough skin), that affects skin, GIT and central nervous system. The symptoms of pellagra are commonly referred to as **three D_S** as **dermatitis, diarrhea** and **depression** or **dementia.** Dermatitis occurs usually on the areas exposed to sunlight such as neck, face, ankle and feet etc. GIT disorders include diarrhea, which if prolonged leads to weight loss. Dementia is associated with CNS lesions causing irritability, memory loss, insomnia, depression, delirium etc.

Niacin deficiency disease pellagra is commonly seen amongst people whose staple diet is corn or maize, that lack free vitamin niacin (present as bound, unavailable form), as well as it lacks amino acid tryptophan, which in body is normally metabolized into niacin and its coenzyme production. Approximately 1 mg niacin is synthesized in the body from 60 mg of tryptophan provided in the diets.

Sorghum (Jawar) is rich in amino acid leucine, that inhibits conversion of tryptophan into niacin and NAD^+, causing pellagra like condition, if is taken as staple diet.

Other conditions leading to pellagra include B_6 deficiency caused by tuberculosis drug isoniazid and genetic disorder **Hartnup's disease** in which tryptophan reabsorption from renal tubules is impaired.

Therapeutic Use of Niacin

Niacin has been used clinically for lowering blood cholesterol levels. Lipoproteins VLDL and LDL are also lowered by the therapeutic use of niacin and thus is used in treatment of hyperlipoprotenemia type II to prevent risks of cardiovascular diseases. This action of niacin is because it inhibits lipolysis in adipose tissue and reduce triglyceride and cholesterol synthesis in the liver and thus VLDL and LDL serum levels are decreased.

Megadose of niacin however can result in liver damage, so it should be used therapeutically in low pharmaceutical doses.

Dietary Sources and RDA

Rich sources of niacin are yeast, whole grain cereals, pulses, beans, peanuts etc. Moderate sources include milk, fish, egg and vegetables. Good quality proteins rich in amino acid tryptophan, such as poultry and milk proteins reduce dietary requirement of niacin.

RDA—13-17 mg/day for adults

PYRIDOXINE (VITAMIN B_6)

Vitamin B_6 consists of three closely related pyridine derivatives, known as **pyridoxine, pyridoxal** and **pyridoxamine,** which are interconvertible in the body.

Chemistry—Vitamin B_6 compounds are substituted pyridine derivatives. They differ from each other in having a different functional group at position 4th of pyridine ring, as pyridoxine has a primary alcoholic group, pyridoxal has an aldehyde group and pyridoxamine has an amine group as shown in the Fig. 25.4.

Coenzyme B_6 PO_4—The active coenzyme of B_6 is **pyridoxal phosphate (PLP),** which is synthesized in body in the presence of enzyme **pyridoxal kinase.** The phosphate group is donated by ATP, that forms an ester with primary alcoholic group of the side chain present at 5th position of pyridine as shown in the Fig. 25.4.

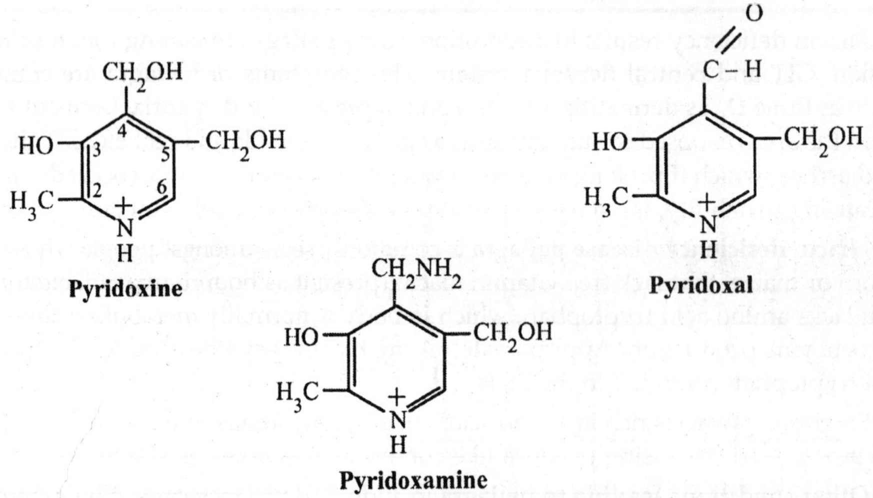

Fig. 25.4 : Structures of B_6 vitamins, pyridoxine, pyridoxal, pyridoxamine and coenzyme pyridoxal phosphate (PLP)

Biochemical Functions

Coenzyme pyridoxal phosphate participates actively in most of the enzyme catalyzed reactions of **amino acid metabolism.** Important examples are given below—

1. Transamination—All amino acids are metabolized by transamination reactions, in which the α-amino group of amino acid is transferred mainly to α -keto glutarate or pyruvate and is concentrated in glutamic acid or alanine. These amino acids then in liver, get deaminated and the ammonia (toxic) liberated is converted into **urea** (less harmful) and excreted in the urine. The role of B_6 PO_4 (PLP) in the transamination reaction is illustrated in Fig. 25.5.

During transamination reaction, the amino group of α-amino acid forms a schiff base with CHO group of PLP, that is changed to aldimine and is converted into pyridoxamine phosphate (Fig. 25.5). Then this amino group of pyridoxamine phosphate is transferred to a keto group of another acceptor (α keto glutarate or pyruvate) to form glutamate or alanine and the coenzyme PLP is generated back. The two most common transaminases are alanine aminotransferase and aspartate aminotransferase (Chapter 15).

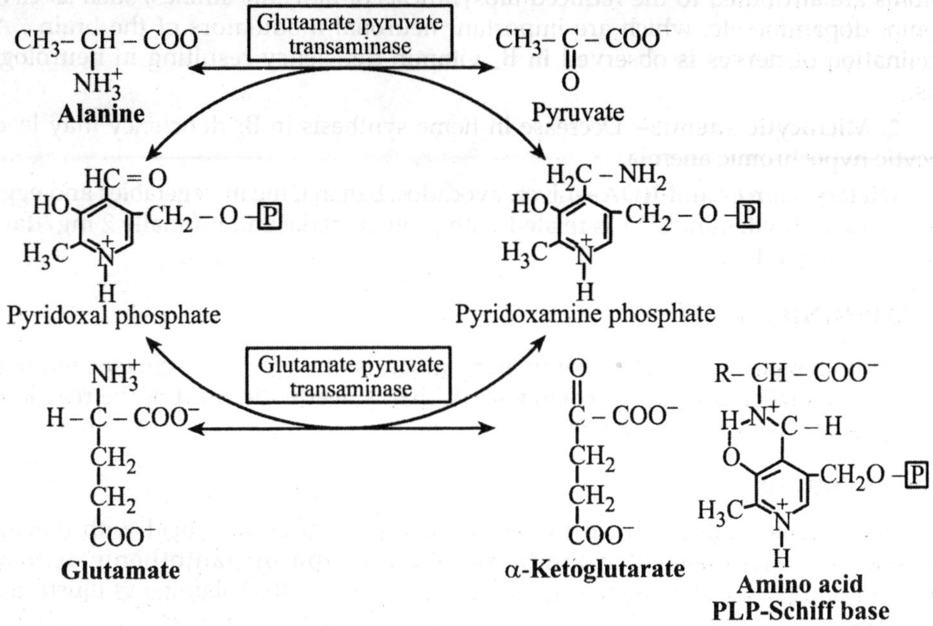

Fig. 25.5 : Role of pyridoxal phosphate (PLP) in transamination reaction

2. Decarboxylation—Many α -amino acids undergo decarboxylation in the presence of specific **decarboxylase enzymes**, where **PLP** is an essential coenzyme and are converted into some **very active primary amines,** that function as **potent hormones** in the body. For instance, synthesis of catecholamines from tyrosine, serotonin from tryptophan, GABA from glutamate, histamine from histidine amino acid etc. all require decarboxylation and PLP as coenzyme. (details are given in Chapter 16).

3. Synthesis of heme—Pyridoxal phosphate (PLP) essentially required as a coenzyme of key enzyme of heme synthesis, called **amino levunic acid synthase (ALA synthase).**

4. Trans sulfuration of S-containing amino acids methionine, cysteine and cystine is an important reaction in their metabolism, where pyridoxal PO_4 plays a key role. (details in chapter 16). After deamination and desulfuration, these amino acids are converted to pyruvate, that enters gluconeogenesis process to synthesize glucose.

5. Other important coenzyme function of PLP include-1) Deamination of serine and threonine, 2)Tryptophan metabolism via kynurenine pathway to vitamin niacin synthesis, 3) Serine synthesis from glycine and 4) Glycogen metabolism—Key enzyme phosphorylase of glycogenolysis, require PLP as an activator.

Deficiency manifestations—Dietary deficiency of B_6 vitamin is rare as it is widely distributed in nature. Deficiency manifestations are usually associated with chronic alcoholism (increase demand of B_6) and intake of anti tuberculosis drug isoniazid (it forms hydrazone with pyridoxal and inactivates the vitamin) and may result in deficiency state. The clinical symptoms include—

1. Neurological symptoms such as depression, irritability, mental confusion and in severe cases e.g. in infants, convulsions and peripheral neuropathy are observed. These symptoms are attributed to the reduced biosynthesis of biogenic amines, such as GABA, serotonin, dopamine etc. which are important neuronal modulators of the brain . Also demyelination of nerves is observed in B_6 vitamin deficiency resulting in neurological lesions.

2. Microcytic anemia—Decrease in heme synthesis in B_6 deficiency may lead to microcytic hypochromic anemia.

Dietary sources and RDA—Liver, avocados, banana, meat, vegetables and egg are good sources of B_6 vitamin. RDA is related with protein intake and normally 2 mg/day for adults is recommended.

PANTOTHENIC ACID

Pantothenate, a vitamin B complex member, is of universal occurrence in nature and forms essential constituent of **coenzyme A,** that is **needed** in most of the reactions of fatty acid metabolism as an acyl carrier.

Chemical Nature

Pantothenic acid is formed by the condensation of pantoic acid with β-alanine. Coenzyme A is synthesized in the body by activation of pantothenic acid with adenine nucleotide and then by binding wih SH-containing ethanolamine as illustrated in the Fig. 25.6.

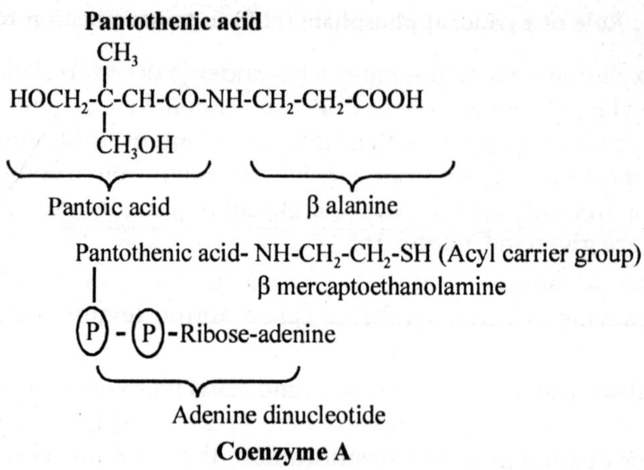

Fig. 25.6 : Chemical structures of pantothenic acid and coenzyme A

Biochemical Functions

The thiol group (SH) of β mercaptoethanolamine of coenzyme A forms a high energy bond on which acyl groups are attached and are activated to participate in various metabolic pathways.

Some important coenzyme A functions are described here.

1. Active acetate—Metabolic end product of carbohydrate, fatty acid and amino acid metabolisms is acetyl CoA, that plays central role in the final oxidation in citric acid cycle as well as in numerous biosynthetic pathways as illustrated in the Fig. 25.7.

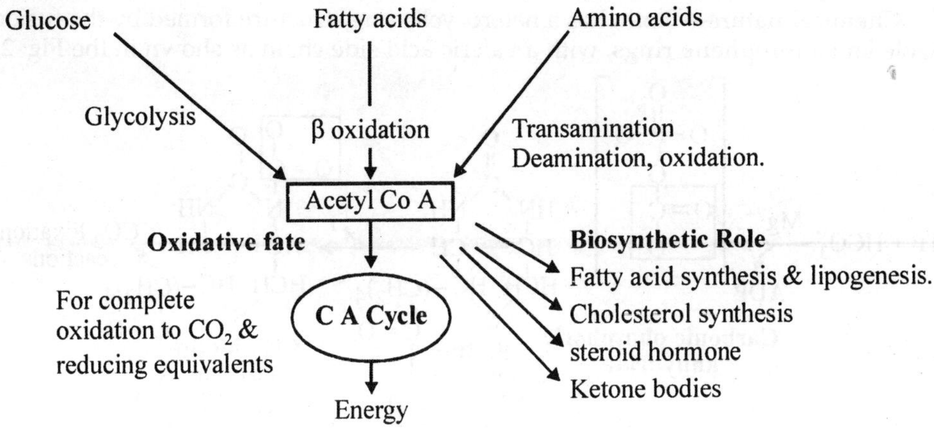

Fig. 25.7 : Formation and utilization of active acetate

2. Succinyl Co A—As a carrier of succinate, succinyl CoA plays key role in some major metabolic and biosynthetic pathways as shown in the Fig. 25.8.

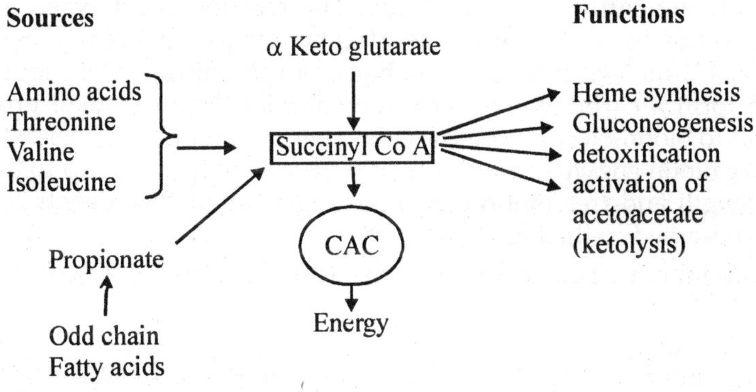

Fig. 25.8 : Sources and functions of succinyl CoA

3. Activation of fatty acids—Fatty acids are activated as acyl CoA before they can enter the metabolic cycle i.e. β oxidation pathway. For biosynthesis of long chain fatty acids, cholesterol and amino acids, the activation of fatty acids as acyl CoA is necessary.

Deficiency manifestations of pantothenate are rare in humans as it's occurrence is widespread in the nature. However severe deficiency state, as seen in extreme famine conditions causes burning foot syndrome, loss of body movement coordination and abnormal gait etc.

In experimental animals pantothenate deficiency resulted in microcytic anaemia (due to reduced heme synthesis), dermatitis, adrenal necrosis and fatty liver.

Dietary sources and RDA—Pantothenate is widely distributed in nature, as well as it is synthesized by intestinal bacteria, so RDA is not well documented.

BIOTIN

Biotin, also known as **anti egg white injury factor,** is a sulphur containing B complex vitamin. It directly performs coenzyme role in CO_2 fixation and carboxylation reactions to increase C-chain length of a metabolite in a variety of biosynthetic pathways.

Chemical nature—Biotin has a heterocyclic ring structure formed by the fusion of imidazole and a **thiophene rings,** with a valeric acid side chain as shown in the Fig. 25.9.

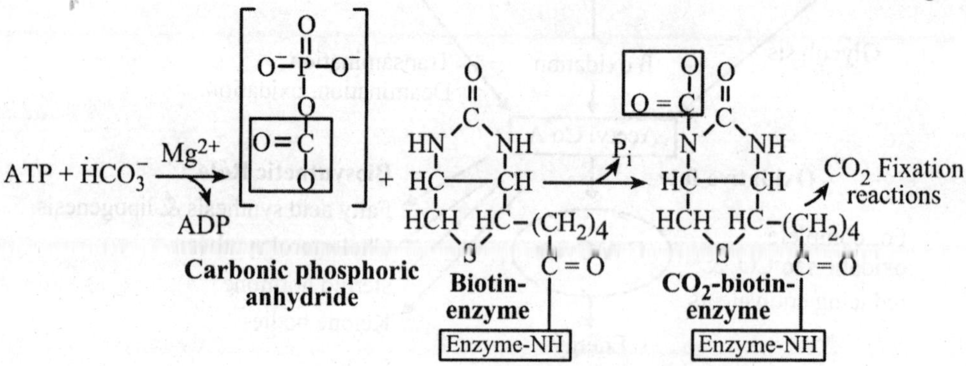

Fig. 25.9 : Structure of biotin and its coenzyme role in CO_2 fixation reaction

Coenzyme Role—Biotin plays central role as conenzyme for many important **carboxylation** (CO_2 fixation) reactions to increase the C-chain length of a metabolite in a variety of biosynthetic pathways in our body. The reaction is catalyzed by a multiunit **carboxylase** enzyme to which biotin is attached through lysine residue. During carboxylation, first a molecule of CO_2 gets bound to the nitrogen of biotin (attached to carboxylase) to form a carboxybiotin enzyme complex. The carboxyl group here gets covalently bound to biotin-enzyme complex and energy required is provided by ATP. This activated carboxy biotin enzyme complex then transfers carboxyl group to the substrate to increase its C- length and free biotin-enzyme is regenerated. The events of CO_2 fixation reactions are represented in the Fig. 25.10.

Some important carboxylation reactions (CO_2 fixation) are described here under.

1. Acetyl CoA Carboxylase is the key enzyme of **fatty acid synthesis** and **steroidogenesis** to increase C-chain length. Biotin plays an important role in CO_2 fixation reaction as a cofactor of the enzyme acetyl CoA carboxylase, as shown in the reaction given below (details in the Chapter 11).

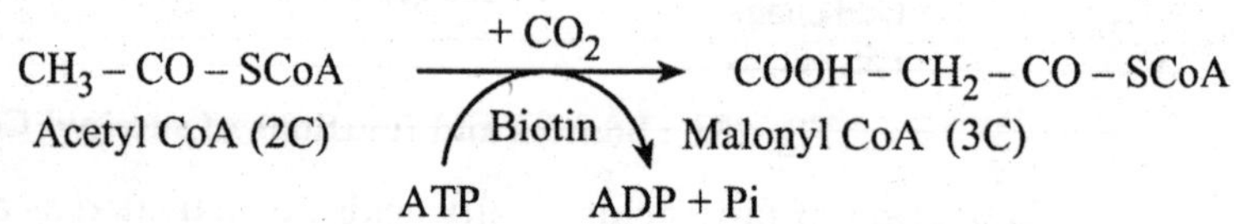

2. Pyruvate Carboxylase enzyme catalyzes the fixation of CO_2 to carboxylate pyruvate (3C- compound) to produce oxaloacetate (4C- compound). This is a key reaction to provide oxaloacetate as the starter of TCA cycle (Chapter 7) and secondly to convert lactate and pyruvate to oxaloacetate, the key reaction of the gluconeogenesis process (Chapter 7 Fig. 7.1).

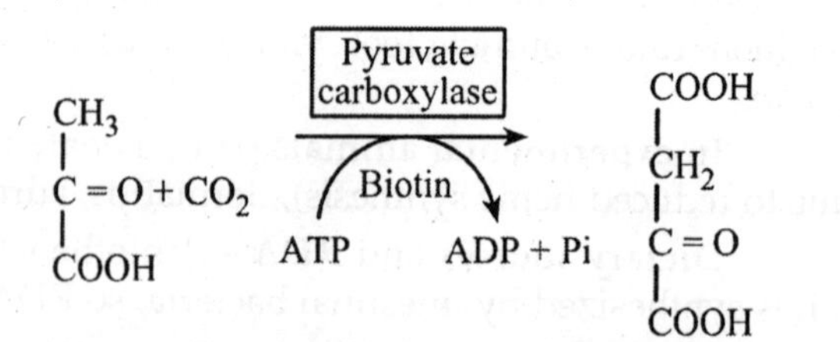

3. Propionyl CoA Carboxylase—It is an important reaction through which propionate, the end product of β oxidation of odd number fatty acids, can enter the gluconeogenesis pathway (Chapter 7) as described below—

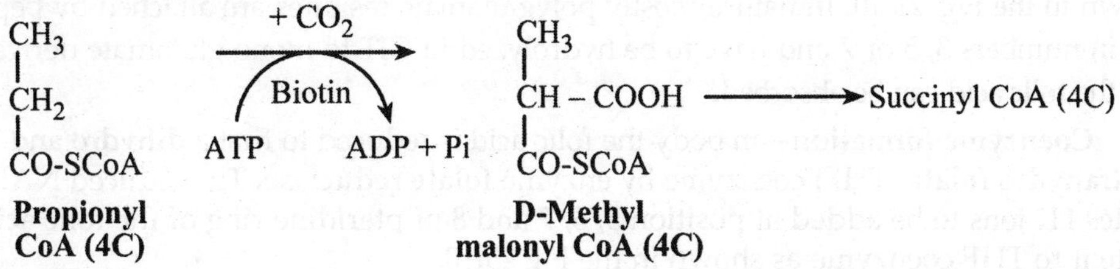

During this carboxylation reaction, first a D isomer of methylmalonyl CoA is formed, that is racemized to L form and then is isomerized to succinyl CoA by a vitamin B_{12} dependent enzyme (Chapter 7). The succinyl CoA then enters gluconeogenesis pathway.

4. Leucine Metabolism—In amino acid leucine metabolism, the intermediate β–methyl crotonyl CoA is carboxylated (Co_2 fixation) in the presence of biotin dependent carboxylase enzyme for further catabolism. (Chapter 16).

Deficiency Manifestations

Dietary deficiency of biotin is rare as it is synthesized by intestinal bacteria. However deficiency of biotin may be caused by eating raw egg or by the prolonged use of antibiotic drugs. Raw egg contains a protein **avidin,** that tightly binds biotin and impairs its absorption from GIT, resulting in biotin deficiency. The deficiency symptoms include dermatitis, depression, hallucinations and muscle pain etc.

Infants with persistent diarrhea may develop biotin deficiency (Leiner's disease)

FOLIC ACID

Folic acid or folacin (folium means leaf) is a vitamin B complex member and is widely distributed in green leafy vegetables. The vitamin is water soluble and is destroyed rapidly on exposure to light.

Chemistry–Folic acid consists of **three components,** a heterocyclic **pteridine ring** linked to para amino benzoic acid and glutamate residue

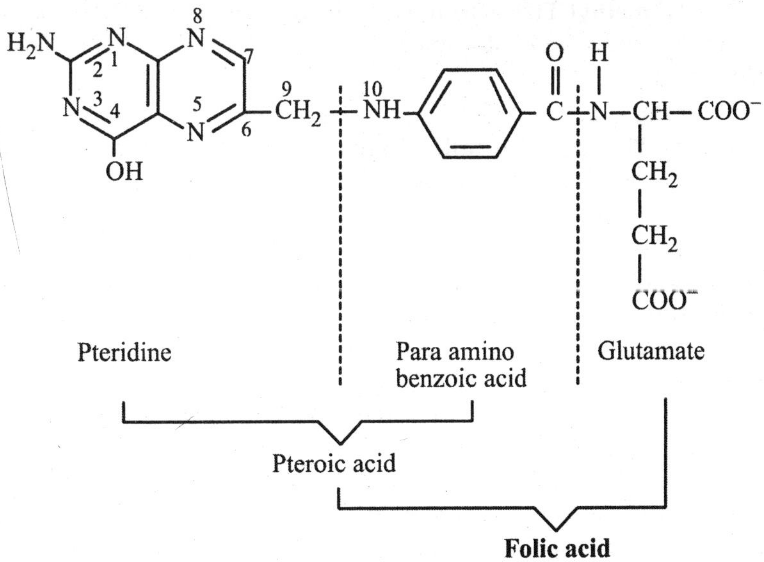

Fig. 25. 10 : Chemical structure of Folic acid

as shown in the Fig. 25.10. In nature mostly polyglutamate residues are attached by peptide bonds in numbers 3, 5 or 7 and have to be hydrolyzed in GIT to mono glutamate derivative before the folic acid can be absorbed.

Coenzyme formation—In body the folic acid is reduced to first a **dihydro** and then to a **tetrahydro folate** (THF) coenzyme by enzyme **folate reductase.** The reduced NADPH provides H^+ ions to be added at position 5, 6, 7 and 8 of pteridine ring of the folic acid to activate it to THF coenzyme as shown in the Fig.25.10.

Biochemical Functions

Tetrahydrofolic acid coenzyme (THF) plays central role in **one C-atom metabolism** and is necessary for the biosynthesis of **purine and pyrimidine nucleotides** required for DNA and RNA synthesis. Cell division, protein synthesis and growth process are all dependent on THF that is required for DNA replication as well as for RNA formation. Other important compounds synthesized by tetrahydro folate(THF) are amino acid serine, glycine, choline and creatine phosphate etc.

One C*Atom Metabolism

Tetrahydro folate is the carrier of 1C* units, such as formyl, formimino, methenyl and methyl groups. They are all **interconvertible** and are in various stages of oxidation/ reduction states. This is important to continue the 1C atom metabolism.

Sources of 1C* atom units—Amino acids **serine, glycine, tryptophan and histidine** are **major** sources of 1C atom units as formyl, formimino and methylene groups that are accepted by THF at **N^5, N^{10}** positions and play central role in 1C atom metabolism as shown in the Fig. 25.11.

Conversion of Homocysteine to Methionine

N^5 methyl THF is formed by the reduction of N^5, N^{10} CH_2- THF and has important role in the methylation of homocysteine to active methionine. Vitamin B_{12} is necessary as cofactor.

One C-atom pool—The generation and utilization of 1C-atom units is under the control of THF and are linked with methionine metabolism through cobamides, the coenzymes of vitamin $B_{12.}$ Methionine so produced (Fig. 25.11) is activated as S-adenosyl methionine (with high energy S~ CH_3 bond), and participates in transmethylation reactions for the biosynthesis of several important physiologically active compounds (Chapter 16). A brief account of 1C-atom metabolism and biological functions of THF and B_{12} coenzymes are illustrated in the Fig. 25.11.

Terahydro Folate Trap—THF (FH_4) is mostly trapped as N^5 methyl- THF during 1C-atom moiety metabolism as shown in the Fig. 25.11 Vitamin B_{12} is necessary for the utilization of N^5 CH_3 -THF for the methionine formation from homocysteine and for the liberation of free THF (FH_4) that can be again used to continue the 1C atom metabolism (trapping and utilization). In B_{12} deficiency, liberation of THF from N^5 methyl THF is blocked leading to folic acid deficiency state.

This is the major route by which N^5 CH_3 –THF donates its methyl group to homocysteine in the presence of B_{12} dependent methyl transferase enzyme. B_{12} deficiency results in lowering of free THF body pool and impairs purine nucleotide and DNA and

RNA biosynthesis. Thus both THF and B_{12} are necessary for DNA and RNA synthesis and for cell division and growth process.

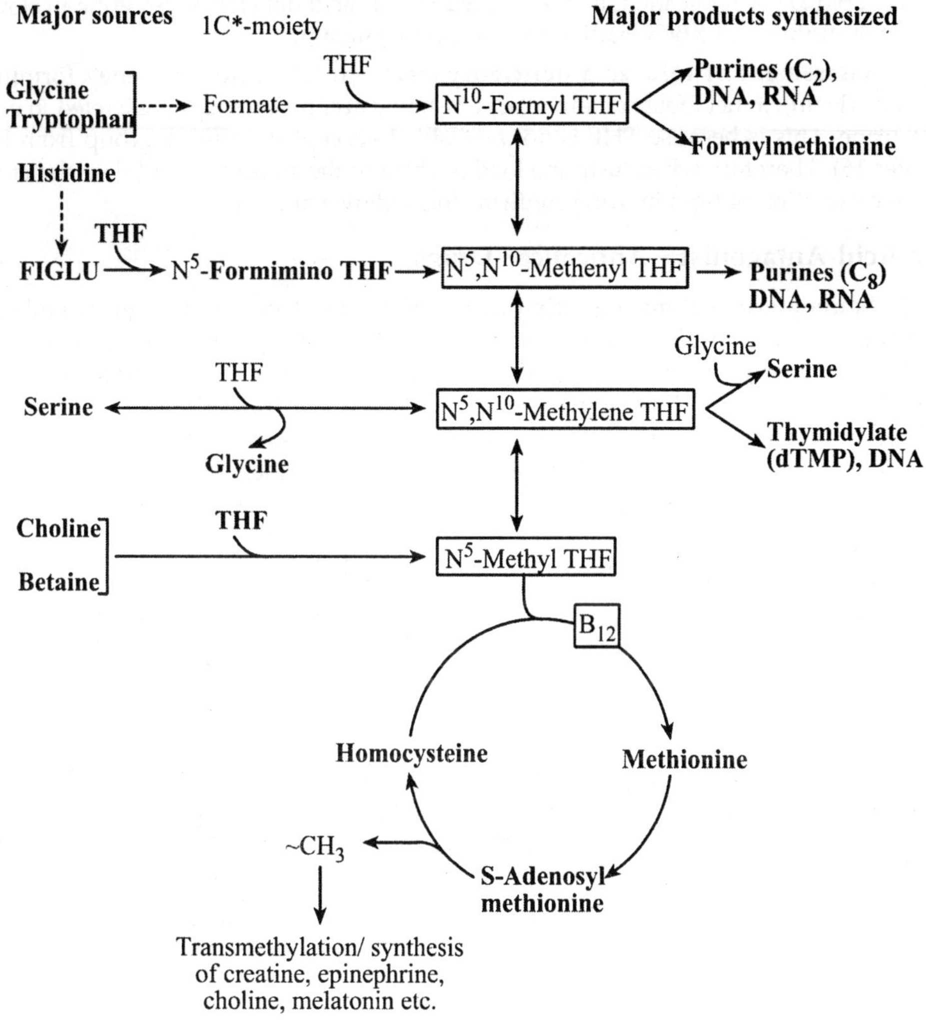

Fig. 25.11 : Sources and utilization of 1C-atom units and role of THF in important biosynthesis processes

Dietary sources and RDA—Folic acid is widely distributed in nature and is also synthesized by intestinal bacterial flora. Rich sources are green leafy vegetables, whole grain cereals, egg, yeast and liver. RDA is 100 μg/day for adults. During pregnancy folic acid requirement increases for fetal development. Around 300 μg/day are recommended for pregnant women.

Deficiency Manifestations

Folic acid deficiency is most common vitamin deficiency observed **during pregnancy,** where its demand is increased for the fetal development. Folic acid deficiency causes **megaloblastic (macrocytic) anemia.** This is because, deficiency of folic acid causes **reduction** in purine nucleotide synthesis as well as decreased d-TMP production, which are necessary for DNA replication. Rapidly multiplying tissues, for example erythropoiesis in bone marrow requires rapid DNA replication, that is blocked in folic acid deficiency, resulting in slowing down of maturation of RBCs leading to **macrocytic anemia.**

Assessment of folic acid deficiency state—In folic acid deficiency formimino glutamate (Figlu), a metabolite of amino acid histidine accumalates and is excreted in excess in the urine. This is because THF is not available to accept formimino group from Figlu (Chapter 16). Therefore when histidine load is given to the subjects with folate deficiency, excessive excretion of figlu in urine confirms folate deficiency state.

Folic Acid Antagonist - Anticancer Drugs

Aminopterin and **methopterin** (Methotrexate) are structure analogues of folic acid and **competitively inhibit** enzyme folic **acid reductase,** that activates folic acid to THF coenzyme. In absence of THF, biosynthesis of purine nucleotide and dTMP are decreased resulting in a **block in DNA synthesis** and thus **cell multiplication** (proliferation) is impaired in cancerous tumor. Thus **aminopterin** and **methotrexate** are successfully used as **anticancerous drugs** in the treatment of leukemia and other cancers where radiation therapy cannot be used.

VITAMIN B_{12}

Vitamin B_{12} is also known as cyanocobalamine and is a heat stable vitamin, that is red in colour. It contains cobalt as essential constituent and a complicated ring structure called corrin ring resembling to that of porpyhrins.

Chemistry

Vitamin B_{12} has a complex chemical ring structure called **corrin** ring, which is similar to porphyrin nucleus. But unlike heme it has **cobalt** instead of iron (Fe^{++}) at the center, attached to the nitrogen of 4 pyrrole rings. To the cobalt ion (Co) at the center of corrin ring, one cyanide group is attached and by its 6th valency it is attached to (below the corrin plane) a different nucleotide structure having **dimethyl benzimidazole (DMB)** instead of adenine, linked to **ribose** 5-phosphate and is attached to **aminoisopropanol** as illustrated in the Fig. 25.12.

Unique vitamin—Vitamin B_{12} is an unique vitamin as it is **not** synthesized either by plants or by animals. It is synthesized **exclusively** and only by **microorganisms.** Thus vitamin B_{12} is absent from plant foodstuffs, but it is synthesized by **intestinal bacterial flora.**

Absorption, Transport and Storage

The absorption of vitamin B_{12} (also called as **extrinsic factor of castle**) from intestine require a specific **glycoprotein** secreted by parietal cells of gastric glands into gastric juice, known as **intrinsic factor (IF).** This intrinsic factor is a dimer and binds 2 molecules of vitamin B_{12} per mole. This binding protects B_{12}, as intrinsic factor is resistant to digestive enzymes. This bound complex (B_{12}-IF) then binds to specific **receptor** sites on the surface of mucosal lining of ileum and via a Ca^{++} mediated transport system, the B_{12} -IF complex is absorbed.

After absorption B_{12} is separated from IF and is transported to the liver via blood, bound to a plasma protein called **transcobalamin.** In liver B_{12} is stored as three derivatives, methylcobalamin (CN group is replaced by CH_3), adenosylcobalamin (CN group is replaced by adenosine) and hydroxycobalamine(OH group instead of CN).

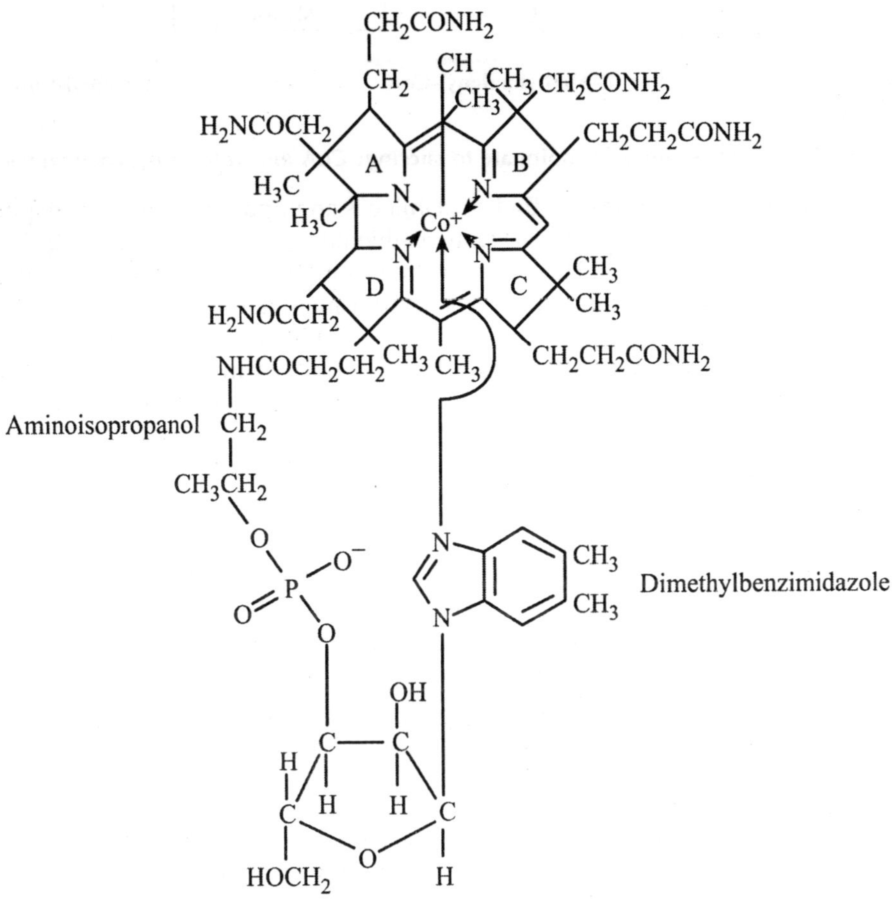

Coenzymes of B_{12} vitamin—In mammals two active coenzymes of vitamin B_{12} are found to occur. They are **methylcobalamin** and **deoxyadenosyl cobalamin.** These coenzymes of B_{12} are essential component of enzyme systems required for two important metabolic pathways as given below—

1. **Deoxyadenosylcobalamin** is required for the metabolism of propionate. A key reaction in propionate metabolism is conversion of methyl malonyl CoA into succinyl CoA (Chapter 7). Succinyl CoA is a member of citric acid cycle and can be utilized for denovo synthesis of glucose. The main fate of propionate (3 C atom compound), that is left after β oxidation of odd C-number fatty acids is to enter gluconeogenesis process. Propionic acid is the major end product of microbial fermentation in the GIT of **ruminants,** that is absorbed and is converted into glucose. Details of the reactions are given in the Fig. 25.13.

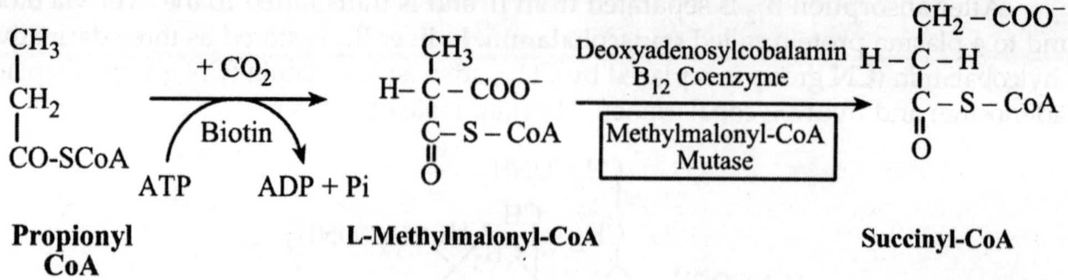

Fig. 25.13 : Conversion of propionate to succinyl CoA and role of B_{12} coenzymes

2. **Methylcobalamine**—Methylcobalamine is an important coenzyme required for the conversion of homocysteine to methionine, as well as in liberating THF from N^5-methylTHF trap. This coenzyme role of **methylcobalamine** is very important in maintaining methionine levels for transmethylation reactions, as well as for the release of THF (FH_4), form the N^5 CH_3 - THF trap. So that it is made available for its repeated use in accepting 1C-atom units form various donors, required essentially for purine nucleotide, d-TMP and DNA synthesis as shown in the Fig. 25.14.

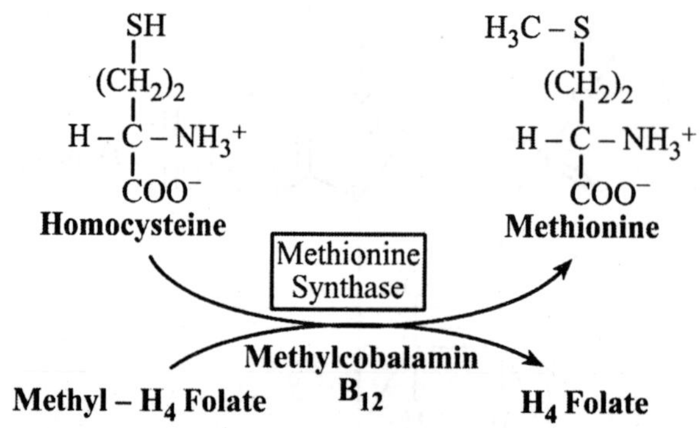

Fig. 25.14 : Coenzyme role of methylcobalamin in libration of THF from methyl trap during conversion of homocysteine to methionine

Dietary sources and RDA—Food stuffs of animal origin that store $B_{12,}$ are the major sources of vitamin $B_{12.}$ Rich sources are liver, egg, tish, pork, milk and curd. RDA is 3 ¼g/ day for adults.

Deficiency Manifestations

Vitamin B_{12} deficiency mainly occurs when its absorption is impaired due to the lack of intrinsic factor. The insufficiency of intrinsic factor can occur either due to autoimmune destruction of gastric glands accompanied by **achlorhydria** or due to surgical removal of stomach . B_{12} deficiency leads to following disorders—

1. Pernicious anemia—It is characterized by **megaloblastic immature RBCs** in bone marrow as well in the blood circulation, due to decreased DNA synthesis. This is caused by unavailability of free THF for purine and pyrimidine nucleotide synthesis, since it remains trapped as N^5 methyl THF (folate trap) in vitamin B_{12} deficiency as explained above (Fig. 25.14).

2. Neurological symptoms—Vitamin B_{12} deficiency also results in **neuronal degeneration** and **demyelination of nerves.** This may be secondary to relative deficiency of methionine, resulting in impaired choline and phospholipid synthesis. **Neurological lesions** include parasthesia, confusion, loss of memory and in severe cases even psycosis results. Neurological symptoms are absent in folic acid deficiency megaloblastic anemia.

Methylmalonic aciduria and **homocystinuria** occur in vitamin B_{12} deficiency due to the impairment of propionate and methionine metabolism as descibed above. This can be taken as an indication of vitamin B_{12} deficiency state.

Interaction and complexities of B_{12} and folic acid deficiencies—

1. Megaloblastic anemia results in both folate as well as in B_{12} vitamin deficiencies. This is due to their common role in methionine synthesis (Fig. 25.14) and due to **folate trap** in B_{12} deficiency, that results in a block in DNA replication leading to impairment of maturation of RBCs.
2. Neurological defects occurs **only in vitamin B_{12} deficiency** in **pernicious anemia** and not in folic acid deficiency megaloblastic anemias.
3. Thus megaloblastic anemia caused by B_{12} deficiency may be alleviated and the blood picture improves (cured) by treatment with folate, but it will not cure neurological disorders of B_{12} deficiency and the neuronal degeneration will progress to **irreversible state**, leading to severe clinical complications.
4. Megaloblastic anaemias are thus generally treated with combined B_{12} and folate therapy to prevent clinical complication.
5. **Proper assessment of folate** and **B_{12} deficiency** is the key aspect in prevention of **clinical complications.**
6. **Folic acid deficiency** is assessed by **figlu test** in face of histidine load.
7. **B_{12} deficiency** should be assessed by methylmalonyl aciduria.

VITAMIN C (ASCORBIC ACID)

Vitamin C is also known as ascorbic acid. It is a water soluble, versatile function vitamin and plays significant role in the maintenance of normal health in humans. Recently its role has been implicated in preventing and /or curing a variety of diseases ranging from common cold to cancers in higher doses.

The deficiency disorder of vitamin C, the **scurvy** has been known for centuries and was called **sailor's** disease. In olden times, sailors on long sea voyages did not have fresh fruits or vegetables and suffered from scurvy. Active principle isolated from the **citrus fruits** that cured scurvy was called vitamin C or as **anti scorbutic factor.**

Chemistry Nature

Vitamin C or ascorbic acid closely resembles in its chemical structure with hexose sugar **glucose,** from which it is derived (synthesized) in **lower mammals** via **uronic acid** pathway (Chapter 7). Higher mammals such as primates including humans, guinea pigs and bats cannot synthesize ascorbic acid probably due to the loss of a gene during evolution. This gene is responsible for the synthesis of enzyme L **gulonolactone oxidase** required for the vitamin C synthesis. Thus humans are unable to synthesize vitamin C and it has to be supplied in diet to maintain normal health.

Chemically vitamin C is a hexose derivative of L-gulonolactone with an enediol group at C-2 and C-3 positions. Strong reducing properties of vitamin C are attributed to this enediol group as shown in the Fig. 25.15.

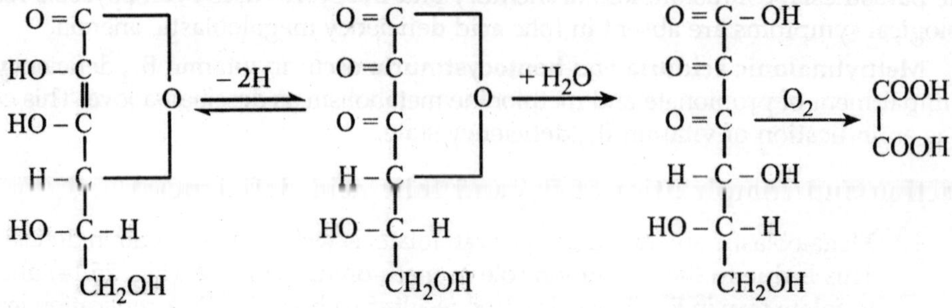

Fig. 25.15 : Structure of ascorbic acid and reversible oxidation/reduction properties

Strong reducing properties of Vitamin C—Due to its enediol group, ascorbic acid serves as a strong **reducing agent** and after oxidation (donating H^+), it is converted to dehydroascorbic acid. Because of this **reversible** reaction, both ascorbic and dehydroascorbic acids are biologically active vitamin C. In body vitamin C mainly exist in reduced form with a ratio of 15 : 1 to oxidized form. The dehydroascorbic acid in body is irreversibly hydrated rapidly to form diketogluconate, that is further oxidized to oxalic acid and excreted in urine.

Biochemical Functions

Most of the biochemical functions of ascorbic acid are due to its **strong reducing properties.** It functions as strong reducing agent in the body capable of reducing compounds such as **molecular O_2, nitrates, cytochrome a and** c and **metals** like iron and copper etc. These metals are important for the physiological activity of **hemoglobin** and many enzymes like **mono** and **dioxygenases.**

Hydroxylation Reactions

Vitamin C is involved in specific reactions in which one atom of O_2 is utilized in **hydroxylation** of some important compounds, that is **significant** for their biological activities while the other O_2 atom is reduced to water by vitamin C. Important biochemical functions of vitamin C are described here.

1. **Post translational modification of procollagen to collagen**—Vitamin C is necessary as cofactor for the **hydroxylation** of proline and lysine residues of procollagen (collagen precursor) for its posttranslational modification to functional collagen fiber. Hydroxyproline and hydroxylysine so formed are essential for collagen fiber cross-linking to form a fine mesh in which ground cementing substance, chondriotin sulfate is laid down to provide **strength to the connective** tissues of blood capillary walls, osteiod (bone) and other tissues. Thus vitamin C is essential for maintaining structural integrity of connective tissues of the body organs.

2. **Tyrosine metabolism**—Ascorbic acid is essential cofactor for **hydroxylation** and **oxidation of** pOH phenyl pyruvate to homogentisic acid.(Chapter 16, Fig. 16.16).
3. **Synthesis of epinephrine**—Vitamin C is required for the **hydroxylation** of dopamine to produce norepinephrine, which is then converted to hormone epinephrine (Chapter 16).
4. **Synthesis of serotonin**—Vitamin C is required as cofactor for **hydroxylation** of tryptophan to form 5 OH-tryptophan which is then carboxylated and converted to hormone serotonin. (Chapter 16).
5. **Bile acid synthesis**—Vitamin C is required at the **initial** step i.e. **7-α** hydroxylation of cholesterol for the synthesis of bile acids and bile salt.
6. **Steroid hormone synthesis**—Adrenal gland has rich amount of vitamin C. When the gland is stimulated, vitamin C is used in specific hydroxylation reactions necessary for steroid hormone synthesis (Chapter 30).
7. **Iron absorption**—Ascorbic acid increases absorption of iron from GIT by reducing ferric iron into ferrous state, in which form it can cross mucosal cell membrane for its absorption.
8. **Antioxidant properties**—Because of its strong reducing action, Vitamin C spares oxidation of tocopherol (Vitamin E) by reducing oxidized tocopherol in biomembranes and thus prevents damage (lipid peroxidation) caused by oxyfree radicals. (Chapter 26).
9. **Anti-carcinogenic functions**—Vitamin C **inhibit** formation of nitrosamine (carcinogen) in GIT during digestion process and thus provides protection from certain cancers.
10. **Immune response and effect on phagocytosis**—Ascorbic acid has stimulating effect on phagocytosis and antibody production, thus have beneficial effects in inflammatory diseases.

Deficiency of vitamin C causes scurvy

Scurvy is vitamin C deficiency disease. Clinical sign and symptoms are related to defective post translational modification of procollagen to collagen fibers, thereby damaging connective tissue integrity of capillary walls and other tissues. Scurvy is characterized by spongy, swollen and bleeding gums, loosening of teeth, swollen and painful bone joints, subcutaneous and internal hemorrhages due to fragile blood capillaries. Scurvy is cured by consuming citrus fruits and fresh vegetables that contain rich amount of vitamin C.

Dietary sources and RDA—Citrus fruits, amla, guava and green leafy vegetables are good sources of vitamin C. RDA is 50 to 60 mg/day in adults. The vitamin is highly water soluble and is easily destroyed by heat, alkali and by storage. Thus fresh fruits (citrus) and raw salad are necessary in the diet to supply vitamin C.

Therapeutic uses—Recently the therapeutic use of vitamin C as an adjuvant (beneficial effects) has been reported in treatment of a variety of diseases, such as common cold, trauma, wound healing, ulcers, bone fracture healing etc. It is believed that vitamin C enhances the phagocytic action of leucocytes and synthesis of immunoglobulins and thus is beneficial in higher doses in inflammatory conditions. As an antioxidant, ascorbic acid is beneficial in preventing risk of heart diseases and certain types of cancers.

FAT SOLUBLE VITAMINS

The fat soluble vitamins A, D, E and K, have unique biochemical and physiological functions in the body. These vitamins must be supplied by diet, as they cannot be synthesized by animals. Chemically they are all **isoprene (hydrocarbon)** derivatives, insoluble in water but soluble in fats, oils and fat solvents. Thus their major dietary sources are fats and oils taken in the diet and their digestion and absorption in GIT is associated with lipid digestion and absorption.

VITAMIN A

Three active compound with vitamin A activity are present in our body. These are **Retinol, (vitamin A alcohol) retinal (vitamin A aldehyde)** and **retinoic acid** and collectively they are known as **retinoids.**

Chemistry—Retinoids are polyisoprenoid derivatives with a cyclohexenyl ring called β ionone ring and a side chain of two isoprene units with a primary alcoholic group to form retinol, or with an aldehyde group (retinal) and with a carboxylic group to produce retinoic acid. The retinoids are interconvertible in the body, however once produced, retinoic acid cannot be converted back to retinal or retinol. Their structure and interconversion is shown in Fig. 25.16.

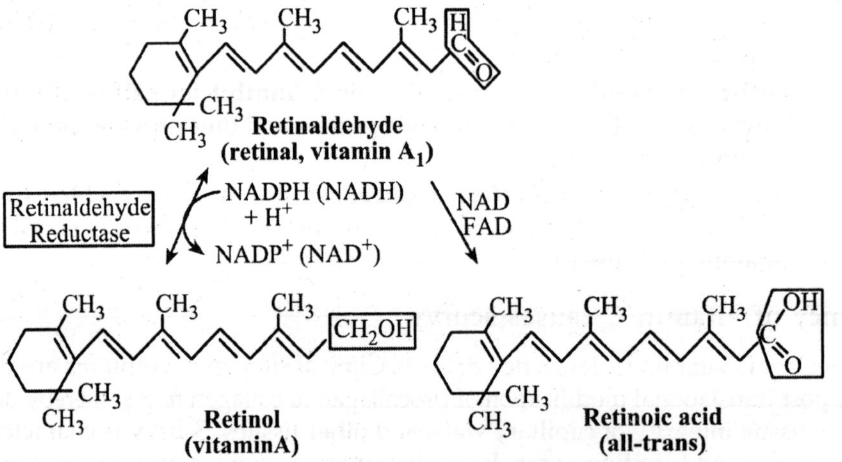

Fig. 25.16 : Chemical structures and interconversion of retinoids

Sources—Animal sources, that contain preformed vitamin A (retinol), are fish liver oils, cod liver oil, egg, milk and butter.

Plant sources—In plants vitamin A exists in **precursor** form called **provitamins** that are associated with yellow pigments called **carotenoids. β -carotene** is the main **provitamin A** present in rich amount in **yellow fruits and vegetables,** such as carrots, papaya, mango, pumpkin etc.

Daily requirement (RDA) for vitamin A is expressed in retinal equivalents where 1RE=1 microgram of retinal or 6 microgram of β-carotene. RDA for children is 500 microgram/day and for adults it is 750 microgram/day.

Digestion and absorption—β -carotenes contain two molecules of retinal joined at the aldehyde end of their carbon chain. Dietary carotenoids are oxidatively cleaved by the

GIT enzyme β -**carotene dioxygenase** into two molecules of retinaldehyde, that can be reduced to retinol with an alcoholic group by enzyme retinal reductase as illustrated in the Fig. 25.17.

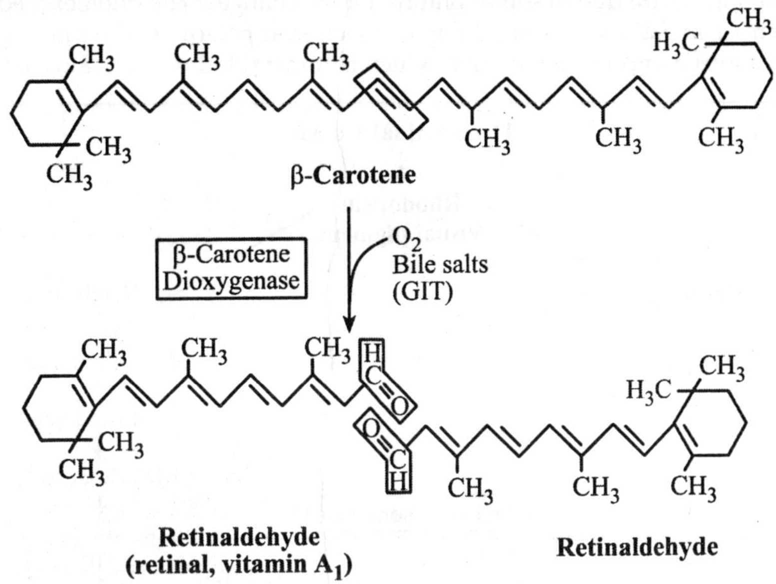

Fig. 25.17 : Cleavage of β-carotene to two molecules of retinal in GIT by a dioxygenase enzyme

Absorption and Transport

Dietary animal sources contain vitamin A mainly as retinylesters with fatty acids and are hydrolyzed in GIT to free retinol that is absorbed. In intestinal mucosal cells, retinol is reesterified with fatty acid to retinylesters and along with lipid digestion products are secreted as chylomicrons into lymphatics, that open into the thoracic duct.

Storage in liver—Chylomicron remnants are taken up by the liver which bring dietary vitamin A and is stored as retinylesters in liver as shown in the Fig. 25.18.

Transport in blood—In liver retinylesters are hydrolyzed to free retinol for its transportation to tissues. **Retinol** is transported in blood, bound to a specific **retinol binding protein (RBP)** to various tissues (Fig. 25.19), where through receptor mediated transport, retinol is taken inside and gets bound to cellular retinal binding protein.

Biochemical Functions

1. Role in Vision—Vitamin A aldehyde retinal plays an important role in the normal vision of a person. The pigment **rhodopsin,** present in rod cells of **retina** of eye is responsible for the vision in the dim light. Rhodopsin is made up of a protein **opsin** bound to **11-cis retinal** (vitamin A aldehyde). When bright light strikes retina of eye, rhodopsin is bleached and dissociated into opsin and **all-trans retinal** through various steps. Scientist Wald (Nobel prize, 1967) proposed the whole events as a **visual cycle** which is illustrated in the Fig. 25.18.

When bright light strike retina, **rhodopsin** absorb light radiations since 11 cis-retinal is a photosensitive chromophore. The primary event in this photo excitation is isomerization of 11-cis retinal to **all trans retinal** and the various photolytic intermediates of rhodopsin are produced (Fig. 25.18) due to some conformational changes. The photoexcited rhodopsin causes, a decrease in Na^+ ionic permeability and increased polarity of the rod cell membrane, that triggers a **nerve** impulse generation which is transmitted to the brain carrying vision message.

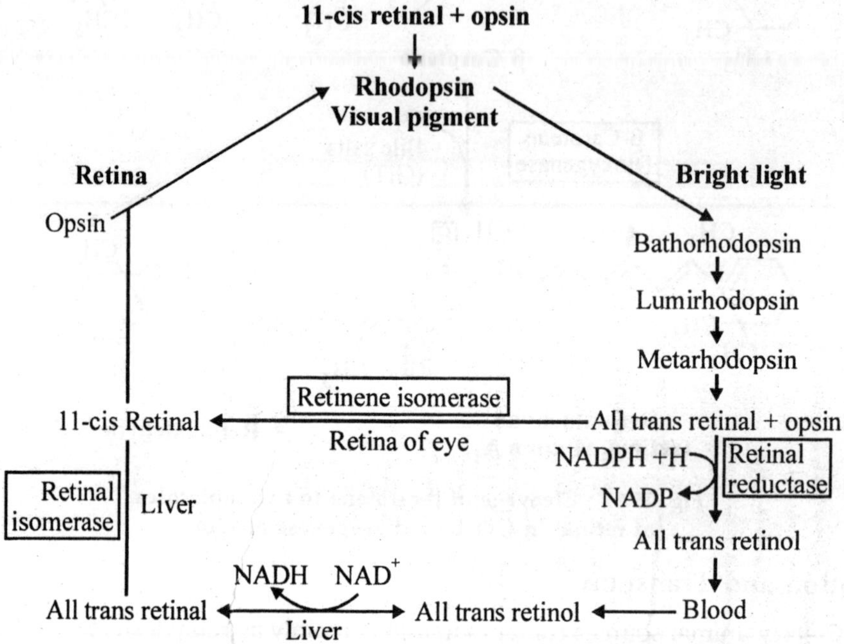

Fig. 25.18 : Wald's visual cycle and regeneration of rhodopsin

Dark Adaptation—Bright light depletes the rhodopsin stores in the rod cells of retina of eyes. Therefore when a person from bright light comes in a dim lighted room, there is difficulty in vision for some time, till the pigment rhodopsin is regenerated and vision in dim light improves. This period is referred to as **dark adaptation.** The vision in dim light improves quickly when vitamin A is available in sufficient amounts to ensure ready supply of 11 cis-retinal to retina of eyes to regenerate rhodopsin rapidly. In vitamin A deficiency, dark adaption is prolonged and **night vision is impaired**. The condition is known as **night blindness** or **nyctolopia.**

Regeneration of Rhodopsin—All trans retinal formed by the dissociation of rhodopsin when exposed to bright light, cannot recombine with opsin to form visual pigment rhodopsin. It has to be first isomerized to 11-cis retinal, but this takes some time.

Retina of eye have enzyme **retinene isomerase** that has **limited activity** to convert all trans retinal to cis-retinal. However most of all trans-retinal is reduced to vitamin A alcohol retinol, that is released in blood and goes to **liver,** where it is first oxidized to trans retinal and then is **isomerized to 11-cis retinal.** The whole events are summarized in a cyclic from known as Wald's visual cycle, that is illustrated in the Fig. 25.18.

The mechanism of absorption of vitamin A, its transport and important biological functions in various body organs is summarized in the Fig. 25.19.

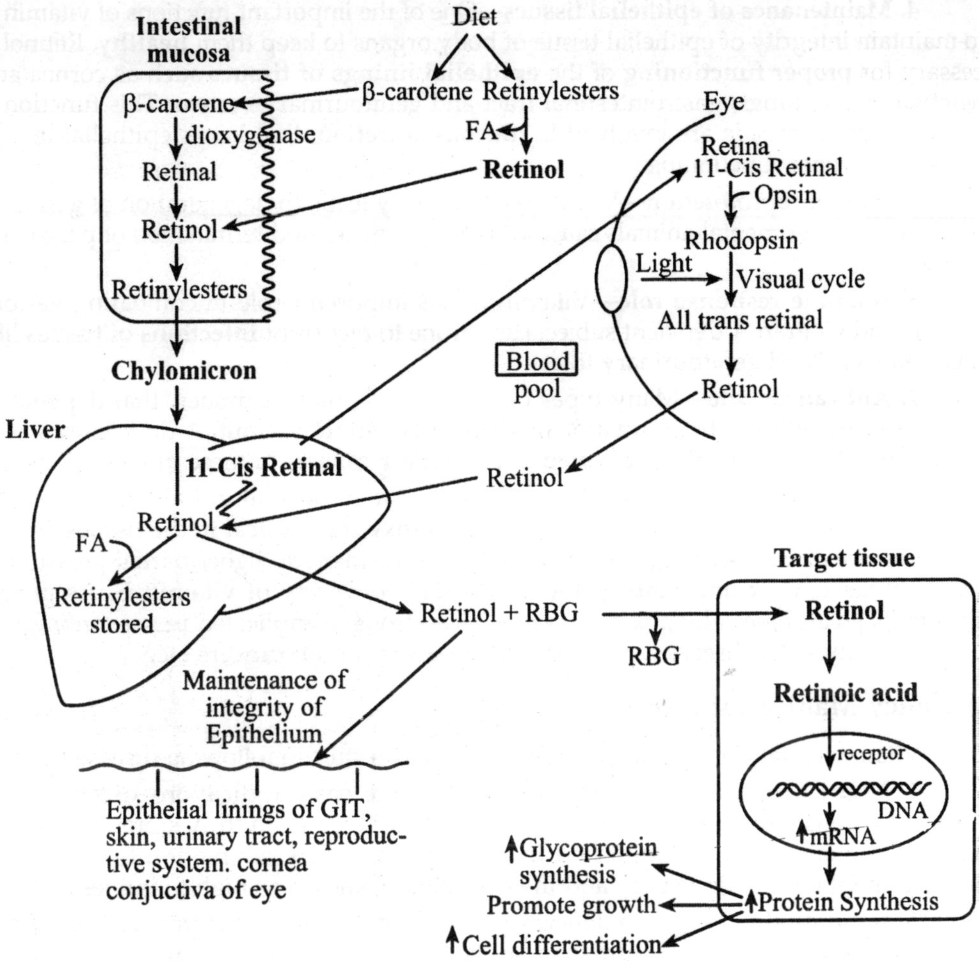

Fig. 25.19 : Diagrammatic representation of absorption, transport and biochemical functions of retinoids in various body organs (RBG—Retinol binding globulin)

Other important biochemical functions of vitamin A and retinoids include—

2. Retinol and retinoic acid act like steroid hormones—In target tissues, retinal is oxidized to retinoic acid and as they are lipid soluble they cross lipid bilayers of plasma membranes of the cell and nucleus. Retinol and retinoic acid act like steroid hormone and combine with **nuclear receptors** that directly stimulate **gene expression** to increase general protein synthesis to **promote growth.** Retinoic acid plays important role in **cellular differentiation** during embryonic development of the fetus.

3. Glycoprotein synthesis—Retinoic acid participates in the glycoprotein synthesis to facilitate growth process. Retinoyl phosphates functions as carrier of oligosaccharides across lipid bilayer of the cell membrane, that are required for **mucopolysaccharide** synthesis. Retinoic acid is involved in the synthesis of chondroitin and glycoprotein sulphates by stimulating gene expression that may account for its role in promoting **skeletal growth** and **tissue differentiation** during fetal development.

4. Maintenance of epithelial tissues—One of the important functions of vitamin A is to maintain integrity of epithelial tissue of body organs to keep them **healthy.** Retinol is necessary for **proper functioning** of the **epithelial linings of tissues** such as cornea and conjuctiva of eye, lungs, gastrointestinal tract and genitourinary system. This function is mainly because retinoids are involved in **mucous secretion**, that keeps epithelial linings wet for their proper functioning.

5. Role in reproduction—Vitamin A deficiency leads to degeneration of **germinal epithelium** in experimental animals causing sterility in males and termination of pregnancy in the females.

6. Immune response role—Vitamin A has important role in combating various infections and vitamin A deficient subjects are prone to recurrent **infections** of tissues like cornea, skin, GIT and genitourinary tract.

7. Anticancer role—Many types of cancers of epithelial tissues, that depend on retinoids for the cellular differentiation, are very frequent in vitamin A deficiency states. Thus Vitamin A is very much implicated in preventing cancers of these tissues.

8. Antioxidant functions—β-carotenes are important natural antioxidants. The isoprenoid units of their chemical structure **trap peroxy free radical** in the tissues, at low partial pressure of O_2. Since vitamin E is effective antioxidant at higher partial pressure of O_2, thus β carotenes **complements** the antioxidant property of vitamin E. As potent antioxidants β-carotenes along with vitamin E are strongly implicated in preventing risk factors for many killer diseases, such as **heart diseases** certain **cancers** etc.

Deficiency Manifestations

Vitamin A deficiency causes tissue lesions and results in following disease states—

1. Night Blindness (Nyctolopia)—One of the early indications of vitamin A deficiency is defective night vision that occurs when liver vitamin A stores are nearly exhausted. Visual acuity is diminished in dim light and dark adaptation time is much prolonged in vitamin A deficiency and the condition is known as night blindness.

2. Xerophthalmia—Severe deficiency of vitamin A results in dryness of conjuctiva and cornea of eye, due to diminished mucous secretion. This is accompanied by **keratinization** of epithelial linings of the tissues of eye, which thickens and become opaque (loss of transparency), while cornea become dry and luster less. Recurrent infections of cornea result in formation of **Bitots spots** (white triangular plaques) on conjuctiva due to bacterial infections and ulcerations. This deterioration of eye tissues is known as **xerophthalmia** and may lead to **complete blindness**. Vitamin A deficiency is the most common cause of **blindness** in developing countries like India, in the children under 5 years of age.

3. Skin and mucous membrane manifestations

a. Skin become dry, rough and keratinized in vitamin A deficiency, that is prone to recurrent infections.

b. Keratinzation of mucous membranes of GIT, respiratory tract and genitourinary tract lead to recurrent **bacteria infections** resulting in serious clinical manifestations.

c. **Urinary stone** formation occurs due to keratinization of urinary tract which becomes susceptible to recurrent bacterial infections, causing pH alterations and deposition of insoluble salts to form renal calculi.

4. Growth retardation—Physical growth retardation, specially impairment of skeletal growth is the general manifestation of vitamin A deficiency.

VITAMIN D

Vitamin D is a fat soluble steroid compound, known as **cholecalciferol.** Vitamin D is considered as pro-**hormone,** as in body it is activated as **calcitriol,** which functions like a hormone and plays central role in calcium and phosphate metabolism, as well as in the **mineralization** of **bones** to keep them strong and healthy.

Formation and Activation of Vitamin D

Vitamin D is one vitamin that can be synthesized in our body. Cholecalciferol (Vitamin D) is synthesized in the skin, in presence of sunlight from its **precursor 7-dehydrocholesterol.** This compound is an intermediate of cholesterol biosynthesis and is readily available in the body. Other **precursor** for vitamin D synthesis is a plant sterol called **ergosterol.** The fungus ergot is an important source of ergasterol.

Synthesis of Vitamin D in skin—In presence of sunlight, 7 dehydrocholesterol in the malpighian layer of epidermis of skin, is activated as **cholecalciferol** (vitamin D_3). Vitamin D is formed as a result of photolysis by ultra violet radiation (photons of 290-315 nm) in the B-ring of sterol, that breaks the bond between position 9 and 10. This results in the opening of the B-ring and the compound acquires the vitamin D activity, after isomerization to trans form as illustrated in the Fig. 25.20.

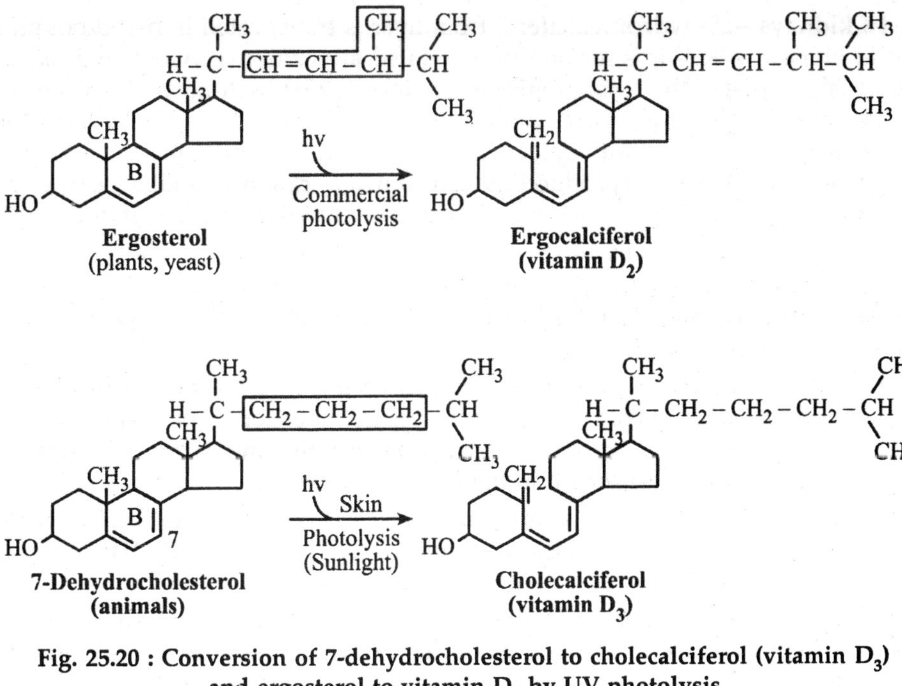

Fig. 25.20 : Conversion of 7-dehydrocholesterol to cholecalciferol (vitamin D_3) and ergosterol to vitamin D_2 by UV photolysis

Sources—Small amounts of vitamin D is found to occur in the food stuffs of animal origin, such as fish liver oil, milk, egg yolk etc. **Cod liver oil** is the richest source. Most of the vitamin D is **synthesized in skin** in the presence of sunlight, as described above.

Recommended Daily Allowance (RDA)

As vitamin D is synthesized in skin in the presence, of sunlight, RDA for vitamin D in adults is not specified. But during growth and pregnancy requirement for vitamin D increases. RDA for children is 400 IU/day, for pregnant women 400 IU/day and for adults it is 200 IU/day.

Commercially the vitamin is prepared from fungus ergot, that contain **ergosterol** (plant sterol) which is very similar to 7-dehydrocholesterol, except that it has a double bond in the side chain, at position 22-23 and an extra methyl group at C-24. The ergosterol when exposed to UV radiation is converted into **ergocalciferol** (vitamin D_2), in a manner similar to the production of cholecalciferol as shown in Fig. 25.20.

Activation of Vitamin D : Synthesis of calcitriol

Vitamin D is activated into calcitriol primarily in the liver and kidney as described here.

In liver—Vitamin D, synthesized in skin and the dietary vitamin D absorbed from GIT are transported via blood, bound to a specific globulin (vit. D binding protein) to the liver. In liver it is hydroxylated at C-25 position by a specific hepatic enzyme, known as **α 25-hydroxylase** to produce **25 hydroxycalciferol,** which is released in the blood or some amount may be **stored** in liver and adipose tissue. Hepatic α-25 hydroxylase is a monooxygenase enzyme and requires NADPH, cytochrome p-450 and molecular O_2 for the hydroxylation reaction. 25-Hydroxycalciferol is the major form of vitamin D in the blood.

In kidneys—25-hydroxycalciferol from liver is transported in blood bound to α-globulin to kidneys where it is further hydroxylated at position-1 by **α 1-hydroxylase**, a mitochondrial enzyme of the renal tubules to produce 1, 25 dihydroxy cholecalciferol.(Fig. 25.20). The enzyme **α 1-hydroxylase** is a mixed function oxidase and requires NADPH, Mg^{++}, O_2 and cofactors, such as cytochrome p 450 and ferrodoxin. **1,25 Dihydroxy cholecalciferol (1,25 DHCC)** so produced is also known as **calcitriol** and is the most potent form of vitamin D in the body. Reactions of calcitriol synthesis are given in the Fig. 25.21.

Regulation of calcitriol synthesis—Like other steroid hormones, **calcitriol** synthesis is subjected to a tight **feed-back regulation.** Rate limiting key enzyme is **α1-hydroxylase** of kidneys and is the major site of **control** of calcitriol biosynthesis. Factors regulating calcitriol biosynthesis includes—

1. Low dietary calcium and hypocalcaemia increase the rate of production of calcitriol while its production decreases in hypercalcaemia. Stimulatory effect of hypocalcaemia on renal **α 1-hydroxylase enzyme** is through parathyroid hormone (PTH) as shown in the Fig. 25.21.
2. When serum calcium level is high (hypercalcaemia), **calcitriol feedback inhibits** its own production and instead stimulates formation of **24, 25 DHCC** (inactivate form).
3. Low phosphate diets and low serum phosphate levels also **induce** (activate) α-1 hydroxylase enzyme to a lesser degree as compared to low serum calcium level.

Major modulators which control rate of production of calcitriol are serum levels of calcium, phosphate, parathyroid hormone (PTH) ant calcitriol itself as illustrated in the Fig. 25.21.

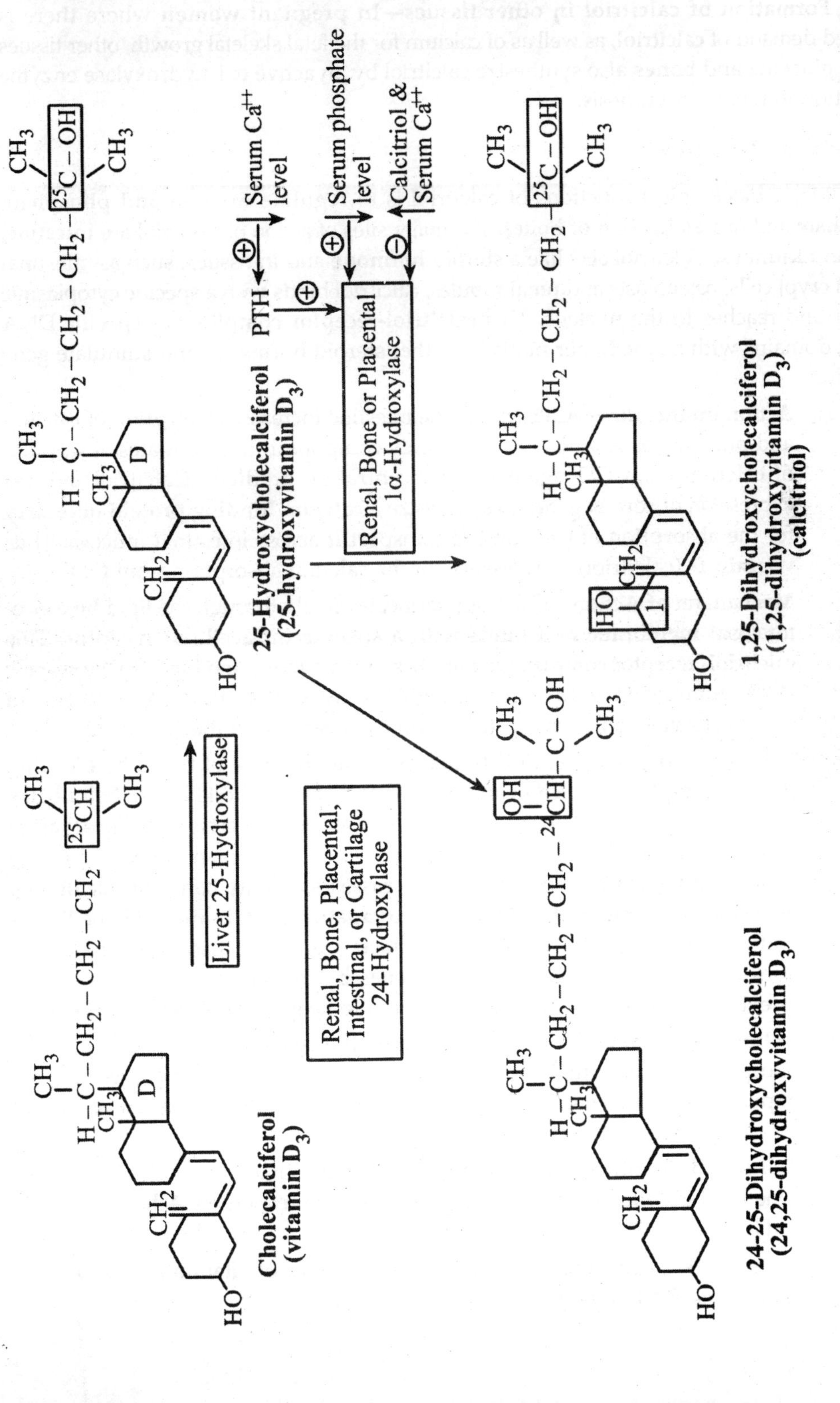

Fig. 25.21 : Conversion of vitamin D to hormone calcitriol in liver and kidneys along with the factors regulating its synthesis at α1-hydroxylase step in renal tissues

Formation of calcitriol in other tissues—In pregnant women where there is increased demand of **calcitriol,** as well as of calcium for the fetal skeletal growth, other tissues such as **placenta** and **bones** also synthesize calcitriol by an active α 1-hydroxylase enzyme in addition to renal biosynthesis.

Biochemical Functions

Major biochemical functions of calcitriol is to **regulate calcium and phosphate metabolism** and **mineralization of bones.** The major sites of action of calcitriol are **intestine, bones and kidneys.** Calcitriol acts like a **steroid hormone** and in tissues, such as intestinal villi and crypt cells, osteoblasts and renal tubules, calcitriol binds with a specific cytoplasmic receptor and reaches to the nucleus. This **calcitriol-receptor complex** has specific DNA binding domains with a zinc finger motif like other steroid hormones and stimulate gene expression.

1. **Action on intestine**—Calcitriol in the intestine increases **absorption** of calcium and phosphates from GIT. It is the only compound that facilitates absorption of calcium from GIT against its concentration gradient. Calcitriol induces expression of certain genes to synthesize a **calcium binding protein necessary** for the absorption of Ca^{++} and to transport it across intestinal mucosa. Thus **vitamin D** (calcitriol) is **necessary** for the calcium absorption from **GIT.**

   Mechanism of Action—Calcitriol permeates freely through the lipid bilayer of mucosal membrane and binds with a specific intracellular receptor. This **calcitriol- receptor complex** then reaches to the nucleus and binds to the specific DNA domains to induce genes (DNA) to synthesize m-RNA and calcium binding protein synthesis is stimulated as illustrated in the Fig. 25.22.
2. **Action on bones**—Calcitriol stimulates **mineralization** of bones by ensuring proper absorption of calcium and phosphate from the intestine and their uptake by **osteoid tissue.** Calcitriol stimulates osteoblast to secrete enzyme **alkaline phosphatase,** that increases local concentration of inorganic phosphate. The ionic product of Ca^{++} and PO_4^{-3} thus crosses solubility constant, facilitating deposition of **calcium phosphate hydroxyapatite crystals** around bone tissues and mineralization of bones takes place. Calcitriol also coordinates the mineralization and remodelling of bones by its action on osteoclasts and osteoblasts along with PTH and calcitonin to keep bones healthy and strong.
3. **Action on kidneys**—Calcitriol increases the reabsorption of calcium and phosphate by renal tubules, thereby decreasing their excretion, as they are important minerals needed for bone formation and mineralization, as well as for other functions (Chapter 24).
4. **Cellular differentiation and immune response**—Vitamin D is also involved at gene level in cellular differentiation and immune response mechanisms of the body.

   The synthesis, activation and biochemical functions of vitamin D and calcitriol are summarized in the Fig. 25.22.

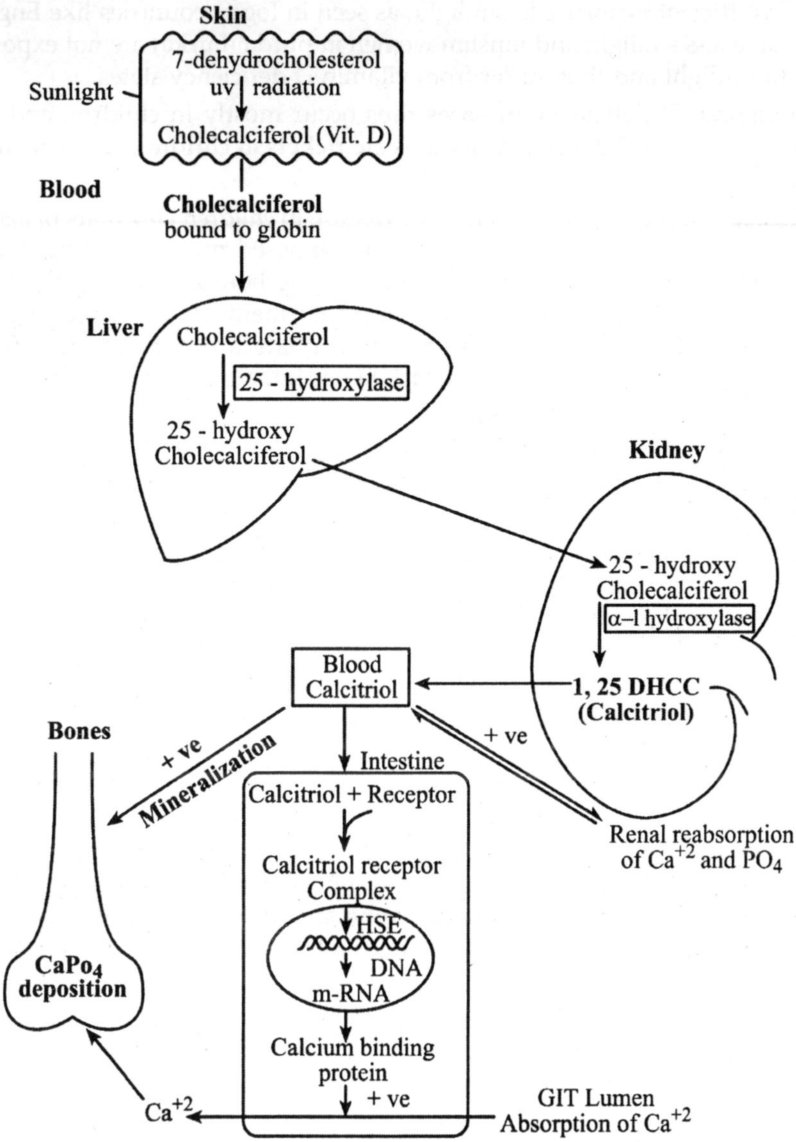

Fig. 25.22 : Diagrammatic representation of activation and biochemical function of vitamin D (calcitriol)

Deficiency Diseases

Dietary deficiency of vitamin D normally does not occur, as it is synthesized in the skin in presence of sunlight. However vitamin D deficiency can occur in following conditions—

a. When requirement of Vitamin D is increased, for instance in growing children and pregnant women for the active skeletal growth and mineralization of bones and in lactating mother for milk secretion.

b. Insufficient exposure to sunlight, as seen in foggy countries like England that have less sunlight and muslim women in purda (burka) are not exposed much to sunlight and thus suffer from vitamin D deficiency states.

The vitamin D deficiency diseases thus occur mostly in children and pregnant women. The two vitamin D deficiency diseases are **rickets** in **children** and **osteomalacia** in adult **women.**

1. Rickets—It is a vitamin D deficiency disease in children (1-3 years of age), caused mainly by consuming poor diets and child is not exposed much to sunlight. Rickets is characterized by **improper mineralization** of growing bones resulting in severe **bone deformities.** Vitamin D deficiency results in impairment of calcium and phosphate absorption from GIT and thus calcium and phosphate are not available for proper mineralization of growing bones, due to which bones remain soft and pliable and **weight bearing bones bend and are deformed.**

Clinical Manifestation—Clinical **symptoms** of rickets are bowed legs, knock knee syndrome leading to difficulty in walking, pigeon chest, rickety rosary on ribs, (beaded ribs), bossing of frontal bone of the skull and enlargement of epiphysis of long bones etc.

2. Osteomalacia—Osteomalacia occurs mostly in pregnant women living in foggy countries or in purdah. Due to the vitamin D insufficiency, plasma calcium and phosphate levels are reduced so the mother's bones gets **demineralized** to provide minerals for fetal skeletal growth (prime importance) resulting in **osteoporosis.** Clincial symptoms include reduced bone density with bone aches and pains and brittle bones that are prone to **fractures.** Softening of bones of pelvic girdle is accompanied with difficulty in the birth of the baby due to fractures.

Biochemical Abnormalities—In Vitamin D deficiency states, blood calcium level may be normal or slightly lowered but **phosphate level is reduced.** This is because PTH secretion is stimulated by low calcium level that causes demineralization of bones to restore serum calcium level. PTH increases excretion of phosphate in the urine resulting in low plasma phosphate levels. **Serum alkaline phosphatase enzyme** is markedly increased as a compensatory measure by bones (osteoblast) to counteract vitamin D deficiency effects.

Management—Rickets and osteomalacia if diagnosed early can very well be managed by giving vitamin D supplements as well as by providing calcium and phosphate rich diets. Cod liver oil is good supplement of vitamin D for growing rachitic children to prevent the irreversible bone deformities.

Hypervitaminosis—As excretion of vitamin D is very less and it is also synthesized in skin, its excessive dietary intake may lead to accumulation and toxicity of vitamin D. Body stores of vitamin D become 10 times than normal. Hypervitaminosis D result in high blood calcium level that may lead to calcification of soft tissues and renal stone formation. High consumption of vitamin D initially causes anorexia, nausea, increased thirst, weakness etc.

VITAMIN E

Vitamin E was named as tocopherol (tokos-childbirth, pheros to bear) because its deficiency in the experimental animals resulted in **infertility.** Presently vitamin E is considered as the most **potent antioxidant** (natural) present in our food that protect tissue damage from oxy-free radicals to maintain good health of an individual.

Chemistry—Eight different tocopherols with vitamin E activity are isolated from the natural fats and oils, all having **6-hydroxy chromane ring** with a **saturated isoprenoid side chain** and were named as α, β, γ, etc. Out of these α-tocopherol (5, 7, 8 trimethyltocol) is of **widest occurrence** in nature and has **greatest biological activity.** The structures of α-tocopherol and some other important vitamin E compounds are shown in Fig. 25.23.

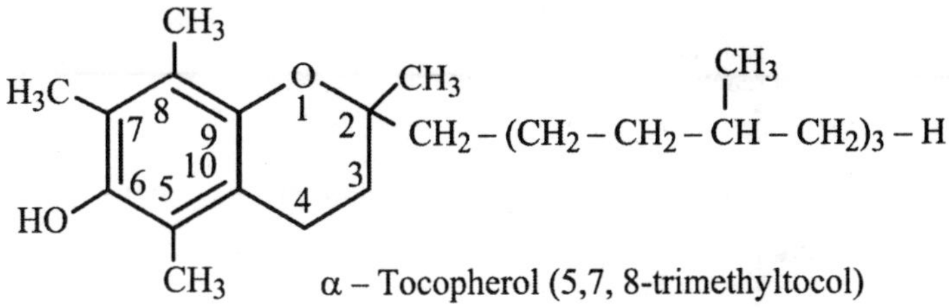

α – Tocopherol (5,7, 8-trimethyltocol)

β – Tocopherol (5, 8-dimethyltocol)

γ – Tocopherol (7, 8-dimethyltocol)

Fig. 25.23 : Chemical nature of tocopherols

Sources of vitamin E—Vegetable oils such as wheat germ oil, sunflower oil, saffola oil, soyabean oil, cotton seed oil etc are rich sources of vitamin E, while fish liver oils are **devoid** of vitamin E. Vitamin E is destroyed by commercial food processing and deep freezing.

RDA - Males – 10 mg/day, while in females it is 8 mg/day

Requirement of vitamin E increase with increase intake of polyunsaturated fatty acids (PUFA), to prevent peroxidation of lipid bilayer of the biomembranes.

Absorption and transport—As vitamin E occurs dissolved in vegetable oils taken in the diet, it is absorbed along with fat digestion products as mixed micelles and secreted as chylomicrons via lymphatics in blood. So active fat digestion and absorption facilitates vitamin E absorption.

Transportation—Vitamin E is transported as lipoproteins complexes first as chylomicrons and then from liver it is transported as **VLDL** and then as LDL fraction to various tissues.

Storage—Vitamin E is stored in adipose tissues along with fats.

Vitamin E deficiency may occur in disorders such as steatorrhea, abetalipoprotinemia, cholestatic liver disease, cystic fibrosis and patients undergoing resection intestinal surgeries.

Biochemical Functions

Most of the functions of vitamin E are due to its powerful antioxidant properties and are as described here under.

1. Vitamin E acts as first line of defence against potential damaging effects of the oxy free radicals. It prevents tissue damage caused by nonenzymatic oxidation of PUFA of lipid bilayers of biomembranes by oxy free radicals such as superoxide anion **(O_2^-)** and hydrogen peroxide etc.
2. Vitamin E is lipophilic in nature and thus has affinity to get concentrated in the lipid bilayers of cell membranes, mitochondrial membranes and endoplasmic reticulum. At these sites vitamin E prevents peroxidation of PUFA

of phospholipids of lipid bilayers and thus protects the integrity of these biomembranes for normal cellular functioning.

3. Vitamin E acts as a **scavenger of oxy free radicals**, and itself gets oxidized to quinone and thus protects PUFA of biomembranes from peroxidation as shown in the Fig. 25.24.

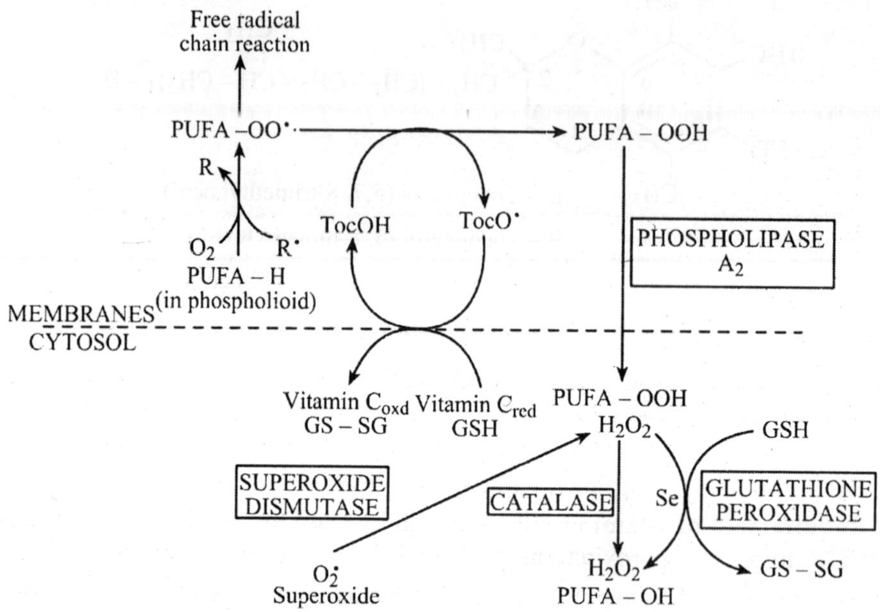

Fig. 25.24 : Role of Vitamin E is scavenging peroxyl free radicals in the lipid phage of cell membranes and interaction and synergism with antioxidant systems operating in the aqueous phase of cytosol (PUFA-OO• peroxyl free radical, TocOH-tocopherol (vit. E), TocO•-free radical of tocopherol, GSH-glutathione, Se-selinium)

4. **Free radical chain breaking action**—Vitamin E is a potent antioxidant mostly because of its **free radical chain reaction breaking actions.** Vitamin E is capable of transferring its phenolic hydrogen to a peroxy free radical of a peroxidized PUFA of the lipid bilayer of biomembranes and thus breaks free radical chain reaction. The peroxy free **Toco·** (free radical) so formed may react with **vitamin C** to regenerate active tocopherol to recycle its free radical chain breaking action as illustrated in the reaction given below.

$$ROO\bullet + TocOH \longrightarrow ROOH + TocO\bullet$$

$$ROO\bullet + TocO\bullet \longrightarrow \underset{\text{Non-free radical product}}{ROOH +}$$

5. Phenoxy free radical (**Toco·**) of tocopherol otherwise can react with yet another peroxy free radical of PUFA to produce an oxidized complex (with oxidized chromane ring) with loss of vitamin E activity and is excreted in bile after conjugation with glucuronic acid in the liver. This results in **net loss** of vitamin E and thus it must be replenished by dietary vitamin E supplementation.

Important Physiological Role of Vitamin E—

1. **Maintenance of biomembrane functional integrity**—Vitamin E is required essentially for the prevention of peroxidation of PUFA of lipid bilayer, to maintain the structural and functional integrity of biomembranes of cell and subcellular components.
2. **Maintenance of RBCs fragility**—Vitamin E along with other dietary and cellular antioxidant protects erythrocytes from free radical injury and thus prevent hemolysis.
3. **Maintenance of reproductive germinal epithelium**—Vitamin E plays important role in protection of structural and functional integrity of the germinal epithelium of the reproductive organs. Vitamin E deficiency in experimental animals results in reproductive failure.
4. Vitamin E protects tissues against damaging action of free radicals thus minimize risk factors for many killer diseases such as **heart diseases, cancers** etc.
5. **Anti aging functions**—As potent antioxidant, vitamin E prevents cumulative damaging effects of oxy free radicals which results in gradual deterioration of the tissues during aging process. Thus vitamin E is strongly implicated as a beneficial adjuvant, preventing **aging process.**

Vitamin E and Selenium—Vitamin E and selenium act **synergistically** and **reinforce** antioxidant actions of each other against **lipid peroxidation. Glutathione peroxidase** (selenium containing enzyme) provides a second line of action against damaging effects of hydrogen peroxide and peroxyfree redicals (ROO•) before they can damage membranes of cell components (Fig. 25.24). In addition, selenium is necessary for the activity of pancreatic enzymes required for fat digestion and absorption in the GIT, including vitamin E absorption. Therefore they complement each other for their antioxidants function as well as they reduce the dietary requirements of each other. Thus selenium has a vitamin E sparing effect.

Deficiency Manifestations

Vitamin E deficiency in experimental animal shows considerable variations in clinical symptoms as listed below.

Rats—Infertility in experimental vitamin E deficient rats is observed due to the deterioration of germinal epithelium of gonads. **Other deficiency manifestations** in animals observed are **renal impairment** due to degeneration of renal tubular epithelium, **hemolysis** of RBCs, acute **hepatic necrosis** etc.

Rabbits are more vulnerable to vitamin E deficiency resulting in **hemolytic anaemia** and **muscular dystrophy**.

Humans—Vitamin E dietary deficiency have not be reported so far in the humans. However vitamin E deficiency in the new born infant has been shown to result in anaemia.

Anaemia in newborn—Insufficient vitamin E supplements to pregnant and lactating women may result in **anaemia** in newborn infants, due to the decreased heme synthesis and increased hemolysis of RBCs.

VITAMIN K

Vitamin K was abbreviated as K because of its important role in blood coagulation. Vitamin K is necessary for the activation of blood clotting factor II- prothrombin, factor VII, IX and X.

Chemistry—Vitamin K group includes many compounds with vitamin K activity. Chemically they are **naphthoquinones** with polyisoprenoid side chains of varing lengths.

Menadione (K_3) is the parent compound of K series (synthetic), but when administered in diets, it is alkylated in body to vitamin K_2. Menadione is water soluble and thus is widely used in clinical practice as vitamin K suppliment. **Phylloquinone** **(K_1)** is the major vitamin K found in plants while animal sources contain menaquinone–7 with unsaturated polyisoprenoid side chain. Chemical structures of important vitamin K members are shown in Fig. 25.25.

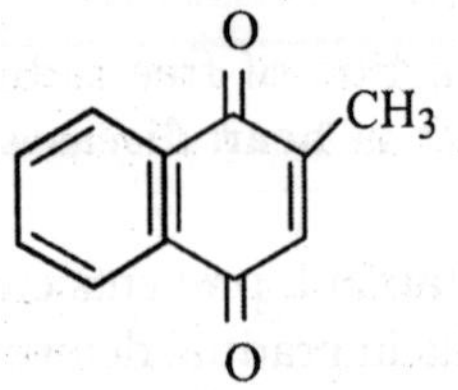

Menadione (vitamin K_3)

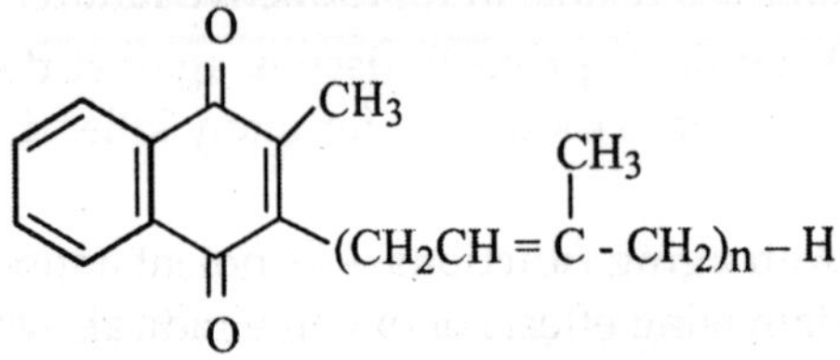

Menaquinone-n (vitamin K_2 n= 6,7 or 9)

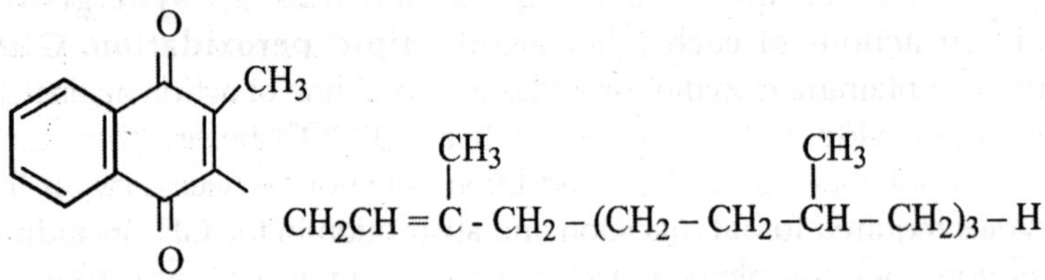

Phylloquinone (vitamin K_1)

Fig. 25.25 : Structures of various members of vitamin K series

Absorption, Transport and Storage

Vitamin K is taken in diet along with dietary fats and oils, as well as it is synthesized by the intestinal bacteria. Vitamin K absorption from the intestine along with lipid digestion products requires bile salts, and then they are secreted as chylomicrons from intestinal mucosal cells into blood via lymphatics, and reach to liver. From liver vitamin K is incorporated into VLDL and later as LDL and is transported to various tissues. Vitamin K is stored mainly in the liver.

Biochemical Functions of Vitamin K

Biosynthesis of blood clotting factors—Main biochemical function of vitamin K is in **biosynthesis** and **maintenance** of normal levels of **blood clotting** factor II prothrombin, factors VII, IX an X, all of which are synthesized as zymogens(precursors, inactive form) in the liver.

Activation of clotting factors—Vitamin K is necessary for the post translational modification of zymogens of clotting factors to produce biologically active clotting factors.

For this activation, a γ -carboxylation of their glutamate residues (prothrombin has 10 glutamate residues) by a **microsomal carboxylase enzyme** is required, that is vitamin K dependent.

Mechanism—This carboxylase enzyme, with **reduced vitamin K** (quinol form) as cofactor, carboxylates the glutamate residues of **precursor proteins** to have an extra carboxyl group at the **γ carboxyl C-atom** (Fig. 25.26) to produce **gamma carboxyl glutamate (gCG)** residues. These are the binding sites which chelate calcium ions in a specific protein calcium-phospholipid interactions, essential for the biological role of clotting factors.

Vitamin K cycle—During this **γ carboxylation**, vitamin K is converted to its epoxide form, that has to be converted back to quinone form of vitamin K, to continue the activation of clotting factors. The epoxide vitamin K is first converted to quinone form by the lipoic acid and then to reduced vitamin K (quinol form) by a **reductase enzyme** to continue the cycle as illustrated in the Fig. 25.26. The process is known as vitamin K cycle, which exists in the endoplasmic reticulum of the liver.

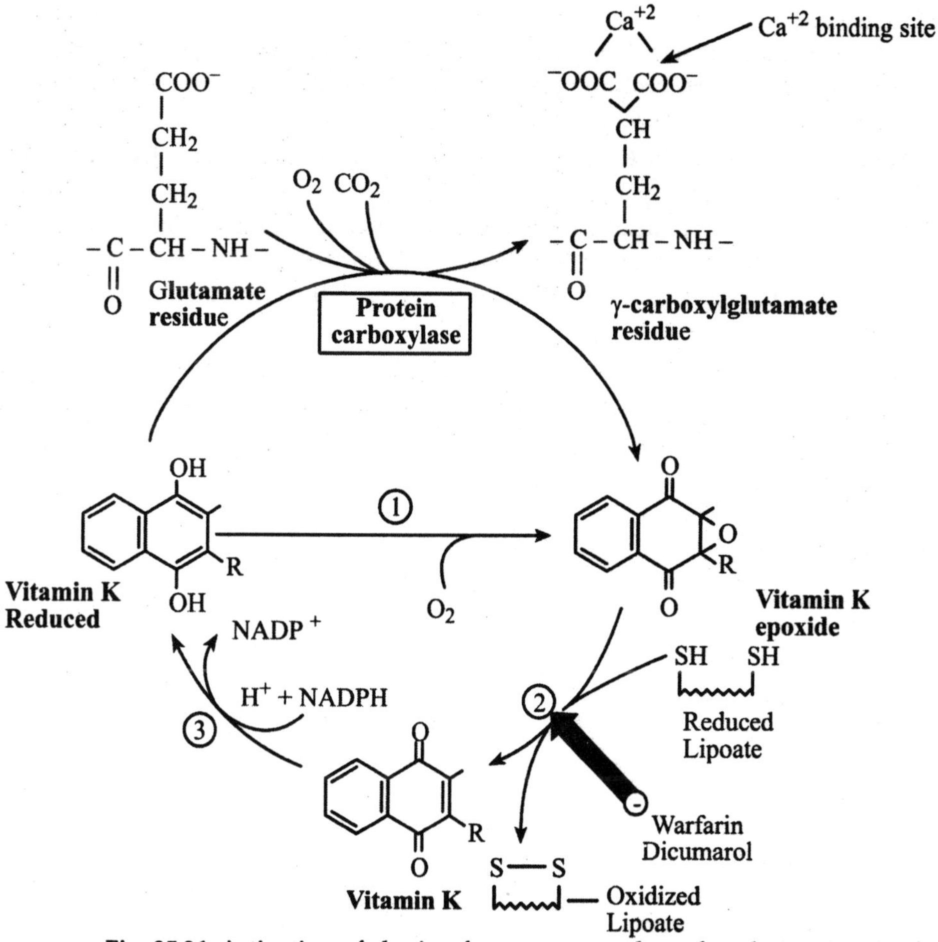

Fig. 25.26: Activation of clotting factor zymogen by reduced vitamin K and carboxylase to generate Ca^{2+} binding sites and regeneration of reduced vitamin K in the vitamin K cycle, (1) monoxygenase, (2) epoxide reductase and (3) reductase enzymes

Inhibition of activation process by dicumarol—Vitamin K antagonists, such as dicumarol and warfarin inhibit reduction of vitamin K epoxide to vitamin K quinone and thus can be used **therapeutically** as **anticoagulant drugs** to prevent intravascular thrombus formation in cardiovascular diseases.

Activation of other important proteins—Recently the role of vitamin K in synthesis and activation (post translational carboxylation) of bone proteins e.g. osteocalcin (for mineralization of bone) and structural proteins of kidney, lungs and spleen has been reported. This role of vitamin K seems to be very significant in the proper functioning of these tissues.

Deficiency Manifestations

As vitamin K is widely distributed in nature and is also synthesized in intestine by the bacteria, normally vitamin K deficiency does not occur in adults. However newborn, particularly preterm infants are vulnerable to vitamin K deficiency, as placenta does not pass vitamin K to fetus and the intestine is sterile at birth. Thus **hemorrhagic disease** may develop in newborn preterm infants. **Hemorrhagic** disease is manifested as bruising patches on skin, mucous membrane hemorrhages, post-traumatic bleeding even by minor injury and internal bleedings. **Prolonged prothrombin time** and **delayed clotting time** are characteristic hallmark of vitamin K deficiency state.

Deficiency of vitamin K may also occur secondary to diseases like steatorrhea, obstructive jaundice, chronic pancreatitis, sprue etc that impair fat digestion and thus absorption of vitamin K is reduced. Therapeutic uses of anticoagulants like dicumarol and warfarin may also result in vitamin K deficiency hemorrhagic condition. An important medicinal use of vitamin K is as **antidote** to the toxic effects of dicumarol and other anticoagulant drugs in the body.

Hypervitaminosis K is caused by administration of excess of menadione as antidote to dicumarol and may results in excessive hemolysis, hemolytic anaemia, hyperbilirubinemia, jaundice and brain damage etc.

❑❑❑

26

OXY FREE RADICALS AND ANTIOXIDANTS

Oxy free radicals are now recognized as one of the most predominant risk factors in the etiology of many diseases. Among the free radicals, the **highly reactive oxygen species (ROS)** play most significant role in causing tissue injury resulting in initiation of disease process. This is becoming an alarming global problem especially because the levels of **proxidants** i.e chemicals capable of generating potentially toxic oxy free radicals are progressively increasing in the highly adulterated food stuffs and in the polluted environment throughout the world.

FREE RADICALS

The free radical is a molecule or molecular fragment that contain one or more **unpaired electrons** in its outer orbit, in contrast to the normal molecule which have paired electrons. Due to this unpaired electron, the free radical has a tendency to acquire an electron from other molecules and this makes the them highly reactive species. Free radicals are represented by a superscript dot or simply a dash and include **superoxide anion ($O_2^{\bullet}$), hydrogen peroxide, peroxyl radical ($RCOO^{\bullet}$), alkoxyl radical ($RO^{\bullet}$) and hydroxyl radical ($OH^{\bullet}$)** etc.

GENERATION OF FREE RADICALS

In most of the oxidation reactions of aerobic organism including humans, it is ensured that O_2 is completely reduced to water. However **partial reductions of O_2** does occur in tissues resulting in the production of oxy free radicals. The oxygen is paramagnetic in nature as it contains two unpaired electron with opposite spin, in the two outermost orbits and thus has significant tendency to produce highly reactive oxy free radicals. When oxygen is reduced to water by cytochrome oxidase system, four electrons are acquired. However the electron and protons can be gained one at a time, successively by univalent reduction of O_2, to generate various oxy free radicals as illustrated in the reactions—

$$O_2 \xrightarrow{+e^-} O_2^{\bullet} \xrightarrow[+H^+]{+e^-} OH^{\bullet} \xrightarrow[H^+]{+e^-} H_2O_2$$

$$O_2 \xrightarrow{+e-} O_2^- \xrightarrow[+H^+]{+e-} H_2O_2 \xrightarrow[H^+]{+e} OH^- \xrightarrow[H^+]{+e} H_2O$$

The intermediate molecules produced in the above mentioned steps of univalent reduction of oxygen, are all highly reactive oxy free radicals and are potentially toxic and damaging to the tissues. These are known as superoxide free radical ($O_2^\bullet$), hydrogen peroxide (H_2O_2) and hydroxyl ($OH^\bullet$) free radical. The last mentioned i.e. hydroxyl radical ($OH^\bullet$) is extremely toxic but has very short half life and therefore it affects tissue components close to the site of its production only.

Productions of free radicals—Free radicals are constantly being produced in the body during aerobic oxidation (biological oxidation) in tissues. Approximately 1 to 4% of O_2 taken up in body is converted to oxy free radicals by following routes—

1. **Autooxidizable flavoprotein enzymes,** such as xanthine oxidase, aldehyde oxidase and peroxidases etc are important oxidase enzymes of the body. When the reduced flavoprotein enzymes are reoxidized by oxygen, superoxide anion is produced as shown in the reaction—

$$EnzH_2 + O_2 \longrightarrow Enz\,H + O_2^\bullet + H^+$$

2. Superoxide anion is generated during oxidation of ferrous iron of hemoglobin to ferric iron and methemoglobin is produced, as given under—

$$Hb\,Fe^{2+} + O_2 \longrightarrow HbFe.^{2+}O_2 \longrightarrow Hb\,Fe^{3+} + O_2^\bullet$$

3. The capacity of H_2O_2 to damage tissues is minimal as compared to other free radicals, such as $O^\bullet_2$, $OH^\bullet$, $ROO^\bullet$ etc. But in presence of iron containing compounds (ubiquitous in body), H_2O_2 can generate hydroxyl free radicals which are the most potent killer agent as illustrated in following reactions —

Fenton and Haber-Weiss Reactions

Hydroxyl free radicals are generated by iron catalyzed break down of H_2O_2. For example in the Fenton and Haber-Weiss reactions as given below—

$$HbFe^{2+} + H_2O_2 \longrightarrow HbFe^{3+} + OH^\bullet + OH^- \text{ and } HbFe^{2+} + O_2 \longrightarrow HbFe^{3+} + O_2^\bullet$$

$$O^\bullet{}_2{}^- + H_2O_2 \longrightarrow O_2 + OH^\bullet + OH^-$$

Beneficial Effects of Free Radicals

Respiratory burst during phagocytosis—Inflammatory cells such as neutrophils, eosinophils, monocytes and others generate superoxide anion during phagocytosis by what is known as respiratory burst. This is a mechanism utilized by phagocytes for the destruction of bacteria and is important beneficial effect of free radicals in the body. During phagocytosis the superoxide is converted to H_2O_2 and then to hypochlorus ion ($HClO^-$) with the help of enzymes **superoxide dismutase (SOD)** and **myeloperoxidase (MPO).** The $O_2^\bullet$, H_2O_2 and $HClO^-$ (hypochlorus ion) are potent bacteriocidal agents and kill bacteria.

Hydroxylation of steroids and other compounds—During the biosynthesis of steroid hormones and other physiological important compounds, superoxide anion is generated in the presence of cytochrome p 450 to add OH group on steroid ring. This is another example of beneficial effect of free radicals as given below.

$$\text{Cyt. p450} + \text{substrate} + Fe^{2+} + O_2 \longrightarrow \text{Cyt. p450 . Subs. } Fe^{2+} + O_2^{\bullet}$$

$$\swarrow \text{Cyt p 450} \qquad \searrow \text{OH . Substrate}$$

Detoxification of certain drugs and xenobiotics—In the body detoxification of xenobiotics and drugs is also achieved by hydroxylation, in the presence of Cyt p450 and superoxide anion. This facilitates their excretion and thus superoxide anion helps in the prevention of their harmful effects in the body.

Exogenous agents generates free radicals—Many exogenous agents can produce free radicals in the body such as exposure to **ionizing radiations,** that generate $O^{\bullet}{}_2$, H_2O_2 and $OH^{\bullet}$ ions and can be extremely damaging to the tissues.

Ultraviolet radiation of a particular wavelength can result in photolysis of oxygen to generate superoxide anion.

Cigarette smoke contains high concentration of various free radicals, which can be damaging to lungs and to respiratory system causing various diseases.

Oxy Free Radical Tissue Injury

Oxy free radicals are highly reactive and thus rapidly cause oxidative damage to the vital cellular components close to where they are generated. Therefore most cellular structural and functional components are vulnerable to the damaging effects of free radicals, including structural proteins, nucleic acids, biomembranes, enzymes etc. resulting in impairment of cellular function with severe tissue damage. Major damages caused by free radicals are described here—

Lipid Peroxidation of Biomembranes

Oxy free radicals cause peroxidation of polyunsaturated fatty acids of lipid bilayer of biomembranes causing severe damage to the cell membrane integrity with the loss of cell membrane functions. This results in multiferous disturbances of electroyte and water balance and may even cause death of the cell.

Free Radical Chain Reaction

Peroxidation of unsaturated double bonds of polyunsaturated fatty acids (PUFA) of lipids of cell membrane results in generation of other free radical, such as $ROO^{\bullet}$, $RO^{\bullet}$ and $OH^{\bullet}$ with severe deleterious effects. Lipid peroxidation may initiate **free radical chain reactions,** that are highly injurious to cellular components. Lipid peroxy free radical chain reaction occurs in three stages as described here—

1. Initiation of free radical chain reaction —

$$ROOH + \text{metal}^{(n)+} \longrightarrow ROO^{\bullet} + \text{metal}^{(n-1)+} + H^{+}$$

$$\text{Or } X^{\bullet} + RH \longrightarrow R^{\bullet} + XH$$

2. Propogation of chain reaction —

$$R^{\bullet} + O_2 \longrightarrow ROO^{\bullet}$$

$$ROO^{\bullet} + RH \longrightarrow ROOH + R^{\bullet} \text{ etc.}$$

3. Termination of free redical chain reaction —

$$ROO^{\bullet} + ROO^{\bullet} \longrightarrow ROOR + O_2$$
$$ROO^{\bullet} + R^{\bullet} \longrightarrow ROOR$$
$$R^{\bullet} + R^{\bullet} \longrightarrow RR$$

The hydroperoxide product ROOH of PUFA is the precursor for initiation process and thus lipid peroxidation is generally a chain reaction with potentially devasting damaging effects. To control and reduce lipid peroxidation, both in humans and in natural food stuffs, the nature has provided various **antioxidants** (Chapter 25). The most important natural antioxidants are vitamin E, selenium, vitamin C and β carotenes etc.

Free radical damaging Effects on Proteins—Free radicals cause oxidative inactivation of sulphhydral group of cysteine residue at the active site of the enzymes, thereby inhibiting key enzymes of the metabolic cycles. Free radicals also modify amino acid residues resulting in structural alteration of protein molecules resulting in impairment of their biological activities.

Acid Mucopolysaccharide damage—Oxy free radicals cause depolymerization of acid mucopolysaccharides of connective tissues, resulting in loss of structural integrity and tissue damage, that may lead to initiation of disease process.

Free radical damaging effects on DNA—Oxy free radicals cause single stranded breaks in DNA double helix, resulting in impairment of gene expression and inhibition of protein and enzyme synthesis. Free radicals also modify the purine and pyrimidine nucleotide structure resulting in mutations of genes causing carcinogenesis.

Carcinogenic radiations and chemicals act via generating free radical superoxide anion and initiating free radical chain reactions causing mutations in the **oncogenes** which may result in cancer development.

Free radicals scavenging systems—To save body tissues from deleterious damaging effect of free radicals, aerobic organisms have effective antioxidant system for scavenging free radicals, as soon as they are generated. The important enzymes which efficiently scavenge free radicals include—

1. **Superoxide Dismutase (SOD)**—Superoxide dismutase enzyme catalyze reduction and removal of superoxide anion to less harmful H_2O_2 as described in the reaction given below—

$$O_2^{\bullet} + O_2^{\bullet} + 2H^{+} \xrightarrow{\boxed{SOD}} H_2O_2 + O_2$$

The superoxide dismutase thus protects the aerobic organisms against potentially deleterious effects of superoxide anion. The SOD enzymes are found to occur in several different compartments of the cells of almost all tissues. Mitochondrial SOD is Mn^{++} dependent enzyme, while cytosolic SOD contains copper as cofactor and both play significant role in scavenging oxy free radicals in the body tissues.

2. **Catalase**—H_2O_2 is generated continuously in aerobic oxidative system. It is acted upon by the enzyme **catalase** and is degraded into water as shown in the reaction given below. This prevents its deleterious harmful effects on cellular structural and functional components.

$$2H_2O_2 \xrightarrow{\boxed{Catalase}} 2H_2O + O_2$$

3. **Glutathione peroxidase system**—Hydrogen peroxide can also be degraded as soon as it is generated, particularly in RBCs by glutathione and glutathione peroxidase system as shown in the Fig. 26.1 (detail in Chapter 7). The detoxification of H_2O_2 by glutathione peroxidase system is highly significant for the maintenance of cell membrane structural integrity and fragility of RBCs, since accumulation of H_2O_2 causes hemolysis of erythrocytes thereby reducing their life span.

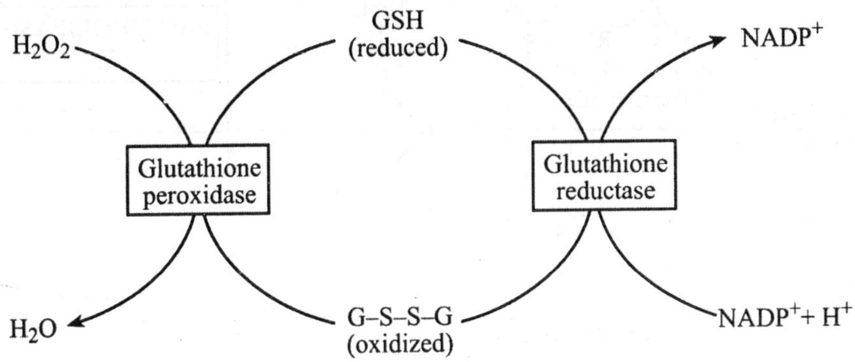

Fig. 26.1 : Detoxification of H_2O_2 by glutathione peroxidase system in erythrocytes

Other Antioxidants—There are other nonspecific natural antioxidants found to be present in food stuffs or produced in the body itself. These antioxidants prevent the formation of free radicals or inhibit free radical chain reaction by breaking or terminating their propagation. Important dietary antioxidants are discussed here.

1. **Vitamin E** is lipid soluble and thus most effective **free radical chain breaking antioxidant** (Chapter 25). Alpha tocopherol reacts with lipid peroxyl free radical and itself is coverted to tocoperoxyl radical, which is relatively stable and prevents propogation of the free radical chain reaction. Vitamin E is lipid soluble and it is found in nature in close association with lipid bilayers and thus prevents lipid peroxidation of biomembranes. If lipid peroxidation is not prevented it may lead to multifarious damage to tissues resulting in initiation of various disease processes. Thus it is very important that **lipid peroxidation stress** must be arrested to prevent clinical consequences and diseases by taking plenty of antioxidants in diet.

2. **Vitamin C** is an important aqueous phase antioxidant and thus prevents tissue damage by oxy free radicals. Vitamin C is implicated in protection against risk of certain cancers as it prevents formation of carcinogens. (Chapter 25).

3. **Ceruloplasmin**—It prevents the formation of free radicals and acts as an antioxidant in blood.

4. **Caffeine and Uric acid** are good antioxidants and prevent formation of oxy free radicals in the body. Uric acid is produced in the body as the end product of nucleic acids and purine nucleotide metabolism, while caffeine is an alkaloid present in tea and coffee.

5. Vitamin A and β carotene are important antioxidants at low pO_2. β carotene is also a chain breaking antioxidant but is less effective than Vitamin E. (Chapter 25).

A summary of highly reactive free radical species and the mechanisms available in tissues for their destruction is given in the Fig. 26.2 along with the interaction of some of the antioxidants.

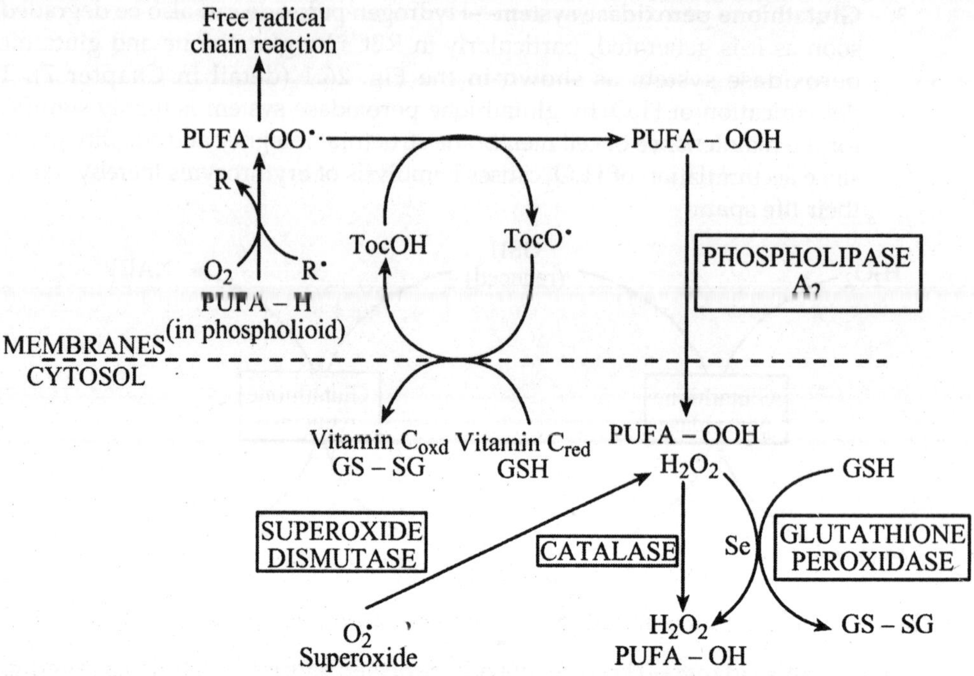

Fig. 26.2 : Role of antioxidants and various enzyme systems in scavenging free radicals in the body

Free Radical Injury—Clinical Implications

There is increasing evidence implicating role of highly reactive species (HRS) and oxy free radicals in the etilogy of disease process. Moreover low levels of antioxidants in plasma or in the diet correlate well with increased incidences of various deadly diseases such as cardiovascular diseases, cancers etc. which further emphasizes the injurious effects of the free radicals in the body. Suggestive role of free radicals in the etiology of some important disease processes is described here under:

1. Chronic Inflammation—As described earlier, at the site of inflammation, activation of macrophages results in increased generation of oxy free radicals. The chronic inflammatory diseases such as rheumatoid arthritis is self perpetuating, due to the release of free radical by neutrophils. Also in these cases there is a decrease in the free radical scavenging enzymes in the joint cavities, which further aggravate the arthritis. Anti inflammatory drugs such as corticosteroids and others, interfere with the generation of free radicals and thus alleviate the disease symptoms.

2. Respiratory diseases—Prolonged breathing of pure oxygen results in destruction of lung endothelium leading to lung edema which can be fatal. **Adult respiratory distress syndrome (ARDS)** is characterized by pulmonary edema, that results from increased vascular permeability due to the release of free radicals by activation of neutrophils.

Cigarette smoke is rich in a variety of highly reactive species and oxy free radicals, which enhance lung damage causing respiratory diseases.

3. Neonatal Free radical injury : clinical aspects—In premature newborn infants, treatment with pure O_2 for prolonged period is responsible for the tissue damage in various organs. This probably is due to the increased production of oxy free radicals causing diseases such as bronchopulmonary dysplasia in lungs, retinopathy in eyes and intracranial haemorrhages etc in the infants. These clinical complications are caused by free radical generation at the site of tissue damage by activated neutrophils causing thromboxane release, sustained vascular contricture and cellular injury etc.

4. Atherosclerosis and Myocardial Infarction—Oxy free radical generation plays significant role in the development of atherosclerotic plaque formation causing clinical vascular complications. Oxy free radicals released by endothelial cells of blood vessels cause **lipid peroxidation** and modify apo LDL, that attracts macro phages which are then converted into the foam cells. This initiates the atherosclerotic plaque formation, leading to cardio vascular diseases.

Low density lipoproteins (LDL) promote atherosclerosis by getting deposited under endothelial cells of blood vessels and are modified due to lipid peroxidation by free radicals released there. This results in elevation of lipid hydroperoxide radicals and leads to decrease in prostacycline—thromboxane ratio causing thrombosis, which may block the blood vessel.

5. Reperfusion injury after myocardial infarction is also caused by free radicals. During ischaemia of myocordium, the concentration of enzyme xanthine oxidase increases and when the myocardium is reperfused, it produces superoxide anions by oxidizing hypoxanthine to xanthine (Chapter 18) and aggravates myocardial injury. Allopurinol, a xanthine oxidase inhibitor, reduces severity of reperfusion injury.

6. Shock related free radical injury—Body ischaemias, during shock also result in oxy free radical release causing lipid peroxidation of the damaged membranes. Leukotrienes released from the platelets and proteases from macrophages cause increased vascular permeability. This results in extravascular exudation of plasma causing tissue edema.

Cerebrovascular damage in hypotensive brain injury and ischaemia, may also be contributed by the free radical generation.

7. Cancer—Free radicals are considered to be main factor contributing in cancer development as they cause DNA damage, resulting in mutations and activation of oncogenes that leads to malignancies. Many carcinogens generate free radicals which are implicated in development of cancer.

8. Age related diseases—Highly reactive oxy free radicals play significant role in degenerative brain diseases such as Parkinsonism, Alzheimer's dementia and multiple sclerosis. Free radical injury also impairs blood- brain barrier resulting in degenerative diseases.

9. Aging Process is thought to be due to cumulative effects of free radical injury to tissues with simultaneous decreased scavenging systems and low levels of antioxidant mechanisms resulting in gradual deterioration of tissues during aging process (old age)

❑❑❑

27

NUTRITION

Optimum nutrition is the prime basis of good health of an organism. Nutrition deals with the quantitative as well as qualitative requirements of various essential dietary principals necessary to maintain health, growth and development of an individual. The **adequate diet** should provide daily energy requirement to carryon life processes, as well as it should supply all the **essential** body building compounds, such as essential amino and fatty acids, minerals, vitamins etc.

In order to plan an adequate diet, one must have the basic knowledge of energy requirements of the body, that is to be met by the diet. Carbohydrates and fats are major source of energy in the diet and to a lesser extent protein are present in varying amounts in foodstuff as energy source.

Calorific value—The energy contents of the dietary components of food are determined from the amount of **heat liberated** on complete oxidation of the nutrient in an atmosphere of oxygen in **bomb calorimeter** and is expressed as **calories. Calorie** is an unit of heat and is defined as—the amount of heat required to raise temperature of 1 gram of water by 1° C (i.e. from 15° C to 16° C). Since it is a very small unit, in medical sciences, the energy requirement and the fuel value of food components are expressed in kilocalories (kcal or simply Cal) that is equal to 1000 calories.

The fuel value or calorific value of three proximate principals of our diet i.e. carbohydrates, fats and proteins as determined in the bomb calorimeter and produced in body is given in the Table 27.1.

Table 27.1 : Calorific values of nutrients

Food components	**Fuel value in K cal per gram of compound**	
	In Bomb calorimeter	**In Body**
Carbohydrates	4.1	4
Fats	9.4	9
Proteins	5.6	4

The calorific values of the carbohydrates and fats is almost same as determined in bomb calorimeter and produced in the body as they are completely oxidized to CO_2 and water in both the cases. However proteins in the body cannot be completely oxidized, rather amino group of proteins has to be converted into urea, that requires extra energy (Chapter 15). Therefore the calorific value of proteins in body is less as compared to that observed in the bomb calorimeter as is evident from the Table 27.1.

Energy Requirement of a Normal Person

Recommended daily allowance of calories for an individual (RDA) is defined as—It is the sum total of basic energy demand, just to sustain life (BMR) plus extra energy needed for the day to day physical work. Also extra provisions should be made for the short durations of leanness e.g. fasting, illness etc.

Caloric Balance—A normal healthy adult, who is maintaining his body weight constant, is said to be in perfect **caloric balance,** that is the amount of daily caloric intake in diet is equal to the total energy expenditure per day.

Basic components of diet—There are six basic components of the diet, the carbohydrates, lipids, proteins, minerals, vitamins and water.

The daily calories intake in the diet is provided by the three major components of the diet i.e. carbohydrates, fats and to some extent by proteins. While the other three dietary essentials i.e. **vitamins, minerals** and **water,** though do not produce energy but are essential in diet to provide for building up of body tissues and for the establishment of biochemical mechanisms and metabolic cycles in which nutrients are oxidized to liberate energy to carry out life processes of an organism.

Recommended Daily Allowance (RDA) for Calories

While calculating the daily caloric requirement, in order to maintain caloric balance of an individual and for good health we must know the **energy expenditure** by an individual for the following processes—

1. Maintenance of basal metabolic rate (BMR)—BMR is the energy expenditure to maintain basic physiologic functions of the body organs to sustain life.
2. Specific dynamic action or thermogenic effects of the food.
3. Energy requirement for the daily physical activities of an individual and
4. Extra energy needed during physiological states such as pregnancy, growth, lactation etc.

Basal Metabolic Rate (BMR)

It can be defined as the basal or minimum amount of energy required just to sustain life for the maintenance of the life processes such as pumping of heart, respiration, working of brain, liver, kidney etc. BMR is measured under certain standard conditions i.e. the subject should be at complete physical and mental rest, but should be awake. He should be in post absorptive state and in comfortable environmental conditions.

Measurement of BMR

Previously BMR determination was an important physiological analytical parameter to assess the thyroid functions, but now a days, it is more of a theoretical and experimental importance. BMR was determined previously either **directly** by measuring amount of heat liberated by a subject in bomb calorimeter or **indirectly,** by the amount of O_2 consumed and CO_2 liberated i.e. by finding **respiratory quotient.** However a simple and convenient procedure to measure BMR is the **closed circuit method,** in which the amount of **oxygen consumed** by the subject under basal conditions is determined using **Benedict Roth's** apparatus as described here under;

Preparation of the subject—The person is asked to come after overnight fast. In morning he is made to lie down for half an hour after giving him a mild tranquilizer for complete relaxed rest. Room temperature should be around 25° C. The subject should be fully awake, as sleep lowers down BMR.

Benedict Roth's Apparatus mainly consists of an inverted jar filled with oxygen, which is suspended in water filled in the space between double walls of another container. This provides a water seal, that prevents leakage of gases (Fig 27.1). The inverted jar is connected through a pulley to a writing pen, which would move, to write on a revolving drum of kymograph, when O_2 is consumed by the subject and inverted jar comes down. The apparatus has two outlets, one for breathing O_2 with one way valve and the other for exhaled CO_2 to be absorbed by a soda lime jar fitted in the center of the double walled tank as shown in the Fig. 27.1.

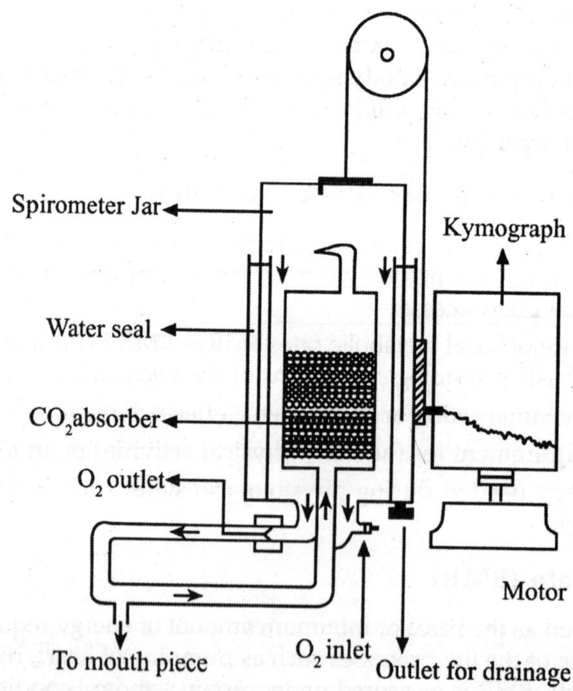

Fig. 27.1 : Diagrammatic representation of Benedict Roth's apparatus

Procedure—The subject is asked to breath O_2 for a 6 minutes period through a mouth piece of the inlet tube (nose is clipped) and exhaled CO_2 goes directly to the soda lime jar and is absorbed. As O_2 is consumed by inspiration, the inverted spirometer jar comes down and the writing pen would record the amount of O_2 consumed on a graph paper pasted on the rotating drum of the kymograph (Fig. 27.1).

Calculation of BMR—BMR is calculated by taking into account the fact, that 4.8 kilo calories are produced per liter of O_2 consumed during metabolism. Suppose 1.2 L of O_2 is consumed in 6 minutes by the subject:

Therefore the heat produced in one hour would be - 4.8 × 1.2 × 10 = 55 Kcal/ hour. BMR or the heat liberated in 24 hours would be - 55 × 24 = 360 Kcal/hour.

Since BMR is affected by the body surface area, it is usually expressed as kilocalories per hour/sq.m. body surface area. **Body surface area** can be found out by measuring the body weight and height of the subject and applying Dubois formula as given below—

$$A = H^{0.725} \times W^{0.425} \times 71.84$$

where A is body surface area in sq. m, H is height in cm. and W is body weight in kilograms.

Normograms are now available showing body surface area from the height and body weight. Suppose the body surface area of the subject comes to be 1.7 sqm, then BMR can be expressed as –

$$\text{BMR} = \frac{55 \text{ Cal}}{1.7} = 32 \text{ Kcal / hour / sqm.}$$

Normal BMR for adult men is 34-37 Kcal/hour/sqm and for women it is 32-35 Kcal/ hour/sqm.

Factors affecting BMR—Several factors affect basal metabolic rate, which should be taken into consideration and are as listed below—

1. **Age**—BMR is high during the period of active growth. So children have high BMR, which reaches to maximum by 5 years of the age, then it decline gradually with lower BMR in old age.
2. **Sex**—Males have higher BMR than females as they have more muscle mass.
3. **Environmental temperature**—BMR increases in cold climate as a compensatory mechanism to maintain body temperature at 37° C. Thus Eskimos have higher BMR than persons living in hot climates.
4. **Exercise**—During exercise BMR increases, thus athletes have higher BMR due to increased muscle mass and increased cardiac output.
5. **Emotional conditions** such as anxiety, nervousness etc increases BMR as brain metabolism increases during stressful conditions.
6. **Sound sleep** lowers BMR, but it may be high during disturbed sleep.
7. **Smoking** increases BMR by about 10%.
8. **Starvation**—Prolonged fasting reduces BMR, probably as a compensatory measure to decrease caloric expenditure.
9. **Fever**—During disease states associated with fever, the BMR is raised as all body functions are enhanced, for example the pulse rate increases with tachycardia. BMR in febrile conditions increases by about 12% per degree rise of body temperature.
10. **Thyroid hormones**—Thyroid hormones have a stimulatory effect on general metabolism and heat production in the body. Thyroxine has calorigenic effect and BMR is increased in **hyperthyroidism** while it is lowered in **hypothyroidism**.

Specific Dynamic Action of Food

It is observed that following intake of food (i.e. after meals), the heat production in body is increased. This is known as **thermogenic effect** or **specific dynamic action** (SDA) **of food.**

Specific dynamic action is defined as the **extra heat** generation over and above the **actual fuel value** of the food, when it is metabolized in the body. For instance an amount of protein with fuel value of 100 Kcal when metabolized in the body, extra 30 calories from the body stores are expended with total heat production of 130 Kcal. The proteins have maximum SDA while carbohydrates have lowest SDA. The values of SDA of the dietary components as observed are—for protein it is 30%, for lipids it is 15% and for carbohydrate it is 5%, extra than the actual fuel values.

Cause of SDA of food components—The extra energy production by consuming food is attributed to following factor—a) Energy expenditure by the body during digestion and absorption of food stuffs, b) Energy expenditure to establish various metabolic cycles and for providing various enzyme, cofactors etc. to metabolize the nutrient and c) For proteins the SDA is **maximum** because body has to expend extra energy during conversion of NH_2 group of amino acids into urea for its excretion (Chapter 15).

SDA of mixed diet—It has been observed that by consuming a mixed diet, the SDA of proteins is compensated by taking more carbohydrates in diet (they have low SDA). This is because the carbohydrates spare the proteins and amino acids for body building up and glucose is preferably oxidized to produce energy. Thus the SDA of normal mixed diet is less (10% only). Therefore it is customary to add 10% extra calories while calculating RDA for calories to account for SDA of the food stuffs.

Physical Activity

Another important factor to be considered while calculating RDA for caloric requirement is the daily physical activity of a person. This is an **highly variable factor** and varies from person to person. The extra caloric requirement for physical activity would depend on **occupation** of the person, his **life style** and the amount of **physical work** he performs everyday.

To simplify and for the sake of convenience, the physical activity level of people may broadly be grouped into three categories as; **sedentary, moderate** and **heavy work.** Thus depending on level of physical activity a person is involved with, additional calories are to be added while calculating RDA for caloric intake as shown in Table 27.2.

Table 27.2 : Extra caloric requirements depending on person's physical activity

S.No.	Type of work	Extra caloric requirement
1.	**Light work** (sedentary work) e.g. clerks, teachers, house wife etc.	- 30-40% of BMR
2.	**Moderate work** e.g. students, industry worker, doctor, engineers etc.	- 40-50% of BMR
3.	**Heavy work** e.g. labourers, athletes, farmers etc.	- 50-60% of BMR

The amount of energy required for performing physical work depends mainly on the duration and intensity of the muscular activity. The expenditure of energy for doing various physical activities has been determined experimentally and is summarized in the Table 27.3.

Table 27.3 : Energy requirements during performing physical work

S.No.	Physical Activity	Energy required in kcal/hour
1.	Sitting, writing, standing, eating, reading etc.	30-40
2.	Typing, driving vehicle, doing household work.	60-80
3.	Walking slowly, going to college	100
4.	Cycling (speed 2 km/hr)	150
5.	Running	300-400
6.	Swimming	500-550
7.	Stairs climbing up	800

Daily Energy Requirement for an Adult

To calculate daily energy requirement, three factors as mentioned above should be taken into account. Thus the BMR, specific dynamic action of food and the kind of physical work the person is involved with, would determine the daily energy requirements of the person. For instance, the daily energy requirement of a person doing **moderate work** would be as given below—

The energy requirement for BMR is	55 x 24	=	1320 Cal/day
Extra calories needed for moderate work (50% of BMR)		=	660 Cal
One hour games/exercise		=	300 Cal
	Total	=	2280
SDA of food (10% of total)		=	228 Cal
	Total	=	2508 Cal

Thus total energy requirement (rounded off) would be = 2500 Cal/day.

Depending on the kind of physical work and occupation of a person, an approximate recommended daily allowance for calories (RDA) for an adult person is as given under —

For light worker	–	2200-2500 Cal/day
For moderate worker	–	2500-2900 Cal/day
For heavy worker	–	2900-3500 Cal/day

Dietary Requirement of Nutrients

The principal components of a diet, that would meet our daily requirement for calories, as well as would supply the dietary essentials to maintain good health are—

1. **Macronutrients**—The three major (proximate) components of our diet, the carbohydrates, lipids and proteins are the main source of energy (calories), as

well as they provide some essential food factors such as essential amino acids, essential fatty acids etc. needed for body building up and to maintain good health.

2. **Micronutrients**—The three essential components of our diet are vitamins, minerals and water, which though do not produce energy, are essential for the life. They are necessary for establishing various biochemical mechanisms and metabolic cycles for the liberation of energy to carry on life processes, as well as for biosynthesis of various physiologically important compounds to maintain normal health and growth of a person.

Distribution of Calories in Diet

Carbohydrates are the most economical and readily available source of energy, while fats are richest and concentrated form of energy in our diet. Where as proteins in our diet mainly provide essential amino acids needed for protein synthesis and for anabolic purposes.

Distribution of calories in the dietary components of food varies from place to place in the world, because of the kind of availability of food stuffs and varying food habits, life style of the people and cultural and social status. A comparison of caloric distribution in diets of Western and Oriental regions of world is given in Table 27.4.

Table 27.4 : Comparison of distribution of calories in dietary components of Western and Indian diet

Source of calories	Western diet	Indian/occidental diet	Ideal distribution according to dietations
Carbohydrates	40-50%	70-80%	55-70%
Fats	20-30%	15-20%	15-20%
Proteins	20-40%	10-15%	10-15%

Recommended Daily Allowance for Nutrients (RDA)

The RDA for various dietary components facilitates planning of an adequate or balanced diet required for the optimum nutrition, good health and welfare of an individual. With understanding of the nutritional aspects of various dietary ingredients and RDA for important nutrients as recommended by nutrition experts, a balanced diet can be planned from the **"Basic seven"** that form the common food stuffs taken in our diet.

BALANCED DIET

The basic composition of balanced diet varies from country to country in the world depending on the food habits, life style, availability of the kind of food stuff etc. The nutrition expert group of Indian council of medical research (ICMR) has recommended the RDA for various nutrients and composition of balanced diet for Indians.

A Balanced diet is that which contains all the nutrients in **proper proportion** and of **good quality,** so as to meet the daily caloric requirement, as well as to provide all the essential dietary components, such as vitamins, minerals, essential amino acids, essential fatty acids etc to maintain good health and development of an individual. Extra provision of nutrients should be made in balanced diet for the physiological states, such as pregnancy, lactation, growth and convalescence period after illness.

To simplify the concept of balanced diet, the various food stuffs that form our diet are grouped as **"Basic Seven"**. The international association of biochemists has prescribed RDA for all the nutrients as per recommendation of WHO and FAO and is as summarized in the Table 27.5.

Table 27.5 : Basic Seven and RDA of major dietary components of food stuffs in a balanced diet

S.No.	Basic Seven	Servings/day	RDA/day	Provide/ Importance
1.	Cereals, enriched whole grain flour	2-3 servings	300-400g	Major source of energy (calories)Bran comprises dietary fiber, vitamins, minerals
2.	Milk and milk products paneer, cheese, curd etc.	1-2 servings	280g	Excellent quality proteins Rich in essential amino acids, calcium, phosphate, other minerals & vitamins.
3.	Vegetables—	2-3 serving		
	Green leafy		110 g	Vitamins, minerals, fibers
	Tuber root		80 g	Carbohydrate, proteins
	Others & raw salad		80 g	Dietary fiber, vitamins
4.	Fruits—	1-2 serving	80 g	Vitamins, minerals
	Seasonal			Vitamin C, antioxidants
	Citrus			Dietary fiber,
	Fruit juice			Carbohydrates
5.	Meat/Egg for non veg. for vegetarians - pulses, legumes, beans, paneer, cheese etc.	1-2 serving	80 g	Good quality proteins Fair quality proteins Variety of food stuffs supply all essential amino acid.
6.	Fats, oils, butter & ghee	Cooking medium	40 g	Fat soluble vitamins Antioxidants Essential fatty acids
7.	Sugar and jaggary	Sweetner	30-40g	Readily available caloric source, Jaggary is rich in iron.

Nutritional Importance of Carbohydrates

Major carbohydrates taken in the foodstuffs are polysaccharide, **starch** and disaccharides such as sucrose (cane sugar) and lactose (milk sugar). Carbohydrates are most **economical** and **major source of calories** in our diet. Approximately 70% of total caloric intake in Indian diet is supplied by carbohydrates. A minimum of 5g/100 calories in diet should be taken as carbohydrates to prevent ketosis. This is because the fats burn in the common furnace, the citric acid cycle, that is properly operative in the presence of carbohydrates (Chapter 11). In addition to calories, the carbohydrates in diet also provide dietary fibers, which are very important for good health.

Dietary fibers—Certain complex plant polysaccharides, such as cellulose, hemicellulose, lignins, pectins etc, that form the cellular frame work of the vegetable food stuffs, are not digested in our intestine and are referred to as **dietary fibers.**

These undigestible dietary fibers are very important in our diet with many beneficial effects and functions and now a days are considered as **essential dietary components.** In general the dietary fibers are divided into two major groups—

1. Insoluble dietary fibers include cellulose, hemicellulose, lignins, pectins etc and constitute mainly the seed coats and the cellular frame work of the vegetables, for example bran, rice polishing, outer coat of vegetable, fruits etc. Major function and beneficial effects of the insolube dietary fibers are enumerated below —

a. The dietary fibers imbibe water, swells up and **prevent constipation** by increasing bowel motility
b. Promote **intestinal peristalsis** and regulate GIT motility.
c. Increase **bulk of diet** and thus **provide satiety** with low caloric diet to maintain caloric balance and thus obesity is prevented
d. Prevent **colic diseases, abdominal pain, diverticulosis** and **cancer of colon** etc.

2. Soluble dietary fibers—The soluble dietary fibers include plant polysaccharides such as mucilages, galactomannans, pectins, rhamnogalacturans, arabinoglycans etc. and are found to occur in the vegetables such as legumes, beans, soyabean, guar gum etc.Their major beneficial effects include —

Hypocholesterolemic effect—The soluble fibers lowers blood cholesterol level by acting as sequestrial agents and bind bile acids and bile salts in the intestine and thus prevent their reabsorption. This blocks enterohepatic circulation of bile acids and sterols and thus enhance their excretion in the feces.

Hypoglycaemic effects—The soluble dietary fibers are beneficial in reducing postprandial rise in blood sugar level by slowing gastric emptying time and thereby reduce the rate of glucose absorption from the intestine. Thus the dietary fibers improve glucose tolerance and are highly beneficial for diabetic patients.

Anticancerous effects—In addition to above mentioned effects the dietary fibers are implicated in prevention and lowering the incidences of gastrointestinal cancers and colic diseases.

Nutritional Importance of Lipids

Fats and oils are main cooking medium in the Indian diet, which improves taste and palatability of the food stuffs. The lipids in diet are **concentrated source of calories** and are major source of **fat soluble vitamins** and **essential fatty acids.**

The fats and oils, chemically are triglycerides, which are esters of glycerol with saturated, monounsaturated and polyunsaturated fatty acids (PUFA).

Essential Fatty Acids

The polyunsaturated fatty acids (PUFA) are considered as essential fatty acids as they cannot be synthesized in body and have to be taken in the diet to maintain normal health and growth of an individual. The chemical nature, major dietary sources and important functions of essential fatty acid are described in the Chapter 11.

The essential fatty acids (PUFA) play very important role in our body that include— a) They are essential constituent of biomembranes and lipids of brain and CNS, b) They play important role in maintenance of epidermal linings and their proliferation. Deficiency of PUFA in experimental animals results in severe scaling of skin, loss of hair and impairment of reproduction. c) Essential fatty acids are precursor for the synthesis of physiologically potent compounds, such as prostaglandins, leukotriens, thromboxanes etc. and d) PUFA lowers blood cholesterol, VLDL and LDL cholesterol levels. Thus PUFA are protective against cardiovascular diseases.

Animal fats are primarily saturated fats, such as butter, ghee and invisible fat contents of meat, chicken, egg etc. as they consists of predominantly saturated fatty acids and are considered to be **atherogenic** with high risk of cardiovascular diseases. The polyunsaturated fatty acids are present in vegetable oils and fish oils. More than half of the PUFA are provided in the invisible fats of seeds, nuts, almond etc. The PUFA are considered to be **antiatherosclerotic** as they lower blood cholesterol level.

Balanced Intake of PUFA, MUFA and Saturated Fatty Acids

However it has been observed that **excessive consumption** of **PUFA** can be **injurious** to body tissues. Presently the nutrition expert groups recommend, that not more than 30% of total fat should be taken as unsaturated fats with PUFA and a serious concern is being expressed regarding undue importance given to the excessive consumption of PUFA in the diet by health freaks.

This is because, it has been observed that excessive PUFA in diet may lead to generation of **peroxy free radicals** due to the peroxidation of unsaturated fatty acids of the body tissues that may cause free radical chain reaction which can be highly injurious to tissues (Chapter 26), unless enough antioxidants are available. Thus recent nutritional recommendations favour a **proper ratio** and a **balanced intake** of saturated, monounsaturated and PUFA in diet for good health and welfare of the persons.

Nutritional Importance of Proteins

Proteins are building blocks of body tissues and are of prime importance as functional components necessary for the life. A living cell synthesizes thousands of different proteins with characteristic and unique biological functions necessary for performing life processes.

In a normal **Indian diet,** protein form 10-15% of total caloric intake. When sufficient amount of carbohydrates are taken in the diet (as in Indian diet), the amino acids are not much utilized for the oxidation and energy production purposes, rather they are spared to be utilized for body protein synthesis and for the anabolic purposes. This is known as **protein sparing action of carbohydrates.**

Essential Amino Acids

An appropriate amount of **good quality** proteins in diet is **essential** as they are the only source of **essential amino acids** and major source of N_2 and sulphur required for various biosynthetic processes in the body for maintenance of good health and growth of a person.

A regular and daily supply of all **essential amino acids** is necessary for two reasons—i) there is no reserve store of proteins or essential amino acids in the body and ii) All proteins of the body are in dynamic state i.e. everyday an appreciable amount of body proteins (400g/day) are degraded and re-synthesized to keep body tissues healthy and functioning optimally. So essential amino acids must be supplied regularly in the diet for the daily protein synthesis. Deficiency of even a single essential amino acid for some time leads to negative N_2 balance, hypoproteinemia, edema etc. causing disease state.

There are several indices which determine the nutritional quality and biological value of proteins. These were developed mainly by experiments on animals. Weaning albino rats were mostly used as experimental animals in these nutrition studies. The four important indices to asses the nutritional value of proteins are described here—

1. Protein efficiency ratio (PER)—It is the weight gain by an animal per gram of protein provided in the diet for a standard time period of 4 weeks and is calculated by the formula as given under—

$$\text{PER} = \frac{\text{Gain in body weight (g)}}{\text{Per gram protein ingested}}$$

Protein efficiency ratio of major proteins is summarized in the Table 27.6.

2. Biological value of proteins—The biological value (BV) of proteins is the ratio between the amount of N_2 retained in the body and N_2 absorbed from the diet by an animal and is determined by the formula—

$$\text{BV} = \frac{\text{Nitrogen retained}}{\text{Nitrogen absorbed}} \times 100$$

Biological value of proteins is determined by feeding protein to an animal as the only source of N_2 and the excretion of N_2 waste products in urine and feces were measured. The biological value provides reasonably good index for nutritive value of the proteins. However it does not make any allowance for the N_2 lost during digestion.

3. Net protein utilization (NPU)—This is a better nutritional index for proteins than the biological value (BV), since it takes into account the digestibility of proteins in GIT. Net protein utilization is determined from the ratio between the amount of nitrogen retained to that of taken in diet (ingested) as given below —

$$\text{NPU} = \frac{\text{Nitrogen retained (g)}}{\text{Nitrogen ingested (g)}} \times 100$$

Chemical score—The essential amino acid contents (EAA) of a protein are expressed in terms of chemical score. It is defined as—mg of EAA per gram of protein as compared to egg protein, taken as the reference protein. Chemical score can be found out by the formula as described below —

$$\text{Chemical score} = \frac{\text{mg of EAA / g test protein}}{\text{mg of EAA / g reference protein}} \times 100$$

The relative nutritional values and different experimental indices to assess the nutritive value of various proteins are summarized in the Table 27.6.

Table 27.6 : Nutritive values of various proteins

Source of protein	BV	NPU	PER	Chemical score
Egg	90	91	4.5	100
Milk	84	75	3.0	65
Meat	80	76	2.8	70
Rice	64	57	2.0	60
Wheat	58	47	1.7	42
Bengal gram	58	47	1.7	44
Dried peas	56	45	1.6	42
Ground nut	54	45	1.7	44
Soyabean	64	54	2.0	57
Gelatin	-	-	0	0
Zein	-	-	0	0

Limiting Essential Amino Acid

The data in Table 27.6 shows that the nutritive value of egg and poultry proteins is highest, followed by milk and meat proteins. On the other hand, proteins from plant food stuffs have lower nutritional value. This is because certain proteins of plant origin were found to be deficient in one or more essential amino acids. When these proteins were fed as the only source of N_2 to experimental animals, they failed to grow. The deficient EAA in that protein is called the **limiting amino acid** and proteins were considered as **fair quality** (nutritional value) proteins.

Mutual supplementation of proteins—The problem of limiting essential amino acid in vegetarian proteins can be solved by taking a mixture of variety of proteins in the diet. The mutual supplementation of proteins to the experimental animals resulted in increased growth rate. This is because the deficiency of an EAA in one protein is being overcome by giving another protein rich in that particular EAA. For instance, rice and wheat proteins lack lysine and threonine (EAAs), while pulse protein (Bengal gram) is limited in S-containing methionine and cystine amino acids. Indian diet with cereals (wheat, rice) is taken along with pulses (dal) and thus the limiting EAAs lysine and threonine in cereals is compensated by their supplementation from the pulses and at the same time the limiting amino acid methionine of pulses is compensated by the cereal proteins which are rich in methionine.

The nutritive value of food stuffs is enhanced by proper combination of variety of pulses, legumes and other food stuffs and thus with the consumption of mixed diets, the requirement of all essential amino acid is fulfilled ultimately.

NUTRITIONAL DISORDERS

Protein Energy Malnutrition

The protein energy malnutrition is the most prevalent nutritional disorder of the children in developing countries. The prevalence varies depending on the socioeconomic status, literacy and awareness of nutritional importance of proteins in the diet.

Isolated deficiency of proteins can occur due to the consumption of staple diet lacking in good quality proteins. Deficiency of one or the other essential amino acid will result in **negative Nitrogen balance,** hypoproteinemia with edema, anemia and other clinical consequences. The major malnutrition disorders are described here—

Kwashiorkar

It is a protein malnutrition disorder. The term kwashiorkar was introduced by Cicely Williams (1933) to a nutritional protein deficiency disorder, commonly affecting the children of gold coast (modern Ghane) of Africa. The term kwashiorkar means "sickness of the older child, weaned away from mother's milk, when next child is born early.

Cause—Kwashiorkar mainly is expressed in children of 1-5 years of the age, primarily due to the lack of good quality proteins in the diet. The diet of weaned child from mother's milk, predominantly consists of cereals (carbohydrates) only and lacks good quality proteins due to poor socio economical status. Typically the child has very thin extremities and protruded belly due to enlargement of liver.

Biochemical abnormalities include negative nitrogen balance and decreased plasma albumin levels. Hypoproteinemia is a characteristic finding (<2g/dl) resulting in water retention and **edema**. Other clinical complications are fatty liver with liver function impairment, low hemoglobin level with mild anaemia, low plasma immunoglobins with decreased resistance to infection etc.

Clinical manifestation—The major clinical symptom are retardation of growth, edema, diarrhea, anaemia, apathy, child is lethargic, skin rough with dermatitis, hair sparse, angular stomatitis, muscle wasting , Jaundice, renal failure etc.

Management—Supplementation of good quality proteins in the diet and control of infection would alleviate clinical manifestations. A mixture of vegetable proteins (3 part) and milk proteins (1 part) is very effective in management of the patients. The monitoring of the treatment is done by improvement in weight gain and elevation of serum albumin level with disappearance of edema.

Marasmus

It is a combined **protein energy malnutrition syndrome.** Marasmus is the most widespread nutritional disorder in developing countries. It results from severe deficiencies of both proteins and caloric dietary intake. Marasmus is a severe condition affecting the children below one year age due to early weaning away from mother's milk and non availability of proper diet. This is mostly because of poor socio economic status as well as due to frequent adverse natural calamities such as famine etc.

Clinical symptoms of marasmus—Marked retardation of growth, severe anemia, diarrhea, dry atrophic skin, muscles weak and atrophic, CNS development is also retarded, decreased resistant to infections etc. Main difference in clinical symptoms of marasmus from that of kwashiorkor is that **edema** is **not present** in the case of marasmus.

Obesity

Obesity is the most prevalent nutritional disorder in the prosperous developed countries. It is a great health hazard and can occur at all ages. Obesity in recent times is implicated as root cause of many dreaded diseases, such as atherosclerosis, cardiovascular diseases, type II diabetes mellitus and certain types of cancers etc.

Obesity is an abnormal increase in body weight due to excessive fat deposition in adipose tissues. It is basically a **nutritional disorder** caused primarily by excess caloric intake than being consumed, due to the lack of physical exercise. Obesity is more commonly considered to be a behavioral and wrong eating habit disorder.

Major Causes of Obesity Include

1. Excess intake of caloric rich food stuffs with behavioral habits of always nibbling something with sedentary life style.
2. Genetic predisposition to heredity has also been correlated with obesity.
3. Endocrine disorders such as hypothyroidism, hypogonadism, Cushing's syndrome etc. are associated with obesity.
4. Psychiatric disorders some times lead to obesity, because patient has compulsive craving to consume excessive food.
5. Lack of brown adipose tissue in obese persons. This establishes the importance of brown adipose tissue in preventing obesity. Some adult individuals are fortunate to possess brown adipose tissue and can burn **extra calories ingested** and would not become obese (Chapter 6), inspite of ingesting excessive calories in food.

Adverse Effects of Obesity

Clinical manifestations of obesity include—a) The major ill effects of obesity are reduced life span, b) hyperlipidemia, with associated atherosclerosis leading to hypertension and c) increased risk of coronary heart diseases and high incidences of cerebrovascular diseases.

- Obesity is associated with development of **type II diabetes mellitus,** as fat ladden peripheral cells have less number of spare insulin receptors and develop resistance against insulin action resulting in non insulin dependent diabetes mellitus (NIDDM).
- Cancer of breast, colon and prostate are correlated with high fat intake and hyperlipidemia in obese persons.

Management of Obesity

1. Low caloric diet with lots of vegetables, salad and dietary fibers that gives satiety at less calories in diet.
2. Regular physical exercise to control body weight.
3. Restriction of refined carbohydrates and sugar, that stimulate lipogenesis and fat gain.
4. Low carbohydrate-low fat with more PUFA in diet to control blood cholesterol and triglyceride levels to prevent risk of cardiovascular diseases.

❑❑❑

28

HORMONES

A living organism possesses an highly efficient **communication system,** that coordinates various body organ functions and metabolism in an integrated manner to maintain its normal health, growth and welfare. This is achieved by **hormones,** the **chemical mediators** which are the major controlling factors to regulate the body functions and metabolism according to the immediate need of a living organism.

Hormones (Greek—**hormo-to excite or to arouse)** are **defined as chemical messengers, produced in small amounts by endocrine or the ductless glands, that are released in the blood circulation to control and regulate metabolic and physiological functions of the target tissues in a coordinated manner.** However certain hormones may have **paracrine** functions i.e. they act on adjacent cells, close to the cells from which they are released, for instance prostaglandins and polypeptide growth factors. Some **autocrine** hormones act in same cells where they are synthesized e.g. interleukin-2, produced by T-cells stimulates the proliferation of T-lymphocyte itself.

Classification of hormones—**Hormones are broadly classified into two groups based on their mechanism of action as given below—**

1. Hormones- acting through intracelluar receptors—Hormones belonging to this group are lipophilic in nature and can permeate through the lipid bilayer of biomembranes. These hormones bind to specific intracellular receptors and then mediate their physiological and metabolic functions inside the target cells. Hormones belonging to this group are mainly steroid hormones. Examples are —

(a) Corticosteroid hormones—such as glucocorticoids, mineralocorticoids etc.

(b) Male and female sex hormones.

(c) Calcitriol (active vitamin D, DHCC) and (d) thyroid hormones, T_3 and T_4.

2. Hormones-acting through cell membrane receptors—These hormones being hydrophilic in nature, cannot cross biomembranes and thus bind to **specific receptors** on the **cell membrane** (plasma membrane) of the target tissue. The hormone-receptor complex then stimulates release of **II messengers,** through which they mediate their action inside the target cells. The hormones belonging to this group are mostly **protein or polypeptide hormones.** Example are—

(a) Protein hormones such as insulin, glucagon, growth hormone, PTH etc.

(b) Peptide hormones such as ADH, vasopressin, LH, prolactin etc.

(c) Catecholamines, epinephrine and norepinephrine.

Mechanism of action of hormones—Group I

These hormones have **specific intracellular receptors** located either in cytoplasm or in the nucleus, to which they bind after **recognition** to form an **hormone-receptor complex** (Fig. 28.1) and then bring about their diversified **short term** and **long term effects** inside the target cells as described here.

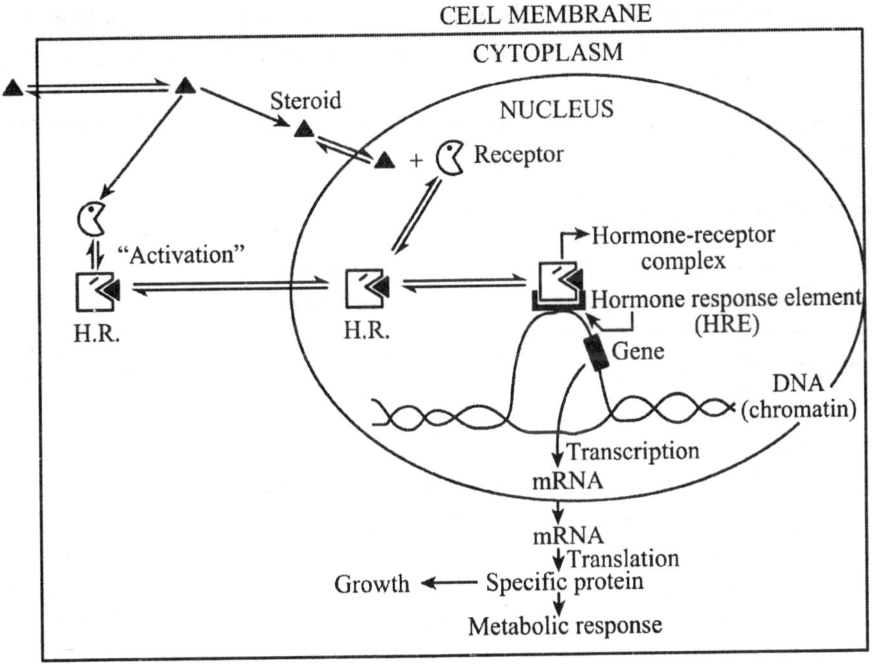

Fig. 28.1 : Mechanism of action of steroid hormones H- Hormone, R-receptor

Effects on Gene Expression—Long term effects

The **hormone-receptor complex** so formed has affinity to bind to specific regions on the DNA, called **hormone responsive elements** (HRE) and increase **expression of specific genes** to stimulate transcription for the synthesis of proteins in general and certain key enzyme as shown in the Fig. 28.1. The enzyme synthesis takes some time i.e. in hours or may be days and thus these effects are known as long term effects to increase or decrease a metabolic process.

Short term effects—These hormones directly activate/ inactivate the key enzymes **of metabolic process by** covalent modifications. This action is rapid and brings about immediate response to regulate a metabolic process.

Mechanism of action of Protein Hormones—Group II

The hormones belonging to II category cannot permeate through the plasma membranes of target tissue and act via binding on specific **cell membrane receptors.** These hormones exert their wide range of action inside the cell through the release of certain potent mediary compounds, collectively known as **second messengers.**

Concept of II Messengers

The term II messenger was coined by scientist Earl Sutherland (1961, Nobel prize 1971), when he discovered **cyclic AMP,** while working on glycogen metabolism in hepatic cells. This was a major break through in our understanding of the mechanisms of hormone action.

Cyclic AMP (c-AMP, cyclic adenosine monophosphate) is a physiologically potent nucleotide, of universal occurrence that is released in the target cell, when some protein hormones bind on the cell membrane receptors. The c-AMP so produced then acts as a **second messenger** to bring about the diversified actions of the hormone (I messenger) inside the cell at various sites.

Chemical nature of c-AMP

Chemically c-AMP is a purine nucleotide, consisting of adenine, ribose sugar and a phosphate group, which froms an extra intramolecular ring(cycle) by forming a **3′-5′ diphosphoester** linkage at position 3rd and 5th C-atom of ribose sugar as shown in Fig. 28.2.

c-AMP is synthesized from ATP and the reaction is catalyzed by a membrane bound **enzyme adenyl cyclase** that is specific in hydrolyzing ATP in a different manner to introduce an intramolecular cycle in AMP moleule to produce 3′-5′ cyclic AMP (Fig. 28.2). After bringing about hormone actions inside the cells, c-AMP is inactivated by enzyme **phosphodiesterase** to form plain AMP (Fig. 28.2). This mechanism regulates the c-AMP concentration inside the tissues.

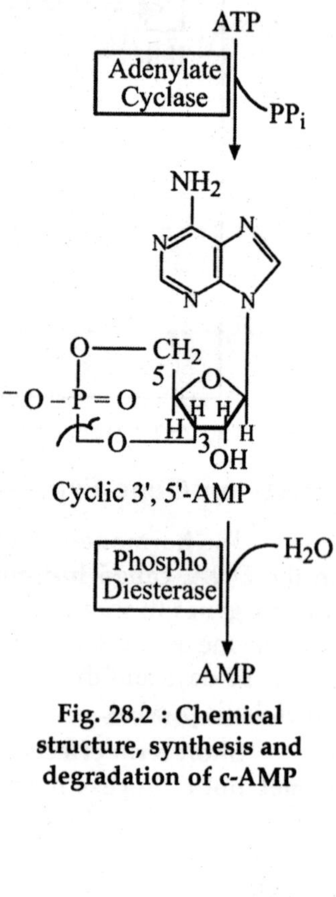

Fig. 28.2 : Chemical structure, synthesis and degradation of c-AMP

Adenyl Cyclase System—Mechanism

When protein hormones (II category) bind on specific cell membrane receptors, a series of events occur leading to the synthesis of c-AMP and the process is mediated through a membrane bound **G-protein.** These events are as described here —

1. The hormone (H) binds to a stimulatory receptor (R) on cell membrane of the target tissue and

activates **G-protein,** a membrane bound specific protein, that has an **affinity** to bind to **guanine nucleotide.**

2. The G-protein is a **trimer,** consisting of α, β and γ-subunits. The α subunit is bound to **GDP** in the native G-protein (Fig. 28.3).
3. The binding of hormone to specific receptor on plasma membrane causes **exchange** of GDP from α subunit of G-protein with **GTP,** and this dissociates the α subunit, now bound to GTP, from the trimeric G-protein to liberate β-γ **dimer.**
4. The separated α subunit bound to GTP stimulates the membrane bound enzyme **adenyl cyclase** to synthesize c-AMP by specific hydrolysis of ATP molecule as shown in the Fig. 28.3.
5. Simultaneously GTP bound to α subunit is hydrolyzed to provide energy for the formation of c- AMP. The α subunit-GDP so produced has affinity to recombine with β-γ dimer to regenerate α-β-γ trimeric G-protein that is reutilize to continue the process. The whole sequence of events is illustrated in the Fig. 28.3.

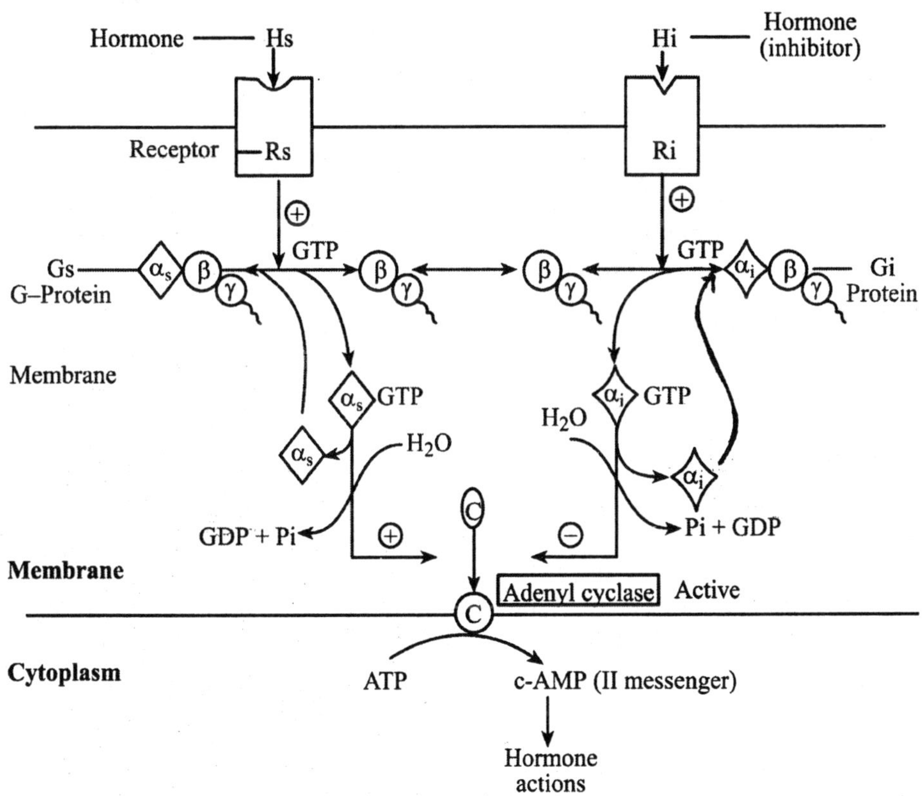

Fig. 28.3 : Mechanism of G-protein mediated stimulation of adenyl cyclase for the synthesis of c-AMP by hormone -receptor complex formed on the cell membrane

Some hormones inhibit the adenyl cyclase activity. This effect is mediated through inhibitory G- protein (Gi). The structure of α subunit of Gi is different (mol wt. 41 kDA) while β and γ subunits are similar to the stimulatory G- protein (Gs). The dissociation of trimeric Gi is similar to that of Gs, however the α subunit released, inhibits adenyl cyclase instead of activating it. The hormones that inhibit adenyl cyclase and decrease c-AMP level include somatostatin, α_2 adrenergics and angiotensin II etc.

Action of II Messenger—c-AMP

The protein hormones that act through c-AMP as II messenger include hormones glucagon, growth hormone, epinephrine and pituitary hormones such as ACTH, FSH, LH, TSH, ADH etc.

Once produced c-AMP mediates the wide range of diversified biochemical and physiological actions of hormones inside the target cells. c-AMP activates and act through an enzyme **protein kinase.** This kinase enzyme is a tetrameric protein having two regulatory (R) and two catalytic (C) subunits as **R_2C_2 inactive form**. c-AMP binds to the regulatory subunit and dissociates the protein kinase into R and C subunits as shown in the reaction.

$$4\text{ c-AMP} + \underset{\substack{\text{Protein kinase} \\ \text{inactive}}}{R_2C_2} \longrightarrow R_2\text{-(4c-AMP)} + \underset{\substack{\text{Catalytic subunit} \\ \text{active}}}{2C}$$

The catalytic subunit so separated is now free to activate/inactivate the key enzymes of metabolic cycles by phosphorylation reactions. The catalytic subunit catalyze the phosphorylation of OH group of serine residue at the calatytic site of key enzymes by ATP, to either activate or inactivate them rapidly to stimulate or inhibit a metabolic cycle as required. Details of this covalent modification of key enzymes by c-AMP-protein kinase system is given in the Chapter 5 along with the reciprocal regulation of glycogenolyis/ glycogenesis and glycolysis/ gluconeogenesis metabolic processes.

Other II Messengers

Some hormones act through different II messengers, other than c-AMP. Some important other II messengers and their mechanism of action is described here.

Ca^{++}–calmodulin System

Hormones such as α adrenergic catecholamine, TRH, vasopressin etc, after binding on cell membrane receptors stimulate **Ca^{++} ion release** in cytosol by either activating Ca^{++}/ H^+ ATPase pump to uptake Ca^{++} inside the target cell or by stimulating release of **bound or stored** intracellular calcium. The Ca^{++} ions so released act as II messenger to bring about hormone action inside the cell by binding with a specific protein known as **calmodulin** as shown below.

$$4Ca^{++} + 4\text{ sites of calmodulin} \longrightarrow 4\ Ca^{++}\text{–calmodulin complex}$$

↓ +ve(activate several fold)

Certain kinases

↓ Activate/inactivate

Many enzyme systems

Phosphoinositol – II Messenger System

Certain hormones mediate their action through activation of phosphoinositol as second messenger. The binding of hormones such as **serotonin** on cell membrane receptor activates phosphotidylinositol (PLP_2) cascade system, by triggering activation of enzyme **phospholipase C** which hydrolyzes PLP_2 to diacylglycerol (DAG) and release **inositol triphosphate** (II messenger). The inositol triphosphate (IP_3) elevates intracellular calcium concentration, which in turn contributes to a variety of actions inside the target cell as mentioned above. The IP_3 and DAG both act sinergestically and phosphorylate the target enzymes to activate/ inactivate them, for example HMG CoA reductase, tyrosine hydroxylase, cytochrome p450 etc. Both IP_3 and DAG increase affinity of protein kinase-C for calcium. The calcium is derived by IP_3 action that releases calcium from the endoplasmic reticulum stores.

Tyrosine Kinase like Activity

Certain hormones such as **insulin** after binding on cell membrane receptor, activate tyrosine kinase bound to the β unit of receptor that transduce transmembrane signals inside the cell to bring about the diversified of actions insulin at various levels (Chapter 8).

A brief account of II messengers of some important hormones is summarized in the Table 28.1.

Table 28.1 : II Messengers of hormones acting through cell membrane receptors

	II messenger	Hormones	Endocrine gland
A	c-AMP	Growth hormone, ACTH, FSH, LH, TSH, ADH	Pituitary
	c-AMP	Protein hormone glucagon,	α cells of Islets of Langerhan's of pancreas
		PTH	Parathyroid gland
	c-AMP	Epinephrine Nor epinephrine	Adrenal medulla
B.	Phosphatidyl inositol/Ca^{++}	Serotonin	-
C	Ca^{++} calmodulin	α-Adrenergic catechol-amines, TRH, ADH etc.	-
D	Tyrosine kinase	Insulin	Islets of Langerhan's of pancreas

c- GMP as II Messenger

In recent years it has become clear that c-GMP also acts as II messenger and phosphorylates its target proteins to mediate action of certain hormones and drugs inside the cells.

c- GMP, synthesized from GTP in an analogous manner to the formation of c-AMP as described above. Here the substrate GTP is acted upon by a membrane bound enzyme **guanyl cyclase** to synthesize c-GMP Atriopeptins are certain peptides produced by atrial cardic tissue cause vasodilation, diuresis and inhibit aldosterone secretion. These actions of atriopeptins are mediated through the II messenger c-GMP that activates the c-GMP dependent protein kinase, which phosphorylates myosin of smooth muscles, leading to relaxation and vasodilation of the tiusse.

PITUITARY AND HYPOTHALAMIC HORMONES

The pituitary is a small endocrine gland (1g in weight) and is situated below the hypothalamus region of the brain. The pituitary gland has two distinct parts, the anterior and posterior lobes. The anterior pituitary is under the control of hypothalamus and secretes many hormones that regulate growth and functions of other endocrine glands of the body and are called tropic hormones. The posterior pituitary secretes hormones that regulate water balance of the body and milk ejection from lactating mammary gland.

Hypothalamus—Master Coordinator

Hypothalamus is a specific center in the brain, that along with its other diversified physiological functions, **coordinates the secretions of many endocrine hormones** with that of the **central nervous system (CNS)** and thus can be regarded as **master coordinator.** In response to certain signals from brain, the hypothalamus secretes certain **releasing hormones (factors)** from the nerve fiber endings, around the blood capillaries of the hypothalamus-hypophysial axis in the pituitary stalk. The releasing factors so liberated then reach to the anterior lobe of pituitary gland and control the secretion of tropic hormones from it. The tropic hormones on their part then regulate the secretion of hormones from their target endocrine gland.

The whole sequence of events mentioned above is termed as hypothalamus-hypophysial (pituitary) endocrine axis and is under control by **tight feed back mechanisms** that ordinary form an integrated feed back regulatory system.

Hypothalamic Releasing Factors—Hypothalamus secretes around six important releasing factors that are listed in the Table 28.2, along with their influence on pituitary tropic hormones secretion and the target endocrine glands that affect hormones produced by them.

Table 28.2 : Hypothalamus – Hypophysial target endocrine acid gland axis

Hypothalamic hormones	Pituitary hormones affected	Target gland hormones affected
Corticotropin releasing hormone (CRH)	ACTH (LPH, MSH, endorphins)	Adrenocorticoid hormone hydrocortisone
Thyrotropin releasing hormone(TRH)	TSH	Thyroid hormones, T_3 and T_4
Gonadotropin releasing hormone (GnRH)	LH, FSH	Androgens, estrogens, progestins
Growth hormone releasing hormone (GRH)	GH	1GF-1
Growth hormone inhibitory hormone (GRH In)	GH(TSH,FSH, ACTH)	IGF-1, T_3 and T_4
Prolactin release/inhibitory hormone (PRH In)	PRL	Neurohormones.

HORMONES OF ANTERIOR PITUITARY

The anterior pituitary lob secrets several tropic and other hormones which serve as main controllers and coordinators of endocrine glands and a variety of physiological and biochemical processes in the body.

The hormones of anterior pituitary can be classified broadly into three categories—

(1) The growth hormone- prolactin-somatomammotropins.

(2) The glycoprotein tropic group which include TSH, LH, FSH and chorionic gonadotropins (CG)

(3) The pro-opiomelanocortin (POMC) peptide hormones.

GROWTH HORMONE-PROLACTIN GROUP

The growth hormone, prolactin and chorionic somatomammotropins constitute a family of protein hormones with considerable **structural similarities** in the amino acid sequence in their molecules and have **common antigenic determinants.** All these hormones **promote growth** and have lactogenic function and probably have evolved by duplication of a single ancestral gene.

GROWTH HORMONE

The human growth hormone is a single polypeptide protein hormone with 191 amino acids (mol. wt. 22,000) and is produced from the acidophilic cells called somatotropes of anterior pituitary, that are the most abundant cells of the anterior lobe. Growth hormone is essential for postnatal growth of the infant, since it stimulates transcription (gene expression) and general protein biosynthesis and has profound effect on metabolisms of carbohydrates, lipids, proteins and minerals in the body.

Secretion of Growth Hormone (GH)

The secretion of growth hormone is under the control of two hypothalamic factors. The GH releasing hormone (GHRH) and GH release inhibiting hormone (GHR In) which in turn are regulated by **tight feedback** mechanisms. Secretion of growth hormone is periodical with large diurnal variations. Peak secretion of GH occurs during sleep i.e. 2 hours after sleeping (sound sleep period), which explains the fact that growth process in the child mainly takes place during sleep. The concentration of GH in the pituitary gland is largest (5-15 mg/g) as compared to other hormones. Daily output of GH is 0.2 to 1 mg/day with plasma half life of about 6-20 minutes.

Physiological and Biochemical Functions

Effect on Growth—The growth hormone is essentially required for the **postnatal growth** of the infant. It promotes growth of the body tissues by stimulating cell division, as well as by promoting **increase in size** and development of cells of a number of tissues such as **cartilages, bones** and **muscles** etc.

The growth promoting actions of GH are mediated through insulin like growth factors (IGFs) which are also known as somatomedins. Individuals fail to grow normally and remain dwarfs or pygmies if IGF-I is lacking in them. Another less important IGF-II has also been reported through which GH may mediate its growth promoting actions.

Effects on protein synthesis—The growth hormone promotes growth of body tissues by **enhancing protein synthesis** through directly stimulating gene transcription and RNA synthesis. GH also facilitates amino acid transport (uptake) into the muscles to be utilized rapidly for the protein synthesis during growth process.

Growth hormone causes **positive nitrogen balance** and **gain in body weight** by enhancing DNA replication, RNA synthesis and generalized protein biosynthesis to promote growth. The anabolic and growth promoting actions of GH resembles some of the actions of the Insulin.

Effects on Carbohydrate Metabolism—Growth hormone increases blood sugar level by antagonizing (opposite) the effects of insulin. GH decreases peripheral utilization of glucose by stimulating mobilization of fats from the adipose tissues and causing release of fatty acids in blood. The fatty acids (FFA) are preferably oxidized in muscles for energy generation, thus **sparing glucose** that results in **hyperglycaemia.** GH also stimulates hepatic gluconeogenesis and increase liver glycogen stores. Overproduction of GH, as seen in hyperpituitarism or chronic administration of GH causes β cells exhaustion and results in diabetes like condition.

Effects on lipid metabolism—GH promote lipolysis in adipose tissues by activating enzyme **hormone sensitive lipase** (HS lipase) via c-AMP-protein kinase cascade activation system (Chapter 11).This results in increased release of FFA in blood, which are oxidized instead of glucose in muscles and other tissues, thereby indirectly increasing blood sugar level. Increased oxidation of FFA may also cause **ketogenesis**.

Effects on mineral metabolism—GH through IGF-I promotes a **positive calcium, magnesium** and **phosphate balance** by increasing their reabsorption from the renal tubules (less excretion). Growth hormone promotes growth by stimulating formation of cartilages and growth of long bones in growing children. GH also causes retention of electrolytes such as Na^+, K^+ and Cl^- in the body.

Lactogenic Effects—GH has actions similar to prolactin and binds to lactogenic receptors and stimulates lactogenesis in mammary gland.

Abnormalities of Growth Hormone—Clinical Aspects

Hyposecretion of GH—Deficiency of growth hormone can occur either due to panhypopituitarism or due to congenital defective GH production and cause serious clinical manifestations in the growing children and infants. GH deficient infants fail to grow resulting in **dwarfism**. The GH deficient dwarfs respond positively and growth is stimulated by administration of growth hormone to the affected subjects.

Hypersecretion of GH—Overproduction of GH mostly occur due to the acidophilic cell tumors of pituitary resulting in clinical syndrome **gigantism** in children and **acromegaly** in adults **Gigantism** in children is caused by GH excess, that occurs **before the closure of epiphysial plate,** leading to excessive growth of long bones with extraordinary height and size of the body.

Acromegaly in adults, on the other hand is caused by GH excess occuring after the closure of epiphysial plate of long bones. It is characterized by very **large hands, feet** and **skull,** along with abnormal facial changes, such as protruding **big jaw, enlarged** nose etc. Other abnormalities include enlargement of vicera, thickening of skin, excessive hairs etc, along with metabolic abnormalities resulting in a diabetes like condition.

PROLACTIN HORMONE

Prolactin is a protein hormone with stricking structural and amino acid sequence similarities to that of growth hormone. Prolactin is secreted from the lactotropes (acidophilic cells) of anterior pituitary. During pregnancy the number and size of lactotropes increases several fold to increase prolactin and luteinizing hormone (LH) synthesis.

Physiological Functions

Lactation—Prolactin and other lactogenic hormones primarily promote initiation and maintenance of lactation in mammals. In rodents, prolactin maintain corpus luteum and functions as luteotropic hormone and stimulates progesterone production, necessary for the maintenance of pregnancy.

Protein Synthesis—In mammary gland, prolactin stimulates transcription of genes, responsible for the synthesis of milk proteins and several enzymes needed for the milk production and synthesis of lactose.

Hypersecretion of Prolactin : Abnormalities—Overproduction of prolactin, caused by the tumors of lactotropes, results in **amenorrhea** (cessation of menses) and enlargement of breasts with **galactorrhea** (milk discharge without pregnancy) in the females and impotency in males.

GLYCOPROTEIN HORMONES OF ANTERIOR PITUITARY

The glycoprotein group of hormones secreted by anterior pituitary are the most complex proteins with remarkable structural similarities, but with diversified physiological actions. This group include: (a) Thyroid stimulating hormone (TSH), (b) Luteinizing hormone (LH), (c) Follicle stimulating hormone (FSH) and (d) Chorionic gonadotropin (CG) which is produced from the placenta.

Mechanism of Action—All these protein hormones act through specific cell membrane receptors and via G-protein adenyl cyclase activation, increase c-AMP concentration, which acts as intracellular messenger to bring about their physiological actions in the target tissues.

Chemical Nature—These hormones consist of two subunits called α and β subunits. The α subunit of these hormones is identical in a particular species, while the β subunits are structurally different and are responsible for the specific physiological actions of these hormones.

Thyroid Stimulating hormone (TSH)—Anterior pituitary secretes TSH in response to the hypothalamic thyroid releasing factor (TRF) and both are under tight feedback control by the thyroid hormone (T_3 & T_4) blood levels.

Physiological functions—TSH hormone stimulates the synthesis and secretion of thyroid hormones from the thyroid gland. There are specific TSH receptors on the thyroid acinar cells, on which TSH binds and **stimulates all the steps of T_3, T_4 synthesis** by activating key enzymes via the release of c-AMP as mentioned above (Chapter 29). The TSH also stimulates the secretion and release of thyroid hormones by the gland.

Gonadotropic hormones (FSH, LH and CG)—These hormones are primarily involved in proper functioning of gonads during **reproduction** and **stimulate gametogenesis** and synthesis of female sex hormones.

Follicular stimulating hormone (FSH)—The FSH binds to the specific receptors on its target cells i.e. the follicular cells in the ovaries and the **sertoli** cells in the testes, to stimulate ovulation and spermatogenesis respectively.

FSH also stimulates **follicular growth,** increases **ovarian size** and **enhance estrogen production** in the females. In males FSH stimulate synthesis of male sex hormone testosterone and promotes **spermatogenesis** by stimulating growth of the seminiferous tubules.

Luteinizing Hormone (LH)—LH binds on specific cell membrane receptors of corpus luteum and promotes production of hormone progesterone in females to maintain pregnancy. In males LH stimulates testosterone production in the **leydig cells** by activating key enzymes of steroidogenesis via increasing c-AMP concentration in the target cells.

Human chorionic gonadotropin (hCG)—It is a glycoprotein hormone produced mainly from the **placenta** and closely resembles in its structure (amino acid sequence) to that of LH. The physiological actions of hCG are not definite. However its plasma levels and **excretion in urine increases after implantation** of fertilized ovum in the uterus. This forms the basis of diagnostic tests for the early confirmation of pregnancy by the detection of hCG in the urine.

Physiological Functions—Human chorionic gonadotropin has been reported to have lactogenic and luteotropic activities along with similar metabolic effects to that of growth hormone.

THE PRO OPIOMELANO CORTIN (POMC) HORMONES

The anterior pituitary secretes a group of pro-opiomelano cortin hormones. The POMC family consists of a group of peptide hormones namely (a) Adrenocorticotropin hormone (ACTH), (b) Lipotropin hormone (LPH), (c) Melanocyte stimulating hormone (MSH) and (d) Endorphins and other several neuromodulators.

Synthesis and structure of POMC peptides—The most stricking feature in the production (synthesis) of the several peptide hormones of POMC family is that they are all produced by a single gene in the anterior and intermediate lobes of pituitary gland. This single gene is expressed in a single precursor large polypeptide chain with 285 amino acids. From this large **POMC precursor,** all the peptide hormones of this family are produced by a complex post- translational processing, that is different in different regions of pituitary gland.

Salient features of POMC hormones—The most conserved sequences between species reside within amino terminal fragment, the ACTH region and the β endorphin region.

POMC and related products are found to occur in several other tissues such as brain, lungs, placenta, GIT, reproductive tract and lymphocytes, where they are synthesized by different genes.

The intermediary lobe of pituitary gland is rudimentary in adult humans but is active in fetus and pregnant women. Different peptide hormones are generated from the precursor POMC polypeptide as a result of difference in the site of cleavage. There are three broad cleavage regions in POMC as described below—

(1) ACTH region at the amino terminal end, that can further be cleaved to produce MSH and corticotrophin like intermediate lobe peptide (CLIP).

(2) Lipotropin region (β-LPH) which can be cleaved to produce γ-LPH, β-MSH and β endorphin as shown in Fig. 28.5.

(3) A large N-terminal peptide that can produce γ-MSH.

The various specific cleavage sites of the precursor POMC have cluster of basic amino acids peptide bonds, that are cleaved by trypsin like enzymes. The major cleavage sites and different peptide hormones formed are illustrated in the Fig. 28.5.

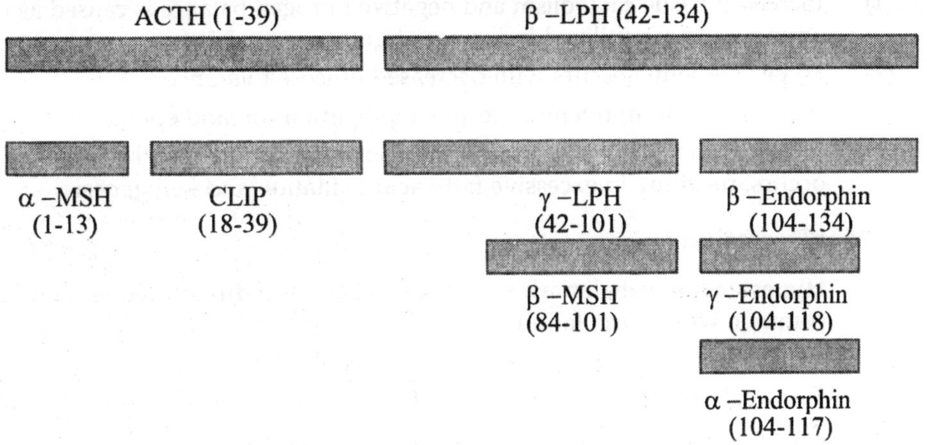

Fig. 28.5 : Products of POMC cleavage at specific site along with the number of amino acids mentioned in bracketm

The POMC in the anterior lobe is first cleaved into two fragments namely ACTH from the amino terminal end with 1-39 amino acids and β LPH with 93 amino acids (42-134 amino acid residues).

In the intermediary lobe of pituitary β LPH is further cleaved into γ LPH (42-101 residues) and β endorphins (104-134 residues) as shown in Fig. 28.5.

Extensive additional modifications of these peptide products occurs, such as glycosylation of ACTH, acetylation of β MSH etc. β- Endorphins are **rapidly acetylated** and **are inactivated** in the pituitary gland, but in hypothalamus, the endorphins are not acetylated and remain in active form with potent physiological functions.

Adrenocorticotropic hormone (ACTH)—ACTH is a single polypeptide hormone consisting of 39 amino acids, produced by specific cleavage from NH_2 terminal end of POMC as mentioned above. Its main function is to stimulate growth and functions of adrenal cortex. The first 24 amino acids from amino terminal end of ACTH are essential for its biological function.

Mechanism of action—ACTH acts via binding to cell membrane receptors and within seconds of its release, it markedly increases c-AMP level in adrenal cortex. ACTH via c-AMP and Ca^{++} ions release, increase gene expression for the synthesis of key enzymes, required in the synthesis of steroid hormones. ACTH increases the synthesis and release of adrenal steroid hormones by enhancing the conversions of cholesterol to pregnanolone, which requires lysis of six carbon side chain with the help of a **cleavage enzymes** (Chapter 30).

Abnormalities of ACTH Secretion—Clinical Aspects

Cushing's syndrome—Excessive production ACTH from the pituitary or ectopic production caused by tumors, results in Cushing's syndrome due to enhanced activity of adrenocorticosteriod hormones (Chapter 30).

Metabolic Manifestations include

(1) Increased tissue catabolism and negative nitrogen balance is caused as more amino acids are utilized in increased gluconeogenesis process.

(2) Hyperglycaemia occurs with decreased glucose tolerance.

(3) Increased sodium retention results in hypertension and edema.

(4) Increased mobilization of fats and release of fatty acids from the adipose tissue occurs, resulting in excessive fatty acid oxidation and ketogenesis.

Clinical Manifestations

(1) **Hypertension** is hallmark symptom caused by sodium retention. Edema is common with muscular atrophy.

(2) Peculiar redistribution of fats results in **trunkal obesity.**

(3) The weak MSH like activity of ACTH results in **hyperpigmentation of skin.**

(4) Decreased circulating eosinophils and lymphocytes with increased polymorphonuclear leukocytes count is observed in blood.

Decreased ACTH secretion due to infection or infarction of the pituitary results in opposite manifestations to that of Cushing syndrome (Chapter 30).

β LPH and Endorphins—Biological function of β LPH (42-131 residue) produced from POMC (Fig. 30.5) is limited and is rapidly cleaved in pituitary gland into γ LPH and β endorphins.

Endorphins—β Endorphin cleaved off from β LPH has 31 amino acids. The α and γ endorphins are modification of β endorphin, by the specific cleavage of 15 and 14 amino acid residues respectively as shown in Fig. 30.5. These α and γ endorphins are acetylated and inactivated in the pituitary gland.

However at other sites, such as neurons of central nervous system, they are not modified or inactivated and hence serve as neurotransmitter or neuronal modulators. Endorphins bind to same receptors, as do the morphine opiates and may play a role in **endogenous control of pain perception.** They have higher **analgesic potencies (18-30 times) than morphine.** Human POMC do have amino acid sequence for enkephalins, but it is not preceded by a sequence of basic amino acids, so are not cleaved and expressed in humans.

Melanocyte stimulating hormone (MSH)—The POMC consists of amino acid sequence of α, β and γ-MSH, but are expressed and cleaved differently in different animal species. α and β MSH are secreted in some animal species, where they stimulate melanogenesis (synthesis of melanine) and cause intracellular dispersion of melanine granules resulting in darkening of the skin.

In humans the actual circulating MSH activity remains contained within larger β or γ LPH molecules. ACTH possesses α MSH sequence (1-13 residue) but it is acetylated and thus exhibits weak MSH activity.

Patients with Addison's disease have hyperpigmentation which could be due to the compensatory increase in ACTH or more likely due to increased secretion of β or γ LPH with their associated MSH activity.

POSTERIOR PITUITARY HORMONES

Posterior pituitary contain two active hormones, called vasopressin or **antidiuretic hormone** (ADH) and **oxytocin.** Both these hormones are nonapeptides (nine amino acid polypeptide) and are produced from specific cells in hypothalamus and are transported by axoplasmic flow to nerve endings in the posterior pituitary lobe. These hormones are then released upon proper stimulation, into the blood circulation. This arrangement of secretion of these hormones is probably to escape restriction of blood-brain barrier. ADH is mainly synthesized by supraoptic nucleus while oxytocin is synthesized by paraventricular nucleus and are transported through axons, bound to specific carrier proteins, called neurophysin I and II respectively (synthesized along with their specific peptide hormones). They are released at the nerve endings into the posterior pituitary.

The two hormones ADH and oxytocin are secreted separately into the blood stream upon appropriate stimulus, from the pituitary depots. In blood, they circulate in unbound form and have a very short half life, of about 2-4 minutes only.

Chemical nature—Both ADH and oxytocin are nonapeptide hormones with cystein amino acids at position one and six from the carboxyl terminal that are linked by intramolecular disulphide linkage as shown in Fig. 30.6. Humans and most animals have arginine ADH, while pigs and other animals have lysine ADH.

Cys-Tyr-Phe-Gln-Asn-Cys-Pro-Arg-Gly-NH_2

Arginine vasopressin

Cys-Tyr-Phe-Gln-Asn-Cys-Pro-Lys-Gly-NH_2

Lysine vasopressin

Cys-Tyr-Ile-Gln-Asn-Cys-Pro-Arg-Gly-NH_2

Oxytocin

Fig. 30.6 : Structures of ADH and oxytocin

Because of the close structural similarities, ADH and oxytocin exhibit some of the effects of each other. ADH was originally called vasopressin because it increases blood pressure, which is one of its important functions in the body.

OXYTOCIN

Regulation of Secretion—The primary stimulus for oxytocin release from posterior pituitary is the neural impulse that results from stimulation of breast nipples by infant sucking. Secondary stimuli include vaginal and uterine distentions due to the pregnancy. Estrogens stimulate production of oxytocin and neurophysin I, while progesterone inhibit its production.

Physiological Function

1. Uterine contraction—Oxytocin causes contraction of uterine smooth muscles and induce labour during child birth and thus is used as drug to induce labour at the time of delivery.

2. Milk ejection—The most important function of oxytocin appears to be, to stimulate milk ejection by contracting the myoepithelial cells surrounding the mammary gland alveoli.

Mechanism of Action

Cell membrane receptors for oxytocin are found in uterine and mammary tissues, which increase in numbers by estrogens and decreased by progesterone. The concomiant elevation of estrogens and fall in progesterone, occurring immediately before parturization, promotes the lactation before the child birth.

ANTIDIURETIC HORMONE (ADH)

ADH secretion is controlled primarily by the change in osmolality of plasma, which is sensed by the osmoreceptors located in hypothalamus and by baroreceptors present in the heart and blood vessels. Increased osmolality of plasma (low blood volume) stimulates ADH secretion while reduced osmolality inhibit its secretion. Other factors which stimulate ADH secretion are emotional and physical stress and pharmacological agents such as nicotine and morphine etc. Epinepherine and certain chemical agents that expand plasma volume, inhibit ADH secretion. Alcohol intake also inhibit ADH secretion.

Physiological Action—The primary action of ADH in the mammals is on distal convoluted tubules (DCT) and collecting tubules of nephrons of the kidneys. These ducts pass through renal medulla, where solute concentration is 4 times that of plasma. These ducts are thus relatively impermeable to water and if plasma volume is more (hemodilution) with decreased osmolality, ADH will not be secreted and more water is excreted in the urine to regulate plasma volume and to maintain osmolality, as well as blood pressure.

In cases of hemoconcentration with increased osmolality, ADH is secreted and it increases the water reabsorption by increasing its permeability from DCT and collecting ducts. Thus water is retained to restore blood volume and urine is concentrated, resulting in urine volume as low as 0.5 to one l/day.

Mechanism of Action—Specific ADH receptors (V_2 type) are present on renal epithelial surface that are linked to adenyl cyclase and ADH actions are mediated through c-AMP in DCT and collecting ducts. Inhibitors of phosphodiestrase enzyme e.g. caffiene mimic the action of ADH. Extrarenal receptors of ADH are V_1 type and ADH acts through activation of phospholipase C and increase of intracellular Ca^{++} ions. Major effects of ADH binding on V_1 receptors, present in vascular system is to cause vasoconstriction and increase in blood pressure. Hence the name vasopressin was originally given to ADH.

Abnormalities of ADH Secretions—Impairment of ADH secretion results due to the destruction of hypothalamus-hypophysial tract, such as in cases of basal skull fracture, tumors or infections etc. Insufficiency of ADH results in excretion of large volumes of dilute urine leading to severe clinical manifestation.

Diabetes insipidus—In an heriditary congenital disorder called nephrogenic diabetes insipidus, the ADH secretion is normal, but the target renal tubules do not respond to ADH action due to a receptor defect (V_2 type) in the renal tubules.

❑❑❑

29

THYROID HORMONES

The thyroid gland is situated in the neck region, on the either side of the trachea and produces two iodoamino acid hormones called **3,5,3' triiodothyronine** (T_3) and **3,5,3'5' tetraiodothyronine** (T_4) which is also known as **thyroxine**. Thyroid hormones have long been known for their potent biological effects on—1) **Gene expression** to increase protein synthesis, **general growth** and **development,** as well as they have profound effects on fetal growth and tissue differentiation. 2) Thyroid hormone increases general metabolism and are essential for the normal functioning of almost all cells of the body and 3) They play central role in the regulation of **BMR** by increasing O_2 consumption, as well as Na^+/K^+ pump and ATPase activity.

CHEMICAL NATURE AND BIOSYNTHESIS

Thyroid hormone biosynthesis requires trace element **iodine** and a tyrosine amino acid rich colloidal protein known as **thyroglobulin.** Only function of iodine in our body is in the synthesis of thyroid hormones and thus I_2 metabolism is dealt along with thyroid hormones in this section.

I_2 is scarce element in most part of world

In most part of the world, iodine is a scarce component of soil, so very less amount of it is present in the food stuffs.

Sources of iodine in diet is mainly the **drinking water** and trace amounts of I_2 present in the cereals and vegetables. **Sea foods** and fish are good sources of I_2. Major source of iodine, now a days is **iodized salt.** Most of the countries have made it compulsory to supplement I_2 in **common salt** to prevent hypothyroid diseases, which are the most common endocrine disorders in the world.

Unique Feature of Thyroid Hormone Synthesis

As iodine is a scarce element in nature, our body has developed a **complex mechanism** to trap and **retain iodine** for thyroid hormone synthesis and once trapped as iodotyrosine residues, it is not allowed to be lost out of the gland normally.

Biosynthesis of thyroid hormone requires thyroglobulin

The production of T_3 and T_4 and all the steps of biosynthesis of thyroid hormone takes place on tyrosine residue, while it remains attached to the thyroglobulin molecule present in the lumen of the thyroid gland. Thyroglobulin is a large protein molecule with approximately 5000 amino acids (mol. wt 660 kDa) and consists of about 115 tyrosine residues. It is synthesized by the thyroid follicular cells and is secreted in the **lumen** of the **acini** where it remains stored.

Thyroid stimulating hormone (TSH) secreted from the pituitary gland is **necessary** for T_3 and T_4 production and all the steps of biosynthesis and release of thyroid hormones are stimulated by TSH.

I_2 metabolism and T_3, T_4 synthesis—Iodine metabolism and biosynthesis of thyroid hormone comprises the following discrete steps—

1. Uptake and concentration of iodide (I^-)—The thyroid gland, salivary glands and stomach epithelial linings are able to uptake and concentrate iodide against strong concentration gradient by an **active transport mechanism,** linked to the Na^+/K^+ pump and requires an **iodide transport carrier system** as shown in the Fig. 29.3. Through this transport mechanism, thyroid gland can takeup and concentrate I^- to about **10,000 times** than the ordinary cells.

This **iodide transport** system is inhibited by **perchlorate** and **thiocyanate ions** that **compete** with I^- for the **carrier** which transport iodide inside the thyroid gland.

2. Oxidation of iodide (I^-)—The thyroid gland is the only tissue, that can oxidize I^- to an **highly reactive, high valency state iodine** ions **(I^+)**. This is **prerequisite** for the **organification** and iodination of tyrosine residues to synthesize thyroid hormones. Oxidation of I^- requires a specific, heme containing enzyme called **thioperoxidase** that is present in the thyroid follicular cells at their **luminar surface.**

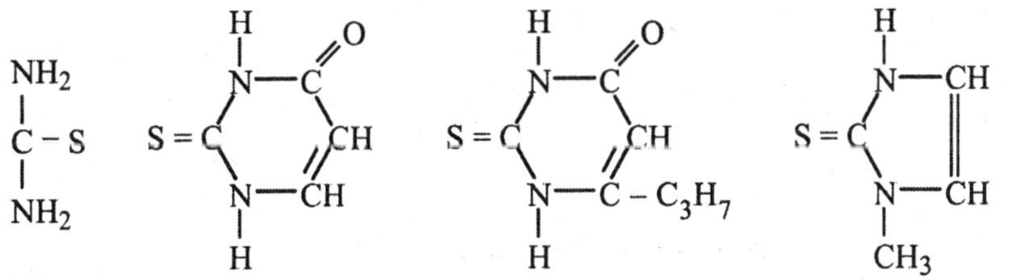

Fig. 29.1 : Thiourea Thiouracil Propylthiouracil Methimazole

Thioperoxidase—It is a tetrameric enzyme (mol wt. 60 k Da), containing heme as cofactor and requires H_2O_2 as oxidizing agent. Hydrogen peroxide (H_2O_2) is produced by an NADPH dependent enzyme resembling **cytochrome c-reductase** present in the acinar cells. Many compounds inhibit oxidation of I^- by the enzyme thioperoxidase, for example **thiourea, thiouracil, methimazole** etc and have great therapeutic uses as described below.

Biomedical importance—The compounds thiourea, thiouracil and methimazole are used as **antithyroid drugs** for the treatment of **hyperthyroid** diseases and **thyrotoxicosis.** These compounds are potent inhibitors of thyroid hormone biosynthesis and thus their use prevents excessive production of T_3, T_4 in the hyperthyroid diseases. Chemical structures of these antithyroid drugs are given in Fig. 29.1.

3. Iodination of tyrosine residues : Organification—In the 3rd step the highly reactive oxidized iodide (I^+) reacts, within seconds of its production with tyrosyl residues of thyroglobulin molecule by an oxidation reaction. The reaction is catalyzed probably by the same **thioperoxidase** enzyme. In this organification reaction, I^+ is added (by displacing H^+ ion) first at position 3rd and then on 5th position of phenyl ring of tyrosine residue to form **monoiodotyrosine** (MIT) and **diiodotyrosine** (DIT) respectively as shown in the Fig. 29.2. Once iodination **(organification of iodide) occurs, iodine is trapped inside the thyroid gland and is stored.**

4. Coupling of iodotyrosine residues—In the next step condensation or coupling of two molecules of DIT to form T_4 (thyroxine) and one DIT and one MIT to produce T_3 (triiodothyronine) takes place, while they remain attached to the thyroglobulin molecule. This condensation reaction also requires oxidation, so the same enzyme thioperoxidase catalyzes the coupling of DIT and MIT also for the thyroid hormone biosynthesis. This is further supported by the fact, that the same thiourea drugs inhibit coupling of MIT and DIT also that inhibit thioperoxidase enzyme. The coupling reaction of DIT and MIT is given in the Fig. 29.2.

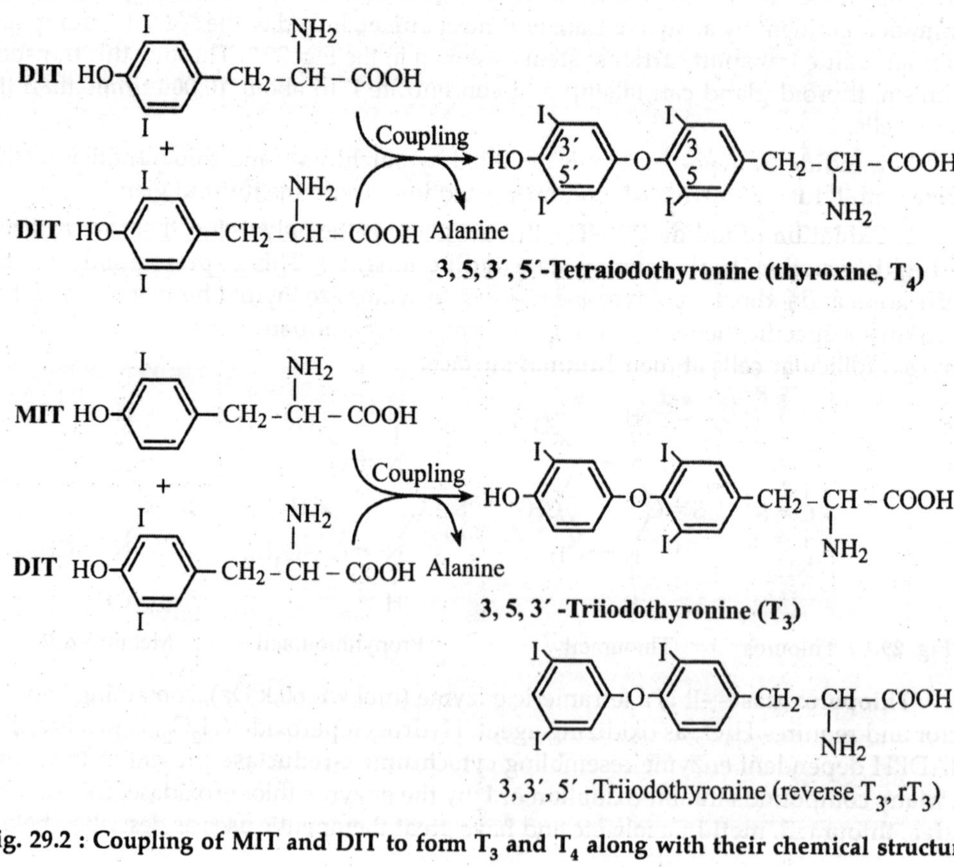

Fig. 29.2 : Coupling of MIT and DIT to form T_3 and T_4 along with their chemical structures

5. Storage of T_3, T_4—The thyroglobulin containing iodinated tyrosyl residues serves as storage form of T_3 and T_4 with several week's supply of thyroid hormones. Thyroid gland is the only endocrine organ which can store appreciable amount of thyroid hormones.

6. Hydrolysis and release of T_3 and T_4—Upon proper stimulation by TSH, secretion of T_3 and T_4 from the thyroid gland occurs, within minutes of TSH release from the pituitary gland. The thyroglobulin containing **iodothyronines** is taken inside the follicular cells from the lumen, by the process of **pinocytosis.** Inside acinar cells, the phagosomes containing thyroglobulin, fuse with lysosomes and by specific proteolysis, the various iodothyronines are liberated. T_3 and T_4 so produced are then released into to the blood for their transport to the target tissues. The whole events involved in the synthesis and secretion of T_3 and T_4 are illustrated in the Fig. 29.3

Salvation of iodide from MIT and DIT

When thyroglobulin is hydrolyzed into T_3 and T_4, appreciable amount of MIT and DIT are also released inside the acinar cells (about 70%). The iodide from MIT and DIT is scavenged (salvaged) by the enzyme **deiodinase** and is reutilized for the thyroid hormone synthesis (Fig. 29.3), and thus iodine is not allowed to be lost or excreted out of the body. This salvation of iodide is very important to retain iodine in the thyroid gland as it is a scarce element and not easily available in food stuffs.

Potassium iodide (KI) depresses hydrolysis of thyroglobulin into T_3 and T_4 and thus is used as **adjuvant** in the treatment of hyperthyroid diseases.

The major events of iodine uptake, thyroid hormone biosynthesis on the thyroglobulin molecule in the lumen of thyroid acini, mechanism of T_3, T_4 secretion and salvation of iodide are illustrated in the Fig. 29.3.

Transport of T_3, T_4

In blood T_3 and T_4 are transported bound to two specific proteins called **thyroxine binding globulin** (TBG) and thyroxine binding **prealbumin.** TBG quantitatively bind T_4 and T_3 100 times more than prealbumin. A very small fraction of T_3 and T_4 (0.03%) is free and is the biologically **active form.** Total protein bound iodine (PBI) is 10 μg/dl of plasma and in this fraction T_4 is 70% more in quantity than T_3. **Plasma half life** of T_4 is 4 to 7 days, while that of T_3 is 1 day only.

Regulation of biosynthesis and secretion of thyroid hormones

The synthesis and secretion of thyroid hormones is under strict **feed back regulation** at pituitary (TSH) and hypothalamus (TRF) level, as shown in the Fig. 29.4. T_3 appears to be more actively involved in the regulation of thyroid hormone production than T_4. Excess of T_3 and T_4 inhibits production of TSH from the pituitary and TRF from hypothalamus by feed back mechanism, while increase in synthesis of TSH and TRF occurs in response to low levels of T_3 and T_4 in blood. The regulation of the synthesis and secretion of thyroid hormones at various levels of hypothalamus - hypophysial - thyroid axis is illustrated in the Fig. 29.4.

Metabolism of T_3, T_4—Catabolism of thyroid hormones is achieved by total deiodination and inactivation by deamination or decarboxylation in the liver. **Hepatic conjugation** with glucuronic acid or sulphate of final product occurs, that are excreted via bile into GIT.

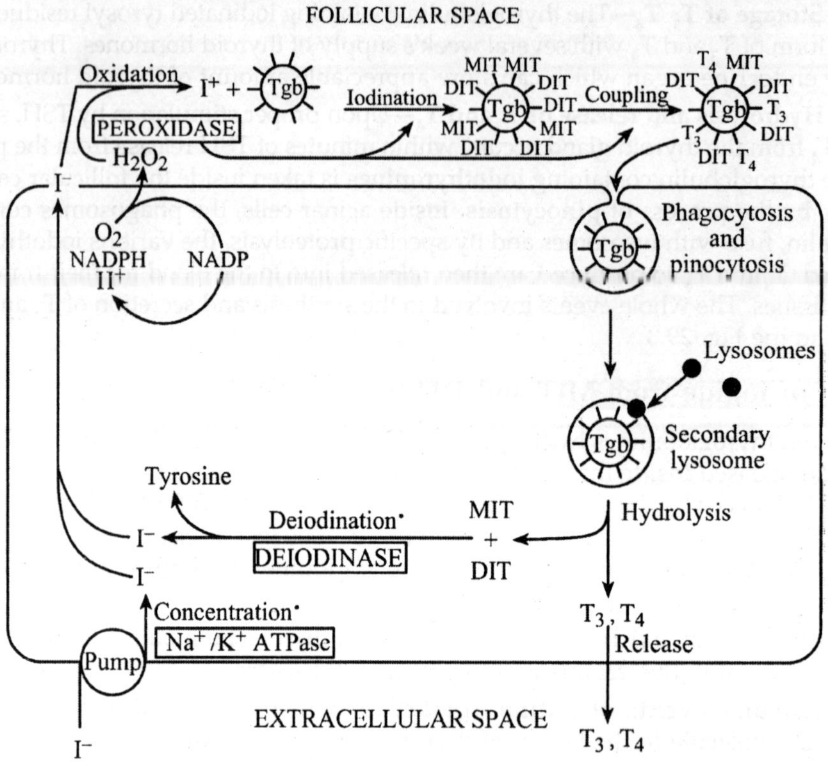

Fig. 29.3 : Diagrammatic representation of I_2 metabolism and T_3 T_4 biosynthesis, their secretion in blood and salvation of iodide by deiodinase enzyme

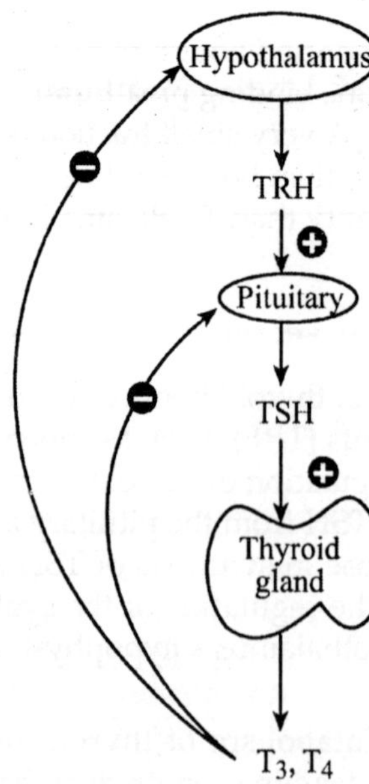

Fig. 29.4 : Regulation of thyroid hormones synthesis and representation of hypothalamus- pituitary- thyroid tract and feed back inhibitions

Mechanism of Action

In peripheral tissues, first a deiodination reaction converts T_4 into T_3 since T_3 binds to the thyroid receptors in the target cells with 10 times more affinity than that of T_4. Thus triiodothyronine **(T_3) is the predominant biological active form** of thyroid hormones. About 80% of T_4 is converted into T_3 in peripheral tissues or into reverse T_3 that is a weak agonist. Thus **T_4 is considered as prohormone** that is deiodinized to T_3 to bring about its biological action inside the target cells.

Thyroid hormones act through nuclear (intracellular) receptors. T_4 is first deiodinized into T_3 that binds with specific nuclear receptors. The **T_3 -receptor complex** then binds specifically on **hormone sensitive elements** on DNA and enhance transcription of certain genes to increase protein synthesis to bring its biological effects in the target tissues.

Biological Functions of Thyroid Hormones

Major functions of T_3 and T_4 are mediated through nuclear receptors in the target cell, that cause a variety of physiological and biochemical effects as described here.

1. **Effects on Gene Expression**—Thyroid hormones like steroid hormones bind to high affinity receptors in nucleus and enhance gene expression to increase protein synthesis in general and thus they **promote growth**, cause **positive nitrogen balance** and are **anabolic** in nature.
2. **Fetal Development and Tissue Differentiation**—A direct association exists between thyroid hormones and growth hormone. Thyroxine and cortisol first enhance gene transcription in pituitary gland to promote synthesis of growth hormone. Then along with growth hormone, T_3 and T_4 modulate fetal development, tissue differentiation and organogenesis during embryonic growth.
3. **Thyroid Hormones** are essentially required for the normal functioning of all cells of the body and thus are considered as **pleotropic hormones.**
4. **Effects on Basal Metabolic Rate**—Thyroid hormones increase BMR by enhancing general metabolism and by increasing oxygen consumption.
5. **Thyroid hormones** enhance functioning of Na^+/ K^+ pump by increasing the number of pump units as well as by stimulating ATPase activity.
6. **Thermogenesis and Calorigenic Effect**—Thermogenesis is one of the major effect of thyroid hormones and by their potent calorigenic effect they increase BMR. One mg. of thyroxine produces an excess of 1000 kilo calories in the body. This calorigenic effect of T_3 and T_4 is mediated through uncoupling of oxidative phosphorylation on electron transport chain, resulting in liberation of heat energy. Thus thyroxine is a physiological uncoupler and increases BMR as observed in hyperthyroidism.
7. **Effect on Blood Sugar Level (BSL)**—Thyroid hormones increase BSL, as observed in hyperthyroidism, probably by increasing intestinal absorption of glucose as it is associated with Na^+/ K^+ pump and ATPase activity that is enhanced by thyroid hormones. (Chapter 4).

Disorders of Thyroid Functions—Clinical Aspects

Diseases of thyroid gland are most common endocrine disorders worldwide and usually are associated with disturbances in the synthesis and secretion of thyroid hormone and release of TSH from the pituitary gland.

1. Hypothyroidism—It results due to the insufficiency of thyroid hormones and usually is associated with development of **simple goiter.**

Simple goiter is the enlargement of thyroid gland, that develops as a compensatory measure to counter balance reduced concentration of T_3 and T_4 hormones. The low levels of thyroid hormones stimulate TSH secretion from the pituitary, which in turn activates proliferation of the thyroid follicles and results in the development of goiter.

Simple goiter is endemic in the hilly regions at high attitudes, away from the sea coast, where iodine is a scarce element in the soil and in drinking water. Major cause of hypothyroidism and goiter is iodine deficiency in diet. Other causes include a variety of rare **inherited genetic diseases**. Depending on main cause of the defect, hypothyroidism can be of two types as described here under—

Primary Hypothyroidism—It is caused by the thyroid failure resulting due to autoimmune destruction of thyroid gland or due to the congenital defects. The genetic defects are due to mutations in the genes responsible for the synthesis of enzymes catalyzing various steps of thyroid hormone synthesis. These genetic abnormalities results in—(1) Iodine transport defect, (2) Iodination (organification) defect, (3) Coupling defects and (4) Deiodination defects. Important clinical hypothyroid diseases are—

Myxedema in Adults—Myxedema is a common hypothyroid disease, often resulting due to autoimmune destruction of the thyroid gland. Women are mostly affected by this disorder than males. **Clinical symptoms include**—Lethargy, mental dullness, sluggish behavior, sleepiness. weight gain, typical moon face with edema, extreme sensitivity to cold, dry skin with shallow complexion are common findings. Severe clinical manifestations include slow heart rate, bradycardia, diastolic hypertension etc.

Biochemical hallmarks are – Decreased BMR, reduced T_3, T_4 with increased TSH plasma levels. Typically high blood cholesterol levels are observed in hypothyroid diseases.

Cretinism in Children—In children retardation of physical growth without mental retardation occurs in hypothyrodism resulting in **dwarfism.**

Neonatal Cretinism—It results due to maternal hypothyroidism usually because of dietary deficiency of iodine in the pregnant women, that cause reduced T_3, T_4 synthesis, which are necessary for the production of growth hormone. Thus along with low T_3, T_4 as well as reduced growth hormone levels, the fetal development is impaired and the child born may have **multiple congenital deformities**. After birth, **all mile stones** of growth are delayed. Cretin child is of **short stature** with **severe mental retardation** and multiple physical deformities. The condition is irreversible.

Secondary Hypothyroidism—This may results from pituitary or hypothalamic defects that impairs secretion of TSH or TRH. This may also be due to failure of feed back regulatory mechanisms. Clinical symptoms are as mentioned in the myxedema.

Hyperthyroidism-Thyrotoxicosis

Thyrotoxicosis is a clinical syndrome caused by excessive production of thyroid hormones. Most common hyperthyroid diseases are **Grave's disease** and **toxic goiter.** Rarely it can also be caused by TSH secreting pituitary tumors.

Grave's Disease—It is caused by over production of **thyroid stimulating immunoglobulins IgG**(TSI), that binds and activates the TSH receptors on the plasma membranes of thyroid acini, even without TSH interaction. This causes toxic enlargment of thyroid gland (toxic goiter) with excessive production of thyroid hormones. Here the feed back regulation fails since TSI (immuno globulins) are not under T_3, T_4 feed back control.

Clinical Manifestations—Major clinical symptoms include—

Tachycardia with rapid heart rate and widened pulse pressure.

Nervousness, anxiety, insomnia, body weight loss in spite of increased appetite.

Excessive sweating, red moist skin with increased sensitivity to heat.

Characteristic exopthalmia is observed

Bio Chemical hallmarks are – Increased BMR, high T_3, T_4 and TSH plasma levels.

Treatment—Hyperthyroidism is treated by blocking T_3, T_4 synthesis and hormones production with **antithyroid drugs**, or by ablating the thyroid gland with radioactive isotope of iodide (I^{131}).

Assessment of thyroid function is important for the diagnosis as well as for monitoring the patients receiving treatment for thyroid diseases. Previously measurement of BMR was used as one of the important criteria for the diagnosis of thyroid disorders. Estimation of protein bound iodide (PBI) and thyroid binding globulin (TBG) were also done to assess thyroid functions.

However with the advancement of biochemical techniques such as radioimmuno-assay and ELISA, the thyroid hormones T_3 T_4 and TSH estimations in plasma have become highly important analytical procedures in the diagnosis, as well as in monitoring and prognosis of thyroid diseases. Plasma cholesterol level though is not a diagnostic parameter, but is an useful index in the prognosis and monitoring the effectiveness of therapy in hypothyroid diseases.

Thyroid Function Tests

Recent clinical thyroid function tests commonly done in the clinical biochemical laboratory are as given below—

1. Determination of T_3 T_4 and TSH plasma levels by **radioimmuno assay** or by **ELISA** techniques (details in the Chapter 37) are accurate and important, both in the diagnosis and differential diagnosis as well as in the prognosis of thyroid diseases.
2. **TRH Response Test**—Thyroid releasing hormone (TRH) assay is the most dynamic and useful parameter, particularly in the differential diagnosis of **secondary hypothyroidism**, that results from the defect in hypothalamus-pituitary- thyroid axis. TRH administration will stimulate TSH production and T_3, T_4 synthesis and secretion would be elevated markedly, if the whole axis is normal. But there will be no response if the defect lies in hypothalamus pituitary tract as is evident by the finding in the Table. 29.1.

Thyroid function tests findings in various thyroid diseases are summarized in the Table 29.1.

Table 29.1 : Thyroid function tests findings in thyroid diseases

Thyroid Diseases	Plasma T_3 T_4	Plasma TSH`	Serum Cholesterol	TRH Response Test
Primary hypothyroidism	Decreased	Increased	High levels	Exaggerated
Secondary hypothyroidism	Decreased	Decreased	High	No Response
Hyperthyroidism				
Grave's disease Thyrotoxicosis	Increased markedly	Decreased or normal	Normal or Decreased	-

PARATHYROID HORMONE

Parathyroid hormone (PTH) is the most important hormone involved in the maintenance of calcium homeostasis and plays central role in the regulation of serum calcium level. PTH is secreted directly in response to low serum calcium level (hypocalcaemia) and immediately restores it to normal by its major action on bone demineralization causing mobilization of calcium from bones to be added to blood.

A more rapid effect of PTH is to retain calcium by decreasing its excretion in the urine, for immediate correction of serum calcium level. PTH increases calcium absorption from intestine, indirectly through stimulating calcitriol synthesis (Chapter 25).

Chemical Nature

Parathyroid hormone chemically is protein in nature, having a single polypeptide chain with 84 amino acids (mol.wt 9.5 kDa). Core biological activity of PTH resides in $1/3^{rd}$ of its molecule starting from NH_2 terminal end i.e. 1-34 amino acid residues.

Synthesis of PTH

Parathormone is synthesized by 2 pairs of parathyroid glands, that are present embeded in the thyroid tissue. The chief cells of parathyroid gland synthesize PTH as a **large precursor polypeptide** with 115 amino acids, known as **preproparathormone** at the rough endoplasmic reticulum. This prepro PTH lateron is activated as PTH in the golgi appratus by specific proteolysis as shown in the Fig. 29.5. The rate of formation of PTH from its precursor **depends on serum calcium level,** which is the most important factor **regulating PTH secretion.**

The PTH so produced is released in the secretory vesicles and has three fates—1) Storage in the gland, 2) degradation into amino acids if not required or 3) immediate secretion in blood if serum calcium level is low. (Fig. 29.5).

Synthesis and secretion of PTH are finely regulated processes

Serum calcium level plays major role in controlling PTH synthesis and regulating the rate of its degradation in the gland itself, or its secretion in blood when needed, as illustrated in the Fig. 29.5. The various factors affecting PTH secretion are described below.

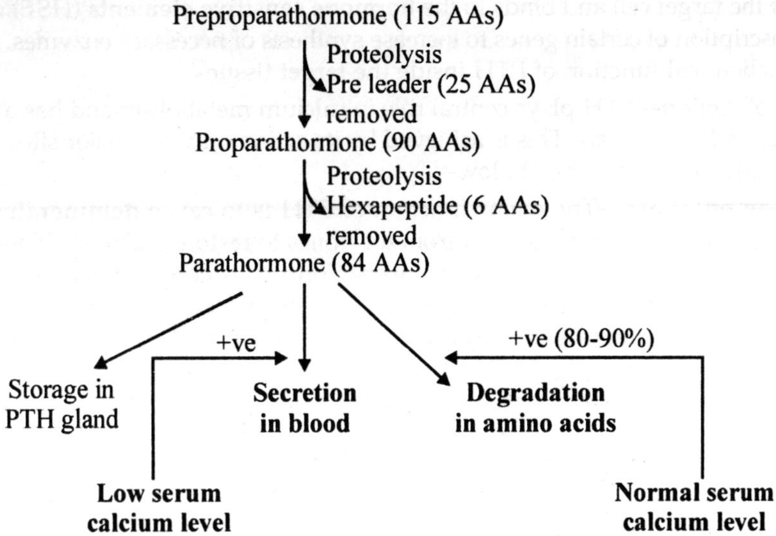

Fig. 29.5 : Synthesis and regulation of secretion of PTH depending on serum calcium status(AAs- amino acids)

1. Secretion of PTH—The release of PTH from parathyroid glands is controlled by an highly complex regulatory process and is under tight negative feed back regulation by serum calcium level.

2. Sensing of **serum calcium status** is accomplished by a calcium- receptor (via G-protein) located on the plasma membranes of the parathyroid glands.

3. At normal serum calcium level degradation of most of (80-90%) the pro PTH and PTH begins within 20 minutes after its synthesis, while hormone is still in the secretory vesicles. This shows that regulation of PTH secretion is achieved mainly by its degradation in gland itself, if it is not required, for instance in case of normal serum calcium status.

4. Low serum calcium level enhances PTH activation and its release rapidly from the parathyroid glands. At the same time the **degradation** of PTH **is inhibited.** The whole events of the regulation of PTH formation, its degradation or its release from the gland are illustrated in the Fig. 29.5.

Biochemical Functions of PTH

PTH is secreted directly in response to low serum calcium level (as mentioned above), that can occur either due to a dietary deficiency of calcium and/or due to deficiency of vitamin D. Low calcium levels can also be caused by inadequate absorption of Ca^{++} from the intestine.

The prime function of PTH is to increase and restore normal serum calcium level to prevent the adverse effects of hypocalcaemia that can be life threatening. (Chapter 24).

Mechanism of Action

Main target tissues of PTH are **bones** and **kidneys**. PTH is a polypeptide hormone and it binds to a specific membrane receptor protein and **initiate** an **activation** cascade system. This activates adenyl cyclase to synthesize c-AMP. The c- AMP so produced reaches to nucleus of the target cell and binds to the **hormone sensitive elements (HSE)** on DNA to enhance transcription of certain genes to increase synthesis of necessary enzymes, that finally mediates biochemical function of PTH inside the target tissues.

Site of Action—PTH plays central role in calcium metabolism and has a significant effect on calcium homeostasis. This is achieved by its action on three major sites, the bones, kidneys and intestine as describe below—

Action on bones—The major function of PTH is to cause **demineralization** and **resorption of bones** to mobilize calcium from the bones to restore normal ECF ionized Ca^{++} level. PTH causes resorption of bones by its action on osteoclasts by enhancing gene expression for the synthesis of enzymes that dissolves the inorganic calcium hydroxyapatite crystals as well as they degrade the organic matrix as illustrated in the Fig. 29.6.

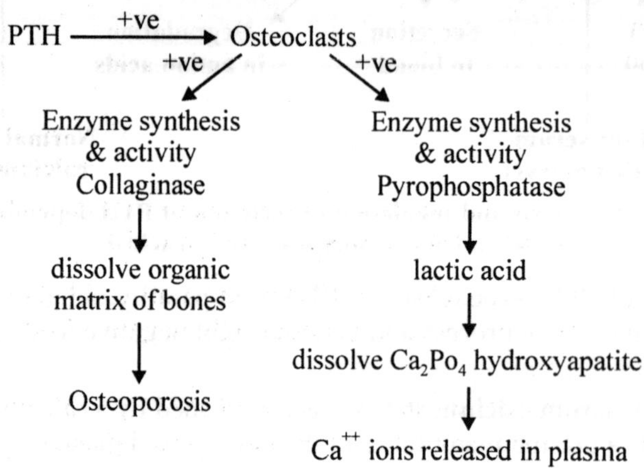

Fig. 29.6 : Mechanism of PTH action on bone resorption

PTH Action on bones is significant

The action of PTH on bone resorption and demineralization is quantitatively highly significant to restore normal serum calcium level, even at the cost of bones. This is necessary since severe hypocalcaemia may result in adverse clinical manifestations which can be life threatening (Chapter 24).

PTH maintain **calcium homeostasis** even during prolonged calcium deficiency at the **expense** of **bones** resulting in loss of calcium from the bones (demineralization), making them weak, brittle and easily fracturable. Clinically the condition is termed as **osteoporosis.**

Action on kidneys—PTH has following important effects on kidneys—

1. **PTH enhances Ca^{++} retention**—PTH increases reabsorption of calcium from the glomerular filtrate and thus reduce calcium excretion in urine. This action of PTH is very rapid and thus it quickly add calcium back to ECF and plasma to restore normal calcium levels.
2. **It increases phosphate excretion**—PTH increases phosphate excretion by reducing its reabsorption by renal tubules. By this action PTH lower serum phosphate levels that is necessary to keep Ca^{++} x PO_4 product in serum below its saturation point, to prevent calcification of soft tissues and renal stone formation.

3. PTH **promotes calcitriol synthesis in kidneys**—PTH stimulates activity of renal enzyme α-1 hydroxylase which synthesize **calcitriol** by α 1- hydroxylation of 25 hydroxy calciferol, a vitamin D derivative. (Chapter 25).

Action on intestine—PTH enhances intestinal absorption of calcium **indirectly** through stimulating calcitriol synthesis in kidneys.

All these actions of PTH i.e. at bones, at renal level and on intestine, ultimately result in increasing ECF and serum calcium levels.

Abnormalities of parathormone production : Clinical Aspects

Hypoparathyroidism—PTH insufficiency occurs usually due to the accidental removal of PTH glands during neck surgery (secondary hypoparathyroidism), or due to autoimmune destruction (rare) of parathyroid glands causing primary hypoparathyroidism. In hypoparathyroid cases, there is insufficient production of PTH, causing reduction of serum calcium level (**hypocalcemia).**

Biochemical hallmarks of hypoparathyroidism are—low ionized serum calcium level and high serum phosphate levels

Clinical menifestation—In mild **hypocalcemia**, there is **increased neuromuscular irritability** and muscular weakness with severe muscle cramps. **In severly low serum calcium level** (< 7 mg/dl) a clinical syndrome **tetany** develops. Clinical manifestation include **spasmic muscular contractions** such as **facial muscle paralysis, carpopedal spasms, paralysis of respiratory muscles,** bronchospasm, laryngospasms, resulting in respiratory arrest, cyanosis and severe convulsions leading to fatal consequences.

Hyperparathyroidism—Excessive production of PTH can be caused due to (a) **parathyroid adenomas** (cancerous tumor of glands), (b) **parathyroid hyperplasia** and (c) ectopic increased production of PTH or related polypeptides in various carcinomas.

Biochemical hall marks are hypercalcemia with elevated ionized serum calcium levels and decreased serum phosphate levels.

Clinical manifestations

Hyperparathyroid cases are associated with excessive demineralization of bones and bone resorption leading to **osteoporosis. Renal manifestations** include kidney stones, nephrocalcinosis, frequent urinary tract infections and renal function impairment.

Osteoporosis—It is characterized by bone resorption and demineralization causing progressive decrease in bone density. The bones become weak, fragile and easily fracturable and may result in bone deformities. Osteoporosis is also an age related problem in elderly people particularly in post menopausal women.

CALCITONIN

Calcitonin is a polypeptide hormone having opposite effects to that of PTH on calcium metabolism, and thus they are directly antagonist to each other. Calcitonin decreases serum calcium level by increasing bone mineralization and renal calcium excretion.

Chemical nature and synthesis

Chemically calcitonin is a polypeptide hormone containing 32 amino acids and is synthesized by the parafollicular 'C' cells arranged near basement membrane of the thyroid glands. The entire calcitonin molecule with 32 amino acids and a **7 member N-terminal loop,** formed by a cys-cys bridge is necessary for its biological activity.

Secretion of Calcitonin

Secretion of calcitonin is stimulated after a calcium rich diet, when serum calcium level tend to increase. Other factors which stimulate calcitonin secretion are hormones gastrin and glucagon that are also secreted after meals.

Biochemical Functions of Calcitonin

Role of calcitonin in human calcium homeostasis is still not very clear. However it is proposed that calcitonin lowers serum calcium level by its actions opposite to that of PTH on bones and kidneys. At bone level calcitonin reduces demineralization and resorption of bones by inhibiting osteoclastic activity and it promotes mineralization of bones by increasing number as well as the activity of osteoblasts.

At kidney level, calcitonin increases phosphate excretion (action similar to PTH) as well as it increases calcium excretion in the urine (action opposite to PTH). By their opposite effects on bones, calcitonin and PTH **coordinate bone remodelling** to keep bones healthy and strong. Calcitriol(activated vitamin D) is also important factor that facilitates bone growth, bone mineralization and remodelling of bones along with calcitonin and PTH.

❑❑❑

30

Hormones of Adrenal Gland

The **adrenal glands** are a pair of small organs (of approximately 10g each) located above the kidneys and each gland has two distinct tissues – an outer **cortex** and inner **medulla.**

ADRENAL CORTEX

The adrenal cortex produces many (about 50) steroid compounds, but only few of them are **biologically active.** The biologically active **corticosteroid** hormones are broadly categorized into three groups, based on their predominant physiological action, though there is some overlapping in their functions.

Three distinct kind of cortex hormones

The adrenal cortex has three **distinct zones** (layers) which secrete three different corticosteroid hormones.These are as described here—

1. **Zona glomerulosa**—It is the subcapsular layer of gland and is associated with the production of **mineralocorticoids,** such as **aldosterone.**
2. **Zona fasiculata**—It is the next zone and produces mainly **glucocorticoids** such as cortisol and cortisone.
3. **Zona reticularis**—Thus is the innermost layer of adrenal cortex and produces small quantities of **androgens.**

The Glucocorticoids

These are **21-carbon steroid** hormones, produced from middle layer zona fasiculata of adrenal cortex and have a variety of functions, the most important of which is to stimulate **gluconeogenesis** to maintain blood sugar level (BSL) during fasting (hypoglycemic) conditions. **Cortisol** is the predominant glucocorticoid hormone in **humans,** while corticosterone is main glucocorticoid in rodents (rats).

Mineralocorticoids

These are also **21 carbon-steroids,** synthesized in zona glomerulosa and predominant action of these hormones is to retain sodium and excrete out potassium and H^+ ions in urine to maintain electrolyte and water balance of the body fluids and tissues. **Aldosterone** is the most potent mineralocorticoid and is produced exclusively in zona glomerulosa.

Androgens

The innermost layer of adrenal cortex, the zona reticularis produces significant amounts of the **androgen precursor, dehydroepiandrosterone** and a weak androgen **androsterone.** These androgens are **19 carbon steroids** and after their release in blood, they reach to other tissues and are converted into more potent androgens and also into estrogens, but only in post menopausal women.

Biosynthesis of corticosteroid hormones—Various steps involved in the steroid hormone biosynthesis are as described here along with the salient features.

1. Cholesterol is the precursor of all steroid hormones synthesis. It is derived from plasma or synthesized in adrenal gland itself from acetyl CoA and stored as cholesterol esters in the gland for steroid hormone synthesis.
2. ACTH (or c-AMP) stimulates adrenal cortex to **activate** an **esterase** enzyme that hydrolyze cholesterylesters into free cholesterol, which is then transported inside mitochondria for steroid hormone biosynthesis.
3. **Cleavage of side chain of cholesterol**—Inside mitochondria, a **cytochrome p450 side chain cleavage enzyme (p450 SCC)** oxidatively cleaves a 6 carbon fragment from the 8 carbon side chain of cholesterol to leave behind a 21 carbon steroid **pregnenolone** as illustrated in the Fig. 30.1.

   Pregnenolone is the common precursor for the synthesis of all the three categories of corticosteroids. An ACTH dependent regulatory protein **(STAR)** is necessary to transport cholesterol to **p450 SCC** through inner mitochondrial membrane. This transport of cholesterol inside mitochondria is a **rate limiting step** in the biosynthesis of steroid hormones.
4. **Progesterone** is the first steroid hormone synthesized from pregnenolone by oxidation of 3 β OH group of the A ring of the steroid nucleus to a carbonyl group in the presence of a **dehydrogenase** enzyme, then the double bond is shifted from 5, 6 position to 4-5 position in the presence of an **isomerase** enzyme as shown in the Fig. 30.1.
5. **Cortisol synthesis**—Cortisol is synthesized from **progesterone** by stepwise action of **17 α hydroxylase, 21 hydroxylase** and **11 β hydroxylase** enzymes which add hydroxyl groups at the respective positions as illustrated in the Fig. 30.1, reaction pathway -1. The 17 α hydroxylase and 21 hydroxylase are **microsomal** cytochrome p450 and NADPH dependent enzymes, while **11 β hydroxylase** is a **mitochondrial enzyme.**

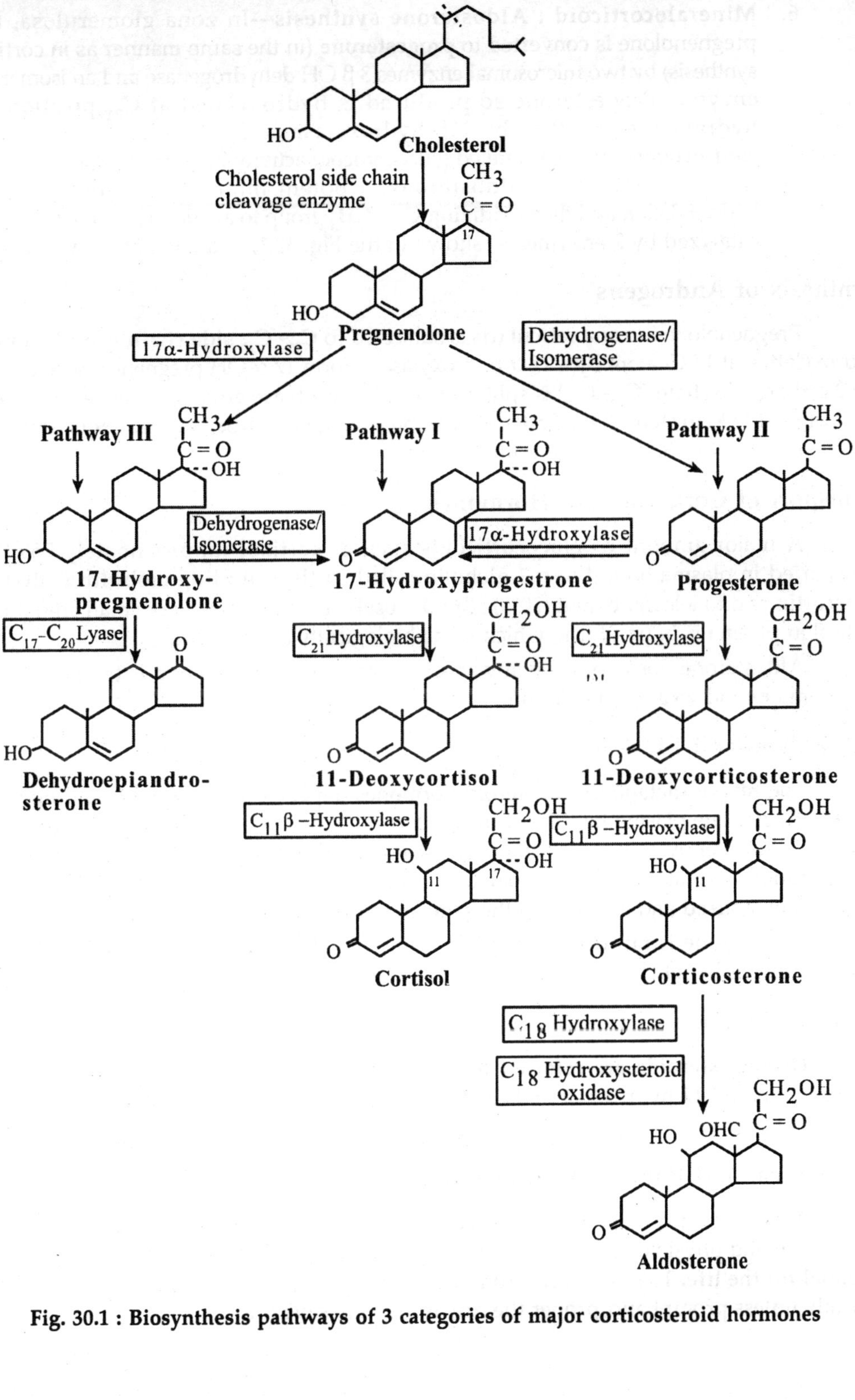

Fig. 30.1 : Biosynthesis pathways of 3 categories of major corticosteroid hormones

6. **Mineralocorticoid : Aldosterone synthesis**—In zona glomerulosa, the pregnenolone is converted to **progesterone** (in the same manner as in cortisol synthesis) by two microsomal enzymes 3 β OH dehydrogenase and an isomerase enzyme. Progesterone so produced is hydroxylated at C_{21} position by **hydroxylase** and then by **11β hydroxylase** at C_{11} position to synthesize **corticosterone,** that has limited glucocorticoid activity in humans. Corticosterone is then converted to **aldosterone,** a potent mineralocorticoid, by first hydroxylation and then **oxidizing C_{18} - CH_3** group to an **aldehyde (CHO)** group catalyzed by 2 enzymes as shown in the Fig. 30.1, reaction pathway – II.

Synthesis of Androgens

Pregnenolone in **zona reticularis** is subjected to **C_{17}- C_{20} side chain lysis,** by first a hydroxylation at 17 C- atom by a 17 α-hydroxylase to form 17 α-OH pregnenolone and then the 2 carbon side chain (C_{19}- C_{20}) is split off from C_{17} by a **lyase enzyme**. This results in the production a **17 keto steroid,** the first androgen precursor, **dehydroepiandrosterone (DHEA)** as shown in the Fig. 30.1 reaction pathway III.

Transport of Corticosteroid Hormones

A major amount of glucocorticoids such as **cortisol** (approximately 75%) is transported in plasma bound to α-1 globulin called **cortisol binding globulin (CBG)** or **transcortin** and to a lesser extent (20%) bound loosely to albumin. A very small amount of cortisol in plasma is in free form, which is the biologically active form.

Aldosterone does not have any specific transporter protein and is transported in the plasma bound weakly to albumin.

Metabolism and Excretion

The rate of metabolism of steroid hormones depends on their tight binding with **carrier** protein. The plasma half life of the glucocorticoids is about two hours and they are metabolized in liver by reduction to tetrahydrosteroids and then **conjugation** with **glucuronic acid** or **sulhates** to make them water soluble for their excretion in bile. Other steroid hormones are also metabolized and excreted in the same manner.

Aldosterone is **rapidly** removed from the blood by liver for its metabolism and excretion, as it is loosely bound to albumin and has very short half life (a few minutes only).

Excretory Products of Steroid Hormones in Urine

Urinary steroids are referred to as—(a) **17 keto steroids** mainly derived from androgens and (b) 17-hydroxy steroids which are directly derived from glucocorticoids and mineralocorticoids and are excreted in conjugation with glucuronic acid as mentioned above.

Biochemical and Physiological Functions

Adrenal corticosteroid hormones have a wide variety of numerous, highly important biochemical and physiological functions in the body. The corticosteroid hormones are very **essential for the life.** This is evident from the fact that loss of corticosteroid activity results in death, unless prompt and proper therapy is given to save life.

Mechanism of Action

All steroid hormones are lipophillic in nature and can easily permeate through lipid bilayer of biomembranes. Inside the target cell, they act through highly specific **intracellular receptors** and directly **modulate gene expression** to increase **transcription process** (details in the Chapter 28, Fig. 28.1). Their specific actions are mediated by inducing genes to stimulate transcription for the synthesis of various important proteins, certain key enzymes of metabolic cycles or some carrier binding proteins. For instance Na^+ binding protein synthesis is induced by aldosterone through enhancing gene expression.

Glucocorticoid Functions

The glucocorticoid hormones affects intermediary metabolism to a great extent. In addition they are very important in host defense mechanisms and play central role in the maintenance of blood pressure, cardic output and management of stress by the body as described here.

(A) Effects on intermediary metabolism

Gluconeogenesis—Glucocorticoids increase **denovo synthesis of glucose** mainly from glycogenic amino acids (from peripheral tissues), to maintain blood sugar level in fasting and **hypoglycemic** states. They elevate rate of **gluconeogenesis** by directly enhancing **gene expression** to increase synthesis of key **enzymes** of gluconeogenesis (Chapter 7 Fig. 7.1).

Glycogen metabolism—Glucocorticoids increase glycogen synthesis and its deposition in liver by stimulating key enzyme **glycogen synthase.**

Promote lipolysis—Glucocorticoid hormones stimulate **hormone sensitive lipase** enzyme in adipose tissue, via c-AMP- protein kinase system to enhance hydrolysis of TAG (stored lipids). Thus they increase the release of fatty acids in blood, that are preferably oxidized for energy production, thereby sparing glucose to increase BSL.

Protein metabolism—In welfed state, glucocorticoids act as **anabolic hormones** and promote **protein synthesis** and **growth.** But during prolonged fasting, they cause tissue protein catabolism to provide amino acids needed for the elevated rate of **gluconeogenesis** to maintain blood sugar level .

(B) Effects on host defense mechanism

1. The glucocorticoids **suppress immune response** by causing cell type specific lysis of lymphocytes. These effects are brought about by **high doses** of steroid hormones used as drugs in inflammatory conditions.
2. They **suppress inflammation** of tissues by decreasing the number of leucocytes, by inhibiting **fibroblast proliferation.** These effects may also be due to inhibition of synthesis of prostaglandins and leukotriens by corticosteroid hormones (Chapter 11).

(C) Other effects of glucocorticoids include—

1. Glucocorticoids maintain **blood pressure** and regulate **cardiac output**
2. They regulate **electrolyte** and **water balance** through reducing ADH release from pituitary (water balance) and via increasing angiotensin production they cause retention of sodium. These effects contribute to maintain normal blood pressure and regulate cardic output.

3. Along with epinephrine, steroid hormones are necessary to **counteract adverse situations,** such as **stressful** and **emotional conditions.** This is achieved by their wide range of diversified physiological and biochemical actions as mentioned above.

Functions of Mineralocorticoids

Aldosterone is the **most potent mineralocorticoid,** 1000 times more potent than corticosterone and 50 times more potent than 11-deoxycorticosterone. It acts on renal tubules to stimulate **active Na^+ transport** from glomerular filtrate to retain Na^+ and excrete K^+ in exchange in the urine. The effect is probably through increasing gene expression to stimulate the synthesis of a **Na^+ binding protein (carrier)** and **ATPase** enzyme, required for the active transport mechanism for sodium retention.

Aldosterone Synthesis Regulation

The production of aldosterone by glomerulosa cells of adrenal cortex is regulated by a different mechanisms. The primary regulation of aldosterone synthesis is though **Renin angiotensin** system (Chapter 35) of kidneys. To a lesser extent, its synthesis is stimulated by increase in K^+ plasma levels, **decreased Na^+ level** and through **ACTH** secretion. The factors regulating aldosterone synthesis described below—

Renin–Angiotensin System

Regulation of blood pressure and electrolyte balance is mostly under the control of renin angiotensin system of the body. Liver synthesize **angiotensinogen,** a globular protein, that is a substrate for enzyme renin. Renin is synthesized by kidney and it is a proteolytic enzyme. It cleaves angiotensin at a specific site to liberate a decapeptide (10 amino acids) **angiotensin** I. This is further changed, by a **converting enzyme** to an octapeptide **angiotensin** II (8 amino acids) as shown in the Fig. 30.2.

Angiotensin II stimulates aldosterone production, and is a potent vasoconstritor, and increases blood pressure (Fig. 30.2).

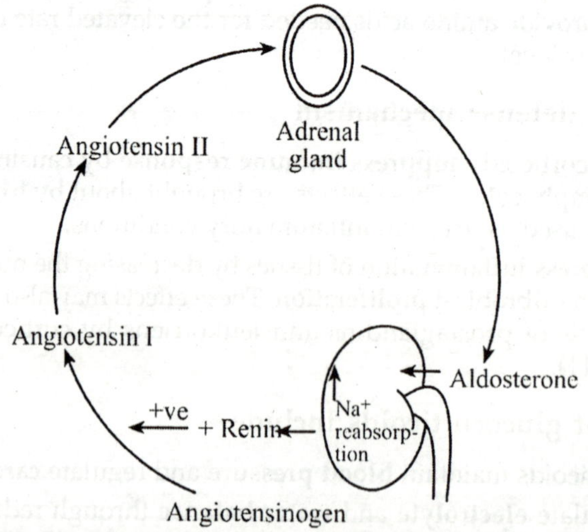

Fig. 30.2 : Diagrammatic representation of coordinated regulation of plasma volume, sodium level and blood pressure by various factors

Plasma Potassium Level

Increase in plasma potassium level immediately stimulates aldosterone production while a decrease in potassium level inhibits its synthesis. K^+ affects some enzymatic steps as does angiotensin II in the synthesis of aldosterone. Actual mechanism is still not clear.

Plasma sodium and ACTH—Under some special circumstances Na^+ and ACTH are also involved in regulation of aldosterone synthesis.

Abnormalities of Adrenocorticoid Hormones : Clinical Aspects

Primary corticosteroid insufficiency—Addison's Disease

Addisons's disease is caused by the adrenal glucocorticoid insufficiency and leads to diversified clinical manifestations. The major alterations are as described here under—

Biochemical abnormalities include—Hypoglycemia with extreme sensitivity to insulin, low plasma Na^+ level along with elevated plasma potassium. Blood picture shows increase in lymphocytes and eosinophil count.

Physiological manifestations are—Anorexia, nausea, body weight loss, weakness and low blood pressure. Patients with Addison's disease show **intolerance to stress.** Renal **functions** alterations include decreased GFR, decreased ability to excrete out water load. Patients often have craving for salt.

Patients with Addison's disease often have increased **pigmentation** of skin and mucous membranes because of exaggerated compensatory secretion of ACTH, associated with increased production of **pro-opiomelanocortin (POMC)** - which possesses weak MSH activity.

Glucocorticoid Excess – Cushing's syndrome

Cushing's syndrome is caused usually due to excessive use of steroids as drugs, but it may also result due to malignant tumors, such as ACTH secreting pituitary adenomas, adrenal carcinomas or due to ectopic production of neoplasmic ACTH.

Clinical manifestations

(a) **Hyperglycemia** and decreased glucose tolerance due to enhanced rate of **gluconeogenesis** process.

(b) Severe **tissue protein catabolism** resulting in loss of body weight, **muscle wasting** and **negative nitrogen balance.**

(c) **Osteoporosis** occurs due to demineralization and resorption of bones.

(d) **Redistribution of fats** in body and deposition in trunk (trunkal obesity) and typical **"buffalo hump"**

(e) **Impaired resistance** to **infection** and **inflammation,** poor wound healing.

Biochemical hall marks include—Increased plasma Na^+ level, decreased K^+ plasma level, **alkalosis, edema** and **hypertension.**

Abnormalities due to Mineralocorticoid Excess

Conn's syndrome—Primary aldosterone excess results due to adenoma of glomerulosa cells of adrenal cortex. Excessive aldosterone production results in abnormality, known as **Conn's syndrome**

Clinical manifestations include **hypertension, hypokalemia, hypernatremia** and **alkalosis.** In this syndrome hyperglycemia is not observed but plasma renin-angiotensin II levels are suppressed.

Assesment of adrenal cortical function—Adrenal cortex function is assessed by measuring plasma cortisol and ACTH levels, and by determining urinary 17-keto steroid levels.

HORMONES OF ADRENAL MEDULLA

Adrenal medulla is associated with sympathetic nervous system and its hormones, known as catecholamines, are secreted mainly as a result of sympathetic stimulation during adverse situation such as acute and chronic stressful conditions. Adrenal medulla produces two important catecholamine hormones.

a. Epinephrine also known as adrenaline

b. Norepinephrine, previously called as noradrenaline.

Dopamine is another catecholamine hormone produced as an intermediate during synthesis of epinephrine, but it is mainly active in brain and CNS. Norepineprine and dopamine are important neurotransmitters, synthesized in brain and autonomic nervous system These catecholamines are so called, because they have a **catechol ring** in their chemical structures (Fig. 30.3).

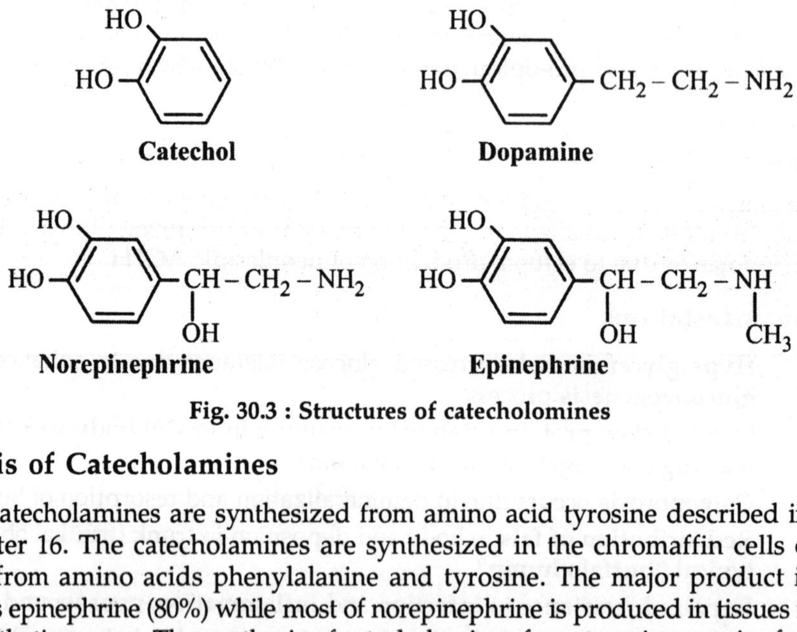

Fig. 30.3 : Structures of catecholomines

Synthesis of Catecholamines

Catecholamines are synthesized from amino acid tyrosine described in detail in the Chapter 16. The catecholamines are synthesized in the chromaffin cells of adrenal medulla from amino acids phenylalanine and tyrosine. The major product in adrenal medulla is epinephrine (80%) while most of norepinephrine is produced in tissues innervated by sympathetic nerves. The synthesis of catecholamines from tyrosine require four steps as illustrated in the Fig. 16.14 Chapter 16, where the first enzyme tyrosine hydroxylase is the rate limiting key enzyme.

Regulation of Catecholamine Biosynthesis

Biosynthesis of catecholamines is regulated at the rate limiting step catalyzed by **tyrosine hydroxylase enzyme** as given under—

(a) By feed back inhibition—Excess of catecholamines decrease the activity of this enzyme.

(b) Tyrosine derivates act as competitive inhibitors of this enzyme, e.g. α methyl tyrosine

(c) Other enzymes of synthetic pathway can also be regulated by drugs.

Catecholamines cannot cross **blood-brain barrier**; hence they are synthesized in brain tissue itself. In **Parkinson's disease** there is a local deficiency of **dopamine** in the brain (Chapter 16).

Secretion of catecholamines occurs as a result of stimulation of splanchnic nerve, which causes **storage chromaffin granules** to fuse with plasma membranes and by the process of **exocytosis**, the hormones are released in the blood. This process is calcium dependent and it is stimulated by cholinergic and β adrenergic agents, while it is inhibited by α adrenergic agents.

Metabolism of Catecholamines

The adrenal medulla, unlike sympathetic nerves does not have a mechanism for **reuptake** and **storage** of discharged catecholamines and thus once released from medulla they are taken up by liver and skeletal muscles from blood and are rapidly metabolized. Two important enzymes are involved in cateholamine metabolism, as described below—

1. **Catechol-O-methyl transferase** (COMT)—It is a cytosolic enzyme present in many tissues and inactivates catecholamines by adding a methyl group at meta position (3rd carbon) of the phenyl ring. The S-adenosylmethionine (SAM) is the methyl group donor. The methylated products are the **normetanephrine, metanephrine** and **homovanillic acid** as illustrated in the Fig. 30.4.

2. **Monoamine oxidase (MAO)**—It is an oxidoreductase enzyme and deaminates various monoamines. This enzyme is found in many tissues and has great role in metabolism of biological amines. The action of these two enzymes on catecholamines is shown in Fig. 30.4. The final metabolic products of **diagnostic** significance are **metaneprines** and **3 methoxy, 4 hydroxy-mandelic acid,** also known as **vanillylmandelic acid (VMA).**

Biomedical importance—Clinical aspects

The concentration of metanephrines and VMA in urine is elevated in patients of **pheochromocytoma** along with elevated levels of catecholamines.

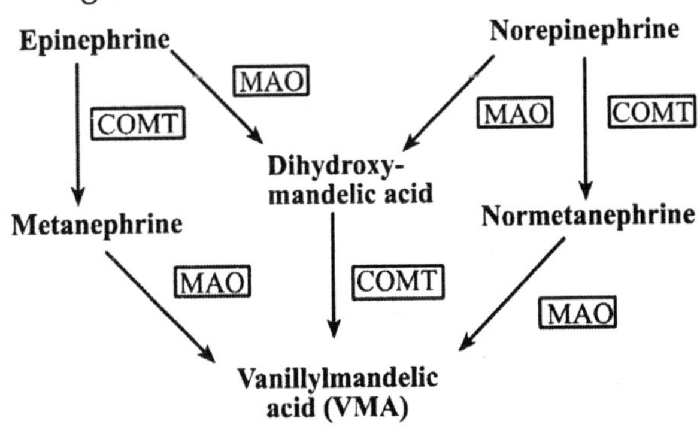

Fig. 30.4 : Metabolism of epinephrine and norepinephrine (COMT—Catechol-O-methyltransferase; MAO—Monoamine oxidase)

Nerve Impulses Regulate Catecholamine Synthesis

Stimulation of splanchnic nerve, that supplies the preganglionic fibers in adrenal medulla, results in both increased synthesis, as well as release of catecholamines. Norepinephrine synthesis increases after **acute stress.** Prolonged stress is accompanied by chronic sympathetic nerve activity resulting in increased catecholamines biosynthesis. Release of catecholamines prepares the subject to adapt itself to physiological stressful conditions.

Mechanism of action of catecholamines—

Catecholamines acts through two major classes of **receptors** as given below—

(a) α adrenergic with two subclasses - α_1 and α_2

(b) β adrenergic with two subclasses- β_1 and β_2.

Epinephrine binds to and activates both α and β receptors. So its biological functions in the target tissue depends on the type of receptor present and its affinity to the hormone.

Norepinephrine at physiological concentration, primarily bind to α-adrenergic receptors. The catecholamine receptors are **G-protein** linked and via adenyl cyclase, a c-AMP – proteinkinase system is activated, that mediates hormone action inside the cell (details in Fig. 27.2).

Biological Function

Catecholamines have diversified biological effects in body, with ultimate goal is to prepare the individual to meet the stressful situations such as shock, emotional conditions, hypoglycemia etc.

1. **Effect on blood sugar level (BSL)**—Catecholamines in general increase BSL by stimulating **glycogenolysis** and **gluconeogenesis** processes. Catecholamines increase blood sugar level to provide more glucose to brain during stressful condition.
2. **Effects on lipolysis**—Catecholamines enhance lipolysis of TAG in adipose tissue by activating hormone sensitive lipase via c-AMP. This results in elevated levels of free fatty acids (FFA) in blood, which are preferably oxidized in skeletal and heart muscle as fuel source. Thus they spare glucose, to make it available for oxidation in brain and CNS (The metabolism of brain is increases during stress).
3. **Physiological effects**—Catecholamines, predominantly epinephrine is a vaso-constrictor and **increase blood pressure** and thus blood supply to tissues is increased. It also increase **cardic output** by increasing cardic muscle contraction, both **force** and **heart rate are increased.**
   - They cause **smooth muscle relaxation** of **bronchi**, GIT and blood vessels supplying skeletal muscles.
   - On the otherhand, catecholamines stimulate **smooth muscle contraction** of blood vessles supplying skin (flushing of face, skin) and kidneys.
   - They also inhibit platelet aggregation.

Abnormalities of Catecholamines Production : Clinical Aspects

1. Pheochromocytomas —These are tumors of adrenal medulla, which are usually not detectable, unless they secrete excess of catecholamines. This excessive production of catecholamines results in severe **hypertension syndrome.** The ratio of norepinephrine to epinephrine is increased in the affected individuals. The VMA excretion in urine is very high (normal < 8 mg/day) and its estimation is of **diagnostic value** in this disease.

❑❑❑

31

Hormones of Gonads

The ovaries and testes produce germ cells and sex hormones **sinergestically,** since high local concentration of sex hormones are required for the germ cell development. The ovaries produce ova and steroid hormones estrogens and progesterone in females while testes produce spermatozoa and male sex hormone, testosterone.

The steroid sex hormones are responsible for **growth, development** and **maintenance** of **reproductive system**. In addition, the sex hormones perform several other important functions in the body, for instance they are **anabolic in nature** and are required for the maintenance of metabolic processes in the tissues such as skin, bones, muscles etc.

ANDROGENS-MALE SEX HORMONES

Testosterone (C-19 steroid) is the primary male sex hormone produced by the **leydig cells** of the testes. To a minor extent the androgens are also produced by the adrenal gland of both the sexes (Chapter 30). The leydig cells are scattered in the connective tissue between coiled seminiferous tubules and secrete testosterone in response to luteinizing hormone (LH) of the pituitary gland. The **sertoli cells**, which form the basement membrane of the seminiferous tubules, provide the environment necessary for germ cell differentiation and maturation. Spermatogenesis is stimulated by FSH and LH and requires a localized high concentration of testosterone for germ cell differentiation and maturation. This requirement is met easily as leydig cells are present in close proximity to seminiferous tubules.

Biosynthesis of Testosterone

Testicular androgens are synthesized in the interstitial leydig cells from cholesterol, which is the precursor for all steroid hormone synthesis in the body. Synthesis of testosterone involves following discrete steps—

Formation of Pregnenolone—The cholesterol is first converted into pregnenolone as in the adrenal gland by the side chain cleavage, which is a **rate limiting step**. The **side**

chain cleavage enzyme, known as **p450 SCC** is required to produce pregnenolone from cholesterol and the reactions are similar as in adrenal gland (Chapter 31). However in ovaries and testes, the reaction is stimulated by the hormone LH rather than ACTH.

Transport of cholesterol across mitochondrial membrane seems to be rate limiting step as in adrenal gland and requires a carrier protein called **steroidogenic acute regulatory (STAR)** protein. This is because the side chain cleavage enzyme is a mitochondrial enzyme and cholesterol has to be transported inside mitochondria.

Conversion of pregnenolone to testosterone—Androgens can be produced from the pregnenolone by two pathways, the **progesterone pathway,** or by the **dehydroepiandrosterone pathway** (details in the Chapter 31). The progesterone route appears to be the main pathway in human testes. The reactions and important enzymes involved in the synthesis of testosterone are illustrated in the Fig. 31.1. The conversion of pregnenolone to testosterone requires the action of 5 enzymes. These enzymes are—a) 3β-hydroxy dehydrogenase, b) $\Delta^{4,5}$ isomerase; c) 17, β-hydroxylase; d) $C_{17\text{-}20}$ lyase and e) 17 β-hydroxy steroid dehydrogenase. The two pathways, the progesterone and the dehydroepiandrosterone pathways are illustrated on right and left hand side of the Fig. 31.1 respectively.

Age Related Changes in Testosterone Production

Testosterone is the dominant hormone in the fetal and neonatal stage in most of the animal species, including humans, but after the birth, testes produce only androsterone. Testosterone production by testes is **restored at puberty** and then the production continues throughout the life.

Dihydrotestosterone is active form—Human testes mainly secrete testosterone (5 mg/day) but 50-100 μg of dihydrotestosterone (DHT) is also released in the blood daily. However in the peripheral tissues **most of the testosterone is first converted to DHT** which is the **active form,** by reduction of the A ring of steroid nucleus by the action of enzyme **5 α-reductase.**

Transport—Testosterone is transported in plasma bound to a specific protein, β–globulin which have high affinity to bind testosterone but have limited capacity. These carrier proteins are synthesized by liver and they also transport estrogens. They are known as sex hormone binding globulin (SHBG) or testosterone- estrogen binding globulins (TEBG).

Since SHBG and albumin bind 97-99% of circulating testosterone, very less amount is in free form. SHBG has high affinity for testosterone than the estradiol.

Mechanism of Action—In peripheral tissues, such as prostate, external genitalia and skin, the testosterone is converted into an highly potent metabolite, DHT, by reduction of the A ring of steroid nucleus by enzyme **5 α- reductase.** It seems that in these tissues DHT is the main active hormone. Testosterone is thus considered as **prohormone** produced by testes, that is converted into a potent hormone DHT, which then bring about physiological functions inside the target tissues (Fig. 31.2). The DHT and testosterone, like other steroid hormones act via nuclear receptors to stimulate gene expression and promote protein synthesis and growth.

Metabolism of testosterone—Testosterone is metabolized in the liver by oxidation at 17th position to produce 17 ketosteroids, that are inactive metabolites and are excreted in the urine. The major 17 ketosteroid metabolites, the androsterone and etiocholanone are conjugated with glucuronic acid and sulphate in the liver to make them water soluble for their excretion in urine.

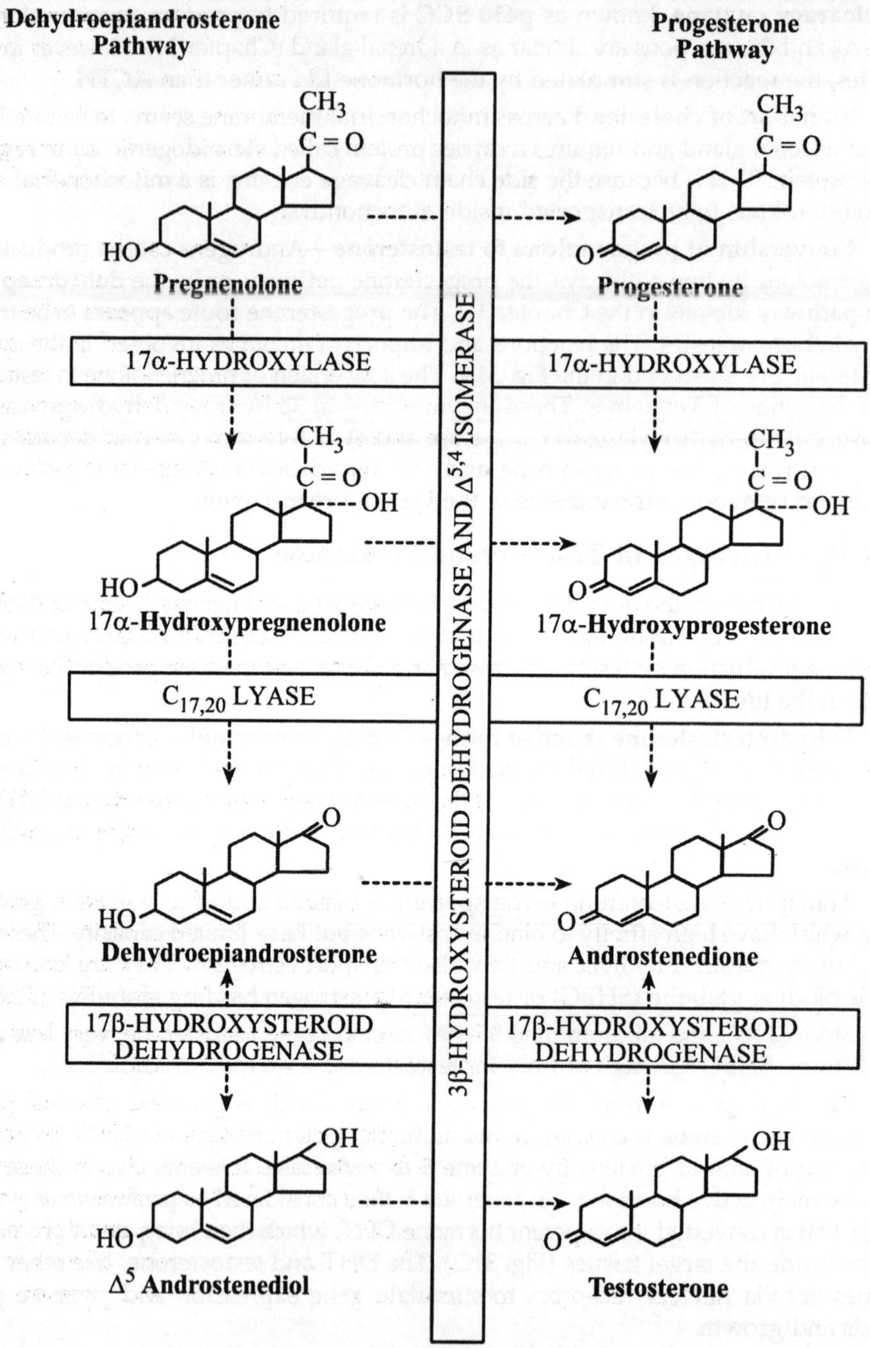

Fig. 31.1 : Pathways of testosterone biosynthesis

Physiological Function of Androgens—The principal androgens, testosterone and dihydrotestosterone (DHT) bring about several complex physiological functions to promote growth and development. The major functions are listed below—

1. Sexual differentiation during fetal growth.
2. Development of secondary sexual organs and sexual characteristics at puberty.
3. Anabolic metabolism and general growth.
4. Growth, development and maintenance of male reproductive system.
5. Promote spermatogenesis and
6. Male pattern of behaviour

Target Tissues and Mechanism of Action

The classic target tissues for DHT (with high 5α reductase activity) are prostate, seminal vesicles, external genitalia and genital skin, where as the main target tissues of testosterone include the embryonic wolffian structure, spermatogonia, muscles, bones, kidney and brain.

Physiological action of androgens inside the target tissues are brought about via nuclear receptors. The testosterone from plasma enters the target cell and some amount of it combine with intracellular receptor to bring about its action inside the target tissues. The rest of testosterone inside the target cell is reduced to Dihydrotestosterone (DHT), a potent androgen which combine with nuclear receptors. This **DHT-nuclear receptor complex** then binds with higher affinity than the testosterone to certain **hormone sensitive elements on DNA** and stimulates gene expression and promote protein synthesis. Testosterone stimulates protein synthesis in male accessory organs along with RNA synthesis including m-RNA, t-RNA and r-RNA. More specific example is activation of ABP gene by testosterone-receptor complex to promote synthesis of ABP. A brief account of mechanism of action and diversified physiological actions of androgen is represented diagrammatically in the Fig. 31.2.

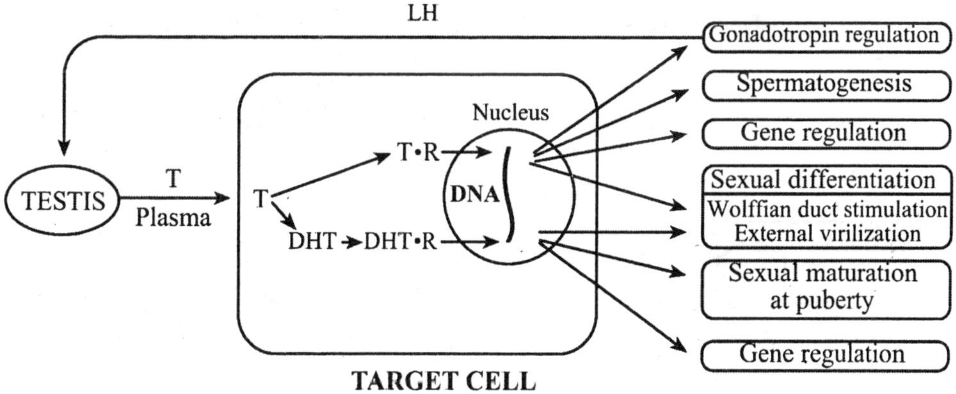

Fig. 31.2 : Mechanism of Action and fuctions of Androgens (T-testosterone, R-receptor, DHT-dihydrotestosterone, LH-Luteinizing hormone)

Abnormalities of Testosterone Production : Clinical Aspects

Hypogonadism—The lack of testosterone production results in **hypogonadism**. If this occurs before puberty, secondary sexual characteristics fail to develop and if it occurs in adults, many of the sex characteristics regress. **Primary hypogonadism** is due to testicular failure to synthesize testosterone because of genetic defect in testes. **Secondary hypogonadism** occurs due to defective secretion of gonadotropins such as LH from pituitary gland because of some genetic defect.

Receptor defect—Certain genetic defects may cause lack of receptors for testosterone or DHT in the target tissues leading to abnormalities in sexual behavior. For instance, persons who completely lacks receptors for androgen, appear to be phenotype female but have an XY (male) genotype.

FEMALE SEX HORMONES

The ovaries produce the female sex hormones, estrogens and progesterone which are responsible for the maintenance of menstrual cycle and reproductive system of females. Estrogens comprises a family of hormones produced by ovaries and number of other tissues, and include **estradiol, estrone and estriol.** 17β-estradiol is the primary estrogen produced from ovaries, while in pregnancy, relatively more estriol is produced and is synthesized by placenta. In some species, estrone is more abundant female sex hormones, that is synthesized in numerous tissues.

Biosynthesis of Estrogens

Cholesterol is the precursor for estrogen synthesis and the general pathway and enzymes involved in the early steps of estradiol synthesis are same as in androgen synthesis. Estrogens are then formed by the **aromatization** of the **A ring** of androgen by a complex process, that involves three hydroxylation steps which require oxygen, NADPH +H^+ and enzyme **aromatase.**

Aromatase is a complex, **mixed function oxygenase enzyme** and probably is cytochrome p450 dependent. Estradiol is produced by aromatization of A ring of the testosterone, whereas estrone is formed by the aromatization of androstenedione. The main reactions and enzymes involved in estrogen biosynthesis are summarized in the Fig. 31.3.

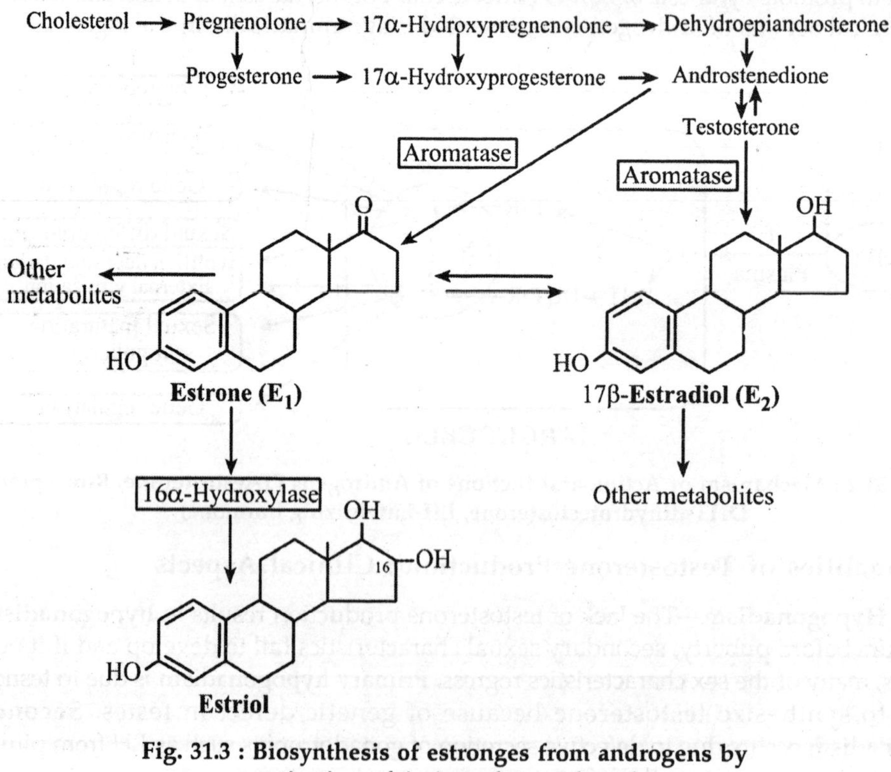

Fig. 31.3 : Biosynthesis of estronges from androgens by aromatization of A ring of steroid nucleus

Site of Synthesis

Progesterone is produced from corpus luteum and some amount is converted into estradiol. **Theca cells** of the ovaries produce androstenedione and testosterone and then in the **granulosa cells** they undergo **aromatization** to **estrone** and estradiol respectively. Significant amount of estrogens are produced by peripheral aromatization of androgens. In males 80% of estradiol is produced by peripheral aromatization of testosterone. In females adrenal steroids are important precursors and during pregnancy about 50% of estradiol is produced by aromatization of androgens. In postmenopausal women, the main source of estrogens is by peripheral aromatization of androstenedione.

Transport of estrogens and progestins—Estrogens are transported in plasma, bound to sex hormone binding globulin (SHBG) and progesterone is bound to cortisol binding globulin (CBG) in the blood. These binding proteins have a large capacity and thus provide for circulating reservoir of these hormones.

The metabolic clearance rate of the sex hormones depends on their affinity for binding proteins in the plasma. Therefore estrone(less affinity) is cleared off rapidly from plasma as compared to estradiol, which in its turn gets cleared more rapidly than testosterone or DHT as testosterone has highest affinity for SHBG.

Metabolism of Estrogens and Progesterone

In liver the estradiol and estrone are first converted into estriol (Fig. 31.3) and then they are conjugated with glucuronic acid or sulphates to make them water soluble for their excretion in the urine. In humans, the conjugation enzymes in liver are less active and thus estrogens administered orally are effective and thus are used as **oral contraceptives.**

Liver actively metabolize progesterone to several steroid compounds. Major progestin metabolite is sodium pregnanediol glucuronide, that is excreted in the urine. Oral progesterone is ineffective as it is rapily metabolized in liver.

Certain **synthetic steroids,** such as derivatives of 17 α-hydroxyprogesterone and 17 α-alkyl nortestosterone possess progesterone functions and escape hepatic metabolism and thus are widely used as oral contraceptives.

Physiological Functions of Ovarian Hormones

The female sex hormone, estrogens and progesterone are primarily involved in the maturation and the maintenance of female reproductive system. These hormones prepare the structural and functional components of female reproductive organs for proper reproduction. Their major functions are—

1. Maturation of primordial germ cells during ovulation and provide hormonal timing for ovulation.
2. Development of uterine tissue for implantation of the blastocyte.
3. Provide proper conditions necessary for the maintenance of pregnancy.
4. Provide the hormonal influences for parturition and lactation.
5. Maintain menstrual cycle if implantation fails.
6. Development and maintenance of secondary sexual characteristics during puberty.

Estrogens : Physiological Functions

Anabolic effects—Estrogens stimulate the **development** of the **female reproductive tissues** by increasing size and number of the cells. In general, the estrogens enhance the rate of synthesis of proteins, r- RNA, t -RNA, m- RNA and DNA replication to promote growth and development.

Estrogens stimulate proliferation of vaginal epithelium and uterine endometrium. They stimalate gland hypertrophy and elongation of acinar cells.

Under estrogen stimulation, uterine myometrium develops an **intrinsic rhythmic motility.**

Breast duct proliferation is caused by estrogens.

Estrogen have **anabolic** and **growth stimulating** effects on **bone and cartilages** and thus promote growth. In postmenopausal women, cessation of estrogen secretion leads to **osteoporosis.**

Estrogens cause vasodilation of peripheral blood vessels resulting in heat dissipation.

Progesterone : Physiological Functions

Progesterone on the other hand reduces the proliferative action of estrogens on vaginal epithelium and convert uterine endometrium from proliferative to secretory state by increasing size and activity of secretory glands and thus prepares endometrium for the implantation of fertilized ovum.

- Progesterone enhances development of breast glands, after estrogens have stimulated the ductal development.

- Progesterone decreases peripheral blood flow and hence reduce heat loss, so that the body temperature is raised during luteal phase of menstrual cycle at the time of ovulation.

- Progesterone action requires **estrogen priming,** because estrogen is necessary for the production of **progesterone receptors** in the target tissue. Thus the estrogens and progesterone **act synergistically,** though they have **opposite effects** as mentioned above.

Age Related Changes in Ovarian Hormone Production

The secretion of ovarion hormones is very low during childhood. These low levels of sex hormones inhibit gonadotropin production in prepubertal girls. At puberty the pulsating release of gonadotropin releasing hormone (Gn RH) from hypothalamus stimulates LH secretion from pituitary, that causes a dramatic increase in the secretion of ovarian hormones. FSH greatly stimulates estrogen production for the development of secondary sexual characters and reproductive system and enhance maturation of follicles for ovulation.

THE MENSTRUAL CYCLE

In humans, the menstrual cycle is maintained by complex interaction between hormones of hypothalamus, pituitary and ovaries. The cycle normally varies from 25 to 35 days in length with an average of 28 days and consists of a follicular phase and a luteal phase which finally terminates into menstruation.

Follicular phase—The Follicular stimulating hormone (FSH) secreted from the pituitary stimulates **enlargement** of **an ovarian follicle** and increases estrogen production. The estrogen levels increase progressively as the follicle enlarges. Estradiol (E_2) level reaches to a **peak (maximum),** 24 hours before the LH (FSH) peak and sensitize pituitary to gonadotropin releasing hormone (Gn RH) and LH is released. The **LH surge** or **peak** then cause the **end of follicular phase** and **ovulation ensues** with the release of ovum from the ruptured follicle. The progesterone levels are low during follicular phase.

The Luteal Phase—After ovulation, the granulosa cells of the ruptured follicle luteinize and form **corpus luteum**. This corpus luteum then begins to produce progesterone and some estradiol. The estradiol level then peaks about midway through luteal phase and then declines to a very low level. The progesterone then becomes the major hormone of the luteal phase, which is required for the preparation and maintenance of secretory endometrium that provide early nourishment for the implanted fertilized ovum.

LH is required for the early maintenance of the corpus luteum for about 10 days and if implantation occurs (days 22-24 of average cycle), the LH function is taken over by the placental hormone, **chorionic gonadotropin.** The human chorionic gonadotropin (hCG) is very similar to LH and it stimulates progesterone production by corpus luteum until the placenta start synthesizing progesterone in large amounts.

In the absence of implantation the **corpus luteum regresses** and the uterine endometrium is **shed** causing **menstruation**. After complete shedding of the endometrium, (which takes normally 3-5 days), a new cycle begins. The luteal phase is always of 14±2 days in length. The variations in the menstrual cycle length of individual females is mostly because of the changes in length of follicular phase.

Placental Hormones Maintain Pregnancy

The implanted blastocyte is subsequently converted into placenta which provides connection between maternal blood and the growing embryo for its nutritional and other needs. Placenta produces a number of hormones, necessary for the maintenance of the pregnancy. The main placental hormones are hCG, progestins and lactogens. Human chorionic gonadotropin (hCG), as mentioned before is a glycoprotein and is similar to LH in its structure. Its primary function is to stimulate the progesterone production from the corpus luteum until the placenta start synthesizing progesterone in amounts large enough to maintain pregnancy. The human chorionic gonadotropin (hCG) can be detected within a few days of implantation of fertilized ovum and thus provides a method for early detection of pregnancy. Peak hCG levels are reached in the middle of the first trimester of the pregnancy after which its production is reduced and its functions are taken over by large amounts of progesterone produced from the placenta.

Estrogens—Plasma concentrations of estradiol, estrone and estriol gradually increase and remain high throughout the pregnancy. Estriol is produced in largest amount and has a number of fetoplacental functions during pregnancy.

MENOPAUSE

The fertility period in females begins after puberty with the commesement of menstruation cycle and it continues till the age of 45 to 50 years. After this the cessation of menstrual cycle occurs along with loss of ovarian functions and estrogen production, causing **menopause** in the women. The progesterone and estrogen decline to very low levels. Weak estrogen, the estrone though is produced by peripheral aromatization of androstenediol, but is not sufficient to suppress pituitary gonadotropins production. Thus marked increased levels of LH and FSH are characteristic of postmenopausal women.

Clinical Aspects—Postmenopausal women are susceptible to certain clinical complications due to the insufficiency of estrogen hormones, which are protective against these clinical manifestations in the fertile women before menopause. The weak estrogen i.e. estrone is not sufficient to prevent tissue catabolism resulting, in various clinical complications as listed below—

1. **Atrophy** of epithelial linings of female reproductive organs, particularly the lower urinary tract and vagina occurs.
2. **Osteoporosis** is a major health problem in post menopausal older women. Bone mass is reduced due to bone resorption and demineralization, resulting in reduced density of the bones which become weak, fragile and easily fracturable.
3. Estrogens during fertility age are protective against heart disease and thus postmenopausal women are susceptible to heart and cardiovascular diseases.

Abnormalities of Sex hormone—There are several disorders that affect the female reproductive system and it is not possible to discuss all of them here. However a few of them are described below.

Primary hypogonadism is due to ovarian dysfunction with decreased hormone production and decreased ovulation. For instance **gonadal dysgenesis** or **Turners syndrome** is relatively frequent genetic disorder in which the individuals have XO karyotype female internal and external reproductive organs with several developmental abnormalities and delayed puberty.

Secondary hypogonadism is due to the lack of pituitary gonadotropins with severc sexual characteristic abnormalities.

Secondary hypergonadism is associated with overproduction of androgens resulting in hirsutism, obesity, irregular menses and impaired fertility.

❑❑❑

32

PLASMA PROTEINS, IMMUNOGLOBULINS AND BLOOD COAGULATION

The blood circulates throughout the body in virtually a closed system of blood vessels. The plasma is the liquid medium of the blood, in which red and white blood cells and platelets are suspended. Total blood volume in an adult human is about 4.5 to 5 liters, in which packed cell volume or **hematocrit** is about 45% while the rest is **plasma.** The plasma contains many solid constituents dissolved in water.

Important functions of blood—Blood performs many important functions, which are absolutely essential for the life. The fundamental role of blood is in the **maintenance of homeostasis,** (internal milieu) for optimum functioning of the body organs. The principal and diversified functions of blood includes: 1) **Transport** of gases, metabolites, electrolytes, minerals, vitamins, hormones etc. to each and every tissues of the body, 2) **Excretion** of waste products and unwanted substances out of the body, 3) **Maintenance** and homeostasis of acid base, electrolyte and water balance, 4) **Immunity**—The white blood cells and gamma globulins provides immunity for host defense against invading microorganism and pathogens and 5) **Blood** coordinates functions of different body organs and regulates body temperature.

PLASMA PROTEINS

Plasma contains a complex mixture of proteins, that constitutes the major part of solutes of plasma. Total protein concentration in human plasma is approximately 7-7.5 g/dl. The plasma proteins include—**simple proteins,** such as albumin, globulins and fibrinogen and **conjugated proteins** like glycoproteins, lipoproteins, metalloproteins etc.

Characteristics of plasma proteins—Most of the plasma proteins are synthesized in the liver except immunoglobulins, which are synthesized in the plasma cells. Almost all plasma proteins are glycoproteins, with the exception of albumin. The plasma proteins are separated into main components by either **salting out process,** or more accurately by **electrophoresis.**

Electrophoresis is the most widely used analytical technique for the separation of various components of plasma proteins. In plasma, proteins exist as **charged particles** and they will move in an electric field either towards cathode or anode, at different rates, depending on the particle size and their affinity to the support medium. The support medium commonly used is either a filter paper or as in advance technology, various types of gels.

Electrophoresis system basically consists of a tank to hold buffer and is fitted with electrodes and a power supply, that provides a constant electric current and voltage (Fig. 32.1A). The plasma sample is applied in the middle of solid or gel support and electric current is passed for about one and half hour to two hours.

Fixing of electrophoretogram—After completion of electrophoretic run, the separated protein **electrophoretogram** is fixed on the solid support by treatment with acetone or methanol etc. and then it is stained by using suitable dyes. Excess dye is then washed out thoroghly. Coloured bands corresponding to various plasma protein components appear on the support as shown in the Fig. 32.1B. The electrophoretogram can be scanned by densitometer to estimate each band quantitatively.

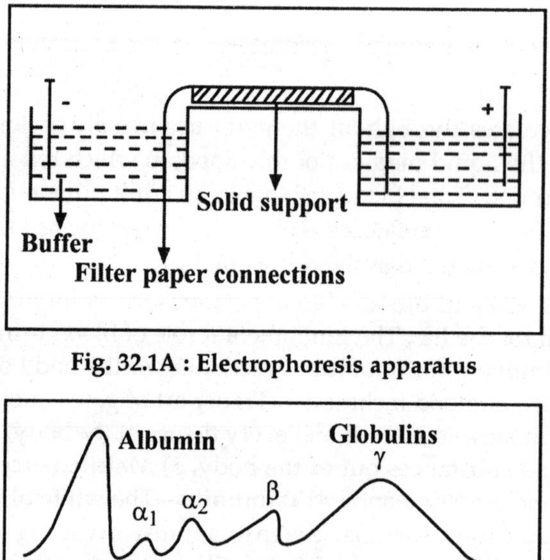

Fig. 32.1A : Electrophoresis apparatus

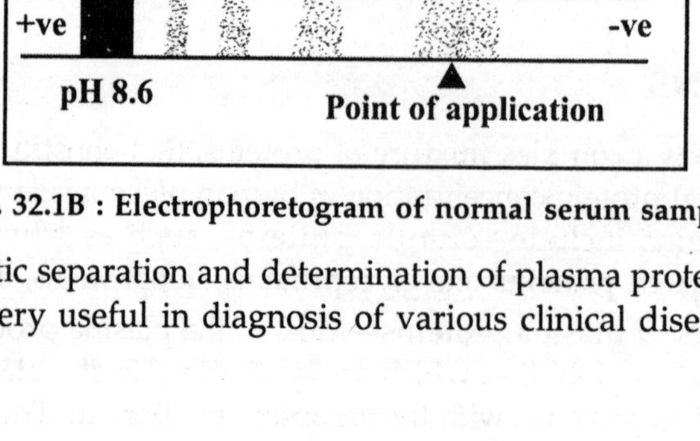

Fig. 32.1B : Electrophoretogram of normal serum sample

Electrophoretic separation and determination of plasma proteins has great clinical significance and is very useful in diagnosis of various clinical diseases (details later in Chapter).

Components of Plasma Proteins

The major fractions of plasma proteins, as evident by electrophoretogram (Fig. 32.1B) are **albumin, prealbumin, α globulin, β globulin** and **γ globulins. Fibrinogen** is another important plasma protein which forms **blood clot** in bleeding conditions. A brief account of major plasma protein fractions along with their biological functions is given in the Table 32.1.

Albumin

Albumin is the major protein fraction of plasma and constitutes approximately 60% of total plasma proteins with a concentration of about 3.5-5.0 g/dl. Its molecular weight is 69 kDa and is made up of 585 amino acid in a single polypepide chain with 17 disulphide linkages.

Albumin molecule is ellipsoidal in shape and has low molecular weight as compared to the globulins. It is synthesized in liver, at the rate of about 12g/day. In liver diseases, the synthesis of albumin is decreased, resulting in lowered plasma albumin concentration and decreased albumin/globulin ratio (A/G ratio).

Functions of Albumin—The major functions of albumin include—

Osmotic pressure functions—Albumin contributes to 75-80% of the plasma colloidal osmotic pressure because of its high concentration and low molecular weight as compared to other protein fractions. Thus albumin has predominant role in regulating blood volume and water balance between extracellular and intracellular fluid compartments. Decrease in plasma albumin concentration results in water retention in the tissue causing **edema,** as observed in nephrotic syndrome, cirrhosis of liver, protein malnutrition diseases etc. (details in Chapter 27).

Transport functions—Another important function of albumin is to transport hydrophobic substances by its ability to bind various ligands, such as free fatty acids (FFA), calcium, steroid hormones, bilirubin, copper and tryptophan etc. The albumin has specific affinity sites (e.g. for bilirubin) to bind these substances and thus facilitates their transport to various tissues for further metabolism. Many drugs such as sulfonamides, penicillin, dicumerol and aspirin etc. also bind to albumin for their transport to the tissues and to liver for their metabolism and excretion.

As buffer—Albumin has maximum buffering capacity amongst the plasma proteins but low in comparison to the bicarbonate buffer. Proteins can neutralize some amount of H^+ ions or alkalies, because they have both positive or negative ionic charges on their molecules.

Globulins

The globulins exist in several different forms in the plasma, that are separated into 4 distinct bands as α_1, α_2, β and γ as is evident from electrophoresis pattern (Fig. 32.1B). Globulins in general are of bigger molecular size and weight as compared to the albumin. The primary functions of globulins are—1) **transport** of a variety of compounds to various tissues and 2) to **provide immunity** to body against invading microorganisms and pathogens.

The important fractions of plasma proteins along with their specialized functions are summarized in the Table 32.1.

Table 32.1 : Important plasma proteins with their major functions

Plasma Protein	Plasma level & mol. weight	Ligand binding	Biological functions
Albumin	3.5-5.0 g/dl 69 kDa	Binds unconjugated bilirubin, fatty acid, steroid hormones, T_4, Calcium & drugs	Osmotic pressure, buffering action Transport of bilirubin, FFA & others
Pre Albumin	20-30 mg/dl 61 kDa	Retinal Steroid hormone & Thyroxine	Transportation of steroids, retinal and thyroxine
α_1 Globulins	0.2-0.3 g/dl		Inhibitor of protease,
α_1 Antitrypsin	54 kDa		trypsin, elastase
α_1 Lipoprotein (HDL)			Transport cholesterol and phospholipids
α_1 Globulin		Thyroxine binding globulin (TBG), retinol binding globulin (RBG), cortisone, binding globulin (CBG)	Transport T_4, Transport Retinol Major transporter of steroid hormones.
α_2 Globulins	0.4-0.8 g/dl		Antiprotease
α_2 Macroglobulin	800 kDa		Anti plasmin
α_2 Haptoglobulin	90 kDa	Free Hemoglobin	Prevent excretion of free hemoglobin
Prothrombin	63 kDa	Form thrombin	**Blood clotting factor II**
Ceruloplasmin	20-30 mg/dl	Binds Cu++	**Ferroxidase** oxidize ferrous to ferric iron & facilitates iron transport.
β Globulins	0.6-1.1 g/dl		
β Lipoprotein (LDL)		Binds cholesterol, TAG,	Transport cholesterol
Transferrin	76 kDa	Binds Fe^{+++}	Transport iron
Hemopexin		Binds free heme in plasma	Prevent heme excretion in urine.
Plasmalogen		Forms plasmin	Participate in fibrinolysis.
γ Globulins	0.8-1.8 g/dl 140 kDa	Bind antigen	Antibody and immune functions
Fibrinogen	0.2-0.4 g/dl 130 kDa	Forms fibrin	Blood coagulation

Acute Phase Proteins

The concentration of certain proteins in plasma increases several folds (sometimes 100-1000 fold) in various inflammatory and neoplastic conditions. These proteins are called **acute phase proteins** and include C-reactive protein (CRP), α_1 antitrypsin, haptoglobulin, α_1-acid glycoprotein and fibrinogen.

These acute phase proteins play important role in body's response to inflammation, for example **C-reactive protein** stimulates **complement** system. It is so named as it reacts with the C-polysaccharide of pneumococci. α_1 Antitrypsin can inhibit certain proteases released during inflammation by microorganism. **Interleukin-1**, a polypeptide released from mononuclear phagocytes, is the main stimulant for the synthesis of acute phase proteins by the liver. After the inflammation subsides, CRP quickly are cleared from the plasma.

Haptoglobin—Haptoglobin is a glycoprotein with mol. wt of 90 kDa. It binds extra-corpuscular free hemoglobin in plasma and thus prevent its excretion by kidneys, that is important to retain iron in the body for its reuse. The binding of free hemoglobin with haptoglobin is as a tight non covalent complex (Hp-Hb) with a capacity of 40-180 mg/dl. The free hemoglobin in plasma is released every day by the lysis of RBCs. The Hp-Hb complex is promptly taken up by the liver and is degraded to release iron to be reutilized. Haptoglobin is an **acute phase protein,** since its concentration increases in several inflammatory conditions.

α_1 Anti protease—This plasma protein is also known as α_1 antitrypsin. It is the major component (>90%) of α_1-globulin fraction of plasma. It is synthesized by liver as a single polypeptide protein with 394 amino acids attached with three oligosaccharide chains. α_1-Antitrypsin is the major **serine protease inhibitor** as it inactivates **trypsin, elastase** and other proteases released in the tissues during infections. It exists in at least 75 polymorphic forms, major genotype form being MM and its phenotype product is PiM.

Clinical Significance

1. α_1-Antitrypsin deficiency is implicated in the lung disease called **emphysema,** which is an abnormal distension of lungs, associated with infectious disease pneumonia. Normally the α_1 antitrypsin counter checks the damage caused to lung tissue by proteolytic enzymes elastase produced in large amounts by white blood cells during infections. But in this genetic abnormality, due to insufficiency of α_1 antitrypsin, the proteolytic elastase cannot be inactivated, which in excess cause damage to the lungs.
2. **Liver disease**—Deficiency of α_1 antitrypsin (α_1-AT) is also implicated in a liver disease (α_1 antitrypsin deficiency liver disease). In this condition due to mutation in gene, abnormal α_1-AT are produced which aggregate and form polymers, causing liver damage leading to **cirrhosis** and severe Jaundice.

α_2-Macroglobulin—It is an high molecular weight tetrameric glycoprotein and constitutes major portion of α_2 globulins of the plasma. α_2 Macroglobins inhibit several proteases and serve as anti plasmin, (that are released in plasma) by binding with them to from complexes which are rapidly cleared and degraded in the tissues.

α_2-Macroglobulin binds many **cytokines** e.g. platelet derived growth factor etc and target them towards specific tissues for their functions, such as cellular growth etc.

Ceruloplasmin—It is a glycoprotein, synthesized by liver and is associated with α_2-globulin fraction of the plasma proteins. Ceruloplasmin is a copper containing protein (mol. wt. 16 kDa) and is blue in colour. It binds copper and 90% of plasma copper is tightly bound to ceruloplasmin (6 Cu atom/mole). While the rest 10% of serum copper is loosely bound to albumin and can easily be transferred to the tissues for various functions (Chapter 24). Albumin bound copper is thus the main transport form of copper in the blood. Ceruloplasmin is not the transport protein; rather it functions as an enzyme ferroxidase and facilitates iron transport (Chapter 24).

Normal plasma level of ceruloplasmin is approximately 30 mg/dl (range 20-50 mg/dl). Its level is markedly decreased in **Wilson's disease.**

Wilson's Disease—Hepatolenticular Degeneration

This congenital disease is inherited as autosomal recessive trait, in which due to genetic defect ceruloplasmin is not synthesized. This results in the accumulation of free copper in plasma, that gets deposited in the tissues causing copper toxicity. Excessive deposition of copper in liver leads to **hepatocellular degeneration** and **cirrhosis.** Copper deposition in brain basal ganglia causes **lenticular degeneration**, resulting in **neurological lesions.** A typical sign in Wilson's disease is a green or golden brown ring around cornea of eyes due to the copper deposits (details in the Chapter 24).

Transferrin is associated with β globulin fraction and is primarily responsible for the transport of iron to various tissues (Chapter 24). Transferrin is a glycoprotein (mol.wt 76 kDa), synthesized by the liver and binds two ferric ions per mole. Free iron is toxic to tissues and its toxicity is prevented by binding with transferrin, that transport iron to the tissues for active heme synthesis and thus plays a central role in iron metabolism.

IMMUNOGLOBULINS

γ Globulins—Plasma γ globulins play major role in body's defense system against invasion of microorganisms and foreign substances, collectively known as pathogens. The defense mechanisms of body are known as **immune system** and consists of two major components—

a) **Humoral immune system**—The humoral (ancient term humor means fluid) immune system depends on B lymphocytes, that are formed in the bone marrow and synthesize soluble proteins (γ globulins) called antibodies. These antibodies are specific for a pathogen that recognize and bind to the invading microoraganism. This binding of antibody to the antigen then triggers pathways that will destroy invading microorganism and thus protect our body from the infections and various disease causing bacteria, protozoa, microorganism, viruses and other foreign pathogen.

b) **Cellular immune system**—This is mediated by T lymphocytes, so called, because their maturation from stem cells occurs in the thymus. The T-lymphocytes are mainly associated with a variety of cell mediated immunological processes. Here the intact T cells themselves are responsible for the recognition and killing (destroying) of foreign invaders. The T cells are cytotoxic (CTL) and thus are also called killer T cells. The T lymphocytes are our primary defense against malignant cells, viruses, hypersensitive reaction against allergens and graft rejection etc.

Some other T lymphocytes provide essential help to B lymphocytes to produce antibodies against pathogens and killer viruses and are known as T helper cells e.g. CD_4 cells.

Immunoglobulins—Structural Aspects

Immunoglobulins are mostly associated with **gamma globulin** fraction of the plasma proteins, however they are not synonymous, rather the term immunoglobulin is a functional term. All immunoglobulins (Ig) basically consists of a similar three dimensional structure of **Y** shape, formed by two identical light chains (L) and two identical heavy **chains** (H) held together as a **tetramer** (L_2 H_2), bonded by **disulphide** bonds as shown in Fig. 32.2. The heavy chains are bigger with 450 amino acids forming the basic Y shape, to which smaller L chains with 212 amino acids are attached to the two arms of Y shape, linked by disulphide bonds. The heavy chains of Ig are linked to carbohydrate unit and thus they are glycoprotein in nature.

Domains of Light and Heavy Chains

Each chain of Ig has two main domains called constant regions and variable regions. **In light chain the amino terminal half** is the **variable region (V_L)** and the **carboxyl terminal half** is **constant region (C_L).** While in heavy chains, the amino terminal **one quarter** is **variable region**, which in size is equal to the variable region of light chain and both are held together by disulfide bonds to form the two arms of Y shape Ig molecule. The rest of three quarter of heavy chain is the constant region as illustrated in the Fig. 32.2.

In other words the amino terminal one half of the two arms of Y shaped immunoglobulin molecule having both light and heavy chain fractions, forms the **variable arms** which recognize and specifically bind **antigen** to form an **antibody-antigen complex.** This complex is then destroyed to eliminate the pathogen from the body to protect us from the infections and diseases.

Hypervariable regions—It has been observed that some **hypervariable** regions exists interspread between the variable regions of the Ig molecule. The light chain has 3 hypervariable regions while the heavy chains have four such regions and are responsible for more specific recognition and binding to antigens.

Fab and Fc fragments—Enzyme papain hydrolyzes and cleaves Ig molecule, specifically into two **antigen binding** fragments (Fab) which corresponds to the two arms of Y shaped molecular structure of Ig, each containing one light chain and a portion of heavy chain (Fig. 32.2).While the rest of the Ig molecule left behind is known as crystalizable fragment (Fc) as shown in Fig. 32.2.

Hinge region—The hinge region is the area of immunoglobulin molecule, on which enzyme papain attacks Ig to split it into 2 Fab and 1 Fc pieces. This hinge region confers the flexibility for the free movement of two arms of Y shape Ig molecule. This facilitates the two Fab arms to move independently in different directions to bind two antigen molecules separately, which may be at some distance away.

Epitope—The site on the antigen to which an antibody binds (Fab arm) is termed as **antigenic determinant** or **epitope.** An antigen can have more than one epitopes for specific binding to different antibodies.

Types of light chains—There are two general types of light chains depending on the basic structure (amino acid sequence) in their constant regions and are accordingly classified as kappa (k) and lambda (λ) types. A given immunoglobulin always contains either two k or two λ chains i.e. identical light chains in one type of immunoglobulin molecule exist.

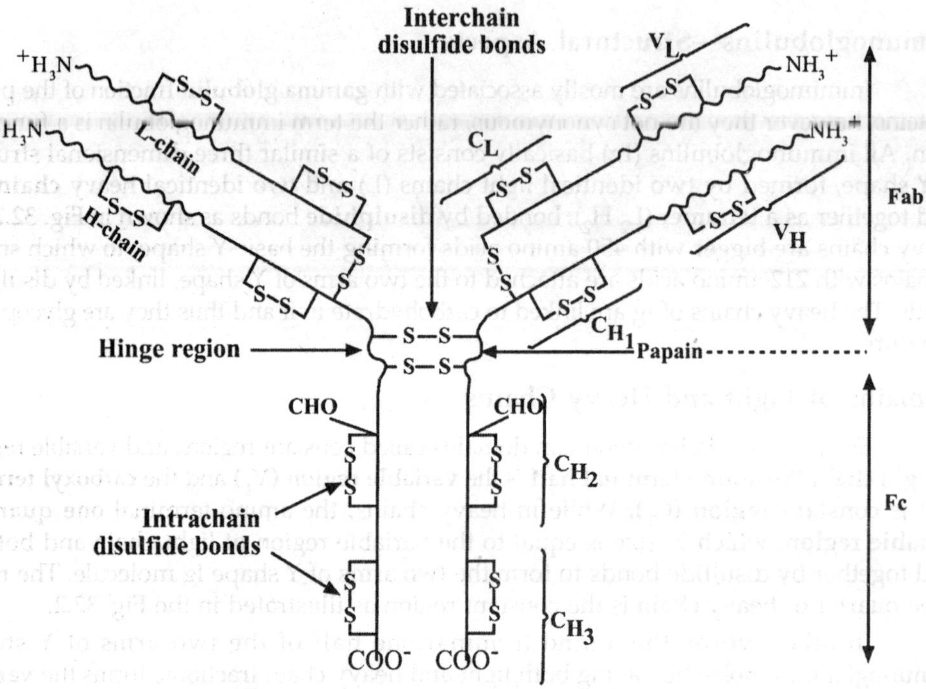

Fig. 32.2 : Diagrammatic representation of human IgG molecule with two light and heavy chains with variable (V), constant (C) and hinge, Fab and Fc regions

In addition to the inter chain disulphide bonds, there are some intrachain disulphide linkages which forms loops in the peptide chains to form **specific domains** (loops). These domains of V_L and V_H are responsible for **antigen binding** where as C_{H2} domain binds the complements, while the C_{H3} domain helps in adherence to antigen or to the cell surface of bacteria or virus.

Hypervariable Regions of Immunoglobulins

The variable regions of two arms of Ig molecule formed by the amino terminal portions of light and heavy chains of immunoglobulin have **quite different** and **specific primary** structures i.e. with a specific **amino acid sequences.** Thus no two variable regions from different human immunoglobulins have been found to have identical amino acid sequence in the hypervariable regions. Because of this, the variable regions are also termed as **complementary determining regions (CDRs).** About 5-10 amino acid in a specific sequence in hypervariable region form CDRs and contribute to specific antigen binding sites.

Five Major Types of Immunoglobulins

Depending on 5 different types of heavy chains with specific primary structure (amino acid sequences) in their constant regions, there exist five major types of immunoglobulins. These are designated as **IgG, IgA, IgM, IgD** and **IgE** and their specific heavy chains are called gamma (γ) alfa (α) mu (μ) delta (δ) and epsilon (ε) respectively. Biological functions of the five type of immunoglobulins are described here.

Characteristics of Different Types of Ig

1. IgG—This is the most predominant class (70-80% total Ig) of immunoglobulins of the plasma. It is a **monomer** with single Y shaped unit, secreted as a **secondary response antibody.**

- IgG can cross **blood vessel walls** readily and can cross placenta to provide mother's immunity to fetus and thus it is the major protective antibody in the newborns.
- IgG can traverse blood vessels and come out into extra vascular space and binds to microorganism and **activates complement components** which triggers the process that lyses (destroy) the foreign organism. By Fc component, IgG gets fixed on the macrophages to facilitate phagocytosis and the process is called **antibody dependent cell mediated cytolysis** (ADCC). Only IgG is active in ADCC mechanisms. IgG scavenges bacterial toxins and viruses also by forming antibody-antigen complex, that is then degraded to clear the toxins from the body.

2. IgM—These are largest (**macro**) **immunoglobulins,** as they are **pentamers** having **five subunits** (Y shaped monomers) joined together by a **joining chain** polypeptide. It can thus combine with 5 different antigenic determinant sites simultaneously and thus IgM is the most effective and highly **efficient** in **agglutination** of bacteria and **lysis** of microorganism to destroy them.

- IgM is the first antibody to be produced in response to invading bacteria and is thus the main **primary response** antibody.
- It fixes complements and acts as an antigen receptor on the B lymphocyte cell surface.
- It can not cross placenta and does not traverse the walls of the blood vessels, because of its large molecular size.

3. IgA—It occurs as single monomer Y shaped Ig (in serum) or as a **dimeric** unit, held together by a joining chain. IgA is secreted mostly in the **body secretions** such as saliva, tears, sweat and mucous of intestine. It inhibits attachment of bacteria and viruses on the mucous membranes and thus prevents their entry inside body.

- IgA is the most predominant antibody present in the **colostrum,** the initial milk secreted by mother (mainly in cattles) to provide immunity to the new born babies.

4. IgD—It is a monomeric Y shaped unit found to occur in serum in low concentrations attached onto B lymphocytes surface. Their function is still not clear and it is presumed that it acts as B cell membrane receptor, for the binding to antigen or microorganisms, which enter in the blood.

5. IgE—This is also a monomeric Y shaped immunoglobulin present in minute amounts normally in the serum, but its levels increase several times in **allergic** conditions, as it is associated with allergic response of the body.

- It causes hypersensitive reactions by stimulating release of mediators from the mast cells. IgE binds tightly to the mast cells and basophils which release **histamine** that is responsible for the allergic reactions on exposure to certain allergens.
- IgE stimulates release of enzymes from **eosinophils.** The eosinophil count increases in allergic conditions.
- IgE does not fix complement, but is mainly responsible for the host's defense against helminthic infections.

Constant regions that constitute Fc portion of immunoglobulins, are responsible for the **class specific functions** of different immunoglobulins, such as **complement fixation, trans placental** passage etc.

Immunoglobulins are Encoded by Multiple Genes

Different segments of light chain of Ig are encoded by three structural genes—A variable gene (V_L), a joining (J_L) gene and a constant region gene (C_L) as described in detail in Chapter 22. While heavy chains of Ig are produced by at least four different genes namely a variable gene (V_H), diversity gene (D_H), a joining (J_H) gene and a constant (C_H) gene.

Antibody Diversity by Gene Rearrangement

A human being is capable of producing specific antibodies, may be **millions** of **them** against unlimited range of antigens to defend our body by binding and destroying them. Humans are thus endowed with a **highly efficient** defense mechanism called **immune system.** This is achieved by the **combination and gene rearrangement** of **various types of structural genes** and by **somatic mutations.**

A person can produce vast number of specific antibodies against any antigen by specific gene rearrangement (Chapter 22). There are approximately 400 different V gene (for variable region of Ig), 6 types of J gene and around 20 constant region genes with specific genetic codes. So when an antigen finds it way in our body, it **induces** production of a specific antibody by gene rearrangement in such a way, that one particular V gene (out of 400) and joining gene (out of 6) and one constant gene (out of 20 variety) come close together and are transcribed into a m-RNA to produce an highly specific antibody (Ig) to bind a particular antigen for its destruction

Gene Switching—Production of Different Ig

An antigen induces production of different class of Ig in a chronological order by the phenomenon of **gene switching.** For instance, the first antibody produced in response to an invading antigen is always IgM type and is called as **primary response antibody**. IgM is a pentamer with many variable arms and several hypervariable regions. One of its hypervariable region recognizes the specific epitope of the antigen and binds to it tightly. After binding it stimulates the B lymphocyte to **switch over** to produce a specific IgG (by gene switching) by another gene rearrangement that moves C (for IgG) next to the recombined V_{D1} (of IgM) and delete the intervening C μ (of IgM) segment. The newly recombined genome retains the specific hypervariable region of IgM, specific for binding to epitope of the invading pathogens, but contain the constant region of IgG. When this gene would be expressed on exposure to the same pathogen again, it would synthesize IgG in large quantities to specifically bind the antigens and trigger the mechanisms to destroy it.

Immunization—Vaccination

When some antigen finds its way into the body or is purposely injected in to body, primary response antibody IgM with specific hypervariable region to bind that antigen are produced. This triggers the memory in B lymphocytes endowed with the gene arrangement (as described above) and the specific antibody gene code is retained in the memory. By this mechanism the subject gets immunized against this pathogen.

Secondary Response—When the person encounters the same pathogen (actual infection) later on in the life, the body is prepared to fight the infection efficiently. The immune response second time is immediate, powerful and more avid. Large amounts of IgG as secondary response antibodies with specific hypervariable region are synthesized to combate the infection efficiently by destroying and killing the microorganisms rapidly and the person would not suffer form the disease.

This forms the **basis of vaccination** of the subject against a variety of dreaded diseases. Vaccination is widely used in modern clinical medicine to protect us from several killer diseases, such as diphtheria, tetanus (DPT), polio, hepatitis B, small pox, tuberculosis etc.

COMPLEMENT SYSTEM

After the specific antibody recognizes and binds to an antigen, it triggers activation of a complement system, that would ultimately cause lysis of the invading microorganisms to destroy them. The complement system comprises about 20 protein factors present in the blood and ECF that facilitates and complement the functions of antibodies in destroying pathogens to protect our body against infections.

The complement proteins are heat labile and are inactivated by heating serum at 56°C for 30 minutes. These proteins are referred to as C-proteins and designated as $C_{1\text{-}9}$ and so on.

Mechanism of Action

Immunoglobulins after binding to invading pathogens activate the C_1 component of complement system by its Fc fragment. C_1 thus activated, acquires a protease activity and cleaves and activates next components **triggering a multienzyme cascade system of activation,** leading to a **chemical amplification** of the original message, similar to that of blood clotting factors activation.

Membrane Attack Complex Formation

Components C_3 of complement system is the key factor, which after cleavage and activation stimulates the other factors. The activated complex so formed **punches microscopic holes (pores)** on the invading bacterial surface membrane causing **osmotic lysis,** thereby killing the bacteria. This is known as classic pathway.

Inflammatory Reaction

The activation of some other components of complement system, generate certain small polypeptide fragments by proteolysis, that trigger local inflammatory reaction, attracting phagocytosis. The various leukocytes (WBC) are activated which phagocytose the pathogens to destroy them.

Alternative pathway of complement activation also exists, which is activated directly by the polysaccharides of the cell wall of foreign invader, even in the absence of antibody-antigen reaction and destroys the microorganisms. This alternative pathway is the first line of defense against invading microorganism and defends the body against infection, at an early stage before the immune response that augments the effects of classic pathway.

Abnormalities Associated with Immunoglobulin Production

Disturbances in the production of immunoglobulins results in various disease states. Abnormalities in the immunoglobulin production can be caused by several factors, such as genetic, malignancies, secondary to certain diseases or as acquired immunosuppression due to infections. Some important disorders of Ig production are described here.

Multiple Myeloma—It is a plasma cell cancer due to a malignant transformation (tumor) of **single clone** of plasma cell. This neoplasmic tumor results in the overproduction of abnormal immunoglobulins, mostly IgG type (70%) and 25% IgA paraproteins. However useful antibodies may be very low with depressed general immunity and thus recurrent infections are common. The patients are usually 50-60 years of the age and females are affected more than males.

Clinical manifestations—Multiple myeloma is characterized by **paraproteinemia,** anemia, osteolytic bone lesions and abnormal proteinuria. Common complaints are bone pains and tenderness. Spontaneous pathological fractures of weight bearing bones, ribs and vertebrae may occur.

Diagnosis is done by the detection of abnormal proteins in the urine (Bence Jones proteins, Chapter 36). Serum electrophoresis shows common M band of myeloma globulins. Diagnosis may be confirmed by bone marrow examination, where large number of malignant plasma cells are seen. Prognosis, generally is good and average survival is more than 5 years with adequate chemotherapy and local radiation therapy.

Bence Jones proteinuria

Approximately 20% multiple myeloma patients have excessive production of abnormal light chain paraproteins due to **asynchronous** production of light chains of Ig only or due to the deletation of a portion of L chain, so that they cannot combine with H chains in a normal way. These abnormal L chain paraproteins were discovered by Bence Jones (1948) and hence the name. These Bence Jones proteins are excreted in urine in excess and are characterized by their unique property of precipitating (coagulating) at low temperature (40-45° C) and dissolving at 80°- 100° C. This coagulum reappears again on cooling. These abnormal L chain fragments get deposited in renal tissues (amyloidosis) and may block renal tubules causing renal failure.

Amyloidosis is a clinical condition characterized by the deposition of L chain fragments in various tissues causing disfunction of the tissue involved, such as liver, kidney, intestine etc of multiple myeloma patients.

Detection of Bence Jones proteins in urine can be done by following methods—

1. By electrophoresis of concentrated urine.
2. By coagulation test at 40-45° C
3. Brad shaw's test which involves formation of a ring of white precipitate, when urine sample of multiple myeloma patients is layered on concentrated hydrochloric acid in a test tube.

Hypergammaglobulinemia

The excessive production of globulin antibodies can occur in clinical conditions associated with diseases such as: a) leprosy, tuberculosis, syphilis, malaria, typhoid and in other chronic infections, b) Aberrant immune reaction in conditions such as rheumatoid arthritis, collagen disorders, glomerulonephrits and autoimmune disorders and c) paraproteinemias by malignancies of cells producing immunoglobulins.

Hypogammaglobulinemia and Agammaglobulinemia

Decreased or complete absence of immumoglobins are observed in certain genetic inherited diseases or in acquired conditions. Agammaglobulinemia is a condition where IgG production is markedly diminished, resulting in impairment of body's defense against infection that may result in fatal conceguences. Agammaglobulinemia may occur due to a primary failure in the production of globulins, such as congenital X-linked disorder called Bruton's disease. **Hypogammaglobulinemia** may also occur secondary to the diseases like myeloma, leukemia or it can be drug induced.

Acquired immuno deficiency syndrome (AIDS) is a severe killer disease caused by human immuno suppressant viral (HIV) infections. The HIV infections specifically attack and destroy CD_4 T-helper cells and thus result in severe immunosuppression, which invariably leads to death of the infected subject. (Detail in Chapter 38).

BLOOD COAGULATION

Blood clotting or coagulation is body's major defense mechanism against blood loss in case of a cut or severed blood vessel. The term **hemostasis** is referred to the cessation of bleeding from a cut or severed vessel, where as **thrombosis** occurs when the endothelial lining of the blood vessel is damaged, for example upon rupture of an atherosclerotic plaque.

Blood clotting or hemostasis involves around **20 different factors** or components present in the plasma that play a crucial role in preventing blood loss during injury by forming a clot. Most of these clotting factors are **glycoprotein** in nature and are synthesized **in the liver.** The blood clotting factors are designated by the Roman numeral in the order of their discovery, except the two factors (Table 32.2).

The mechanism of blood clotting involves a series of reactions with **activation cascade of the clotting factors.** These clotting factors are present in the plasma as **inactive zymogens,** that are converted to their active forms only in case of a cut or tissue injury, when blood clotting is required. This is very important to prevent intravascular thrombosis, that may block an artery with severe clinical consequences.

The activation cascade system of blood clotting factors causes **chemical amplification of a stimuli into several 1000 folds** to activate millions of final active clotting factors within seconds to promptly stop bleeding by forming a stable blood clot at the site of the injury.

Table 32.2 : Blood coagulation factors in humans

Factor Number	Common name(s)	Subunit molecular weight
I	Fibrinogen	340,000
II	Prothrombin	720,000
III	Tissue factor, thromboplastin	370,000
IV	Calcium (Ca^{2+})	—
V	Proaccelerin, labile factor	330,000
VII	Proconvertin, serum prothrombin-conversion accelerator (SPCA)	50,000
VIII	Antihemophilic factor A, antihemophilic-globulin (AHG)	330,000
IX	Christmas factor, antihemophilic factor B, Plasma thromboplastin component (PTC)	56,000
X	Staurt-Prower factor	56,000
XI	Plasma thromboplastin antecedent (PTA)	160,000
XII	Hageman factor	80,000
XIII	Fibrin-stabilizing factor (FSF), fibrinoligase, Liki Lorand factor	320,000
—	Prekallikrein	88,000
—	High molecular weight kininogen (HMK)	150,000

The blood coagulation or hemostasis involves four main phases—

1. Constriction of blood vessel—Initially a vasoconstriction of the injured blood vessel occurs to restrict the blood flow, distal to the site of injury.

2. Formation of loose platelet plug—The platelets bind to collagen fibers at the site of injury and are activated either by thrombin formed at the site or by ADP released from other activated platelets. Upon activation the platelets change their shape and aggregate along with fibrinogen to form a hemostatic plug to prevent bleeding. This plug is loose and temporary.

3. Formation of stable blood clot—Fibrinogen, a plasma protein is then converted into fibrin threads, that cross link to form a fine fibrin mesh in which blood cells and other components are entangled to form a stable blood clot. This clot then binds on platelet aggregate plug firmly to stop bleeding. The conversion of fibrinogen to fibrin is triggered by an activation of a cascade system of blood clotting factors, as described below.

4. Lysis of blood clot—After healing of the injured tissue, a partial or complete dissolution (lysis) of blood clot or thrombus occurs by an activated protease **plasmin** and the process is known as plasminolysis.

Blood Clotting Pathways

Blood clotting process involves broadly two major pathways known as **intrinsic** and **extrinsic** pathways, which finally lead to the formation of **fibrin clot.** The two pathways in strict sense are **not independent** and are **coupled together** in the **final common clotting**

process. The extrinsic pathway is the initial process in the clotting of the blood and it is very rapid. It involves certain tissue factors which are not present in the blood. While the intrinsic pathway is slow and involves participation of many factors present in the blood. Blood clotting process may broadly be divided into three stages—

Activation of factor X—The primary event in the blood clotting process is the **activation of factor X to Xa** that leads to final common pathway of blood coagulation (Fig. 32.3). Activation of factor X is accomplished by both the intrinsic and extrinsic pathways independently and involves different mechanisms as shown in the Fig. 32.3.

1. **Conversion of prothrombin to thrombin**—The zymogen prothrombin is activated to thrombin by a cascade system of activation of various clotting factors. The active clotting factors are all **serine proteases.** The activation of prothrombin occurs on the platelets and requires the presence of active factors **Va and Xa besides Ca^{++} and phospholipids.** (Fig. 32.3).
2. **Conversion of fibrinogen to fibrin**—Thrombin is a strong proteolytic enzyme and converts fibrinogen to fibrin fibers (threads) that form a fine mesh in which the blood cells get entangled to form a stable blood clot. Overview of blood clotting steps are summarized in the Fig. 32.3.

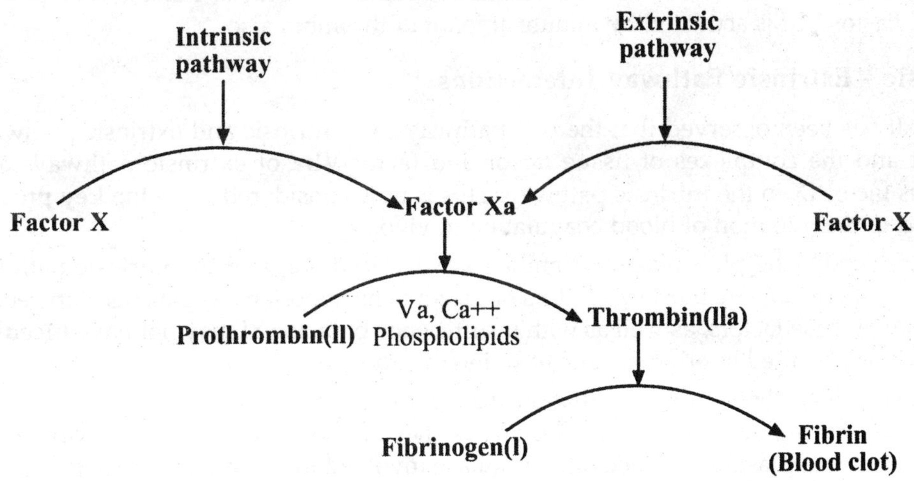

Fig. 32.3 : Overview of blood clotting with the final common pathway

Extrinsic Pathway—Mechanism

The extrinsic pathway is very rapid and is initiated in response to a tissue injury with the expression of **tissue factor (Tf),** on the endothelial cells. This pathway essentially comprises the conversion of **proconvertin (factor VII)** to its active form VIIa, by the interaction of the tissue factor. The active VIIa then activates factor X to Xa. Here tissue factor acts as cofactor, enhancing the enzymatic activity of VIIa, that cleaves an Arg-Ile bond of factor X to activate it to Xa (Fig. 32.4).

Intrinsic Pathway—Mechanism

This pathway commences with the **contact phase,** where the clotting factors of blood, namely prekallikrein, factor XII and factor XI are exposed to a **negatively charged activating**

surface, for example glass tube in the laboratory. In vivo these protein factors probably assemble on the endothelial cell membrane **(wounded surface),** where factor XII is activated to factor XIIa upon proteolysis by activated **kallikrein**. This active XIIa attacks prekallikrein to generate more kallikrein (Fig. 32.4). Active factor XIIa then activates factor XI to XIa and also releases bradykinin (nonapeptide), a potent vasodilator factor.

Role of Calcium—Several of the clotting factors require Ca^{++} ions for their activation. The Ca^{++} ions are chelated by the gamma carboxyl group of glutamate residues of the factors prothrombin, VII, X, XI and XII as described here (details in the Chapter 25, Fig. 25.26).

Factor XIa in presence of Ca^{++} activates factor IX (a zymogen containing vitamin K dependent γ carboxyl residues) to active IXa, a serine protease. This in turn cleaves an Arg-Ile bond in factor X to activate it into Xa, which is a **two chain serine protease.** This reaction requires the assembly of components called the **tenase complex** on the surface of activated platelets and comprises Ca^{++}, factor VIIIa as well as IXa and X as illustrated in the Fig. 32.4.

Tenase complex—For assembly of tenase complex, the platelets first must be activated to expose acidic (anionic) phospholipids (e.g. phosphotidyl serine and phosphotidyl inositol), which normally remains on the internal side of the platelet membrane. Factor VIII is a cofactor that serves as a receptor for factor IXa and X on platelet surface. Factor VIII is activated by minute amount of thrombin also.

Intrinsic - Extrinsic Pathway Interactions

It has been observed that the two pathways, i.e. intrinsic and extrinsic pathways interact and the complexes of tissue factor and factor VIIa of extrinsic pathways also activates factor IX in the intrinsic pathway. This is now considered to be the **key process** involved in the initiation of blood coagulation in vivo.

Secondly the physiological significance of **initial steps** of the intrinsic pathway (involving XII, prekallikrein & HMW) does not seem to be important, as patients with genetic deficiency of these factors, as well as with insufficiency of factor XI may not **have bleeding problems** as observed in other genetic bleeding disorders.

Thus it has been proposed that intrinsic pathway may actually be more important in **fibrinolysis** rather than in coagulation, as these factors (mentioned above) cleave plasminogen into plasmin and activate urokinase involved in **plasminolysis** (details given below).

Final Common Pathway of Blood Coagulation

The final common pathway of blood clotting involves the following stages—

1. Activation of prothrombin to thrombin—This activation also occurs on the surface of activated platelets and requires an assembly of a **prothrombinase complex** consisting of Ca^{++}, platelet acidic phospholipids, factor Va and factor Xa (Fig. 32.4). Factor Va is a cofactor like VIIIa of tenase complex and binds to specific receptors on the platelet membrane.

Prothrombin, the factor II (mol. wt. 70 kDa) is a single chain glycoprotein synthesized in the liver. Upon binding to the complex of factors Va and Xa on the platelet membrane, prothrombin is cleaved by the **activated factor Xa** at two sites, producing an active, two chain thrombin molecule which is released from the platelet surface. The A and B chains of the thrombin are held together by **disulphide bonds.**

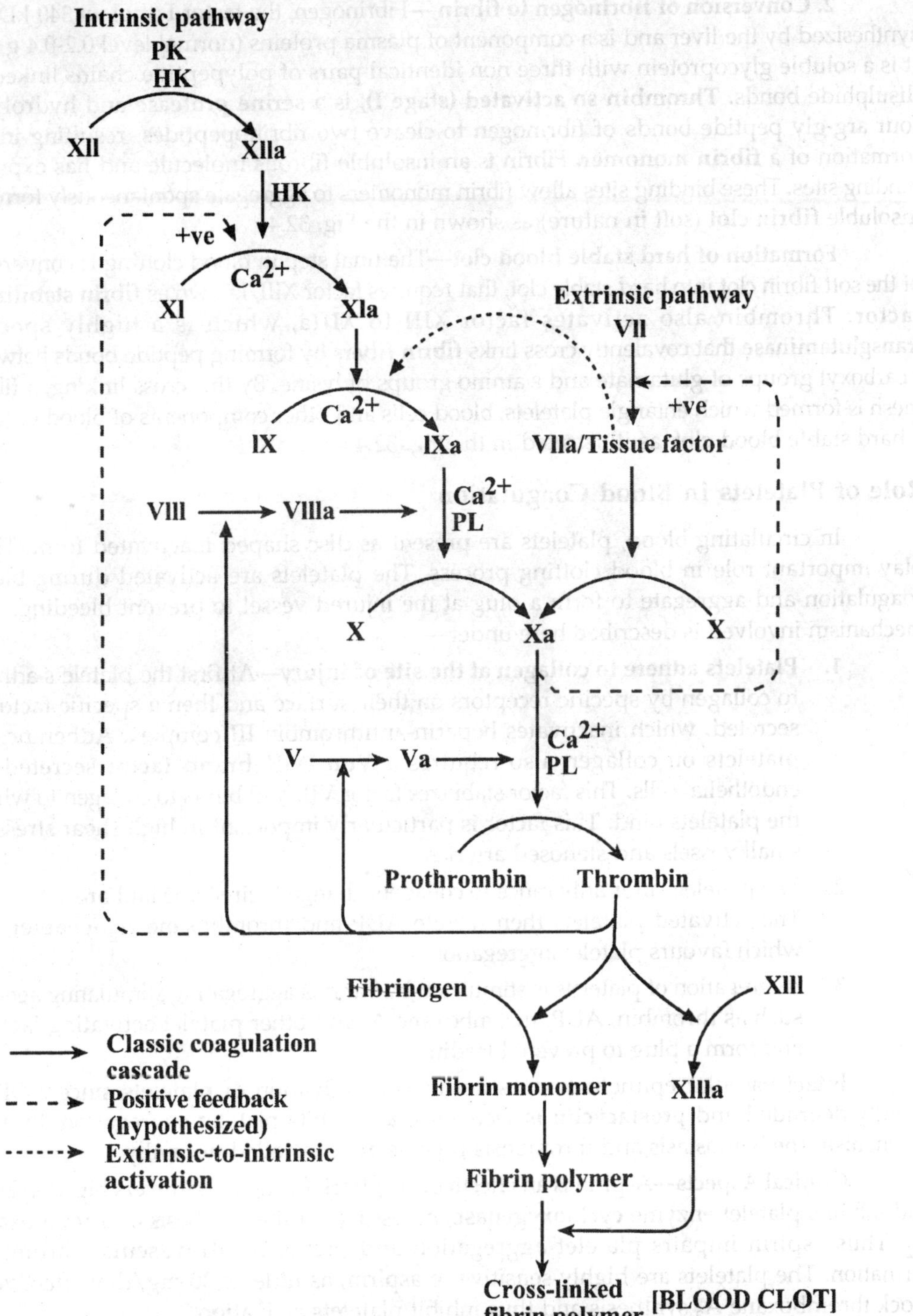

Fig. 32.4 : Intrinsic and Extrinsic pathways of blood clotting factors activation cascade system and final common pathway to produce blood clot

2. Conversion of fibrinogen to fibrin—Fibrinogen, the factor I (mol wt 340 kDa) is synthesized by the liver and is a component of plasma proteins (normal level 0.2-0.4 g/dl). It is a soluble glycoprotein with three non identical pairs of polypeptide chains linked by disulphide bonds. **Thrombin so activated (stage I),** is a **serine protease** and hydrolyzes four arg-gly peptide bonds of fibrinogen to cleave two fibrinopeptides, resulting in the formation of a **fibrin monomer.** Fibrin is an insoluble fibrous molecule and has exposed binding sites. These binding sites allow fibrin monomers to aggregate spontaneously forming insoluble **fibrin clot** (soft in nature) as shown in the Fig. 32.4.

Formation of hard stable blood clot—The final step in blood clotting is conversion of the soft fibrin clot into hard stable clot, that requires factor XIII, known as **fibrin stabilizing factor.** Thrombin also activates factor XIII to XIIIa, which is a highly specific **transglutaminase** that covalently cross links **fibrin fibers** by forming peptide bonds between γ carboxyl groups of glutamate and ε amino groups of lysine. By this cross linking, a fibrin mesh is formed which entangle platelets, blood cells and other components of blood to form a hard stable blood clot, as illustrated in the Fig. 32.4.

Role of Platelets in Blood Coagulation

In circulating blood, platelets are present as disc shaped inactivated form. They play important role in blood clotting process. The platelets are activated during blood coagulation and aggregate to form a plug at the injured vessel to prevent bleeding. The mechanism involved is described here under—

1. **Platelets adhere to collagen at the site of injury**—At first the platelets adhere to collagen by specific receptors on their surface and then a specific factor is secreted, which inactivates heparin-antithrombin III complex. Adherance of platelets on collagen also requires a **Von Willebrand factor** secreted by endothelial cells. This factor stabilizes factor VIII and binds to collagen to which the platelets bind. This factor is particularly important in high shear stress of small vessels and stenosed arteries.
2. The platelets after adherance to collagen, change their shape and are activated. The activated platelets then secrete ADP and thromboxane A_2 (chapter 11) which favours platelet aggregation.
3. Aggregation of platelets is stimulated by various aggregating stimulating agents, such as thrombin, ADP, thromboxane A_2 and other platelet activating factors and form a plug to prevent bleeding.

Intact vascular epithelium does not favour activation of platelets since ADP is rapidly degraded and **prostacyclin** is secreted that inhibits platelet aggregation. By this mechanism the hemostasis and thrombosis process are controlled normally.

Clinical Aspects—Aspirin is an effective **antiplatelet** drug as it irreversibly acetylates and inhibits platelet enzyme **cyclooxygenase,** necessary for the synthesis of thromboxane A_2. Thus aspirin impairs platelet aggregation and prevents intravascular thrombus formation. The platelets are highly sensitive to aspirin, as little as 30 mg/day effectively block thromboxane A_2 synthesis and thus inhibit platelets activation.

Aspirin is thus recommended in the management of heart diseases involving myocardial infarction and transient cerebral ischemic attacks caused due to thrombi formation.

Prevention of Intravascular Thrombus Formation

Active **thrombin,** formed in the course of **hemostasis** or thrombosis in response to tissue injury, is a **very potent protease** and its level in blood has to be controlled to prevent intravascular fibrin formation or platelet activation, which may result in thrombosis.

Factors Decreasing Thrombin Concentration

Feed back mechanism—The levels of circulating thrombin is efficiently and effectively controlled by **feed back mechanisms,** at each point of clotting factors activation cascade (Fig. 32.4).

Antithrombin III—The second means of controlling thrombin activity is its inactivation by **circulating inhibitors.** The most prominent inhibitor is **antithrombin** III which contributes approximately 75% of antithrombin activity. **Antithrombin III** also inhibits activities of factors IXa, Xa, XIa , XIIa and VIIa

Anticoagulants

Heparin present in plasma increases the activity of antithrombin III and is a potent inhibitor of thrombin. Heparin by binding and **inducing** conformational change in the antithrombin III **facilitates** its binding to thrombin to inactivate it. **α_2 Macroglobulin** contributes most of the remaining antithrombin activity along with heparin, cofactor II and α_1 antitrypsin acting as minor inhibitors under physiological conditions.

In vitro anticoagulants—These anticoagulants prevent clotting of blood sample in laboratory by removing Ca^{++} ions, that is essential for several steps of activation cascade of blood coagulation factors. Examples are oxalates, citrate, EDTA etc.

Biomedical Importance of Anticoagulants

1. Heparin is used therapeutically as an anticoagulant during dialysis process in cases of renal failure, in thromboembolic diseases and in the case of intravascular thrombosis formation.
2. Dicoumarol, coumarin, warfarin etc are antagonist to vitamin K and are used therapeutically as anticoagulants in intravascular thrombosis and in pulmonary embolus deseases. These drugs inhibit vitamin K dependent γ carboxylation of the factor II, VII, IX and X, which is necessary for chelation of Ca^{++} ions needed for their activation.

Genetic Defects of Blood Coagulation

Genetic inherited deficiencies of clotting factors leads to **bleeding disorders** in humans. The most prevalent defect is deficiency of factor VIII causing **hemophilia A.**

Hemophilia A—It is a X-chromosome linked genetic disease due to the lack of clotting factor VIII. It affects males and has played a major role in the history of royal families of Europe. Clinical manifestations are prolonged clotting time and hence even a minor cut or wound, for instance tooth extraction will result in excessive bleeding. Patients are prone to internal hemorrhages into joints and gastrointestinal tract with severe clinical consequences.

Treatment—Administration of **anti hemophilic globulin (AHG)** i.e. factor VIII, prepared from the pooled sera would help in management of this bleeding disorder. AHG is now produced by recombinant DNA technology for the therapeutic uses.

Hemophilia β or Christmas disease—This bleeding disorder is caused by a genetic defect in the production of factor IX. The IX factor is named Christmas factor, after the name of first patient reported with this disease. The clinical features are identical to those of hemophilia A, but somewhat mild in nature.

Fibrinolysis

The dissolution of blood clot is known as **fibrinolysis.** It has been observed that the coagulation system of body is in a **dynamic state** i.e. blood clots are constantly being laid down and dissolved and the two processes are in a state of equilibrium. This is highly significant in tissue wear and tear process.

Plasmin causes Fibrinolysis

Plasmin is mainly responsible for dissolving fibrin clot, as well as degrading circulating fibrin if present. Plasminogen is a zymogen (precursor) of **plasmin** and is present in plasma at a normal level. It is activated only when it binds to fibrin blood clot. Any amount of active plasmin formed in liquid phase of the blood is inactivated immediately by the fast acting plasmin inhibitor, called the **α_2 antiplasmin** present in the plasma.

Mechanism—Plasminogen binds to fibrin and gets incorporated into the blood clot, as they are being produced. Bound to blood clot, plasminogen is activated to plasmin (bound in blood clot, it is protected from α_2 antiplasmin action) by proteolytic cleavage of arg-val bond by certain activators (present in most of body tissues) to produce the two chain **plasmin,** which is a potent **serine protease.** The tissue plasminogen activators (alteplase – tPA) are released from vascular endothelium in cases of tissue injury. It is active only when bound to fibrin where it cleaves plasminogen within clot to produce plasmin, which in turn digests and dissolves the fibrin clot.

Urokinase is another plasminogen activator produced by cell types such as monocytes, macrophages, epithelial cells etc. Its main action is in degradation of extracellular matrix by converting plasminogen to plasmin that brings about fibrinolysis.

Recombinant tPA and streptokinase are used therapeutically as fribrinolytic drugs in **thrombosis** and related diseases.

❑❑❑

33

HEMOGLOBIN STRUCTURE AND FUNCTIONS

Hemoglobin and myoglobin are conjugated proteins with heme as prosthetic group, containing ferrous iron that provides them the ability to **bind** and **transport oxygen** to the tissues and **store** it, to be utilized for biological oxidation and energy generation purpose.

The biological functions of heme proteins, hemoglobin and myoglobin are related predominantly to their molecular structure and provides an insight to the mechanism of structural relationship with physiological functions of the proteins. The importance of the influence of subunit interactions, as the main determinant of the physiological functions of these hemoproteins is clearly illustrated by a comparison of their molecular structure.

STRUCTURE AND PROPERTIES OF HEMOGLOBIN

Hemoglobin is present in the erythrocytes and contains heme as prosthetic group. The **heme is a cyclic tetrapyrrole porphyrin** derivative containing **iron**, red in colour and is responsible for the characteristic red colour of the blood. Heme contains one atom of **ferrous iron** at the center of its planar ring, attached to the N_2 of pyrrole rings of the prophyrin.

Hemoglobin is a **tetrameric** conjugated protein, consisting of four polypeptide chains with different primary structures. The adult type of hemoglobin (HbA) has 2 alfa chains and 2 β chains. The α chain has 141 amino acid residues with a specific amino acid sequence while the β chain consists of 146 amino acids with different primary structure. These α and β chains are encoded by different genes and have different primary structures i.e. the specific sequence of amino acids in the α and β polypeptide chain is different. However the primary structures of γ and δ chains of other human hemoglobins, for instance Fetal hemoglobin (HbF) and HbA_2 are more similar to β chain, with **some** highly conserved amino acid sequences. The tetrameric structure of the common types of hemoglobin is as follows—HbA_1 has $\alpha_2\beta_2$, HbF (Fetal hemoglobin) has $\alpha_2\,\gamma_2$, and HbA_2 (minor adult Hb) has $\alpha_2\,\delta_2$ tetrameric polypeptides chains.

Tetrahedron Structure of Hemoglobin

Both α and β chains of hemoglobin have 75% alpha helical secondary structure with 7 and 8 segments respectively. Each polypeptide chain of hemoglobin is bound to one heme to form one subunit. The four subunits of hemoglobin at quaternary level form a tetrahedron structure as shown in the Fig. 33.3. The subunits are bonded together by certain **salt bridges** that stabilize the tetrameric hemoglobin structure.

Hemoglobin Subunit Interactions

The heme group is located in a cleft between E and F helices of globin chain and is guarded by a valine residue. The nonpolar vinyl groups of heme are burried deep in the hydrophobic interior of the cleft while the charged polar propionate groups are oriented towords outer hydrophilic surface of the subunit. The propionic acid residues of porphyrin ring form electrostatic bonds separately with arginine residues of the globin polypeptide.

Histidine residues—Specific histidine residues of the polypeptide chains of hemoglobin binds ferrous iron of heme and are designated as **proximate histidine** residues, as they are present near to the heme, at position 87 of α chain and 92 of β chain. While the **distal histidine** residues are far away from the heme and are present at position 58 of α chain and 63 of β chain.

Position of ferrous iron—The ferrous iron is placed in the center of porphyrin ring of the heme and has six valencies (ferric iron has only 5). In hemoglobin, the iron is always in ferrous state, irrespective of whether it carries oxygen or not. The ferrous iron is linked to the nitrogen of 4 pyrrole rings by its four valencies and by its fifth one to the imidazole nitrogen of the proximate histidine. In deoxy Hb, 6th valency is occupied through water with **distal histidine,** but in **oxy hemoglobin the 6th valency** of ferrous iron **binds oxygen** as shown in the Fig. 33.1.

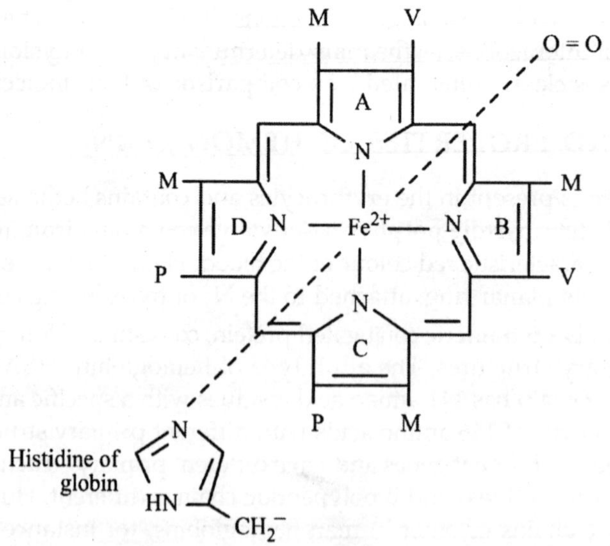

Fig. 33.1 : Molecular structure of heme with ferrous iron in the center of porphyrin ring, bonded to 4 N_2 of pyrrol rings, and to N_2 of histidine residue of globin. Ferous iron of heme binds molecular O_2 by its 6th valancy

Hemoglobin Transport Oxygen, CO_2 and Protons

Two major functions of hemoglobin present in the erythrocytes are—1) It binds and transport O_2 from lungs to the tissues and 2) It transport CO_2 and protons from peripheral tissues to the lungs for their elimination out of the body in exhaled air.

Oxygen Binding by Hemoglobin is Cooperative

Hemoglobin binds four molecules O_2 per tetramer molecule i.e. one molecule of O_2 is bound to one heme subunit and hemoglobin gets oxygenated (Oxy Hb). It should be made clear here that hemoglobin by binding O_2 is not oxidized, rather it acts as a carrier of O_2 to transport it. Hemoglobin is best suitable as **respiratory pigment** because of its unique molecular structure and properties, such as—a) It can transport large (maximum) quantities of O_2 from lungs to the tissues, b) It is very soluble and thus can easily circulate via blood, c) It can load and unload oxygen at an appropriate partial pressure of O_2 prevailing at lungs and tissue levels respectively and d) It is a powerful buffer and neutralizes H^+ ions and transport CO_2 to lung for its excretion.

Oxygenation of hemoglobin is accompanied by **conformational** changes in its molecular tetrameric structure and shows cooperativity in oxygen binding. This is illustrated best by the oxygen dissociation curve, which is typically **sigmoid in shape.** The **ability** of hemoglobin to load O_2 and release it at the physiological partial pressure of oxygen (pO_2) is described best by oxygen dissociation curve.

Myoglobin Structure and Functions

Myoglobin, present in the red muscle tissues, **stores oxygen** and acts as a **reservoir** of stored oxygen at the pO_2 of 30 mm Hg in tissues, at which oxyHb unload oxygen. **Oxymyoglobin** releases this oxygen later on under the condition of oxygen deprivation (i.e. hypoxia, pO_2-5 mm Hg), when muscles are vigorously exercizing, to be utilized by muscles for biological oxidation to generate ATP.

Myoglobin is a **monomeric hemoprotein** with a single polypeptide chain (mol.wt. 17 kDa) with 153 amino acid residues. It is similar in its molecular structure to that of one subunit i.e. the β subunit of hemoglobin with almost identical secondary and tertiary structures. Myoglobin is **unsuitable** for oxygen transport, but is very effective as an oxygen storage protein. This is because myoglobin has high affinity for O_2 than hemoglobin and it binds oxygen very tightly. The oxygen dissociation curve (ODC) of myoglobin is **hyperbolic** in shape, in contrast to the sigmoid shaped ODC of hemoglobin.

Oxygen dissociation curve—It is the binding capacity of oxygen by heme proteins at different partial pressure of O_2 (pO_2) and is graphically represented as oxygen dissociation curve (ODC). The curves of hemoglobin and myoglobin are depicted in the Fig. 33.2.

Sigmoid Shaped Oxygen Dissociation Curve of Hemoglobin

Sigmoid (S-shaped) dissociation curve of hemoglobin shows the **cooperative oxygen binding** by hemoglobin and is explained as follows—When one O_2 molecule binds to one heme subunit of the tetrameric Hb molecule, the subsequent binding of O_2 molecules is **facilitated** by certain conformational changes in the other three subunits, resulting in **many fold increase in O_2 binding capacity.**

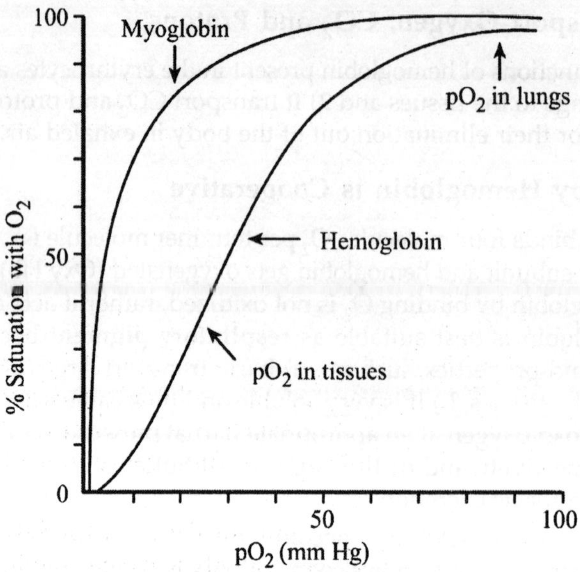

Fig. 33.2 : Oxygen binding curves of hemoglobin and myoglobin at different partial pressure of oxygen

The oxygen dissociation curve of Hb shows that initially at low pO_2, the binding of O_2 by Hb is very less. As pO_2 increases, a steep increase in O_2 binding capacity is observed (e.g. at pO_2 of 50 mm of Hg) and reach to a maximum (100% binding capacity) at higher pO_2 i.e. pO_2 of 100 mm Hg, as it occurs in the lungs. After this, there is no further increase in O_2 binding, as all the Hb molecules become fully saturated with O_2 (Fig. 33.2). This

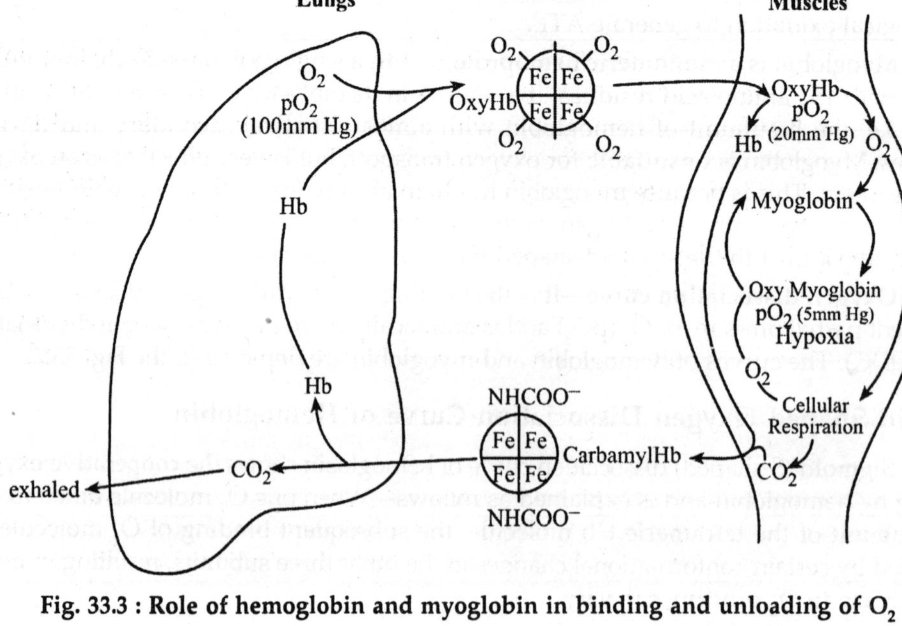

Fig. 33.3 : Role of hemoglobin and myoglobin in binding and unloading of O_2

sigmoid shaped ODC enable hemoglobin to bind maximum quantity of O_2 at lungs and transport it to the tissues, where pO_2 is less (20 mm of Hg) and oxyHb quickly releases (unload) O_2 to be utilized for biological oxidation and generation of ATP.

Oxygen dissociation curve of myoglobin, on the other hand is **hyperbolic** in nature, showing that myoglobin has higher affinity for O_2 than that of hemoglobin. Thus myoglobin **effectively takes up oxygen** released by hemoglobin in tissues, at low pO_2 i.e. around 20 mm of Hg and store it as reservoir of stored oxygen in the muscles. Oxymyoglobin releases O_2 later on under the conditions of O_2 deprivation as occuring in severely exercizing muscles. The various events of O_2 binding, its transport and unloading of O_2 by hemoglobin and role of myoglobin as reservoir of stored O_2 are illustrated in Fig. 33.3.

Conformational Changes in Hb on Oxygenation

Oxygenation of hemoglobin is accompanied by gross conformational changes in its tetrameric molecule as described here under—

a) Breaking of salt bridges between the carboxyl terminal residues of all the four subunits of hemoglobin occurs that greatly alter its secondary, tertiary and quaternary structure.

b) The quaternary structure of deoxy Hb is described as the **tout state (T form)** and that of oxy Hb as **relaxed (R) form**. By binding of O_2 on one subunit of deoxy Hb, due to the rupture of the salt bridges, the tout (T) form of Hb is relaxed and changes to R form as illustrated in the Fig. 33.5.

c) One pair of **α/β subunits rotates** with respect to the other α/β pair by 15 degree, thereby increasing the affinity of the other three heme units for O_2.

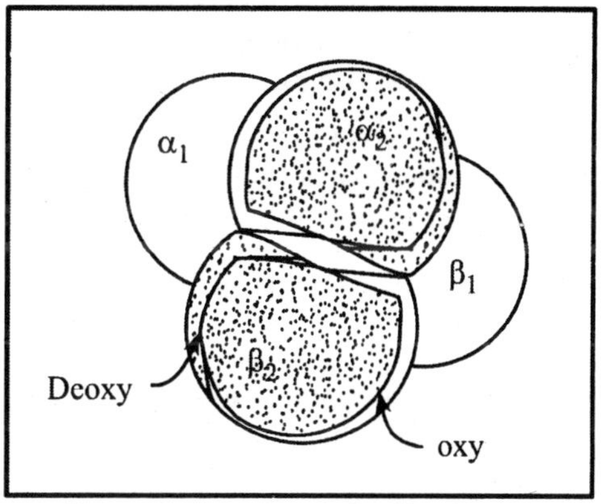

Fig. 33.4 : Rotation of α/β subunits of Hb on O_2 binding

d) In relaxed form of hemoglobin, the **ferrous iron** of deoxy Hb, that was 0.06 mm away from the plane of heme ring, shifts into the plane of porphyrin ring as shown in the Fig. 33.4. This facilitates O_2 binding on the **adjacent** heme groups many folds. This is known as **cooperative O_2 binding** and is responsible for the **sigmoid shape** of oxygen saturation curve of hemoglobin (Fig. 33.2).

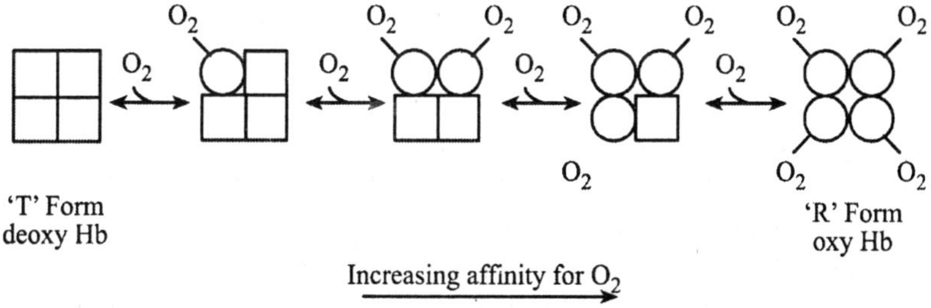

Fig. 33.5 : Transition of the T form of deoxyHb to R form on oxygenation of hemoglobin

Allosteric proteins in general undergo conformational alteration in successive ligand binding. That is, the **cooperativity** increases with successive O_2 binding on the other subunits of hemoglobin. The binding of O_2 to one heme unit increases the affinity of O_2 binding by the remaining heme residues of other subunits of tetrameric hemoglobin in a manner as depicted below.

$$\underset{+\atop O_2}{HB} \xrightarrow[\text{1 time}]{\text{affinity}} \underset{+\atop O_2}{HbO_2} \xrightarrow[\text{2 times}]{\text{affinity}} \underset{+\atop O_2}{HbO_4} \xrightarrow[\text{4 times}]{\text{affinity}} \underset{+\atop O_2}{HbO_6} \xrightarrow[\text{18 times}]{\text{affinity}} HbO_8$$

Heme–Heme interactions

As mentioned above, the conformational change during oxygenation of hemoglobin causes shifting of ferrous iron into the plane of heme ring, that facilitates binding of O_2 strongly by the 6th valency of ferrous iron. The shifting of ferrous iron in the plane of heme ring is accompanied by pulling of **proximal histidine** residues, which seems to be the **primary event in heme – heme interactions** and is transmitted to the **adjoining subunits** resulting in conformational change from T (tout) to R (relaxed) forms of rest of the Hb subunits, and the binding of subsequent O_2 occurs more rapidly.

Hemoglobin thus exhibits **cooperative binding kinetics**, that permits it to load maximum quantity of O_2 at lungs (pO_2 >100 mm of Hg) and to deliver maximum quantity of O_2 at the peripheral tissues, at their prevailing pO_2 of around 20mm of Hg.

Different Hemoglobins have Different Affinity for O_2 Binding

By comparing relative affinities of various types of hemoglobin to bind O_2 at their p50 i.e. partial pressure of O_2 that half saturate them with O_2, wide variations were observed. Human fetal Hb illustrates the variation in affinity for O_2 well, at p50. For HbA, p50 is at 26 mm of Hg while for HbF, p50 is at 20mm Hg. This difference in relative affinity for O_2 permits **HbF to extract oxygen from HbA of placental blood** from the mother to be utilized for fetus. However after birth, the HbF is unsuitable for the infants, since its high affinity for O_2 would diminish its capacity to release O_2 at tissue level. Thus HbF of infants is replaced by adult type hemoglobin (HbA), some weeks after postpartum by **gene switching mechanism** (details in Chapter 22).

Bohr's Effect

During aerobic metabolism in the tissues, CO_2 is released and pCO_2 is elevated with simultaneous increase in H^+ ion concentration. The pH of tissues is reduced due to the formation of metabolic acids, such as carbonic and lactic acids. At low pH, the affinity of Hb for oxygen is decreased and oxygen dissociation curve (ODC) shifts to right with rapid unloading of O_2 by oxyHb in the tissues. This is known **as Bohr's effect.** Reverse condition prevails in the lungs, where pCO_2 is low and pH is high, so the affinity of Hb to bind O_2 increases. With binding of O_2 on one heme, the conformational changes occurs and the tetrameric hemoglobin molecule changes from T form to relaxed (R) form, increasing cooperativity in the O_2 binding capacity to maximum.

The increased oxygenation of hemoglobin in lungs and rapid unloading of maximum quantity of O_2 in tissues under the effects of pH and pCO_2 is referred to as Bohr's effect which shift oxygen dissociation curve to right, as illustrated in Fig. 33.6.

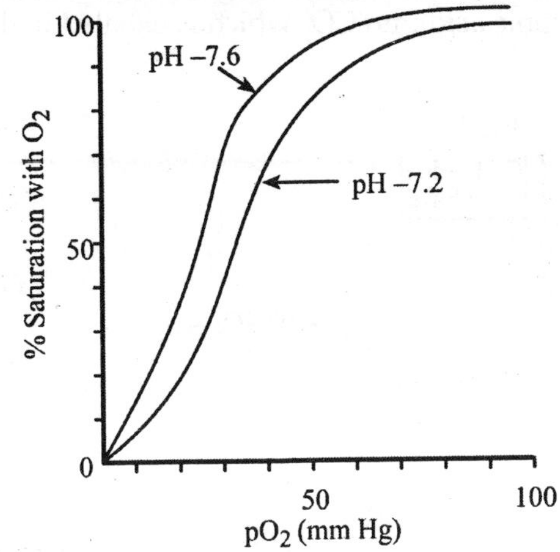

Fig. 33.6 : Bohr's effect due to change in pH on oxygen dissociation curve of hemoglobin

Transport of Carbon dioxide by Hemoglobin

Hemoglobin actively participates in the transport of CO_2 from tissues to the lungs. Hemoglobin binds CO_2 in the tissues after unloading O_2 and can carry 15% of total CO_2 transported in blood, directly on its molecule. Carbon dioxide reacts with amino terminal α-amino groups of globin polypeptides forming a **carbamino group** and protons are released, which produce Bohr's effect causing further oxygen unloading by oxyHb as shown below.

$$CO_2 + Hb\text{-}NH_3^{+} \longrightarrow 2H^{+} + Hb\text{-}NH\text{-}\overset{\overset{\displaystyle O}{\|}}{C}\text{-}O^{-}$$

This change in positive charged amino terminal end to the negative charge favours salt bridges formation between subunits of hemoglobin and relaxed form of Hb changes to "T" form of deoxyHb. The H^+ ions liberated due to CO_2 binding by hemoglobin are immediately buffered by deoxy Hb to form HHb (Fig. 33.7).

Rest of CO_2 is transported as bicarbonate—In addition to carrying CO_2 directly on its molecule as **carbamino Hb,** hemoglobin also facilitates transport of major portion of CO_2 from tissues as bicarbonate in blood to the lungs. When pCO_2 in tissues is high, CO_2 diffuses into the erythrocytes where the enzyme **carbonic anhydrase** catalyzes the formation of carbonic acid, which spontaneously dissociates into H^+ and bicarbonate ions.

$$CO_2 + H_2O \underset{}{\overset{\boxed{\text{Carbonic anhydrase}}}{\rightleftharpoons}} H_2CO_3 \xrightarrow{\text{Spontaneous}} H^{+} + HCO_3^{-}$$

Buffering Action of Hemoglobin

The protons so released would tend to decrease pH of RBCs, which is prevented by **potent buffering action of hemoglobin.** Hemoglobin **bind 2 protons** for every four O_2

molecules released in the tissues and thus contributes significantly to the buffering capacity of blood. In lungs, the process is reversed i.e. as oxygen binds to deoxy Hb, the protons are released which combine with bicarbonate forming carbonic acid. Here carbonic anhydrase catalyzes breaking of carbonic acid into CO_2 which is exhaled by lungs and is excreted out as illustrated in the Fig. 33.7.

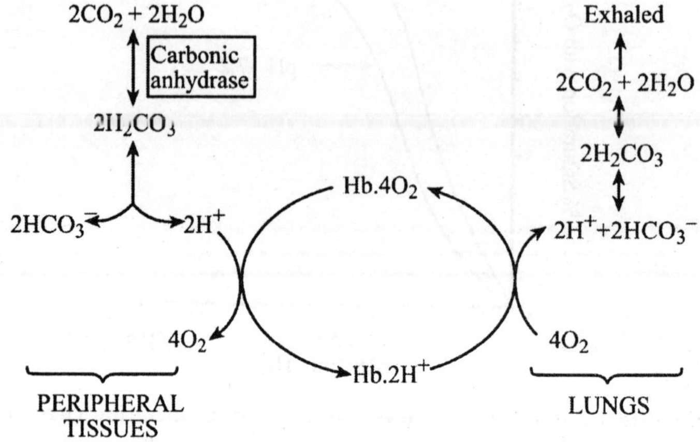

Fig. 33.7 : Role of hemoglobin as potent buffer to eleminate protons liberated in tissues

This reversible phenomenon is called **Bohr's effect,** which shift ODC at tissues to right facilitating unloading of O_2 and shifting of ODC in lungs to left, facilitating maximum binding of oxygen. The events occuring are illustrated in Fig. 33.7.

Chloride Shift Facilitates Unloading of OxyHb

When HCO_3^- concentration in erythrocytes increases as shown above, it diffuses out into the plasma. To maintain electric neutrality, chloride ions from plasma shift inside RBC in exchange with HCO_3^- and bind to deoxyHb, as chloride ions have more affinity to bind tightly to deoxyHb (Fig. 33.8). This further facilitates O_2 release from oxyHb. The HCO_3^- ion released in the plasma are transported to lungs where they bind with protons to form carbonic acid as mentioned above.

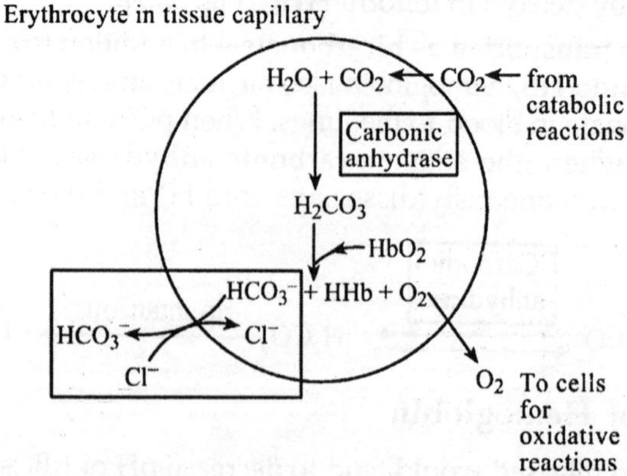

Fig. 33.8 : Mechanism of chloride shift to facilitate unloading of O_2 by oxy Hb

Oxygenation of Hemoglobin Generate Protons

At lungs, the binding of O_2 to T form of deoxy hemoglobin causes breaking of the salt bridges with release of two protons responsible for the Bohr's effect (Fig. 33.7). The protons are released from the nitrogen atoms of the histidine residues of β chain and combine with bicarbonate ion (brought by plasma as mentioned above) to form carbonic acid. The carbonic acid is then degraded by **carbonic anhydrase** into CO_2 in alveolar blood to be exhaled out by the lungs and water is formed.

On the other hand, the generation of protons (metabolic acids, lactate and carbonic acid, Fig. 33.7) in tissues favours formation of salt bridges by protonating the histidine residues of the β chain and this facilitates the release of O_2 from OxyHb with simultaneous change of R from to T form of deoxyHb. Thus an increase in proton concentration causes rapid unloading of O_2 from OxyHb in tissues and shifts ODC to right (Bohr's effect).

Effect of 2, 3 bisphosphoglycerate (2, 3 BPG)—

2, 3 BPG is the most abundant organic phosphate present in the erythrocytes. It is produced by a bypass reaction (alternative) cycle of main **glycolysis process** of glucose oxidation (Chapter 5 Fig. 12.5.) and plays very important role in facilitating unloading of O_2 by oxyHb in the tissues. 2, 3 BPG has more affinity to bind with deoxyHb than with oxyHb, and thus its binding to Hb results in the release of O_2 from oxyHb as shown below—

$$\underset{\text{oxyHb}}{HbO_2 + 2,3\ BPG} \longrightarrow \underset{\text{deoxyHb bound to 2, 3 BPG}}{Hb\text{-}2,3\ BPG} + O_2$$

Binding of 2, 3 BPG to deoxyHb—One molecule of 2, 3 BPG bind per molecule of tetrameric hemoglobin, in the **central cavity** formed by all the 4 subunits. This cavity, with a size sufficient to fit 2, 3 BPG molecule is formed when the space between H helices of β–globulin chain widens due to the change of relaxed form of hemoglobin into T form of deoxyHb. 2, 3 BPG binds strongly in this central cavity by forming salt bridges between its negatively charged phosphate group with the positive NH_3^+ groups of the amino terminal ends of both the β chains, as well as with the positively charged free amino groups of basic amino acid lysine (EF_6) and with histidine H_{21}. By forming these **strong salt bridges**, BPG **stabilizes the T structure** of deoxyHb, since these **salt bridges are extra** and must be broken prior to the formation of relaxed form of oxyHb.

Allosteric effectors—The four allosteric effectors, namely BPG, protons, CO_2 and chloride ions interact strongly with deoxy hemoglobin and thus facilitate release of O_2 at the tissue level by shifting oxygen dissociation curve to right i.e. Bohr's effect.

Biomedical Significance

All physiological and clinical conditions associated with oxygen shortage and deprivation stimulate increased production of 2, 3 BPG in the erythrocytes, which is required essentially to increase the release of O_2 from oxyHb. Thus 2, 3 BPG has great clinical significance in various diseases associated with **hypoxia**. The concentration of BPG is closely related to the tissue demand for oxygen in various conditions as described here under—

Hypoxia—In hypoxic state, the BPG concentration in RBCs is elevated by increasing its production as a bypass of glycolysis. Increased levels of BPG causes more binding to deoxyHb, thereby decreases affinity of Hb to O_2 and thus facilitates O_2 release by oxyHb in the tissues. This is an example of a method of **adaptation to hypoxia** by body to survive in the adverse situations. Examples of clinical conditions with chronic hypoxia are chronic pulmonary diseases, cyanotic cardic diseases etc in which 2, 3 BPG production in RBCs is elevated.

Anaemia is a clinical condition associated with decreased hemoglobin blood levels. Due to the reduced Hb concentrations, the oxygen supply to the tissues by blood is considerably decreased. In anaemias, the 2, 3 BPG concentration in RBCs is elevated to increase oxygen unloading in the tissues. It has been observed that in most of the cases, BPG level varies **inversely** to the Hb concentration in anaemias.

High Altitude Adaptation—People living in high altitude regions, where pO_2 is very low, body undergoes various physiological changes in order to adapt to the conditions of oxygen deprivation. These changes include—polycythemia, hyperventilation and increased production of 2, 3 BPG in the erythrocytes. Increased concentration of BPG shifts oxygen dissociation curve of oxy Hb to right, ensuring maximum unloading of O_2 in the tissues.

Fetal hemoglobin (HbF) has more affinity for O_2 than adult HbA. This is because 2,3 BPG binds weakly to fetal Hb than HbA. Fetal hemoglobin unlike HbA has two α and two gamma (γ) chains instead of two β chains of the adult type hemoglobin. The gamma chain, though is similar in size to β chain with 146 amino acid, the primary structure is different. The gamma chains of HbF have serine residue instead of histidine H_{21} residue of β chain of HbA and thus cannot form salt bridge with 2, 3 BPG. Thus BPG has less ability to stabilize T form of HbF so it easily changes to R form to bind O_2 showing more affinity of HbF for O_2.

Hemoglobin Derivatives

Various types of hemoglobin derivatives are produced by binding of Hb with different ligands on heme units in place of O_2 or by oxidation of ferrous iron to ferric state. These hemoglobin derivatives show **characteristic absorption spectrums** and can be identified by comparing their absorption spectra.

OxyHb and deoxyHb are common derivatives of hemoglobin present in the blood. Normally concentration of deoxyHb in blood is less than 5% of total Hb and if its concentration increases, cyanosis results. OxyHb is dark red in colour while deoxyHb is bluish purple. Other Hb derivatives are carbonmonoxy Hb (cherry red coloured), met Hb (dark brown in colour) while sulph Hb is greenish brown coloured.

Carboxy hemoglobin (COHb)—Carbon monoxide has very high affinity to bind to hemoglobin than O_2 (200 times more) and so if it is inhaled, CO replaces O_2 from hemoglobin to form carboxy hemoglobin (COHb). The carboxy Hb will not transport O_2 to tissues resulting in carbon monoxide poisoning which results in death of the affected subjects.

Methemoglobin (metHb)—When ferrous iron of hemoglobin is oxidized to ferric state, metHb is formed. MetHb cannot transport O_2, as ferric iron has only 5 valencies (ferrous Fe^{++} has 6) and cannot bind to O_2. Normally the auto oxidation of ferrous iron is prevented by its position in hydrophobic heme pockets. Small quantities of metHb formed in RBCs by oxy free radicals are immediately reduced back to ferrous state by glutathione peroxidase enzyme system of the RBCs (Chapter 26).

Oxidizing agents such as ferricyanide can oxidize ferrous iron of Hb and convert it to metHb. The colour of blood changes to dark brown and it shows characteristic extra absorption band in the red region of the spectrum in addition to the bands of oxyHb in yellow green region.

ABNORMAL HEMOGLOBIN - HEMOGLOBINOPATHIES

Several mutations in genes that code for alfa or beta chains of hemoglobin have been reported that cause production of **abnormal hemoglobins,** resulting in inherited genetic diseases called **hemoglobinopathies.** The abnormal Hb so formed has impaired physiological functions and altered properties. Several hundred mutations with resultant Hb variants have been identified. Most of them are extremely rare and benign with no severe clinical consequences. However there are several genetic disorders with abnormal hemoglobin accompanied by serious clinical manifestations, which can be life threatening due to the altered functions of hemoglobin variants. Some important hemoglobinopathies are described here.

Methemoglobulinemia—In this genetic abnormality, because of the point mutation in the gene encoding β chain, tyrosine amino acid replaces histidine of β chain (details in the chaper 21) due to which ferrous iron of heme is oxidized to ferric state. The ferric iron of metHb is stabilized by forming tight ionic complexes with phenolate ion of tyrosine of β chain. The metHb with ferric iron cannot bind O_2 resulting in fatal consequences.

Sickle Cell Anaemia—Hemoglobin 'S' (HbS)

The phenomenon of **sickling of RBCs** in an inherited genetic disease, known as sickle cell anaemia is because of the synthesis of abnormal hemoglobin. In this congenital disease due to a **point mutation** in the gene encoding β chain of Hb, the **polar glutamic** acid is replaced by non polar amino acid **valine** at 6th position, producing abnormal hemoglobin, HbS (details in the chapter 21). This single amino acid change in the primary structure of β chain causes alteration in the properties of **hemoglobin,** such as change in electrophoretic mobility and markedly **reduced solubility** of HbS in **deoxygenated T state.**

Sickle cell anaemia is most prevalent genetic hemoglobinopathy in the black population of tropical countries of Africa (frequency rate of 1 in 500 births). This disease is so named because the discoidal shape of normal erythrocytes is altered to **crescent** or **sickle shape** (Fig. 33.9) at low oxygen concentration and they are prone to excessive hemolysis causing sickle cell anaemia.

Normal Crescent

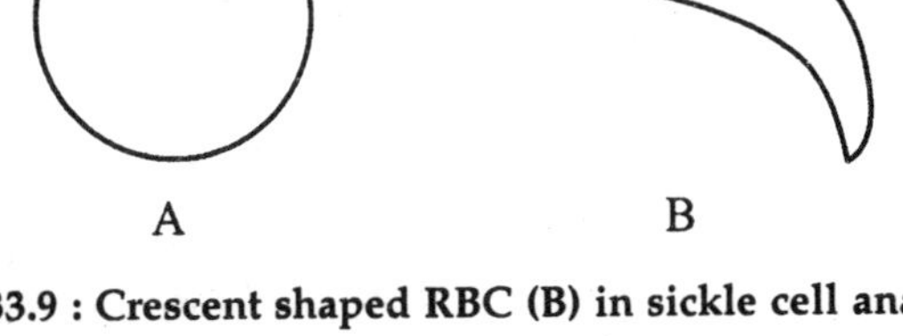

A B

Fig. 33.9 : Crescent shaped RBC (B) in sickle cell anaemia

Pathophysiology of sickle cell disease is due to the single nonpolar valine amino acid replacement on β chain located on the surface of hemoglobin. This change leads to **abnormal polymerization** of **deoxy HbS** forming intracellular fibers, which cause **distortion of RBCs** into **crescent** (sickle) **shape** with increase vulnerability to hemolysis.

Sticky patches on HbS—The substitution of nonpolar valine on the surface of β chain of hemoglobin in place of hydrophilic glutamate in HbS, results in localized stickiness with formation of **sticky patches** on the sickle cell hemoglobin. These sticky patches on HbS have **complementary** receptor site **only on deoxyHbS** as shown in the Fig. 33.10.

Polymerization of deoxyHbS—The sticky patch of one deoxyHbS bind with the complementary receptor of another deoxyHbS and the process continues to form long aggregates of HbS molecules in form of fibers which lateron hardens into long stiff fibers, distorting the shape of RBCs to crescent shape.

Normal oxyHbA does not have sticky patches on its surface but **deoxy** adult HbA do possess complementary receptor sites. However the binding of **sticky HbS** to **normal deoxy HbA** cannot extend polymerization, since the later does not have a sticky patch and thus polymerization is terminated immediately as is illustrated in the Fig. 33.10.

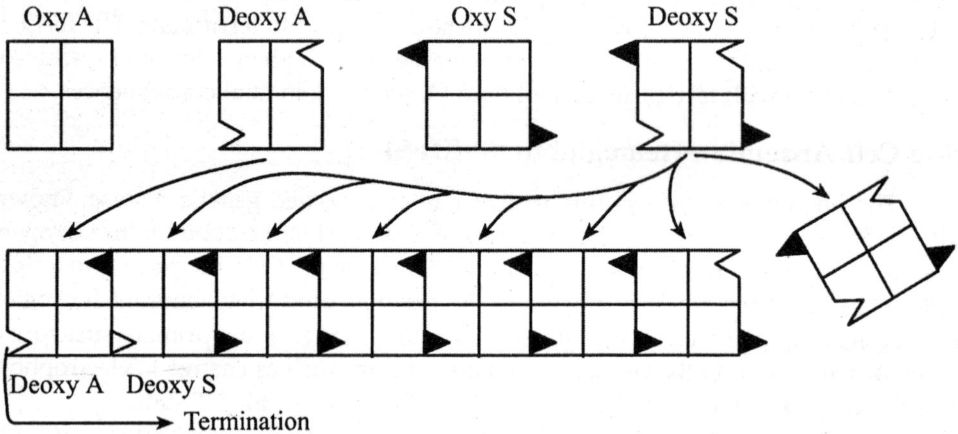

Fig. 33.10 : Diagrammatic representation of polymerization of deoxyHbS to form a fiberous structure

Oxygenated Hemoglobin S cannot polymerize to form fiber as it is in **relaxed form** and the complementary sites for sticky patches **are masked** in R form, and though sticky patch is present on **oxyHbS**, they cannot bind to each other or with deoxy HbS. Thus **oxygenation** of **hemoglobin HbS** prevents sickling of RBCs. However it is observed that once polymerization of deoxy HbS has occurred, a vicious cycle starts and it becomes irreversible, causing sickling of erythrocytes. The crescent shape RBCs have increased tendency to adhere to endothelial cells and form plug in the capillaries, blocking flow of blood and RBCs are excessively hemolyzed.

Sickle Cell Trait provides Resistance to Malaria

The incidences of sickle cell anaemia are more prevalent among the black population of tropical areas of world, such as Africa, which are **highly infested** with **malaria.** This disease is caused by plasmodium falciparum and this malarial parasite's life cycle partially occurs inside erythrocytes. In order to survive by this deadly disease, the expression of genes has altered in this population, due to **induced mutations** in the genes, to **aquire sickle cell trait.** With this aquirement of sickle cell trait, the increased hemolysis of erythrocytes becomes beneficial as it will **interrupt** the parasite's life cycle, thereby preventing multiplication of parasite and infection will be cured.

Thus sickle cell trait appears to be an **induced gene mutation** in African population as a means to adapt and survive in the malaria infested regions.

Homozygous and Heterozygous HbS

The sickle cell trait is inherited as autosomal recessive trait and thus only homozygous offsprings would suffer from severe sickle cell anaemia with serious clinical consequences, resulting in death of the affected individual around 20 years of the age.

However the heterozygous infant carrying only one recessive mutant gene from the parent and other gene being normal, would contain both the types of hemoglobins in the RBCs i.e. normal HbA as well as HbS. These subjects can lead a normal life, as the condition is mild and usually do not show severe clinical symptoms of the disease. This heterozygous sickle cell trait is more common, almost 1 in 10 are affected in the black population and is beneficial in protection from malarial infection as described above.

Thalassemias

The thalassemias are a group of inherited genetic diseases, in which mutations in genes encoding alfa or beta chains of hemoglobin results in reduction or no synthesis of the α or β-polypeptide chains of globin. The disease is characterized by excessive hemolysis of RBCs. Greek world **thalassa** means sea and the disease is so named, as it was most prevalent in the population of the coastal regions of Mediterranean sea.

In thalassemia, mutation in gene results in either α or β chains of globin are not synthesized causing imbalance in the tetrameric hemoglobin formation. Reduction in α–chain synthesis is called **alfa thalassemia,** that is **rare** but very severe disorder and is **incompatible with life.**

Beta thalassemia is more common and is characterized by decreased or absence of β chain synthesis which may be compensated by increasing gamma and delta chain synthesis and tetrameric Hb with $\alpha_2\ \gamma_2$ (HbF) or $\alpha_2\ \delta_2$ (HbA_2) chains may be formed. These hemoglobins can transport O_2 but not as efficiently as HbA.

Mutations in thalassemia can occur due to—a) an operator gene defect or b) due to deletation or insertion of a nucleotide in the genetic code for α or β chain, which results in frame shift mutations, where whole genetic code is altered. So α or β chain will not be synthesized. Other causes are—c) Impairment in m-RNA synthesis due to defect in initiation of transcription process or d) premature termination of m-RNA synthesis.

Thalassemia Syndrome is characterized by overall decreased synthesis of normal adult type HbA leading to excessive hemolysis of erythrocytes, resulting in severe hemolytic anaemia.

Thalassemia minor—This is a heterozygous trait with mutation in only one gene of the two β chain gene allele pair. So some amount of β chain synthesis does occur and some normal tetrameric HbA are produced, but quantity is less. The clinical signs and symptoms are minimal and is usually asymptomatic.

Thalassemia Major—This **homozygous** trait is caused by mutation and defect in both the alleles of genes encoding the β chain. The infants born with thalassemia major are healthy at birth, since during fetal development HbF with $\alpha_2\,\gamma_2$ tetramer is produced and the β chains are not needed. After some weeks the child becomes severely anaemic and die within 1-2 years due to excessive hemolysis of RBCs.

In these cases, synthesis of unaffected α chains continues but since they donot have complementary β chains to bind, the α chains aggregate and precipitate within the erythrocytes. These precipitates, as **inclusion bodies** (abnormal) cause cell membrane damage, resulting in destruction of RBCs. This is also responsible for ineffective erythropoiesis in β thalassemia major. There is increased oxidation stress in HbA deficient red blood cells, causing production of superoxide free radicals and consequent destruction of RBCs.

Management—Frequent blood transfusions is the only available treatment to sustain life in affected children. This may lead to iron overload with severe clinical consequences (Chapter 24). Bone marrow transplantation has been successfully tried in a few cases. Azacytidine drug temporarily activates dormant gene encoding gamma chain, that can be expressed to synthesize gamma chains which combine with α chain to form fetal Hb ($\alpha_2\,\gamma_2$), to reduce severity of disease to the some extent.

Gene replacement therapy seems to be the best possible treatment in management of the thalassemia major, but it is a very costly treatment as highly advanced technology is needed.

❑❑❑

34

HEMOPROTEINS, HEME SYNTHESIS AND CATABOLISM

Heme containing proteins are **hemoglobin, myoglobin, cytochromes** and certain enzymes, such as catalase, tryptophan pyrrolase etc. The hemoproteins are essential for the life as they play central role in binding and transport of oxygen to tissues, and transport of electrons during biological oxidation for the generation of ATP to carry on life processes. Hemoglobin and myoglobin are **conjugated proteins,** with **heme** as non-protein prosthetic group and they possess the ability to bind **reversibly** with molecular oxygen and thus transport O_2 to tissues for the final oxidation of reducing equivalents to generate energy.

Heme is a **metalloporphyrin** containing iron while **chlorophyll** of green plants is a magnesium containing porphyrin derivative that plays important role in photosynthesis process.

Hemoglobin

Hemoglobin is a conjugated **metalloporphyrin protein** having **heme** as prosthetic group and **globin,** the protein part. It is a **chromoprotein**, red in colour and imparts characteristic red colour to the blood. It is present exclusively in **erythrocytes** (greek, erythros -red colour) of the blood. Normal level of hemoglobin in males is 14-16 g/dl of blood and in females it is 13-15 g/dl.

Myoglobin

Myoglobin is a monomeric hemoprotein found mostly in the red skeletal muscles and heart muscles. It is a conjugated protein, with a single polypeptide chain (152 amino acids) and one **heme molecule** attached to the 92nd histidine residue of the globin polypeptide chain. Myoglobin structure resembles to one subunit of the tetrameric hemoglobin and its molecular weight is 17,500 Da.

Cytochromes are components of electron transport chain (ETC),that is established on the inner mitochondrial membrane. The cytochromes are heme containing conjugated proteins with a oxidation/reduction system of ferrous iron of heme to ferric state (Chapter 6). The cytochromes facilitate quick transport of electrons and reducing equivalents, removed in various metabolic cycles to the final acceptor i.e. oxygen during biological oxidation to produce water. The energy of biological oxidation on ETC is captured as high energy phosphate compound ATP, by coupling oxidation with phosphorylation (Chapter 6).

Heme Enzymes—Certain enzymes such as **catalase** and **tryptophan pyrrolase** are hemo proteins, which play important role in degradation of H_2O_2 and tryptophan metabolism respectively (details in Chapters 16 & 26).

Structure of Heme

Heme is an iron containing porphyrin derivative called **protoporphyrin IX** (III) with ferrous iron in the center of the porphyrin ring. The **porphyrins** are cyclic compounds formed by the fusion of four pyrrole rings linked together by methelene bridges. Usually the four pyrrole rings of porphyrins are substituted with side chains (Fig. 34.1).

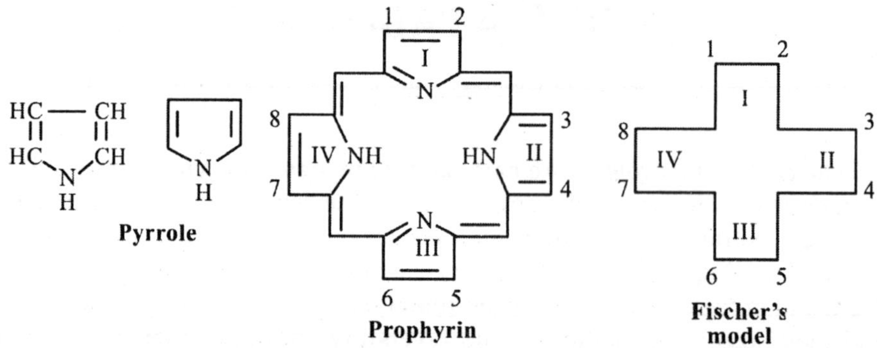

Fig. 34.1 : Structures of pyrrole and porphyrin derivatives and Fischer's model of representation of porphyrin nucleus

Pyrrole ring is a heterocyclic, 5 membered ring with 4 C-atoms and one N_2 atom (Fig. 34.1). The four pyrrole rings in the porphyrin molecule are numbered as I, II, III and IV and the methelene bridges, which link the pyrrole rings, as α, β, γ and δ. The possible sites of the side chain substitution in the pyrrole rings are denoted as 1-8 C-atom in the porphyrin nucleus (Fig. 34.1). Hans Fischer proposed a shorthand model for the presentation of the porphyrin structures, in which each pyrrole ring is represented as a bracket, while omitting the methelene bridges as shown in Fig. 34.1.

Types of Porphyrins

The natural porphyrins have substituted side chains on the positions 1-8 of porphyrin nucleus and depending on the types of side chains and their arrangement, they are classified in following major types—

1. Porphyrins type I—In type I porphyrins, the substituted side chains on 1-8 positions of porphyrin nucleus are **symmetrically** arranged. For example uroporphyrin I has four acetate (A) and four propionate (P) side chains arranged symmetrically as shown in the Fig. 34.2.

2. Porphyrins type III—The type III porphyrins have side chains substituents arranged **asymmetrically on IV pyrrole ring.** For example, uroporphyrin type III has side chains acetate (A) and propionate (P) on IV pyrrole ring arranged asymmetrically, as the expected order of A and P groups is reversed as illustrated in the Fig. 34.2. Type III porphyrins are of most common occurrence in the biological systems.

3. Heme is a type III porphyrin known as **protoporphyrin III,** also referred to sometimes as **protoporphyrin** IX, as it was ninth (IX) in the series of possible isomers as designated by Hans Fischer, the pioneer scientist in the field of porphyrin research. The protoporphyrin III (heme) has side chain substituents **as 4 methyl (M), 2 propionyl (P)** and **two vinyl (V)groups** on position 1-8 of porphyrin ring, but the arrangement of M and P groups on pyrrole ring IV is asymmetric as shown in the Fig. 34.2.

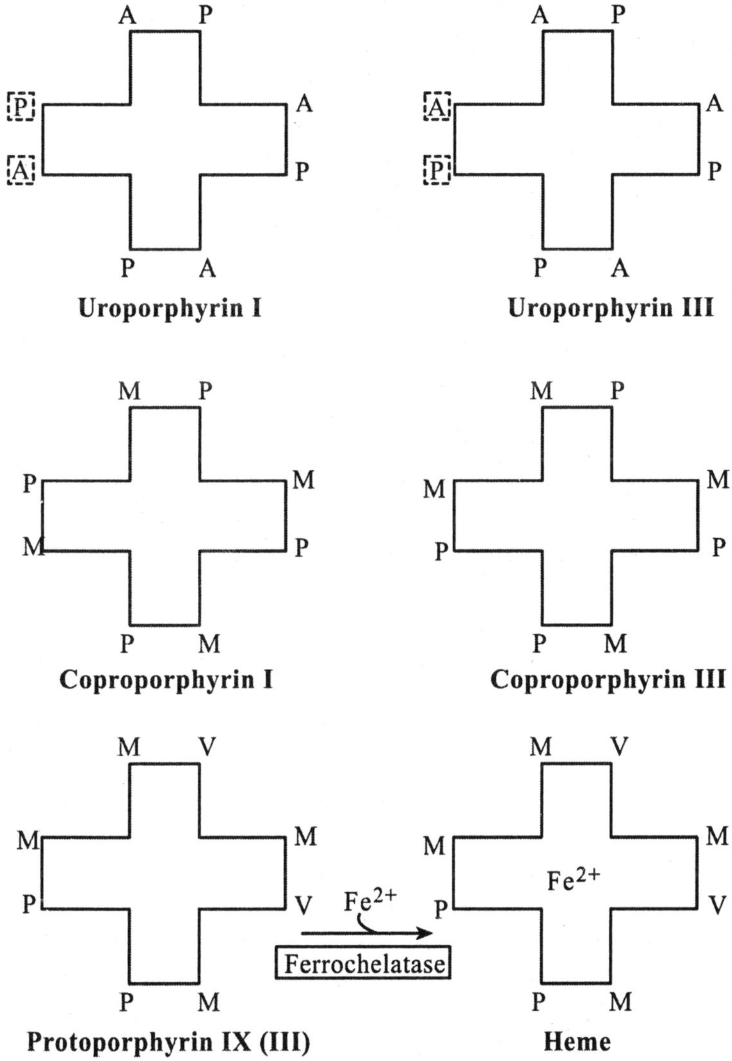

Fig. 34.2 : Fischer's model representation structures of physiologically important porphyrins

Protoporphyrin III – the precursor of Heme

When an iron atom is added to a protoporphyrin III nucleus, it forms **heme, a metallo porphyrin derivative.** The iron atom of heme occupies the central position of the porphyrin ring and in hemoglobin, it is always in **ferrous** (Fe^{++}) state, irrespective of whether it is oxy hemoglobin or deoxy hemoglobin. The ferrous iron has six valencies and is held in the center of heme linked by 4 coordinated bonds to the nitrogen of four pyrrole rings. The other two bonds are formed on **either side of the plane of** heme ring, between Fe^{++} and the imidazole nitrogen of the **proximate histidine** residues of α (87 position) and β chain (93 position) as shown in Fig. 34.3.

In oxyhemoglobin the 6th valency of ferrous iron binds the O_2, but in deoxy Hb it is bound via water molecule to the **distal histidine** residue. The 58th residue in α-chain and 63rd residue in β, γ, and δ chains are called distal histidine residues as they are far away from the heme group. The structure of heme with ferrous iron is shown in the Fig. 34.3.

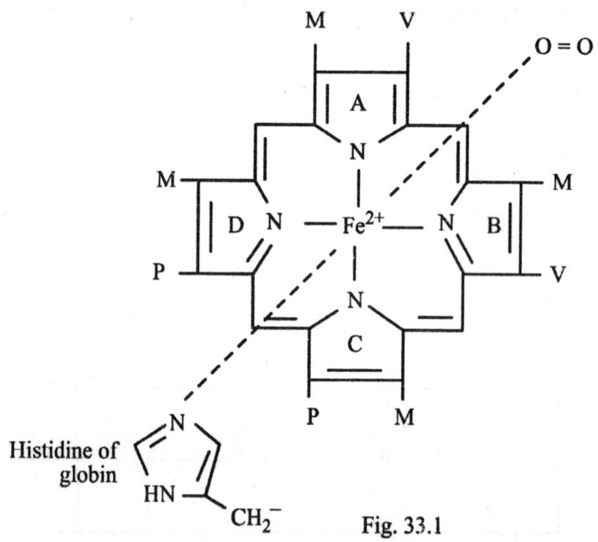

Fig. 34.3 : Molecular structure of heme with ferrous iron in the center of porphyrin ring, bonded to 4 N_2 of pyrrol rings, and to N_2 of histidine residue of globin. Ferous iron of heme binds molecular O_2 by its 6th valancy

Methemoglobin

If the ferrous iron in hemoglobin is oxidized to ferric state, it looses two of its valencies and will not bind with molecular oxygen to transport it to the tissues, resulting in fatal consequences. The hemoglobin with oxidized ferric iron is called **methemoglobin.** The oxidation of ferrous iron of hemoglobin in the blood is normally prevented by various mechanisms, such as—1) by its position in hydrophobic heme pockets, 2) Reduced glutathione present in rich amount in RBCs keeps iron always in ferrous state (chapter 7 and 3) Many other factors, antioxidants and HMP shunt going on actively in RBCs prevent oxidation of ferrous iron of hemoglobin to ferric state by degrading oxy free radicals as soon as they are produced.

BIOSYNTHESIS OF HEME

Heme is the most important porphyrin present as nonprotein prosthetic group in hemoproteins, such as hemoglobin, myoglobin, cytochromes and enzymes catalase, tryptophan pyrrolase etc. Heme is synthesized primarily in the bone marrow, during erythropoiesis, for the synthesis of hemoglobin. It is also synthesized in the liver and in other tissues for the production of various enzymes.

Precursors of Heme Synthesis—Succinyl CoA and amino acid glycine are main precursors needed for heme biosynthesis. Succinyl CoA is derived from the citric acid cycle, as it is an important intermediate of CAC, while glycine is a non essential amino acid, which can easily be synthesized in tissues by glycine synthase complex (Chapter 16, Fig. 16.3). **Pyridoxal phosphate** (PLP) is essentially required as coenzyme in heme synthesis and it activates glycine amino acid to enter the biosynthesis process. **Heme synthesis** takes place partly inside mitochondria and partly in the cytosol and involves following steps—

1. **Condensation of glycine with succinyl CoA**—In the first step succinyl CoA condenses with amino acid glycine, catalyzed by key enzyme **ALA synthase** with PLP as coenzyme to form **α-amino β-keto adipic acid**, which is rapidly decarboxylated to produce δ **amino levulenic acid (δ ALA)** as shown in Fig. 34.4. The mitochondrial **ALA synthase** is the key regulatory, rate limiting enzyme and the reaction takes place inside mitochondria. The δ ALA so formed then comes out in the cytosol.
2. **Formation of porphobilinogen**—In the cytosol, two molecules of ALA condense together in the presence of enzyme **ALA dehydratase,** which removes two molecules of water and the condensation complex so formed **cyclizes** to produce a **heterocyclic porphobilinogen (PBG)**. It is a pyrrole derivative with three substituted side chains, namely acetate, propionate and methylamine (Fig. 34.4). ALA dehydratase is a zinc containing enzyme and is inhibited by lead, as seen in lead poisoning cases where heme synthesis is impaired.
3. **Production of cyclic porphyrins**—The porphyrins are formed by the condensation of four molecules of porphobilinogens, in head to tail manner, to form first a **linear tetrapyrrole derivative** by deamination reactions. The reaction is catalyzed by enzyme **PBG deaminase** which is also known as **uroporphyrinogen synthase.** Four molecules of ammonia are removed from the **methylamine side chain of the porphobilinogens** during this condensation reaction and the 4 PBG get linked together by methylene bridges (Fig. 34.5). This linear tetrapyrrole derivative so produced then may cyclize spontaneously to produce uroporphyrinogen type I.

Production of type III Porhyrins

However in body tissues another enzyme, **uroporphyrinogen III cosynthase** is present, which facililates the cyclization of the linear tetrapyrrole derivative formed described above, in a different manner and **isomerizes** the compound to uroporphyrinogen III, with asymmetric arrangement of substituted side chains on IV pyrrole ring as shown in Fig. 34.5.

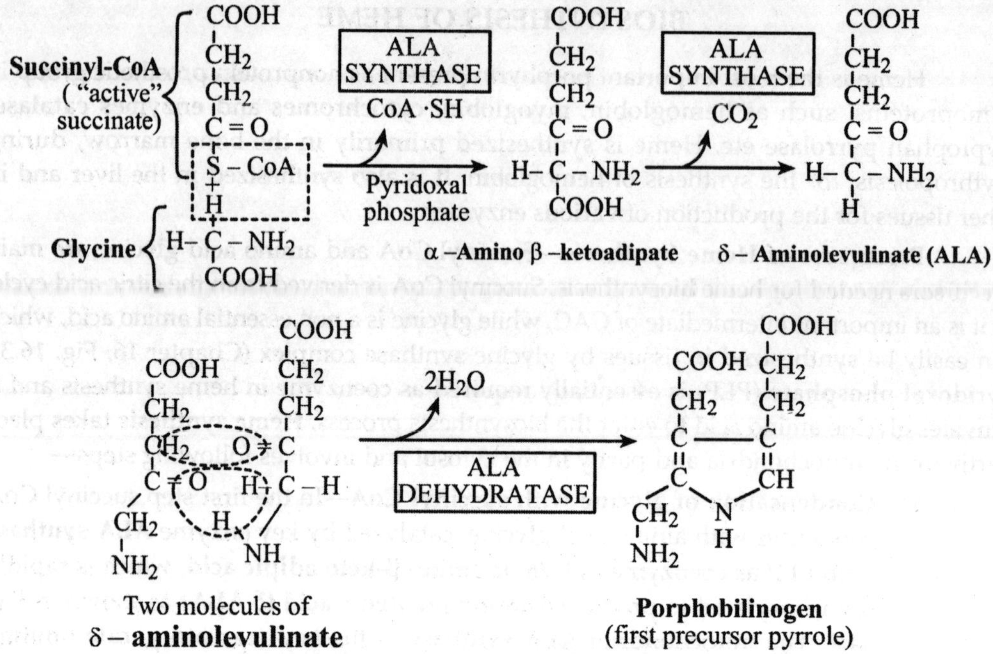

Fig. 34.4 : Formation of a pyrrole derivative (PBG) by condensation of two δ amino Levulenic acid (ALA)

Thus normally the **cyclization of 4 pyrroles,** linked by methylene bridges is almost exclusively is to form **type III uroporphyrinogen** that is facilitated by **cosynthase III.** However in certain genetic **congenital porphyria disorders,** in which this cosynthase III enzyme is lacking, type I isomers are produced in excess. These type I porphyrins cannot be utilized in heme synthesis and are excreted in urine in excess causing **porphyrias.** (details later in the Chapter).

Production of Heme

Uroporphyrinogen III produced as described above, has 4 acetate (A) and 4 propionate (P) as side chains substitutions on position 1-8 of porphyrin nucleus (Fig. 34.5). In next steps the 4 acetate groups are decarboxylated to form 4 methyl (M) groups and two propionate residues of two pyrrole rings I and II are decarboxylated and dehydrogenated to produce two unsaturated vinyl (V) side chains respectively (Fig. 34.6) as described in the following steps

1. Production of coproporphyrinogen III—First the 4 acetate groups of uroporphyrinogen III are decarboxylated in the presence of a **decarboxylase** enzyme to form 4 methyl groups and **coproporphyrinogen III** is produced, which is then transferred into the **mitochondria** for further reactions.

2. Formation of protoporphyrinogen III—Inside mitochondria the two propionate side chains (pyrrole I and II) of coproporphyrinogen III are first decarboxylated and then oxidized in the presence of enzyme **coproporphyrinogen oxidase** to two unsaturated vinyl groups to produce **protoporphyrinogen** III as shown in the Fig. 34.6. All the intermediates in heme synthesis are **porphyrinogens** which have methylene bridges with six extra hydrogen atoms (instead of methelene CH= bridges of heme) and are **colourless compounds.**

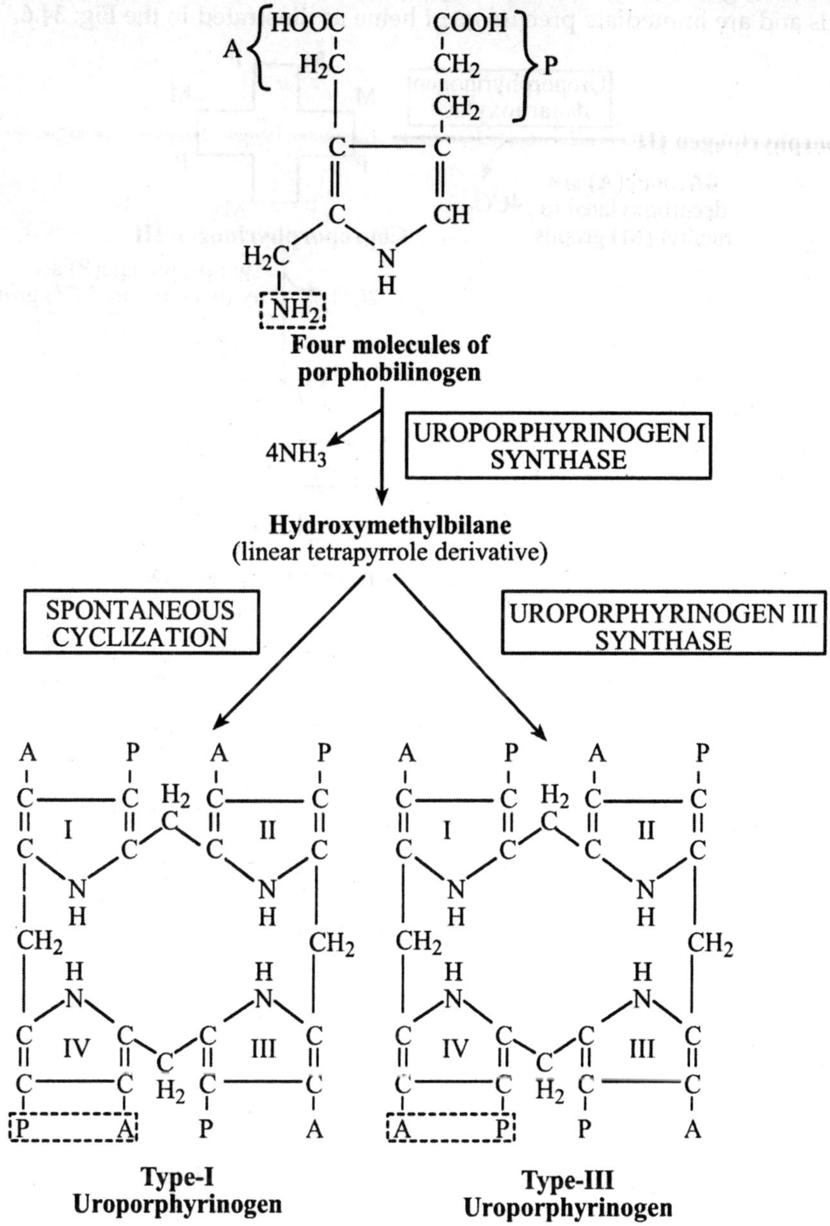

Fig. 34.5 : Condensation of 4 PBGs to produce type III porphyrin, uroporphyrinogen III

3. Formation of Protoporphyrin III

The oxidation of protophyrinogen to **protophyrin** III occurs in the mitochondria and requires molecular oxygen. In this conversion the four methylene bridges of protoporphyrinogen are oxidized to **methelene (-CH=) bridges** in the presence of enzyme **protoporphyrinogen oxidase.** The protoporphyrin III so produced are red coloured compounds and are immediate precursors of heme as illustrated in the Fig. 34.6.

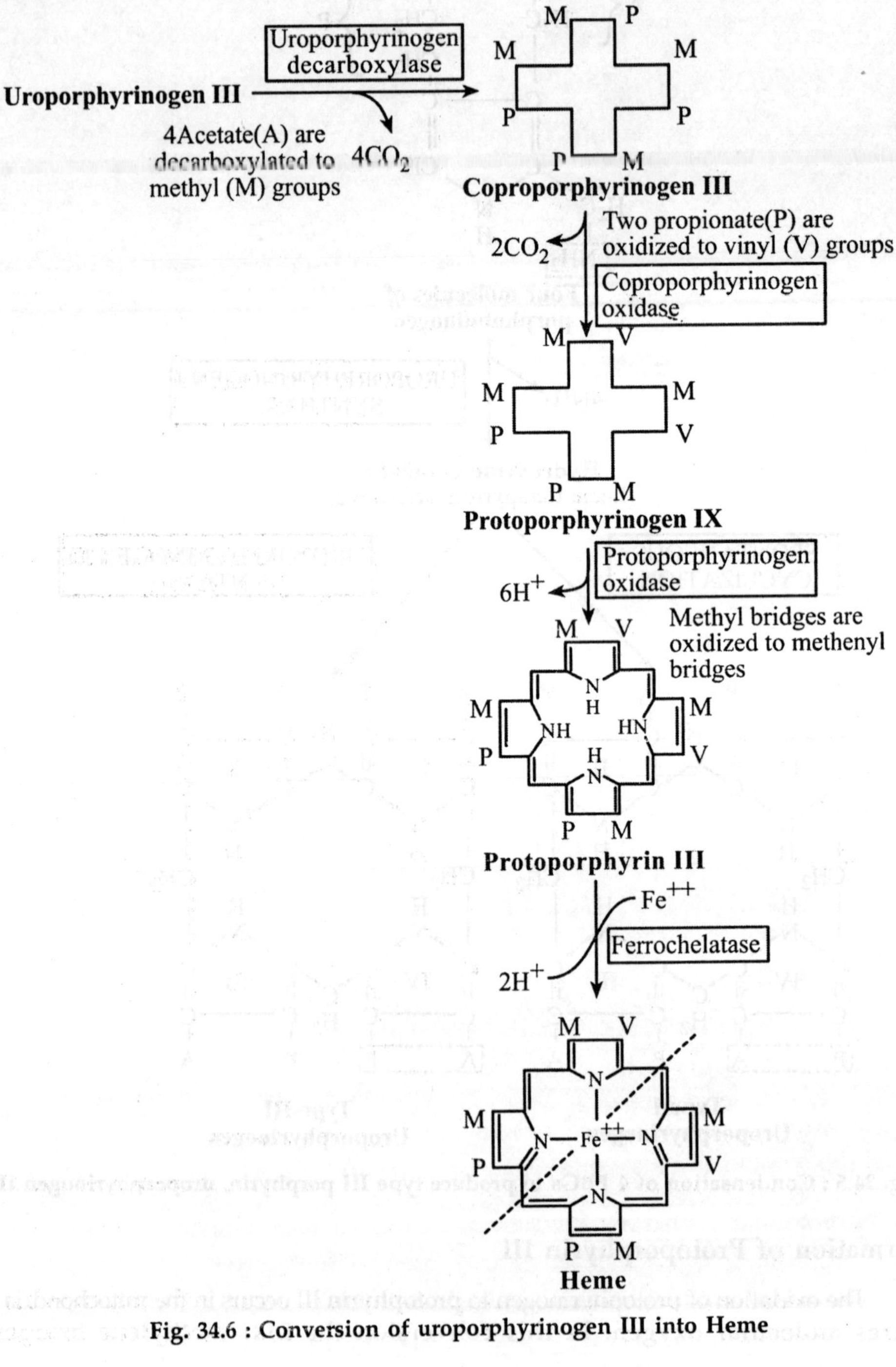

Fig. 34.6 : Conversion of uroporphyrinogen III into Heme

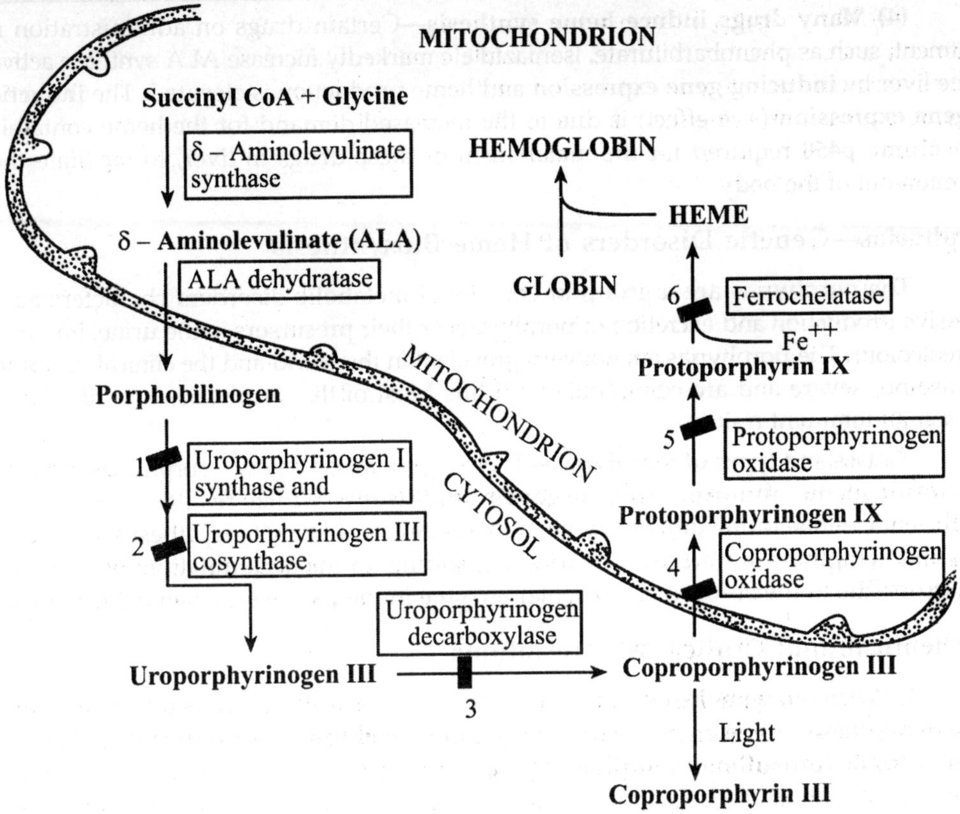

Fig. 34.7 : Overview of Heme biosynthesis occuring partly inside mitochondria and partly in cytosol (block show enzyme defects in congenital porphyrias)

Regulation of Heme Biosynthesis

The major site of regulation of heme synthesis is at the rate limiting initial step, catalyzed by **ALA synthase** and is under tight **feedback inhibition** by the accumulation of heme in excess than is needed for its utilization. The various mechanisms of the control of heme biosynthesis are described here—

(1) Repression of gene expression—ALA synthase formation is regulated at the gene level by the **repression** of gene expression due to the accumulation of heme. Heme in excess inhibits the synthesis of ALA synthase by acting as a **corepressor** and inside the nucleus, it binds with an **aporepressor** molecule (inactive) to produce a **functional repressor.** This active repressor specifically binds on DNA, at the **promoter region** with a regulatory **gene (operator) to suppress** the **transcription process,** and the biosynthesis of ALA synthase is reduced. (details in the Chapter 22).

(2) Allosteric inhibition by hematin—Excess of free heme (not bound to protein part) is oxidized to hematin, in which ferrous iron is oxidized to ferric state. Hematin allosterically inhibits ALA synthase enzyme and thus block further heme synthesis.

(3) Hematin impairs translocation of ALA synthase from cytosol to mitochondria where it is needed for the heme synthesis.

(4) Many drugs induce heme synthesis—Certain drugs on administration for treatment, such as phenobarbiturate, isoniazid etc markedly increase ALA synthase activity in the liver by **inducing gene expression** and heme production is elevated. The **induction of gene expression** (+ve effect) is due to the increased demand for the heme containing **cytochrome p450** required for the metabolism of these drugs in liver, to facilitate their excretion out of the body.

Porphyrias—Genetic Disorders of Heme Biosynthesis

The porphyrias are a group of congenital metabolic disorders characterized by excessive production and excretion of porphyrins or their precursors in the urine, imparting it a red colour. The porphyrias are not very prevalent in the world and the clinical symptoms are also not severe and are compatible with life. Most of the porphyrias are inherited as autosomal dominant trait.

Causes and types of porphyrias—There are six major types of porphyrias produced as a result of the **abnormalities in genes (mutations)** which results in the lack or insufficiency of one or the other enzyme of the pathway of heme biosynthesis. The clinical signs and symptoms of porphyrias varies, depending on the accumulation of the type of precursors due to the absence of particular enzyme in the pathway of heme biosynthesis.

Biochemical and Clinical Manifestations

1. When **enzyme lesion** due to mutations occurs **at the initial steps** (early) in the heme biosynthesis, for example insufficiency of **uroporphyrinogen I synthase** (Table 34.1), i.e. **prior to the formation of porphyrinogen,** the precursors **ALA** and **PBG** accumulate in excess. Accumulation of ALA and PBG causes **toxic** effects on **abdominal nerve** and to **CNS** resulting in severe **abdominal pain** and **neuropsychiatric** symptoms but the urine voided is not red coloured.

2. When enzyme defect occurs in the **later stages** in heme synthesis, it results in the accumulation of **porphyrinogens** and their oxidation products **porphyrins** are produced that are excreted in excess in urine, imparting a typical red colour to the urine.

Clinical Manifestations of Porphyrias

1. Photosensitivity—The phorphyrins on accumulation are deposited under skin and cause photosensitivity. The porphyrins in skin on exposure to sunlight (photons of 400nm wave length), get excited and react with molecular O_2 to generate **free radicals.** The free radicals cause injury to lysosomes resulting in the release of degradative enzymes, that cause skin damage, leading to itching and scarring of skin, skin rashes, eczema etc.

2. Neuropsychiatric symptoms—Some patients with porphyrias develop neurological manifestations, such as sensory and motor disturbances, confusion, agitation and psychiatric problems such as maniac (abnormal) behavior etc. The causes of neurological symptoms proposed are—a) deficiency of heme in neuronal cells, b) toxicity caused by excess of ALA and PBG in the brain and c) decreased tryptophan pyrrolase activity in the brain.

Historical importance—It has been speculated that King George III suffered from **variegate porphyria** (Table 34.1). which may explain some of the characteristic clinical manifestations and accounts for his abnormal maniac behavior.

A brief account of major types of porphyrias along with the biochemical and clinical manifestations is summarized in the Table 34.1.

Table 34.1 : Major types of porphyrias with biochemical and clinical manifestations

Type of porphyria	Enzyme defect	Biochemical Abnormalities	Clinical signs & symptoms
Acute intermittent porphyria	Uroporphyrinogen synthase	Excretion of ALA & Porphobilinogen. Urine darken on exposure to air	—Abdominal pain Neuropsychiatric symptoms No photosensitivity
Congenital erythropoietic porphyria	Uroporphyrinogen III cosynthetase	Increase excretion of uroporphyrinogen and uroporphyrins (red coloured urine)	–Photosensitivity with itching, burning, scarring etc. of skin
Porphyria cutanea tarda (hepatic) **most common**	Uroporphyrinogen decarboxylase	Urinary excretion of uroporphyrin I and III. Red coloured urine	—Photosensitivity, liver damage
Hepatic coproporphyria	Corproporphyrinogen oxidase	Urinary excretion of ALA, PBG & coproporphyrinogen III	—Photosensitivity Abdominal pain Neuropsychiatric symptoms.
Variegate porphyria hepatic	Protoporphyrinogen oxidase	urinary excretion of PBG, coproporphyrin and protoporphyrin increases (Red urine)	—Photosensitivity Abdominal pain Neuropsychiatric symptoms
Erythropoietic protoporphyria	Ferrochelatase	Protoporphyrins excreted in urine and feces (Red urine)	—Photosensitivity

Management of Porphyrias

Porphyrias are usually treated and managed by infusing hematin, which inhibits **ALA synthase,** the key enzyme of heme biosynthesis and thus reduces production of various **precursors** responsible for causing clinical symptoms.

Acquired porphyria (Toxic) is due to the toxicity caused by certain compounds, such as hexachlorobenzene, griseofulvin drug and due to heavy metal poisoning (lead) etc. These compounds inhibit many enzyme of heme synthesis, such as ALA dehydratase, uroporphyrin - I synthase and ferrochelatase etc. causing porphyrias.

HEME CATABOLISM

After completing their life span (120 days), the senescent erythrocytes are lysed liberating hemoglobin. The daily turnover rate of hemoglobin is 6 g. i.e. around six gram of hemoglobin is degraded per day in an adult human and same amount is resynthesized. The RBCs are destroyed by the macrophages of reticuloendothelial (RE) system of the spleen, liver and bone marrow. Hemoglobin is broken into the globin, which is proteolyzed to amino acids and heme, that is degraded to form **bile pigments,** the end **product of heme catabolism.** During catabolism of heme, the porphyrin ring is broken and iron is liberated, that enters the plasma pool to be reutilized for heme synthesis, while the broken porphyrin nucleus is converted to bile pigments, mainly into **bilirubin in humans.**

Heme Catabolism : Mechanism

Heme liberated from the degradation of various hemoproteins, is catabolized in microsomal fraction of the cells by a complex enzyme system, known as **heme oxygenase** and the process involved is as described here.

Microsomal Heme Oxygenase—This enzyme is induced by heme (substrate inducible) and requires molecular O_2 and reduced NADPH coenzyme. The component cytochrome c of electron transport chain is necessary for the activity of heme oxygenase, to regenerate reduced NADPH $+H^+$ to continue the cycle as shown in the Fig. 34.8.

Mechanism of Oxidation of Heme

During oxidation of heme, as catalyzed by heme oxygenase, one atom of O_2 is added to **hydroxylate** the **methelene bridge** between pyrrole rings I and II of the porphyrin nucleus. While the other atom of O_2 is reduced to water by reduced NADPH$+H^+$ and simultaneously ferrous iron is oxidized to ferric state as illustrated in the Fig. 34.8.

Further oxidation of hydroxylated methelene bridge between **pyrrole** rings I and II by molecular O_2 results in its **cleavage to remove it as carbon monoxide,** which is exhaled out by the lungs and porphyrin ring is broken to liberate ferric iron and a linear tetrapyrrole compound is produced, known as biliverdin IX. In birds, the end product of heme catabolism is biliverdin, a green bile pigment that is excreted in the feces. In mammals, the green biliverdin is further converted into **bilirubin** by reduction in the presence of an enzyme **biliverdin reductase,** by addition of $2H^+$ ions at the methelene bridge between pyrrole rings III and IV to form methylene group. The bilirubin so produced is a yellow coloured bile pigment and is excreted via bile into GIT.

Bilirubin is Major bile Pigment in Humans

Bilirubin, the end product of heme catabolism is a **waste product and is excreted** via bile into intestine. Approximately 35 mg of bilirubin is formed from the degradation of one gram of heme. The daily production of bilirubin in an adult human is around 250-300 mg, that is derived mainly from hemolysis of RBCs (80% from hemoglobin degradation), while some bilirubin is also produced from ineffective erythropoiesis (10%) and rest 10% from the catabolism of myoglobin and other heme containing enzymes.

Transport of Bilirubin—The bilirubin produced by heme catabolism in the reticuloendothelial system is **insoluble** in water and therefore it is transported in plasma, bound to albumin. One molecule of albumin binds two molecules of bilirubin and it appears to have one high affinity site and one low affinity site for bilirubin. In plasma, approximately 25 mg of bilirubin is tightly bound to albumin. Bilirubin in excess than this amount is bound loosely and can easily be detached and diffuses into the tissues, where it **gets deposited as yellow pigment** causing clinical condition known as **jaundice.**

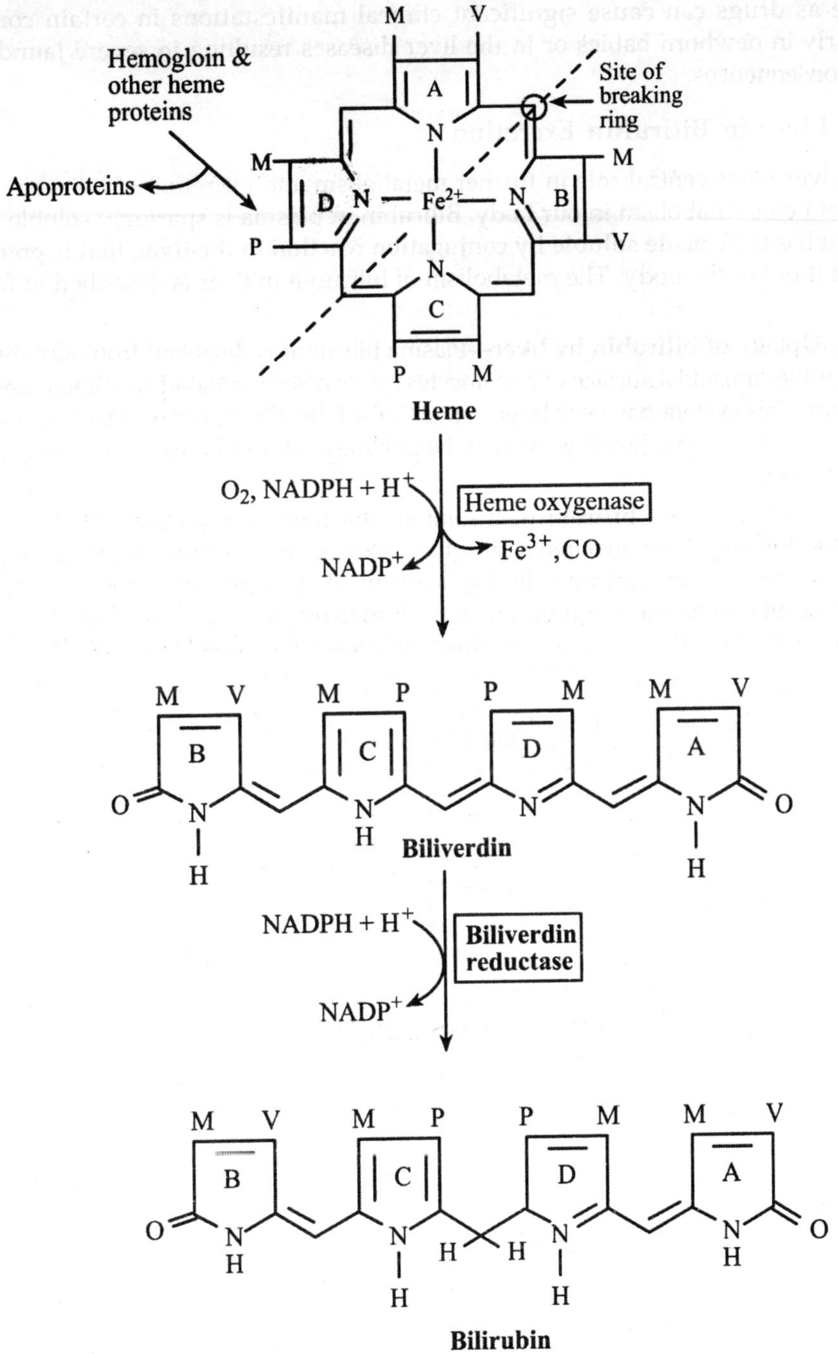

Fig. 34.8 : Degradation of heme to produce bilirubin

Certain drugs, such as aspirin, sulphonamide, penicillin etc have affinity to bind on high affinity site of albumin and thus displace bilirubin from the albumin. Therefore, their use as drugs can cause significant clinical manifestations in certain conditions, particularly in newborn babies or in the liver diseases resulting in severe Jaundice with serious consequences.

Role of Liver in Bilirubin Excretion

Liver plays central role in further metabolism and excretion of bilirubin, the end product of heme catabolism in our body. Bilirubin in plasma is sparingly soluble in water and thus it has to be **made soluble by conjugation reaction in the liver, that is prerequisite** to excrete it out of the body. The metabolism of bilirubin in liver is described in following steps—

1. Uptake of bilirubin by liver—Plasma bilirubin is detached from albumin and is taken up at the sinusoidal surface of hepatocytes by a **carrier mediated facilitated absorption mechanism**. This system has very large capacity for bilirubin transport inside the liver and thus does not become rate limiting, even in the pathological conditions associated with severe hyperbilirubinemia.

2. Conjugation of bilirubin—Insoluble bilirubin is made water soluble in liver by conjugation (adding polar groups) with glucuronic acid. The conjugation of bilirubin is catalyzed by an hepatic enzyme called **glucuronosyl transferase,** present in the smooth endoplasmic reticulum and requires active UDP-glucuronic acid. This UDP-glucuronate is derived from the hepatic uronic acid pathway of glucose metabolism (Chapter 7).

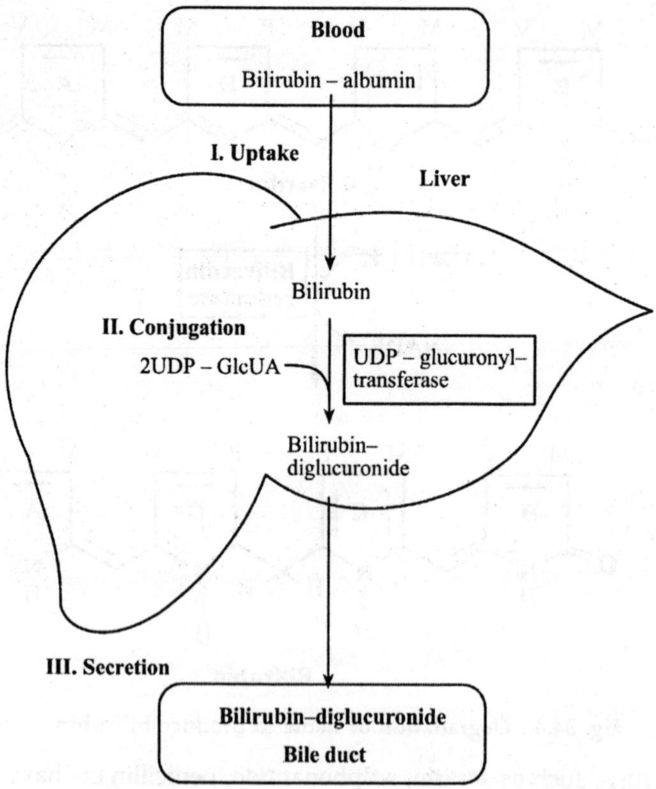

Fig. 34.9 : Conjugation of bilirubin with glucuronic acid in liver

Bilirubin is conjugated with UDP-glucuronate, first to produce a monoglucuronide and finally to bilirubin diglucuronide, that are water soluble and can easily be excreted via bile into GIT and finally in the feces. The conjugation of bilirubin takes place at the carboxyl groups of two propionyl side chain substituents of pyrrole rings III and IV with glucuronyl groups. UDP-glucuronyl transferase activity can be induced by a number of clinically useful drugs including phenobarbital.

3. Secretion of bilirubin in bile—Secretion of bilirubin diglucuronide by hepatocytes into the bile takes place by an **active transport mechanism** requiring ATP as energy source. This step seems to be **rate limiting step** in the entire process of hepatic bilirubin metabolism. In liver diseases associated with liver dysfunction, the secretion of bilirubin conjugates into bile is severely impaired. This active transport mechanism of bilirubin secretion in the bile is also **inducible** by same drugs that **induce conjugation steps of bilirubin** and thus it seems that both processes behave as a **well coordinated functional unit**. Under normal physiological conditions, almost all bilirubin secreted in bile is in conjugated form.

Fate of Bilirubin in Intestine

The bilirubin diglucuronide secreted through bile in the intestine is acted upon by intestinal bacteria to metabolize it further into various other bile pigments. It is first **deconjugated** and the glucoronide groups are removed by specific bacterial enzyme called **β-glucuronidase.** Free bilirubin so liberated is then reduced by the fecal flora to a group of colourless tetrapyrrole compounds called **urobilinogens (UBG)** by successive addition of H_2 ions through various intermediates such as mesobilirubinogen, uroblinogen and then finally to stercobilinogen as shown in the Fig. 34.10.

Approximately 20% of urobilinogen is reabsorbed via portal blood and is re-secreted in the bile (enterohepatic circulation). A small fraction of the urobilinogen from the blood is also excreted in the urine (less than 4 mg/day).

Reduction of the vinyl side chains of mesobilinogen at the pyrrole rings I and II produce urobilinogen and stercobilinogen, that are excreted finally in feces. The urobilinogen and stercobilinogen are colourless but are oxidized on exposure to air to yellow orange coloured pigments, urobilin and stercobilin respectively which impart normal colour to the feces. On long standing, the colour of the feces turns to dark brown black due to further oxidation of these pigments. The major steps of bilirubin metabolism in the liver and in intestine are summarized in the Fig. 34.10.

Plasma Bilirubin levels

Normal plasma **total bilirubin level** ranges from 0.2-0.8 mg/dl. Major fraction of total bilirubin is **unconjugated,** that is bound to albumin and normal level is 0.2-0.6 mg/dl. While **conjugated bilirubin glucuronide** level in plasma is very less normally and ranges from 0.0 to 0.2 mg/dl.

Abnormalities associated with Bilirubin Metabolism—

Hyperbilirubinemia—Clinical Syndrome Jaundice

In conditions where plasma bilirubin levels are elevated and exceeds 1 mg/dl, it is referred to as hyperbilirubinemia. When plasma bilirubin levels exceeds 2 mg/dl, all the

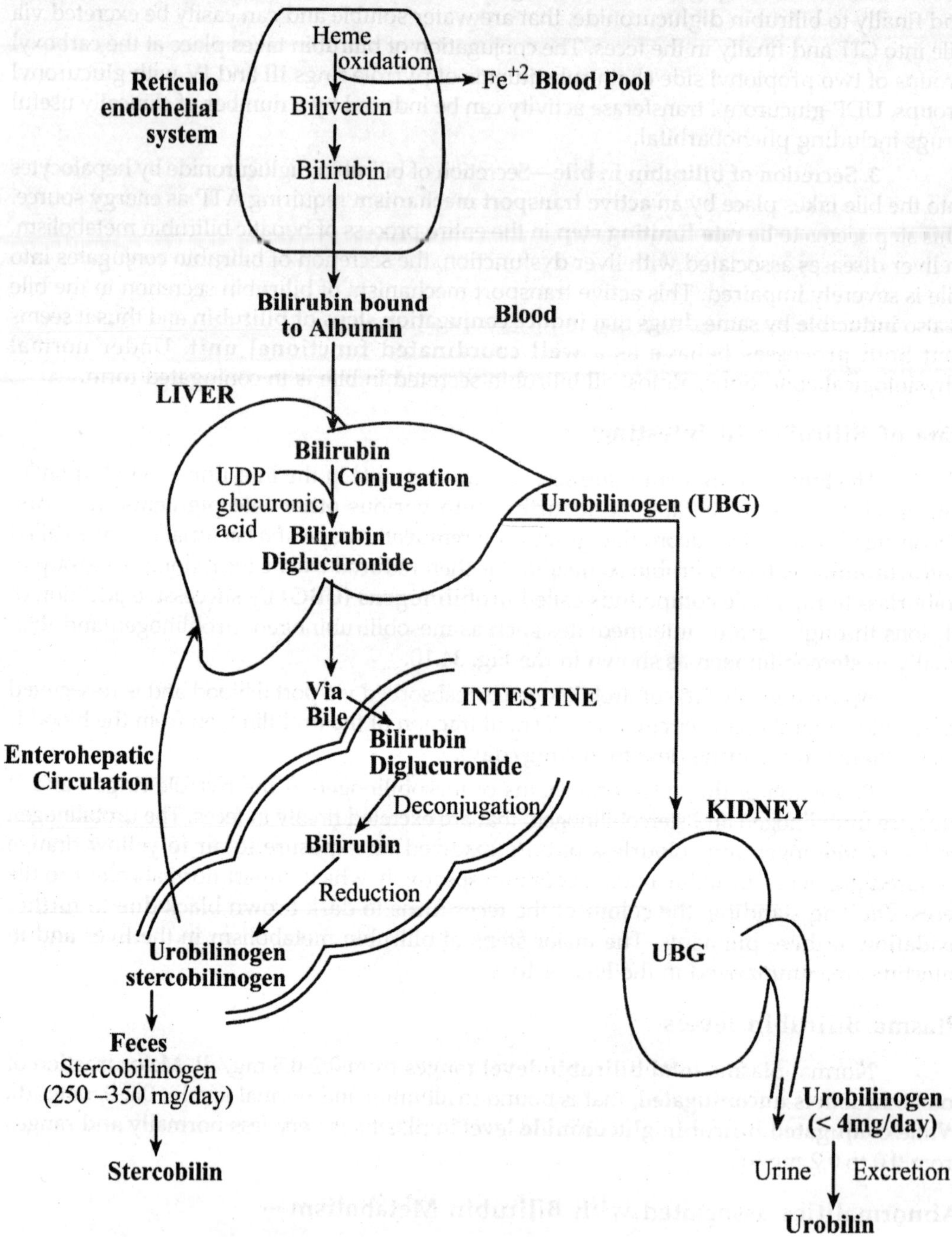

Fig. 34.10 : Diagrammatic representation of bilirubin metabolism in liver and its fate in intestine

high affinity sites on albumin becomes saturated and the extra bilirubin (loosely bound) easily gets detached and diffuses into the tissues, such as under skin, cornea of eyes, nails etc. Excess of bilirubin gets deposited in these tissues as yellow pigment and is excreted in urine, imparting it a **deep yellow colour** and the condition is known as **Jaundice** or **Icterus.**

Clinical Assay Procedures

Measurement of plasma levels of bilirubin is of great clinical significance in the diagnosis and prognosis of various types of jaundice. Van den Bergh was first to devise the test for the detection of bilirubin in urine sample by using **Ehrlich's diazo reagent.** The test is based on coupling of diazosulphanilic acid (Ehrlich reagent) with bilirubin to produce a **reddish purple** coloured **azo** compound. Conjugated bilirubin diglucuronide being water soluble gives direct Van den Bergh test positive (purple colour), without adding alcohol and is called **direct reacting bilirubin.** While the unconjugated protein bound bilirubin (insoluble in water), first has to be solubilized in **methanol**, and then it will give Van den Bergh's test positive, hence it is called as **indirect reacting bilirubin**. Most of the frequently used clinical estimation procedures for assaying plasma levels of total and conjugated bilirubin are based on Van den Bergh's test.

Causes of hyperbilirubinemia—Types of Jaundice

Elevated serum bilirubin levels can be caused by several factors and depending on the cause and the type of bilirubin elevated, the clinical syndrome jaundice is differentiated into following major types—

1. Hemolytic Jaundice—This type of jaundice occurs due to the excessive hemolysis of RBCs and **overproduction** of **bilirubin** than the amounts liver can excrete out of the body. Here the active transport mechanism of the secretion of bilirubin in bile becomes **rate limiting** step leading to hyperbilirubinemia and clinical manifestation known as hemolytic jaundice.

Causes—The hemolytic jaundice is associated with diseases with excessive rate of hemolysis of red blood cells resulting in excessive production of bilirubin and hence the name hemolytic jaundice. This type of jaundice usually occurs in the conditions, such as uncompatible blood transfusions, congenital sickle cell anaemia, glucose 6-PO_4 dehydrogenase deficiency, thalassemias and other hemolytic anaemias. Elevated bilirubin is mostly the **uncojugated indirect reacting bilirubin** in the plasma.

2. Hepatic Jaundice—In liver diseases, damage caused to hepatic tissue results in impairment of liver functions, with drastic reduction in bilirubin conjugation and its excretion capacity. The most common causes are **viral infective hepatitis, cirrhosis** of liver and damage caused by certain toxins. The characteristic feature is a biphagic Van den Bergh's test due to the increase in both conjugated as well as unconjugated bilirubin in the plasma.

3. Obstructive Jaundice—This is caused by obstruction in biliary tract, because of which the bilirubin cannot be excreted out and **regurgitates** back to liver and in the blood causing severe **hyperbilirubinemia** and **jaundice.** The obstruction in the biliary tract might be caused by gall stones, tumors etc. The elevated bilirubin is mostly **conjugated directly reacting type** in the plasma.

A summary of important **liver function tests** and clinical findings in the three types of jaundice, that facilitates the differential diagnosis also, is given in the Table 34.2.

Table 34.2 : Liver function tests in differential diagnosis of various types of jaundice

Type of jaundice	Type of bilirubin in plasma	Van den Berg's test	Enzyme assay in plasma			Urinary findings	Faecal findings
			SGOT & SGPT	Alkaline phosphatase	5′ Nucleotidase		
Hemolytic	Unconjugated Bilirubin increased	Indirect Positive	Usually normal	Normal	Normal	Urobilinogen +ve/no bilirubin	Urobilin +ve Stercobilin +ve dark brown coloured feces
Hepatic	Both conjugated & unconjugated Bilirubin eleveated	Biphasic +ve	SGPT, SGOT markedly increased	Slight increase	Slight increase	Bilirubin +ve deep yellow coloured urine	Stercobilinogen Urobilin +ve
Obstructive Jaundice	Conjugated Bilirubin increased	Direct reacting bilirubin +ve	SGPT, SGOT elevated	Markedly increased (30 to100 KA units)	Markedly increased	Bilirubin increased, deep yellow urine	Decreased/ or absent stercobilin, clay coloured stool

Hyperbilirubinemia is also observed in association with certain conditions caused by several other factors as described below—

Neonatal Jaundice—A physiological transient mild type of jaundice is observed commonly in newborn babies. It results due to increased hemolysis of RBCs in new borns coupled with undeveloped hepatic mechanisms to excrete out bilirubin. In newborn babies the UDP-glucuronyl transferase activity is low. Normally the condition disappears within 4-5 days of the birth of child.

Kernicterus—Certain cases of incompatibility between maternal and fetal blood groups (Rh -ve mother), known as **Rh incompatibility,** results in severe jaundic in newborn baby, that can lead to fatal consequences. This abnormality is usually expressed in the second pregnancy, as in first pregnancy, the Rh +ve fetus induces production of antibodies in Rh -ve mother. Therefore in the second pregnancy the Rh antibodies of mother, mostly IgG type are passed on to the fetus, which **destroy fetal RBCs** and child born is with severe hemolytic jaundice. The condition is referred to as **erythroblastosis fetalis.** The blood level of bilirubin is very high (> 20 mg/dl.) and it crosses **blood-brain barrier** which is under developed in newborns. Excess of bilirubin gets deposited in brain, especially in the basal ganglia, causing CNS damage and **encephalopathy** that can be fatal. Otherwise it may result in **spasticity** and mental retardation in the surviving infants. The condition is referred to as **kernicterus.**

Management—Kernicterus patients do not respond to usual treatment with drugs like phenobarbital but **phototherapy** does help to some extent. Blue light (400 nm wave length) isomerizes the insoluble bilirubin (ZZ type) to more soluble form (ZE & EE type), that can be excreted in urine even without conjugation and thus is beneficial in reducing severity of jaundice.

Congenital Genetic Jaundice

These are inherited metabolic disorders due to genetic abnormalities leading to congenital hyperbilirubinemia and jaundice. These genetic disorders are caused by mutations in genes resulting in insufficiency or lack of enzymes involved in conjugation of bilirubin in liver or its secretion in the bile. Some important genetic disorders are described here.

Crigler-Najjar Syndrome Type-I—It is a rare, autosomal recessive disorder caused by mutant gene, encoding hepatic enzyme β **glucuronyl transferase** that leads to defect in conjugation of bilirubin in the liver. This results in impairment of the excretion of bilirubin causing hyperbilirubinemia and sever jaundice with fatal consequences. Death occurs within five to six months of the age of infant. Unconjugated bilirubin levels are usually very high (> 20 mg/dl) and leads to **kernicterus,** with severe clinical manifestations as described above.

A benign type II Crigler - Najjar syndrome due to the mild conjugation defect, has also been reported. In this case the serum bilirubin levels are not very high and bile secreted does have some **bilirubin monoglucuronides.** This disorder is probably due to the defect in addition of second glucuronide molecule that impairs production of more soluble bilirubin diglucuronides. The children suffering from this defect respond to the treatment with **phenobarbital,** which induces both conjugation and secretion of bilirubin in bile for its excretion.

Gilbert Syndrome

This syndrome includes a heterogenous group of genetic metabolic disorders, many of which are due to compensated hemolysis that is associated with unconjugated hyper bilirubinemia. The condition is harmless, as the plasma bilirubin level is not very high and is around 3 mg/dl with mild jaundice. This genetic abnormality seems to affect at three levels of bilirubin metabolism in liver i.e. in uptake of bilirubin (carrier defect), in conjugation impairment due to enzyme lack or due to reduced hepatic secretion (transport defect). High carbohydrate diets are beneficial in decreasing bilirubin levels in the plasma.

Toxic hyperbilirubinemia—This is an acquired condition in which the unconjugated hyperbilirubinemia is caused by certain **toxin induced liver dysfunction** due to hepatocellular damage. The toxins causing hepatic damage include chloroform, carbon tetrachloride, acetaminophan and amanitus mushroom poisoning etc.

❑❑❑

35

Water, Electrolyte and pH Maintenance

The maintenance of the "**internal milieu**" (environment) of the body fluids and tissues constant, within narrow physiological limits, is of paramount importance, for the proper functioning of the life processes and to maintain good health of an individual. This **homeostasis** or maintenance of composition of the body fluids constant, is achieved by regulating distribution of compounds, water and electrolytes between various body compartments and by maintaining water, electrolyte and acid base balance of the body through various mechanisms.

A living organism has tremendous capability to maintain these homeostasis mechanisms, irrespective of wide variations in the dietary habits and in adverse environmental conditions. The maintenance of extracellular and intracellular fluid volume, as well as electrolyte and pH balance are closely interrelated and thus are considered together here.

WATER BALANCE

Water is **essential** for life and provides the aqueous medium for all biochemical reactions going on in the body. It is an excellent neucleophil and serves as reactant in several metabolic reactions. Water is the principal end product of biological oxidation of nutrients for the generation of energy to carry on life processes.

The maintenance of fluid volume and electrolyte balance within narrow limits is essential and any alteration between intra and extracellular fluid volume and electrolyte balance would adversely affect cellular functions and health of an individual.

Regulation of water balance and **plasma volume** is a complex process and is achieved primarily by following mechanisms—

1. **Hypothalamic mechanisms** to control the fluid intake or thirst of an individual according to the need of the body.
2. **Antidiuretic hormone** that regulates retention of water by kidneys or to excrete water depending on the osmolality status of plasma.
3. The states of water depletion and of excess of body water are quite common and in most of the cases, they are associated with **sodium** and other **electrolyte imbalances,** which finally are regulated by hormones, such as aldosterone.

Disturbances in plasma volume may be caused by following conditions.

1. Depletion of water or decreased fluid intake
2. Increased loss of water due to vomiting, diarrhea etc.
3. Excessive water loss by skin as profuse perspiration during hot climate
4. Water loss by lungs due to excessive evaporation during respiration in hot weather.
5. Renal water loss due to excessive urination (polyuria) in diseases such as diabetes mellitus, diabetes incipidus etc.

Water balance is normally maintained by controlling the amount of daily water intake and regulating daily water excretion by various homeostasis mechanisms, as described above and are summarized in the Table 35.1.

Table 35.1 : Maintenance of water balance by regulating daily water gain and losses

Water intake /day	Average	Water loss / day through	Average
Water content of food stuffs.	1200ml	Urine	1500ml
Water produced by oxidation	300ml	Skin	500ml
Drinking water / fluid intake	1200ml	Lungs	700ml
		Feces	500ml
Total	2700ml	Total	2700ml

ELECTROLYTE BALANCE

Electrolytes are readily dissociable positively and negatively charged ions and their concentration in body fluids is expressed as milliequivalents per liter (m. Eq./L), by taking into account their atomic weight and valencies as given below —

$$\text{m. Eq./L} = \text{mg per liter} \times \text{valency}$$

Electrolyte Composition of Body Fluids

Electrolytes are appropriately distributed between **intracellular fluid (ICF)** and **extracellular fluids (ICF)** to maintain their osmolality and hence their **fluid volume is maintained.**

Osmolality of body fluids—Osmolality of a fluid is due to the concentration of electrolytes present in it and is the major determinant for regulating fluid volume and water balance between intra and extracellular compartments. These two compartments are separated by a semipermeable cell membrane. Crystalloids and water can diffuse across the membrane freely, but if an **impermeable, nondiffusible colloidal ion,** such as **protein** is present in one compartment, the diffusion of permeable ions becomes **unequal** and this is known as **Donnan's effect.**

An osmotic gradient develops between ECF and ICF due to the osmotic pressure exerted by the nondiffusible protein particle (ion) on one side of the membrane. This colloidal osmotic pressure exerted by the proteins is the major factor, that maintains the intracellular and extracellular fluid volumes and water balance. For instance, in cases of hypoproteinemia i.e. low plasma albumin level, the osmotic pressure of plasma is reduced and the water extravasate and accumulates in the interstitial space of cells of the tissues, causing **edema.** Normally the fluid of blood vessels is retained within the vascular compartments due to the osmotic gradient exerted by the plasma proteins, primarily by albumin, because of its high concentration and low molecular size as compared to the other plasma proteins.

Composition of Body Fluids

The composition of intracellular (ICF) and extracellular fluid (ECF) compartments differ from each other due to the difference in the composition of electrolytes and other solute particles and this is achieved primarily by Donnan's effect as mentioned above and by various mechanisms described below. This difference in composition of ICF and ECF is necessary for carrying out various physiological and biochemical processes in a living cell. The electrolyte composition of body fluids is summarized in the Table 35.2.

Table 35.2 : Electrolyte composition of body fluids in m. Eq./L

Anion/cation	Plasma	ECF	ICF
Na^+	140	146	12
K^+	4	5	160
Ca^{++}	5	3	-
Mg^{++}	1.5	1	34
Chloride	105	117	2
Bicarbonate	24	27	10
Phosphate	2	2	140
Protein	15	8	55
Other anions	5	6	-

The electrolyte compositions of ICF and ECF is maintained by various **active transport mechanisms,** that utilize energy as ATP, for instance Na^+/K^+ pump and through various carriers, hormones etc. These carriers and hormones facilitate transport of substances across biomembranes according to the need of the body tissues.

As evident from the composition summarized in the Table 35.2, the sodium and calcium are the major cations of ECF and plasma, while potassium and magnesium are principal cations of ICF. Where as chloride is the major anion of ECF and plasma and the phosphates constitute major anion of ICF. In each fluid compartment, the electrical neutrality is maintained essentially, inspite of differences in the electrolyte composition.

Osmolality of Plasma

Since osmolality of a fluid compartment is dependent on the number of solute particles present, the major determinant factor is sodium concentration in ECF. The sodium concentration and water balance are dependent on each other. Osmolality of plasma can be measured directly using osmometer or indirectly as the concentration of effective osmoles as described below.

The osmolality of plasma is normally maintained in the range of 285-295 milliosmols/Kg and since plasma Na^+ is predominant contributor to osmolality, it can be calculated as plasma osmolality in m. osmol/kg. = 2× plasma Na^+ (m.mol./kg)

Osmotic equilibrium—In health state, the osmotic pressure of ECF and ICF is equal and **osmolality** of both fluid compartment is **identical**. As such there is no net passage of water across membranes and ICF and ECF water balance is maintained.

Regulation of Sodium and Plasma Volume

The major regulatory factors that maintain sodium and water balance are hormones, namely **aldosterone, antidiuretic hormone** (ADH) and **renin-angiotensin system**. By regulating sodium and water balance, these three hormones play major role in regulation of plasma volume and blood pressure.

Aldosterone is the main mineralocorticoid hormone, secreted form the adrenal cortex. It increases sodium reabsorption from the glomerular filtrate by renal tubules in exchange of K^+ causing sodium retention, according to the need of the body.

Antidiuretic hormone (ADH)—When osmolality of plasma rises (due to Na^+ retention), it is sensed by the **osmoreceptors** of the hypothalamus and ADH secretion is stimulated. ADH effectively increases water reabsorption by renal tubules (DCT and collecting tubules), which otherwise are impermeable to water. Thus by simultaneous actions of aldosterone and ADH, proportionate amounts of Na^+ and water are retained to maintain osmolality of plasma and thus its volume is also maintained to regulate blood pressure.

In the case of decreased osmolality, the plasma volume expands (hemodilution) and the secretion of ADH is reduced by **feed back inhibition**. This results in excretion of more water in urine and water balance is restored.

Renin-Angiotensin System

The regulation of the plasma volume, sodium concentration and blood pressure is simulataneously achieved by well controlled and well coordinated mechanisms, through **renal renin-angiotensin** system, as summarized in Fig. 35.1 and is described below.

When there is a fall in plasma volume (ECF volume), the renal plasma flow is decreased causing stimulation of secretion of **renin** by **juxtaglomerular** cells of kidneys. In addition to renal plasma flow, other factors which stimulate renin secretion are: a) decreased blood pressure, b) sodium depletion, c) prostaglandin secretion and d) beta adrenergic antagonists.

The renin-angiotensin system on **activation** is a major stimulus for the aldosterone secretion from the adrenal cortex, in the cases of low plasma sodium level and reduced plasma volume. The renin is a **potent enzyme** and acts on angiotensinogen. Angiotensinogen is synthesized in liver, and is a large globulin protein with 400 amino acids. Renin specifically cleaves and splits off a small decapeptide called **angiotensin** I (with 10 amino acids) from angiotensinogen which undergoes further proteolysis by a **converting enzyme** to remove a dipeptide (2 amino acid) leaving behind a potent **octapeptide** called **angiotensin** II as illustrated in the Fig. 35.1.

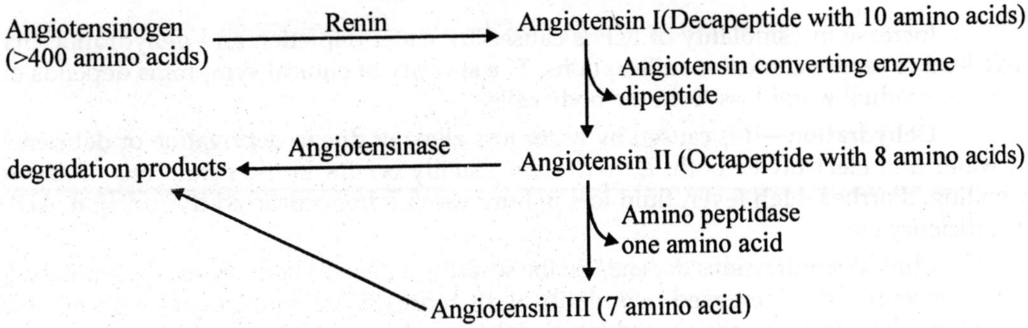

Fig. 35.1 : Formation and metabolism of angiotensins

Functions of Angiotensin II

Angiotensin I is an inactive peptide in humans, while the angiotensin II and III are active, but the plasma levels of the angiotensin II is 4 times more than angiotensin III. Thus quantitatively angiotensin II is the most active peptide hormone involved in the regulation of plasma fluid volume, osmolality and blood pressure.

The important functions of angiotensin II include—(a) It increases blood pressure by causing vasoconstriction of the arterioles, (b) It stimulates aldosterone secretion from the adrenal cortex, that enhances sodium reabsorption by renal tubules along with water (obligatory reabsorption) and thus restores normal plasma volume and osmolality and the blood pressure is regulated.

Angiotensin II also inhibits further release of renin. The events leading to co-ordinated control and the regulation of plasma volume, sodium level and blood pressure are summarized in Fig. 35.2 (details in the Chapter 30).

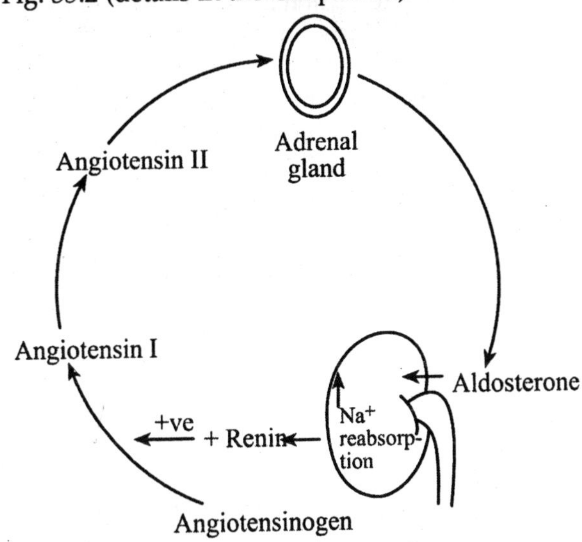

Fig. 35.2 : Diagrammatic representation of coordinated regulation of plasma volume, sodium level and blood pressure by various factors

Disturbances in Fluid and Electrolyte Balance

Increase in osmolality of ICF is caused by water depletion and dehydration that may lead to severe clinical manifestations. The severity of clinical symptoms depends on acute or gradual water loss from the body cells.

Dehydration—It is caused by water loss alone or due to deprivation or deficiency of water and electrolytes both. Dehydration usually occurs in the conditions of severe vomiting, diarrhea, high fever, fluid loss in burn cases, adrenocorticoid dysfunction, ADH insufficiency etc.

Clinical manifestions depends on the severity of dehydration. The symptoms include hemoconcentration, increased osmolality with reduced ICF and plasma volume and shrunken (dehydrated) cells. A sudden shrinkage of brain cells may result in decreased intracranial pressure leading to hemorrhages in the brain. Depression of mental state, seizures, tremors, muscular rigidity and spasticity are observed in the dehydration and in severe cases, patient may become comatose and it can be life threatening.

ACID BASE BALANCE

The maintenance of acid base balance or pH of body fluid is one of the most important homeostasis of the body and is effectively achieved by minutely regulated mechanisms.

In normal circumstances the pH of the body fluids is maintained within narrow physiological limits of 7.35 to 7.45. The **blood pH homeostasis** is highly efficient and is not allowed to alter much. This is essential, since any change in pH would severely affect metabolic processes resulting in impairment of functioning of the body organs. This is because pH alterations distort and denaturate protein molecular structure, causing loss of biological activity of proteins of the body. For instance the enzyme catalysis is reduced, muscle contractility is impaired, CNS disfuctioning etc. leading to severe clinical manifestations.

Acidosis and Alkalosis—A reduction in blood pH below 7.35 is known as **acidosis** and the pH above 7.45 is called as **alkalosis.** Severe reduction in pH (<7.25) and unmanaged **acidosis can be life threatening** and patient may go into coma and finally results in death. A living organism can **tolerate** and may **survive** the pH alteration of body fluids within narrow range of 7.0 to 7.6 only, beyond which life is not possible.

Normally the maintenance of pH of blood is efficiently achieved and is minutely regulated by three major mechanisms, which involve: 1) Efficeint body buffer systems, 2) the lungs and respiratory system and 3) The kidneys and renal mechanisms. Before discussing the acid base homeostasis, it is important to consider some fundamental concepts.

Acids and Bases

Acids are the compounds, that are capable of donating hydrogen ions (H^+), while bases are the compounds that accept H^+ or protons. For example

Acids	**Bases**
$HCl \longrightarrow H^+ + Cl^-$	$NH_3 + H^+ \longrightarrow NH_4^+$
$H_2CO_3 \longrightarrow H^+ + HCO_3^-$	$HCO_3^- + H^+ \longrightarrow H_2CO_3$

Depending upon the degree of ionizations, the acids can be grouped as—a) **Strong acids** which ionize rapidly and completely, for example mineral acids, such as hydrochloric and sulfuric acids. b) **Weak acids** are those which ionize slowly and incompletely, for instance organic acids, such as carbonic and lactic acid. **Dissociation constant** (Ka) of an acid can be determined by the formula as given under —

$$Ka = \frac{[H^+][A^-]}{[HA]}$$

Where $[H^+]$ is concentration of H^+ ions, $[A^-]$ is the concentration of anion or conjugated base and [HA] is the concentration of undissociated acid. The acidity of a solution is determined by calculating $[H^+]$ concentration as follows.

$$[H+] = Ka\frac{[HA]}{[A^-]}$$

Concept of pH

Sorensen in 1909 introduced the **concept of pH** for the sake of convenience and practical purposes by expressing H^+ concentration as negative log and designed it as pH. In other words, pH of a solution is inversely proportional to $[H^+]$ and it can be expressed as:

$$pH = -\log[H^+] \text{ or } pH = \log\frac{1}{[H^+]}$$

It is evident from the above formula that pH value of a solution is **inversely proportional** to the $[H^+]$ concentration. Thus higher the $[H^+]$ or acidity, lower will be the pH of the solution. At the pH value 7 the medium is **neutral,** i.e. it is neigher acidic nor alkaline. The total pH range of the medium is 1- 14 and is depicted as follows:

$$\text{pH range} = 1 \xrightarrow{\text{Acidic pH}} \underset{\text{Neutral}}{7} \xrightarrow{\text{alkaline pH}} 14$$

Determination of pH

The pII of a solution can be measured directly by determining voltage difference between a glass electrode and a calomel (reference) electrode in a instrument called pH meter.

Henderson-Hasselbalch Equation

The relationship between pH, pKa and the concentrations of an acid and its conjugated base can well be expressed by Henderson-Hasselbalch equation as given below. This equation also explains the effect of salt upon the dissociation constant of the acids.

$$pH = pKa + \log\frac{[base]}{[acid]} \quad \text{OR} \quad pH = pKa + \log\frac{[salt]}{[acid]}$$

The formula shows that pH and pKa will have same values when concentrations of acid and base or salt are equal or when acid is half ionized in the medium.

Buffers

The buffers are the system that resists any change in pH of the medium on addition of an acid or alkalies. Walter and Nernst were first to introduce the concept of buffers in biological reaction medium, that are required essentially for pH maintenance and they were awarded Noble prize in the year 1920.

A Buffer system usually contains a mixture of weakly ionizable acid and its salt with a strong base or a mixture of weak base and its salt with strong acid. For instance—

1. **Bicarbonate buffer** ($H_2CO_3/NaHCO_3$) consists of a weak acid, carbonic acid and its salt with strong base i.e. sodium bicarbonate.

2. **Phosphate buffer** (Na_2HPO_4/NaH_2PO_4), consists of disodium hydrogenphosphate (basic salt) with sodium dihydrogen phosphate (acidic salt)

Buffering Capacity—Efficiency of the buffer system is determined by the capacity or the ability of the buffer to resists the change in pH on adding acids or alkalies.

For example, when hydrochloric acid is added to the bicarbonate buffer, it is immediately neutralized by its component salt, forming a weak acid (carbonic acid) which do not ionizes much or contributes to H^+ ion concentration. So pH of medium is not affected much. The reaction involved is as given below:

$$NaHCO_3 + HCl \longrightarrow H_2CO_3 + NaCl.$$

Similarly when an alkali (NaOH) is added to the bicarbonate buffer, it immediately is neutralized by its component carbonic acid, without altering pH of the medium as shown below:

$$H_2CO_3 + NaOH \longrightarrow NaHCO_3 + H_2O$$

The pH of a buffer solution is affected by certain factors, such as—a) The value of pKa i.e. lower the pKa, the lower will be pH of the buffer solution and b) **The ratio of salt** (the conjugated base) to that of **the acid** concentration.

Buffering capacity, as mentioned above depends very much upon the ratio of acid to its salt concentration and higher the ratio of salt (**alkali reserve**) to that of the acid, more will be the efficiency of the buffer system.

The Henderson-Hasselbalch equation is important in determining the buffering capacity, or the efficiency of a buffer system, as it can depict the ratio of base (alkali reserve) to the acid. According to this equation, a **buffer is most effective** when—a) pH is equal to pKa and 2) when concentration of salt and acid are equal and ratio is 1 : 1. By measuring the pH of the solution, the concentration of the salt or acid and their ratio can be found out by this equation. Therefore the Henderson-Hasselbalch equation has great practical application in clinical practice, for assessing the acid base status and knowing the **compensation limits of** the blood and body buffers, to meet an acid or alkali challenge.

Production of Acids during Metabolism

In the body during various metabolic processes, large quantities of acids are produced in tissues and are being added to blood, that would tend to lower the pH of the blood. These acids have to be effectively buffered by the body buffers, in order to maintain the pH of blood in the physiological range of 7.35 to 7.45. This is essential to prevent severe clinical manifestations due to the adverse effects of acid base disturbances as mentioned earlier.

The acids produced during metabolism may be volatile in nature or nonvolatile, fixed acids. The most predominant acid produced is the **volatile acid,** carbonic acid. Nearly 20000 m.Eq. of carbonic acid are produced per day during metabolism. The non volatile or the fixed acids, such as lactic acid, keto acids, sulphuric and phosphoric acids etc. are produced in lesser amounts (60-80 m.Eq.) per day during various metabolic processes.

Production of alkalies during metabolism is normally **negligible** in the body. Small amounts of bicarbonate may be produced during metabolism of the organic acids. The variation in diet has a tendency of a net production of some bases and thus has an alkalizing effect, while a diet rich in animal proteins may result is more acid production.

MAINTENANCE OF BLOOD pH

In normal circumstances, the acids or alkalies produced during metabolism are effectively buffered by buffer systems of the body and are **eliminated** out of the body to maintain blood pH efficiently. Body has three major mechanisms to regulate the acid base balance and pH of the body fluids—1) the blood buffers, 2) respiratory system to eliminate CO_2 and 3) the renal mechanisms to finally eliminate acids or alkalies out of the body.

Blood Buffers

The buffer system of the body fluids is the **first line of defense** against acid or base challenge and effectively resists any alteration in the pH of blood, as long as the acid load is **not excessive** and **the alkali reserve** of body is not completely **exhausted.** Once the acid is neutralized by the existing buffer base (alkali reserve), the base has to be replenished to meet the further acid load to maintain blood pH.

Buffer Systems of Blood

The **extracellular** buffers of blood includes—1) The bicarbonate buffer, 2) phosphate buffer and 3) the protein buffer. While the **intracellular buffers** are—1) Hemoglobin buffer of erythrocytes and the by carbonate, phosphate and protein buffers of intracellular fluid.

Bicarbonate Buffer System

The most efficient and effective buffer of the plasma is the **bicarbonate buffer system.** It is quantitatively the most significant and predominant buffer of the body fluids. It is an open buffer system, as its components- bicarbonate and carbonic acid are freely diffusible across cell membranes and the vascular epithelium.

When the Henderson - Hasselbalch equation is applied to bicarbonate buffer system at pH of blood 7.4, it is observed that the ratio of bicarbonate to carbonic acid was 20 : 1 under normal conditions. The normal bicarbonate level of plasma varies from 22-26 m mol/ L with an average of 24 m. mol and the level of carbonic acid is 1.2 m. mol/L while the pKa of carbonic acid is 6.1. By substituting these values in Henderson-Hasselbalch equation, the pH of blood obtained is 7.4 which is the actual pH of the blood.

$$pH = pKa + \log \frac{[HCO_3^-]}{[H_2CO_3]}$$

$$\begin{aligned} 7.4 &= 6.1 + \log 24/1.2 \\ &= 6.1 + \log 20 \\ &= 6.1 + 1.3 \\ &= 7.4 \end{aligned}$$

Thus it is evident from the above equation, that the ratio of bicarbonate to carbonic acid at the pH of blood (7.4) is 20 under normal conditions. This is much higher than the theoretical value of 1, that ensures maximum buffering capacity. Thus HCO_3^- ratio of 20 as compared to carbonic acid represent high **alkali reserve** of this buffer system. This is highly beneficial to meet an appreciable acid load produced in the body during metabolism. The bicarbonate buffer system of blood accounts for 40-50% of total buffering capacity of the body buffers.

Phosphate Buffer System

The phosphate buffer is primarily an intracellular buffer, as its concentration in the plasma is very low and thus is less important in plasma. Its pKa value is 6.8, which is very near to the pH of blood i.e. 7.4, so it would have been most effective buffer in plasma if its concentration was more.

By applying Henderson-Hasselbalch equation to phosphate buffer by substiting pKa value of 6.8 —

$$pH = pKa + \log \frac{[Na_2HPO_4]}{[NaH_2PO_4]}$$

$$7.4 = 6.8 + \log \frac{[Na_2HPO_4]}{[NaH_2PO_4]}$$

$$7.4 - 6.8 = \log \frac{[salt]}{[acid]}$$

$$0.6 = \log \frac{[salt]}{[acid]}$$

Anti log of 0.6 is 4, hence the ratio of phosphate buffer is 4 : 1, which is found to be the actual ratio of its components i.e. the basic Na_2HPO_4 to the acidic NaH_2PO_4 of phosphate buffer in the body fluids.

Protein buffer system

The buffering capacity of the plasma protein buffer depends on the pKa value of its ionizable side chains of amino acid residues. The most effective ionizable group is the imidozole N_2 of histidine with a pKa value of 6.1. The protein buffer system of the plasma is effective to some extent and accounts for 1-2% of the total buffering capacity of plasma.

Hemoglobin Buffer System

The intracellular buffers are also quantitatively very significant and account for 50-60% of total buffering capacity of the blood. The role of hemoglobin buffer is described in detail in Chapter 33 and shall be considered here along with the respiratory regulation of pH. The hemoglobin buffer plays highly significant role in regulation of blood pH, as along with the transportation of CO_2 from tissues to lungs, it also transports protons (H^+) and help in their elimination. (Chapter 33).

The buffer systems immediately neutralize the acids or bases produced during metabolism and thus resist any significant change in pH of body fluids. But for **final elimination** of acids and for the **replenishment of the alkali reserves**, the respiratory and renal mechanisms are involved, which play highly significant and central role, that is essential for the maintenance of the acid base balance of the body.

Respiratory Regulation of pH

Respiratory mechanisms provide rapid regulation of pH and acid base balance and are body's **second line of defense** against acid or alkali challenge. The respiratory regulation of pH is achieved by eliminating CO_2 in exhaled air at a rate, required to keep the bicarbonate-carbonic acid ratio at 20 : 1 by regulating the rate of respiration.

During metabolism of nutrients and biological oxidation to generate ATP, large quantities of CO_2 are produced in the tissues, where it is trapped as carbonic acid in the presence of enzyme **carbonic anhydrase** by the reaction as given under—

$$CO_2 + H_2O \xrightarrow{\text{Carbonic anhydrase}} H_2CO_3$$

The carbonic acid so formed would tend to lower ICF pH and would disturb the acid base equilibrium of the blood. Therefore in normal circumstances, carbonic acid is buffered by the blood buffers and is then brought to lungs, where it is eliminated out as CO_2 by **increasing the rate of respiration** i.e. **hyperventilation** as described here under.

The rate of respiration is controlled by the **chemoreceptors** present in the medulla of the brain, that are highly sensitive to any change in the pH of blood. Thus any fall in blood pH will stimulate chemoreceptors to increase the respiratory rate and cause hyperventilation to excrete out excess of CO_2. This would decrease carbonic acid level of the plasma and simultaneously the H^+ ions are eliminated as water. This results in restoration of bicarbonate-carbonic acid ratio to 20 : 1 to maintain blood pH. (details in Chapter 33).

Hemoglobin buffer is very important and acts concomitantly along with the respiratory mechanisms to transport and eliminate CO_2 and H^+ ions. It acts as an efficient buffer and help in the transport of CO_2 as bicarbonate ions and protons as HHb with minimum change in the pH of blood. Simultaneously to this, hemoglobin facilitates generation of HCO_3^- to **replenish the alkali reserve** of plasma as illustrated in the Fig. 35.3 (details in Chapter 33).

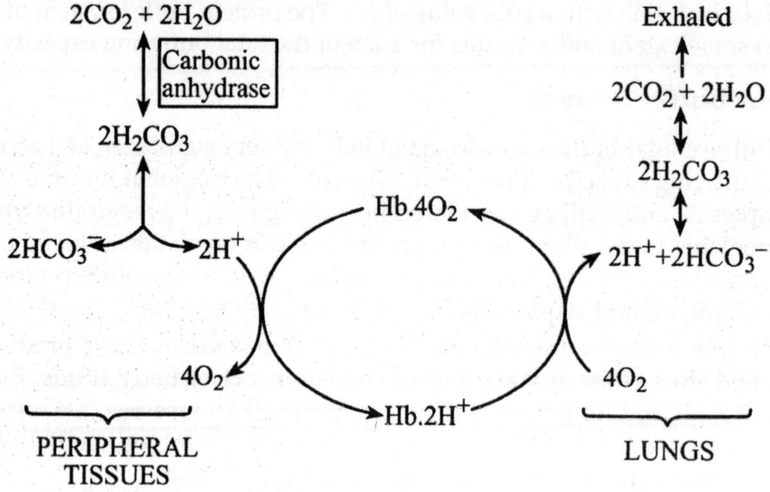

Fig. 35.3 : Role of hemoglobin as potent buffer to eleminate protons liberated in tissues

Mechanism

As evident from the Fig. 35.3, CO_2 from the tissue comes in plasma and diffuses into RBCs, where it combines with water to form carbonic acid, in the presence of enzyme **carbonic anhydrase.** In RBCs, the carbonic acid dissociates into H^+ and HCO_3^-. The H^+ ions are buffered by hemoglobin to from HHb, while the bicarbonate ions diffuse into plasma in exchange with chloride ions (chloride shift) to maintain electric neutrality. The net result is replenishment of bicarbonate ions into plasma to keep bicarbonate-carbonic acid ratio to 20 to regulate blood pH at 7.4, inspite of excess of CO_2 being produced during metabolism in the tissues and is added to the plasma.

The reverse occurs in the lungs during oxygenation of hemoglobin and CO_2 is excreted in excess by increasing the rate of respiration as described earlier (details in Chapter 33).

The respiratory regulation of blood pH is **immediate and rapid** and is achieved by increasing or depressing respiratory rate, in response to acidosis or alkalosis respectively. But respiratory mechanism is a short time regulation, as hyperventilation cannot proceed for long.

RENAL REGULATION OF pH

Kidneys play highly significant role in maintenance of acid base balance and pH of the body fluids. Kidneys are the primary organs that **finally** excrete out excess of acids or alkalies in urine to regulate blood pH and the acid base equilibrium of the body is maintained. Kidneys provide the only route through which H^+ ions and acids can be excreted out from the body.

During metabolic processes in body, large quantities of acids are produced, that diffuse into the blood, efficiently buffered by the buffer system and are brought to kidneys.

The Kidneys excrete H^+ ions (acidification of urine) and simultaneously reabsorb the HCO_3^- ions from the glomerular filtrate to recover and replenish **alkali reserve** of the blood to regulate acid base equilibrium and blood pH.

Urinary pH is lower than ECF

Normally the pH of urine is acidic, average being 6.0, which is lower than pH of the blood (7.4). This indicates that kidneys excrete H^+ ions and a net acidification of urine occurs in renal tubules. One of the major functions of kidneys is to eliminate and excrete the acids produced during metabolic processes of the body to maintain the acid base balance.

Normal pH of urine varies and ranges between 4.8 to as high as 9.8 depending on the amount of acids or alkalies produced during the metabolic processes and being excerted. However in normal circumstances, the pH of urine is acidic with an average pH of 6.0.

Renal Mechanisms to Maintain Blood pH

The major renal mechanisms by which kidneys excrete out acid load of the body to maintain acid base equilibrium and pH of body fluids are described here.

1. Excretion of H^+ ions—The proximate convoluted tubules of nephrons secrete H^+ ions and excrete acid load of the body and along with it, bicarbonate (HCO_3^-) ions generated in tubular cells are added to plasma to replenish the alkali reserve to meet further acid load.

Mechanism of H^+ secretion—In the renal tubular cells, the enzyme carbonic anhydrase (CA) catalyzes condensation of CO_2 and H_2O to produce carbonic acid, that dissociates into H^+ and HCO_3^- ion. The H^+ so produced are excreted into tubular lumen in exchange of Na^+ ions (important cation of ECF), which is reabsorbed from the glomerular filtrate. The sodium ions, then along with HCO_3^- formed in the tubular cells, are added back into plasma as illustrated in the Fig. 35.4 to replenish the alkali reserve of the plasma.

This renal mechanism is very effective in regulating blood pH, especially when excess of acid load has to be eliminated. This mechanism serves to excrete acids with net excretion of H^+ ions (acid) and reabsorption of Na^+ and HCO_3^- ions into the blood to replenish the alkali reserve of plasma.

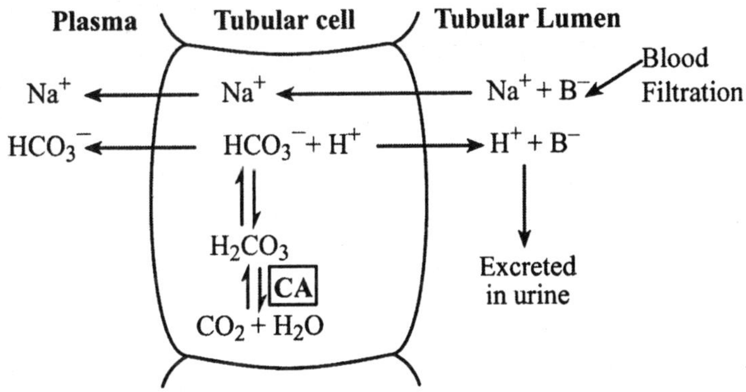

Fig. 35.4 : Excretion of H^+ in exchange to Na^+ in the urine and replenishment of plasma HCO_3^- by renal tubules

2. Reabsorption of sodium bicarbonate—This renal mechanism is primarily for the reabsorption of sodium bicarbonate of the plasma, filtered in the glomerular filtrate to

recover the alkali reserve i.e. bicarbonate ion back in plasma. This ensures **no net loss and excretion** of sodium bicarbonate in the urine, so as to maintain bicarbonate/carbonic acid ratio as 20 : 1 to regulate blood pH. In this renal mechanism, there is no net H^+ excretion and is primarily for recovering the HCO_3^- as shown in the Fig. 35.5.

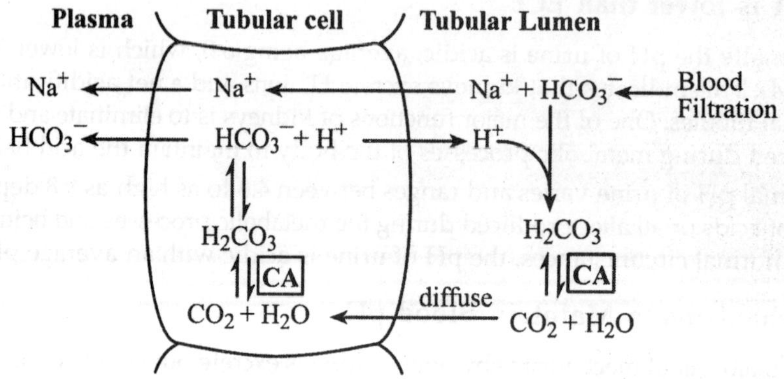

Fig. 35.5 : Reabsorption of bicarbonate and sodium from glomerular filtrate by renal tubules

In this renal mechanism the bicarbonate ions filtered by glomerulus is almost **completely** reabsorbed by renal tubules and the urine is normally free from bicarbonate ions. As evident from the Fig. 35.5, in the renal tubular lumen, the H^+ secreted by the tubular cells combine with bicarbonate, filtered from the blood in glomerulus to form carbonic acid, that is degraded to CO_2 and water by the enzyme carbonic anhydrase of tubular cell membranes. As the CO_2 concentration builds up in the lumen, it diffuses into the tubular cells and form H_2CO_3 to continue the cycle as shown in the Fig. 35.5.

3. Excretion of titratable acids in urine—The acidic pH of the urine (pH<6.0) is due to the excretion of titratable acids and secretion of H^+ ions in urine. The **term titratable acid** of urine is the measure of net acids excreted in urine and is defined as the number of milliliters of N/10 NaOH required to neutralize the acid contents of one liter of urine to pH 7.4.

The major titratable acid of urine is due to the excretion of **acidic sodium dihydrogen phosphate** that lowers the pH of urine to approximately 6.0. Initially the pH of the glomerular filtrate in proximate convoluted tubular lumen is 7.4 due to the presence of **basic disodium hydrogen phosphate,** filtered from the blood. As the tubular fluid passes down the tubules, more and more H^+ ions are secreted by tubular cells to eliminate the acid load of the body, in exchange to Na^+ ions to form the **acidic NaH_2PO_4** and the pH of fluid steadily falls. As the tubular fluid reaches to distal convoluted tubules, urine becomes acidic with pH 4.8, due to complete conversation of basic phosphates to acidic salts (Fig. 35.6).

During the excretion of titratable acids, the renal tubules excrete H^+ ions in exchange to Na^+ from disodium hydrogen phosphate (basic) to convert it to the acidic sodium dihydrogen phosphate (pH<5.0) by the renal mechanism as illustrated in the Fig. 35.6.

The maximum limit of acidification of urine is to pH 4.5 at which all the phosphates of urine are in form of acidic NaH_2PO_4 and further acidification would lead to sodium depletion and normal ratio of phosphate buffer (Na_2HPO_4/NaH_2PO_4) as 4 : 1 in plasma would be disturbed.

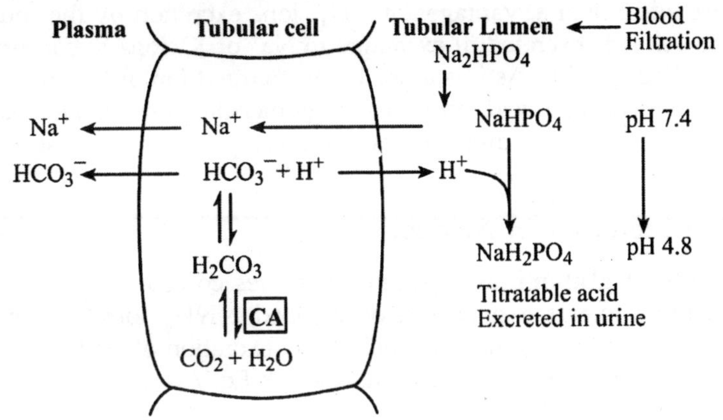

Fig. 35.6 : Acidification of urine in renal tubules by excretion of acidic sodium dihydrogen phosphate

4. Excretion of ammonium ions—This is another important renal mechanism to excrete out H^+ ions (acid load) as well as to excrete ammonia, brought to kidneys in form of glutamine (Chapter 16). This renal mechanism facilitates excretion of large quantities of acids in urine, in the face of excessive acid load. This is achieved by excretion of H^+ and NH_3 in form of well buffered ammonium ion (NH_4^+) with minimum alteration in urinary pH.

This is very important, because normally the urine formed is a **well buffered solution** with minor pH alternations, inspite of excretion of large quantities of acids. This is necessary so as not to affect solubility of various salts and waste products excreted in the urine, such as sodium urates, oxalates etc. to **prevent renal stone formation** and other clinical manifestations (Chapter 18).

Mechanism of NH_4^+ ion excretion—The distal convoluted tubular (DCT) cells have active enzyme **glutaminase** that hydrolyzes glutamine to glutamate and ammonia. The NH_3 so liberated diffuses into the tubular lumen and combine with H^+ secreted in exchange with Na^+ ions by the tubular cells as shown in the Fig. 35.7 to form NH_4^+ ion. This ammonium ion cannot diffuse back into tubular cells and is excreted in the urine.

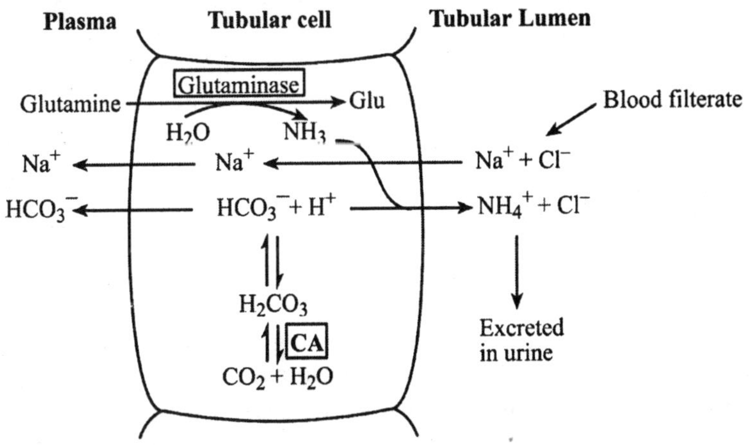

Fig. 35.7 : Excretion of NH_4^+ (H^+ trapped) to excrete out excess of acid load, as well as NH_3 out of body

There are additional advantages of NH_4^+ ions excretion by the renal tubules as enumerated here—a) It is excreted in exchange to Na^+ or K^+ ions to conserve important electrolytes of the body and b) As it is a positively charged ion, it forms **neutral soluble salts** with negatively charged acid anion of the glomerular filtrate, such as Cl^- ions (as NH_4Cl) and is excreted without much alteration in the pH of urine as shown in the Fig. 35.7.

NH_4^+ Excretion Increases in Acidosis

The activity of **glutaminase enzyme** increases considerably in the conditions associated with acidosis to excrete out excess of acid load as NH_4^+ ions (H^+ trapped) without much alternation in the pH of urine. Normally daily excretion of acids (trapped as NH_4^+ ions) is about 70 m.Eq., which rises to as high as 400 m.Eq. / day in conditions associated with acidosis.

Intracellular Cytoplasmic Buffers

The regulation of ICF acid base equilibrium and pH is as important, as the maintenance of ECF and plasma pH, since pH disturbances would adversely affect the cellular functions resulting in severe clinical consequences. These can be life threatening and are prevented to a great extent by ICF buffers. The major tissues involved in regulation of pH alterations are bones and skeletal muscles, which have large exchangeable pool of Na^+ or K^+ alkali reserve to buffer H^+ ions and various acids produced during metabolism.

The intracellular cation, potassium plays important role and diffuse in plasma in exchange of H^+ ions in acidosis conditions, when H^+ concentration in ECF is high. The net effect is an apparent increase in ECF potassium level. Also at renal level in acidosis, there is exchange of H^+ with Na^+ ions instead of normal mechanism of excretion of K^+ ions in exchange to Na^+. This would result in retention of K^+ ions. Thus **hyperkalemia** is often found to be **associated with acidosis,** that can be life threatening, if it is not corrected immediately.

Disorders of Acid Base Balance

In normal circumstances the acid base equilibrium and the blood pH is not allowed to alter much from 7.4, by the efficient pH maintenance mechanism of the body as described above. However disturbances in acid base balance are often found to occur in several conditions and disease states. Clinically the term **acidemiaor acidosis** refers to an increase in H^+ ion concentration (low pH) due to excessive production of acid in the tissues.

The body can **tolerate** and **survive** only minor alterations of pH within the narrow limits of 7.0 to 7.6, beyond which the life is threatened with fatal consequences. This is because minor changes in pH are accompanied by gross increase of H^+ ion concentration, that adversely affects the functioning of tissues leading to server clinical manifestations

As described earlier, at normal pH of blood i.e. 7.4, the bicarbonate-carbonic acid ratio is 20 : 1 and this ratio would be disturbed during the disorders of acid base balance. Accordingly the disturbances of acid base equilibrium are classified as follows—

1. Acidosis—A decrease in blood pH (<7.35) can be due to respiratory or metabolic causes and are referred to as—a) **Respiratory acidosis** due to **primary excess** of **carbonic acid** as seen in the lung diseases associated with hypoventilation and impairment of excretion of CO_2 and b) **Metabolic acidosis** due to **primary deficiency of bicarbonate** (exhaustion of alkali reserve) as a result of excessive acid load, which occurs in conditions such as ketoacidosis, lactoacidosis etc.

2. Alkalosis—Increase in blood pH (>7.45) is caused by: a) **Respiratory alkalosis** due to primary deficiency of carbonic acid, as a result of hyperventilation conditions and b) **Metabolic alkalosis** which occurs when there is primary excess of bicarbonate.

Acid Base Disorders—Clinical Aspects

In actual clinical conditions, mixed disturbances of acid base balance are seen, as the patients are in different states of compensations by the respiratory and renal mechanisms as described earlier. Normally the compensatory measures by lungs and kidneys would try to restore the pH of body fluids to normal. However depending on the severity of acid load and functional states of respiratory and renal mechanisms (that may be affected by diseases), the clinical states associated may be divided as—a) Fully compensated, b) partially compensated and c) uncompensated.

Anion Gap

For the better understanding of the disorders of acid base balance, a clear concept of anion gap is necessary to assess the clinical manifestations. Normally the electric neutrality is maintained in the ECF and ICF and the sum of cations and anions in body fluids is always equal. The Na^+ and K^+ ions account for 95% of the cations of ECF, where as chloride and bicarbonate ions account for only 85% of total anions of ECF. These are the commonly measured electrolytes in clinical laboratory while the **unmeasuranble anions** are the proteins, sulphates, phosphates and organic acids etc. These unmeasured anions of body fluids constitute the **anion gap**.

Thus **anion gap** can be defined as the difference between the measured cations i.e.(Na^+ and K^+ (95%) and measured anions i.e. Cl^- and HCO_3^- (86%) of the plasma. Normally this difference is 12± m.mol/L and is designated as anion gap. This anion gap is due to A^-, i.e. the unmeasured anions of ECF as illustrated below—

ECF cations in m. Eq./L		ECF anions in m. Eq./L
$Na^+(136) + K^+ (4)$	=	$Cl^- (100) + HCO_3^- (25) + A^-$

Where $A^- = 15$ m. Eq./L and is approximate measure of the anion gap.

The clinical states of acid base disorders are often associated with differences in the anion gap and thus its determination is extremely useful in the clinical assessment and management of the patients with acidosis or alkalosis conditions.

Metabolic Acidosis—Primary Bicarbonate Deficit

The metabolic acidosis is usually due to the excessive production of acids in abnormal conditions, such as **ketoacidosis** and **lactoacidosis**. In these cases the bicarbonate ions are used up to neutralize the excess acid load and the anion gap rises to a very high level and is referred to as high anion gap acidosis.

High Anion Gap Acidosis

In the metabolic acidosis conditions such as **diabetic ketoacidosis** and **hypoxic lactoacidosis,** the anion gap increases due to the deficit of bicarbonate ions and are known as **high anion gap acidosis.** In the renal acidosis associated with kidney diseases, the excrĕtion of H^+ (acids) is decreased and along with this the generation and reabsorption of bicarbonate is also reduced to a great extent. This results in loss of bicarbonates in the urine causing high anion gap acidosis in renal diseases.

Normal Anion Gap Acidosis—This occurs in the cases of excessive vomiting, diarrhea etc., where there is loss of both anions as well as cations, along with water loss. Restoration of ECF volume and electrolytes is essential to manage this type of acidosis. **Hyperchloremic acidosis** sometimes occurs in renal tubular acidosis due to the failure to excrete H^+ ions or to reabsorb bicarbonate ions.

Respiratory compensation of acidosis—The acute metabolic acidosis is usually compensated by stimulating respiratory rate and hyperventilation, leading to increased elimination of CO_2. This reduces carbonic acid level and restores bicarbonate-carbonic acid ratio to 20 : 1 to maintain blood pH, though the concentration of both is lower than normal. However this is a short term compensation, as hyperventilation cannot proceeds for long, as it would cause respiratory muscle fatigue.

Renal Compensation of Acidosis—The final compensation of metabolic acidosis occurs at the renal level. The renal compensatory mechanisms (as described earlier) set within 3-4 days of acidosis to excrete out excessive acid load by secreting H^+ and NH_4^+ ions as described earlier.

The renal compensation is very effective to excrete acid load as well as to restore bicarbonate ions (alkali reserve) to normal ratio of 20.

Respiratory Acidosis—Primary Excess of Carbonic Acid

The respiratory acidosis is characterized by the primary excess of carbonic acid due to the failure to eliminate and excrete CO_2. The acute respiratory acidosis is caused by the lung diseases associated with hypoventilation as seen in bronchiopneumonia and asthma etc. In the lung diseases, the respiratory rate decreases leading to respiratory acidosis. Another cause of acute respiratory acidosis can be due to overdose of **sedatives or narcotics,** that depresses the respiratory centers in the brain causing hypoventilation.

Chronic obstructive lung diseases leads to chronic respiratory acidosis conditions, where fall in blood pH is minimal and the patient is mostly in the partially compensated state.

The renal compensation is the major factor to restore the blood pH to normal in the respiratory acidosis. The renal compensatory mechanisms are very effective and include a) activation of the enzyme carbonic anhydrase, b) to generate and reabsorb more bicarbonate and c) to excrete more acid load by secreting H^+ and NH_4^+ ions.

Important causes of acidosis in various disease states are summarized below—

1. Diabetic ketoacidosis
2. Lactoacidosis of hypoxic states associated with various diseases.
3. Reduced GFR in renal diseases.
4. Inherited renal tubular acidosis where H^+ ions are not secreted.
5. Fanconi's syndrome where HCO_3^- is nor reabsorbed properly.

6. Addison's disease, in which there is decreased reabsorption of sodium and bicarbonate ions.
7. Respiratory acidosis in lung diseases, such as pneumonia, asthma etc.

Metabolic Alkalosis—Primary increase in HCO_3^-

The metabolic alkalosis, characterized by primary excess of bicarbonate, results due to excessive loss of acids as seen in severe vomiting (HCl loss) causing hypochloremic alkalosis.

In Cushing's syndrome (excess of aldosterone), there is excessive retention of sodium and excretion of potassium. Hypokalemia is closely associated with metabolic alkalosis. This is because renal tubules attempts to retain H^+ ions in alkalosis, and in exchange potassium is excreted out in the urine. The increase in ECF pH depresses the respiratory centres and the respiratory mechanisms initiate the compensation by hypoventilation to retain CO_2, thereby increasing carbonic acid to restore bicarbonate/carbonic acid to 20 : 1, though the concentration of both is higher than the normal values.

Renal Compensation—The renal mechanisms are more effective to compensate metabolic alkalosis. The renal tubules conserve H^+ and more HCO_3^- ion are excreted in the urine to regulate the blood pH

Management—However the complete correction of alkalosis is possible only by correcting hypokalemia by administration of potassium and removing the cause of alkalosis.

Respiratory Alkalosis—It is characterized by the primary deficiency of carbonic acid. This can occur as a result of **hyperventilation,** where excess of CO_2 is eliminated out. Causes of hyperventilation include—hysteria, raised intracranial pressure, brain stem injury and excessive artificial ventilation etc.

Renal mechanisms tries to compensate respiratory alkalosis by increasing urinary excretion of HCO_3^- to restore carbonic acid-bicarbonate ratio normal to regulate blood pH, however the concentration of both is lower than the normal.

❑❑❑

36

CLINICAL BIOCHEMISTRY I—KIDNEY FUNCTIONS

The major function of kidneys is to excrete metabolic waste products and unwanted substances out of the body in form of urine. The kidneys are vital organs of body and perform several important functions which include—

1. Excretion of nitrogenous waste products such as urea, uric acid, creatinine, ammonia etc. and other unwanted and toxic substances out of the body.
2. Maintenance of homeostasis—The kidneys are primary organs that are responsible for the maintenance of **"internal milieu"** of the body constant. The regulation of water, electrolyte and acid-base balance of the body is effectively and ultimately is achieved by kidneys.
3. Hormonal functions—The kidneys have endocrine functions and produce hormones such as—a) **erythropoietin** that stimulates erythropoiesis, b) **calcitriol** to enhance calcium and phosphate absorption and c) **renin-angiotensin system** that regulates electrolyte balance, blood volume and blood pressure.

Formation of Urine—The **functional unit** of the kidneys is nephron and approximately one million nephrons are present in each kidney. The nephron consists of a glomerulus apparatus formed by the tuft of afferent and efferent blood arterioles, surrounded by the Bowman's capsule that collects ultra filtrate of blood and continues in the proximate convoluted tubule (PCT), loop of henle, distal convoluted tubule (DCT) and collecting tubule as illustrated in the Fig. 36.1.

Renal Blood Flow (RBF) through kidneys is very large, i.e. about 1200 ml of blood per minute, which is ultra filtered in the glomerulus apparatus. During ultrafiltration, blood corpuscles and plasma proteins are retained in the blood, while crystalloids (small mol. wt substances) dissolved in water as glomerular filtrate is collected in the Bowman's capsule. **Glomerular filtration rate (GFR)** is about 120-125 ml per minute in a person with 70 kg body weight. The total glomerular filtrate formed in a day is around 170 to 175 liters; out of which 99% is reabsorbed back by the renal tubules and only 1.5 liters of urine (average) is excreted per day normally.

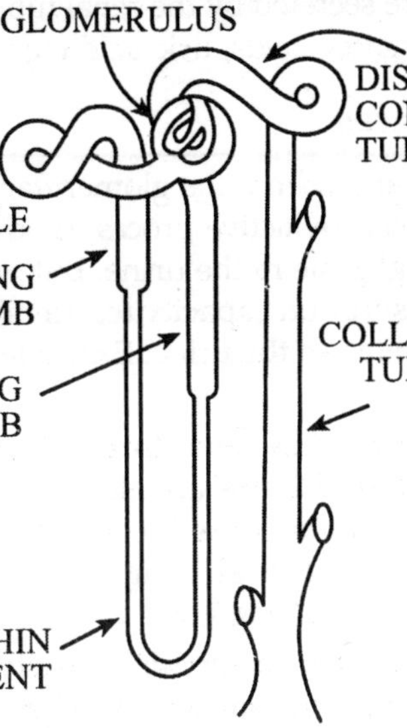

Fig. 36.1 : Structure of nephron

Glomerular Function—When the blood is perfused through the closed system of a tuft of arterioles of glomerulus, an ultrafiltration of blood occurs in which blood cells and plasma proteins (mol. wt > 69,000) are retained in the blood. The sieves (pore size) of glomeruli are such that the substances of molecular weight less than 69 kDa can pass through them. For example hemoglobin with mol. wt of 67 kDa can pass through the pores of glomerulus but albumin (mol. wt. 69 kDa) is retained back. Thus glomerular filtrate consists of all the constituents of plasma with mol.wt less than 69 kDa dissolved in water and is collected in the Bowman's capsule, that acts as a funnel and continues into the renal tubules.

Functions of Renal Tubules

Selective reabsorption—As mentioned above, the glomerular filtrate contains all the crystalloids of plasma including important nutrients, electrolytes etc. which cannot be allowed to be lost in the urine. Thus in proximate convoluted tubules, selective reabsorption occurs. The nutrients such as glucose, amino acids and potassium are reasbsorbed back completely (100%) by an **active transport mechanism**, i. e. mediated by specific carrier system, in which ATP is utilized as the energy source. In proximate convoluted tubules 70% of water is reabsorbed along with Na^+, bicarbonate and Cl^- ions as an obligatory process.

Loop of Henle (thin segment) of nephron is mainly involved with the absorption of electrolytes such as Na^+, Cl^-, Ca^{++} and Mg^{++} ions.

Facultative reabsorption—In distal convoluted tubules, under the action of hormones, such as **aldosterone** and **ADH**, rest of the sodium and water is reabsorbed, depending upon the need of the body to maintain water and electrolyte balance, as well as to regulate blood volume and blood pressure. This is known as facultative reabsorption.

Secretion of substances—In addition to the reabsorption of important constituents of blood from the glomerular fitrate, the renal tubules play very important role to secrete H^+ ions and NH_4^+ ions to eliminate acid load and to maintain acid base equilibrium of the body. Other substances that are secreted by the renal tubules to be excreted in urine include K^+ (in exchange with Na^+), organic acids, uric acid etc.

Renal Threshold Substances

Some important substances from the glomerular filtrate are reabsorbed completely (100%) at the normal blood levels by active process mediated through specific carriers and normally their excretion is negligible in the urine. But when their concentrations in blood are elevated, the tubular reabsorption capacity (carriers) becomes saturated and excess of that compound would be excreted in the urine. Examples are glucose, amino acids, ketone bodies etc.

These substances are called as **renal threshold compounds**, as their excretion in urine occurs only when their blood levels are elevated and **cross renal threshold levels.** The renal threshold blood level of glucose is 180 mg/dl (normal 70-110 mg/dl), for ketone bodies it is 1.5 to 3 mg/dl (normal level 1 mg/dl), for calcium it is 10 mg/dl and for bicarbonate it is 30 mg/dl (normal 24 mg/dl).

Tubular Maximum—It is the maximum capacity of the renal tubules to reabsorb a substance, above which the carriers become saturated and the compound is excreted in the urine. The **tubular maximum for glucose (TMG)** is 360 mg/minute.

NORMAL URINE

The urine is formed in kidneys by the mechanism described above and serves as a main route to excrete out nitrogenous and other metabolic waste products and unwanted substances dissolved in water. The physical characteristics and chemical constituents of normal urine are described here.

Physical Characteristics

In clinical biochemistry, the following physical characteristics are usually considered when analyzing an urine sample to assess the normal kidney functions. The urine (freshly voided) has following physical characteristics—

Volume—The average output of urine by an adult per day is 1 to 1.5 liters and the volume is affected by several factors. For instance, urinary volume is altered and depends on the amount of fluid intake, climatic conditions, loss of water in excessive perspiration, vomiting, diarrhea and in pathological states, fever etc.

Appearance—Normal urine is perfectly clear, transparent fluid when freshly voided. It may become turbid on standing due to the ammonical fermentation and pH alterations, resulting in precipitation of calcium and magnesium salts.

Colour—Normal urine has a pale amber yellow tinge due to the presence of a pigment urochrome. Depth of the colour depends on the density of urine or its concentration. Abnormally, urine may be deep yellow coloured in case of jaundice, due to the excretion of excess of bilirubin or some times after ingesting B complex capsule or certain drugs. Urine may acquire a red tint in cases with hematuria or porphyrias due to the excretion of hemoglobin or porphyrins.

Odour—Normal urine has a faint aromatic odour when freshly voided, due to the presence of volatile organic acids. On standing, bacterial fermentation occurs resulting in typical ammonical smell. Abnormally urine may have a fruity odour in **ketosis** due to the excretion of volatile acetone or it can have a foul smell in some pathological conditions.

Specific gravity—The specific gravity (sp. gr.) of the normal urine ranges from 1.015 to 1.025 and is affected greatly by the amount of fluid intake or water loss, which may alter urinary volume or its concentration. The sp.gr. of urine is abnormal in disease states such as diabetes, nephritis etc. The specific gravity of the urine is measured with the help of urinometer.

Reaction—The pH of normal urine is acidic, with an average pH of 6.0 (range is 5.5 to 7.5). The acidic pH of urine is because, kidneys excrete and eliminate acid load of the body by secreting H^+ ions to maintain acid base balance of the body (Chapter 33).

NORMAL CONSTITUENTS OF URINE

The composition of the normal urine varies widely as it is influenced greatly by dietary habits, **environmental** conditions and other factors. The normal urinary constituents are broadly grouped as (a) Inorganic constituents comprising soluble salts, such as chlorides, phosphates and sulphates of sodium, potassium, calcium and magnesium etc. and (b) The organic constituents, which mainly include the nitrogenous waste products of metabolism, such as urea, uric acid, creatinine etc. The daily excretion of normal constituents of urine is summarized in the Table 36.1.

Table 36.1 : Composition and daily excretion of solutes in normal urine

Organic constituents	Daily excretion g/day (average)	Inorganic constituents	Daily excretion g/day (average)
Water	1200.00	Sodium chloride	12.00
Solids	60.00	Sodium	04.00
Urea	30.00	Potassium	02.00
Uric acid	00.70	Calcium	00.20
Creatinine	01.20	Magnesium	00.15
Ammonia	00.70	Phosphates	01.00
Others- Indican, hippuric acid, oxalic acid, phenols etc		Sulphates	01.00

Inorganic Constituents of Urine

Sodium chloride is the major inorganic salt excreted in the normal urine. About 12 g of NaCl are excreted per day, and is the second largest component of the urine after urea, which is the largest solute excreted. The other inorganic constituents of the normal urine are phosphates and sulphates of major cations of the body as is evident from the Table 36.1.

Organic Constituents of Urine

Urea—The urea is the major nitrogenous waste product of the body and approximately 30g of urea (half of total solutes of urine) are excreted daily in the urine. Urea is the end product of protein and amino acid metabolism in the humans. The amino group of amino acid is highly toxic to body tissues and thus it is converted into less harmful, water soluble compound urea in the liver, to be excreted in urine (details in the Chapter 15).

Urea is freely filtered by glomerulus but about 40% of it is reabsorbed back by the renal tubules and some amount is re-secreted into tubular lumen. Rate of urea reabsorption is increased by tubules when GFR is decreased, as observed in kidney diseases. Thus **blood urea level increases** excessively in **renal impairment**, associated with kidney diseases.

Urea is detected by following tests

Hypobromite test—Sodium hypobromite (freshly prepared) decomposes urea present in the urine sample to CO_2, which is absorbed by the alkali of the reagent and N_2 gas, that evolves as copious effervescence, indicating a positive test. The reaction of the test is as given under—

$$\underset{\text{Urea}}{CO(NH_2)_2} + \underset{\text{Sodium hypobromite}}{3NaOBr} \longrightarrow \underset{\text{Sodium bromide}}{3NaBr} + 2H_2O + CO_2 + N_2$$

The test is nonspecific but significant as other interfering substances are in negligible amounts as compared to the amount of urea present in the urine. Sodium hypobromite is prepared by mixing 3ml of 40% NaOH with 2ml of bromine water.

Urease Test—The enzyme urease is specific for urea and decomposes it into carbonate and ammonia, which change the pH of the medium from acidic to alkaline (detected by indicator phenol red). Or in other way, on slight heating the mixture of urease powder with urine sample, ammonia gas evolves which turns a moist red litmus paper kept on the mouth of test tube to blue, indicating a positive test.

Creatinine

It is an important nitrogenous waste product excreted in the normal urine. Creatinine is an anhydride of creatine present in muscles as creatine phosphate, the storage form of high energy compound ATP. During muscle contraction, creatine phosphate is converted into creatinine, a waste product to be excreted in the urine.

Amount of creatinine excreted per day in urine (Table 36.1) is quite constant for an individual and depends on his muscle mass. Creatinine can be detected in the urine by Jaffe's test.

Jaffe's Test—Creatinine in urine sample is allowed to react with alkaline picric acid (3 ml of staturated picric acid solution with 0.5 ml of 10% NaOH) to produce a deep yellow orange colour, due to the formation of creatinine picrate. The Jaffe's test is utilized frequently in the estimation procedures for creatinine in urine as well as in the plasma.

Uric acid

The uric acid is the end product of nucleic acids and purine catabolism in humans, to be excreted in urine. Uric acid is sparingly soluble in water and is mostly excreted as comparatively more soluble sodium urates in the urine. The excretion of sodium urates is influenced greatly by the pH of urine, as urates are less soluble at the acidic pH of urine which is about 6.0 (details in Chapter 18).

The excretion of uric acid, therefore is problamatic and is influenced by several factors. Renal impairment results in elevation of serum uric acid levels with adverse consequences, which may cause gout (Chapter 18).

Murexide Test—Uric acid can be detected by murexide test in normal urine sample. The uric acid, in the residue left after evaporating the urine sample by boiling in a porcelain dish, is heated with a few drops of concentrated HNO_3. A red orange murexide is formed by oxidation of uric acid, which turns to purple red on adding a few drops of 40% NaOH solution.

ABNORMAL CONSTITUENTS OF URINE

Many substances, considered as pathological or abnormal constituents of urine, are normally not excreted and may be present only in trace amounts which cannot be detected by usual analytical tests in the normal urine.

These **abnormal constituents** are excreted excessively in urine in certain **pathological or abnormal conditions** and can be detected in urine by specific chemical tests and are thus helpful in diagnosis of several diseases. The abnormal constituents usually analyzed in pathological states are proteins (albumin), sugar, blood, ketone bodies and bile pigments and bile salts.

Proteins

The sieves of glomeruli of kidneys are impermeable to plasma proteins (mol. wt > 69,000) and thus normally they are not excreted in detectable amount in urine. Traces of protein (<100 mg/day) may be present in normal urine.

In renal diseases, such as glomerular nephritis, the glomeruli are damaged resulting in increased pore size of their membranes and cause excretion of proteins in urine. The albumin (low mol. wt plasma protein) is excreted in urine more readily in abnormal conditions as compared to globulins whose molecular size and weight is much higher.

Proteinuria (albuminuria)—The excretion of albumin in the urine is called albuminuria and may be associated with physiological states such as strenuous exercise (transient) and in pregnancy occasionally. In pathological diseases of kidneys, such as nephritis, nephrotic syndrome and eclampsia of pregnancy, albuminuria occurs and is of great clinical significance in the diagnosis and monitoring of the patient's response to the treatment. Albumin is detected in pathological urine by following tests.

Heat coagulation test—The upper column of 10 ml of the urine sample in a test tube is heated to boiling point, cooled and 5 drops of dilute (2%) acetic acid are added (to dissolve phosphates if precipitated). The top layer is then compared with the lower portion of the sample, that acts as a control test. A white coagulum in the upper layer of sample indicates the presence of heat coagulable proteins i.e. albumin in the abnormal urine sample.

Sulphosalicylic acid test—To 3 ml of urine sample, a few drops of 3% sulphosalicylic acid solution are added. Formation of a white precipitate shows the presence of albumin or globulin in the urine sample.

Bence Jone's Proteins—In patients suffering from multiple myeloma, some abnormal immunoglobulins are produced in excess. Usually the light chain immunoglobulins of low molecular weight are produced excessively by plasmacytomas and excreted in the urine. These are called as Bence Jone's proteins after the scientists who discovered them in the urine of patients with multiple myeloma.

Bence Jone's proteins are detected in urine by their characteristic property of precipitating at low temperature by heating to 45° to 50°C, which start redissolving at 80° C and the coagulum disappears completely at 100°C. The coagulum reappears on cooling the sample to 40°C.

This is in contrast to albumin and globulins which are coagulated **irreversibly** at 100°C and will not redissolve on cooling.

Microalbuminuria—It is a clinical condition in which micro (small) amounts of albumin is excreted in the urine. In microalbuminuria about 150 mg of albumin is excreted in urine per day and the amount is sufficient to be detected by some specific tests.

The **microalbuminuria** is **very significant clinically** as it is the **earliest indication of renal function impairment.** The microalbuminuria is associated with the advanced stages of diabetes mellitus, in which the unmanaged and persistant hyperglycaemia damages nephrons of kidneys causing nephropathy (details in Chapter 8).

Blood and Hemoglobin

The blood is normally not excreted in the urine, as blood corpuscles (containing hemoglobin) are retained back by glomeruli during ultrafiltration. The pathological conditions in which blood is excreted in urine is known as **hematuria.**

Hematuria—It occurs in renal disease, such as nephritis, nephrotic syndrome etc. in which glomeruli are damaged extensively and pore size of glomerular sieves increases very much to allow blood along with RBCs to pass through them to be excreted in urine. Hematuria also occurs in **post renal hemorrhages** caused by **renal stones,** trauma or mechanical (accidents) injury caused to urinary tract.

Hemoglobinuria is the condition in which free hemoglobin (RBCs are absent) is excreted in the urine. This occurs in the diseases associated with excessive hemolysis of the RBCs, such as hemolytic anaemias (genetic disorders), malaria, typhoid etc.

Benzidine test is performed to detect blood as well as hemoglobin in the urine. This test is very sensitive and is given positive even with the presence of trace amounts of blood or hemoglobin in the urine. However it is not specific for hemoglobin, as it is given positive by any catalyst or enzyme that can split H_2O_2 to nascent oxygen.

Procedure—To the urine sample, 2 drops of benzidine solution (in glacial acetic acid) and 2 drops of 6% H_2O_2 solution are added and mixed. A blue or green colour indicates the positive test. Mechanism and reactions of the test are explained as follows—The heme catalyzes splitting of H_2O_2 to nascent oxygen, which oxidizes the colourless benzidine to a blue or green coloured compound. Pus cells or bacteria present in the urine contain the enzyme peroxidase, that can split H_2O_2 to nascent oxygen and thus will also give the test positive. Therefore the urine sample should be boiled before performing the test, so that bacteria and enzymes, which interfere in the test are destroyed.

Reducing Sugars

Normally the glucose or other sugars are not excreted in urine, though trace amounts may be present, but that will not give usual chemical tests positive. Glucose filtered by glomeruli is reabsorbed completely (100%) by the renal tubules from the glomerular filtrate as mentioned earlier in the section. However when blood sugar level is elevated, as in diabetes mellitus, it may cross renal threshold value (> 180 mg/dl) and the extra glucose is excreted in urine. The condition is known as **glucosuria.**

Benedict's test—Glucose and other reducing sugars in urine are detected by Benedict's test. This test is very sensitive for the reducing sugars and is not affected by other reducing substances present in the urine, as the reagent contains a mild alkali. When the test is performed in the standard manner, it serves as a **semi quantitative** test to assess the approximate amount of sugar excreted in the urine.

Procedure—To 5ml of Benedict's reagent add 8 drops (0.5 ml) of urine sample and boil for 2 minutes, cool and observe the colour of the precipitate formed. The colour produced depends on the extent of reduction of the cupric ions of the reagent in hot alkaline medium, by the amount of sugar present in the sample.

The Benedict's test is semi quantitative, as the colour of precipitate obtained would give us the approximate concentration of sugar in the urine sample. The approximate amount of sugar present in the urine would be—0.2% when colour of precipitate is green, 0.5% when colour is yellow, 0.8% when orange precipitate is formed and it is more than 1 gm% when colour of the precipitate is brick red.

Other sugars may also be excreted in urine in certain conditions such as lactose (during lactation), fructose, galactose and pentose sugars and will also give Benedict's test positive.

The fructose is excreted in urine in genetic disorder called **essential fructosuria** and galactose in a severe inborn error of metabolic disorder called **glactosemia** (details in the Chapter 7). Pentose sugars sometimes may appear in urine after ingestion of fruits rich in pentoses, or in case of **essential pentosuria**, a genetic disorder. (Chapter 7).

All these are reducing sugars when present in urine will give Benedict test positive. The differential diagnosis may be facilitated by the osazone test or by other specific colour tests for the type of sugar present in the abnormal conditions. **Clinistics** with specific enzyme for the detection of glucose in urine are now available commercially (Chapter 8).

Ketone Bodies

The ketone bodies include acetone, acetoacetate and β hydroxy butyric acid, that are excreted in urine in the condition known as **ketosis**. The excessive production and excretion of ketone bodies in urine occurs in the abnormal and pathological states, such as diabetes mellitus (uncontrolled or type I), starvation etc. (details in the Chapters 8 & 11).

Rothera's test—This test is given positive with acetone and acetoacetate but not with β OH- butyrate present in the urine. The test is performed by saturating 2 ml of the urine sample with solid ammonium sulphate and then 2-3 drops of sodium nitroprusside solution are added and mixed, followed by 1-2 ml of ammonia solution to form an upper layer. Appearance of a violet ring at the junction of the two layers indicates the presence of the ketone bodies in the urine sample.

Bile Pigment and Bile Salts

In normal urine, small amounts of urobilinogen may be present, but bilirubin (bile pigment) and bile salts are not excreted normally. In abnormal pathological conditions, excess of bile pigments i.e. bilirubin, biliverdin and urobilinogen are excreted in urine imparting it a deep yellow colour. The clinical condition is known as Jaundice and it is associated with liver dysfunction and pathological diseases (details in the Chapter 34).

Gmelin's test for bilirubin—Three ml of the urine sample is layered on the top of 3 ml of the concentrated HNO_3 in a test tube. Various coloured rings i.e. green, blue, violet and red rings appear in rapid succession at the junction of the two layers, due to the oxidation of bilirubin to various coloured bile pigments indicating a positive test.

Hay's surface tension test for bile salts—Bile salts are excreted in the urine in obstructive jaundice, that may be caused by the obstruction of biliary tract due to gall stones or in case of malignancies etc. (Chapter 34).

Procedure—Hay's test is performed by sprinkling a few flowers of sulphur powder on the surface of the urine sample and the sinking of sulphur flower gradually to the bottom of the test tube is observed, indicating a positive test. This is because, the bile salts lower the surface tension of urine and thus light sulphur flowers sink to the bottom of the test tube (compare the test with plain water).

ASSESSMENT OF RENAL FUNCTIONS

Kidneys are important organs of the body and perform several sensitive vital functions, that are absolutely essential for the life as well as to maintain good health. Thus it is clinically very significant to assess renal functions for early detection of any abnormality or the impairment of kidney functions. In kidney diseases, early screening of renal function is necessary to prevent clinical complications, which can be life threatening and may result in fatal consequences.

Renal Function Tests

Several tests are employed in clinical biochemistry laboratory to assess renal functions. Early screening is done by routine urine examination, and determination of blood and serum parameters as described here under—

1. **Urine Examination**—Preliminary assessment of the renal function is done by complete urine examination, which include **physical characteristics** such as volume, pH, specific gravity etc. and **microscopic examination** to observe any abnormality such as the presence of bacteria, pus cells, RBCs and salt deposits etc. Presence of **abnormal constituents** as described above will help in diagnosis and detection of the diseases associated with kidney function impairment.
2. **Analysis of blood or serum**—Since the most important function of the kidneys is to excrete out nitrogenous waste products in urine, the estimation of blood urea and serum creatinine levels are extremely useful and clinically significant in renal function assessment. As blood urea level is influence by other factors also (Chapter 15), serum creatinine level is a better indicator to assess renal function. However elevated levels of both, serum urea and serum creatinine in a ratio of increasing order, clearly indicate renal impairment and the severity of the renal failure can be assessed by their increased blood concentrations.

Estimation of blood urea—Urea is the major nitrogenous waste product of the body to be excreted in urine by kidneys. The normal blood urea level ranges between 15 to 40 mg/dl and is estimated in laboratory by the methods, such as—(1) Urease nesslerization method of Van Slyke or by (2) Diacetyl monoxime method.

Diacetyl monoxime method is very simple and convenient and is widely used in clinical biochemistry to estimate serum urea concentration. In this method urea is allowed to react with diacetyl monoxime at 50° C, in the presence of a strong acid to produce a rose pink colour which is estimated colorimetrically (Chapter 15).

High blood urea levels may occur in various conditions, that can be conveniently classified as prerenal, renal and post renal.

Pre-renal conditions associated with elevated blood urea levels include dehydration, that may occur in conditions such as severe vomiting, intestinal obstruction, severe diarrhea, fever etc.

Renal diseases are associated with increased blood urea levels. In acute glomerular nephritis urea level may be as high as 300 mg/dl. In chronic nephritis, blood urea level increases along with increasing renal failure. In cases of malignant hypertension and chronic **pyelonephritis blood** urea level may reach to as high as 500-700 ml/dl, that may lead to fatal consequences.

Post renal conditions in which blood urea level rises, are obstruction in urinary tract due to renal stones, stricture of urethra, enlarged prostrate, tumors of bladder etc and may cause acute renal failure with severe clinical consequences.

Serum Creatinine—Creatinine is a nitrogenous waste product formed by creatine phosphate, which is the storage form of energy (ATP) in the muscles. The production of creatinine is nonenzymatic and amount formed depends on total muscle mass of the body and is not affected by diet, age etc. Increase in serum creatinine level along with high blood urea is indicative of renal impaiment and serum levels of creatinine and urea usually parallels the severity of renal failure.

Estimation of Serum Creatinine—Creatinine estimation procedures are based on Jaffe's reaction. The creatinine in deproteinized serum is made to react with alkaline picrate to give an yellow orange colour which is estimated colourimetrically.

Normal serum creatinine level in adult male is—0.7-1.4 mg/dl, in females it is 0.6 to 1.3 mg/dl and in children it is 0.4 to 1.2 mg/dl, depending on the muscle mass.

Assessment of Glomerular Function

The measurement of **glomerular filtration rate (GFR)** is the most useful index for the assessment of renal function and to assess the severity of kidney damage in renal diseases. The excretion of nitrogenous waste substances is the most important function of the kidneys and can be expressed quantitatively by using the **concept** of **clearance.**

Clearance Test—The clearance tests are mainly done to measure GFR, for assessing renal function and include—creatinine, inulin and urea clearance tests.

Clearance is defined as the quantity in milliliters of plasma,that would be completely cleared off from a substance per minute by the kidneys. In other words clearance of a substance is referred to the milliliters of plasma which contain the amount of the substance, that is excreted by kidneys per minute and can be expressed as—

$$\text{Clearance} = \frac{\text{mg of substance excreted per minute}}{\text{mg of substance per ml of plasma}}$$

Clearance of a substance by kidneys is calculated by using the formula

$$C = \frac{U \times V}{P}$$

Where C is clearance in ml/minute, U is concentration of the substance in urine, P is concentration of substance in plasma and V is the ml of urine excreted per minute.

Maximum clearance—The maximum rate at which the plasma can be cleared off from a substance is when the substance is freely filtered by the glomerulus, but neither reabsorbed nor secreted by the renal tubules. This clearance value would be equal to GFR, for instance, inulin clearance test.

Inulin clearance test—Inulin is a plant polysaccharide formed by the condensation of many fructose units. Inulin clearance satisfies the above mentioned criteria and thus inulin clearance test can be performed to measure GFR for the assessment of renal functions. Inulin clearance value is about 125 ml/minutes which is very near to GFR and thus can be taken as measure of GFR. But inulin (exogenous) has to be administered intravenously to estimate GFR and thus this clearance test is not done routinely.

Renal plasma Flow is another factor which would influence GFR and other kidney function tests. Diodrast (diiodopyridone acetate) and para amino hippuric acid (PAH) are peculiar substances, as they are completely and entirely excreted by a single passage of blood through kidneys and thus their clearance value is taken as equivalent to renal plasma flow.

Diodrast is also used as a contrast medium to take urinary tract X-rays, for the detection of any obstruction or renal stones in urinary tract.

Creatinine Clearance Test—It is the most sensitive test for assessing kidney function and for GFR measurement. Creatinine is an endogenously produced nitrogenous waste product, derived from creatine phosphate of the muscles. Creatinine excretion rate is relatively constant in an individual and production rate is not influenced by the diet, metabolism or hormonal factors. Thus creatinine clearance is a good measure for assessing glomerular functions and GFR determination.

Creatinine in the glomerular filtrate is only minimally absorbed and is not secreted by the renal tubules and thus is **ideally suitable** for the clearance test to measure GFR. The value of creatinine clearance is very near to the measure of GFR, hence creatinine clearance test is better indicator to assess glomerular function. Creatinine clearance is defined as the volume (ml) of plasma that would be cleared off from the creatinine per minute by kidneys.

Procedure—Instead of a 24 hour urine collection, the procedure is modified to collect urine sample for 1 hour time period. After giving 500 ml of water to promote good urine flow, the urine sample is collected after 1 hour and volume is measured. Blood sample is withdrawn and both plasma and urine (1 in 100 dilution) samples are analyzed for creatinine concentrations. Creatinine clearance is then calculated by substituting the values in the formula—

$$\text{Creatinine clearance in ml / minute } C = \frac{U \times V}{P}$$

Creatinine clearance value normally ranges between 110 to 145 ml/min for adult male. For females it is slightly lower, as muscle mass is less and thus should be corrected in terms of body surface area.

Clinical significance—A decrease in creatinine clearance value (75% of normal) is very significant and indicates decreased GFR due to acute or chronic glomerulonephritis involving renal damage. However other prerenal and postrenal conditions, that affect urea and creatinine excretion would also influence creatinine clearance.

The creatinine clearance estimation is important in early detection of functional impairment of kidneys, where clinical symptoms are still not apparent. For instance, with a rise in serum creatinine to 2 mg/dl, the creatinine clearance value will fall to 50% of the normal, that is highly significant clinically.

Creatinine coefficient—Since creatinine excretion is relatively constant for an individual, creatinine coefficient is determined and expressed in mg/kg body weight. The normal values for creatinine coefficient in males is 20-28 mg/kg while in females it is 15-20 mg/kg body weight. The value is elevated in muscular dystrophy and muscle wasting diseases.

Urea Clearance Test—Urea is the main nitrogenous waste product in humans and is excreted out by the kidneys. The urea clearance values are less than GFR, since urea is partially reabsorbed by the renal tubules from the glomerular filtra . Urea clearance is influenced by several factors, such as hypertension and abnormality in renal reabsorption etc. Because of the wide fluctuations in urea clearance, the creatinine clearance test is preferred to assess renal functions.

Urea clearance is defined as the number of milliliters of plasma which is cleared off completely from urea per minute by the kidneys. Procedure of determination of urea clearance is similar to that of creatinine clearance estimation. The patient is given a glass of water at 9 AM in morning to ensure good urine flow and the urine sample is collected after one hour and volume is measured. The blood is withdrawn and the blood and urine (1 in 100 dilutions) samples were analyzed for urea concentration.

The urea clearance is calculated by substituting values of blood urea, urinary urea and urine volume in the formula as given below.

$$\text{Urea clearance in ml / min.} \quad C = \frac{U \times V}{P}$$

Where C is ml of plasma cleared off urea per minute, U is urinary urea, P is plasma urea level and V is urine volume. The urea clearance value is greatly affected by urinary flow and thus it is expressed in two ways (a) the maximum urea clearance and (b) standard urea clearance.

Maximum urea clearance is when urine volume is more than 2 ml per minute and it is calculated by the formula as given above. The maximum urea clearance values normally found to be around 75 ml/minute (average).

Standard urea clearance—The urea clearance values are found to be decreased when urine volume per minute is less than 2 ml and is called as standard urea clearance. The standard urea clearance is calculated by a modified formula, in which under root of the value V (urine volume) is substituted as given below—

$$C = \frac{U \times \sqrt{V}}{P}$$

The standard urea clearance value is found to be about 54 ml/minute (average).

Significance of urea clearance—When urea clearance value is below 75% of the normal, it is considered to be abnormal and indicates renal function impairment. The values of urea clearance fall progressively with severity and extent of renal failure. The decrease in urea clearance value is observed at an early stage of renal impairment, even when plasma urea level is within normal range. It has been observed that the plasma urea values start rising only when urea clearance values are below 50% of the normal.

Tubular Function Tests

Other important kindney tubular function tests include—

(a) **Measurement of Specific gravity** of urine which is elevated considerably in the case of proteinuria associated with the kidney diseases.

(b) **Concentration test** is significant as the earliest manifestation in kidney function impairment. This test assess the difficulty in concentrating the urine by the kidneys. The concentration tests is done by restricting the fluid intake (< 200 ml/day) and measuring specific gravity of 24 hour urine sample collected next day. When the specific gravity is more than 1.022, the renal function is considered to be adequate.

(c) **Dilution test** on the other hand assess the amount of urine volume, sp.gr. and osmolality of the urine on the face of water load. After giving 1200 ml of water in 20 minutes time period, hourly urine samples are collected for the next 4 hours separately and the volume, sp.gr. and osmalality (by osmometer) of each sample are measured. The normal renal function is indicated if the patient excretes almost all water load within 4 hours and specific gravity of atleast one urine sample should fall to 1.003.

❑❑❑

37

Clinical Biochemistry II—Analytical Procedures

The rapid advancement in analytic procedure technology in clinical biochemistry has great impact on medicine. In modern clinical practice, the physicians are getting increasingly dependent on the laboratory analytical methods and advanced technology, for the diagnosis of diseases as well as to monitor the patient's response to the treatment. Doctors are now a days very much concerned about screening and early detection of several dreaded killer diseases, so as to manage the patients to prevent irreversible clinical complications, that can be life threatening.

However proper care should be taken to ensure quality control of the clinical biochemical analytic procedures. This is must to ensure correct interpretation of the analytical findings, for the diagnosis of diseases.

Sample Collection—Body fluids and tissue samples should be properly collected by trained technicians under the guidance of physicians to ensure correct sampling and to avoid any error in sample collection.

Blood samples for clinical biochemical analysis may be venous, arterial or capillary. Venous blood is usually a preferred specimen for laboratory analytical methods. Arterial blood is used mainly for blood gas analysis, while the capillary blood is used primarily in hematological analysis. During blood sample collection, proper care should be taken to prevent hemolysis of blood cells, which interfere in the analytical procedures.

Some important measures to prevent hemolysis are listed below—

1. Dry, sterilized syringe should be used while drawing blood sample.
2. Forceful suction of the blood by syringe or forceful transfer of blood from syringe to collecting tube may cause hemolysis of RBCs. Before transferring the blood from syringe, needle should be removed. Even traces of hemolyzed blood would interfere in the analytical procedure, particularly in electrolyte (potassium) estimations.

Anticoagulants—When whole blood or plasma is required in an analytical procedure, an anticoagulant is added in the withdrawn blood specimen. **Heparin** is the best anticoagulant, as it least interferes in the analytical procedure. However it is very costly, so is not used frequently in routine analytical procedures.

Anticoagulants widely used in clinical biochemistry are EDTA (ethylene diamine tetra acetate), sodium citrates, oxalates etc. EDTA is a chelating agent and acts by chelating calcium ions to prevent clotting of the blood. It is particularly used in hematological procedures as it does not much affect the osmolality of plasma and thus preserve the integrity of cellular components.

Potassium oxalate (1-2mg/ml blood) is widely used as anticoagulant. It prevents clotting by forming insoluble salts with calcium ions. Sodium fluoride is used as anticoagulant particularly in blood sugar estimation, as it inhibits glycolysis by RBCs in the collected blood sample, otherwise the estimated level for sugar may be quite low.

However now a days, serum sample is the specimen of choice for most of the clinical biochemical analytical procedures, where anticoagulants are not required.

Serum Collection—After withdrawal, the blood is allowed to clot at room temperature (30° C) for half an hour and the supernatant fluid oozed out is separated from the blood clot and is centrifuged to obtain the serum sample.

Separation and storage of Samples—Plasma or serum should be separated from the blood cells as soon as possible by centrifugation and should be stored in tightly capped tubes at 4°C. If a specimen need storage for more than 2 or 3 days, it should be kept in freezer at minus (-) 4° C.

The **basic principles** of some widely used analytical methods in clinical biochemistry are described here—

Photometry

The photometry is the basis of most of the analytical procedures for accurate measurment of a wide range of biochemical indices and parameters in biological specimens.

The photometric analytical procedures are primarily based on the absorption of light radiation of a specific wave length by a photosensitive compound, to be estimated in a sample. Technically a solution is considered to be coloured, when it contains a photosensitive compound, that is capable of absorbing light radiations of some definite wave length from the light spectrum, while the rest of the radiations are transmitted through it.

Colorimeter and **spectrophotometers** are the main instruments used in clinical biochemistry laboratory for the photometric procedures to estimate the chemical compounds in biological specimens.

Principle of photometry—The amount of light absorbed or transmitted by a coloured solution is proportional to the intensity of colour, which in turn depends on the concentration of the photosensitive compound present in the solution.

The photometry is based on the principles of **Lambert-Beer's law** which states that amount of the light absorbed or transmitted by a coloured solution and the intensity of the colour is directly proportional to the concentration and length of the coloured solution through which the light is passing. The Lambert-Beer's law can be expressed by the formula as given under—

$$OD \propto C \times \ell$$

Where OD is optical density, C is concentration and ℓ is length of the coloured solution through which light is passing.

Optical density (OD)—The optical density is defined as negative log of the transmittance and is a commonly used term to express the colour intensity or density of the coloured solution.

It has been observed that the plotting of a graph of concentrations of coloured solutions against transmittance is not always linear, but a graph of concentration verses absorbance or against optical density obtained is linear. Thus in photometric procedures, usually optical density is measured with the help of instruments such as colorimeter or spectrophotometer.

The colormetric estimation procedures are simple and very convenient. The length or thickness of the colored solutions to be compared in the colormeter is kept constant by using cuvettes (special test tubes) of same diameters. So that the colour intensity would depend only on concentration of the photosensitive compound to be estimated.

Procedure—A coloured derivative of the compound to be estimated in a biological specimen is prepared with the specific reagents and its optical density or absorbance is measured by a photoelectric colorimeter. This value is then compared with a standard solution of known concentration of the compound, treated similarly with the colour producing reagent. The concentration of the compound in the biological sample then can be calculated by the formula as given below—

$$Ct = \frac{ODt}{ODs} \times Cs \times 100 = \text{mg of compound /dl of Serum}$$

Where C_t is concentration of compound in the sample to be calculated, ODt is optical density of test solution, ODs is optical density of standard (known concentration) and Cs is the concentration of standard solution

Colorimeter—The basic components of a photoelectric colorimeter are described here—

1. **Light source**—A filament lamp is normally used to provide light radiation.
2. **Filters**—The filters are used to obtain a monochromatic light. Filters are usually coloured glass or gelatine papers, that would select a light radiation of small range of wave length to pass through the coloured solution.
3. **Cuvettes**—The cuvettes (sample holders) are special glass tubes of fixed constant diameter, so as to keep the thickness of the coloured solutions to be same, that are compared in the colorimeter.
4. **Photo electric cell**—A photoelectric detector called photocell consists of a selenium coated (photo sensitive material) plate, that converts the light stricking on it (on selenium) into a flow of electrons (electric current), which would deflect the needle of a galvanometer with a displaying scale to give directly the transmittance or optical density of the coloured solution (Fig. 37.1).

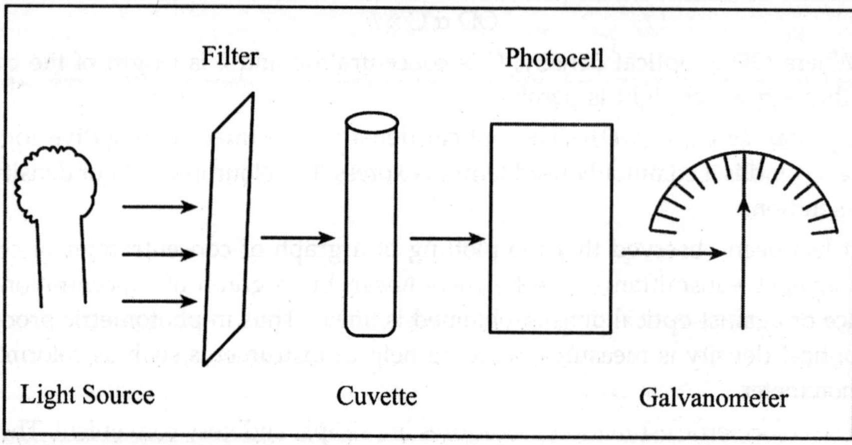

Fig. 37.1 : Basic components of a colorimeter

Monochromatic light

Monochromatic (mono means single, chrome is colour) light is the light radiation of single colour or wave length and is obtained by using appropriate light filter. The filters would absorb light of unwanted wave lengths and would allow only the desired light radiation to pass through a coloured solution. Usually the **complementary (opposite)** coloured lights are absorbed maximally by a coloured solution. Therefore contrast or complementary light is used in a colorimetric estimation to get a linear graph when concentration verses absorbance of coloured solution is plotted.

The coloured filters of desired wave lengths, normally used in the colorimeters along with the complementary colours of solution is summarized in the Table 37.1.

Table 37.1 : Coloured light filters and complementary colours of solution to be estimated in the colorimeter

Filters to obtain mono chromatic light	Wave length (Median)	Complementary colour of solution to be analyzed
Violet	420	Brown
Blue	470	Yellow
Green	520	Pink, Red
Yellow	580	Purple
Red	680	Green, Blue

When monochromatic light falls on the coloured solution, a portion of this light would be absorbed by the photosensitive (coloured) compound to be estimated, depending on its concentration in the solution. The rest of the light would be transmitted through the coloured solution, that is allowed to fall on the selenium coated photocell. Depending on the amount of light striking on photocell, an electric current would be generated, that would deflect the needle of the galvanometer scale, which is calibrated to give directly the optical density of the coloured solution.

Spectrophotometer

Spectrophotometer is the instrument that measures the absorbance or transmittance of light radiations by a photo-sensitive compound, from the ultraviolet and infra red regions of the light spectrum along with the visible range. The photosensitive compound that absorbs light radiations from ultraviolet or infra red regions of light spectrum, would appear colourless to the naked eye, but technically it is coloured solution and its concentration can be measured by the use of spectrophotometers.

The spectrophotometers have all the basic components of the colorimeter with some extra sophisticated and advanced additional devices to make them more sensitive in their operativity. For **precision operation,** a very narrow range of wave length (sharp) of monochromatic radiations are obtained in spectrophotometers by means of **prisms** and **diffraction grating device.** By these devices, desired monochromatic raditions from UV and infra red regions of the spectrum can be obtained. In addition, cuvettes used are of specialized **quartz glass,** which do not absorb UV radiation, in contrast to the ordinary glass tubes, that absorb UV radiations partially.

Light source to obtain UV radiation in the spectrophotometer, is a special gas discharge tube connected to a power supply. The electric power supply must provide stable voltage, as spectrophotometer is highly sensitive to even minute fluctuations in the voltage and would cause exaggerated effect.

The spectrophotometers are thousand times more sensitive than colorimeters and therefore even minute quantities of the substance can be estimated in the biological specimens. The procedure of the estimations is otherwise same in both the instruments.

Flame Photometry

The flame photometers are the instrument used for the accurate quantitative estimations of **electrolytes in biological samples.** Frequently used flame photometric procedures for the estimation of electrolytes in serum samples are for the determination of levels of **sodium, potassium, calcium, lithium,** etc.

Principle—Flame photometric analysis is based on the property of **emission photometry** rather than on optical absorption principles used in photometric analysis. It is observed that when these electrolytes (mention above) in solutions are sprayed on a flame, they get excited by absorbing heat energy and start emitting a light radiation of characteristic wave length which is specific for that particular electrolyte.

The intensity of this characteristic light of specific wave length emitted by the electrolyte sprayed in a flame, is proportional to its concentration in the solution and can be measured by allowing it to fall on a photocell in the same way as in colorimeter.

Components of Flame Photometers

The basic components of a flame photometer are (a) an atomizer with an compressor pump to draw and spray electrolyte solution to be estimated, (b) Blue flame is obtained by a gas burner of LPG (Fuel petroleum gas), (c) light filter to allow only characteristic radiation emitted by the electrolyte to pass through and (d) photoelectric selenium coated cell with a display galvanometer.

Procedure

When the electrolyte solution is sprayed on a blue flame, the electrolyte gets excited by heat to an high energy level and start emitting a characteristic light radiation of specific wave length and the flame acquires a different colour (e.g. golden yellow colour in case of sodium). This emitted light is focused on the photosensor through optical lenses and specific light filters (different for different electrolyte), that would select only the emitted light by electrolyte to be estimated. This selected characteristic emitted light is then allowed to fall on a photocell to generate an electric current which would deflect the needle of a galvanometer to give the optical density of the emitted light. This OD is then compared with a series of standard electrolyte solutions with known concentrations, in the range of the serum levels in the flamephotometer, to determine the concentration of electrolyte in the sample.

Ion selective electrodes

More sensitive equipments with **Ion selective electrodes** are now available to estimate sodium, potassium and other electrolytes in the serum samples. These instruments are highly sensitive and give accurate concentrations but are very costly.

Autoanalyzers

With advancement in the technology in the field of clinical biochemistry, a remarkable expansion has taken place in respect to the range, volume etc. in the speedy yet complex procedures, primarily because of the development of computerized autoanalyzers. It is now possible to analyse a wide range of biochemical parameters **at a time simultaneously** in large number of specimens in an autoanalyzer. Some important frequently used autoanlayzers in the clinical laboratories are described here.

Automated Photoelectric Analyzers

These are commonly known as semi autoanalyzers, since some of the steps involving pipetting of sample and reagents, incubation etc. are done manually. These autonalyzers operate on single step reagent methods. The reagents are now available as ready made kits to prepare coloured solutions of the compound to be analyzed in a biological specimen. The colour intensity of the test sample is then compared with standard solution (known cocentration) by aspirating them into these semi autoanalyzers (computerized equipment), that automatically measure, calculate and display the results directly or give print out of the results.

The semi autoanalyzers analyze one parameter at a time and thus are meant for parameter oriented analytical procedures. However, many samples can be analyzed at a very fast rate with the help of these autoanalyzers.

Batch and Selective Multichannel Analyzers

Fully automated analyzers are highly sophisticated and have computerized hard and software to estimate simultaneously 40 or more parameters at a time in a sample. At the same time more than 50 to 100 different specimen of different patients can be analyzed simultaneously, for a wide range of biochemical parameters or indices. All steps involved in the procedure of the analytic techniques are carried out by automated instruments. The laboratories in big hospitals and organizations are now equipped with fully autoanalizers for speedy estimations of several biochemical parameters in 50 to 100 specimens simultaneously to facilitate quick reporting of the tests.

Dry Chemistry System

The expression, dry chemistry is referred to the technique in which all the reagents and auxiliary substances required in the reactions of an analytical procedure are in dry powder form, embedded in the matrix of paper or plastic sticks called **clinisticks.**

The method of estimation is very simple and convenient and involves application of the sample over the reagent matrix on the clinistick. The colour produced by the reaction in a standard time period is then measured by **reflectance spectrophotometry**.

Reflectance photometer equipments, based on dry chemistry systems are now available commercially and have great impact in clinical practice, particularly in monitoring response to the treatment to control and manage several dreaded diseases which require periodic checkups. For example diabetes mellitus patient can check up his own blood sugar levels at home with the help of glucometer, an instrument based on the reflectance photometry with use of clinisticks.

Quality Control

Regular quality control programmes are necessary and should be an integral part of an organization having clinical biochemistry laboratory. The quality control is a **must** to ensure the reliability of the analytical procedure of a sample to enable doctors to correctly interpretate the test reports (results) for the diagnosis of diseases as well as in monitoring the patient (control of the diseases) under treatment.

Control Reference Laboratories operates in different parts of the country that send serum samples with known quantities of various biochemical parameters commonly assayed in clinical practice. This specimen is analyzed in the peripheral labs and if results obtained are comparable with that of the reference laboratory, then we can take the procedure in the clinical lab as reliable.

Accuracy and precision are the two most important aspects of quality control programmes and should be checked periodically.

Radioimmuno Assay—RIA

The technique of radioimmuno assay (RIA) was developed by Rosalyn Yalow (Noble price 1977) and Samuel Berson. It is a highly sensitive and specific analytical procedure that can accurately measure several physiologically important compounds present in trace amounts, even in nanogram or picogram in a biological sample.

Principle—The basic principle involved in RIA combines the specificity of an antigen- antibody reaction and radioactivity to estimate certain biochemical indices present in very low concentration in the blood. The RIA technique is based on the competition between the labelled (radio active isotope incorporated) antigen to that of unlabelled natural protein (present in serum) to be estimated, to bind with a specific antibody.

The more is the concentration of the unlabelled normal antigen (protein) present in the serum, less amount of labelled antigen will combine with antibody. This is because the unlabelled normal (natural protein) antigen has more affinity to bind with antibody than the radioactive labelled antigen.

Procedure—First an antibody against the normal protein (antigen) to be estimated is produced by injecting it in an animal, such as goat or rabbit. Such specific antibodies against a protein to be estimated in serum are now available, commercially for the ready use in RIA procedures. The major steps involved in RIA technique are summarized here.

(1) The specific antibody (Ab) is first incubated with a labelled antigen (e.g. insulin, I^{131}) in a test tube (A) and after a standard time period, the excess of radioactive insulin is washed away thoroughly. The radioactivity of the complex left in the tube is then determined by Gieger Muller counter.

(2) Now the serum sample containing unlabelled natural (insulin) antigen to be estimated is added in the test tube (A) containing antibody bound to labelled antigen and

incubated for standard time period. The natural unlabelled antigen present in the serum will compete and displace the radioactive antigen from the antibody complex and the amount of displaced labelled antigen would depend on concentration of natural antigen in the serum.

(3) The supernatant solution (containing displaced labelled antigen) is centrifuged and separated. The antibody-antigen complex left in the test tube is washed thoroughly and radio activity is determined. From the loss of radioactivity, estimated by Gieger Muller counter (difference from 1st and 2nd reading), the amount of normal level of unlabelled antigen (insulin) in the serum can be calculated.

Or in other way the radioactivity determined in the supernatant solution containing labelled antigen, separated from the antibody-antigen complex in the test tube can also be taken as measure of displaced labelled antigen. This will be equal to the concentration of unlabelled antigen in the serum.

Enzyme linked Immuno Sorbent Assay (ELISA)

ELISA is a **non radioactive immuno assay** technique and has advantage over RIA procedure, as there are no risk of radiation hazards. Moreover costly instruments such as Gieger Muller counter, as well as special laboratories and RIA trained technicians are not required. RIA is a tedious radius and time consuming process and need specialized training, equipment and laboratories.

ELISA has further advantage of a highly sensitive and specific technique, comparable to the RIA to measure physiologically important compounds accurately in biological specimens present even in nanograms.

In ELISA, specific antibodies are used to measure the amount of the corresponding antigen (protein) in a biological specimen. In this technique an enzyme is utilized to tag the antigen- antibody complex that facilitates the production of a coloured solution, hence the name Enzyme linked immuno sorbent assay (ELISA) is designated to the analytical procedure.

Procedure

In ELISA specific antibody is allowed to bind on a solid support, such as a plastic tray with moulded wells (a microtiter tray) or to plastic beads. Sample to be assayed is then added to the antibodies bound to the solid support. The specific antigen (protein) to be estimated, present in the sample is recognized by the antibody and gets bound on its specific epitope (Fig. 37.2).

The epitopes are the specific domains or sites of antigens, which are recognized by specific antibody. They are also called antigen determinants and an antigen can have many different epitopes to bind to different antibodies.

After the antigen gets bound to antibody on solid support, the wells of the tray are washed thoroughly to remove unbound antigens (proteins) and then the antibody- antigen complex is incubated with a second antibody (tagged with an enzyme) that recognizes and binds to the different epitope of the antigen than the first antibody as illustrated the Fig. 37.2.

In the next step the excess of second antibody tagged with the enzyme from the wells of the microtiter plate is washed off and colour producing reagent with a specific substrate (colourless compound) is added in the well. The enzyme linked to the antigen-antibodies complex would catalyze the reaction to produce a coloured solution. For instance, peroxidase is the enzyme commonly used in ELISA technique, that produces coloured compound by catalyzing degradation of H_2O_2 to liberate nascent oxygen, that oxidizes a colourless diamino-benzene of reagent to a coloured compound as described in the reaction given below—

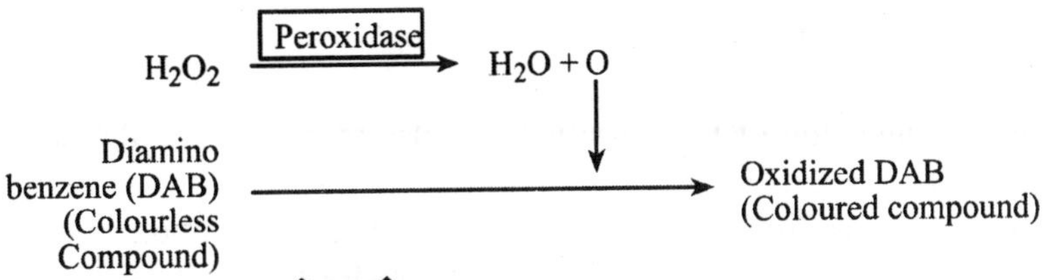

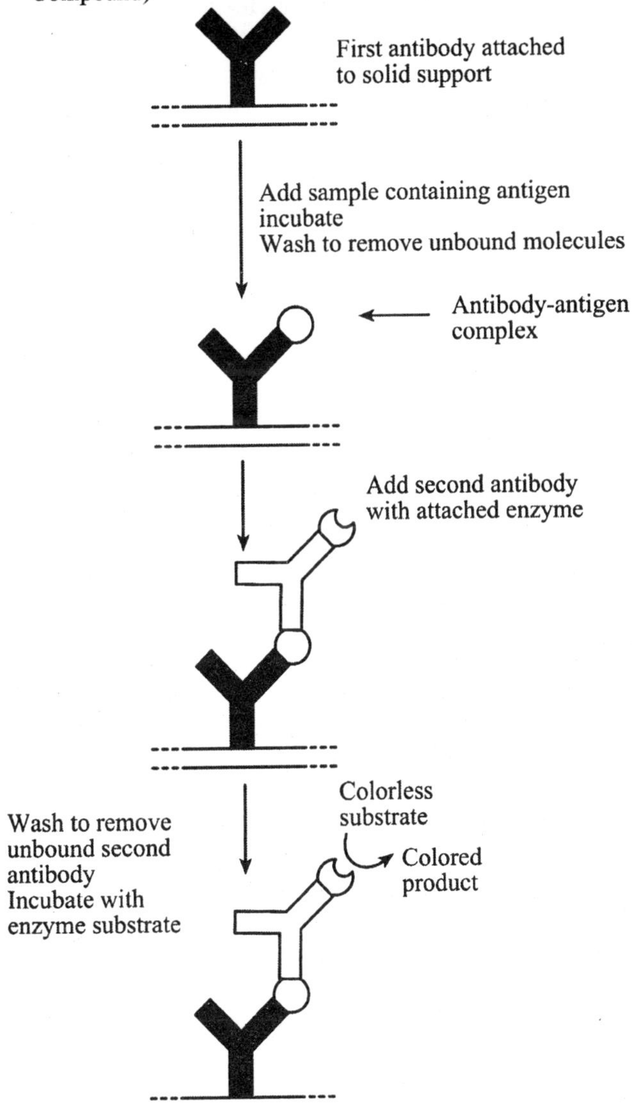

Fig. 37.2 : Diagrammatic representation of ELISA technique to estimate an antigen in a biological specimen

Several equipments (ELISA scanners) are now available that scan the coloured solution in the wells of the microtiter plate to facilitate the determination of the amount of the antigen (protein) to be estimated in the biological specimens.

HIV detection by ELISA technique

ELISA has a highly significant applications in the medical practice and has proved very useful in the detection of HIV positive samples to diagnose AIDS and other killer diseases.

ELISA is very useful in detecting small quantities of antibodies produced by an organism against specific antigen, for instance gp-120 of HIV in the sero-positive HIV infected AIDS patients (Chapter 38).

Procedure—HIV antigen (virus with gp-120) is added in the wells of microtiter plate containing serum sample of the suspected HIV infected patient. If antibodies are present (as in HIV infected seropositive patient) in serum, it will bind to specific epitope of HIV antigen. Next a second antibody (IgG) attached with an enzyme (peroxidase) is added which will bind to different eiptope of HIV antigen. The procedure of ELISA as mentioned above is then followed to produce a coloured solution, which is then scanned.

❑❑❑

38

BIOCHEMISTRY OF AIDS

Acquired immunodeficiency syndrome (AIDS) is the most dreaded disease, discovered towards the end of 20th century and is a serious burning problem world over. AIDS was first reported and confirmed in 1981 in USA in the homosexual men suffering from pneumocystitis carinii pneumonia that infects only AIDS patients. Based on the clinical manifestations, the disease was named as **acquired immunodeficiency syndrome** and is represented by the acronym of **AIDS.**

AIDS—THE KILLER DISEASE

AIDS is caused by a retrovirus, named as **human immunodeficiency virus (HIV)** and the disease is characterized by severe immunodeficiency with serious clinical manifestations with **fatal consequences.** AIDS invariable results in death of the affected individuals as there is **no cure of the disease till date.**

Brief History—AIDS was prevalent in Africa as early as 1970, which was proved by retrospective analysis of serum samples of African patients kept in deep freeze for the past 35 years. When those blood samples were tested by modern ELISA method, they were found to be HIV positive. It was seen that the **brunt of the attack by AIDS** was mainly on **young adults** and in certain African villages the young adult population was completely wiped off by the killer disease. Only old persons and small children survived and were found to be living in the villages.

Discovery of AIDS—The first human retrovirus (HIV) was isolated in 1984 from the lymphocytes of the AIDS patients independently by Gallo of USA and Montagniers in France. The virus was originally designated as HTLV-III, but later on was re-designated as HIV (human immunodeficiency virus) in 1986.

With the discovery of AIDS, it attracted great attention world over. USA has majority of reported AIDS cases. The prevalence of the killer disease (AIDS) has been since reported from most of the countries of world and approximately 5 to 10 crores persons are found to be HIV positive (WHO 1992). The number is rapidly rising and it has doubled in 2000 AD.

It is expected that more than 5-10 crore persons would be under attack by AIDS in this decade. Thus AIDS has become a **serious global problem** with an **alarming increase** in number of HIV positive patients in almost every country of the world.

Transmission of HIV

The human immunodeficiency virus is transmitted by the affected individuals to another person, **essentially through the exchange of body fluids,** primarily by three major routes as listed below.

1. **Unsafe Sex**—75% of the AIDS patients get the infection through unsafe sex, particularly in the homosexuals.
2. **Through blood**—About 15% of AIDS patients get the infection through blood exchange, for instance, the drug abusers use same needle (syringe) without proper sterilization or the persons receiving blood transfusions. This is a major risk, if blood is donated by HIV infected subject.
3. **From Mother to Fetus**—The virus (HIV) may be transmitted from HIV infected mother to fetus through placenta. About 30% of infants born from HIV positive mothers may get AIDS infection.

AIDS is not Transmitted by Touch

AIDS is not transmitted by casual personal contact, such as by touching, kissing or hugging the AIDS patient at home or at the working place. This awareness is very important in taking care and dealing with the HIV infected patients.

Characteristics of HIV

AIDS is caused by human immunodeficiency virus (HIV), which is a type of retrovirus. The **retrovirus** typically contain **RNA as genetic material** instead of DNA and are endowed with an enzyme **reverse transcriptase** (hence the name retrovirus), or the RNA dependent DNA polymerase. When the HIV infect the host cell, the RNA of the virus acts as a template strand and transcribes the genetic code (of HIV RNA), to synthesize a complementary copy of DNA (c-DNA) strand in presence of enzyme reverse transcriptase. This proviral c-DNA then replicates to produce double helix DNA which is then integrated into the genome of the infected host cell. When the host cells multiply, the pro-viral DNA is also replicated. After some time a large population of viral RNA are produced inside the host cell utilizing host cell enzymes and high energy nucleotides.

Two types of HIV have been isolated from the serum of AIDS patients and are called as HIV-I and HIV- 2. The type HIV- I is more common and shall be described here—

Structure of HIV

The HIV is spherical in shape with 110 nm diameter. It has a **cylindrical core** containing two strands of genome RNA, surrounded by a capsid with major antigen protein p24. The proteins of the virus are numbered after the molecular weight in kDa. There are four core proteins in the virus namely p24, p15, a reverse transcriptase enzyme (p66/51) and an endonuclease (p32).

The core RNA of virus with capsid is surrounded by a **lipid envelope derived from the host plasma membrane.** The lipid membrane is **studded** with **two HIV specific glycoproteins** called gp120 and gp 41. This surface antigen gp120 is characteristic of HIV and is very important for the viral infection as well as for the detection of HIV seropositive patients by modern ELISA technique. The HIV structure is illustrated in the Fig. 38.1.

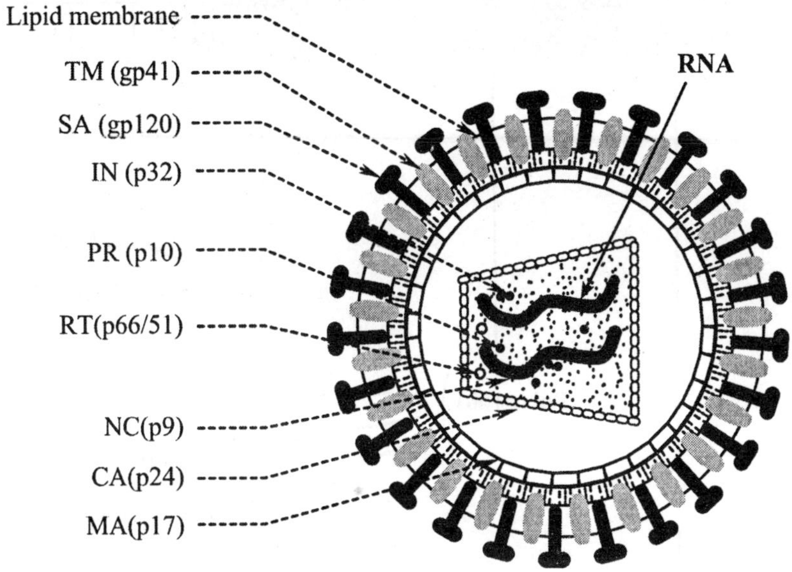

Fig. 38.1 : Structure of human immunodeficiency virus HIV (gp 120-glycoprotein with mol. wt. 120 k Da, RT-reverse transcriptase, IN-integrase, PR-protease, SA-surface antigen, TM-trans-membrane protein, MA-membrane antigen, CA-capsid antigen, NC-nucleocapsid)

HIV Genome and gene products

The HIV genetic matter consists of following genes—

1. **Three structural genes** are present in HIV RNA called gag, pol and env that encode for core proteins, reverse transcriptase (RT) and envelope proteins respectively.

   The gag and pol genes overlap by 241 bp (Fig. 38.2) and are transcribed in such a way that structural gag proteins are made in very large amounts, while enzymatic pol proteins (RT, p66/51 & INP32) are produced in relatively smaller amounts.
2. **Regulatory genes**—HIV contain three regulatory genes called tat, rev and nef which control transcription of m-RNA, from genome RNA for the synthesis of core proteins.
3. **Accessory Genes**—There are five accessory genes called vif, vpr vpu, vpt and tev present in HIV RNA genome.
4. **LTRs**—On either side of HIV genome, two **long terminal repeats (LTRs)** are present as shown in the Fig. 38.2.

Importance of LTRs

The LTRs consists of a **promoter region** with TATA box where the RNA polymerase binds to start transcription process for the synthesis of core viral proteins. In addition to this, LTRs contain certain conserved consensus regions called **transcription control**

elements. The LTRs are also responsible for the **integration of proviral DNA** into the host cell DNA genome as described earlier.

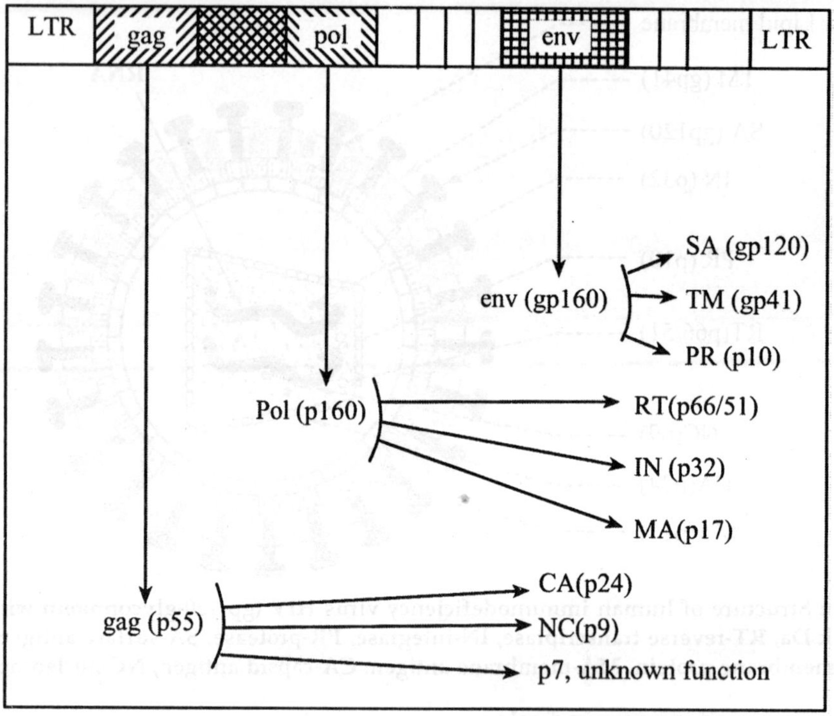

Fig. 38.2 : The HIV genome and the gene products

Maturation of HIV

The gag and pol genes are transcribed into m-RNAs, which in cytoplasm produce precursor protein p55 and p160. These are then incorporated into viral assembly and during or after budding process, are cleaved by viral protease (p10) into mature active proteins as illustrated in the Fig. 38.2. These mature proteins then get organized in the structure of virus as shown in the Fig. 38.1.

HIV Infection—Mechanism

HIV primarily affects the cell mediated immune system, that protects the body from the intracellular parasites, such as protozoa, mycobacteria, viruses etc. HIV infections severely suppress immune system by **reducing the number of CD_4 (Cluster determinant 4) cells of T lymphocytes (T- helper cells),** as well as it grossly impairs functions of CD_4 Cells. These CD_4 (T-helper) cells are the predominant cells responsible for providing the cell mediated immunity to our body. They are also important in the body as they produce cytokinines, macrophage chemotactic factors, hemopoietic growth factors etc.

HIV enters the CD_4 lymphocytes by binding to a specific receptor, the CD_4 protein molecules, through its membrane **(surface) glycoprotein gp120.** The CD_4 molecule has affinity for gp120 and thus acts as a receptor for HIV and therefore the CD_4 T- helper cells are attacked by HIV maximally. The monocytes and other macrophages are also susceptible to HIV infection and its propagation. In fact monocyte/ macrophage system acts as the **reservoir of the virus** in the body for **long periods (may be years)** and are store houses of HIV to be spread to other organs of the body.

After binding of HIV through gp120 on CD_4 receptors, the other membrane proteins gp 41 of HIV produce **fusion between viral envelope** and the **T- helper cell membrane.** The viral RNA is then internalized into the host cell and multiply there. The genome RNA of HIV acts as template strand and in the presence of enzyme reverse transcriptase, the genetic code of virus is transcribed as such into a complementary strand of DNA (c-DNA), which then replicated to produce a **proviral double helix DNA**. This proviral DNA is then integrated into the host cell DNA, by the action of viral integrase enzyme (p32). The viral genome terminal regions contain two long invert repeats (LTRs, Fig. 38.2) which are necessary for viral DNA integration into the host DNA.

The proviral DNA once integrated into the host DNA of T- helper cell (CD_4 cells), it replicates and multiply along with activated helper T-cells which are actively undergoing cell division. Retrovirus in general do not multiply in the non dividing cells or quiescent T-cells (inactive). During this period, a large population of human immunodeficiency **viral RNA** are transcribed from the proviral DNA integrated in the host genome **inside CD_4 cells.**

Stages of HIV Infection

Natural course of AIDS can broadly be divided into three stages—

Stage 1. Window Period or Latent Period—As mentioned above, when the HIV enters into CD_4 T - lymphocytes, it multiplies there and may remain **locked in the host cell genome** for **weeks** or may be for **months (up to 6 months).** Therefore during this period, HIV is not released in blood and so it cannot be detected in the blood and this period is known as **window period.**

Stage 2. Seropositive Period—After a few weeks or months, in any case by 6 months, the newly synthesized HIV by multiplication (as mentioned above) leave the host CD_4 cells by **erupting** as **buds on the surface** membrane of the host cell and are released in the blood with **simultaneous lysis of CD_4 cells.** The HIV in blood can now be detected by ELISA test and patient is called as **HIV positive** and the period as **seropositive period.**

Detection of Seropositive (HIV +ve) Patients

Detection of HIV in blood is done by ELISA technique, which is quite sensitive procedure for the detection of HIV infection. The HIV infected patients produces antibodies against viral gp120, which can be detected by ELISA (details in the Chapter 37). As only one antibody is detected by ELISA against HIV gp120, thus there are some chances of false +ve results in ELISA technique. So HIV should be confirmed by Western blot technique (Chapter 23.) which detects 6 different antibodies against different component antigens of the virus. However this Western blot technique is very costly and time consuming process, so it is not done in routine for HIV detection.

During this seropositive period (HIV +ve), which is the longest stage (5 to 10 years), the person is completely normal without any clinical manifestations of AIDS and is completely unaware of the fact that he has HIV infection unless his blood is tested for HIV. However this HIV infected person is a carrier of the disease (virus) for long period and unknowingly can transmit and infect other persons through body fluid exchange (3 routes) as mentioned earlier. This is the major cause of spread of AIDS worldover at an alarmingly fast rate.

Stage 3. AIDS Disease : Clinical Manifestations—This is the III and last stage of HIV infection when the full fledged AIDS developes. About 10% of HIV +ve persons will develope AIDS disease within 5 years of getting infection, about 50% will get AIDS in 10 years and rest in 15 years. This **long gap** between HIV infection and actual development of clinical manifestations of AIDS is responsible for the spread of the disease worldover with fast rate. This is because the HIV +ve person seems to be very normal, without clinical symptoms for long period (5-10 years) and no one will know that he is HIV +ve, unless blood test is done. Thus due to unawareness, that he has HIV infection, the patient can infect other persons or his wife through unsafe sex and thus unknowingly is responsible for the spread of HIV infection.

Clinical Manifestations of AIDS—The 3rd stage in the natural course of the disease is **severe immunosuppression** in the HIV +ve patients, which develops within 5-10 years of the infection. During this long seropositive period, the extensive budding of virus on the cell membranes of helper T- lymphocytes cause rupture of their plasma membranes with the lysis and degradation of host cells. The released viral particles in blood then infect other CD_4 T- lymphocytes and start multiplying inside the newly infected cells. The whole process is repeated several times, which **takes many years and this propagation of virus** ultimately results in **extensive lysis of CD_4 T- helper cells,** thereby reducing their count grossly in the blood, which finally leads to AIDS.

By this time the patient enter the 3rd stage and the immune status of the patient is severely depressed. A failure of defense mechanisms due to severe immunosuppression results in full-fledged clinical manifestations of AIDS, which are listed below—

1. Plasma levels of HIV are tremendously increased.
2. CD_4 T- lymphocyte count is drastically decreased (below <200/cmm).
3. Common microbial infections develop due to severely low resistance against infections, resulting in serious clinical complications.
4. Patient is unable to produce antibodies against HIV as well as against any other pathogens, due to the suppression of immune system resulting in severe infections and clinical manifestation, which can not be controlled or cured even by strong antibiotic drugs.
5. Patient ultimately and invariably **succumbs to death** within two years of this critical 3rd stage of AIDS.

Clinical Presentation of AIDS

Lymphoadenopathy, constant fever and **weight loss** are observed by the end of the 2nd stage. CD_4 T-helper cell count is decreased and the reduction is proportional to the severity of AIDS.

AIDS related symptoms (ARS) are wide ranged and beyond the scope of this book. Due to severe immunosuppression, non pathological comers microbes infect and multiply causing lesions in skin, GIT, lungs, urinary tract and brain.

- Gastroenteritis and tuberculosis are predominant infections caused in the AIDS patients in India. While in the Western countries, Kaposi's sarcoma is commonly associated with AIDS.

- CNS damage and neurological lesions results due to HIV infection in brain. Monocytes (reservoirs of HIV) enter brain and release virus which infect CNS, where HIV multiply and cause neurological damage.

Immunology of AIDS

1. As the name suggest the primary lesion in HIV infection is the severe immunosuppression. This is mainly because the HIV infects CD_4 T-helper lymphocytes, which are destroyed and their count decreases grossly in the blood as described earlier.
2. Another important mechanism for immunosuppression is the release of gp-120 molecules by the destruction of infected CD_4 cells. These liberated gp-120 molecules in the circulation have affinity and may get attacted to the nearby normal T-helper cells. Body tries to contain HIV infection, so the immunological system of body i.e. T Killer cells and antibodies against gp120 interpretate these normal T-lymphocytes with gp-120 on their surface, as HIV infected cells and attack these normal CD_4 T-helper cells and kills them. This further aggravates the condition by decreasing the number of circulating CD_4 T-helper cells drastically.
3. Tat protein released by HIV, further inhibit antigen induced proliferation of lymphocytes causing generalized immunosuppression.
4. Monocytes and macrophages infected with the HIV, store the virus as reservoir. These cells (store house of virus) may enter various organs such as brain, CNS etc. causing HIV infection and neurological lesions.
5. HIV infected patient's immune system produces antibodies against various viral proteins, particularly against gp-120 surface antigen, which is detected by ELISA test to identify HIV in the infected subjects.
6. HIV infection severely impairs functioning of T-helper cells and T-killer cytotoxin production is reduced drastically. Antibody dependent cell mediated cytolysis (ADCC) is thus severely inhibited causing severe immunosuppression.
7. Macrophage activity become sluggish and monocyte count is also reduced.
8. Generalized antibody response against foreign antigen is grossly reduced.
9. Delayed type hypersensitivity is decreased.
10. Production of lymphokinins, such as interferon, interleukin-2 etc. are decreased.

The effective immune mechanisms of body's defense system are thus paralyzed and AIDS patient is infected with all kinds of infections with severe clinical complications and ultimately succumbs to death.

Anti-HIV Drugs Used in AIDS Treatment

At present there is **no cure for AIDS** and once infected, the HIV +ve patient will ultimately succumbs to severe clinical manifestations of AIDS, that finally results in death.

However the **dideoxy nucleoside analogues** used in AIDS therapy, are effective in suppressing HIV multiplication and are able to prolong the life span of the AIDS patients for atleast some years. Chemically the dideoxy nucleoside drugs have only hydrogen atoms instead of normal OH groups at position 2nd and 3rd of ribose unit of the nucleoside.

Zidovudine or AZT

Zidovudine abbreviated as AZT is a commonly used and effective anti HIV and AIDS drug. Chemically it is **azido 2,3 dideoxy thymidine** and is a structure analogue of deoxy thymidine. It is the drug of choice and its mechanism of action is as described here— AZT inactivates reverse transcriptase by acting as a competitive inhibitor. It may be **incorporated in proviral DNA nucleotide sequence** in place of dTMP and would block formation of 3-5 diphosphoester bond between the incoming nucleotide and proviral DNA replication would be terminated prematurely. This would block viral multiplication and its spread to other CD_4 cells and thus life span of AIDS patients may be prolonged for some time.

Other dideoxy drugs used in AIDS treatment are dideoxy citidine (ddC), dideoxy inosine (ddI) etc.

Recently combination of AZT, ddI, ddC, saquinavir and ritonavir were shown to eliminate HIV from the body tissues. The last mentioned two drugs are inhibitors of viral protease enzyme. These drugs are very costly, as well as they are highly toxic to body and cause bone marrow suppression. Moreover drug resistant mutations may develop in viral genome within 6 months of the use of these drugs, making them ineffective. All these factors create great problems in effective AIDS therapy.

Another experimental approach to contain check HIV infection, is to inject micro beads coated with CD_4 proteins which absorbs all the free HIV from the blood and thus protects new CD_4 T-helper cells from being infected with HIV.

Prospects of Vaccine Against AIDS

Both HIV-1 and HIV-2 exhibit remarkable microheterogenicity due to the hyper variations in certain regions of viral genome. Such mutations or the "hot spots" accumulate to produce various types and strains of the virus, which would make it highly difficult to develop an effective vaccine against AIDS.

The most divergent part is the envelope gene. About 15 % of the amino acid sequence of gp120 are highly variable, so antigenicity of virus also varies and escapes from host's immune attack. The hypervariability in the gp120 also makes it difficult for our immune system to produce an effective antibody against gp120. This antibody may be unable to impart complete immunity against AIDS because of the mutations occurring in virus continuously.

However encouraging results have been obtained from the animal experiments and trials are being conducted with recombinant gp-120. But these are in early stage of testing and may take another 10 years to perfect an efficient vaccine against HIV and AIDS.

Other Types of AIDS Viruses

Other types of AIDS viruses exists in animals. For instance, **Simian virus** (present in monkeys) would not produce disease in monkeys because of the perfect symbiosis which has come in existance through evolution. But these viruses when transmitted into the new hosts (e.g. humans) would cause AIDS. Similarly human **herpes simplex (HHV-6) virus,** which may infect T-helper cells and rapidly kills them, may cause severe immunosuppression syndrome (AIDS) in humans.

Prevention of AIDS is the Only Cure

Since there is no cure for AIDS, preventive measures are best to protect oneself from this deadly killer disease. Public education and awareness about major risk factors which may cause AIDS, are the predominant means to limit the spread of HIV infection. Persons should be educated to take following precautions to protect themselves from HIV infection.

- Since major route of transmission of HIV is through sex with affected patients, persons should refrain from extramarital and unsafe sex.
- Before use, the needles and syringes should be properly sterilized or only disposable needles should always be used.
- Before blood transfusion, each sample should be tested for HIV by ELISA technique.
- Autoclaved surgical instruments and sterilized gloves should always be used.
- Some chemicals such as sodium hypochlorite (1%) or 2% glutaraldehyde (cidex) decontaminate heat sensitive surgical appliances and thus should be used in sterilization procedures.

❑❑❑

39

BIOCHEMISTRY OF CANCER

Cancer is the most dreaded killer disease prevalent world wide and is the most common cause of mortality in the developed countries, second only to heart diseases. There is no cure of cancer till date, and the patients ultimately die due to malignant tumors. It is becoming a serious global problem and challenging to the physicians world over.

Cancer is characterized by **uncontrolled proliferation of** cellular growth, which invades the local tissue causing tumors and may spread to other parts (metastasis) of the body. These characteristics differentiate **cancerous cells** from the normal cells, which multiply and proliferate under **strict control,** according to the physiological demands of the organ involved.

The **unrestrained** cell multiplication and the proliferation of the cellular growth results in tumor formation or cancerous growth. International union against cancer (IUAC) has defined cancer as—a disturbance of normal cell multiplication, characterized by excessive proliferation of the cellular growth with no relation to actual demand of the organ involved. **Oncology** deals with the etiology, diagnosis, treatment, prevention and research aspects of cancer.

Tumors broadly can be of two types—a) benign tumors and b) malignant tumors.

Benign tumors—The cells of benign tumors also show diminished control on cell multiplication with slow proliferation of cellular growth. But they do not invade local tissue or spread to other parts of the body, as they remain encapsulated by a layer of connective tissue. These tumors are not much harmful.

Malignant tumors—On the other hand, the malignant tumors are usually fast growing cancers and spread to other parts of the body. The process is referred to as **metastasis** and is invariably associated with severe clinical manifestations, ultimately resulting in death of the patient.

Most human cancers are caused spontaneously, without any particular cause which could be pinpointed. The incidence of the cancer increases with advancement of the age. More than 70% of the new cancer cases occur in the persons above 60 year of the age.

Etiology of Cancer—Multiple factors are responsible for the causation of cancer and they can broadly be grouped as: (a) Physical agents, (b) Chemical agents, (c) Biological agents, such as viruses etc. and (d) the genetic factors.

Most of the cancers (90%) are caused by chemical carcinogens, short wave length radiations and viruses. These agents damage DNA, resulting in mutations or interference with its replication or repair process.

Genetic Factors—Mutations

Spontaneous mutations, some of which predispose to cancer development, occur at the rate of 10^{-7} to 10^{-6} per cell per generation. These mutations result in the generation of potential cancerous cells from the normal stem cells, resulting in uncontrolled proliferation of cellular growth causing tumor formation. Oxy free radicals, such as hydroxyl ions ($OH^{\bullet}$) released from the mitochondria, cause oxidative damage to DNA causing mutations, some of which may be important etiological factors in the generation of cancerous cells.

Physical Agents—Radiation energy

The radiations such as X-rays, ultraviolet rays and γ-rays etc are **mutagenic** in nature and can be **carcinogenic**. These rays damage DNA in various manner. For instance, UV rays cause formation of pyrimidine dimers. The short wave length radiations can bring about single or double stranded breaks in DNA or may cause cross linkages formation. Radiations are also responsible for the insertion or deletation of a purine or pyrimidine nucleotide in the genetic code on DNA, resulting in mutations some of which may be carcinogenic.

Damage to DNA is presumed to be the basis of carcinogenic effects of the radiations (Chapter 19). In addition to directly damaging effects on DNA, the short wavelength radiations, such as X-rays and, γ-rays can cause generation of **oxy free radicals** in tissues, such as superoxide anion, $OH^{\bullet}$ ions etc. These oxy free radicals are potentially damaging to DNA and other macromolecules, causing **molecular damage,** which may result in carcinogenic effects.

Chemical Carcinogens

Many human cancers are caused by a variety of chemical agents. It is estimated that approximately 80% of cancers in humans are caused by environmental chemicals. Exposure to such compounds can occur by means of following routes—(a) subject's occupation,(e.g. asbestos, benzene etc),(b) through diet (aflatoxin fungus contamination), (c) life style, such as smoking etc. and (d) certain therapeutic drugs can also be carcinogenic in nature.

Chemical carcinogens act cumulatively and are present in tobacco, food additives, colouring agents and food preservatives etc. Aflatoxins as fungal contamination are common carcinogens in our environment.

Thousands of chemicals have been reported to be carcinogens, as they are mutagenic in nature. These carcinogenic chemicals may be organic or inorganic compounds. Some important inorganic metals which are implicated in carcinogenesis are asbestos, arsenic, cadmium etc.

Organic Carcinogenic Chemicals

1. Methyl cholanthrene is a powerful carcinogen, as only nano grams can cause tumor growth in the mice. It is observed that intestinal bacteria can synthesize cholanthrene derivates from cholesterol that may be involved in causing colon cancers.

2. Aflatoxins are synthesized from fungi and are powerful carcinogens, mainly acting on liver to produce **hapatomas**. The fungi contamination in grains, nuts or vegetable etc. may enter body through diet.

3. Cigarette and tobacco contain many carcinogens, such as nicotine, carbon mono oxide, nitrogen dioxide, and benzopyrenes etc. Thus cigarette smoking and chewing tobacco are implicated strongly in the development of lung, throat and oral cancers.

The **diversity in the chemical structures** of these compounds (carcinogens) shows that they do not possess one common group or radical responsible for their carcinogenicity and mutagenic effects on DNA.

Mechanism of Action

Direct carcinogens—Some of the organic carcinogens, such as methyl cholanthrene and β propiolactone have been found to interact directly with target cellular components or molecules to cause mutations in DNA and are considered as direct carcinogenic compounds.

Procarcinogens—However most of the carcinogenic chemicals require prior modification by metabolism in the body to become active carcinogens and thus are called **procarcinogens.**

Metabolic Activation of Procarcinogens

The activation of procarcinogen requires action of certain enzymes, such as monoxygenase, which basically is cytochrome p450 dependent and are also involved in the metabolism of xenobiotics and drugs in our body.

The ultimate (active) carcinogen binds covalently to the cellular macromolecules, such as DNA and interacts with purine (guanine) or pyrimidine nucleotides or phosphodiester groups and cause damage to the DNA. Some of these persistent unrepaired lesions in DNA molecule result in mutations causing carcinogenesis.

Mutagens—Most of the chemical carcinogens are thus mutagens and in laboratory can be detected by **Ames assay**. This assay is rapid, economical and has proved to be very useful in screening for the presence of potential carcinogens in biological samples. This assay uses a special strain of bacteria, the salmonella typhimurium (His), which has a mutation in its gene and cannot synthesize histidine. This amino acid has to be added in the culture media for growth of the bacteria. Interestingly it has been observed that some potential carcinogens are able to **reverse this mutation** by attacking on abnormal site of the gene to **restore** its histidine sythesizing ability (reverse mutation). The ames assay can detect about 90% of the chemical carcinogens in the biological samples and has become a routine preliminary screening procedure for the drug trials.

Promoters—Most of the chemical carcinogens require certain promoters for their action to promote cancer development. The various stages in the production of a malignant tumors by certain mutagens, which need promoters can be sequenced as follows—

Normal cell $\xrightarrow[\text{Initiation}]{\text{Carcinogens}}$ Mutations in DNA $\xrightarrow{\text{Promoter}}$ Pre-cancer $\xrightarrow[\text{Immunity}]{\text{Progression}}$ Malignant tumor

For instance, when the carcinogen benzopyrene is applied on the skin, it does not produce cancers, but when it is followed by an application of **croton oil,** tumor developed. Here the croton oil is termed as promoter. The active agent in the croton oil is **phorbol ester,** most active being **tetradecanoyl phorbol acetate (TPA).**

Mechanism of Promoter Action

Protein Kinase C enzyme present on the surface of skin act as the receptor for TPA. When TPA binds on protein kinase C, the enzyme is activated and phosphorylates the membrane proteins leading to the triggering of carcinogenesis causing malignancy. These events indicate that carcinogens can cause mutations in the genes, but it is the promoter which gives the drive for uncontrolled cell multiplication leading to malignancies.

Several compounds that act as **promoters in triggering carcinogenesis** in various body organs, along with chemical carcinogens have been identified. These include saccharin, phenobarbital, phorbol esters (TPA) etc.

Carcinogenic Viruses

Oncogenic viruses contain either DNA or RNA as the genetic material i.e. genome. It has been observed that under certain circumstances infection of appropriate cells with polyoma virus or SV 40 viruses can result in their malignant transformation in some animal species. For example, SV-40 (simian virus) that causes lytic infections in monkeys, when introduced in the mice or hamster, **produces sarcomas.** These viruses are small in size (genome of 5 kbp) and their circular DNA (genome) code for only about 5 to 6 proteins. In the case of SV-40, these proteins are known as T and t and in polyoma virus as T, mid T and small t. These T antigen proteins have a affinity to bind tightly to the host DNA and alter gene expression resulting in carcinogenesis.

Transformation to Cancerous Cell

Another probable etiological factor of carcinogenesis is the integration of viral genes into the host DNA (animal and humans). When the oncogenic virus infects the animals, its circular DNA is integrated into the host DNA and became its part (as in chimeric DNA) and replicates along with host DNA replication. However the **powerful drive of multiplication** by **viral genome overrules** the **regulatory checks and control of host DNA** replication, (that balances cellular multiplication) and results in uncontrolled multiplication of cells causing production of tumors and malignancies. Here the oncogenic virus infection in the host, **transforms** the **normal stem cell into a cancerous cell.**

Examples of **oncogenic viruses** are **adenovirus,** such as Epstein Barr virus which is associated with Burkitt lymphoma and nasopharangeal carcinoma in humans. Hepatitis B virus is probably the major etiological factor involved in liver cancers.

Retroviruses—Many oncogenic viruses containing RNA as genome, called retroviruses have been identified. Much of our understanding about oncogenes in recent years has been derived from these viruses. Retrovirus R type is implicated in the development of mammary gland tumors in mice studies and retrovirus C is associated with cancers of blood, such as leukemia and sarcomas etc.

Molecular Basis of Cancer—Oncogenes

Oncogenes are genes which are capable of causing cancer and play an important role in the process of carcinogenesis. Their discovery has had a major impact in our under standing the etiology of carcinogenesis. Oncogenes were first discovered as **unique genes** of **tumor causing viruses.** A definite proof for oncogene was first demonstrated in **Rous sarcoma virus** that produces sarcoma (cancer arising from connective tissue) in avians. However the evidence was little blurred, when it was shown by recombinant DNA hybridization techniques, that normal host avian cells also contain DNA sequences similar to that of viral oncogenes, which are now recognized as **protooncogenes.** In recent years about 80 such protooncogenes or cellular C-oncogens in humans have be identified.

Presently the protooncogenes are considered to be **important regulatory genes** which encode for some growth regulatory proteins such as tyrosine kinase, protein kinase C etc.

The C-oncogenes on expression generate protein factors that are essential for the normal function and cell division but they are under the control of regulatory genes and are expressed only when required physiologically. However these protooncogenes can be activated to oncogenes by various carcinogenic factors for example oncogenic viruses etc.

Activation of Protooncogenes

The activation of C-oncogenes or protooncogenes into oncogenes is a crucial step in the process of carcinogenesis. The C- oncogenes may be activated by several factors. A few important mechanisms are described here.

1. Viral insertion as promoter—When certain retroviruses, such as avian leukemia virus with RNA as genetic material, infects host cell, a complementary DNA copy(c-DNA) of their RNA genome is synthesized by the enzyme **reverse transcriptase**. This c-DNA then is replicated to produce a **proviral DNA** and is **integrated** into the **host genome near inactive protooncogene (e.g. myc gene).** The myc gene gets activated by this proviral insert, that acts as a promoter of transcription, and takes over the control of the transcription of myc gene. Due to this the normal regulatory check and control on cell division is lost which transforms the normal cell into a cancerous cell. The activation of myc gene (protooncogene) by proviral DNA insert, that acts as promoter is shown in the Fig. 39.1.

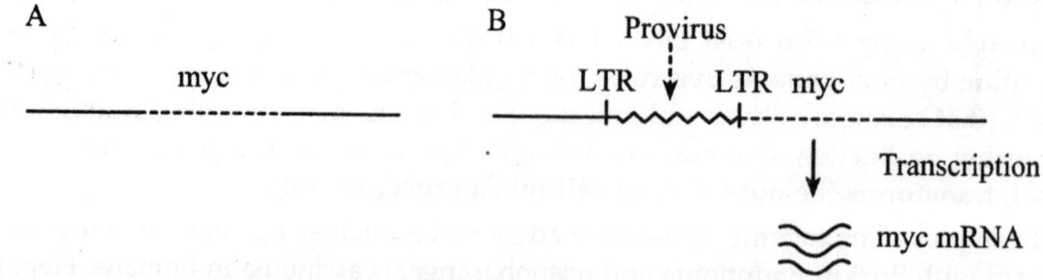

Fig. 39.1 : Diagrammatic representation of provirus insertion to activate (A) protooncogene (myc) to increase its expression (B)

2. Enhancer insertions—Some proviral DNA so produced (as described above) are inserted downstream (away) from the myc gene but can activate it to oncogene by acting as

enhancer elements on DNA (Chapter 20). This results in uncontrolled gene expression, that transforms the normal cell into cancerous cell.

3. Chromosome translocation—Certain proviral DNA insertions cause rearrangement of genetic material by chromosomal translocation and activate protoonocogenes to become oncogenes. For example in Burikitt's lymphoma of humans, a fragment from chromosome 8 is split off and gets joined (integrated) to chromosome 14 containing inactive myc gene by the viral infection. This translocation results in activation of myc gene leading to its increased transcription and synthesis of certain growth factors or proteins in excess, which transform the cell to become cancerous or malignant in nature.

4. Gene amplification—Amplification of certain genes is also implicated in the development of several types of malignant tumors (Chapter 22). Prolonged administration of anticancer drug, methotrexate makes the cancerous cell resistant to it. This drug acts by inhibiting enzyme dihydrofolate reductase (details in Chapter 22). After sometime the gene encoding the dihydrofolate reductase becomes amplified several folds resulting in increased synthesis and activity of this key enzyme to about 400 times than the normal, making the anticancer drug ineffective.

Certain cellular oncogenes can also be amplified in similar manner and are activated to produce excess amounts of certain protein factors, such as c-ras which may cause progression of tumor to become malignant as describe below.

5. Point Mutations—The ras protooncogenes produce a protein termed p21, (mol. wt. 21 kDa) that has GTPase like activity that suppresses the activity of adenyl cyclase after its action. This ras protooncogene is activated to ras oncogene by a point mutation with single base change in a triplet (coding for specific amino acid) leading to the synthesis of an abnormal p21 protein with single amino acid change, at position 12th resulting in loss of its activity. Due to this the **adenyl cyclase activity remains continuous (abnormal)** resulting in increased action of certain growth factors that transform normal cells into cancerous cells. Example is the c-ras oncogene produced in human bladder cancer.

However it is important to know that the activation of oncogenes is not the sole factor in carcinogenesis. There are other genes which protect individuals from cancers. These are called antioncogenes or oncosupressor genes.

Antioncogenes or Oncosuppressor Genes

Antioncogenes normally protect the individuals from getting cancer. The products of these cancer suppressor genes **restrain** and **apply brake to control and regulate normal cell division.** When this gene is deleted or mutated, it result in the loss of this check on the cell multiplication, resulting in unrestrained cell division, cellular growth and proliferation leading to cancerous tumor.

It has been shown that along with the **activation of oncogenes** simultaneous **inactivation** of **oncosuppressor genes** is involved in the development of most of the cancers. The oncosuppressor genes are abbreviated by capital letters, for instance for retinoblastoma (RB), Wilm's tumor (WT), familial adenomatus polyposis (FAP), deleted in colon cancer (DCC) etc.

The oncogenes can be activated by a variety of factors with simultaneous deletation or mutation of antioncogene, for example by viruses, chemical carcinogens, chromosome translocation, short wave length radiations and spontaneous mutations etc. This provides explanation for the multifactorial origin of cancer by the oncogene activation with simultaneous inactivation of oncosuppressor genes, that is later on manifested in malignancies.

In case of retinoblastoma when both the alleles of the RB genes are mutated, then only the cancer is caused.

A part of short arm of chromosome 17 was shown to be deleted in various human cancers and it is this region which **contain oncosuppressor gene p53.** Most tumors have a complete absence or mutant p53 gene.

Mechanism of Action of Oncogenes

General mechanisms through which the products of oncogenes stimulate excessive or uncontrolled cell proliferation resulting in cancers are described here. The oncogenes encode for certain **oncoproteins, which are altered (abnormal) form** of the normal protein factors involved in the check and control of cell division and growth according to the physiological demand of the tissue.

These oncoproteins act on and alter key intracellular pathways involved in the control of cell multiplication and growth according to the normal physiological demands. Examples are—(a) src oncoprotein acting as a tyrosine kinase, (b) c-ras oncoprotein causing continuous adenyl cyclase activity and (c) activation of myc gene to produce DNA binding proteins. Each of these oncoproteins affect the cell cycle, resulting in **unrestrained cell proliferation** and malignancies.

Growth Factors

Several growth factors which stimulate the cell division (mitosis), cellular differentiation and proliferation have been identified (Table 39.1). They regulate cell division by transmitting the message across the plasma membrane by transducing transmembrane signals, mostly through protein kinases. These protein kinases (e.g. tyrosine kinase and protein kinase C) phosphorylate intracellular target proteins to activate them, which in turn via intracellular messengers reach to nuclear DNA to stimulate cell division.

Now it has become clear that the oncogene products are altered version of normal growth factors and cause uncontrolled cell proliferation resulting in malignancies. For example the transforming growth factor (TG F α) is normal growth factor required for the epithelial cell growth. It has been shown that TGF- α is produced in high concentration in individuals suffering from skin cancer psoriasis. Some important growth factors of our body along with their functions are summarized in the Table 39.1.

Table 39.1 : Some Important Growth Factors along with their functions

S.No.	Growth Factor	Source	Functions/Stimulate growth of
1.	Epidermal growth factor (EGF)	Salivary gland fibroblasts	Epidermal and Epithelial growth
2.	Platelet derived growth factor (PDG-1)	Platelets	Mesenchymal cell growth & accelerate wound healing

contd. ...

... contd.

S.No.	Growth Factor	Source	Functions/Stimulate growth of
3	Transforming growth factor (TGF α)	Tumor cells placenta	similar to EGF
4	Erythropoietin (EP)	Kidneys	early phase of erythropoiesis
5	Insulin like growth factors (IGF-1)	serum	Incorporation of sulphate in cartilages
6	Interleukin 1 & Interleukin 2	Monocytes Lymphocytes	Synthesis of IL-2 Growth and maturation of T-cells

Growth factor receptors—Some oncogenes encoding growth factor receptors have been identified. On activation, the oncogenes produce **structural alteration in the growth factor receptors** of the target tissues, resulting in over expression of the action of growth factor causing carcinogenesis. For instance oncogene erb-B on activation causes EGF- receptor alteration, resulting in over expression of EGF that leads to the lung cancer.

GTP binding products of ras oncogenes is the single most dominant cause of many human cancers. The ras protein p21(mol.wt.21kDa) is involved in control of cell multiplication as described before. A point mutation in ras gene results in the production of an altered form of p21, resulting in continuous activations of enzyme adenyl cyclase, that leads to carcinogenesis.

Cancerous Tumors—Salient Features

In the normal tissues, only 1% cells are in the dividing state while in cancerous tumor 2% are in cell division stage (mildly growing tumor) and in some case 5% are in multiplying stage as in the case of aggressively growing cancers. This demarcation guides the clinician to select and plan treatment to control cancerous growth.

When a normal cell acquires cancerous characteristics, the morphological and biochemical changes occur and it is referred to be transformed. Renato Dulbecco was pioneer in transformation studies in cancerous tissue culture (Noble Prize 1975) and reported characteristic features of cancerous growth that are listed here under—

1. Cancerous cells show multilayered appearance where as normal cells are monolayered (Fig. 39.2)
2. **Loss of contact inhibition**—It is characteristic of the normal cells to have contact inhibition. For instance when skin is cut, the cells from both the sides of cut, starts multiplying till the cut is filled (wound healing). When they come in contact, the multiplication is stopped and tight junctions are formed for cell to cell communication. But in case of cancer cells, **tight junctions are rare, contact inhibition is lost** and adjacent cells continue to multiply to form a multilayer **jumbled structure** as is evident from the Fig. 39.2.

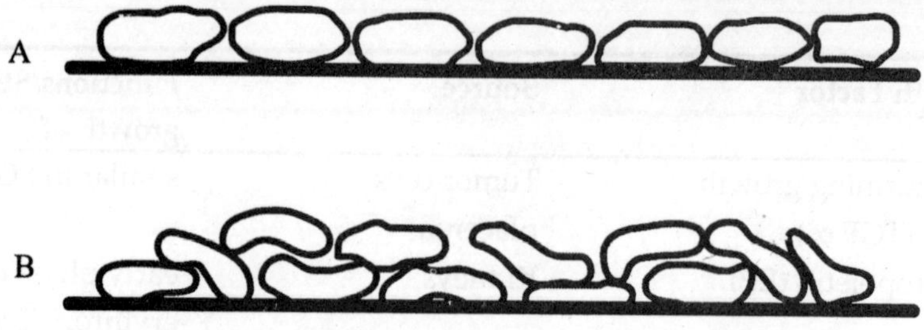

Fig. 39.2 : Normal (A) and cancer cell multiplication (B)

3. **Loss of anchorage dependence**—The normal cells adhere firmly on to the tissue surface and normally do not get disintegrated from each other or from the basement membrane through an adhesive protein called **vinculin**. In cancerous tissue, an oncogene product, tyrosine kinase phosphorylates vinculin (abnormal) and it looses its adhesion character. This result in the loss of anchorage dependence of the cancerous cell and it may migrate to invade and spread to other tissues. Many cancer cells possess more negative charges on their surface than normal cells, due to the higher N-acetylneuraminic acid (NANA) contents on their membranes. Because of high negative charges, the cancer cells repel each other resulting in less adhesiveness. This results in their moving away from the primary site to invade other tissues causing **metastasis.**

Metastasis

The spreading of cancerous cell from the primary site to the other parts of body organs, where they proliferate and cause secondary tumors is referred to as metastasis.

The factors which contribute to the tendency of a cancerous cell to form secondary tumors (metastasis) are the loss of anchorage dependence to each other or to the surface of tissues and their high negative charges. Because of these factors the cancer cells disintegrate from the main tumors mass and migrate to invade nearby tissue or to the distant organs to spread the cancer through out the body, causing secondary tumors or metastasis.

Metastasis or secondary tumors affecting other body organs is the major cause of morbidity and mortality in cancer patients.

Anticancer Role of Vitamin A

The enzyme collagenase and stromolysin released by most of the disintegrated and spreading cancer cells help in their penetration into the surrounding tissue and distant body organs, causing metastasis. Retinoic acid is shown to decrease activity of these enzymes produced by cancer cells, that may explain the anticancer activity of vitamin A.

Biochemical Alterations in Cancer Cell

1. Loss of normal antigens—Abnormalities in glycoproteins on the surface layer of cancer cell results in loss of normal antigens and appearance of new antigens.
2. Many cancer cells thrive on minimal enzyme requirements. For instance many tumor cells prefer anaerobic glycolysis (excessive glucose consumption) and

thus delete citric acid cycle enzymes as well as **aspargine synthase** (see anticancer drugs) in certain lymphomas.

Apoptosis—In normal healthy tissues, the old sensate cells are lysed and new cells are produced in equal numbers. This programmed death of cells after completing their life span (e.g. RBC, 120 days) is known as **apoptosis** and is regulated by certain genes, such as c- fos, p53, Rb etc. Certain cancers are produced by preventing programmed cell death. For instance the bcl 2 prevents apoptosis in B lymphocytes causing B cell lymphomas.

Tumor Markers

The diagnosis and early detection of cancer is the key factor that would help in its effective treatment and to prevent severe clinical complications. Certain tumor markers, which are produced abnormally in case of cancerous growth can be detected in blood and urine and are utilized as biochemical indicators for the early diagnosis and more importantly in prognosis of cancer treatment.

Some important tumor markers include—oncogene products, abnormal proteins, enzymes, growth factors and surface antigens etc. Some of these clinically useful markers can be measured in plasma and can support screening of cancer as well as can be useful in monitoring the response to the therapy and for early detection of recurrence of cancerous growth after surgery. Some selected most commonly used tumor markers for the diagnosis and prognosis of cancer are listed in the Table 39.2.

Table 39.2 : Important tumor markers useful in screening of cancers

Name/ Class of tumor markers	Plasma level increased in cancers of
Oncofetal antigens—	Colon, rectal, stomach, lung, pancreas &
-Carcino embryonic antigen (CEA)	—breast cancers.
-α fetoprotein (AFP)	—Hepatomas, germ cell cancer of testes.
-Cancer antigen 125 (CA-125)	Ovarian cancer
Hormones—	
Human chorionic gonadotropins (hCG)	Chorionic carcinoma
Epinephrine & metabolites, VMA.	Pheochromatomas & neuroblastomas
Enzymes & proteins—	
Prostate specific antigen(PSA) & -Acid phosphatase	Prostate cancer
Abnormal immunoglobulins—	
Bence Jones proteins	Multiple myeloma
Tissue catabolic products—	Argentaffinoma
Hydroxy indole acetic acid (OH IAA)	Cancer of serotonin producing cells.
Hydroxy proline	Bone malignancies

Anticancer Drugs

Prime modalities used in cancer therapy are surgery, radiotherapy and chemotherapeutic agents.

Anticancer drugs that check cancerous growth by killing cancer cells are effective but they should not be excessively toxic to the normal tissues. Some widely used anticancer drugs are listed in the Table 39.3. Many of these drugs act by inhibiting DNA synthesis and thus cell multiplication would be blocked and cancer growth is checked. They would also damage normal body tissues which are rapidly multiplying, such as bone marrow and GIT etc.

Tumors with high growth fraction (rapidly growing) are usually more responsive to chemotherapy than the slow growing cancers.

Cell cycle specific anticancer drugs, such as methotrexate, flourouracil and cytorabine act primarily on proliferating cells (multiplying cells), while the nonspecific drugs such as alkylating agents and cisplastin act irrespective of the state of proliferation i.e. all kind of cells in a tumor would be affected and killed.

High Dose - Intermittent Therapy

In effective cancer therapy, it is vital to detect and treat tumor at an early stage. In general **high-dose intermittent therapy** is more effective, as it exposes all tumor cells to a higher level of the drug used. Secondly during **intermittent no dose time**, the RBC level is restored to normal to prevent clinical complications.

Combination chemotherapy with 3 or 4 agents has proved successful in some cancers, as these drugs act **synergistically** and thus are more effective. Secondly the onset of resistance against the drugs may be delayed for some time. In the combination therapy the toxic effects are also less severe.

Table 39.3 : Important chemotherapeutic drugs used in cancer treatment

Name/class of drug	Treatment in Cancer	Mechanism of Action
Anti metabolites—		
Mercapto purine	Acute lymphatic leukemia	Prevents amination of IMP to AMP, so impairs DNA replication
Fluoro uracil	Adenocarcinoma of colon & breast.	Thymine analogue Inhibits thymidylate and DNA synthesis
Methotrexate	Chorionic carcinoma, Acute Leukemia	Inhibit enzyme dihydrofolate reductase, prevents puirne nucleotide & DNA synthesis
Alkylating agents—		
Melphalan	Lymphomas, multiple myeloma, myeloid leukemia	Alkylates DNA, forms cross linkages, so DNA strands are not separated for replication.
Antitumor antibiotics—		
Mitomycin C	Adenocarcinoma of stomach, breast, colon.	Forms cross linkages, so inhibits DNA replication.
Actinomycin D	Wilm's tumor Chorionic carcinoma	Blocks transcription and m- RNA synthesis.

contd. ...

... contd.

Name class of drug	Treatment in Cancer	Mechanism of Action
Plant compounds— Vineristine, Vinblastine are alkaloids isolated from vincarosea	Lymphomas, leukemia & solid tumors.	Arrest cancer cells in metaphase, so cell division is inhibited.
Others— Azaserine & 6- azauridine.	Anti cancer drugs effective in several cancers	Inhibit purine and pyrimidine nucleotide & DNA synthesis.

Drug Resistance—A major problem in cancer chemotherapy is, that cancerous tumors develop resistance against drugs after some time. The cancerous cells acquire the resistance by several mechanisms which render the chemotherapeutic agents ineffective later on during the treatment. The mechanisms of drug resistance reflects high spontaneous mutation rates in the DNA of the cancer cells.

Some important mechanisms of drug resistance by cancerous cells are as described below—

1. **Gene amplification** to increase synthesis of an enzyme needed for the proliferation by several hundred folds, for example methotraxate resistance by increasing dihydrofolate reductase synthesis (Details in Chapter 22).
2. **P-glycoprotein : Multidrug resistance**—The cancer cells which exhibit multi drug resistance often have increased amount of p-glycoprotein on their plasma membranes. This protein acts as an **energy dependent pump** which **eject out** a **wide variety of anticancer drugs** from the cancer cells, to be metabolized in liver and excreted out of the body.
3. Drug uptake is reduced by deleting or mutations in genes encoding carrier proteins.
4. Increased drug catabolism by increasing cytochrome p450 synthesis which is responsible for degrading the drug in the liver to make it ineffective.

❑❑❑

Abbreviations

A	adenine, adenosine
AA	Amino acids
Ab	antibody
ACP	acyl carrier protein
ACAT	acyl CoA-Cholesterol acyl transferase
ACTH	adrenocorticotropic hormone
Acyl CoA	fatty acid derivative of coenzyme A
ADA	adenosine deaminase
ADH	alcohol dehydrogenase
ADH	antidiuretic hormone
ADP	adenosine diphosphate
AFP	α-fetoprotein
Ag	antigen
A/G	albumin/globulin (ratio)
AIDS	acquired immunodeficiency syndrome
Ala	alanine
ALA	δ-aminolevulinic acid
ALP	alkaline phosphatase
ALT	alanine transaminase
AMP	adenosine monophosphate
Apo-A	apoprotein-A
Arg	arginine
Asn	asparagine
Asp	aspartic acid
AST	aspartate transaminase
AT	α_1-antitrypsin
ATCase	aspartate transcarbamoylase
ATP	adenosine triphosphate
BAL	British antilewisite
BAO	basal acid output
BHA	butylated hydroxyanisole
BHT	butylated hydroxy toluene
BMR	basal metabolic rate
bp	base pair
BP	blood pressure
BPG	bisphosphoglycerate (2, 3-BPG &1,3-BPG)
BSP	bromosulphthalein
BUN	blood urea nitrogen
BV	biological value

C	cytosine, cytidine
CA	carbonic anhydrase
Cal	calorie
Cam	calmodulin
cAMP	3', 5'-cyclic adenosine monophosphate (cyclic AMP)
CAP	catabolite activator protein
CBG	corticosteroid binding globulin
CCK	cholecystokinin
CCCP	Chlorocarbonyl cyanide Phenylhydrazone
cDNA	complementary DNA
CD_4	cluster determinant antigen 4
CDP	cytidine diphosphate
CEA	carcinoembryonic antigen
cGMP	3', 5'-cyclic guanosine monophosphate
C_H	constant heavy chain
CHD	coronary heart disease
ChE	choline esterase
Chl	chlorophyll
C_L	constant light chain
CLIP	corticotropin like intermediate lobe peptide
CMP	cytidine monophosphate
CNS	central nervous system
CoA or CoASH	coenzyme A
COHb	carboxyhemoglobin
COMT	catechol-o-methyltransferase
CoQ	coenzyme Q (ubiquinone)
CPK (CK)	creatine phosphokinase (creatine kinase)
CPPP	cyclopenteno perhydrophenanthrene
CPS	carbamoyl phosphate synthase
CRH	corticotropin releasing hormone
CRP	C-reactive Protein
CS	chorionic somatomammotropin
CSF	cerebrospinal fluid
CT	calcitonin
CTP	cytidine triphosphate
Cys	cysteine
d	day
D	dextrorotatary
dA	deoxyadenosine
dADP	deoxyadenosine monophosphate
DAG	diacylglycerol
DAM	diacetyl monoxime
dAMP	deoxyadenosine monophosphate
dATP	deoxyadenosine triphosphate
dCMP	deoxycytidine monophosphate
DCT	distal convoluted tubule
DEAE	diethyl aminoethylamine
DFP or DIFP	diisopropyl fluorophosphate
dGMP	deoxyguanosine monophosphate
DHAP	dihydroxyacetone phosphate
DHCC	dihydroxycholecalciferol (1, 25-DHCC; 24, 25-DHCC)
DHEA	dehydroepiandrosterone
DHF	dihydrofolate
DHT	dihydrotestosterone
DIT	diiodotyrosine
dl	deciliter
DMB	dimethyl benzimidazole
DMS	dimethyl sulfate
DNA	deoxyribonucleic acid

DNase	deoxyribonuclease
DNP	2, 4-dinitrophenol
DOPA	dihydroxy phenylalanine
DPG	diphosphoglycerate
DPP	dimethyl allyl pyrophosphate
dTMP	deoxythymidine monophosphate
E_O	redox potential
EC	enzyme commission
ECF	extracellular fluid
EDRF	endothelium-derived releasing factor
EDTA	ethylene diamine tetraacetate
EF	elongation factor
EFA	essential fatty acids
EGF	epidermal growth factor
ELISA	enzyme-linked immunosorbent assay
EM	Embden-Meyerhof
ER	endoplasmic reticulum
ES	enzyme-substrate complex
ETC	electron transport chain
FA	fatty acid
Fab	antigen binding fragment
FAD	flavin adenine dinucleotide
$FADH_2$	reduced FAD
FAS	fatty acid synthase
Fru.1,6-BP	fructose 1, 6-bisphosphate
Fru.2,6-BP	fructose 2, 6-bisphosphate
Fc	crystalline fragment
FDNB	1-fluoro 2, 4-dinitrobenzene
FFA	free fatty acid
FGF	fibroblast growth factor
FH_4	tetrahydrofolate
FIGLU	formiminoglutamic acid
fMet	N-formylmethionine
FMN	flavin mononucleotide
$FMNH_2$	reduced FMN
Fru.l-PO_4	fructose 1-phosphate
Fru.6-PO4	fructose 6-phosphate
Fp	flavoprotein
FSH	follicle stimulating hormone
FTM	fractional test meal
G	guanine, guanosine
g or gm	gram
ΔG	free energy change
GABA	γ-aminobutyric acid
GAG	glycosaminoglycans
Gal-Cer	galactocerebroside
GAR	glycinamide ribotide
GDH	glutamate dehydrogenase
GDP	guanosine diphosphate
GFR	glomerular filtration rate
GGT (GT)	γ-glutamyl transpeptidase
GH	growth hormone
GHRH	growth hormone releasing hormone
GIP	gastric inhibitory peptide
GIT	gastrointestinal tract
Gla	γ-carboxy glutamate
GLC	gas liquid chromatography
Gln	glutamine
Glu	glutamic acid
Glu-Cer	glucocerebroside
Gly	glycine
GLUT	glucose transporters
GN	glucose-nitrogen (ratio)
GMP	guanosine monophosphate
GnRH	gonadotropin releasing hormone
GRH	growth hormone releasing hormone
GRIH	growth hormone release-inhibiting hormone

Glu.6-PO_4	glucose 6-phosphate
Glu.6-PD	glucose 6-phosphate dehydrogenase
GPP	geranyl pyrophosphate
GSH	glutathione (reduced form)
GSSG	glutathione (oxidized form)
GTP	guanosine triphosphate
GTT	glucose tolerance test
ΔH	change in enthalpy
h	hour
Hb	hemoglobin
HbA_1	adult hemoglobin
HbA_{1c}	glycosylated hemoglobin
HbF	fetal hemoglobin
HbO_2	oxyhemoglobin
HbS	sickle-cell hemoglobin
hCG	human chorionic gonadotropin
HDL	high density lipoproteins
HGPRT	hypoxanthine guanine phosphoribosyltransferase
HIAA	hydroxy indole acetic acid
His	histidine
HIV	human immunodeficiency virus
HMG CoA	p-hydroxy p-methylglutaryl CoA
HMP	hexose monophosphate
hnRNA	heterogeneous nuclear RNA
Hp	haptoglobin
HPLC	high performance liquid chromatography
HRE	hormone responsive element
5HT	5-hydroxytryptamine
Hyp	hydroxyproline
ICD	isocitrate dehydrogenase
IDDM	insulin dependent diabetes mellitus
IDL	intermediate density lipoproteins
IDP	inosine diphosphate
IF	initiation factor
Ig	immunoglobulin
IgG	immunoglobulin G
IGF	insulin-like growth factor
Ile	isoleucine
IMP	inosine monophosphate
INH	isonicotinic acid hydrazide (isoniazide)
$InsP_2$ (IP_2)	inositol 1,4-biphosphate
$InsP_3$ (IP_3)	inositol 1,4,5-triphosphate
IPP	isopentenyl pyrophosphate
IR	infrared
ITP	inosine triphosphate
IU	international unit
IV	intravenous
K	dissociation constant
KA	King Armstrong
K_a	dissociation constant of acid
Kbp	kilo base pair
KDa	kilodalton
Keq	equilibrium constant
α-KG	α-ketoglutarate
Ki	inhibition constant
Km	Michaelis constant
KJ	kilojoule
LATS	long acting thyroid stimulator
LCAT	lecithin cholesterol acyltransferase
LDH	lactate dehydrogenase
LDL	low density lipoproteins
Leu	leucine
LFT	liver function tests
LH	luteinizing hormone

LPH	lipotrophic hormone (lipotropin)
LT	leukotrienes
Lp-a	lipoprotein-a
LSD	lysergic acid diethylamide
Lys	lysine
M	molar
MAO	maximal acid output
MAO	monoamine oxidase
Mb	myoglobin
MbO_2	oxymyoglobin
MCAD	medium chain acyl CoA dehydrogenase
MDH	malate dehydrogenase
Met	methionine
mEq	milliequivalents
mg	milligram
MI	myocardial infarction
MIT	monoiodotyrosine
mol	mole(s)
mM	millimolar
mol.wt.	molecular weight
mRNA	messenger RNA
MSH	melanocyte stimulating hormone
mtDNA	mitochondrial DNA
myc	melanocyte oncogene
MW	molecular weight
NAD^+	nicotinamide adenine dinucleotide
NADH	reduced NAD^+
$NADP^+$	nicotinamide adenine dinucleotide phosphate
NADPH	reduced $NADP^+$
NAG	N-acetylglutamate
NANA	N-acetylneuraminic acid
NDP	nucleoside diphosphate
NE	niacin equivalents
NEFA	non esterified fatty acid
ng	nanogram (10^{-9} g)
NGF	nerve growth factor
NIDDM	non-insulin dependent diabetes mellitus
NMP	nucleoside monophsphate
NMR	nuclear magnetic resonance
NPN	non-protein nitrogen
NPU	net protein utilization
OAA	oxaloacetate
OD	optical density,
OMP	orotidine monophosphate
Osm	osmoles
PABA	para amino benzoic acid
PAF	platelet-activating factor
PAGE	polyacrylamide gel electrophoresis
PAH	para amino hippurate
PAPS	phosphoadenosine phosphosulfate
PBG	porphobilinogen
PBI	protein bound iodine
PCM	protein-calorie malnutrition
pCO_2	partial presence of CO_2
PCR	polymerase chain reaction
PCT	proximal convoluted tubule
PDGF	platelet derived growth factor
PDH	pyruvate dehydrogenase
PEG	polyethylene glycol
PEM	protein-energy malnutrition
PEP	phosphoenol pyruvate
PER	protein efficiency ratio
PEST	proline, glutamine, serine, threonine
PFK	phosphofructokinase
PG	prostaglandins
PGA	pteroyl glutamic acid

pH	negative log of $[H^+]$
Phe	phenylalanine
PI	phosphatidyl inositol
Pi	inorganic phosphate
pI	isoelectric pH
PIF	prolactin inhibitory factor
PIP_2	inositol 4, 5-bisphosphate
pKa	negative log of Ka
PKU	phenylketonuria
PL	phospholipid
PLP	pyridoxal phosphate
pO_2	partial pressure of O_2
POMC	pro-opiomelanocortin
PPi	inorganic pyrophosphate
ppm	parts per million
PRIH	prolactin release-inhibiting hormone
PRL	prolactin
Pro	proline
PRPP	5-phosphoribosyl 1-pyrophosphate
PT	prothrombin time
PTH	parathyroid hormone
PTH	phenyl thiohydantoin
PUFA	polyunsaturated fatty acids
QPRT	quinolinate phosphori bosyltransferase
RAIU	radioactive iodine uptake
ras	rat sarcoma
RBC	red blood cells
RBP	retinol binding protein
RDA	recommended dietary allowance
RE	retinol equivalents
RER	rough endoplasmic reticulum
RF	releasing factor
Rf	ratio of fronts
RFLP	restriction fragment length polymorphism
R-form	relaxed form
RIA	radioimmunoassay
RNA	ribonucleic acid
RNAP	RNA polymerase
RNase	ribonuclease
R 5-P	ribose 5-phosphate
RQ	respiratory quotient
rRNA	ribosomal RNA
RT	reverse transcriptase
rT_3	reverse T_3
SAM	S-adenosyl methionine
SCID	severe combined immunodeficiency
SOD	Superoxide dismutase
SDA	specific dynamic action
SDS	sodium dodecyl sulfate
Ser	serine
Sf	Svedberg floatation
SGOT	serum glutamate oxaloacetate transaminase
SGPT	serum glutamate pyruvate transaminase
SHBG	sex hormone binding globulin
SIDS	sudden infant death syndrome
sn	stereospecific number
snRNA	small nuclear RNA
snRNP	small nuclear ribonucleoprotein
SrcsRNA	rous Sarcoma virussoluble RNA
SRS	slow reacting substance
T	thymine, thymidine
T	thymus (T-lymphocyte)
T_3	3, 5, 3' -triiodothyronine
T_4	3, 5, 3',5'-tetraiodoth-yronine (thyroxine)
TBG	thyronine binding globulin

TBPA	thyroxine binding prealbumin
TCA	tricarboxylic acid
TF	tissue factor
T-form	taut or tense form
TAG	triacylglycerol
TGTgb	triglyceridethyroglobulin
TGF	transforming growth factor
THF	tetrahydrofolate
Thr	threonine
TIBC	total iron binding capacity
TLC	thin layer chromatography
Tm	tubular maximum
TMP	thymidine monophosphate
TNF	tumor necrosis factor
TPP	thiamine pyrophosphate
TRH	thyrotropin releasing hormone
tRNA	transfer RNA
Trp	tryptophan
TSH	thyroid stimulating hormone
Tyr	tryrosine
TX	thromboxane
μm	micrometer
UBG	urobilinogen
UDP	uridine diphosphate
UDPG	uridine diphosphate glucose
UDP-Gal	Uridine diphosphate galactose
UDP-Glc	Uridine diphosphate glucose
UDP-GlcUA	Uridine diphosphate glucuronic acid
μl	microliter
μM	micromoles
UMP	uridine monophosphate
UTP	uridine triphosphate
UV	ultraviolet
Val	valine
V_H	variable heavy chain
VIP	vasoactive intestinal peptide
V_L	variable light chain
VLDL	very low density lipoproteins
VMA	vanillyl mandelic acid
V_{max}	velocity maximum
WBC	white blood cells
XMP	xanthosine monophosphate
Xyl	xylose

□□□

Clinical Biochemical Indices

BLOOD/SERUM

1)	Bilirubin Total	:	0.2 - 1.2 mg/dl	3.5 - 20.5 µmol/L
	- Direct (Conjugated)	:	0.1 - 0.4 mg/dl	<7.0 µmol/L
	- Indirect (Unconjugated)	:	0.2 - 0.8 mg/dl	<12.0 µmol/L
2)	Calcium Total	:	9-11 mg/dl	2.1 - 2.6 mmol/L
	Calcium Ionized	:	4.5 - 5.5 mg/dl	2.1 - 2.6 mEq./L
3)	Ceruloplasmin	:	25-43 mg/dl	1.7 - 2.9 µmol/L
4)	Chloride (S/P)	:	96-106 m.Eq./L	96-106 mmol/L
5)	Cholesterol Total	:	150-220 mg/dl	3.9-5.72 mmol/L
	- HDL fraction	:	30-60 mg/dl	
	- LDL fraction	:	80-150 mg/dl	
	- VLDL fraction	:	20-40 mg/dl	
6)	Copper	:	100 - 200 µg/dl	16-31 µmol/L
7)	Creatinine	:	0.7 - 1.5 mg/dl	62-132 mmol/L
8)	Glucose	:	70-110 mg/dl	3.6 - 6.1/ mmol /L
9)	Hemoglobin	:		
	- Male	:	14-16 g/dl	
	- Female	:	13-14 g/dl	
10)	Iron	:	50-175 µg/dl	9-31.3 µmol/L
11)	Magnesium	:	1.8 - 3.0 mg/dl	0.75 - 1.25 mmol/L
12)	Phospholipids	:	145 - 200 mg/dl	
13)	Phosphate (Inorganic)	:	3-4.5 mg/dl	1-1.5 mmol/L

14)	Potassium	:	3.5 - 5 mEq/dl	3.5-5mmol/L
15)	Proteins (Total)	:	6-8 g/dl	
	- Albumin	:	3.5-6.5 g/dl	
	- Globulin	:	2-3.6 g/dl	
	- Fibrinogen	:	0.2 - 0.6 g/dl	
16)	Sodium	:	135-145 mEq/L	135-145mmol/L
17)	Transferrin	:	200-400 mg/dl	23-45 μmol/L
18)	Triglycerides	:	75-150 mg/dl	
19)	Urea Nitrogen (BUN)	:	8-25 mg/dl	2.9-8.9mmol/L
20)	Urea (S)	:	15-40 mg/dl	
21)	Uric acid			
	- Male	:	3-8 mg/dl	0.18 - 0.54 mmol/L
	- Female	:	2.5-6 mg/dl	0.15 - 0.45 mmol/L

ENZYMES

1)	Acid Phosphatases	:	0-3 KA Units/dl KA- (King Armstrong Units)
2)	Alanine aminotransferase (ALT/SGPT)	:	0-45 IU/dl at 37ºC
3)	Alkaline Phosphatase (ALP)	:	3-13 KA units/dl
4)	Amylase	:	80 - 180 Somogyi Units/dl
5)	Aspartate Transaminase (AST,SGOT)	:	5-41 IU/L at 37ºC
6)	Creatine Kinase	:	10-50 IU/L at 30ºC
7)	Creatine Phosphokinase MB	:	0-3 IU/L
8)	Lactate Dehydrogenase (LDH)	:	50 - 200 IU/L
9)	Lipase	:	<150 U/L

HORMONES

Cortisol (P)	:	5 - 25 μg/dl at 9 AM
Epinephrine	:	Supine < 100 pg/ml
Insulin (S)	:	6-25 μU/ml

Testosterone –

Adult male – 300-800 ng/dl

Adult Female - 25-50 ng/dl

Triiodothyronine (T_3)—110 - 180 ng/dl

Thyroxine/Tetraiodothyronine (T_4)—5-12 μg/dl

Thyroid Stimulating Hormone—5-10 μU/ml (TSH)

CEREBROSPINAL FLUID

Chloride	:	120 - 130 mEq/L
Glucose	:	50 - 75 mg/dl
Proteins	:	15-40 mg/dl

❑❑❑

INDEX

A

B

C

D

E

F

H

I

M

N

O

P

Q

R

S

T

U

V

W

X

Z